(machine tool technology and manufacturing processes)

C. Thomas Olivo

C. Thomas Olivo Associates

DEDICATION

This book is dedicated to my wife Hilda for life-time interest, continuing inspiration, understanding, and helpfulness at each successive stage of development from research, to design, to technical writing, to editing, to production and, finally, to publishing.

Library of Congress Cataloging-in-Publication Data

Olivo, C. Thomas.
   Machine tool technology and manufacturing processes.

   Includes index.
   1. Machine-tools.   2. Manufacturing processes.
3. Machine shop practice.   I. Title.
TJ1185.039   1987   621.9′02   86–18216
ISBN 0–938561–08–1
ISBN 0–938561–09–X (laboratory manual)
ISBN 0–938561–10–3 (instructor's guide)
ISBN 0–938561–11–1 (set: Vol. I, II, III)

Cover credits: Cutting tools: Universal Houdaille
Laser process: Control Laser Corporation

PREFACE

Machine Tool Technology and Manufacturing Processes is a refreshingly new, comprehensive, authoritatively-written, highly-practical book. The organization, scope, sequence, and contents are the end product of extensive analyses of industrial needs and other studies of "how-to-do" skill and "why-to-do" technology requirements.

APPLICATIONS

The book meets the needs of . . . students and trainees in trade, technology, industrial-technical, and engineering programs . . . apprentices . . . employed individuals participating in upgrading, retraining, or occupational extension training . . . and others engaged in self-study.

The content is designed for use in regular and area vocational-technical schools, technical institutes, postsecondary and college-level institutions, engineering technology programs, training-within-industry programs, military occupational specialty training, and other programs.

MAJOR FEATURES

The book is divided into sixteen *Parts* and 64 instructional *Units*. While the units are arranged in a functional progression for presenting technology and processes, the sequence may be altered to meet specific program needs.

There are over 650 selected line drawings and industrial photos that complement the content. The drawings appear in two-color format in order to highlight important design features and operations.

Customary and SI metric units and hard and soft conversions are used throughout the book to the same degree that each system of dimensioning and measurement is applied in industry. Personal and machine safety and product protection are stressed in each unit. Review and self-test items for each unit provide practical problems by which skill and technology competency may be measured.

Directly-referenced tables are applied in most units. Related mathematics, applications of physical science principles, the interpretation of blueprints, and the making of technical sketches are interlocked within each unit with required shop and laboratory skills and technology.

The Appendix includes 29 selected industrial *Handbook Tables*, a functional *Index* for rapidly locating significant items in the book, and a *Glossary of Technical Terms*.

A companion **Laboratory Manual** contains supplemental industry-related *Review and Test Items* and *Unit Objectives*. All problems are directly correlated with a specific Unit in this textbook.

The **Instructor's Guide** provides an *Outline of each Unit* as an instructional guideline. *Answers* are given for all Review and Self-Test Items in the textbook and the complementary problems in the **Laboratory Manual**.

CONTENTS OF EACH PART

- **Part 1** provides a brief *historical overview* and an insight into *multiple-job opportunities* and levels of employment.

- **Part 2** introduces *conventional* and *SI metric measurement systems*, line-graduated and precision measuring tools, quality control principles, and surface finish characteristics and measurement.

- **Part 3** deals with basic *hand tools* and cutting and noncutting processes *and layout tools* as applied to *bench and surface plate work*.

- **Part 4** describes *cutoff power saws* and *bench and floor (hand) grinding* machines and processes.

- **Part 5** relates to *drilling machines*, accessories, hole-forming tools, and machine processes.

- **Part 6** introduces basic and advanced *turning machines:* engine lathes, production turret lathes, and automatic screw machines. Machine tools, accessories, tooling, setups, and internal and external processes are included.

- **Part 7** covers *horizontal milling machines* and *vertical milling machines.* Accessories, dividing heads, cutting tools, cutting fluids, and basic and advanced milling processes up through helical gearing and cam milling are treated.

- **Part 8** deals with *vertical band machining* processes and technology related to sawing, filing, grinding, and polishing.

- **Part 9** presents current industrial practices for *shaping, slotting,* and *planing machines;* setups and tooling, processes, and technology.

- **Part 10** relates to *abrasive machining.* Abrasive wheel properties, systems, and selection; machine tools, setups with accessories; and grinding technology and processes are applied to surface grinding, tool and cutter grinding of milling cutters and other cutting tools, and cylindrical grinding.

- **Part 11** introduces *precision production tooling* and machining processes using jig borers, jig grinders, and universal measuring machines.

- **Part 12** covers *numerical control (NC)* and *computer-assisted numerical control (CNC) machine tools;* including programming systems and applications of computer-aided drafting and design (CADD), CADAM, and other computerized input.

- **Part 13** presents the production of *industrial and cutting tool materials*, including ferrous and nonferrous metals, alloys, cemented carbides, ceramic cutting tools, and diamonds.

- **Part 14** covers principles of and practices relating to *basic metallurgy, heat treatment, and hardness testing* of metals, alloys, and nonmetallic materials.

- **Part 15** deals with new *nontraditional machine tools, tooling, and machining* processes such as: electrical discharge, abrasive flow, chemical, laser, and others.

- **Part 16** concludes with a treatment of *automated, flexible, multimachine manufacturing systems (FMS) and robotics.*

Machine Tool Technology and Manufacturing Processes is a fascinating study of skills and technology required to develop human resources and to maintain national growth. This book displays the state-of-the-art, starting with Wilkinson's chipped and filed lathe bed and patented lathe carriage of 1794. The expansion of the mid-1800 concept relating to the manufacture of interchangeable parts and "sustained, incremental contributions of successive generations of machine tool and accessory builders, tool designers, materials producers, systems and programming planners, craftspersons, and technicians, have resulted in the present level of highly sophisticated manufacturing and productivity.

About the Author

Dr. C. Thomas Olivo ("Dr. Tom") is one of the nation's foremost authorities with lifetime experiences in machine tool technology, tooling, and metal products manufacturing. He served an apprenticeship and achieved journeyman and master status as a tool and die maker.

Entering teaching from industry, Dr. Olivo served in successively responsible leadership positions as state supervisor, bureau chief of vocational-technical curriculum planning and instructional materials development, bureau chief of industrial teacher training, and State director of industrial-technical education. Dr. Olivo is also known for pioneering post-secondary vocational-technical institute and area vocational school movements.

The thread of continuity to Dr. Olivo's writings is the continuous interlocking of his work within industry and the development of progressive training programs.

CONTENTS

Preface iii

Acknowledgments xx

Part 1 MACHINE TOOL TECHNOLOGY, MACHINING, AND MANUFACTURING OCCUPATIONS

Section One INTRODUCTION

UNIT 1 HISTORICAL OVERVIEW AND CAREER OPPORTUNITIES

A. Impact of Early Machine Tools and Concept of Mass Production 1

Interchangeable Parts in Manufacturing 3 • End Products of Manufacturing Enterprise 3 • Nontraditional Machine Tools and Machining 3

B. Career Opportunities: Machine Tool, Machining, and Manufacturing Industries Occupations 4

Bases of Job Qualifications 4 • Apprenticeship and Other Job Training Systems 4 • Classification of Job Titles by Level and Area 5 • Job Mobility and Advancement 7 • Review and Self-Test 7

Part 2 MEASUREMENT: TECHNOLOGY AND PROCESSES

Section One MEASUREMENT AND MEASUREMENT SYSTEMS

UNIT 2 MEASUREMENT: CUSTOMARY AND SI METRIC SYSTEMS 9

Dimensional Measurements 9 • Measurement and Industrial Drawings 11 • Limitations of the Inch and Metric Measurement Systems 13 • Common Linear Measuring Tools 13 • Conversion Tables 14 • Review and Self-Test 14

Section Two MEASUREMENT AND LAYOUT TOOLS AND INSTRUMENTS

UNIT 3 DIMENSIONAL MEASUREMENT: INSTRUMENTS AND LAYOUT TOOLS

A. Line Graduated Rules 15

Discrimination and Measurement Scales 15 • Flexible Steel Tapes 16 • Errors in Reading Measurements 16 • *How to Accurately Measure with a Steel Rule* 17

B. Layout and Transfer Measuring Tools 18

How to Measure with a Depth Gage 19 • The Combination Set 19 • *How to Measure and Lay Out Angles with a Protractor Head 21* • Transfer Instruments 21 • The Surface Gage/Surface Plate 24 • Safe Practices 25 • Review and Self-Test 26

UNIT 4 PRECISION INSTRUMENTS: MICROMETERS, VERNIERS, INDICATORS, AND GAGE BLOCKS

A. Standard and Vernier Micrometer Principles and Measurements 27

Major Design Features of the Micrometer 27 • Micrometer Principle and Measurement 28 • *How to Measure with the Inch-Standard Micrometer 29* • The Vernier Micrometer and 0.0001″ (0.002mm) Measurement 30 • The Metric Micrometer 31 • *How to Measure with the Depth Micrometer 33* • The Inside Micrometer 34

B. Vernier Measuring Instruments, Dial Indicating Instruments, and Gage Blocks 34

Vernier Principle and Vernier Caliper Measurement 36 • Measuring with the Vernier Depth and Vernier Height Gage 38 • The Universal Vernier Protractor 39 • Solid and Cylindrical Squares 40 • Test Indicators and Dial Indicating Instruments 40 • Precision Gage Blocks 43 • Precision Measurement of Angles and Tapers 44 • Safe Practices 46 • Review and Self-Test 47

UNIT 5 ADVANCED PRECISION MEASUREMENT, QUALITY CONTROL, AND SURFACE TEXTURE

A. Measurement with Optical Flats and Micrometers 48
Shop and Toolmaker's Microscopes 48 • Optical Height Gage and Industrial Magnifiers 50 • Precision Measurement with Optical Flats 51 • Measurement of Surface Flatness 52 • Interference Band Conversion Table 52

B. Measurement with High Amplification Comparators 52
Mechanical Comparators 53 • Mechanical-Optical Comparators 54 • Operation of Flow and Pressure Types of Pneumatic Comparator 57 • Electrical and Electronic Comparators 58

C. Quality Control Principles and Measurement Practices. 59
Fixed-Size Gages 60 • Adjustable Thread Ring Gages 62 • Allowances for Classes of Fits 62 • Interchangeability of American National and Unified Form Screw Threads 63 • Quality Control Methods, Plans, and Charts 63

D. Surface Texture Characteristics and Measurement. 66
Representation of Lay Patterns and Symbols on Drawings 68 • Measuring Surface Finishes 68 • Surface Roughness Related to Common Production Methods 69 • Surface Texture Measuring Instruments 69 • Safe Practices 73 • Review and Self-Test 73

Part 3   BENCH AND SURFACE PLATE WORK: TECHNOLOGY AND PROCESSES

Section One   LAYOUT AND INSPECTION PROCESSES

UNIT 6 PRECISION ACCESSORIES FOR SURFACE PLATE WORK

A. Linear and Angular Layouts 77
Parallels and Other Forms of Granite Accessories 77 • Precision Angle Measurements and Accessories 78 • Table of Constants and Sine Bar, Sine Block, and Sine Plates 79

B. Techniques for Checking Parallelism, Squareness, Roundness, and Concentricity . . . 81
General Precision Layout and Measurement Practices 82

C. Preparation of Workpieces 83
Work Habits 84 • Safe Practices 84 • Review and Self-Test 85

Section Two   BENCH WORK: TOOLS AND PROCESSES

UNIT 7 BASIC CUTTING AND NONCUTTING HAND TOOLS

A. Noncutting Hand Tools 86
Hammering Tools, Fasteners and Fastening Tools 86 • Screwdrivers 87 • Hand Wrenches 87 • Clamping and Gripping Devices 89 • Industrial Goggles and Face Shields 90

B. Hand Sawing Processes and Technology. 90
Hacksaw Frames 90 • Hacksaw Blades 91 • Cutting Speeds and Force (Pressure) 92 • Holding Irregular Shapes and Thin-Walled Materials 92

C. Chiseling, Punching, and Driving Tools and Processes 92
Chisels 92 • Hand Punches 94 • Drifts 94

D. Hand File Characteristics and Filing Processes. 94
Precision Files 95 • Features of Files 97 • Care of Files 97 • Filing Processes 97 • Burrs and Burring 98

E. **Coated Abrasives and Hand Finishing (Polishing)** 99

Materials and Characteristics 100 • Grain Sizes 101 • Safe Practices 101 • Review and Self-Test 103

UNIT 8   HAND REAMERS AND REAMING PROCESSES 105

Features 105 • Basic Types of Hand Reamers 106 • Chatter Marks Resulting from Reaming 108 • Cutting Fluids 108 • Safe Practices 108 • Review and Self-Test 109

UNIT 9   CHARACTERISTICS, TOOLS, AND HAND TAPPING PROCESSES OF INTERNAL THREADING 110

American National and Unified Screw Threads 110 • ISO Metric Threads 111 • Cutting Internal Threads with Hand Taps 111 • Tap Features 111 • Other Types of Hand and Machine Taps 113 • Flute Depth and Tap Strength 114 • Tap Drill Sizes and Tables 114 • *How to Tap a Through Hole 115* • Tap Extractors 116 • Safe Practices 116 • Review and Self-Test 117

UNIT 10   DRAWINGS, DIES, AND EXTERNAL THREADING 118

Measuring Screw Thread Pitch 118 • External Thread Cutting Hand Tools 119 • Screw Extractor 120 • Safe Practices 120 • Review and Self-Test 121

Part 4   BASIC CUT-OFF AND GRINDING MACHINES: TECHNOLOGY AND PROCESSES

Section One   METAL-CUTTING SAWS

UNIT 11   POWER HACKSAWS: TYPES AND PROCESSES 122

Major Machine Parts 123 • Heavy-Duty and Production Hacksaws 123 • Power Hacksaw Blades 124 • Safe Practices 125 • Review and Self-Test 125

UNIT 12   HORIZONTAL BAND MACHINES AND CUTOFF SAWING PROCESSES 126

Dry Cutting and Wet Cutting 126 • Cutting Action and Cutting Fluids 127 • Kinds and Forms of Saw Blades (Bands) 129 • Cutoff Sawing Requirements and Recommendations 129 • *How to Select the Saw Blade and Determine Cutoff Sawing Requirements 130* • Safe Practices 131 • Review and Self-Test 131

Section Two   BENCH AND FLOOR GRINDERS

UNIT 13   HAND GRINDING MACHINES, ACCESSORIES, AND BASIC PROCESSES 132

Bench and Floor Pedestal Grinders 132 • Grinding Wheels 133 • Dressing and Hand Dressers 134 • *How to Grind an Angular Surface 135* • Safe Practices 135 • Review and Self-Test 136

Part 5   DRILLING MACHINES: TECHNOLOGY AND PROCESSES

Section One   DRILLING MACHINES AND ACCESSORIES

UNIT 14   DRILLING MACHINES AND PROCESSES 137

Heavy-Duty Standard Upright 139 • Multiple-Spindle, Way-Drilling, Deep-Hole, and Radial Drill Machines 140 • Turret and Special Hole Drilling Machines 141 • Work-Holding or Clamping Devices 142 • Machine Accessories 144 • Safe Practices 145 • Review and Self-Test 146

Section Two CUTTING TOOLS AND DRILLING MACHINE OPERATIONS

UNIT 15 DRILLING MACHINE TECHNOLOGY AND PROCESSES **147**

Cobalt High-Speed Steel and Tungsten Carbide Drills 147 • Drill Point Shapes 149 • Split-Point Drill 150 • Drill Point Gage 151 • Taper Shank Adapters (Sleeves and Sockets) 151 • The Drill Drift 151 • Drill Point Grinding 152 • Designation of Drill Sizes 152 • Causes and Correction of Drilling Problems 153 (Table 154) • Cutting Speeds 153 • Cutting Feeds 155 • *How to Drill Holes 155* • Safe Practices 156 • Review and Self-Test 157

UNIT 16 MACHINE REAMING: TECHNOLOGY AND PROCESSES. **158**

Jobbers, Shell, Fluted Chucking, Rose Chucking Reamers 159 • Expansion and Adjustable Chucking, Step, and Taper Reamers 160 • Machine Reaming Speeds and Feeds 160 • Reamer Alignment and Chatter 161 • Cutting Fluids 161 • Causes of Reamer Breakage and Wear 161 • *How to Machine Ream Holes 161* • Safe Practices 162 • Review and Self-Test 162

UNIT 17 MACHINE THREADING ON A DRILL PRESS **163**

Reversing and Nonreversing Spindle Tap Driver 163 • Torque-Driven, Micro-, and Heavy-Duty Tapping Attachments 164 • Factors Affecting Machine Threading Efficiency 164 • Analysis of Tapping Problems 165 (Table 166) • Features of Machine Taps 165 • Checking and Measuring a Threaded Part 165 • *How to Machine Tap (Drill Press) 165* • Safe Practices 167 • Review and Self-Test 167

UNIT 18 COUNTERSINKING, COUNTERBORING, SPOTFACING, AND BORING. **168**

Combination Drill and Countersink 168 • Counterbores and Counterboring 169 • Drill/Counterbore Combination 169 • Spotfacers and Spotfacing 169 • Boring and Boring Tools 170 • *How to Counterbore 170* • Safe Practices 170 • Review and Self-Test 171

Part 6 TURNING MACHINES: TECHNOLOGY AND PROCESSES

Section One ENGINE LATHES: DESIGN FEATURES AND OPERATIONS

UNIT 19 ENGINE LATHES: FUNCTIONS, CONTROLS, AND MAINTENANCE

A. Functions, Types, and Features. 172

Lathe Components for Producing Work and Tool Movements 173 • Headstock: Drives, Speeds, Gearing, Spindles, and Nose Types 174 • Cross Slide and Compound Rest 176 • Lathe Accessories 176 • Lathe Attachments: Taper and Threading 178 • Steady and Follower Rests 179 • Power (Rapid) Traverse 179

B. Lathe Controls and Maintenance 181

Machine Controls for Feeds and Threads 181 • Lost Motion 182 • *How to Align the Tailstock 182* • Lathe Maintenance 183 • Factors Affecting Machining Conditions 183 (Table 184) • Safe Practices 184 • Review and Self-Test 185

Section Two LATHE CUTTING TOOLS AND TOOLHOLDERS

UNIT 20 CUTTING TOOLS

A. Single-Point Cutting Tools: Technology and Processes. 186

Toolholders and Carbide Insert End-Cutting Operations 187 • Rake and Side- and End-Cutting Edge Angles 188 • Lead Angle, Chip Thickness, Cutting Force 188

B. Cutting Tool Materials, Forms, and Holders. 189

High-Speed Steel, Cast Alloys, Cemented Carbide Cutting Tools 189 • Ceramic and Diamond Cutting Tools 190 • Cutting Speeds for Single-Point Tools 190 (Table 191) • System for Designating Brazed-Tip Carbide Tool Bits 191 (Table 192) • Chip Formation and Control of Cutting Forces 192 • Quick-Change Tool System 193

C. Off-Hand Grinding HSS Tool Bits 194

How to Grind High-Speed Steel Tool Bits 194 • Safe Practices 195 • Review and Self-Test 196

Section Three   LATHE WORK BETWEEN CENTERS

UNIT 21   CENTER WORK AND LATHE SETUPS **197**

Preparing the Lathe for Turning between Centers 198 • Common Center Drilling Techniques 199 • *How to Drill Center Holes 199* • Safe Practices 200 • Review and Self-Test 200

UNIT 22   FACING AND STRAIGHT TURNING **201**

Depth of Cut Setting 201 • Facing Tools, Setups, and Processes 202 • *How to Face 202* • *How to Do Straight (Parallel) Turning 203* • Safe Practices 204 • Review and Self-Test 204

UNIT 23   SHOULDER TURNING, CHAMFERING, AND ROUNDING ENDS. **205**

Measurement Procedures and Application of the Radius Gage 205 • Representing and Dimensioning Shoulders, Chamfers, and Rounded Ends 206 • *How to Turn a Beveled (Angular) Shoulder 206* • *How to Turn a Rounded End 207* • Safe Practices 207 • Review and Self-Test 207

UNIT 24   GROOVING, FORM TURNING, AND CUTTING OFF **208**

Form Turning 208 • Cutting-Off (Parting) Processes 209 • Problems in Cutting Off Stock 209 • *How to Cut Off (Part) Stock 210* • Safe Practices 211 • Review and Self-Test 211

UNIT 25   LATHE FILING, POLISHING, AND KNURLING. **212**

Filing Technology and Processes 212 • Polishing on the Lathe 213 • Knurls and Knurling 213 • *How to Knurl 214* • Safe Practices 215 • Review and Self-Test 215

UNIT 26   TAPER AND ANGLE TURNING TECHNOLOGY AND PROCESSES **217**

Taper Definitions and Calculations 218 • Measuring and Gaging Tapers 219 • Taper Turning Processes 220 • Angle Turning with the Compound Rest 221 • Computing Compound Rest Angle Setting 222 • Safe Practices 222 • Review and Self-Test 223

Section Four   LATHE WORK HELD IN A CHUCK

UNIT 27   CHUCKS AND CHUCKING TECHNOLOGY AND PROCESSES **224**

Spindle Noses and Chuck Adapter Plates 224 • Collet Chucks 226 • Work Truing Methods 227 • *How to Mount Lathe Spindle Accessories 228* • Safe Practices 228 • Review and Self-Test 229

UNIT 28   CENTERING, DRILLING, COUNTERSINKING, AND REAMING ON THE LATHE **230**

Centering and Center Drilling Work Held in a Chuck 230 • Drilling and Reaming Practices on Lathe Work 230 • *How to Drill Holes 231* • Safe Practices 232 • Review and Self-Test 233

UNIT 29   BORING PROCESSES AND MANDREL WORK. **234**

Boring Tools, Bars, and Holders 234 • Carbide Insert Setups for Boring 236 • Counterboring, Recessing, and Boring Tapers 236 • *How to Bore a Hole 236* • Mandrels and Mandrel Work 237 • Safe Practices 238 • Review and Self-Test 239

UNIT 30   CUTTING AND MEASURING 60° FORM EXTERNAL AND INTERNAL SCREW THREADS   240

Gages, Cutting Tools, and Holders 240 • Thread-Chasing Attachment 241 • Spindle Speeds 242 • Design Features and Thread Form Calculations 242 • Allowance, Tolerance, Limits, and Size 243 • Thread Measurement and Inspection 243 • Leading and Following Side Angles 245 • Depth Settings 245 • *How to Set Up a Lathe for Thread Cutting 245* • *How to Cut an Outside Right- or Left-Hand Thread 246* • *How to Clean Up the Back Side of the Thread 246* • Lathe Setup for Cutting SI Metric Threads 247 • Internal Threading 247 • Machine Setup for Taper Thread Cutting 247 • Safe Practices 247 • Review and Self-Test 248

UNIT 31   ADVANCED THREAD-CUTTING TECHNOLOGY AND PROCESSES. **249**

Production Methods of Making Threads 249 • American Standard Unified Miniature Screw Threads 251 • *How to Cut a Square Thread 251* • *How to Cut a Right-Hand Acme (29° Form) Thread 253* • Multiple-Start Threads 254 • Safe Practices 255 • Review and Self-Test 255

Section Five   PRODUCTION TURNING MACHINES: TECHNOLOGY AND PROCESSES

UNIT 32   TURRET LATHES AND SCREW MACHINES

A. Turret Lathes: Components and Accessories   256

Bar and Chucking Machines 257 • Components 260 • Basic Machine Attachments 261

B. Spindle, Cross Slide, and Turret Tooling and Permanent Setups   262

Universal Bar Equipment for Permanent Setups 263 • Slide Tools for Turrets 265 • Cutting Tool Holders 267 • Cutting-Off Tools 267 • The Planned Tool Circle 268 • Permanent Setups with Chucking Tooling 268 • Tool Chatter Problems 269 • Safe Practices 269 • Review and Self-Test 270

UNIT 33   BASIC EXTERNAL AND INTERNAL MACHINING PROCESSES **272**

Cutting Speeds (sfpm) and Feeds (ipr) for Bar Turners 272 • *How to Sequence External Cuts 274* • Basic Internal Processes 275 • *How to Sequence Internal Cuts 277* • Safe Practices 278 • Review and Self Test 279

UNIT 34   SINGLE- AND MULTIPLE-SPINDLE AUTOMATIC SCREW MACHINES. **280**

Major Components 280 • Tooling 282 • Operation of Fully Automatic Screw Machines 283 • Functions and Design of Cams 283 • Cutting Fluids 284 • Automatic Screw Machine Accessories 284 • Troubleshooting 285 • Safe Practices 285 • Review and Self-Test 286

Part 7   MILLING MACHINES: TECHNOLOGY AND PROCESSES

Section One   HORIZONTAL MILLING MACHINES, CUTTERS, AND ACCESSORIES

UNIT 35   MILLING MACHINES AND ACCESSORIES

A. Functions and Maintenance of Milling Machines and Accessories   287

Industrial Types of Milling Machines 287 • Functions and Construction of Major Units 290 • Maintenance of the Lubricating Systems 292

B. Milling Cutter and Work-Holding Devices. 293

Cutter- and Work-Holding Devices 293 • *How to Mount and Remove a Milling Machine Arbor and a Cam-Lock Adapter 296* • *How to Position Swivel Vise Jaws and Use Work-Holding Accessories 297* • Safe Practices 297 • Review and Self-Test 298

UNIT 36   TECHNOLOGY AND APPLICATIONS OF STANDARD MILLING CUTTERS

A. Features and Applications of Milling Cutters. 299

Basic Forms of Cutter Teeth 299 • Standard Types of Milling Cutters 300

B. Speeds, Feeds, and Cutting Fluids for Milling 304

Feed and Speed Formulas 305 • Feed Mechanisms and Controls 305 • *How to Set the Feed (Table, Saddle, and Knee) 306* • Cutting Fluids and Milling Machine Systems 307 • Kinds of Cutting Lubricants and Application 308 • Safe Practices 308 • Review and Self-Test 309

Section Two   TYPICAL MILLING SETUPS AND PROCESSES

UNIT 37   PLAIN MILLING ON THE HORIZONTAL MILLING MACHINE 311

Conventional and Climb Milling 311 • *How to Mill Surfaces Parallel and/or at Right Angles 313* • Probable Causes and Corrective Steps of Common Milling Problems 313 • Safe Practices 315 • Review and Self-Test 315

UNIT 38   FACE MILLING ON THE HORIZONTAL MILLING MACHINE. 316

Effect of Lead Angle and Feed 316 • Factors Affecting Face Milling Processes 317 • Face and Shoulder Milling with a Solid End Mill 318 • *How to Face Mill with a Face Milling Cutter 319* • Safe Practices 320 • Review and Self-Test 320

UNIT 39   SIDE MILLING CUTTER APPLICATIONS 321

Setups of Side Milling Cutters 321 • Safe Practices 323 • Review and Self-Test 324

UNIT 40   SAWING, SLOTTING, AND KEYSEAT AND DOVETAIL MILLING. 325

Metal-Slitting Saws and Slotting Cutters 325 • Work-Holding Setups 326 • Keyseat Milling 326 • *How to Mill a Keyseat with a Woodruff Cutter 327* • Characteristics of Dovetails 328 • Safe Practices 329 • Review and Self-Test 330

Section Three   INDEXING DEVICES: PRINCIPLES AND APPLICATIONS

UNIT 41   DIRECT AND SIMPLE INDEXING 331

Functions of the Universal Dividing Head 333 • Direct Indexing 334 • Simple (Plain) Indexing and Calculations 334 • Safe Practices 335 • Review and Self-Test 335

Section Four   GEAR AND CAM MILLING

UNIT 42   GEAR PRODUCTION, DESIGN FEATURES AND COMPUTATIONS 336

Generating Machines for Bevel Gears 338 • Gear Finishing Machines 338 • Gear Tooth Measurement 338 • Spur Gear Drawings 340 • Module Systems 341 • Characteristics of the Involute Curve 342 • Rack and Pinion Gears 343 • Helical Gearing 343 • Calculating Helical Gear Dimensions 343 • Bevel Gearing 343 • Worm Gearing 344 • Gear Materials 345 • Review and Self-Test 346

UNIT 43   MACHINE SETUPS AND MEASUREMENT PRACTICES FOR GEAR MILLING. 347

Involute Gear Cutter Characteristics 347 • *How to Mill a Spur Gear 347* • Rack Milling 348 • *How to Mill Helical Gear Teeth 348* • *How to Mill Teeth on Miter Bevel Gears 349* • Worm Thread and Gear Milling 350 • *How to Mill a Worm Gear 350* • Safe Practices 351 • Review and Self-Test 352

UNIT 44 HELICAL AND CAM MILLING **353**

Lead, Angle, and Hand of a Helix 353 • Helix Lead and Gear Ratio Formulas 354 • *How to Mill a Helix on a Universal Milling Machine 356* • *How to Mill a Secondary Clearance Angle 356* • Milling a Helix by End Milling 357 • Functions of Cams in Cam Milling 357 • Cam Motions and Types of Cams and Followers 357 • Basic Cam Machining Processes 359 • *How to Mill a Uniform Rise Cam 359* • Safe Practices 360 • Review and Self-Test 360

Section Five VERTICAL MILLING MACHINES

UNIT 45 MACHINE AND ACCESSORY DESIGN FEATURES **361**

Combination Vertical/Horizontal Milling Machines 362 • Bed-Type and Numerically Controlled Vertical Milling Machines 363 • Components of Standard Machines 363 • Attachments for Spindle Heads: High-Speed (365), Right-Angle and Rotary Cross Slide Heads and Optical Measuring System 366 • Quick-Change Tooling System, Boring Heads, Tool Adapters 367 • Digital Readout 368 • Safe Practices 368 • Review and Self-Test 369

Section Six CUTTING TOOLS AND BASIC PROCESSES

UNIT 46 CUTTING TOOLS, SPEEDS AND FEEDS, AND BASIC PROCESSES **370**

High-Helix, Square and Ball, Keyway-Cutting, Tapered, and Cobalt High-Speed Steel End Mills 371 • Types of End Teeth 372 • Cutting Fluids 372 • Cutting Speed and Feed Data 372 (Table 373) • *How to End Mill a Flat Surface 372, an Angular Surface 373, a Slot or Keyway 374, a T-Slot 374* • Safe Practices 374 • Review and Self-Test 375

UNIT 47 HOLE FORMING, BORING, AND SHAPING **376**

Work- and Tool-Holding Devices 376 • Factors Affecting Hole-Forming Processes 376 • Offset Boring Head and Cutting Tools 376 • Shaping on a Vertical Milling Machine 377 • *How to Machine Holes Using a Conventional Machine Head 378* • *How to Bore Holes with an Offset Boring Head 378* • Compound Angle, Radius, and Stepped Area Boring 379 • Safe Practices 380 • Review and Self-Test 380

Part 8 VERTICAL BAND MACHINES

Section One BAND MACHINE TECHNOLOGY AND BASIC SETUPS

UNIT 48 BAND MACHINE CHARACTERISTICS, COMPONENTS, AND PREPARATION **382**

Sawing Band Recommendations 383 (Table 384) • Major Components 384 • Roller- and Insert-Type Saw Band Guides 386 • Saw Band Welding: Cutoff Shearing and Table 387 • Problems, Causes, Corrective Action 389 • Safe Practices 390 • Review and Self-Test 391

Section Two   BAND MACHINING: TECHNOLOGY AND PROCESSES

UNIT 49   BAND MACHINE SAWING, FILING, POLISHING, AND GRINDING **392**

How to Make External Cuts 393 • Internal Sawing 393 • Table of Sawing Problems, Causes, Corrective Action 394 • Slotting and Slitting 394 • Radius and Other Contour Cutting 395 • Friction Sawing 395 • Friction Saw Band Characteristics 396 (Table 397) • Friction Sawing Setup Procedures 397 • Spiral-Edge Band Sawing 397 • Diamond-Edge Band Sawing 398 • Band Filing 398 (Table 399) • Band Grinding and Polishing 399 • Line Grinding 400 • Knife-Edge Band Processes 400 • Safe Practices 401 • Review and Self-Test 402

Part 9   SHAPING/SLOTTING AND PLANING MACHINES

Section One   MACHINE TOOL TECHNOLOGY AND PROCESSES

UNIT 50   SHAPERS, SLOTTERS, AND PLANERS: MACHINES AND SETUPS **403**

Basic Types of Shapers 403 • Shaper Cuts 404 • Carbide Cutting Tools 406 • Modern Planer Design 406 • Setups in Preparation for Planer Work 407 • Planing Flat Surfaces 408 • Simultaneous Planing with a Side Head 409 • Planer Cutting Tools and Holders 409 • Safe Practices 410 • Review and Self-Test 411

Part 10   ABRASIVE MACHINING: TECHNOLOGY AND PROCESSES

Section One   GRINDING MACHINES, ABRASIVE WHEELS, AND CUTTING FLUIDS

UNIT 51   GRINDING MACHINES AND GRINDING WHEELS

A. Grinding Machines: Functions and Types 412

Surface Grinding Machines 414 • Cylindrical Grinding Machines 415 • Centerless Grinding 417 • Tool and Cutter, Gear Grinding, and Other Abrasive Machines 418 • Microfinishing, Honing, and Lapping 419 • Superfinishing, Polishing, and Buffing 420

B. Grinding Wheel Characteristics, Standards, and Selection 421

Wheel Types, Uses, Marking System, and Peripheral Grinding Wheels 422 • Side and Peripheral Grinding Wheels 424 • Mounted Wheels (Cones and Plugs) 425 • Work Speed and Table Travel 425 • Problems and Corrective Action 427 • Selection of Grinding Wheels (Table) 427 • Safe Practices 427 • Review and Self-Test 428

UNIT 52   GRINDING WHEEL PREPARATION AND GRINDING FLUIDS **430**

Concentric and Parallel Face Dressing (Truing) 430 • Using Diamond Dressers 431 • *How to True and Dress Surface and Cylindrical Grinder Wheels* 431 • Effect of Traverse Rate on Wheel Cutting Action 432 • Properties and Features of CBN and Diamond Wheels 432 • *How to Prepare Borazon (CBN) and Diamond Abrasive Wheels* 433 • Wheel Balancing 434 • Problems Related to Truing, Dressing, and Balancing 434 (Table 435) • Grinding Fluids and Coolant Systems 435 • Safe Practices 437 • Review and Self-Test 437

Section Two   SURFACE GRINDERS AND ACCESSORIES: TECHNOLOGY AND PROCESSES

UNIT 53   DESIGN FEATURES AND SETUPS FOR FLAT GRINDING **438**

Horizontal-Spindle Surface Grinder 438 • Digital Controls 439 • Wheel and Work-Holding Accessories 440 • Flat Surface Grinding Problems 441 • Down-Feeds and Cross-Feeds 442 • Grinding Edges Square and Parallel 442 • Holding Thin Workpieces 443 • Safe Practices 444 • Review and Self-Test 444

UNIT 54   ANGULAR, VERTICAL, AND PROFILE GRINDING AND CUTTING OFF

A. Grinding Angular and Vertical Surfaces   445

Magnetic Work-Holding Accessories 445 • *How to Grind Angular Surfaces 446* • *How to Grind an Angle with a Gage Block Setup 447* • Undercutting; Truing and Dressing Devices 447 • *How to Form Dress a Wheel for Angular Grinding 449* • *How to Grind Shoulders 449*

B. Form (Profile) Grinding and Cutting-Off Processes   449

Wheel Forming Accessories 449 • Diamond Dressing Tools 451 • Form Dressing with a Pantograph Dresser 452 • Forming Crush Rolls on a Microform Grinder 452 • Cutoff Wheels and Cutting-Off Processes 453 • Safe Practices 454 • Review and Self-Test 455

Section Three   CYLINDRICAL GRINDING: MACHINES, ACCESSORIES, AND PROCESSES

UNIT 55   MACHINE COMPONENTS: EXTERNAL AND INTERNAL GRINDING   456

Features of the Universal Grinder 456 • Automated Cylindrical Grinding Machine Functions 460 • Problems and Corrective Action 461 • Wheel Recommendations 462 • *How to Grind Straight Cylindrical Work 462* • *How to Grind Small-Angle and Steep-Angle Tapers 463* • Face and Internal Surface Grinding on a Universal Cylindrical Grinder 464 • *How to Set Up for and Cylindrically Grind Internal Angular Surfaces 465* • Cylindrical Grinder Wheel Truing and Dressing Devices 465 • Safe Practices 465 • Review and Self-Test 466

Section Four   CUTTER AND TOOL GRINDING: MACHINES, ACCESSORIES, AND PROCESSES

UNIT 56   MACHINE FEATURES, SETUPS, AND CUTTER AND TOOL GRINDING

A. Cutter and Tool Grinder Features and Setups   467

Machine Attachments 468 • Centering Gage, Mandrel, and Arbor Accessories 471 • Tooth Rests and Blades 471 • Grinding Considerations and Clearance Angle Tables 473 • Checking Accuracy of Cutter Clearance Angles 475

B. Grinding Milling Cutters and Other Cutting Tools   476

How to Sharpen Peripheral Teeth on a Slitting Saw 476 • Staggered Tooth Milling Cutters 476 • Considerations for Grinding Carbide-Tooth Shell Mills 477 • *How to Sharpen an End Mill 478* • Grinding of Form Relieved and Single- and Double-Angle Cutters 479 • Grinding Machine Reamers 479 • Features of Adjustable Hand Reamers 480 • Grinding Cutting, Clearance, and Rake Angles on Flat Cutting Tools 480 • Sharpening Taps 480 • Safe Practices 480 • Review and Self-Test 481

Part 11   PRECISION PRODUCTION TOOLING

Section One   PRECISION MACHINE TOOLS FOR PRODUCTION TOOLING

UNIT 57   JIG BORING, JIG GRINDING, AND UNIVERSAL MEASURING MACHINES AND PROCESSES   483

Jig Boring Machine Tools, Processes, and Accessories 483 • Design Features of a Locating Microscope 485 • *How to Set Up and Use a Jig Borer 486* • Jig Grinding Machine Tools and Processes 486 • Design Features of Jig Grinders 487 • Dressing Attachments for Jig Grinding Wheels 488 • Wheel and Diamond-Charged Mandrel Speeds 488 • Material Allowance for Grinding 488 • Universal Measuring Machines and Accessories 489 • Safe Practices 490 • Review and Self-Test 491

Part 12   NUMERICAL CONTROL AND COMPUTER-ASSISTED (CNC) MACHINE TOOLS: TECHNOLOGY AND PROCESSES

Section One   NUMERICAL CONTROL MACHINE TOOLS

UNIT 58   NC SYSTEMS, PRINCIPLES, AND PROGRAMMING **492**

Application of NC to Standard Machine Tools 494 • NC Control and Rectangular Coordinate Systems 495 • Binary System Input to NC 497 • Numerical Control Word Language 497 • Positioning the Spindle 499 • NC Tape Programming 500 • TAB Sequential Format Programs 501 • *How to Program a Single-Axis, Single-Machining Process 502* • *How to Program for Two Axes and Tool Changes 503* • *How to Program for Milling Processes Involving Feed/Speed Changes 504* • Word Address and Fixed Block Format Programs 505 • Mirror Image Programming 505 • Safe Practices 506 • Review and Self-Test 506

Section Two   NC APPLICATIONS AND COMPUTER-ASSISTED PROGRAMMING (CAP) FOR CNC MACHINE TOOLS

UNIT 59   BASIC NC AND CNC TOOLS AND MACHINING PROCESSES **508**

NC Data Processing 512 • Computer-Assisted NC Programming 512 • Functions of the Machine Control Unit (MCU) 512 • Computer-Assisted Programming (CAP) 513 • *How to Prepare an APT Program Manuscript 515* • Components and Features of a CNC System 516 • Expanding Productivity/Flexibility of a CNC System 517 • Presetting NC Cutting Tools 518 • Safe Practices 518 • Review and Self-Test 518

Part 13   PRODUCTION OF INDUSTRIAL AND CUTTING TOOL MATERIALS

Section One   STRUCTURE AND CLASSIFICATION OF INDUSTRIAL MATERIALS

UNIT 60   MANUFACTURE, PROPERTIES, AND CLASSIFICATION OF METALS AND ALLOYS . .   **520**

Steel Manufacturing Processes 520 • Chemical Composition of Metals 522 • Mechanical Properties 523 • Classification of Steels 523 • SAE and AISI Systems of Classifying Steels 524 • Aluminum Association (AA) Designation System 526 • Visible Identification of Steels 527 • Characteristics of Iron Castings 527 • Steel Castings 529 • Groups of Nonferrous Alloys 530 • Applications of Industrial Materials 531 • Safe Practices 533 • Review and Self-Test 533

Part 14   BASIC METALLURGY, HEAT TREATMENT, AND HARDNESS TESTING

Section One   METALLURGY AND HEAT TREATING: TECHNOLOGY AND PROCESSES

UNIT 61   HEAT TREATING AND METALS TECHNOLOGY **534**

Application of Iron-Carbon Phase Diagram Information 537 • Liquid Baths for Heating 540 • Quenching Baths for Cooling 540 • Quenching Bath Conditions Affecting Hardening 541 • Interrupted Quenching Processes 541 • Heat-Treating Equipment 542 • Hardening and Tempering Carbon Tool Steels 543 • Heat Treatment of Tungsten High-Speed Steels 545 • Annealing, Normalizing, and Spheroidizing Metals 545 • Casehardening 546 • Special Surface Hardening 548 • Subzero Treatment of Steel 550 • *How to Stabilize Gages and Other Precision Parts 550* • Safe Practices 550 • Review and Self-Test 551

UNIT 62   HARDNESS TESTING. **552**

How to Test for Metal Hardness by Using a Rockwell Hardness Tester 553 • Brinell 554; Vickers, Scleroscope, and Microhardness Testing 556; and Knoop Hardness Testing 557 • Safe Practices 557 • Review and Self-Test 558

Part 15   NEW MANUFACTURING PROCESSES AND MACHINE TOOLS

Section One   NONTRADITIONAL TOOLING AND MACHINES: TECHNOLOGY AND PROCESSES

UNIT 63   NONTRADITIONAL MACHINE TOOLS AND MACHINING PROCESSES. **559**

Surface Integrity 561 • Surface Roughness Ranges 561 • Metal Removal Rates 561 • Nontraditional Processes and Machines 561 • Metal Removing Rates and Surface Finish 562 • EDM Accessories 563 • Traveling Wire (Wire-Cut) EDM System 563 • CRT, MDI, and CNC Systems 563 • Electrochemical Machining (ECM) 564 and Deburring (ECD) 565 • Electrochemical Grinding (ECG) 565 • Chemical Machining (CHM) 565 • Ultrasonic Machining (USM) 565 • Plasma Beam (Arc) Machining (PBM) 566 • Abrasive-Jet Machining (AJM) 566 • Electron Beam Machining (EBM) 566 • Laser Beam Machining (LBM) 567 • Safe Practices 568 • Review and Self-Test 569

Part 16   FLEXIBLE MANUFACTURING SYSTEMS (FMS) AND ROBOTICS

Section One   AUTOMATED FLEXIBLE MULTIMACHINE MANUFACTURING SYSTEMS

UNIT 64   FLEXIBLE MANUFACTURING SYSTEMS (FMS) AND ROBOTICS. **570**

A. Subsystems of Highly-Automated Manufacturing Systems   571

CAD/CADD, CIM/CAM, and CIM/GEN Components 570 • Graphic Display (571) and Functions of Automated Subsystems in Flexible Manufacturing Systems (FMS) 572 • Materials Handling, Tool-Handling and In-Process and Post-Process Gaging Systems 572 • Precision Surface Sensing Probes 572 • Automated Feed, Speed, and/or Adaptative Controls Subsystems 573

B. Robotics in Conventional and Multimachine Manufacturing Systems   574

Functions and Application of Computer Controlled Six-Axis Robot 574 • Factors Affecting Selection 574 • Versatility of Robots 575 • CADD Robot Model 575 • Review and Self-Test 576

APPENDIX

Glossary of Technical Terms   578

Handbook Tables   588

Index.   609

HANDBOOK TABLES

A-1   Decimal Equivalents of Fractional, Wire Gage (Number), Letter, and Metric Sizes of Drills. 588

A-2   Conversion of Metric to Inch-Standard Units of Measure 590

A-3   Conversion of Fractional Inch Values to Metric Units of Measure. 590

A-4   Machinability, Hardness, and Tensile Strength of Common Steels and Other Metals and Alloys 591

A-5   Recommended Saw Blade Pitches, Cutting Speeds, and Feeds for Power Hacksawing Ferrous and Nonferrous Metals. 592

A-6   Recommended Drill Point and Lip Clearance Angles for Selected Materials. 592

A-7   Conversion of Cutting Speeds and RPM for All Machining Operations (Metric and Inch-Standard Diameters). 593

A-8   Recommended Cutting Speeds for Drilling with High-Speed Drills 594

A-9   General NC Tap Sizes and Recommended Tap Drills (NC Standard Threads) 595

A-10   General NF Tap Sizes and Recommended Tap Drills (NF Standard Threads) 595

A-11   General Metric Standard Tap Sizes and Recommended Tap Drills (Metric Standard Threads). 595

A-12   Suggested Tapping Speeds (fpm) and Cutting Fluids for Commonly Used Taps and Materials. 596

A-13   Recommended Cutting Speeds for Reaming Common Materials (High-Speed Steel Machine Reamers) 596

A-14   Recommended Cutting Speeds (in sfpm) for Carbide and Ceramic Cutting Tools. . . . 596

A-15   Recommended Cutting Speeds and Feeds (in sfpm) for Selected Materials and Lathe Processes (High-Speed Steel Cutting Tools and Knurls) 597

A-16   Recommended Cutting Speeds and Feeds for Various Depths of Cut on Common Metals (Single-Point Carbide Cutting Tools) 597

A-17   Average Cutting Tool Angles and Cutting Speeds for Single-Point, High-Speed Steel Cutting Tools. 598

A-18   Recommended Cutting Speeds (sfpm) of Common Cutter Materials for Milling Selected Metals 598

A-19   Recommended Cutting Fluids for Ferrous and Nonferrous Metals 599

A-20   Natural Trigonometric Functions (1° to 90°) 600

A-21   Tapers and Included Angles. 601

A-22   Conversion Tables for Optical Flat Fringe Bands in Inch and SI Metric Standard Units of Measure. 601

A-23   Microinch and Micrometer (μm) Ranges of Surface Roughness for Selected Manufacturing Processes 602

A-24   Constants for Setting a 5″ Sine Bar (0°1′ to 10°60′). 603

A-25   Suggested Starting Speeds and Feeds for High-Speed Steel End Mill Applications. 604

A-26   Allowances and Tolerances on Reamed Holes for General Classes of Fits 605

A-27   ANSI Spur Gear Rules and Formulas for Required Features of 20° and 25° Full-Depth Involute Tooth Forms 606

A-28   Conversion of Surface Speeds (sfpm) to Spindle Speeds (RPM) for Various Diameters of Grinding Wheels (1″ to 20″ or 25.4mm to 508.0mm) 607

A-29   Recommended Heat Treatment Temperatures for Selected Grades and Kinds of Steel 607

ACKNOWLEDGEMENTS

The most gratifying experience in writing this book was the generous, gracious assistance of many individuals and an equally great number of industries. Help was provided throughout successive stages of development, production, and manufacturing by teachers, supervisors, curriculum specialists, and directors. These lead people were active in secondary, post-secondary, and college-level institutions, and instructional materials centers. Counterpart shop foremen, industrial management leaders and trainers, and training directors assisted with in-plant analyses of occupational qualifications and training demands and technical resource materials.

Special recognition is made of the overwhelming support of industry. Industry provided the state-of-the-art know how about machine tools, accessories, tooling and setups, materials, inspection and testing systems, and standard and automated manufacturing processes. Technical information was provided about basic and highly sophisticated technology and processes using conventional and nontraditional machine tools. The wealth of line art and photos supplied by industry is testimony of their concern for developing highly-capable craftspersons and technicians.

A courtesy line appears above each illustration to acknowledge the company source. In addition, a personal "thank you" is expressed to each individual who provided technical assistance from the following companies.

Acco Industries, Inc., Measurement Systems Division
Acme-Cleveland Corporation
AGF Inc. (formerly American Gas Furnace Company)
American Drill Bushing Company
American Iron and Steel Institute
American Precision Museum Association, Inc.
American Tool, Inc.
American Tool & Cutter Company, Inc.
B.C. Ames Company
Apex Tool & Cutter Company, Inc.
Armstrong Bros. Tool Company
Barber-Colman Company, Machines and Tools Division
Bausch & Lomb, Inc.
Bay State Abrasives, Dresser Industries, Inc.
Beech Aircraft Corporation
Bendix Corporation, Automation and Measurement Division
————, Industrial Controls Division
Blanchette Tool & Gage Manufacturing Corporation
Boston Gear, Incom International, Inc.

Bridgeport Machines, Division of Textron, Inc.
Brown & Sharpe Manufacturing Company
Buck Supreme, Inc.
Burgmaster Division, Houdaille Industries, Inc.
Cabot Corporation
Cincinnati Inc.
Cincinnati Milacron Inc.
Clausing Machine Tools Corporation
Cleveland Twist Drill, an Acme-Cleveland Company
Coherent General
Colt Industries, Elox Division
Control Laser Corporation
(The) Cooper Group, Nicholson File Division
Cushman Industries, Inc.
(The) Desmond-Stephan Manufacturing Company
DeVlieg Machine Company, Microbore Division
DoALL Company
Dorsey Gage Company, Inc.
Dunham Tool Company, Inc.

Engis Corporation, Diamond Tool Division
Enco Manufacturing Company
Ex-Cell-O Corporation
Federal Products Corporation
Fellows Corporation
General Electric Company, Carboloy Systems Department
Grinding Wheel Institute
Hammond Machinery, Inc.
Hardinge Brothers, Inc.
Harig Products, Bridgeport Machines, Division of Textron, Inc.
Hitachi Magna - Lock Corporation
Illinois/Eclipse: A Division of Illinois Tool Works, Inc.
Industrial Plastics, Inc.
(The) Jacobs Manufacturing Company
J & S Tool Company, Inc.
Johnson Gas Appliance Company
Jones & Lamson Metrology Products, Waterbury Farrel Division of Textron, Inc.
Kasto-Racine, Inc.
Kearney & Trecker Corporation
Kennametal, Inc.
Landis Tool Company
LeBlond Machine Tool Company
Linde Division, Union Carbide Corporation
Lodge & Shipley Company
Mahr Gage Company, Inc.
McDonnell Douglas Manufacturing Industry Systems Company
Moore Special Tool Company, Inc.
Morse Cutting Tools Division, Gulf + Western Manufacturing Company
National Acme, an Acme Cleveland Company
National Broach & Machine, Division of Lear Siegler, Inc.

National Twist Drill, Division of Lear Siegler, Inc.
W.H. Nichols Company
Norton Company, Abrasives Marketing Group and Knoebel Diamond Tool Division
Pratt & Whitney Machine Tool Division, Colt Industries
Rockford Machine Tool Operations, Ex-Cell-O Corporation
Science Museum of England
Shore Instrument & Manufacturing Company
Simmons Machine Tool Corporation
South Bend Lathe Company, Inc.
Standard Tool Division, Lear Siegler, Inc.
(The) L.S. Starrett Company
System 3 R U.S.A., Inc.
TRW/Geometric Tool Division
TRW/Greenfield Tap & Die Company
TRW/J.H. Williams Division
Taft-Pierce Manufacturing Company
Thermolyne Corporation
Thompson Grinder Company, Waterbury Farrel Division of Textron, Inc.
Tinius Olsen Testing Machine Company, Inc.
United States Steel Corporation
Universal Houdaille
Universal Vise & Tool Company, Schwartz Fixture Division
USM Machinery Division, Emhart Corporation
VR/Wesson Division of Fansteel
(The) Walton Company
Warner & Swasey Company
Waterbury Farrel, Division of Textron, Inc.
White-Sundstrand Machine Tool Company

Recognition is made of Hilda G. Olivo for excellent, thorough editorial and research services. Jim Linville, Editor-Technical Publications, National Fire Protection Association, is recognized for his early assessment of the manuscript and professional assistance.

Credit is given to Jean LeMorta of Thoroughbred Graphics for her expertise in copy planning and unusual typesetting services. Special commendation is made to Judi S. Orozco of The Drawing Board. Layout and design features, certain line art, specifications writing, and outstanding work in preparing camera-ready copy for production reflect her contribution. The cover design is the work of John R. Orozco. Finally, thanks is expressed to Frederick J. Sharer, Director of Manufacturing, Delmar Publishers, for expert technical advice on development and production items.

The quality and contribution of this end product is a tribute to all who gave so generously and graciously.

C. Thomas Olivo

PART 1 Machine Tool Technology, Machining, and Manufacturing Occupations

SECTION ONE

Introduction

UNIT 1

Historical Overview and Career Opportunities

A. IMPACT OF EARLY MACHINE TOOLS AND CONCEPT OF MASS PRODUCTION

This book relates to the "state-of-the-art" of machine tools, tooling, machining, and mass production processes. The contents provide what Edward A. Battison (Director of the American Precision Museum in Windsor, Vermont) identifies as. . ."the story of sustained, incremental contributions. . .by generations of individuals who stand out for their contributions to the American System of Manufacturing." This endless line of contributors had advanced design concepts beyond Leonardo da Vinci's early drawings of machine elements and mechanical movements to sophisticated, computerized and roboticized equipment.

In "this endless line," it is David Wilkinson of Pawtucket, Rhode Island who provides the reference point to modern machine tool technology and manufacturing processes. Wilkinson applied a carriage he had invented in 1794 to a lathe which he patented in 1798. This lathe advanced the concept of using three bearing points on the underside of a slide rest (carriage). The *main guide* or sliding surface became known as

the *front way* on a lathe bed. This *way* provided a precision guide for two bearing points. The *rear guide* (way) served as a mating surface for the third bearing area of the carriage. Figure 1–1 shows this early design feature.

THE LATHE AS THE DOMINANT MACHINE TOOL

The main guides of the Wilkinson lathe were formed by chisel and file! Precision in turning was achieved by "trial-and-fit" filing to "continue" the rear guide to compensate for variations with the front guide. Other features covered by the Wilkinson patent drawing include a lead screw mounted on centers to permit changeover to a different pitch; a split-nut to move the carriage; and a fixed center in an adjustable tailstock.

From these simple beginnings, the lathe became the dominant machine tool in the manufacturing revolution which followed. In this almost 200 year span, machine elements which were once filed are now machined and held to surface qualities which are measured in microinches (μin.), or micrometers (μm) in SI metric.

Figure 1–1 Wilkinson application of three bearing surfaces on a carriage at right angles to the bed ways

Tolerances in dimensional, parallelism, concentricity, runout, and other measurements are held to within a thousandth, ten-thousandth, or millionth part of an inch, and equivalent metric measurements.

"INCREMENTAL CONTRIBUTIONS" BY SUCCEEDING GENERATIONS

A parallel lathe development by Henry Maudslay in England incorporated and improved on Wilkinson's major design elements. The same features displayed in Maudslay's early 1800 lathe (Figure 1–2) are still basic components in modern computer-programmed, automated lathes and other turning machines.

In 1845, Stephen Fitch of Middlefield, Connecticut built an "eight station" turret lathe. The turret held and positioned multiple tooling setups. Then, in 1860 Christopher Spencer added a "brain wheel." This feature provided a cam-activated device for controlling the movements of cutting tools.

Following in the path of measuring instruments, Joseph R. Brown applied the vernier principle to a slide caliper. The new instrument enhanced the capability to measure and control dimensions to ±0.001″ and ±0.0001″ limits of accuracy.

Each new instrument and machine tool added to the capability of its predecessor and spurred still other machine tool developments for drilling

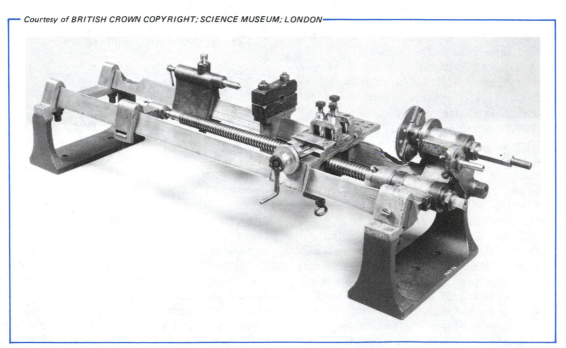

Figure 1–2 Maudslay Lathe of 1800

and hole forming, turning, milling, contouring, boring, grinding, stamping, and other manufacturing processes.

INTERCHANGEABLE PARTS IN MANUFACTURING

Along with machine tool and instrument developments, a concept of using machinery for the manufacture of interchangeable parts emerged. The first successful application of this concept is credited to Simeon North of Middletown, Connecticut. North demonstrated (through on-site examination by government inspectors of his plant, equipment, and manufacturing processes, and the actual fulfillment of a military contract) that it was economically feasible to mass produce interchangeable parts. North used existing machine tools and designed others to replace costly (and comparatively inaccurate) hand processes. North made significant design changes on the milling machine, making it an equally "dominant" production machine to the lathe.

Over the years, the commercial turret lathe of Henry B. Stone; the manufacturing milling machine of Thomas Warner; the profile miller of Frederick Howe; the impact of Thomas Blanchard's cam system for profile turning in woodworking as applied to metalworking machines and processes. . .are examples of. . . "sustained, incremental contributions. . . ." Each generation improved on the work of the preceding generation.

END PRODUCTS OF MANUFACTURING ENTERPRISE

While rooted in the production of guns and military weapons, applications of mass production of interchangeable parts spread rapidly from the mid 1800's to the manufacture of fine parts and mechanisms in watchmaking; business machines like the typewriter; the sewing machine as a household item; and other implements spreading across all economic sectors of society and the world.

NONTRADITIONAL MACHINE TOOLS AND MACHINING

Today's "state-of-the-art" is reflected in the computer-assisted programmed machining center as shown in Figure 1–3 for conventional

Courtesy of DeVLIEG MACHINE COMPANY

Figure 1–3 Machining center with an automatic tool changing system and computer numerical control requires programming by a skilled technician/craftsperson

machining. Concurrently, since 1940 there has emerged a whole new series of nontraditional machine tools and nontraditional material removal processes. These cope with needs for new alloys and materials which are extremely hard and difficult to fabricate and machine.

Nontraditional machining processes and equipment depend on hydrostatic and sound forces; electrical, electronic, and chemical energy; laser (light) beams converted to thermal energy; and other combined energy systems, using nonconventional tools.

B. CAREER OPPORTUNITIES: MACHINE TOOL, MACHINING, AND MANUFACTURING INDUSTRIES OCCUPATIONS

The industrial sector creates wealth through the production of goods which are needed by each person and are essential to the national well-being of all people, and for national security. This creation of wealth by transforming natural resources into useful, needed products requires a labor force with employable skills.

BASES OF JOB QUALIFICATIONS

The qualifications and duties at each job level are established by analyses of tasks, jobs, occupations, and clusters of jobs which require comparable levels of skill and technology competencies. Analyses reveal the sum-total of learning experiences which interlock mathematics; science; drawing, art, design, blueprint reading, and sketching skills, with required shop and laboratory skill competencies.

It is important for each individual who is planning a career to have the desire to prepare for, meet occupational qualifications, and have the capacity to profit by instruction in the occupation, consistent with labor market needs.

PLANNED EARLY LEARNING DECISIONS

The usual beginnings of career preparation in the industrial sector are rooted in planned early learning experiences in schools in industrial arts courses. In dealing with tools, materials, processes, and products, the student develops an *awareness* and receives an *orientation* to occupations. Occupational *exploration* is provided through sampling techniques within generalized groups of industries such as transportation, communications, construction, and manufacturing. Assisted by career guidance, counseling, and self-appraisal measures, these early experiences often lead to the selection of an occupation for which preparation within a secondary or post-secondary program leads to an entry level job.

APPRENTICESHIP TRAINING SYSTEM

One common pathway to job preparation is through an apprenticeship. Apprenticeship means that there is a contract (understanding) between the employer and the individual who is to prepare full-time on-the-job to attain the skills and technology essential for a defined trade or technical occupation. Stated are the conditions of training, time and wage increment schedules, conditions of employment, the assessment of occupational competency, and other responsibilities of each party.

Generally, apprenticeships are four years in duration, based on the skill and technology requirements of the occupation. In-school, military occupational specialty, and other previous training experiences may be credited in partial fulfillment of the requirements. Instruction in related subjects is provided in schools and industry training programs as part of the apprenticeship. This instruction deals with related technical communication skills (blueprint reading and sketching), computational skills (applied mathematics), and related science.

After satisfactorily completing and passing all facets of the program each year, the apprentice advances to the first level of journeyman/journeywoman status. With further experiences and continuous updating, upgrading, and retraining, the skilled worker advances to a second level, then a third level and on to master status as the highest level of employment as a trade craftsperson or technician.

OTHER JOB TRAINING SYSTEMS

Limited numbers of trainees are accepted into apprenticeship. Another common practice used by industry is to accept completers of vocational-industrial-technical programs, military occupational specialty training, and other industry/

training agency programs. In many cases, competencies are validated by the satisfactory passing of theory and/or performance skills of a particular occupation. Such persons enter industry as advanced learners for a specific occupation.

With added years of experience, coupled with updating and upgrading experiences, an individual may achieve the equivalent of an apprenticeship and may advance. Importantly, advancement to higher level jobs requires experiences which take place continuously throughout the life span of each worker.

CLASSIFICATION OF JOB TITLES BY LEVEL AND AREA

The classifications: *beginning, training,* or *entry; intermediate,* and *terminal levels,* are used for identification of *employment levels* (Figure 1–4). Within each of these levels, a number of jobs may be clustered. Each job title in the cluster has common skill and technology requirements.

BASIC JOBS, SUPPLEMENTARY JOBS, AND RELATED JOBS

At each of the three classification levels, there are other job cluster *areas.* For example, in the *basic job area* (in the left column of Figure 1–4), entry into employment and progress vertically is based on preparation for a specific occupation or the jobs within the cluster. An increasingly higher order of skill and technology is required to advance vertically.

Supplementary area jobs, which are identified in the middle column, range throughout the entry, intermediate, and terminal levels. "Supplementary" means that thorough preparation in one occupational field, coupled with additional training, makes it possible to move laterally on a *career ladder* to other jobs that partly depend on the initial skills in the basic area.

Related area jobs within a career ladder are those clusters of jobs that spin off from either basic area or supplementary area jobs and require still further preparation for what may be considered complementary occupations.

In summary, Figure 1–4 shows the structure for classifying jobs at the three levels for each of the three areas. If the nine blocks were filled in with job titles in machine tool technology,

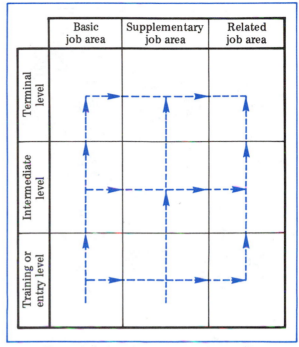

Figure 1–4 Job clusters grouped according to employment levels and areas

machining, and manufacturing, well over 500 different job titles may be included within the career ladder Descriptions, requirements, and other technical information about job titles are provided in standard industrial occupation classification handbooks.

It should be noted that unskilled workers, helpers, single-purpose operators, and low-level skill jobs are not included. These require minimal skills which are usually developed in a short time and on-the-job.

EXAMPLES OF JOB TITLES AT EACH EMPLOYMENT LEVEL

Training, Learner, or *Entry Level.* Job titles at this level relate to *semiskilled* craft and technical workers, operatives, and trainees (including apprentices).

Examples of job titles include: machinist apprentice, tool and die maker apprentice; automatic screw machine operator, milling machine operator; materials tester, production heat treater; machine components assembler, parts inspector; technical office personnel.

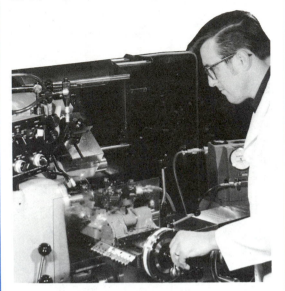

Figure 1–5 Instrument maker operating a Super-Precision® HC-AT chucker and bar machine

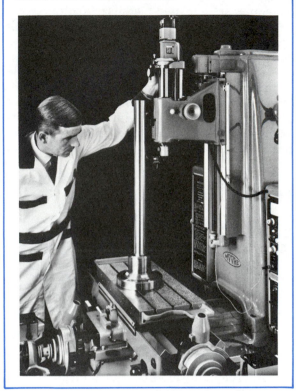

Figure 1–6 Quality control inspector checking the column squareness of a precision machine tool using a cylinder square master

Intermediate Level. This category includes job clusters for *skilled* craftspersons, *semiprofessional* technicians and support persons for engineers; sales and marketing representatives; middle management supervisors; industrial trainers.

Examples of job titles include: machinist (grades I, II, III, IV), toolmaker (grades I, II, III) instrument maker (Figure 1–5); metallurgist; CNC, CAD, CAM, CIM system programmer; machine designer, tool designer; parts and gage inspector grade III (Figure 1–6); production set-up mechanic; heavy-duty machine tool operator specialist; all-round heat treater, laboratory technician; production supervisor, in-plant training instructor; sales operations personnel, marketing analyst, technical operations office manager.

Terminal Level. This category includes job titles which depend on the highest levels of decision making and responsibility. Terminal level is the "top level" for manager, director, engineer, researcher, and scientist positions.

Examples of job titles include: mechanical, electronic, metallurgical, tool, industrial, and production engineers; plant manager, plant supervisor; systemwide training director; overall

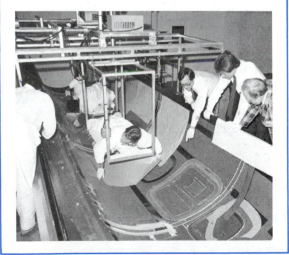

Figure 1–7 Starship component involving a team of workers from three different job levels and responsibilities

computer systems planner or administrator; research and scientific laboratory coordinator.

The relationship of team workers from different levels of job titles and responsibilities is shown in Figure 1–7. Two *CADAM engineers* are checking specifications with a *department chairman* (in sports coat) on a new Starship component. The three *composites technicians* are laying-up epoxy resin-impregnated materials to form the fuselage skin of a new composite airplane.

JOB MOBILITY AND ADVANCEMENT

Job mobility and advancement to intermediate and terminal level positions in the basic, supplementary, and related areas are generally based on the following four groups of qualifications.

- Additional years of successively higher levels of experience and responsibility.
- Continuous updating, upgrading, and extension of skill and technology competencies on-the-job, complemented by special institutes and other training.
- Continuous college level professional experiences (where extablished by occupational analyses). These may be obtained through formalized institutional programs or equivalent advanced degree work through industrial on-the-job professional programs.
- Personal qualities of leadership in group working relationships.

UNIT 1 REVIEW AND SELF TEST

A. IMPACT OF EARLY MACHINE TOOLS AND CONCEPT OF MASS PRODUCTION

1. Explain briefly (a) what the main design features were of David Wilkinson's carriage and (b) the impact it had on lathes becoming the dominant machine tool.

2. Identify four basic inspection and measurement tests which are made today in checking the accuracy of machine tools.

3. State two important outcomes on manufacturing which were brought about by refinements in measuring instruments.

4. Interpret what is meant by "producing interchangeable parts in manufacturing."

5. Indicate what effect the parallel developments of machine tools and the manufacture of interchangeable parts had on the growth of the nation.

6. Name two distinct characteristics or uses of nontraditional machine tools.

B. CAREER OPPORTUNITIES: MACHINE TOOL, MACHINING, AND MANUFACTURING INDUSTRIES OCCUPATIONS

1. Interpret briefly what "job clusters" means.

2. Name three major levels and three area groupings which form the matrix of a career ladder.

3. Identify a sequence of planned early in-school experiences which are designed to guide an individual to select an occupational preparation goal.

4. a. Cite four pathways leading to preparation at an entry job level.
   b. Identify a contract plan between an individual and an employer which, upon satisfactory completion, provides mobility to serve as a beginning-level skilled craftsperson.
   c. State three conditions of the contract.

5. Explain the effect on employment of the following two conditions.
   a. Experiences at successively higher skill and technology levels.
   b. Continuous occupational preparation throughout the life span of the individual.

6. a. Layout and label the framework of a career ladder.
   b. Insert at least two appropriate job titles in each of the job level and job area categories.
   c. Contrast (briefly) the difference in job preparation requirements between the entry level jobs in the basic job area and the terminal level jobs in the supplementary job area.

7. State two sets of qualifications which are required for job mobility and advancement to a higher level of employment.

PART 2 Measurement: Technology and Processes

Measurement and Measurement Systems

Accurate interpretation and measurement of dimensions are skills every individual involved in machine technology must have. A thorough understanding of dimensions, units of measure, measurement systems, and measurement terms is necessary for proficiency in machine technology. This section presents the basic principles and practices that are fundamental to accurate dimensional measurement.

UNIT 2

Measurement: Customary and SI Metric Systems

DIMENSIONAL MEASUREMENTS

Each measurement is written or expressed as a *dimension*. Whoever designs an object uses a dimension to describe or record a distance or a particular feature. *Dimensional measurement* relates to straight and curved lines, angles, areas, and volumes. Plane surfaces, as well as regular- and irregular-shaped objects, are included. Dimensional measurements, therefore, are the foundation of machine technology and of all industry, business, and commerce.

Dimensional measurements are necessary in designing, building, assembling, operating, and maintaining all parts and mechanisms. Some measurements are computed; others are taken directly with direct-reading measuring rules and other measuring instruments. The degree of accuracy establishes the precision requirements of the measuring tool or instrument to be used.

There are three prime reasons for dimensional measurement:

- To describe a physical object;
- To construct a physical object. The object may be a single part, a component in a mechanism, or a complete unit of movable and stationary parts;
- To control the way in which an object is produced by many different people. Controlled measurements are necessary in the manufacture of interchangeable parts.

CHARACTERISTICS OF MEASUREMENT

The most widely used dimensional measurement is *length*, or linear distance. The term used is *linear measurement*. It means that a distance separating two points may be reproduced. Each linear measurement has two parts; a *multiplier*

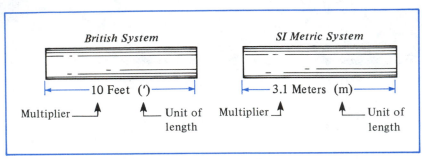

Figure 2–1 Parts of a Linear Measurement

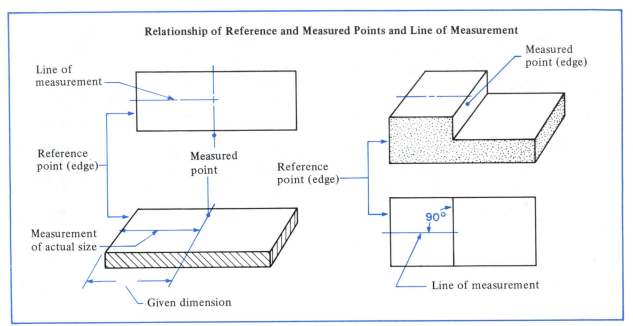

Figure 2–2 Bases of Linear Measurements

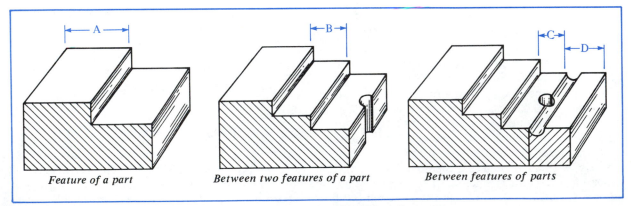

Figure 2–3 Dimensions and Workpiece Features

and the *unit of length* (Figure 2–1). The multiplier may be a whole number, a fractional or decimal value, or a combination of the two.

DIMENSIONS, FEATURES, AND MEASUREMENTS

Every line of measurement must have *direction*. The bases of linear measurements are pictured in Figure 2–2. The measurement of an actual dimension begins at a *reference point*. The measurement terminates (ends) at the *measured point*. These two points lie along the *line of measurement*. The two points are also called *references*. Two such points on an object or part define what are called *features*. The features represent the physical characteristics of the part.

The notes on drawings, as well as lines that are used for laying out or describing features, are bounded by other lines, edges, ends, or plane surfaces. Three examples of features of different parts are given in Figure 2–3. Features are also bounded by areas. Two shop terms that are used to identify such features are *male* and *female*. These features appear in Figure 2–4. Measurements are taken from an end or an inside or an outside edge. Most measurement lines are edges that are formed by the intersection of two planes. Measurements are indicated on drawings as dimensions (A, B, C, and D in Figure 2–3). A *dimension* really states the designer's concept of the exact size of an intended feature. By contrast, a *measurement* relates to the actual size.

Measurements are affected by temperature. Thus, particularly with precise work, a feature may measure larger or smaller than a required dimension. Measurement accuracy depends on the temperature of both the measuring instrument and the workpiece.

PRECISION, TOLERANCES, AND ACCURACY

Three other dimension-related terms must be discussed at this point. *Precision* refers to the degree of exactness. For instance, a part may be dimensioned on a drawing to indicate a precision with *limits* of ±0.001″ or ±0.02 mm. The actual measured dimension, if it *falls within* these limits, known as *tolerances*, may vary by these limits from the size as stated on the drawing. Tolerances thus give the fineness to the range of sizes of a feature. Tolerances are needed to indicate maximum and minimum sizes of parts that will fit and function as designed.

Accuracy deals with the number of measurements that conform to a standard. This number is in comparison to those that vary from the standard. However, in common shop and laboratory practice, the terms *precision* and *accuracy* are used interchangeably.

MEASUREMENT AND INDUSTRIAL DRAWINGS

An object may be described by spoken words or with a photograph or some other form of picture. The sketch of the die block in Figure 2–5 is such a picture. Spoken words, photographs, and simple sketches do not, however, reveal:

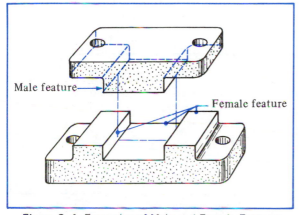

Figure 2–4 Examples of Male and Female Features

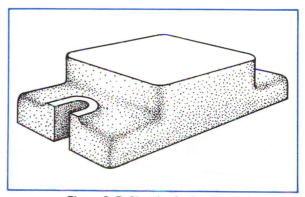

Figure 2–5 Sketch of a Die Block

- Exact dimensions,
- Cutaway sections that show internal details,
- Required machining,
- Additional work processes.

The craftsperson needs to rely on precise technical information. Shape, dimensions, and other important specifications must be furnished on a detailed drawing. Such a drawing conveys identical information to many different people. Some people are involved in the initial design stage; others deal with the same part or mechanism in a final inspection, assembly, or testing stage.

Designers, engineers, technicians, and craftspersons all may work in the same plant or may be separated in many plants throughout the world. They may not speak the same language, except technically. The *industrial drawing* is their universal language. Measurement is an important part of the universal language of an industrial drawing. Measurement is required in the production of every part and device.

INDUSTRIAL DRAWINGS AND BLUEPRINT READING

An *original drawing* (referred to as a *master drawing*) is usually produced with the aid of drafting instruments. While drawings of simple parts sometimes are sketched freehand, other drawings of more complex parts are made on automated drafting machines. Statistical and design data provide input for a computer. The computer then programs and controls automated drafting equipment, and a master drawing may be produced.

Complicated shapes and features may also be drawn directly on a workpiece. The layouts guide the worker in performing hand or machine processes and assembly work.

Original drawings are usually filed as a record. Subsequent design changes are made on these drawings. Reproductions are generally made of original drawings for use in the shop and laboratory and are referred to as *prints of the original*. While few of these prints are *blueprints*, the name still is generally applied among craftspersons.

By developing an ability to *read a blueprint* or sketch, the craftsperson is able to:

- Form a mental picture of a part,
- Visualize and perform different production and assembly processes,

- Make additional drawings that may be needed to machine a part,
- Take necessary measurements.

Reading a print requires an understanding of basic principles of graphic representation and dimensioning. Combined into what is called *drafting*, the principles include:

- The alphabet of lines,
- Views that provide exact information about shape,
- Dimensions and notes (these principles relate to measurements and accuracy),
- Cutting planes and sections (these principles show cross-sectional details of special areas of a workpiece),
- Shop sketching techniques.

THREE BASIC SYSTEMS OF MEASUREMENT

The most important measurement to be considered at this time is the unit of *length* (see Table 2–1). Units of measure for length in the British Imperial System are considered first. Units in the decimal-inch and metric systems follow.

THE BRITISH (U.S.) SYSTEM OF LINEAR MEASUREMENT

The inch is the most widely used unit of linear measure in the *British Imperial System*. The inch is subdivided into a number of equal parts. Each part represents a common fraction or a decimal fraction equivalent. These parts are shown in Figure 2–6.

Inch Common Fractions. The fractions most commonly used on steel rules represent halves,

Table 2–1 Inch-Standard Linear Units of Measure

British Imperial System
Smallest Unit of Measure = International Inch (″)
1 foot (′) = 12″
1 yard (yd) = 3′
1 rod = 5 1/2 yds
1 mile (mi) = 1,760 yds
1 mile (mi) = 5,280 ft
SI Metrics
25.4 millimeters (mm) = 1″

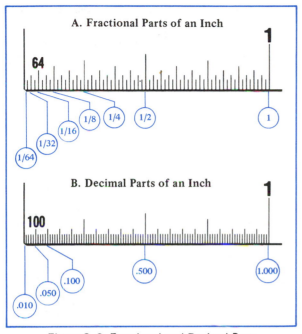

Figure 2–6 Fractional and Decimal Parts
of an Inch on Steel Rules

quarters, eighths, sixteenths, thirty-seconds, and sixty-fourths of an inch. The enlarged illustration in Figure 2–6A shows how these fractions are related.

Inch Decimal Fractions. The inch is also divided into tenths, fiftieths, and hundredths of an inch. In the shop and laboratory, the parts are called *decimal fractions*, or *decimal divisions* (Figure 2–6B). Decimals are applied to both British and metric system basic units of measurement. Fractional divisions that are smaller than 1/64″ or 1/100″ are not practical to measure with a rule.

THE DECIMAL-INCH SYSTEM

In the late 1920s the Ford Motor Company promoted a *decimal-inch system* (Figure 2–6B). This system was based on the decimal division of the inch. Dimensions were given on drawings with subdivisions of the inch in multiples of ten. Parts were measured according to the decimal-inch system. The term *mil* was coined to express *thousandths*.

THE METRIC SYSTEM

The units of linear measure in the metric system are similar to units in the decimal-inch system. The difference is in the use of the *meter* instead of the inch. The meter (m) is the standard unit of measure in the metric system. For all practical shop purposes, the *meter is equal to 39.37″*. All other units in the metric system are multiples of the meter. The centimeter equals 0.01m or 0.3937″. The millimeter equals 0.001m or 0.03937″.

LIMITATIONS OF THE INCH AND METRIC MEASUREMENT SYSTEMS

The metric system has the advantage over the British inch system for ease of computation and communication. However, the inch system provides fractional parts of the basic units that are more functional. For example, in the inch system there is a significant range of finely spaced divisions that can be read on steel rules graduated in inches and fractional parts. Readings of 1/64″ and 1/100″ are common. Micrometer measurements in the British system may be taken with a wide range of discrimination: one one-thousandth and one ten-thousandth part of the inch.

By contrast, the millimeter (milli″ = 1/1,000; meter = 39.37″) is 1/1,000 of 39.37″, or almost 0.040″. The next smaller unit in the metric system, 0.1mm, is one-tenth this size, or about 0.004″. It is almost impossible to graduate or read such fractional measurements on a steel rule. Similarly, for micrometer measurements this same 0.1mm (0.004″) is too inaccurate for most shop measurements.

The next submultiple in SI metrics is 0.01mm (the equivalent of almost 0.0004″). A 0.01mm measurement may require a closer tolerance and more expensive machining than is necessary. In practice, metric measurements are rounded off to the nearest whole or half millimeter value. This rounding-off produces larger measurement errors than the inch system does.

COMMON LINEAR MEASURING TOOLS

Linear measurements may be taken with accuracies of 1/64″, 1/100″, and 1/2mm using steel rules. Micrometers, vernier tools, and dial indicators are used on precision measurements.

These measurements are held to one one-thousandth (0.001) and one ten-thousandth (0.0001) part of the inch, or 0.02mm or 0.002mm (almost 0.001″ and 0.0001″), respectively.

Gage blocks provide accuracies in measuring to within 1/100,000th and 1/2,000,000th part of an inch. Similar accuracies in metric measurements may be made with metric gage blocks. Comparator instruments of high amplification and other measuring devices are used to measure to still finer degrees of accuracy.

CONVERSION TABLES

Most trade handbooks contain conversion tables for ready use by skilled craftspersons and technicians. Practical daily applications require the changing of fractional and metric values to decimal equivalents, and vice versa. Table 2–2 shows selected examples of the kind of information that is provided by conversion tables.

Conversion tables with accuracies of measurements rounded to five or six decimal places are widely used in shops and laboratories. Expanded tables now include fractional, decimal, and metric unit equivalents.

Table 2–2 Partial Conversion Table of Standard Measurement Units

Millimeter	Decimal Inch	Fractional Inch	Equivalent Decimal Inch
0.1			0.00394
0.2			0.00787
	0.01		0.01000
0.3			0.01181
0.397		1/64	0.01563
0.4			0.01575
0.5			0.01968
	0.02		0.02000
0.6			0.02362
0.7			0.02756
	0.03		0.03000
0.794		1/32	0.03125
0.8			0.03149
0.9			0.03543
1.0			0.03937
	0.04		0.04000

UNIT 2 REVIEW AND SELF-TEST

1. a. Cite two examples of direct measurements.
   b. Give two other examples of computed measurements.

2. State three major functions of dimensional measurements.

3. Describe briefly the relationship between producing interchangeable parts and dimensional measurements.

4. Tell how each of the following measurement characteristics relates to dimensional measurement: (a) features, (b) reference point, and (c) measured point.

5. Differentiate between the use of the following measurement terms: (a) precision, (b) tolerance, (c) limits, and (d) accuracy.

6. Give three reasons why a craftsperson must be able to read and interpret industrial drawings.

7. Identify the smallest practical unit of length that it is possible to measure with a steel rule in (a) the British (U.S.), (b) decimal-inch, and (c) metric systems.

8. List three instruments or other measuring tools that are used for layout and measuring to accuracies of 0.000001″ (0.000025mm).

Measurement and Layout Tools and Instruments

The accuracy with which metal parts are shaped and dimensioned depends, in part, on the craftsperson's ability to correctly manipulate and read a variety of semi-precision and precision measuring instruments. This section presents the principles of and correct practices for using semiprecision line-graduated measuring instruments and related transfer measuring tools.

UNIT 3

Dimensional Measurement: Instruments and Layout Tools

A. LINE-GRADUATED RULES

The terms *scale* and *rule* are commonly used in industry. Their difference may be explained by comparing an architectural drafting scale with a machinist's rule. Different divisions (graduations) on the drafting scale may represent different linear measurements. For example, a drawing may show a scale of 1″ = 1′ or 1 cm = 1m. By contrast, the graduations on the rule represent actual measurements. The graduations may be in inches and common or decimal fractions. They may also be in millimeters and fractional parts of millimeters in the metric system.

DISCRIMINATION AND MEASUREMENT SCALES

The terms *scale* and *graduation* are used interchangeably in the shop. There are usually four sets of graduations (scales) on a steel rule, one on each edge, front and back. These scales are 1/64″, 1/32″, 1/16″, and 1/8″. The scale to use depends on the *discrimination* that is required—that is, the degree to which a unit of length is divided and the accuracy of the tool for taking measurements. Figure 3–1 illustrates the discrimination among international inch, decimal-inch, and metric type rules.

For instance, when an inch is divided into 64 equal parts, each division represents 1/64 of an inch. On a metric rule, the centimeter (10mm) is divided into millimeters and half millimeters. The finest reading that can be taken on this rule has a discrimination of one-half millimeter. Thus, common metric rules have a coarser (less accurate) reading than the 1/64″ scale in the inch system.

ALIGNMENT OF GRADUATIONS

The type of rule and the scale to use also depend on the nature of the measurement and the degree of precision required.

The reference points that were described earlier are important in measurement. Measurements with a steel rule must fall along the line of measurement. Either the end of the steel rule or a particular graduation must be aligned with a reference point or edge. *Aligned* means that the

15

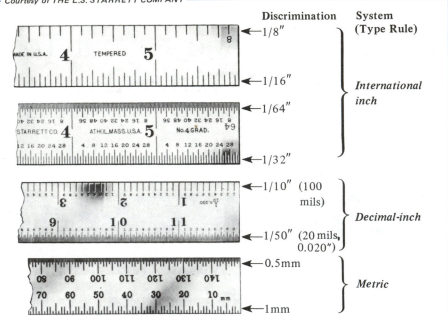

Courtesy of THE L.S. STARRETT COMPANY

Figure 3–1 Discriminations of Three Common Types of Steel Rules

edge or a graduation on a rule falls in the same plane as the measurement reference point or edge. A measurement is then read at the *measured point*.

FLEXIBLE STEEL TAPES

Steel tapes (Figure 3–2) are another form of line-graduated measuring rule. Steel tapes are more practical for making long linear measurements because of their flexibility.

Most *pocket size tapes* are designed with a square head of a specified size. The design permits the tape to be used to measure to a measuring point in a confined area.

When using steel tapes for precision measurements, a flexible tape must be kept on and flush against the line of measurement (Figure 3–2). Compensation must also be made for expansion and contraction caused by differences in temperature.

ERRORS IN READING MEASUREMENTS

Measurements should be taken as accurately as possible using the *best scale*—that is, the scale selected to give the most accurate reading

depending on the requirements of the job. For example, a rule graduated in 20 mils, or 1/64″, or 1/2mm is intended for measurements to the selected degree of accuracy. Errors in reading measurements are caused by (1) inaccurate observation, (2) improper manipulation, and (3) worker bias.

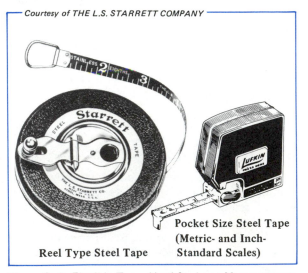

Courtesy of THE L.S. STARRETT COMPANY

Reel Type Steel Tape

Pocket Size Steel Tape (Metric- and Inch-Standard Scales)

Figure 3–2 Flexible Tapes Used for Long Measurements

OBSERVATIONAL ERRORS

Whenever possible, a steel rule should be held at right angles to the workpiece and flush along the line of measurement (Figure 3–3A). The position of the observer, then, will not influence the reading and cause *observational errors*.

Parallax Errors. *Parallax* is a form of observational error that relates to the misreading of a measurement. The misreading is caused by the position or angle from which the graduated division is read in relation to the measuring point. The preferred accurate method of measuring a feature is shown in Figure 3–3A. The parallax conditions, as shown in Figure 3–3B, affects the reading and causes an observational error.

Errors in Reading Graduations. Fine graduations, particularly 1/64″ and 1/100″, are another possible source of observational error. In matching such fine graduations with a reference point on a workpiece, the exact dimension may be read incorrectly.

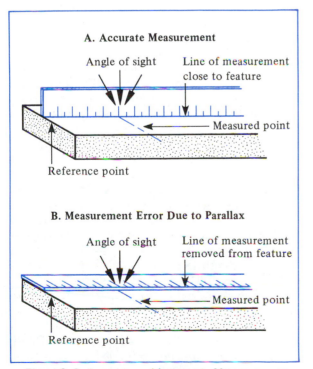

Figure 3–3 Accurate and Inaccurate Measurements

MANIPULATION ERRORS

Errors in measurement also may be caused by *improper manipulation*, or handling of, a steel rule, a workpiece, or both. Sometimes a rule is *cramped*. Cramping is the result of using excessive force. The rule is squeezed, which causes it to bend.

Another common error is to take a measurement when the rule does not lie on the line of measurement.

There is also the possibility of *slippage*, which may occur if a part is heavy or if it takes a comparatively long time to make a measurement. A dimensional error is caused by such slippage, or movement, between the measuring tool and the workpiece.

Good measurement practice requires that a rule be positioned on the line of measurement and held against the reference point using a *light touch*. This light touch is developed with repeated practice.

WORKER BIAS ERRORS

Finally, measurements are affected by *worker bias*, which means that an individual unconsciously reads an error into a measurement. Sometimes this bias is caused by a person's natural tendency to move either the rule or the line of sight to avoid parallax. It is difficult to manipulate a workpiece by hand, keep a rule along the line of measurement and aligned at the reference point, and, at the same time, see the graduation at the measured point.

How to Accurately Measure with the Steel Rule

STEP 1 Select a steel rule with graduations (scale) within the required limits of accuracy.

STEP 2 Remove any rough edges, or *burrs*, from the workpiece. Clean the rule and work. Determine whether the measurement is free of observational, manipulation, and bias errors.

STEP 3 Align the steel rule so that the scale edge is along the line of measurement.

STEP 4 Position the steel rule with the end held lightly against the reference point (Figure 3–4).

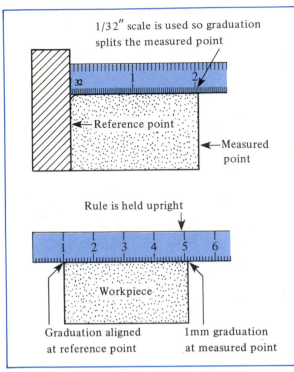

1/32″ scale is used so graduation splits the measured point

Reference point

Measured point

Rule is held upright

Workpiece

Graduation aligned at reference point

1mm graduation at measured point

Figure 3–4 Standard Practices for Taking Accurate Linear Measurements

STEP 5 Read the measurement at the measured point (Figure 3–4). For greater precision in measurement, or if a great number of measurements are required, use a magnifier.

Note: If the measurement point does not align with a graduation, turn the rule to another edge having a finer graduation.

STEP 6 Recheck the measurement using standard practices as illustrated in Figure 3–4.

B. LAYOUT AND TRANSFER MEASURING TOOLS

Layout is a common shop term. Layout refers to the preparation of a part so that work processes can be performed according to specific requirements. A layout identifies the actual features. Lines are scribed on a workpiece:

• To show boundaries for machining or performing hand operations,

• To indicate the dimensional limits to which material is to be removed or a part is to be formed,

• To position a part for subsequent operations.

Layout requires the use of three groups of layout, measurement, and inspection instruments:

• Group 1, the steel rule with attachments that make it more functional and precise for layout as well as measurement operations;

• Group 2, related tools for the transfer of measurements;

• Group 3, precision tools, gages, and instruments.

The basic measuring and transfer tools in the first two groups are covered in this unit. Principles and applications of precision micrometers, verniers, dial indicators, and gage blocks in the third group are covered later.

THE DEPTH GAGE

Many machined surfaces, like the surfaces shown in Figure 3–6, require a depth measurement for a groove, hole, recess, or step. Sometimes it is not practical to use a steel rule. The rule may be too wide, or it may be easy to hold the rule on an angle and read it incorrectly.

One of the common adaptations of the steel rule is the attachment of a *head*. The head spans a groove, slot, or other indentation and rests on the reference surface. The steel rule slides through the head and may be made to *bottom* at the measured point. The combination of head and rule is called a *depth gage*.

The rules for this depth gage are narrow and flat. In other depth gages, rods are used for small holes where a measurement with a narrow rule is not possible. The rods may or may not be graduated. Examples of narrow rule depth gages are illustrated in Figure 3–5.

Some depth gages are marked for angles of 30°, 45°, and 60° as shown in Figure 3–5. The blade may be moved to the right or left. A depth gage with an angle adjustment is useful for making a linear measurement to a surface that is at 30°, 45°, or 60° to the reference surface. Depth gages, however, are *not* intended to be used in making angular measurements.

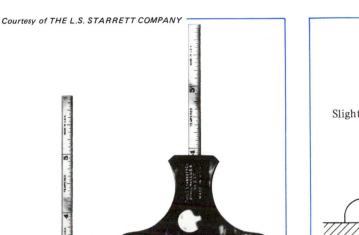

Courtesy of THE L.S. STARRETT COMPANY

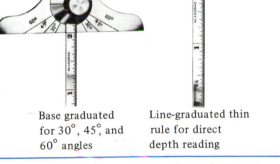

Base graduated for 30°, 45°, and 60° angles

Line-graduated thin rule for direct depth reading

Figure 3–5 Common Types of Line-Graduated Depth Gages

How to Measure with a Depth Gage

STEP 1 Remove burrs and chips from the workpiece and wipe it clean.

STEP 2 Select a depth gage with a blade size that will fit into the groove, hole, or indentation. Then, loosen the clamping screw so that the blade (rule) may be moved by a slight pressure.

STEP 3 Hold the head firmly on the reference surface (Figure 3–6).

STEP 4 Slide the rule gently into the opening until it just bottoms (Figure 3–6). Lock the head and blade.

STEP 5 Release the pressure on the head. Then, try to lightly slide the head on the refer-

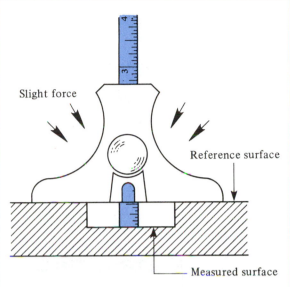

Figure 3–6 Head Held against the Reference Surface

ence surface. *Feel*—that is, sense with a light touch—whether the end of the rule is bottomed correctly.

STEP 6 Remove the depth gage and read the depth. If a nongraduated rod is used, place a steel rule next to the rod and read the measurement.

Note: In good shop practice, the measurement steps are repeated. The rule or rod is checked to see that it is properly bottomed and that the measurement is read correctly.

THE COMBINATION SET

The *combination set* is another adaptation of the steel rule. The combination set includes a *square head*, a *center head*, and a *protractor head* (Figure 3–7). A grooved rule is used with each head. The groove permits the rule to be moved into position and clamped.

THE SQUARE HEAD

One feature of the *square head* is that it provides a right-angle reference and thus extends the

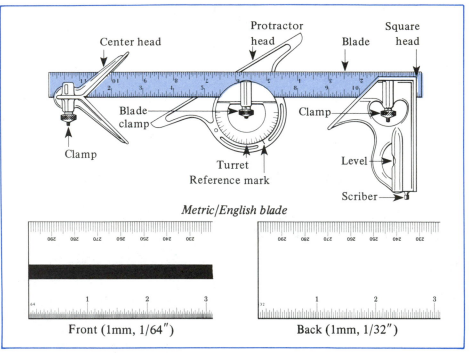

Figure 3–7 Blade and Heads of the Combination Set

use of the steel rule to laying out right-angle or parallel lines. Like the depth gage, the square head and rule may also be used to measure the depth of a hole, indentation, or other linear distance. When a measurement is to be read, the square head covers a portion of one scale beyond the measured point. Although there are four scales on the steel rule, the locking groove on one side permits direct sighting on only two scales. The scale that provides the required discrimination should be selected on the head side. The dimension should be measured at the graduation on the head side.

Another feature of the square head is that one of its surfaces is machined at a 45° angle. This feature permits the layout and measurement of many parts that require a 45° angle. Its application in scribing lines at 45° is illustrated in Figure 3–8.

THE CENTER HEAD

The *center head* is designed so that the measuring or layout surface of the rule falls along the centerline of a diameter. The rule may

be moved in the center head and locked in any desired position.

The center head is used primarily with round stock and cylindrical surfaces. It provides a quick and satisfactory method of measuring a diameter. An application of the center head for laying out centerlines on a round object is shown in Figure 3–9.

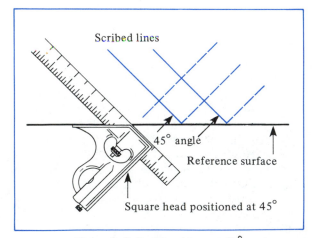

Figure 3–8 Scribing Lines at 45°

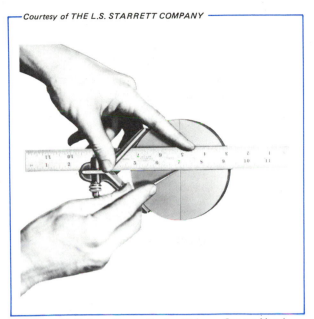

Figure 3–9 Locating a Center Using a Center Head

THE PROTRACTOR HEAD

How to Measure and Lay Out Angles with a Protractor Head

Note: The work must be free of oil, grease, and burrs. All tools should be examined in advance for any imperfections that could impair their accuracy.

Measuring an Angle

STEP 1 Remove burrs and chips. Wipe the protractor head, other layout tools, and the workpiece clean.

STEP 2 Insert the blade (rule) so that it extends far enough to measure the reference surface.

STEP 3 Loosen the turret clamp to permit the turret to be rotated when a slight force is applied.

STEP 4 Hold the protractor base firmly against the reference surface of the workpiece (Figure 3–10). Position the protractor head so that the blade and turret may be moved to the desired angle.

STEP 5 Bring the blade down gently to the workpiece and adjust it. The desired angle is

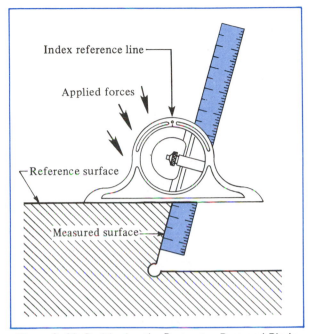

Figure 3–10 Positioning the Protractor Base and Blade

reached when there is no light showing between the measured surface of the workpiece and the blade.

STEP 6 Tighten the clamping screw. Read the graduation on the turret that coincides with the reference mark on the protractor head.

Measurements on an Angular Surface

STEP 1 Hold the protractor head securely against the reference surface.

STEP 2 Extend the rule until the end of it splits the measured point.

STEP 3 Read the measurement directly on one of the scales on the rule (Figure 3–10).

TRANSFER INSTRUMENTS

Technically, the *caliper* is a tool that is applied to mechanically *transfer* a measurement that originates at a reference point or surface and terminates at the measured point or surface. The four basic forms of calipers are the (1) *divider*, (2) *inside caliper*, (3) *outside caliper*, and (4) *hermaphrodite (morphy) caliper*. The hermaphrodite caliper combines some of the features of the divider and the other calipers.

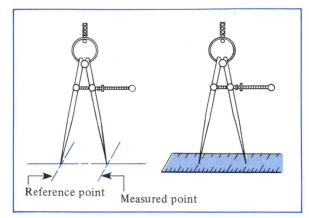

Figure 3-11 Setting the Divider Legs

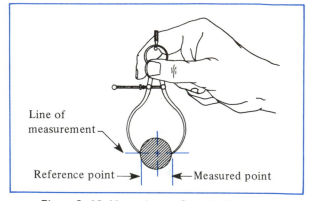

Figure 3-12 Measuring an Outside Diameter

Still other calipers, like the *slide caliper*, are made with scales. Measurements are read *directly*. The slide caliper is described later in this unit. The more precise vernier calipers and instruments are considered in Unit 4.

THE DIVIDER

The divider is a simple, accurate tool for transferring or laying out a linear dimension. Parallax in measurement is at a minimum. The sharp points of the divider legs are accurately aligned with measurement lines. The divider legs may be set to a desired dimension with a rule or they may be adjusted to conform to a measurement. The distance between the divider points is then measured on a rule.

It is common practice when setting the divider legs for a measurement to place the point end of one divider leg on a graduation on a rule. Usually, the 1″ or 1cm graduation on the metric scale is used. The distance to the point end of the other leg is read on the rule (Figure 3-11). The difference between the scale readings represents the measurement.

Circles may be scribed by setting the divider legs at the distance representing the radius. One leg is set in the center point; the other leg is swung around. A full circle or an arc is scribed by applying a slight force.

When a number of parts have the same linear measurement, the divider needs to be set only once to duplicate the measurement.

INSIDE AND OUTSIDE CALIPERS

An important point to remember for measurements that are made with inside and outside calipers is that the actual line of measurement is established by *feel* in using the tools. For example, when the legs of an outside caliper are adjusted so that they just slide over a piece of stock, the measurement taken is on the line of measurement (Figure 3-12). If the outside caliper cannot slide over the workpiece with this feel or if it is set too large, an inaccurate measurement results. Accuracy is developed by correct practice in getting the feel of the instrument. Force applied to a caliper causes the legs to *spring* and produces an inaccurate measurement.

Inside and outside calipers are also used to transfer a measurement to or from a particular feature of the workpiece—that is, either the inside or outside caliper is *set* to a particular size. The setting of an outside caliper may be done by measuring the leg distance with a steel rule (Figure 3-13). An inside caliper leg may be set

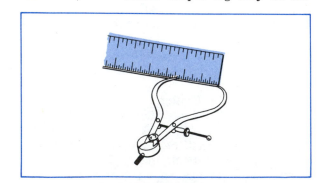

Figure 3-13 Setting an Outside Caliper

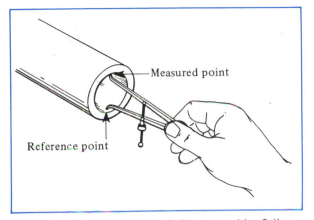

Figure 3–14 Measuring an Inside Diameter with a Caliper

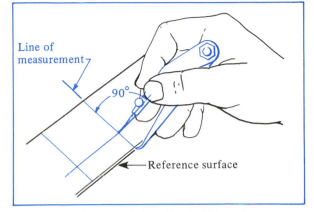

Figure 3–15 Scribing a Parallel Line with a Hermaphrodite Caliper

against another surface feature (reference point, as shown in Figure 3–14), with a micrometer, or with a gage.

If a caliper leg is set against a surface feature, the caliper may then be used to compare a feature against a specific dimension. The size is established by the feel of the caliper legs at the line of measurement. To repeat, if any force is applied, the caliper legs will spring. The lighter the worker feels the caliper measurement, the more accurate the measurement.

Calipers are usually *rocked*—that is, one leg is held lightly or rests on a surface, while the other leg is moved in a series of arcs. The rocking movement helps to establish when the caliper is located at the line of measurement. It is important to be able to manipulate the caliper from a comfortable position.

As a transfer tool, a caliper may be set according to the feature of the workpiece that is

to be measured. The caliper setting then is measured with a measuring tool. The setting also may be compared with a similar feature or part having the desired measurement.

> **Caution:** When measuring a feature of a part which is being machined, stop the machine before calipering the work. There must be no movement of a cutting tool or of the workpiece.

THE HERMAPHRODITE CALIPER

The hermaphrodite caliper (Figure 3–15) is a layout and measurement tool. It has one leg like that of a divider and one like that of a caliper. The caliper leg may be turned 180° for either inside or outside work. The caliper leg usually rests or rides on the reference surface. The divider leg is used to scribe lines parallel to a surface, find the center of a part, or take measurements.

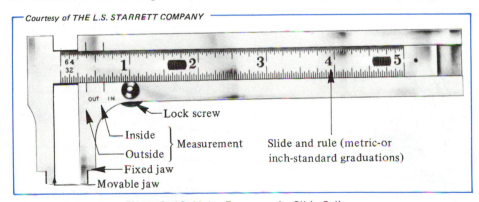

Courtesy of THE L.S. STARRETT COMPANY

Figure 3–16 Major Features of a Slide Caliper

THE SLIDE CALIPER

The slide caliper combines the features of inside and outside calipers and a rule in one instrument (Figure 3–16). The slide caliper has the advantage of providing a direct reading. A measurement can be locked, or *fixed*, at any position. Locking prevents measurement changes caused by handling. Thus, the direct-reading feature eliminates the need to remember a measurement and is an important feature when a number of cuts are to be taken or a series of operations performed.

However, the slide caliper has disadvantages that must be considered at all times:

- The range of use is less than it is with regular caliper and rule combinations,
- The instrument cannot be adjusted to compensate for wear or the springing of the legs,
- The range of discrimination is limited by the 1/64″ or 1/2mm scale.

THE SURFACE GAGE

The surface gage is a layout and transfer measuring tool. Its main parts are illustrated in Figure

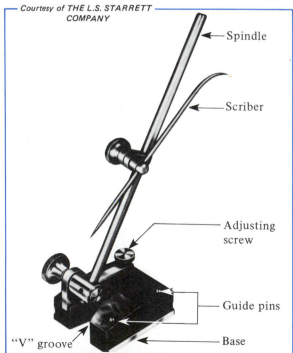

Courtesy of THE L.S. STARRETT COMPANY

Spindle

Scriber

Adjusting screw

Guide pins

"V" groove

Base

Figure 3–17 Main Parts of a Surface Gage

3–17. The base is machined with a "V" groove to rest on a flat or round surface. An adjustable spindle is provided with a clamp to hold and position a scriber.

The scriber is set at the approximate required linear measurement. It is then accurately raised and lowered to the required dimension by the adjusting screw. The base also has four *guide pins*. These are pushed down when the surface gage is to be used to scribe lines parallel to a reference surface (Figure 3–18).

When scribing a line, it is easy to spring the scriber if it extends too far from the spindle. Therefore, both the scriber and spindle must be positioned close to the work.

THE SURFACE PLATE

The surface plate is an auxiliary *accessory*. As an accessory, it is combined with other tools, measuring instruments, and test equipment. It is used universally to provide a reference surface. The surface plate is essential to layout and measurement processes. Figure 3–19 provides examples of popular manufactured surface plates.

A surface plate may be a simple, flat plate that has been accurately machined. The term usually refers to a plate that has been machined and scraped to an extreme accuracy or ground

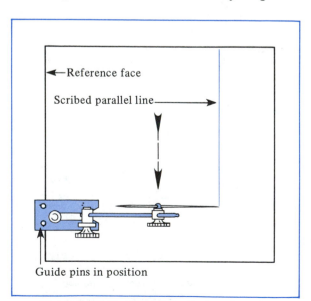

Reference face

Scribed parallel line

Guide pins in position

Figure 3–18 Position of the Surface Gage for Scribing Parallel Lines

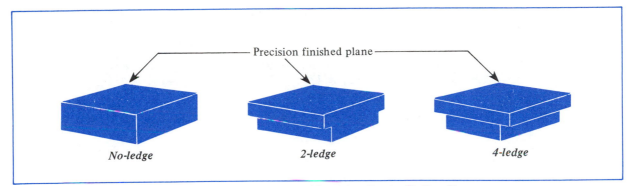

Figure 3–19 General Styles of Precision Granite Surface Plates

to a fine surface finish. The underside is honey-combed with a number of webbed sections that prevent the surface plate from *warpage* and thus provide a permanent flat surface.

The *granite* surface plate is used extensively today and has a number of advantages over the cast-iron type:

- High degree of surface flatness;
- High stability (maintains surface flatness because there is no warpage);
- Capability to remain flat, even when *chipped* (cast-iron surface plates, by contrast, tend to *crater* and require rescraping);
- Simplified maintenance (the surfaces are rustproof and may be cleaned easily with soap and water and other readily available cleaners).

The size, construction, and degree of accuracy of surface plates vary. The size naturally is governed by the nature of the layout, measurement, or assembly operations that are to be performed. The degree of precision required determines: (1) the accuracy to which the flat (plane) surface must be machined and scraped and (2) the finish of the surface. The flat surface is used as a reference point or surface.

A. Safe Practices in Dimensional Measurement

- Remove burrs from a workpiece. Burrs cause both inaccuracies in measurement and unsafe handling conditions. Burrs also affect the accurate fitting and movement of mating parts.
- Recheck all precision measurements.
- Consider the possibility of inaccurate measurement readings. These may result from observational, manipulation, and worker bias errors.
- Make it a practice to check and clean measuring tools and the workpiece surface before taking a measurement.

B. Safe Practices in Using Layout and Transfer Measuring Tools

- Examine the workpiece, layout tools, and measuring instruments. Each item should be free of burrs.
- Use a wiping cloth to clean the workpiece and layout and measuring tools. All chips and foreign matter should be removed.
- Handle all tools and parts carefully. Avoid hitting tools or parts together or scraping them against each other.
- Recheck each layout and measurement before cutting or forming. Each dimension should meet the drawing or other technical specification.
- Use a light touch to feel the accuracy of a measurement at the line of measurement. Force applied to a layout or measuring tool may cause the legs to spread and thus produce an inaccurate measurement.
- Replace each layout and measurement tool in a protective container or storage area.

UNIT 3 REVIEW AND SELF-TEST

A. LINE-GRADUATED RULES

1. State what graduated scale on a steel rule should be used to measure each of the following dimensions: (1) 3 1/64", 2 19/32"; (b) 5.01", 14.08"; (c) 75mm, 10.7mm.

2. Indicate three conditions that produce errors in reading measurements on a steel rule.

3. Define the term *aligned* in relation to a direct measurement.

4. State one general practice followed by the craftsperson to check a measurement that has been taken.

5. Give two reasons for removing burrs before measuring and/or fitting two mating parts.

B. LAYOUT AND TRANSFER MEASURING TOOLS

1. List three functions that are served by layout lines.

2. Identify the tool to use to take the following measurements: (a) the diameter of a hole, (b) the distance from the face to the bottom of a hole, (c) the vertical height to the bottom of a dovetail slot, (d) the length of a part that is machined with an angle to within $1°$, (e) the diameter of a round bar of stock.

3. Tell what function transfer instruments serve.

4. Describe how a measurement on an inside caliper may be accurately read on a steel rule.

5. Explain the meaning of using a *scale that provides the necessary discrimination.*

6. Cite three examples of layout processes that require the use of a hermaphrodite caliper.

7. Describe briefly how to take an inside measurement with a slide caliper.

8. Identify three layout work processes requiring the use of a surface gage.

9. Cite three advantages of granite surface plates over cast-iron surface plates.

10. State three safety precautions to take to ensure accuracy in laying-out processes.

Precision Instruments: Micrometers, Verniers, Indicators, and Gage Blocks

Dimensional measurements that must be more accurate than 1/100″ or 1/2mm require the use of precision measuring instruments. Such instruments include standard and vernier micrometers; vernier calipers, depth gages, height gages, and bevel protractors; test and dial indicators, and gage blocks. The principles of and correct practices for manipulating and reading these instruments are examined in this unit.

A. STANDARD AND VERNIER MICROMETER PRINCIPLES AND MEASUREMENTS

Many parts or features of a part require a degree of accuracy that is ±0.001″ or ±0.0001″ (±0.02mm or ±0.002mm for metric dimensioning). A *standard micrometer* is widely used for precision measurements to within 0.001″ (or 0.02mm). More precise measurements to within 0.0001″ (or 0.002mm) are taken with a *vernier micrometer*. In the shop and laboratory, the term *mike* is used for micrometer and for micrometer measuring practices.

Micrometers are made for direct hand use. They may also be designed as part of another measuring instrument. Their functions parallel the functions of calipers. Of the many different types of micrometers, the three most commonly used types—outside, depth, and inside—are covered in this unit.

Among the advantages of the micrometer are the following:

- Easier and clearer readability than rules or calipers;
- Consistently accurate measurements with close tolerances;
- No observational error due to parallax;
- Portability, comparative ease of handling, fast operation, and reasonable cost;
- Design features that include a built-in adjustment to correct for wear.

MAJOR DESIGN FEATURES OF THE MICROMETER

The major parts of the micrometer are identified in Figure 4–1. The micrometer *anvil* and *spindle*

are made of a hardened alloy steel. The ends, or measuring surfaces, are finely lapped so that they are perfectly parallel to each other. As illustrated, a thread is cut directly on the spindle.

A ratchet stop is attached to the end of the spindle. Spring action in the ratchet provides a constant amount of pressure (force). This action makes it possible to obtain the correct feel for all measurements.

Many micrometers are provided with a spindle lock. Rotating the *lock nut* in one direction *locks* the spindle from turning. Slight rotation in the opposite direction releases the pressure. The spindle is then free to turn. Outside micrometers are available in a wide range of sizes from 1/2″ to 60″ (12mm to 1,500mm) capacity. The maximum range of measurements that may be taken with a micrometer is usually 1″ or 25mm.

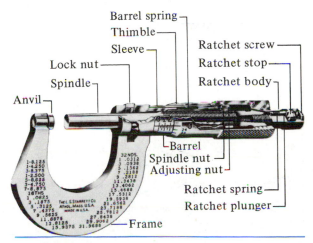

Figure 4–1 Cutaway View of an Inch-Standard Outside Micrometer

The micrometer design provides for wear on the anvil and spindle faces and for adjustments due to frame conditions. The *thimble* and spindle are made in two parts. They may be loosened and the thimble scale turned and reset to zero at the index line. The two parts are then secured as a single part. An *adjusting nut* may be turned to take up the clearance caused by wearing of the threads.

MICROMETER PRINCIPLE AND MEASUREMENT

All micrometers operate on the principle that a *circular movement* of a threaded spindle produces an *axial movement*. Thus, as the spindle assembly is turned, the end (face) of the spindle moves toward or away from the fixed anvil. The distance (axial movement) per revolution (circular movement) depends on the pitch of the threaded spindle. The *pitch* refers to the number of threads either per inch or in 25mm.

For correct micrometer measurement a workpiece is calipered along its line of measurement. The actual measurement is the distance between the reference and measured points. Figure 4–2 shows the positions of the micrometer and workpiece in taking a correct measurement. The accuracy of the measurement depends on the feel—that is, the contact between the instrument, the work, and the operator in a controlled environment.

Inch and metric outside micrometers are used for similar measurements. The instruments are handled in the same manner. Direct readings are made on the *barrel* and *thimble scales*.

THE INCH-STANDARD MICROMETER

The screw thread on the spindle of the inch-standard outside micrometer has 40 threads per inch. Its pitch, then, is 1/40″, or 0.025″. As the spindle is turned one complete revolution, it moves 0.025″ (Figure 4–3A).

Attached to the spindle is the thimble. The thimble is graduated with 25 graduations around its circumference (periphery). By dividing the movement of the spindle (0.025″) for each revolution by 25, the distance represented by each graduation on the thimble is 0.001″ (Figure 4–3B).

For greater readability the lengths of some of the lines that represent the graduations on the thimble and barrel are varied (Figure 4–4). For example, since every fourth graduation on the barrel represents 0.100″ (0.025″ × 4), each 0.100″ is represented by a long graduation line. The long lines on the barrel are numbered 0, 1, 2, 3, and so on to 10. Each graduation on the thimble represents 0.001″. The long lines on the thimble mark every fifth graduation and are

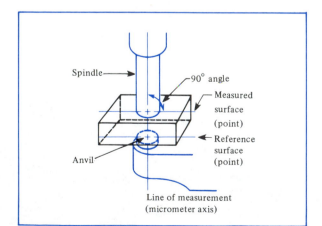

Figure 4–2 Positions of the Micrometer and Workpiece for Correct Measurement

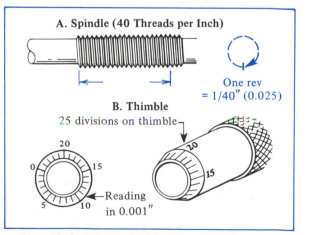

Figure 4–3 Relationships between (A) Micrometer Spindle Pitch (40 threads per inch); Spindle/Thimble Movement (1/40″, or 0.025″, per revolution); and (B) Graduations on the Thimble of an Inch-Standard Micrometer

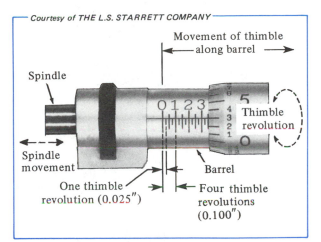

Figure 4–4 Spacing between Graduations on the Micrometer Barrel (each represents 0.025") and on the Thimble (each represents 0.001")

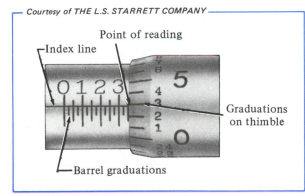

Figure 4–5 Establishing a Micrometer Reading

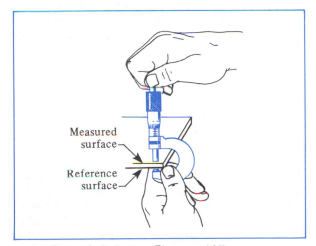

Figure 4–6 Correct Finger and Micrometer Positions for Taking an Accurate Measurement

numbered 5, 10, 15, 20, and 0 (for 25). There usually is a series of smaller-sized numbers for the intermediate graduations.

A micrometer measurement is obtained from the readings on the barrel and the thimble. The thimble reading is taken at the point where the index line on the barrel crosses a graduated (or fraction) line on the thimble (Figure 4–5). The barrel and thimble readings are added and then represent a micrometer measurement.

Through correct practice, micrometer measurements are read directly and almost automatically. Starting at the index line, the readings on the barrel and thimble are read as a continuous numerical value.

How to Measure with the Inch-Standard Micrometer

Taking a Measurement

STEP 1 Adjust the micrometer opening so that it is slightly larger than the feature to be measured.

STEP 2 Hold the frame of the micrometer with the fingers of one hand. Place the micrometer on the work so that the anvil is held lightly against the reference point.

STEP 3 Turn the ratchet (or the thimble directly if there is no ratchet) with the fingers of the other hand. Continue turning until the spindle face just touches the work (measured point). The correct position of the hands and micrometer are illustrated in Figure 4–6.

Reading a Measurement to within 0.001"

STEP 1 Lock the spindle with the lock nut. Note the last numeral that appears on the graduated barrel (A in Figure 4–7).

STEP 2 Add 0.025" for each additional line that shows on the barrel (B in Figure 4–7).

STEP 3 Add the number of the graduation on the thimble (C in Figure 4–7).

STEP 4 Read the micrometer measurement in *thousandths*. The measurement is the *sum* of the barrel and thimble readings (A, B, and C in Figure 4–7).

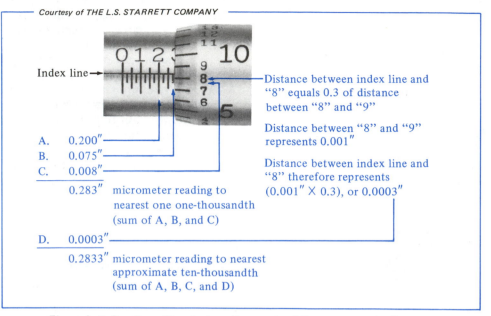

Courtesy of THE L.S. STARRETT COMPANY

Index line →

Distance between index line and "8" equals 0.3 of distance between "8" and "9"

Distance between "8" and "9" represents 0.001″

Distance between index line and "8" therefore represents (0.001″ × 0.3), or 0.0003″

A.   0.200″
B.   0.075″
C.   0.008″

0.283″ micrometer reading to nearest one one-thousandth (sum of A, B, and C)

D.   0.0003″

0.2833″ micrometer reading to nearest approximate ten-thousandth (sum of A, B, C, and D)

Figure 4–7 Reading a Standard, or Nonvernier, Micrometer to the Nearest Approximate Ten-Thousandth of an Inch (0.0001″)

Note: Where a micrometer larger than 1″ is used, the total measurement equals the whole number and the decimal fraction read on the micrometer head. For example, with an 8″ micrometer the reading would be 7″ plus whatever decimal is shown on the micrometer.

Estimating to Finer than 0.001″

STEP 1   Estimate the distance between the last graduation on the thimble and the index line. For instance, in Figure 4–7 the distance between "8" and "9" is about one-third (0.0003″).

STEP 2   Add the decimal fractional value of this distance (D in Figure 4–7) to the thousandths reading (the sum of A, B, and C in Figure 4–7, or 0.283″). The answer gives an approximate four-place decimal value (0.2833″ in Figure 4–7).

THE VERNIER MICROMETER AND 0.0001″ (0.002mm) MEASUREMENT

The vernier micrometer is used to make precise measurements that are within the *one ten-thousandth of an inch*, 0.0001″, range (or the metric system equivalent, 0.002mm). The addition of the vernier scale to the barrel of a micrometer extends its use in making these precise measurements. The vernier principle dates back to 1631 and is named for its inventor, Pierre Vernier.

The vernier principle is comparatively simple. There are ten graduations on the top of the micrometer barrel that occupy the same space as nine graduations on the thimble. The difference between the width of one of the nine spaces on the thimble and one of the ten spaces on the barrel is one-tenth of one space.

Each graduation on the thimble represents one-thousandth of an inch. Thus, the difference between graduations on the barrel and the thimble is one-tenth of one-thousandth, or one ten-thousandth part of an inch (Figure 4–8A).

The vernier reading is established by sighting across the graduated lines on the barrel and the divisions on the thimble. The line that coincides represents the *ten-thousandths vernier reading*. This reading is added to the three-place (one-thousandths) reading. The result is a four-place (ten-thousandths) vernier micrometer reading (Figure 4–8B) of 0.2970″.

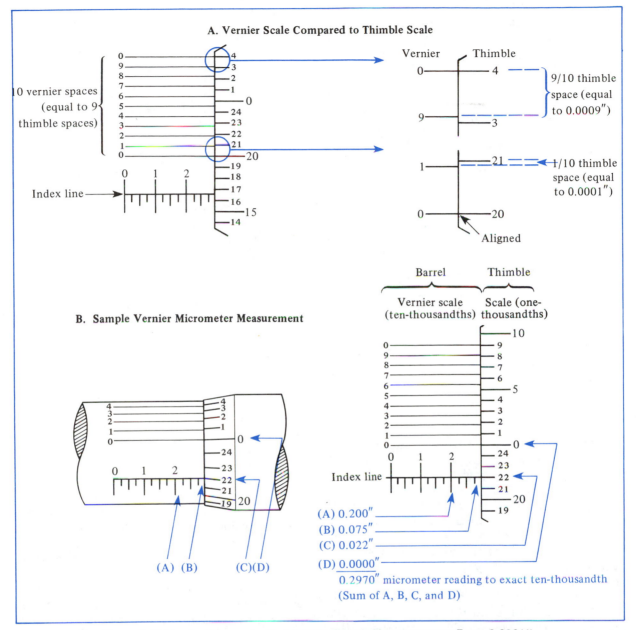

A. Vernier Scale Compared to Thimble Scale

B. Sample Vernier Micrometer Measurement

(A) 0.200″
(B) 0.075″
(C) 0.022″
(D) 0.0000″
0.2970″ micrometer reading to exact ten-thousandth
(Sum of A, B, C, and D)

Figure 4–8 Vernier Micrometer Scales and a Sample Measurement to Exact 0.0001″

THE METRIC MICROMETER

The principles and the applications of inch and of metric micrometers are similar. The difference lies in the basic unit of measure and the discriminating ability of the instruments. For example, standard inch micrometers can be used to measure to .001″ (0.02mm) and vernier micrometers, to .0001″ (0.002mm). By comparison, metric micrometers have discriminations of 0.01mm and 0.002mm. The 0.01 mm measurement is equal to 0.0004″, which is a higher degree of accuracy than is normally required for some work. The 0.002mm measurement (equal

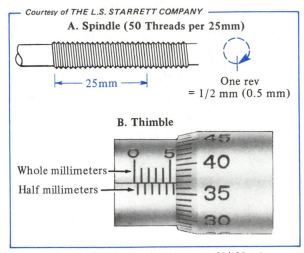

A. Spindle (50 Threads per 25mm)

← 25mm →

One rev
= 1/2 mm (0.5 mm)

B. Thimble

Whole millimeters →

Half millimeters →

Figure 4–9 Relationship between (A) Metric Micrometer Spindle Pitch (50 threads in 25mm) Spindle/Thimble Movement (1/2mm, or 0.50 mm, per revolution); and (B) Spacing between Graduations on the Barrel (whole and half millimeters)

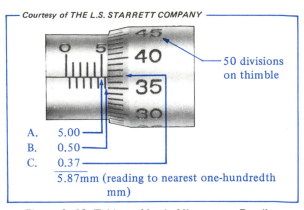

50 divisions on thimble

A. 5.00
B. 0.50
C. 0.37
____
5.87mm (reading to nearest one-hundredth mm)

Figure 4–10 Taking a Metric Micrometer Reading

to approximately 0.00008″) is beyond the range of accuracy of most workpieces and the instrument itself.

METRIC MICROMETER GRADUATIONS AND READINGS

The screw of the metric micrometer has 50 threads in 25mm (Figure 4–9A). Each turn of the spindle produces an axial movement of 1/2mm (about 0.020″). Two turns moves the spindle 1mm (almost 0.040″). In contrast to the inch micrometer, there are two rows of graduations on the metric micrometer barrel (Figure 4–9B). The top row next to the index line represents whole millimeters. They start at 0, and every fifth millimeter is numbered. The graduations on the lower row represent half millimeters (0.50mm).

There are 50 graduations on the thimble (Figure 4–10). One complete revolution of the thimble moves it 1/2mm (or 0.50mm). The difference between two graduations on the thimble is 1/50 of 0.50mm, or 0.01mm.

The metric micrometer reading is established by adding three values (Figure 4–10):

- The last whole number of millimeters that is visible in the top row near the thimble (A in Figure 4–10),
- The half-millimeter graduation in the lower row that shows (B in Figure 4–10),
- The graduation on the thimble at the index line (C in Figure 4–10).

Like the inch micrometer, the metric micrometer may be used to estimate any fractional part that falls between two graduations. Estimating adds to the accuracy of the measurement.

THE DEPTH MICROMETER

The depth micrometer (Figure 4–11) combines the features of a micrometer head and base. The same operations are performed as with a depth gage having a steel rule blade. The precision of the depth micrometer is within the 0.001″ range on the standard inch micrometer and 1/2mm (0.50mm) on the metric micrometer head. While vernier graduations on the head would increase the possible range of accuracy of measurements to 0.0001″, such measurements would not be consistently reliable.

The usual range of measurements for depth micrometers is 1″ or 25mm. This range is increased by the addition of interchangeable *measuring rods*. As with all measuring tools, care must be taken to properly seat the measuring rods. The accuracy of the zero setting of the depth micrometer may be easily checked by holding the base against a flat surface. The face (end) of the measuring rod is then brought into contact with the surface by using the ratchet. Any variation from zero should be adjusted following the steps used with a regular micrometer.

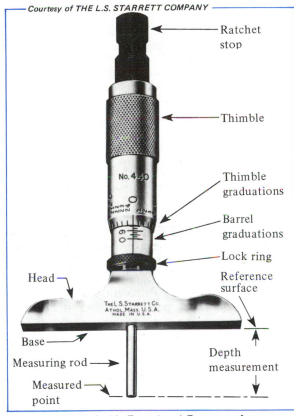

Figure 4–11 Functional Features of the Depth Micrometer

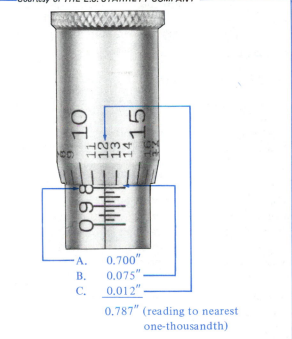

A.	0.700″
B.	0.075″
C.	0.012″

0.787″ (reading to nearest one-thousandth)

Figure 4–12 Reading a Depth Micrometer to the Nearest One-Thousandth of an Inch (0.001″)

Gage blocks of specified sizes may be used to check the accuracy along the depth range of the instrument.

An important point to remember with the depth micrometer is that it measures *in reverse* from other micrometers. (1) The zero reading of the depth micrometer appears when the thimble is at the *topmost* position. (2) The depth micrometer graduations are in *reverse* order. (3) The thimble reads *clockwise*.

How to Measure with the Depth Micrometer

STEP 1 Prepare the workpiece. Select the appropriate length measuring rod. Prepare and test the depth micrometer for accuracy. Then, adjust the measuring rod so that it is almost to the depth to be measured.

STEP 2 Position and apply a slight pressure to hold the micrometer base against the reference surface.

STEP 3 Lower the measuring rod. Use the ratchet to bottom the rod.

STEP 4 Note the numbered graduation that appears in full at the index line on the barrel. The first number of the reading is less than the full numeral that is exposed (A in Figure 4–12, 0.700″).

STEP 5 Read the last (0.025″) graduation that is covered (Figure 4–12). The reading becomes the second and third decimal place reading (B in Figure 4–12, 0.075″).

STEP 6 Read the graduation on the thimble at the index line (Figure 4–12). It represents thousandths of an inch (C in Figure 4–12, 0.012″).

STEP 7 Combine the readings at the index line of the graduations on the barrel and the thimble. The sum represents the depth in thousandths (0.787″ in Figure 4–12).

Note: Where the extension rods are used, the length of the rod is added to the measurement on the micrometer head.

DEPTH MEASUREMENTS IN METRIC

Follow the same procedures with the metric depth micrometer as described for the standard inch micrometer. The measurements will be in millimeters and fractional parts of a millimeter.

THE INSIDE MICROMETER

The inside micrometer (Figure 4–13) consists of a micrometer head, a chuck, and two contact points. Precision extension rods and a spacing collar extend the range of measurements possible with a single micrometer head. A handle is provided for use when small holes or distances are

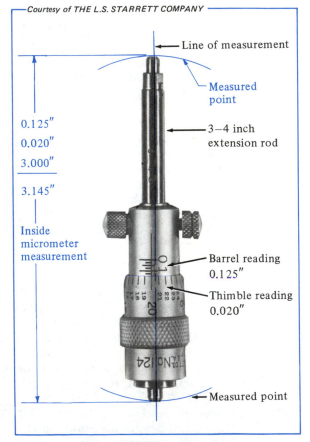

Courtesy of THE L.S. STARRETT COMPANY

Line of measurement

Measured point

0.125″
0.020″
3.000″

3.145″

Inside micrometer measurement

3–4 inch extension rod

Barrel reading 0.125″

Thimble reading 0.020″

Measured point

Figure 4–13 Taking an Inside Diameter Measurement

to be measured. A micrometer head with an extension rod is illustrated in Figure 4–13.

The graduations on the micrometer head may be in thousandths of an inch, in hundredths of a millimeter, or finer. The instrument, like that of all other precision measuring instruments, must be checked against an accurate standard and adjusted for accuracy.

APPLICATION OF THE INSIDE MICROMETER

The features and correct alignment of an inside micrometer for taking an accurate inside measurement are shown in Figure 4–13. A measurement is taken by closing the inside micrometer to almost the size of the feature to be measured. The micrometer thimble is turned until contact is established at the measured point along the line of measurement. The micrometer is removed from the work. The reading is made directly by combining the barrel and thimble readings.

TELESCOPING AND SMALL-HOLE GAGES

The micrometer head size limits its use to minimum internal measurements of close to 2″ or 50mm. For smaller measurements it is necessary to use transfer measuring tools. Among these tools are the *telescoping* and *small-hole gages*. Measurements with these gages extend down to 1/8″ (3mm).

The telescoping gage consists of two tubes (legs) that are spring loaded. The legs are locked at the reference points of a measurement by a locking nut or screw in the handle. The measurement is then transferred to and read on a micrometer or other measuring instrument.

B. VERNIER MEASURING INSTRUMENTS, DIAL INDICATING INSTRUMENTS AND GAGE BLOCKS

Measurements in thousandths of an inch or 0.02mm are the most widely used measurements in industry. Finer measurements are obtained by applying the vernier principle to metric and inch-standard micrometers and other instruments. Vernier scale graduations extend the range of fineness of measurement to 0.0001″ (one ten-thousandth of an inch) and 0.002mm (the metric equivalent of 0.00008″).

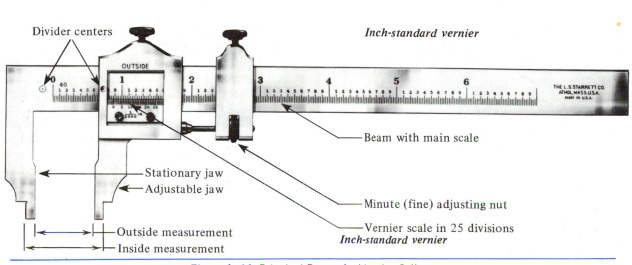

Figure 4–14 Principal Parts of a Vernier Caliper

Vernier measuring instruments incorporate the vernier principle with special features of measuring tools. The vernier measuring instruments described in this unit are the:

- Vernier caliper (inside and outside),
- Vernier depth gage,
- Vernier height gage,
- Vernier bevel protractor.

The vernier principle is also applied to graduated handwheels and tables on machine tools. In each of these applications, the vernier scale increases the precision to which a machine or a workpiece may be set for machining operations.

Two groups of precision measuring and layout instruments also covered in this unit are:

- *Test indicators* and *dial indicators*. These instruments provide for accurate, direct measurements by multiplying minute movements through a system of levers or gears;
- *Gage blocks*. These standards are combined with vernier measuring and dial indicating instruments for both measurement and layout work. Gage blocks are made to extremely close tolerances. These blocks permit measurements to within two one-millionths of an inch (0.000002″) or the equivalent metric value.

PARTS AND SCALES OF VERNIER INSTRUMENTS

In the mid 1800s Joseph R. Brown (a cofounder with his father of the Brown & Sharpe Manufacturing Company) applied the vernier principle to the slide caliper. The series of measuring instruments that resulted from Brown's application are identified by names that incorporate the term *vernier* with the name of the type of instrument. For example, the combination of *vernier* and *height gage* identifies the vernier height gage. Similarly, the names of the vernier depth gage and vernier bevel protractor incorporate the main features of the instruments with the vernier term.

The main parts and scales of a vernier measuring instrument are shown on the vernier caliper in Figure 4–14. A *vernier scale* is generally mounted on the adjustable jaw of the instrument. A *beam*, or frame, with a stationary jaw contains the *main scale*. The vernier and main scales are designed in a straight line, as exemplified by the vernier caliper (Figure 4–14), vernier height gage, and vernier depth gage. These scales also may be imprinted along a curved surface. The vernier bevel protractor is an example.

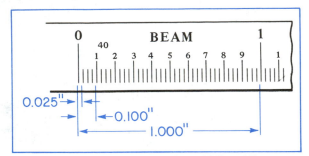

Figure 4-15 Inch, Subdivided into 40 Equal Parts, on the Main Scale of an Inch-Standard Vernier Caliper

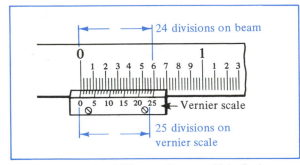

Figure 4-16 Difference between Main Scale Graduations and Vernier Scale Graduations (0.001″)

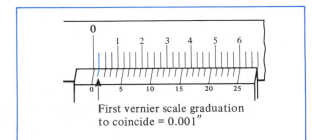

Figure 4-17 Vernier Scale Reading of 0.001″

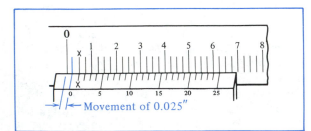

Figure 4-18 Zero (0) on the Vernier Scale Coinciding with a 0.025″ Graduation on the Main Scale

VERNIER PRINCIPLE AND VERNIER CALIPER MEASUREMENT

The vernier caliper differs from the vernier micrometer in construction and in method of taking a measurement. The vernier caliper is used for *internal* (inside) and *external* (outside) linear measurements. The following explanation of the main and vernier scales is made in relation to the inch-standard vernier caliper. The common 25-*division vernier caliper* is used to describe the vernier principle. The vernier principle also applies to the metric vernier caliper. However, measurements are taken in decimal parts of a millimeter.

THE INCH-STANDARD VERNIER CALIPER

The main scale of the beam of the inch-standard vernier caliper is divided into inches (Figure 4-15). Each inch is subdivided into 40 equal parts. The distance between graduations is 1/40″, or 0.025″. A second scale on the vernier caliper, the vernier scale, is imprinted on the movable jaw and therefore moves along the main scale. The 25 divisions on the vernier scale correspond in length to 24 divisions on the main scale (Figure 4-16). The difference between a main scale graduation and a vernier scale graduation is 1/25 of 0.025″, or 0.001″.

When the adjustable jaw of the vernier caliper is moved until the first vernier scale graduation coincides with the first graduation on the beam, the jaw is opened 0.001″. The vernier scale reading is illustrated in Figure 4-17. As the movement is continued until the second graduations coincide, the opening represents a distance of 0.002″. When the jaw has moved 0.025″, the *zero index line* on the vernier scale and the first graduation (0.025″) on the beam coincide (Figure 4-18). This represents a measurement of 0.025″.

A vernier caliper reading consists of (1) the number of inches (for dimensions larger than one inch) and (2) decimal parts of an inch. The measurement is read from the beam and vernier graduations. The decimal values on the beam (0.025″ graduations) and on the vernier scale (0.001″ graduations) are added to any whole-inch values.

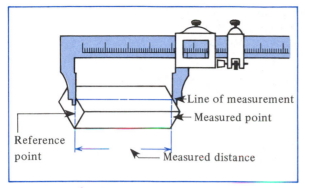

Figure 4-19 Position of the Vernier Caliper for an Accurate Measurement

Two features of the vernier caliper are that (1) inside and outside measurements may be taken directly and (2) an adjustment screw permits the movable jaw to be minutely adjusted to a measurement (the jaw is then locked in position by a clamping screw). Because a vernier instrument is often more difficult to use than a micrometer, the worker must establish which instrument is more practical for the job. Figure 4-19 shows the positioning of a vernier caliper for an accurate measurement. Steps in reading a measurement are illustrated in Figure 4-21.

METRIC VERNIER CALIPER MEASUREMENTS

The applications of metric and inch-standard vernier calipers are the same. The major difference is that measurements are taken in metric units. The beam of the metric vernier caliper is graduated in millimeters. A graduation on the beam, illustrated in Figure 4-20 equals 1mm.

Every tenth graduation is marked with a numeral. The "80" graduation on the beam scale thus represents 80.00mm.

The illustration in Figure 4-20 shows a beam with two scales. The top scale is offset horizontally from the bottom scale to compensate for the width of the vernier caliper jaws. Internal (inside) measurements are read on the top scale. The bottom scale has the same graduations as the top scale has. These graduations are also 1mm apart. The bottom scale is used for direct external (outside) measurements. On the model in Figure 4-20, inside and outside measurements are read on the same side. Some vernier calipers are available with only inch-standard graduations. Others have one scale for metric and a second scale for inch-standard measurements.

On the all-metric vernier caliper in Figure 4-20, the top and bottom vernier scales are graduated in 0.02mm. Every fifth graduation is numbered. The "10" graduation on the vernier scale thus represents 0.10 mm, "20" is 0.20mm, and so on.

The metric vernier caliper is read in the same manner as the inch-standard vernier caliper. The decimal fraction of a millimeter on the vernier scale is added to the whole number of millimeters on the beam. The decimal value is established at the point where a vernier graduation coincides with a beam graduation.

The internal measurement indicated on both the top and vernier scales in Figure 4-20 is 78.08mm. This measurement consists of a beam reading of 70, or 78.00mm. The vernier reading is the "4" graduation × 0.02mm, or

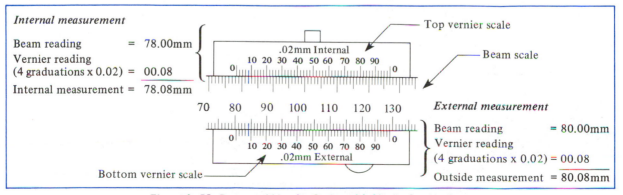

Figure 4-20 Beam and Vernier Scales with Metric Graduations

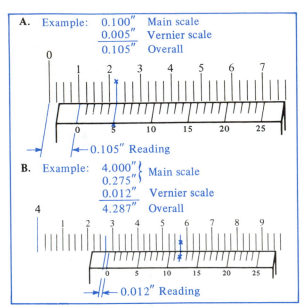

A. Example: 0.100″ Main scale
               0.005″ Vernier scale
               0.105″ Overall

← 0.105″ Reading

B. Example: 4.000″ } Main scale
              0.275″
              0.012″ Vernier scale
              4.287″ Overall

← 0.012″ Reading

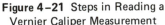

Figure 4–21 Steps in Reading a Vernier Caliper Measurement

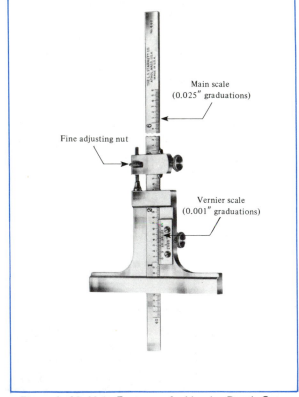

Courtesy of THE L.S. STARRETT COMPANY

Main scale
(0.025″ graduations)

Fine adjusting nut

Vernier scale
(0.001″ graduations)

Figure 4–22 Main Features of a Vernier Depth Gage

0.08mm. The correct measurement is thus 78.08mm. The external measurement, taken from the bottom scale, is 80.08mm.

MEASURING WITH THE VERNIER DEPTH GAGE

The vernier depth gage (Figure 4–22) is essentially a regular depth gage with locking screws, an adjusting nut, a rule, and a vernier scale added. The rule in the inch-standard system is graduated in 0.025″.

Measurements are taken with the same care as for a regular depth gage with a line-graduated rule or a depth micrometer. The adjusting screw permits the blade to be brought against the work surface and locked in position. However, taking a precise accurate measurement is difficult.

MEASURING WITH THE VERNIER HEIGHT GAGE

The vernier height gage (Figure 4–23) is widely used in layout work and for taking measurements to an accuracy of 0.001″ or 0.02mm.

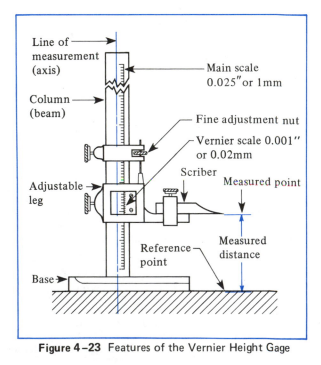

Line of measurement (axis)

Main scale 0.025″ or 1mm

Column (beam)

Fine adjustment nut

Vernier scale 0.001″ or 0.02mm

Scriber

Measured point

Adjustable leg

Measured distance

Reference point

Base

Figure 4–23 Features of the Vernier Height Gage

The three main parts are the base, the column (beam), and the slide arm. The main scale is on the column. The vernier scale is attached to the slide arm.

Measurements are taken in the same manner as with other vernier measuring instruments. Whole-inch and 0.025″ (or centimeter and fractional millimeter) readings are obtained from the graduations on the beam column. The additional decimal values in multiples of 0.001″ (or 0.02mm) are read on the vernier scale. These readings are taken at the two graduations that coincide. The vernier height gage measurement consists of the (1) whole-inch number, (2) 0.025″ graduations to the left of the zero index line on the vernier scale, and (3) vernier scale reading in thousandths (0.001″).

There are a number of attachments used with the vernier height gage. A flat scriber or an offset scriber may be secured to the slide arm. These attachments are used in layout work or for making linear measurements. Depth measurements may be taken with a depth gage attachment or offset attachment. Many other special attachments are used for different measurements.

THE UNIVERSAL VERNIER PROTRACTOR

The vernier principle is also applied in making angular measurements. The vernier bevel protractor is an instrument used to measure angles to an accuracy of 5 minutes (5′), or 1/12 of one degree. The universal vernier protractor includes a number of attachments that make it possible to make a wide range of measurements. All measurements are within the same degree of accuracy as the standard vernier protractor.

Figure 4–24 shows the main features of a vernier bevel protractor. The protractor main scale (dial) is graduated in whole degrees. On some instruments this scale is on the body of the instrument itself. On other instruments the graduations are on the rotating turret. Depending on the manufacturer, the vernier scale is part of either the rotating turret or the body. Some main scales are arranged in four 90° quadrants (0° to 90° to 0° to 90° to 0°).

The vernier scale has 24 divisions; 12 divisions are on each side of the zero index line. The 24

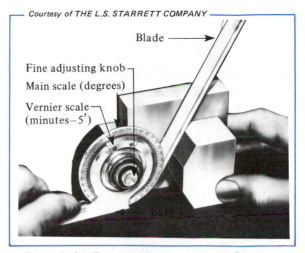

Figure 4–24 Features of a Vernier Bevel Protractor

vernier scale divisions are numbered from 60 to 0 to 60, as illustrated in Figure 4–25. Each vernier scale graduation represents 5′. The graduations on the vernier scale are numbered on *both* sides of the zero index line to make it easy to read the minute values. Some vernier scales are marked at every third graduation (60, 45, 30, 15, 0, 15, 30, 45, 60). Again, each line (graduation) on the vernier scale represents 5′.

In taking angular measurements, it is important to read the vernier scale in the same direction from zero as the main scale reading. When the angle is an exact whole number of degrees, the index line (0) on the vernier scale coincides with

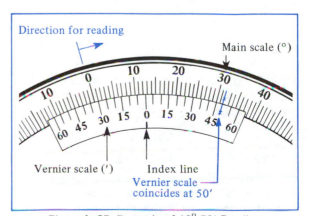

Figure 4–25 Example of 12° 50′ Readings

a whole-degree graduation on the main scale. However, if the angle is more than an exact whole number of degrees, the fractional value (in 5′ increments) is read on the vernier scale. In Figure 4–25 the index line (0) has moved past 12° on the main scale. The 50′ line on the vernier scale coincides with a graduation (32°) on the main scale. When the vernier scale is read in a clockwise direction, the angle represented on the vernier protractor in Figure 4–25 is 12° 50′.

SOLID AND CYLINDRICAL SQUARES

SOLID SQUARE

The solid, hardened steel square is a precision layout and angle-measuring tool. The solid square may have either a flat or a knife edge. This square is widely used for checking the squareness of a part feature that is 90° from a reference surface. The square is brought into contact with the feature. Any variation from a true 90° may be detected by placing a white paper behind the square and observing whether any white is visible.

CYLINDRICAL SQUARE

The cylindrical square, shown in Figure 4–26, is another tool that is used to make extremely accurate right-angle measurements. Markings on the cylindrical square show deviations from a true square condition in steps of 0.0002″.

The cylindrical square and the workpiece usually are placed on a surface plate. The part surface to be measured is moved into contact with the cylindrical square. The cylindrical square is rotated until all light between the part feature and the square is shut out. The deviation from squareness is read directly from the 0.0002″ graduations on the cylindrical square. Figure 4–26 shows a common application of the cylindrical square.

TEST INDICATORS AND DIAL INDICATING INSTRUMENTS

Two groups of indicators are commonly used in shops and laboratories. Test indicators and dial indicators are *comparison instruments*. Unlike direct measuring instruments, which incorporate

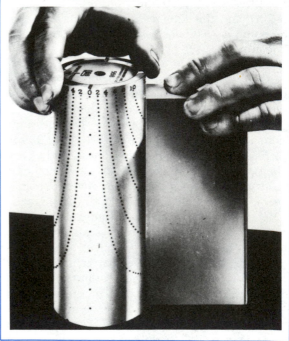

Courtesy of BROWN AND SHARPE MANUFACTURING COMPANY

Figure 4–26 Measuring Squareness with a Cylindrical Square

a standard unit of length, test indicators and dial indicators first must be set to a reference surface. The unknown length of the part to be measured is then *compared* to a known length. Test indicators and dial indicators are positioned from a stand or adjustable holder to which they may be secured.

TEST INDICATORS

Test indicators (Figure 4–27) are used for trueing and aligning a workpiece or a fixture in which parts are positioned for machining. Once the workpiece or fixture is aligned, machining, assembling, and other operations are performed.

Courtesy of THE L.S. STARRETT COMPANY

Figure 4–27 A Test Indicator and Attachment

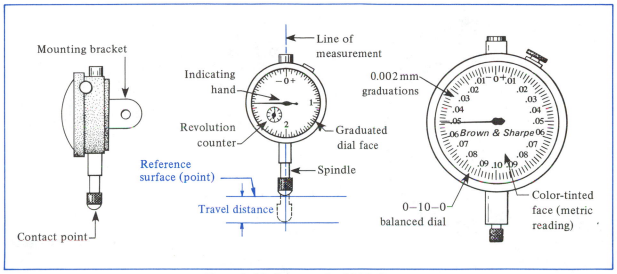

Figure 4–28 Features of a Dial Indicator

Test indicators are also used to test workpieces for roundness or parallelism or for comparing a workpiece dimension against a measurement standard.

The test indicator is held securely in a stationary base (Figure 4–27). The ball end of the test indicator is usually brought into contact with the workpiece surface. As the workpiece is moved, any deviation from the dimension to which the test indicator is set is indicated by movement of the indicator. The range of measurements is limited. Most test indicators employ a lever mechanism that multiples any minute movement of the measuring (ball) end.

DIAL INDICATING INSTRUMENTS

The dial test indicator (Figure 4–28) is commonly called the dial indicator. It contains a mechanism that multiplies the movement of a contact point. This movement is transmitted to an indicating hand. The amount of movement is read on a graduated dial face. Because the dial indicator is more functional and more accurate than the test indicator, it has a wider range of applications.

The dial indicator serves two major functions:

• To measure a length (the distance between a standard dimension to which the dial indicator is set and the length of the part being measured),

• To measure directly how much a feature of the workpiece is *out-of-true*.

Inch-standard dial indicators are commercially produced to measure to accuracies of 0.001″, 0.0005″, and 0.0001″. Metric dial indicators are accurate to 0.02 mm and 0.002 mm. Dial face graduations are marked to indicate the degree of accuracy.

Some dials are graduated to show + and – variations from 0 to 0.025″ or from 0 to 0.0005″. These dials are known as *balanced dials*. Examples of the many types of balanced dials are: 0–5–0 (0–0.0005″–0), 0–50–0 (0–0.050″–0), and 0–100–0 (0–0.100″–0).

Continuous dials are graduated from 0 to 0.100″. Readings beyond 0.100″ are made by multiplying the number of complete dial hand revolutions by 0.100″. The revolutions then are added to the reading of the indicating hand.

Long-range dial indicators are available for measuring dimensions up to 10″. These dial indicators are equipped with revolution counters. The long-range indicator in Figure 4–29 has a 0–2.000″ range, a 0–0.100″ dial, a revolution counter (0–1.000″), and an inch counter (0–1.000″–2.000″).

Common Applications of Dial Indicators. The dial indicator is constructed so that the

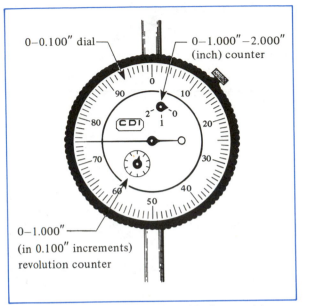

0−0.100″ dial

0−1.000″−2.000″ (inch) counter

CDI

0−1.000″ (in 0.100″ increments) revolution counter

Figure 4−29 Long-Range Dial Indicator with Revolution Counters (Balanced Type Dial)

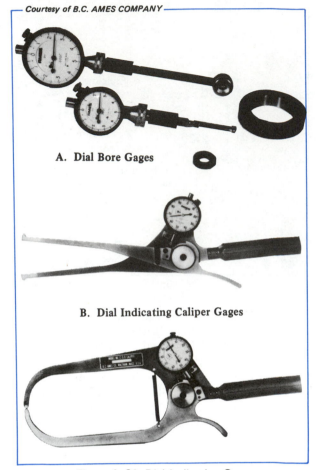

A. Dial Bore Gages

B. Dial Indicating Caliper Gages

Figure 4−30 Dial Indicating Gages

indicator mechanism may be used alone. It may be secured to a holding device and positioned to take a measurement. For example, in applying the dial indicator to check the out-of-roundness of a part on a lathe, the indicator is clamped in a holder. The indicator holder, in turn, is mounted in a lathe tool holder. As the workpiece is slowly rotated, any roundness variation causes the contact end of the indicator to move up or down.

Another common application is checking the accuracy of a lathe spindle. A precision-machined test bar, which is inserted in the taper of the spindle, is tested for runout (levelness) with the dial indicator.

Dial Indicating Gages. The dial indicator often is combined with other gaging devices and measuring tools. A few examples are given in Figure 4−30 to show the versatility of the instrument.

When the dial indicator is combined with a depth gage, it is called a *dial indicating depth gage*. When the dial indicator is adapted to testing holes for size and out-of-roundness or other surface irregularities, the instrument is called a *dial indicating hole gage*, or *dial bore gage*. The dial indicator may also be adapted for taking

inside or outside measurements. Instruments called *dial indicating caliper gages* have revolution counters. The counter permits measurement of dimensions to 3″ (75mm).

Dial indicating snap gages are another adaptation of the dial indicator. These gages are used for linear measurements as well as for measuring diameters.

Dial Indicating Micrometers. Inch-standard *dial indicating micrometers* (Figure 4−31) can measure dimensional variations in one ten-thousandth of an inch (0.0001″) and one one-thousandth of an inch (0.001″). Metric dial indicating micrometers are used to measure to accuracies of 0.02mm and 0.002mm.

The lower anvil of the dial indicating micrometer is a sensitive contact point. It is

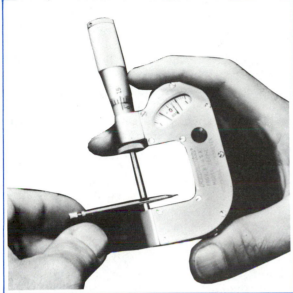

Figure 4–31 Dial Indicating Micrometer (±0.0001″)

connected directly to a dial indicator that is built into the instrument. When used as a micrometer, the thimble is set to the one one-thousandth (0.001″) setting nearest the required measurement and is locked in position. As the dial indicating micrometer is brought into contact with the workpiece, it serves as a gage. Any variation from the required size appears on the dial indicator. The dial is graduated in increments of 0.0001″, with a range up to 0.001″.

Universal Dial Indicator Sets. The term *universal dial indicator set* denotes that there are a number of attachments for holding and positioning the dial indicator. For instance, to check the height of a machined surface, the dial indicator head may be mounted on a T-slot base (Figure 4–32). The indicator is set at a desired height. It is then moved along a surface plate and positioned over the part of the workpiece that is to be measured.

Dial indicator applications include checking the out-of-roundness, runout, parallelism, or face alignment of a workpiece. In these applications the dial indicator may be held in a tool post holder attachment on a lathe or other machine. The dial indicator may be positioned to measure along any one, or combination of three axes (Figure 4–32).

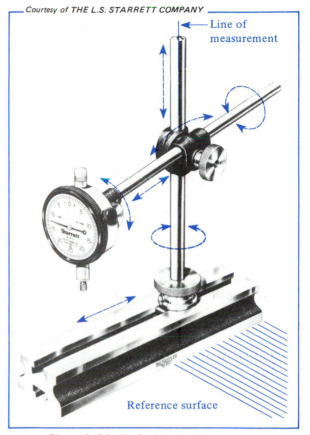

Figure 4–32 Six Positioning Movements of a Dial Indicator

PRECISION GAGE BLOCKS
PARALLEL GAGE BLOCKS

Gage blocks are blocks of steel that are heat treated to a high hardness. Gage blocks are machined, ground, and finished to precise limits of dimensional accuracy. The heat-treating process makes it possible to retain accuracy and dimensional stability. The hardness produced by heat treating helps prevent wear and damage to the block surfaces.

Parallel gage blocks are available in three general grades: (1) *master blocks*, which are accurate to within 0.000002″ or 0.00005mm; (2) *inspection blocks*, which are accurate to within 0.000005″ or 0.0001mm; and (3) *working blocks*, which are accurate to within 0.000008″ or 0.0002mm. Master and inspection blocks are usually used in temperature-controlled laboratories.

The dimensional accuracy of parallel gage blocks relates to flatness, parallelism, and length. Some sets include two additional *wear blocks* (0.050″), which are used as *end blocks* to prevent wear on the other blocks.

PARALLEL GAGE BLOCK SETS AND APPLICATIONS

Practical gage block applications include:

- Checking precision measuring instruments and gages;
- Setting other instruments such as dial indicators and height gages to make comparison measurements;
- Laying out machined surfaces where highly accurate linear measurements are required.

Gage blocks are furnished in many combinations. Two common sets have 83 or 35 gage blocks each. The sizes of the gage blocks in the 83-piece inch-standard set are given in Table 4–1. Note that in the 9-block 0.0001″ series each block increases in size by an increment of 0.0001″. The increment in the 49-block 0.001″ series is 0.001″. In the 0.050″ series there are 19 blocks varying in size from 0.050″ to 0.950″ in increments of 0.050″. There are 4 blocks in the 1.000″ series, with sizes from 1.000″ to 4.000″.

Over one-hundred thousand different measurements may be made with the 83-piece set by combining different gage blocks. The gage block surfaces are so precisely finished that when perfectly clean, they may be carefully *wrung together* to become almost one gage block.

Gage blocks are assembled by overlapping one clean block on another. A gentle force is applied while sliding the blocks together. The gage blocks may be taken apart by reversing the assembly process. In all gage block applications, it is important to use the least number of blocks possible.

Gage blocks are selected starting with a block in the 0.0001″ series which corresponds with the fourth place decimal value in the required measurement. This is followed by one or more blocks in the 0.001″ series, then the 0.050″ series, and the 1.000″ series, as required. Figure 4–33 provides an example of a gage block combination for a measurement of 4.8355″ (±0.000 008″).

PRECISION MEASUREMENT OF ANGLES AND TAPERS

PRECISION ANGLE MEASUREMENT WITH ANGLE GAGE BLOCKS

Angle gage blocks permit the measurement of angles within limits of a fraction of a second

Table 4–1 Series and Sizes of Gage Blocks—83 Piece Set

0.0001 Series (9 Blocks)								
.1001	.1002	.1003	.1004	.1005	.1006	.1007	.1008	.1009
0.001 Series (49 Blocks)								
.101	.102	.103	.104	.105	.106	.107	.108	.109 .110 .111 .112 .113
.114	.115	.116	.117	.118	.119	.120	.121	.122 .123 .124 .125 .126
.127	.128	.129	.130	.131	.132	.133	.134	.135 .136 .137 .138 .139
.140	.141	.142	.143	.144	.145	.146	.147	.148 .149
0.050 Series (19 Blocks)								
.050	.100	.150	.200	.250	.300	.350	.400	.450 .500 .550 .600
.650	.700	.750	.800	.850	.900	.950		
1.000 Series (4 Blocks)								
1.000	2.000	3.000	4.000					
0.050 Wear Blocks (2 Blocks)								

of arc. Angle gage blocks are long, tapered rectangular blocks.

Angle gage blocks are available in three accuracy grades. *Toolroom* angle gage blocks are used for angle measurements within 1-second accuracy. *Inspection grade* angle gage blocks are applied to measurements within 1/2-second ac-

curacy. The most precise set is called the *laboratory master*. Laboratory masters are used for measurements within 1/4-second accuracy. The surface finish of the gaging surfaces also varies from 0.6 microinch for the toolroom grade to 0.1–0.3 microinch for the laboratory grade.

A standard 16 angle gage block set has a range of measurement from 0° to 99° in steps of one second. The number of blocks in this set and the angles of each block appear in Table 4–2. It is possible, with a laboratory grade set of 16 blocks, to measure 356,400 angles in steps of one second to an accuracy of ±1/4 second. In addition to the angle gage blocks, a 6″ parallel and a 6″ knife edge (form of parallel having one edge that is narrowed to a smaller area) are available.

The angle gage blocks are designed to be combined in plus and minus positions. One end of a gage block is marked *plus*. The other end is marked *minus*. For example, Figure 4–34 shows

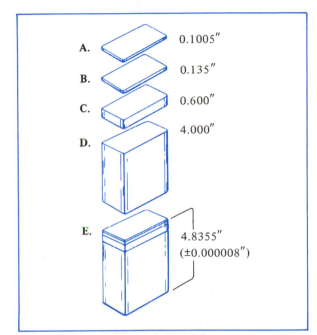

A. 0.1005″

B. 0.135″

C. 0.600″

D. 4.000″

E. 4.8355″ (±0.000008″)

Figure 4–33 Selection of Gage Blocks for a Measurement of 4.8355″

Table 4–2 Angles of the Gage Blocks in a 16-Block Set

Number of Blocks	Angles
6	1, 3, 5, 15, 30, 45 degrees
5	1, 3, 5, 20, 30 minutes
5	1, 3, 5, 20, 30 seconds

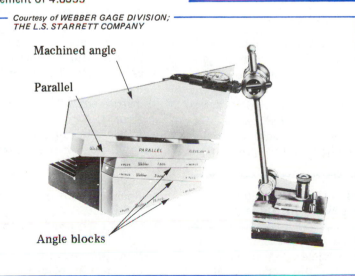

Courtesy of WEBBER GAGE DIVISION; THE L.S. STARRETT COMPANY

Machined angle

Parallel

Angle blocks

Figure 4–34 Angle Gage Blocks and Dial Test Indicator Setup to Measure a Required Machined Angle on a Workpiece

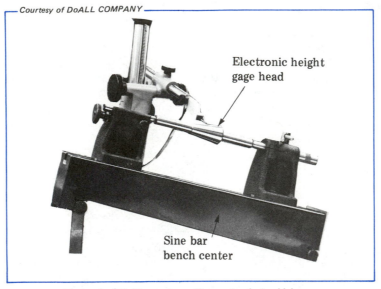

Courtesy of DoALL COMPANY

Electronic height gage head

Sine bar bench center

Figure 4–35 Measuring a Taper Angle by Using a
Granite Sine Bar Bench Center and Electronic Height Gage

a workpiece that is being checked with a dial test indicator and angle gage blocks. The three angle gage blocks produce an angle of 13°. The workpiece is resting on a parallel block that is wrung to the angle blocks. The setup is aligned at a right angle by positioning the parts against an angle plate. The 13° angle requires a 15° angle gage block, minus a 3° block, plus a 1° block.

MEASURING TAPER ANGLES

Tapers that have been machined between centers may be measured with bench centers. These centers are sometimes mounted on a sine bar, as shown in the setup in Figure 4–35. The required gage block height combination is determined for the required angle. The part is carefully inserted between the bench centers. The sine bar is then raised to the height of the gage blocks.

Safe Practices in Using Precision Measuring and Layout Instruments

---

• Check the workpiece to see that it is free of burrs and foreign particles. A clean workpiece helps to ensure the accuracy of all measurements.

• Clean all contact surfaces of vernier measuring and dial indicating instruments and related accessories.

• Check each instrument for accuracy. For example, in the fully closed position, the inside and outside readings of the vernier caliper must be zero (0.000″ or 0.00mm).

• Bring the jaws or other instrument surfaces lightly into contact with the workpiece. Any force applied to the instrument will cause springing and inaccurate measurements.

• Take measurements as close to the instruments as possible. The longer the distance a measurement is taken from the base or frame of the instrument, the greater the probability of error.

• Check the alignment of the instrument, the condition of the reference surface, and the position of the workpiece before taking a measurement.

• Recheck each measurement to ensure that it is accurate.

• Clean each precision tool thoroughly. Instrument moving parts must be lubricated with a thin, protective film. All instruments should be carefully placed and stored in appropriate containers.

UNIT 4 REVIEW AND SELF-TEST

A. STANDARD AND VERNIER MICROMETERS

1. Cite three advantages of using micrometers over steel rules and calipers.

2. Tell why measurements taken for the inch micrometer with a ratchet stop are consistently more accurate than measurements made with a standard micrometer.

3. State three safe practices that must be observed when using a micrometer.

4. Indicate how a 0.001″ micrometer may be used to estimate a measurement finer than 0.001″.

5. Describe briefly the vernier principle as applied to an outside micrometer.

6. State how the spindle graduations of a depth micrometer differ from the spindle graduations on an outside micrometer.

7. Explain how to take an accurate measurement with an inside micrometer.

8. Cite an example of where telescoping and small-hole gages must be used instead of inside micrometers or transfer measuring tools.

B. VERNIER MEASURING AND DIAL INDICATING INSTRUMENTS AND GAGE BLOCKS

1. Indicate the functions of (a) the beam containing the main scale of a vernier measuring instrument and (b) the vernier scale.

2. Use the inch-standard 25-division vernier caliper to explain how to read a measurement of 4.377″.

3. State how the accuracy of a metric vernier height gage may be checked.

4. Give three major differences between a universal bevel protractor and a standard bevel protractor of a combination set.

5. State two distinguishing design features between a solid square and a cylindrical square.

6. Indicate the different purposes that are served by test indicators and dial indicating instruments in relation to vernier instruments.

7. Give four applications of dial indicators.

8. List three examples of instruments that combine the measurement features of dial indicators with other measurement instrument functions.

9. Explain how to build up a gage block measurement of 2.5255″ ±0.000008″.

10. Cite two advantages of using angle gage blocks for laying out and inspecting angle surfaces in preference to regular gage blocks and a vernier height gage setup.

11. State two safety precautions to follow to ensure accurate precision measurements.

Advanced Precision Measurement, Quality Control, and Surface Texture

All of the measuring instruments covered thus far are mechanical. The range of accuracy has been from ±1/64″ (±0.5mm) using a steel rule to ±0.0001″ (±0.002mm) using a vernier-scale instrument. More precise measurements to higher degrees of accuracy are obtainable with *optical flats* and *microscopes*. These instruments depend on physical science principles related to light waves and optics. Still other *high amplification comparators* are treated in this unit. The comparators depend on electrical, electronic, and pneumatic principles, often in combination with mechanics and optics.

Measurements related to flatness, parallelism, precise dimensional, and form differences are treated in terms of quality control and surface texture.

A. MEASUREMENT WITH OPTICAL FLATS AND MICROSCOPES

SHOP AND TOOLMAKER'S MEASURING MICROSCOPES

Two common optical instruments found in the shop and laboratory are the *shop microscope* and the *toolmaker's measuring microscope*. Practical measurement and inspection procedures may be carried on with these microscopes.

OPERATION OF THE SHOP MICROSCOPE

A shop microscope of simple design is pictured in Figure 5-1. The main features include an adjustable ocular lens and tube, a column and stand, an adjusting ring with a graduated scale, and a light source. The microscope is mounted in a stand. The tube may be raised or lowered by a clamp screw.

The ocular lens is adjusted by turning it to the right or left. Adjustment brings the object into proper focus. The light source in the column is directed to the object through the relieved

area in the base. The degree of magnification of the microscope is the product of the magnifications of the ocular lens and the objective lens. For example, the 40X shop microscope has a power of magnification of 40 (10X ocular lens with a 4X objective lens).

A micrometer scale is mounted within the ocular. The 150-division scale represents 0.150″. Measurements may be made to 0.001″. Subdivisions of 0.001″ (0.02mm) may be estimated to within 0.00025″ (0.005mm). The instrument is thus capable of a high degree of accuracy.

One advantage of the shop microscope is that it is small and convenient to carry to a required location. A disadvantage is that it has a limited

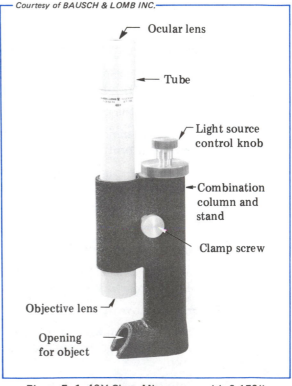

Ocular lens

Tube

Light source control knob

Combination column and stand

Clamp screw

Objective lens

Opening for object

Figure 5-1 40X Shop Microscope with 0.150″ Scale Graduated in 0.001″

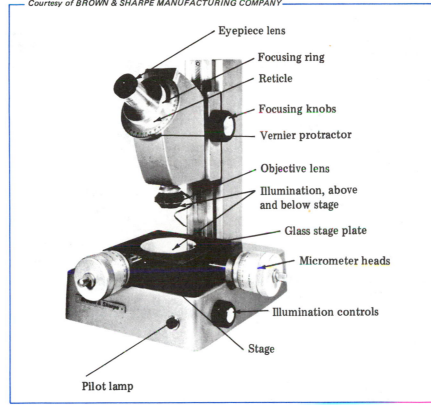

- Eyepiece lens
- Focusing ring
- Reticle
- Focusing knobs
- Vernier protractor
- Objective lens
- Illumination, above and below stage
- Glass stage plate
- Micrometer heads
- Illumination controls
- Stage
- Pilot lamp

Figure 5–2 Main Features of a Toolmaker's Measuring Microscope

field. *Field* refers to the diameter of the area that may be observed in a single position. The model shop microscope, as illustrated in Figure 5–1, has a field of 7/32" (5.5mm) and magnifies 40X.

GENERAL APPLICATIONS OF THE SHOP MICROSCOPE

General applications of the shop microscope are as follows:

- Measuring the diameter of small holes having tolerances of ±0.001" (0.02mm) (such holes often require measurement and inspection during machining);
- Measuring small dimensions;
- Checking the thickness of the case on hardened parts;
- Measuring the diameter of impression made during hardness testing;
- Checking mechanical parts quickly for wear (welds, for example, may be inspected easily);

- Examining surfaces on-the-spot to detect cracks and flaws (the condition of plated, polished, and painted surfaces may be checked quickly).

TOOLMAKER'S MEASURING MICROSCOPE

The toolmaker's measuring microscope is a ruggedly designed microscope with high-power magnification. It is adaptable to taking linear and angular measurements. The microscope is especially valuable for inspecting small parts and tools. The main features of the microscope are labeled in Figure 5–2.

A number of attachments permit the microscope to be used as a comparator. A thread form or other tool contour may be compared with a charted outline that is superimposed directly upon the virtual image of the object. The contour of the object may thus be checked to within 0.0001" (0.002mm).

Angles may be measured to an accuracy of five minutes of arc on the toolmaker's measuring microscope. The microscope in Figure 5–2 has a *protractor eyepiece* (eyepiece and lens assembly, combined with a vernier protractor) that is used to make angular measurements.

The toolmaker's microscope may also be used to measure pitch diameter, major and minor diameters, pitch, and lead. Lead-measuring and thread-measuring attachments extend the practical applications of this microscope. Some advanced models include a projection attachment that permits making contour and other precision measurements normally performed on comparators.

OPTICAL HEIGHT GAGE

The *optical height gage* is another instrument that applies optics. This instrument combines design and measurement features of gage blocks and a measuring microscope. The optical height gage consists basically of a "stack of gage blocks." These blocks are permanently wrung together. The accuracy of the stack is ±0.00005″ per inch (0.0002mm per 25mm) of height. Measurements are read directly in the eyepiece of the microscope.

Some older height gage models are being retrofitted by their manufacturers to replace the optical feature by an easier reading digital readout. One manufacturer's model, the Digi-Chek® II, is shown in Figure 5–3. This particular height gage includes the digital readout component.

The instrument requires the use of a standard reference bar. The reference bar consists of a master stack of alternating 0.300″ gage block jaws and 0.700″ spacer blocks. The gage blocks are permanently wrung and fastened together to form 1″ increments. The master stack has a special bushing arrangement. The bushing allows the stack to conform to thermal conditions under actual use.

The linear accuracy of the master stack of gage blocks is +0.000004/–0.00000″ per inch of length. The accuracy of a master stack in metric units of measure is +0.00001/–0.0000mm per 25mm of length. The parallelism of the gage surfaces to the base and to one another is 0.000015″ (0.0004mm). The measuring system accuracy is

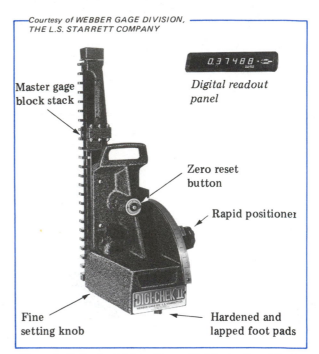

Master gage block stack

Digital readout panel

Zero reset button

Rapid positioner

Fine setting knob

Hardened and lapped foot pads

Figure 5–3 Optical Height Gage with Master Gage Block Stack (Standard Reference Bars) and Digital Readout

0.00005″ (0.00125mm). Most digital systems are equipped with a safety power failure flashing display.

INDUSTRIAL MAGNIFIERS

Industrial magnifiers are used in the shop and laboratory for magnifying lines, surface areas, and design or machining features. Some magnifiers are mounted and provided with an outside source of illumination. They are adjustable. They permit working at a short distance from the workpiece or object. Other magnifiers attach to vernier scales and permit easier and more accurate readings.

Figure 5–4 shows a hand measuring magnifier that has four transparent scales. The magnification of this particular magnifier is 7X. Each transparent scale allows adequate illumination to produce a sharp, flat image. Each of the four scales is contained in a separate mount. The scales are readily interchangeable. The *general-purpose scale* permits measurements in increments of 0.005″, 0.1mm, and 1° and the measurement of 1/16″ to 5/8″ radii and 0.001″, 0.002″, and 0.003″ line thicknesses. A *separate inch scale* has graduations in 0.005″ intervals up to 3/4″. The *metric scale* is graduated in 0.1mm intervals over 20 mm. The *protractor scale* is divided in single degrees.

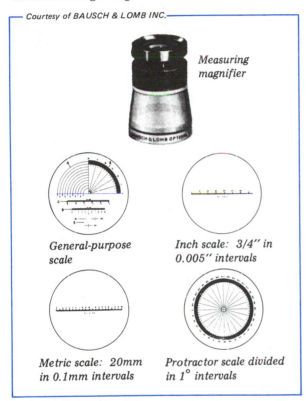

Measuring magnifier

General-purpose scale

Inch scale: 3/4″ in 0.005″ intervals

Metric scale: 20mm in 0.1mm intervals

Protractor scale divided in 1° intervals

Figure 5–4 Hand Measuring Magnifier equipped with Four Transparent Scales

Simple designs of magnifiers with single lenses are identified as hand magnifying readers (magnifying glasses).

PRECISION MEASUREMENT WITH OPTICAL FLATS

CHARACTERISTICS OF OPTICAL FLATS

Optical flats are used for precision checking flatness, parallelism, size, and surface characteristics. The optical flats used for measurement and inspection are made of a high-quality optical quartz. This material has a low coefficient of expansion (approximately 0.32 millionths of an inch for each degree Fahrenheit change). Ultra-low expansion fused silica optical flats are also available. One or two faces may be finished as a measuring surface. A band on the side of the optical flat indicates the measuring surface(s). The accuracy of the flat is marked on the band in millionths of an inch.

Optical flats are manufactured in three grades. *Working flats* are produced with a degree of flatness of 0.000004″ (0.0001mm). No point on a working flat will deviate from any other point by more than 4 microinches (millionths of an inch). *Master flats* are accurate to within 0.000002″ (0.00005mm). *Reference flats*, used primarily under close controlled laboratory conditions, are held to an accuracy of 0.000001″ (0.000025mm).

Optical flats are also furnished with coated surfaces. A thin film of titanium dioxide on the optical flat sharpens the image for greater accuracy.

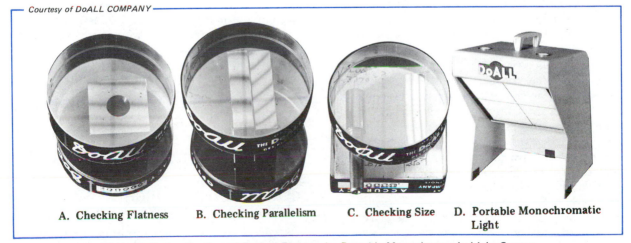

A. Checking Flatness **B. Checking Parallelism** **C. Checking Size** **D. Portable Monochromatic Light**

Figure 5–5 Application of Optical Flats under Portable Monochromatic Light Source

Commercial optical flats are available in sizes ranging from 1″ to 10″ diameter (25 mm to 250 mm). The thicknesses (depending on diameter) vary from 1/2″ (12mm) to 2″ (50mm). The common sizes of square optical flats are 2″, 3″, and 4″ on a side by 5/8″ and 3/4″ thickness. Metric sizes are 50mm, 75mm, and 100mm by 15mm to 18mm thick.

MONOCHROMATIC LIGHT SOURCE

A monochromatic light source, when used with an optical flat, produces a distinct *fringe pattern*. This pattern, consisting of alternately light and dark bands, shows up clearly to establish the flatness, parallelism, and size (over a limited dimensional range), as well as surface characteristics (Figure 5–5A, B, and C). A portable monochromatic light, or monolight, is adaptable for bench-type applications (Figure 5–5D).

Each color in the solar spectrum has a wavelength that tends to blend with the next color. The wavelengths range from 0.0000157″ (0.0003937mm) for ultraviolet rays to 0.0000275″ (0.0006985mm) for red rays. The light emitted from a monochromatic (rays of one definite wavelength) light is 23.1 microinches (millionths of an inch) or 0.0005867mm.

MEASUREMENT OF SURFACE FLATNESS

Thus far, only straight interference bands have been discussed. The straight, parallel, uniform bands relate to a flat surface. Variations in a flat surface produce curved interference bands. Such bands indicate a concave or convex surface, a slight angle, or a radius.

Interference bands that curve slightly at the ends show that the edges of the workpiece are rounded. Interference bands in a series of curved lines that form elliptical circles indicate a high spot on the surface of the workpiece. The pattern may be identified when a slight pressure is applied to the optical flat above the spot. Curvature due to a high spot (hill) points toward the open side of the air wedge. Interference bands as a bent pattern indicate a low spot (valley) in the surface of the workpiece. The curvature points toward the closed side of the air wedge when it is due to a valley.

Table 5–1 Partial Conversion Table for Optical Flat Fringe Bands (Inch and Metric Standard Units of Measure)

Number of Bands	Equivalent Measurement Value		
	Microinches	Inches	Millimeters
0.1	1.2	0.0000012	0.000029
0.2	2.3	0.0000023	0.000059
0.3	3.5	0.0000035	0.000088
. . .	. . .	. . .	. . .
1.0	11.6	0.0000116	0.000294
2.0	23.1	0.0000231	0.000588
3.0	34.7	0.0000347	0.000881
. . .	. . .	. . .	. . .
19.0	219.8	0.0002198	0.005582
20.0	231.3	0.0002313	0.005876

INTERFERENCE BAND CONVERSION TABLE

Part of the full conversion table that appears in the Appendix for converting fringe bands to microinches (or metric equivalents) is reproduced in Table 5–1. Values are given in fractional parts of an inch and of a millimeter. When close approximations are sufficient, one band indicates an accuracy of approximately 10 millionths of an inch (0.00025mm).

B. MEASUREMENT WITH HIGH AMPLIFICATION COMPARATORS

A comparator is highly sensitive instrument that checks the accuracy of a measurement for any variations from a standard. It incorporates within its design:

- A device for holding a workpiece,
- A master against which the workpiece is checked,
- An amplification unit that permits small variations from basic dimensions to be observed and measured.

GENERAL TYPES OF HIGH AMPLIFICATION COMPARATORS

Six general types of high amplification comparators are available:

- *Mechanical comparators* include high amplification dial indicators and bench comparators.
- *Mechanical-optical comparators* are combination mechanical-optical instruments. A light beam casts a shadow on a magnified scale, and dimensional variations are indicated on the scale.
- *Optical comparators* are projection and reflection comparators. Features are measured by casting a magnified shadow of a part on a screen and then using a chart for comparison measurement.
- *Electrical-electronic comparators* are measuring instruments that have the added feature of power amplification.
- *Pneumatic comparators* require a flow of air between a part and a measuring gage. Pressure and flow changes indicate any variation in dimensional accuracy.
- *Multiple gaging comparators* are a combination of mechanical, electronic, optical, and pneumatic comparators. Multiple gaging comparators are used when a number of dimensions are to be gaged at one time.

Principles underlying each type of comparator, with general applications, are covered in this unit.

HIGH AMPLIFICATION MECHANICAL COMPARATORS

HIGH AMPLIFICATION DIAL INDICATORS

The most commonly used dial indicators are graduated for readings in 0.001″ (0.02mm) and 0.0001″ (0.002mm).

Two mechanical comparators, high amplification dial indicators, are shown in Figure 5–6. The dial indicator pictured in Figure 5–6A has a *discrimination* of 20 microinches. Jeweled bearings are used in the instrument to reduce friction and wear and to ensure *repeatability*—that is, the dial indicator reads (after a number of repeated insertions of a workpiece) to a standard of accuracy of ±one-fifth of a division from an original reading. The dial indicator here is graduated so that each division equals 0.00002″ (0.0005mm); the repeatability is 0.000004″ (0.0001mm).

A precision double dial test indicator, which is particularly suited to machine tool applications,

is shown in Figure 5–6B. This indicator has a balanced 0.004″ (0.01mm) dial (0.002–0–0.002″). The graduations are in 0.00005″ (0.00125mm). The repeatability error of the instrument is ±0.00001″ (0.0002mm). The round steel or carbide contact points are available in three diameters: 0.040″, 0.080″, and 0.120″ (1mm, 2mm, and 3mm).

MECHANICAL COMPARATORS

Mechanical comparators consist of a base, column, adjusting arm, and a gaging head. A mechanical comparator works on the same principles as a dial indicator. It is used for taking

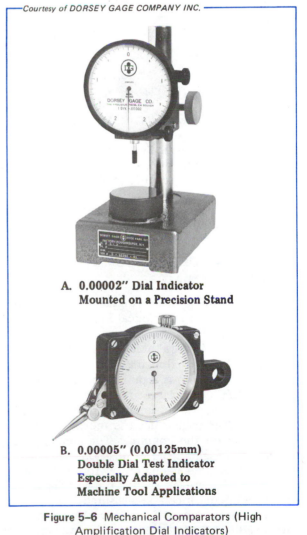

A. 0.00002″ Dial Indicator Mounted on a Precision Stand

B. 0.00005″ (0.00125mm) Double Dial Test Indicator Especially Adapted to Machine Tool Applications

Figure 5–6 Mechanical Comparators (High Amplification Dial Indicators)

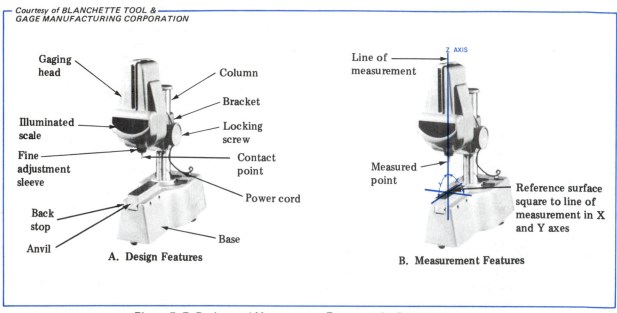

Courtesy of BLANCHETTE TOOL & GAGE MANUFACTURING CORPORATION

A. Design Features

B. Measurement Features

Figure 5–7 Design and Measurement Features of a Reed Comparator

accurate linear measurements by comparison with a standard. Any dimensional variation between the part being measured and the standard against which the instrument is set is shown on a magnified scale. The scale is graduated with plus and minus inch or metric units of measurement. The mechanical comparator in Figure 5–6B provides for extremely high amplification.

MECHANICAL-OPTICAL COMPARATORS

REED COMPARATOR

A *reed comparator* (Figure 5–7) combines the features of a mechanical instrument and an optical instrument. It consists of a base, column, and a gaging head and bracket. The main design features of a reed comparator are shown in Figure 5–7A. A lamp is required in the magnification system. This lamp is housed in the base. The base supports an anvil, a back stop, and a column. The gaging head is moved by a rack and pinion on the column. The locking screw clamps the gaging head in position. The gaging head may also be swung away from the base so that parts that are larger than the working height of the comparator can be measured.

The measurement features are shown in Figure 5–7B. The reference surface for the Z axis is at a right angle to the line of measurement in the X and Y axes. The general range of reed comparators is from 500X to 20,000X.

Reed Comparator Scales. There are two different scales for a simple reed comparator. The note above the inch-standard scale indicates the instrument has a magnification of 1000X. Each whole plus or minus numerical division (0.001″) is divided in tenths. The scale is marked 1/10,000 and indicates that each graduation is equal to 0.0001″. The range of the scale is ±2 thousandths (±0.002″), or 0.004″ overall.

The metric scale also has a magnification of 1000X. Each whole plus or minus numerical division represents 0.01mm. Since each numbered division is divided into five equal parts, each graduation equals 0.002mm. The range of the scale used here is from ±0.05mm or 0.1mm overall. The plus and minus range of the sample comparator scales are ±0.002″ (0.05mm). Reed comparators, having a 10,000 to 1 amplification, are capable of measurement accuracies to 10 microinches (0.000010″) or 0.00025mm.

OPTICAL COMPARATORS

Optical measuring instruments have great versatility, range, precision, and accuracy. One of the

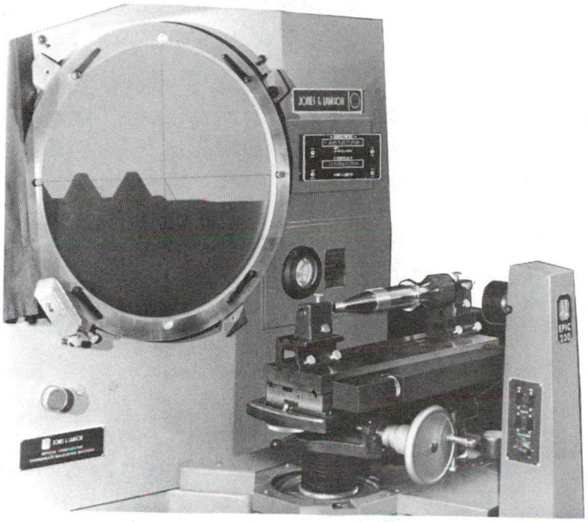

Figure 5-8 Application of Optical Comparator (with Digital Readouts) for Contour and Size Measurements

most popular and widely used optical measuring instruments is the *optical comparator* (Figure 5-8). It is sometimes called a *contour projector*. This projector provides an accurate method of measuring and comparing the contour of irregularly shaped parts. Flat or circular tools, gages, grooves, angles, and radii may be checked for size. Metal, plastic, and soft materials such as rubber may be measured without physical contact. Tolerances within ±0.0005″ (±0.01mm) may be measured accurately.

STAGE AND ACCESSORIES

The stage of an optical comparator may be moved horizontally or vertically. A work table provides a plane surface for accessories and workpieces. The table is provided with a large-diameter micrometer head that permits horizontal movement of the stage to accuracies within 0.0001″ (0.002mm). Vertical travel is obtained with a dial indicator. A work stage that is moved by turning micrometer dials permits the accurate measurement of linear dimensions. Fixture bases with a small rotary vise or adjustable horizontal and center attachments extend the use of the comparator. Fixtures provide an exact location for a part and a master.

CONTOUR MEASURING PROJECTOR

The contour measuring projector is similar in principle and application to the optical comparators previously described. The image is visible in the same position as the operator views the object on the table.

The contour measuring projector has a number of advantages:

- The need to hold the object in special tools or fixtures is eliminated,
- Setup time is reduced,
- The object is placed automatically at a right angle to the beam of light,
- The object may be positioned by hand to permit rapid setup of the image with the lines or outline on the chart.

The table is positioned longitudinally, transversely, and vertically by feed handles. Horizontal measurements are read directly on micrometer dials on each feed screw. Common accessories include center supports and a screw thread fixture. Screens and charts also are used to make linear, angle, radius, and contour measurements.

Today, in addition to establishing the inside size and characteristics of holes, pneumatic comparators are applied in checking outside diameters, concentricity, squareness, parallelism, and other surface conditions.

OPERATION OF PNEUMATIC COMPARATORS

The principle of operation of pneumatic comparators is simple. Dependence is placed on the flow of air between the faces of the jets in the gaging head and the workpiece. The clearance between the gaging head and the workpiece controls both the velocity and the pressure of air. The larger the workpiece is than the gaging head, the greater the flow (higher velocity) of air and the lower the back pressure. Conversely, the smaller the clearance between the gaging head and the workpiece, the slower the velocity and the greater the back pressure.

By knowing this principle, it is possible to make a comparison of either the flow of air (velocity) or its pressure. Flow (column-type) pneumatic gages indicate the velocity of air. Pressure-type pneumatic gages are designed to indicate air pressure. Both types of pneumatic gages are operated from either a portable or a plant supply of compressed air.

PNEUMATIC GAGING HEADS

The common forms of gaging heads include plug, ring, and snap gages. A *gaging head* consists of a flow tube and a gaging spindle. The flow tube provides the channel through which the air supply flows and is measured. The gaging spindle consists of a plug with a central air channel. The channel terminates in two or more jets in the side of the spindle. The air flows through these jets. The form of the spindle is determined by the operation to be performed. When the diameter of a hole is to be measured, a spindle with air jets that terminate in annular (around the circumference) grooves is used. When a hole is to be checked for roundness, bell-mouth, or taper, a spindle with air channels that run parallel to the axis of the spindle is used. Two diametrically opposite jets or more may terminate in the channels. The longitudinal (lengthwise) channels permit a point-to-point check.

ADVANTAGES OF PNEUMATIC COMPARATORS

The major advantages of gaging with pneumatic instruments over other measurement instruments are as follows:

- Amplifications range from 1 to 1000 to 1 to 40,000. Pneumatic instruments provide a simple and direct method of high amplification.
- Minute dimensional variations can be controlled within close tolerances.
- Parts may be gaged without contact between the workpiece and the gaging head. Thus, scoring of highly polished surfaces or damage to soft material parts is prevented.
- Less skill is required in using pneumatic gaging instruments than in gaging processes with precision conventional plug, ring, and snap gages.
- The work life of pneumatic gaging heads is longer than for conventional gages.
- Pneumatic gages may be used at a machine or on the bench. The work can be brought to the instrument, or vice versa.

Courtesy of SHEFFIELD GAGES

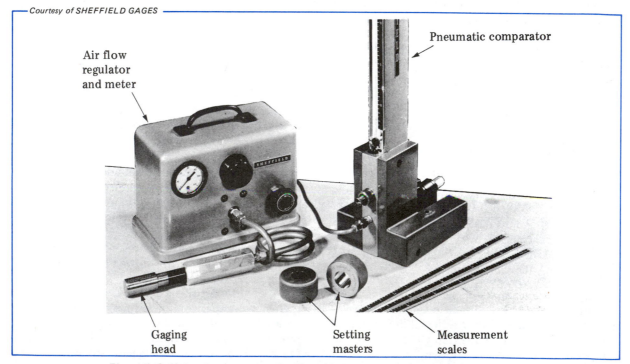

Air flow regulator and meter

Pneumatic comparator

Gaging head

Setting masters

Measurement scales

Figure 5–9 Major Components of a Flow (Column-Type) Pneumatic Comparator

- Multiple diameters may be checked at the same time.
- Pneumatic gages are easier to use than mechanical gages. Holes may be inspected for out-of-roundness, parallelism, concentricity, and other surface irregularities.
- Gaging heads and parts are self-cleaning.

OPERATION OF A FLOW (COLUMN-TYPE) PNEUMATIC COMPARATOR

The flow, or column-type, pneumatic comparator measures air velocity. The major components of such a comparator are shown in Figure 5–9. Air is passed through a filter and a regulator. The air flows through a tapered, transparent tube and reaches the gaging head at about 10 pounds pressure per square inch. The air flow causes a float in the tube to be suspended.

As air flows through a metering gage, it exhausts through the channels of the gaging head and the clearance area between the head and the workpiece.

OPERATION OF A PRESSURE-TYPE PNEUMATIC COMPARATOR

The major features of a pressure-type pneumatic comparator gage system includes a constant air supply which passes through a filter, a pressure regulator, and a master pressure gage. The supply of air is then branched through two channels. The upper branch is identified as the reference channel. The lower branch is identified as the measuring channel. Air escapes from the reference channel to the atmosphere through a zero setting valve. Air from the measuring channel flows through the gage head jets and the part being measured and then into the atomsphere. Note

Courtesy of FEDERAL PRODUCTS CORPORATION

Single setting master

Figure 5–10 Measuring a Diameter with a Pressure-Type Pneumatic Comparator and Gaging Spindle

that the two channels are connected through a differential pressure meter.

SETTING UP THE GAGE SYSTEM FOR MEASUREMENT

The pressure-type pneumatic comparator gage system is set by placing a setting master over the gaging spindle plug. The zero setting valve is adjusted until the gage pointer (needle) reads zero. Any variation between the size of the master gage and the workpiece affects the pressure in the measuring channel. The difference is indicated by the movement of the dial gage pointer.

Table 5–2 Amplification, Discrimination, and Measurement with a Pneumatic Pressure-Type Balanced System

	Ranges	
Amplification	Discrimination	Measurement
1,250:1	0.0001″ (0.0025mm)	0.006″ (0.15mm)
2,500:1	0.00005″ (0.00125mm)	0.003″ (0.075mm)
5,000:1	0.00002″ (0.0005mm)	0.0015″ (0.0375mm)
10,000:1	0.00001″ (0.00025mm)	0.0006″ (0.015mm)
20,000:1	0.000005″ (0.000125mm)	0.0003″ (0.0075mm)

The range of amplification of pressure-type pneumatic comparators is from 1250 to 1 to 20,000 to 1. A common pressure-type pneumatic comparator is pictured in Figure 5–10. The data in Table 5–2 indicates the range of amplification, discrimination, and measurement with this type of instrument in a *balanced system*. With increased amplification, a higher level of discrimination and a lower measuring range result.

ELECTRICAL COMPARATORS

The three major components in an electrical comparator are a gaging head, a power unit, and a microammeter. The microammeter contains a graduated scale. Any movement of the end of the gage head is magnified greatly by the microammeter.

The spindle (lever) of the gage head bears against an armature. The end of the armature is positioned between two electrical induction coils. Current flows between the two coils. When the armature is midway between the two coils, the circuit is balanced. The scale on the instrument reads zero.

Any minute variation in the size of a workpiece causes the spindle to move and change the circuit balance. The amount of movement on the scale is proportional to the movement of the spindle. The magnification factor of the instrument affects the value of each division of the scale. For example, if the magnification of the instrument is 10,000, each division of the scale equals 0.000010″ (10 millionths of an inch) or 0.00025mm (25 hundred-thousandths of a millimeter). The total measurement range on the scale of such an instrument is usually 0.0005″ (0.01mm).

ELECTRONIC COMPARATORS

The versatility of the electronic comparator makes it an ideal instrument for comparison measurements. Standard components may be used in a great number of measurement problems. The absence of mechanical parts makes electronic instruments comparatively rugged as high precision instruments. Also, their sensitivity and speed reduce the time lag during measurements. Electronic comparators are particularly useful, therefore, when measurements of a moving feature are required. Measurements along a strip passing through a rolling mill are a typical example of *dynamic measurement*.

Other considerations for selecting an electronic comparator are as follows:

- Rapidity of operation even at high amplifications,
- Use of the same instrument for multiple amplification ranges,
- Portability of the instruments, which are also self-contained,
- Easily adjusted and understandable controls,
- High instrument sensitivity in all amplification ranges,
- Limited self-checking,
- Instrument accuracy that compares favorably with other instruments.

ELECTRONIC AMPLIFIER AND READOUT

The two major components in an electronic measurement unit are a gage amplifier and one or more gaging heads. The amplifier and a single gaging head serve as a comparator and height gage. In applications where two heads are used, measurements may be taken of differences in size, length, thickness, diameter, flatness, taper, or concentricity. In such applications, the amplifier measures either the difference or the sum of the gaging head outputs.

Electronic gage amplifiers have magnification ranges from ±0.010″ (±0.200mm) to ±0.0001″ (±0.002mm).

Amplifier readouts are provided on either a sensitive meter or a direct digital readout display. The meter version has a precision-calibrated meter dial that is designed for easy readability without parallax error. Both amplifiers contain solid-state circuitry. The result is extreme reliability from variations of temperature, line voltage, and repeatability of measurement.

ELECTRONIC GAGING HEADS

Input into an electronic amplifier is provided by the movement of one or more gaging heads. The two basic types of gaging heads are identified in Figure 5–11 as a *lever (probe) type* and a *cartridge (cylindrical) type*. The gaging heads

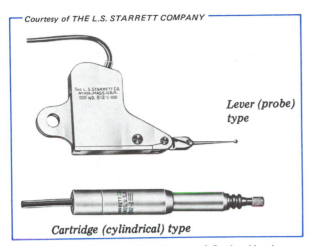

Lever (probe) type

Cartridge (cylindrical) type

Figure 5–11 Two Basic Types of Gaging Heads

are actuated by a contact of 8 to 12 grams of pressure against an object.

Lever (Probe) Type Gaging Head. The four major parts in a gaging head are the outer shell, or housing, the transducer, the probe tip, and the lever assembly. The transducer has matched coils and a core and is designed as a compact unit. The transducer is sensitive to fine measurements. Movements of the probe tip are converted into electrical energy in the transducer. These movements are amplified in the electronic amplifier. Direct readouts are indicated by the pointer movement on the amplifier scale.

C. QUALITY CONTROL PRINCIPLES AND MEASUREMENT PRACTICES

IMPORTANCE OF QUALITY CONTROL

The function of quality control is to ensure that the specifications and standards established by design and engineering departments are maintained. Quality control is essential to the maintenance of the quality of a product and to continuous improvements in the output. Quality control requires dimensional control, the testing of materials, inspection of surface roughness, assembly run-in periods, and other checks.

DIMENSIONAL MEASUREMENT AND CONTROL

The tools and instruments used for dimensional measurement are determined by the spec-

ified accuracy. Nonprecision linear measurements may be taken indirectly by using squares, dividers, calipers, and surface gages. These instruments may be used with line-graduated tools such as steel rules, bevel protractors, and depth gages. Common precision measurements are taken to tolerances of ±0.001″ (±0.02mm) and ±0.0001″ (±0.002mm) with inside, outside, depth, and other forms of micrometers and vernier-graduated instruments.

DIMENSIONAL CONTROL USING GAGES

The line-graduated tools just described are designed for a wide range of measurements and are adjustable. It is more practical and economical in mass production to check parts for size with fixed gages. There is less chance of error when no adjustable parts are involved. Of the many types of gages, six are commonly used. They include the *fixed-size gage*, the *micrometer-feature gage*, a *limit (standard) comparator*, the *indicating comparator*, the *combination gage*, and the *automatic gaging machine*. These gages and instruments are checked against master gages such as precision gage blocks, angle gage blocks, sine bars and plates, and optical flats (all of which were discussed in earlier units). The gages are dimensionally accurate and stable to within a fraction of 0.000002″ (0.00005mm).

DIMENSIONAL TOLERANCES

The permissible variation of a part from its nominal dimension (designated size) is identified by the designer, engineer, and craftsperson. The designer establishes the dimensional limits and surface characteristics of a part and mechanism. The more exacting a dimension, the more difficult and expensive a part is to produce. There is a minute range within which a part will function effectively.

On some dimensions (a shrink fit, for example), only a smaller dimension than the nominal size is permissible. Where a running fit is required, the part may be larger than the nominal size. The amount of variation from the nominal size is called *tolerance*. A tolerance may be *unilateral* (only plus or minus) or *bilateral* (both plus and minus). The amount of tolerance depends on factors such as the function of a part, the kind of material used, the required service life, cost, and production methods and systems.

FIXED-SIZE GAGES

Fixed-size gages are widely used in quality control. Although limited in scope, they are used to check either an internal or external dimension. With use, a plug gage becomes smaller; a ring gage, larger. Therefore, the wear life of fixed-size gages is determined by the number of parts that may be inspected before the gage size exceeds the allowable work tolerance. Constant wear requires that the working gages be checked against a master set of gages. A master gage is easily identified and is used as a dimensional control gage.

In the design of fixed-size gages, a *wear allowance* is provided. For example, if it is established that after 2,000 gagings a 1-1/4″ plug gage wears 0.0001″ (0.002mm), the plug may be finished within the allowable tolerance of the workpiece to 1.2501″. The wear life is thus increased to measure 4,000 workpieces. Wear life may be further increased by chrome plating a worn gage and regrinding and lapping to size. Cemented carbide gages are available for long-run production gaging. Such gages resist abrasive action and wear. Wear life of some carbide gages is increased to almost 100 times longer than wear life of steel gages.

GO AND NO-GO GAGES AND XX, X, Y, AND Z CLASSES OF (GAGE MAKER'S) TOLERANCES

Fixed gages are commercially available in a wide range of sizes for GO and NO-GO measurements. Ring, plug, and thread gages are com-

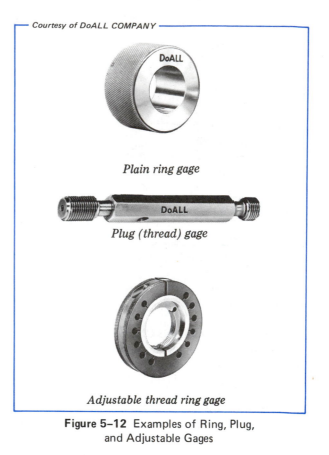

Plain ring gage

Plug (thread) gage

Adjustable thread ring gage

Figure 5–12 Examples of Ring, Plug, and Adjustable Gages

monly used for such measurements. Some examples of these gages are illustrated in Figure 5–12. These gages are available in inch and metric units of measure and within the same tolerance limits. The accuracy of the gages has been standardized. The tolerances are specified in terms of four classes: XX, X, Y, and Z. Table

Table 5–3 Nominal Sizes and Gage Maker's Dimensional Tolerances

Range of Nominal Sizes (inches)	Gage Maker's Tolerance According to Class			
	XX	X	Y	Z
0.029 to 0.825	0.00002	0.00004	0.00007	0.00010
0.826 to 1.510	0.00003	0.00006	0.00009	0.00012
1.511 to 2.510	0.00004	0.00008	0.00012	0.00016
2.511 to 4.510	0.00005	0.00010	0.00015	0.00020
4.511 to 6.510	0.000065	0.00013	0.00019	0.00025
6.511 to 9.010	0.00008	0.00016	0.00024	0.00032
9.011 to 12.010	0.00010	0.00020	0.00030	0.00040

5–3 gives the range of nominal sizes from 0.029″ to 12″ and the four classes of tolerances. These tolerances are known as *gage maker's tolerances*. The tolerance classes are generally defined as follows:

- *Class XX* refers to master and setup standards that are precision lapped to laboratory tolerances;
- *Class X* refers to working and inspection gages that are precision lapped;
- *Class Y* refers to working and inspection gages that have a good lapped surface finish with slightly increased tolerances;
- *Class Z* refers to working gages for short production runs where the tolerances are wide or commercially finished gages that are ground and polished but are not fully lapped.

Gage manufacturers recommend that the gage used should have an accuracy of one-tenth the tolerance of the dimension the gage is to control. Thus, the gage used for a part having a tolerance of 0.001″ (0.02mm) should have a tolerance of 0.0001″ (0.002mm). The surface finish of the gage should also be finer than the workpiece.

FOUR BASIC GAGE DESIGNS

Most types of plug and ring gages are marked for size and tolerance. There are four basic gage designs. Each of the following designs is adaptable within a range of sizes.

A *reversible cylindrical plug gage* has a pin-type gage design with GO and NO-GO members. Gage maker's tolerances are plus on the GO member and minus on the NO-GO member. This type of gage is primarily used for checking hole sizes and depths, gaging slots, and checking locations, threads, and distances between holes. Pin-type gages are available in steel, plated steel, and carbide. A common range of sizes is from 0.005″ to 1.000″ (0.125mm to 25mm). Each size is produced in the four tolerance classes. Plug members are available in increments of 0.001″ and are provided with collet bushings.

A *trilock plug gage* has a body design that permits GO and NO-GO members to be secured to the ends by means of a fastener (Figure 5–13).

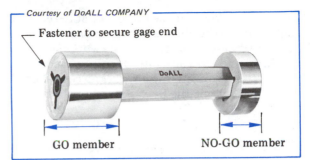

Fastener to secure gage end

DoALL

GO member   NO-GO member

Figure 5–13 Trilock Type of GO and NO–GO Plug Gage

A *taperlock cylindrical plug gage* has plug ends with a taper shank that fits a taper hole in the body (handle) of the gage.

A *progressive plug gage* is a single-end gage. The design permits complete inspection of a hole at one time. The entering portion is the GO gage, which is followed by the NO-GO gage.

While only plug gage and holder designs have been described, many other designs are available. Taper, spline (multiple grooves), flat plug, and special gages are widely used.

Snap Gages. Another design of a progressive (combination GO and NO-GO) gage is the *adjustable limit snap gage* (Figure 5–14). This gage consists of a structurally strong but lightweight and balanced frame. Adjustable gaging pins, buttons, or anvils are brought to size with adjustment screws. They are then secured to and locked in place on the frame with locking screws. A marking disk is stamped. The disk indicates the dimensions and identifies the gage in relation to a production plan.

The GO and NO-GO buttons of an adjustable snap gage are set to gage blocks or other masters. Precision measurements with snap and other gages

Figure 5–14 Adjustable Limit Snap Gage

depend on taking the measurement at the correct angle with respect to the reference planes. Small parts are measured by holding the workpiece with one hand and the gage with the other hand. The gage is then brought to the workpiece and carefully moved along the line of measurement.

ADJUSTABLE THREAD RING GAGES

Adjustable thread ring gages are available for standard metric threaded sizes and pitches and for American National and Unified thread forms. Other gages are available for Acme thread form sizes.

In thread control, *roll-thread gages* are also available. The term indicates that the thread-measuring form is a roll instead of a flat plate. There are even adaptations of roll-thread gages. In some applications, a dial indicator is used to show any variation from the basic size to which the gage is set.

ALLOWANCES FOR CLASSES OF FITS

The term *fit* indicates a range of tightness between two mating parts. This range is expressed on a drawing in terms of the *basic (nominal) size* and the *upper* and *lower (high limit and low limit) dimensions*. Tolerance, to repeat, is the total permissible variation from the basic size. Nominal size is the dimension from which the maximum and minimum permissible dimensions are derived.

CLEARANCE AND INTERFERENCE FITS

Some mating parts must move in an assembly. A *positive clearance* is therefore required. How much clearance to provide depends on the nature of the movement, the size of the parts, material, surface finish, and other design factors. The drawing in Figure 5–15 illustrates limit dimensioning and a positive clearance. There is a 0.002″ (0.05mm) clearance allowance on the punch. The clearance allowance on the block is 0.001″ (0.02mm). The clearance fit has a maximum allowance of 0.0035″ (between 2.0005″ and 1.9970″) and a minimum allowance of 0.0005″ (between 1.9995″ and 1.9990″).

The drawing in Figure 5–16 shows a *negative*, or *interference*, *fit*. The plug is ground to a high limit of 1.253″. The low limit of the mating

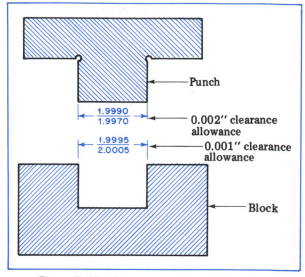

Figure 5–15 Allowance on Each Mating Part for a Clearance Fit

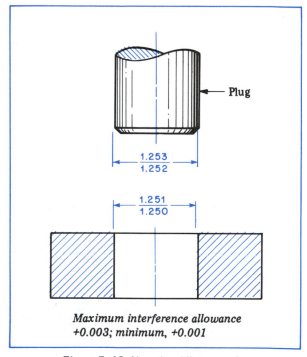

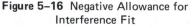

Maximum interference allowance +0.003; minimum, +0.001

Figure 5–16 Negative Allowance for Interference Fit

hole is 1.250″. The size limits of this particular interference fit is 0.001″ minimum and 0.003″ maximum.

ALLOWANCES ON HOLES FOR DIFFERENT CLASSES OF FITS

Standard allowances on holes have been developed for different *classes of fits*. Some allowances relate to tolerances for general purposes. Others relate to *forced*, *driving*, *push*, and *running fits*. The term *class* is followed by a particular letter (or combination of letters) that relates to specific design functions.

Tables of allowances provide a general guide. The material in the mating parts, the conditions under which a mechanism operates, the kind of work to be done, and a host of other factors must be considered. Usually, the designer indicates the ideal allowances through the dimensions on a drawing. There are situations, however, where the machinist, toolmaker, and inspector must refer to tables and make decisions.

INTERCHANGEABILITY OF AMERICAN NATIONAL AND UNIFIED FORM SCREW THREADS

Interchangeability of Unified and American National Form threads is provided through the standardization of thread form, diameter and pitch combinations, and size limits. The general characteristics of the internal and external Unified Form threads are shown in Figure 5–17.

The form is theoretical. Actually, neither British nor American industry has made a complete changeover. The 60° thread angle, diameter and pitch relationships, and limits are accepted. However, the British continue to machine threads with rounded roots and crests. There are production advantages and extended tool wear life in producing a rounded root. American industry continues with flat roots and crests. Fortunately, the manufacturing limits agreed upon permit parts to be mechanically interchangeable.

QUALITY CONTROL METHODS AND PLANS

The control of the quality of a process is agreed upon by the parts designer and industrial management. *Quality control* is the means of ensuring that a standard of acceptance (the product quality) of machined parts is maintained. Further, the inspection of the parts is kept at an economical level. Quality control requires a mathematical approach based on a normal frequency-distribution curve, sampling plans, and control charts.

NORMAL FREQUENCY-DISTRIBUTION CURVE FOUNDATION

Tests carried on over many years indicate that measurement variations occur during manufacturing processes. When they are plotted on a

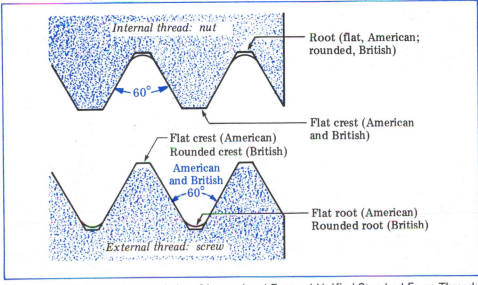

Figure 5–17 General Characteristics of Internal and External Unified Standard Form Threads

graph, the normal size variations follow a curved pattern. The general shape of the pattern is shown in Figure 5–18A. A limited number of the parts are larger or smaller than the basic dimension. The curve formed is known as the *normal frequency-distribution curve.*

This curve is divided into six zones or divisions (Figure 5–18B and C). Each zone is mathematically equal in width. The average on the curve is represented by the centerline. Note that there are three zones on each side of the centerline. The Greek letter sigma (σ) is used with a number to indicate the range of measurements.

For example, the 50 machined lathe parts in Figure 5–18A were measured by micrometer. The diameters varied from 0.9372″ to 0.9378″. Note that the greatest number of parts fall along the centerline (the basic diameter of 0.9375″). The next largest numbers are 0.9376″ and 0.9374″—either +0.0001″ or –0.0001″. These parts are said to fall within plus or minus one sigma ($\pm 1\sigma$).

Statistically, 34% of the workpieces fall within this first sigma. Thus, the diameters of 68% of the parts lie between 0.9374″ and 0.9376″. At $\pm 2\sigma$, the diameters of 95 1/2% of the parts fall between 0.9373″ and 0.9377″. At $\pm 3\sigma$, 99 3/4% of the parts measure between 0.9372″ and 0.9378″. The $\pm 3\sigma$ limit is the *natural tolerance limit.*

REFINING THE MANUFACTURING PROCESSES

Excessive process spread in relation to design tolerance indicates that the manufacturing processes need to be refined. The products fall outside the tolerance requirements. More precise tooling is needed. The machined dimensions must be controlled to produce all parts within the specified tolerance. Only accidental deviations need to be inspected. Where the products fall outside the tolerance, the designer sometimes corrects the problem by redesigning the part or mechanism. The design tolerance is then brought within the limitations of the machine and setup.

SAMPLING PLANS USED IN QUALITY CONTROL

Unless there is automatic inspection, it is costly and not practical to carry on a 100% inspection.

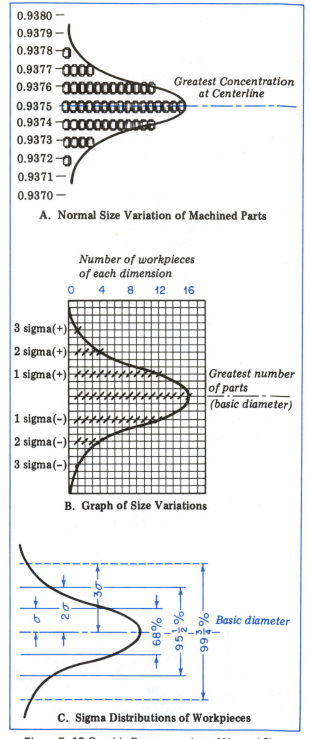

A. **Normal Size Variation of Machined Parts**

B. **Graph of Size Variations**

C. **Sigma Distributions of Workpieces**

Figure 5–18 Graphic Representation of Normal Size Variations of 50 Machined Workpieces

The term *sampling* is used in regard to quantities (batches) of parts that have been machined, cleaned, and are ready for inspection. The plan for sampling is referred to as a *sampling plan.* Such plans are based on the premise that a definite number of defective parts is allowable. This number is identified as the *acceptable quality level (AQL).*

The AQL is usually based on previous trial runs on machined parts. For example, the AQL for a particular part that is machined on a bar chucking machine may indicate that 0.8% to 1.4% are defective. This small percent of defective workpieces is allowable.

The same batch may have an *average outgoing quality limit (AOQL)* of 0.9% to 1.6% defective. This AOQL means that the batch of parts leaving the inspection station will average no more than the fixed percent of defects. The batches (lots) meet the sampling requirements. Any lot that fails to meet the sampling requirements is re-inspected. The defective parts are sorted out before they move out of inspection.

There are several different types of sampling plans. The *single-sampling, double-sampling,* and *sequential-sampling* plans are most common.

SINGLE-SAMPLING PLAN

Sampling plans for the part and lot sizes are shown in Table 5–4. The first column lists the three most common types of sampling plans. If a single-sampling plan is used, a random number of machined parts is required for inspection. In this example, the single-sampling size is 55, as shown in the third column. As a result of inspection, if there are 3 or less parts defective, the lot is accepted. If 4 parts in the lot are defective, the lot is rejected. The acceptance and rejection numbers appear in the sampling plan in the fifth and sixth columns.

DOUBLE-SAMPLING PLAN

Rejecting the lot means that each part in the sample is inspected. The defective parts are replaced. In a double-sampling plan, Table 5–4 shows that a first double-sampling size of 36 is required. Note that if there is more than 1 defective part within the range between the acceptance number (1) and the rejection number (5), a second screening must take place. In the second screening, 72 additional parts are needed. The acceptance number is now 4; rejection, 5. The lot is accepted if there are 4 or less defective parts. If there are 5, the lot is rejected. The lot is subjected to further screening. In this double-sampling plan, the total number of parts checked is 108 when a second sample is needed.

SEQUENTIAL-SAMPLING PLAN

The representative sequential-sampling plan in Table 5–4 for the 800 to 1,300 lot parts lists

Table 5–4 Example of Sampling Plans for Inspecting Lot Sizes Ranging from 800 to 1,300 Parts (1.1 to 2.1 AQL)

Type of Plan	Sample and Sequence	Size of Sample	Combined Samples		
			Cumulative Size of Sample	Lot Acceptance Number	Lot Rejection Number
Single-Sampling	First	55	55	3	4
Double-Sampling	First	36	36	1	5
	Second	72	108	4	5
Sequential-Sampling	First	15	15	*	2
	Second	15	30	0	3
	Third	15	45	1	4
	Fourth	15	60	3	5
	Fifth	15	75	3	5
	Sixth	15	90	3	5
	Seventh	15	105	4	5

*Requires inspection of two samples to permit acceptance

seven samples. The first sample size consists of 15 parts. If 0 or 1 part is defective, a second sample must be inspected. If 2 or more parts in the first sample are defective, the lot is rejected. If the number of defectives in the second sample is between 0 and 3, a third sample of 15 is used. The process continues with additional samples of 15 as long as the combined samples are within the acceptance and rejection numbers given in the table.

SELECTION OF SAMPLING PLAN

Each sampling plan serves a different function. Major factors affecting the choice of plan to use are cost and additional inspection time. Single-sampling plans are used on complicated precision parts that require close dimensional and surface finish accuracy. Sequential sampling is used with homogenous, large lots and provides a quick and comparatively inexpensive check. Any lots that fail to meet sampling requirements are reinspected and sorted (screened). The defective parts are replaced with acceptable parts.

QUALITY CONTROL CHARTS

Control charts show graphically the processes undergoing inspection. There are four common types of control charts:

- c charts are used to plot the number of defects in one workpiece;
- p charts show the percent of defective parts in a sample;
- $\overline{X}$ charts provide a graph of the variations in the averages of the samples
- $\overline{R}$ charts show variations in the range of samples.

D. SURFACE TEXTURE CHARACTERISTICS AND MEASUREMENT

TERMS AND SYMBOLS USED TO SPECIFY AND MEASURE SURFACE TEXTURE

The terms *surface texture*, *surface finish*, *surface roughness*, and *surface characteristics* are used interchangeably in the shop and throughout this text. However, particular American (ANSI) standard terms, ratings, and symbols are described

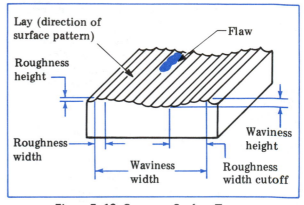

Figure 5–19 Common Surface Texture Characteristics and Terms

and illustrated. Use of these terms, ratings, and symbols permits designers, skilled mechanics, and other workers to communicate. Thus, part specifications may be accurately interpreted to produce a given surface quality. Figure 5–19 identifies some common surface texture characteristics and terms.

Surface Texture. *Surface texture* relates to deviations (from a nominal surface) that form the pattern of the surface. The deviations may be repetitive or random. The deviations may result from roughness, waviness, lay, and flaws.

Surface. According to American (ANSI) standards, an object is bounded by a *surface*. This surface separates the object from another object, substance, or space.

Surface Finish. *Surface finish* is indicated on a drawing by using the modified mathematical symbol $\sqrt{}$ and a microinch measurement. For example, a roughness height of 63 microinches appears on a drawing as $\sqrt[63]{}$.

Profile. A *profile* is the contour of a machined surface on a plane that is perpendicular to the surface. The plane may also be at an angle other than 90°, if it is specified. A short section of a surface profile is illustrated in Figure 5–20.

Centerline. The *centerline* is a line that is parallel to the general direction of the profile. Figure 5–20 shows a centerline of an exaggerated

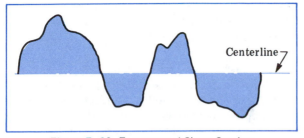

Figure 5–20 Exaggerated Short Section of a Surface Profile

profile. Roughness is measured around the centerline. This line lies within the limits of roughness width cutoff. The centerline in the profile serves as the cutoff line between roughness areas on both sides of the line.

Nominal Surface (Mean Line). A *nominal surface* is considered to be a geometrically perfect surface. Such a surface would result if the peaks were leveled to fill the valleys. The theoretical surface is represented by a *mean line.*

Microinch. A *microinch* is the basic unit of surface roughness measurement. A microinch designates one-millionth of an inch (0.000001″). The numerical value is sometimes followed by the symbol $M\mu$ or μ.

Lay. *Lay* is the direction of the predominant surface finish pattern. Lay patterns, symbols, and applications in drawings are described later.

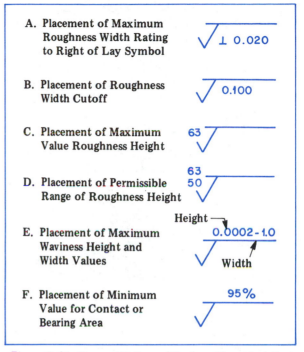

A. Placement of Maximum Roughness Width Rating to Right of Lay Symbol ⊥ 0.020

B. Placement of Roughness Width Cutoff 0.100

C. Placement of Maximum Value Roughness Height 63

D. Placement of Permissible Range of Roughness Height 63 50

E. Placement of Maximum Waviness Height and Width Values Height 0.0002–1.0 Width

F. Placement of Minimum Value for Contact or Bearing Area 95%

Figure 5–21 Representation of Surface Characteristics on Drawings and Specifications

Roughness. The fine irregularities produced by a cutting tool or production process are identified as *roughness (surface roughness).* Roughness includes traverse feed marks and other irregularities.

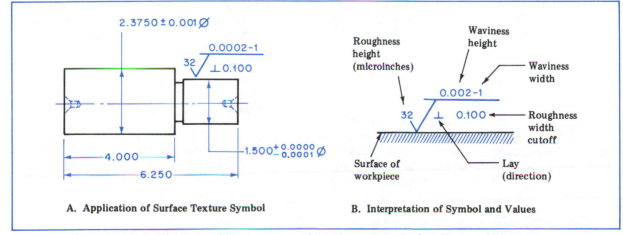

A. Application of Surface Texture Symbol

B. Interpretation of Symbol and Values

Figure 5–22 Application and Interpretation of Lay Symbol and Surface Texture Specifications

Roughness Width. The irregularities produced by production processes extend within the limits of the roughness width cutoff (the distance between successive peaks or ridges). *Roughness width* is measured parallel to the nominal surface. The peaks considered constitute the dominant pattern. The maximum roughness width rating, in inches, is placed to the right of the lay symbol. A rating of 0.020″ appears in Figure 5–21A.

Roughness Width Cutoff. A rating in the thousandths of an inch is used for the greatest spacing (width) of surface irregularities that are repetitive. These irregularities are included in the measurement of average roughness height. The *roughness width cutoff* rating appears directly under the horizontal extension of the surface finish symbol. A rating of 0.100″ is given in Figure 5–21B. If no value is shown, a rating of 0.030″ is assumed.

Roughness Height. *Roughness height* is expressed in microinches. The height is measured normal to the centerline. The height is rated as an arithmetical average deviation. The roughness height value is placed to the left of the surface finish symbol leg. If only one value is given—for example, 63 microinches as shown in Figure 5–21C–it indicates the maximum value. When maximum and minimum values for roughness height are given, they indicate the permissible range. In Figure 5–21D the range indicates 63 microinches maximum and 50 microinches minimum.

Waviness. *Waviness* is a surface deviation (defect) consisting of waves. These waves are widely spaced. Waviness may be produced by vibrations, chatter, or machine or work defects. Heat-treating processes and warping strains also produce waviness.

Maximum Waviness Height. The *maximum waviness height* is the peak-to-valley distance of a wave. The maximum waviness height rating is placed above the horizontal extension of the surface finish symbol, as shown in Figure 5–21E by the 0.0002″ value.

Waviness Width. *Waviness width* is the spacing of successive wave peaks or valleys and is measured in inches. The value also appears above the surface finish symbol but next to the waviness height, as shown in Figure 5–21E by the 1.0″ value.

Flaws. *Flaws* are surface defects. They may occur in one area of the unit surface. Flaws may also appear at widely varying intervals on the surface. Casting and welding cracks, blow holes, checks, and scratches are common flaws. Normally, the effect of flaws is not included in measurements of roughness height.

Contact or Bearing Area Requirement. The *contact or bearing area* is expressed as a percentage value. This value is placed above the extension line of the surface finish symbol, as shown in Figure 5–21F by the 95%. The value is the minimum contact or bearing area requirement as related to the mating part or reference surface.

REPRESENTATION OF LAY PATTERNS AND SYMBOLS ON DRAWINGS

Six lay symbols are commonly used in drawings and parts' specimens. Tool or surface finish patterns are indicated by perpendicular (⊥), parallel (||), angular (X), multidirectional (M), circular (C), and radial (R) symbols.

The manner in which lay symbols and values are applied on drawings is shown in Figure 5–22A.

MEASURING SURFACE FINISHES

A few American standard measurement values of surface roughness height, roughness width cutoff, and waviness height are given in Table 5–5. The full table of surface texture values is included in the Appendix.

FUNCTIONS OF SURFACE COMPARATOR SPECIMENS

Surface comparators (specimens) are produced commercially to measure surface roughness. The comparators have scales that conform precisely with ANSI standards of surface roughness and lay. Surface comparators are generally flat or round. Different scales are designed for gaging common types of machined, ground, and lapped surfaces. The finer microinch-finish

Table 5–5 American Standard Surface Texture Values

Dimensional Measurement	Measurement Values				
	Microinches				
Roughness height	1	13	50	200	
	2	16*	63*	250*	
	. . .				
		Inches			
Standard roughness width cutoff	0.003	0.010	0.100	0.300	1.000
	. . .				
		Inches			
Waviness height	0.00002	0.0001	0.001	0.010	
	0.00003	0.0002	0.002	0.015	
	. . .				

*Recommended values

surface comparator specimens are often used with a *surface finish viewer* (Figure 5–23).

SURFACE MEASUREMENT WITH COMPARATOR SPECIMENS

In actual practice, a number of different machined surfaces may be measured by comparator specimens. Several surfaces are listed in the first column of Table 5–6. General ranges of surface roughness heights and lay appear in the second and third columns respectively. Roughness widths are given in inches in the fourth column for machining processes that produce roughness heights between 2 and 500 microinches. The actual measured values in microinches are given in the fifth column. The letters in the sixth column tell that the measured value was obtained by either a brush analyzer (B) or a profilometer (P).

The skilled mechanic may also use this check to provide information about the material, machine and cutting tools, speeds, and feeds.

SURFACE ROUGHNESS RELATED TO COMMON PRODUCTION METHODS

Different production methods produce variations in surface texture. Figure 5–24 provides ranges of surface roughness for selected manufacturing processes. A more complete table appears in the Appendix. General and nonconventional machining processes and forming methods are considered.

SURFACE TEXTURE MEASURING INSTRUMENTS

SURFACE FINISH ANALYZER

Variations between the high and low points of the work surface may be measured by a *surface finish analyzer*. The three main components of the instrument are a tracer head, amplifier unit, and a recording dial.

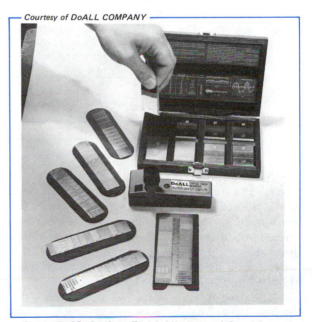

Courtesy of DoALL COMPANY

Figure 5–23 Surface Finish Viewer and Comparator Specimen for Lay, Process, and Surface Roughness

Table 5-6 Machined Surfaces That May Be Measured by Comparator Specimens

Machined Surface	Roughness Height (Microinches)	Lay	Roughness Width (Inches)	Actual Microinch Value	Type* of Instrument
Honed, lapped, or polished	2	Parallel to long dimension of specimen	....	2.4	B
	4		....	3.8	B
	8		....	6.5	B
Ground with periphery of wheel	8	Parallel to long dimension of specimen	....	7.8	B
	16		....	14.	B
	32		....	28.	B
	63		....	64.	B
Ground with flat side of wheel (Blanchard)	16	Angular in both directions	....	17.	B
	32		....	28.	B
Shaped or turned	32	Parallel to long dimension of specimen	.002	31.	B
	63		.005	60.	P
	125		.010	130.	P
	250		.020	250.	P
	500		.030	530.	P
Side milled, end milled, or profiled	63	Circular	.010	57.	P
	125		.020	125.	P
	250	Angular in both directions	.100	235.	P
	500		.100	470.	P
Milled with periphery of cutter	63	Parallel to short dimension of specimen	.050	55.	P
	125		.075	115.	P
	250		.125	240.	P
	500		.250	520.	P

Tolerances

Roughness Height (Microinches)	Tolerance (Percent)
2 to 4	+25 –35
8	+20 –30
16	+15 –25
32 and above	+15 –20

*B represents brush analyzer; P, profilometer.

A diamond stylus is mounted in the tracer head. The tip of the stylus is rounded. The two basic tip radii are 0.0001″ and 0.0005″. The profiles of surface roughness produced by using these two tip radii in four examples of machining processes are presented in Table 5–7. The same wave forms are shown with readings taken using 0.0001″ and 0.0005″ tip radii.

The tracer head is moved across the work surface either manually or mechanically. More precise surface roughness measuring instruments, such as a *brush analyzer*, for example, are moved mechanically. The stylus moves as it follows the surface profile. The minute movements are converted in the tracer head into electrical fluctuations. These fluctuations are magnified in the amplifier. The mechanical movement of the stylus is registered by a needle on the surface recording dial in microinches.

PROFILOMETER

The *profilometer* is another surface roughness measuring instrument. It employs a tracing point

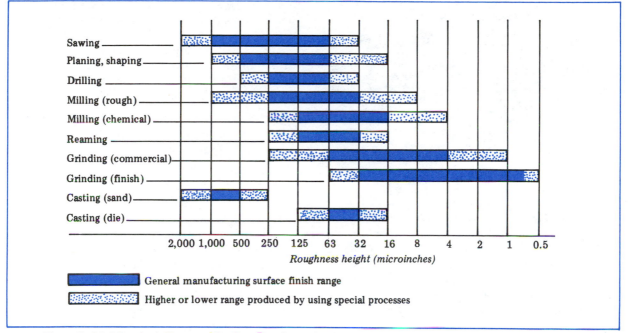

Figure 5–24 Microinch Range of Surface Roughness for Selected Manufacturing Processes (Partial Listing)

Table 5–7 Readings of Machined Surface Profiles
Obtained by Using Two Radii Stylus Sizes

Machining Process	Surface Profiles and Roughness	Stylus Radius (0.0000″)	Average Reading (Microinches)
Shaping		0.0005 0.0001	100–105 110–115
Turning		0.0005 0.0001	45–50 40–44
Grinding		0.0005 0.0001	8–9 10–11
Polishing		0.0005 0.0001	1.4–1.7 1.4–1.7

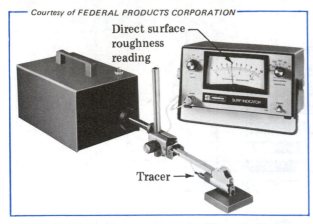

Courtesy of FEDERAL PRODUCTS CORPORATION

Direct surface roughness reading

Tracer →

Figure 5–25 Profilometer Setup for Manually Checking the Roughness of a Plane (Flat) Surface

or stylus to establish the irregularities. These irregularities are greatly magnified and recorded. The profilometer in Figure 5–25 can be set up for manually checking a ground cylindrical surface. The amplimeter pointer shows the roughness on a microinch scale. Direct readings are given as the instrument tracer moves across the work surface.

Another type of profilometer produces a recorded profile on a tape, as shown in Figure 5–26. The height of the horizontal line is 25μ (microinches). Waviness, roughness, and waviness/roughness profiles are recorded.

SURFACE FINISH MICROSCOPE

Surface texture may also be measured with a *surface finish microscope*. This microscope depends on a light wave principle of measurement similar to the principle behind optical flats. The surface to be measured is magnified from 125 to 200 times. The surface finish is then compared

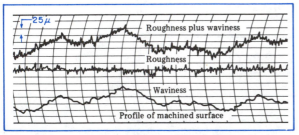

25μ

Roughness plus waviness

Roughness

Waviness

Profile of machined surface

Figure 5–26 Recorded Roughness and Waviness Profiles of a Machined Surface

to a comparator plate of a known standard. The view of one surface is superimposed on the other surface.

The variation indicates the accuracy of the surface finish. Some microscopes have a direct-reading attachment. Variations in surface finish are read on a dial face.

A. Safe Practices in the Use of Industrial Microscopes and Optical Flats

- Normalize the temperature of the measuring instruments, workpiece, and accessory before taking a measurement or carrying on surface inspections. Optical flats have a low conductivity of heat; they normalize at a slower rate than metallic parts.
- Use a fine, soft-bristle brush to wipe over the surfaces of an optical flat or microscope and areas of contact.
- Wipe instrument lenses with a clean, soft cloth or lens tissue. Saturate the cloth with an approved cleansing solution. Any dirt rubbed across a lens will scratch the surface.
- Place a part on the stage of a toolmaker's measuring microscope in order to permit movement over the length or width to be measured.

B. Safe Practices in the Use and Care of High Amplification Comparators

- Investigate the potential sources of error before making high precision measurements. Check for the following conditions:
  - Direct heat from sunlight;
  - Electric lights or drafts;
  - Outside vibrations;
  - Heat transfer due to handling the stand, parts, and gages;
  - Differences in materials, finish, and temperature between the parts and gage blocks;
  - Fastening of the locking nuts for the arm and the gaging head.
- Use the same scale screen on a contour measuring projector that corresponds with the magnification of the lens system.

- Avoid temperatures above 140°F that may concentrate on plastic charts. Distortion may result.
- Measure holes that have a rough surface finish with a pneumatic spindle fitted with a hardened contact blade.
- Place the measuring instrument and setup on a vibration-free surface. Vibration may cause inaccuracies in measurement.
- Avoid handling the stand, parts, and gages. Handling may produce uneven temperatures.

C. Safe Practices in Applying Gages for Quality Control Inspection

- Pay particular attention to the abrasive action of metals on the dimensional accuracy of a gage. Cast iron, cast aluminum, and some of the plastic materials have a higher abrasive action than steels, brass, and bronze.
- Take care when gaging close tolerance holes in soft metals to prevent seizing and loading. Excessive wear may be produced on the gage

and an inaccurate measurement may result. A special gage lubricant is produced to coat the gage.
- Burr, degrease, and thoroughly clean machined parts. Wash all parts in a lot that are to be used for quality control.

D. Safe Practices in Using Surface Finish Analyzers and Instruments

- Set the range switch on a surface finish analyzer to a high setting when the surface roughness is not known. After the initial test, a finer setting may be needed to get an accurate surface reading.
- Allow a surface finish analyzer instrument to warm up before taking a measurement.
- Protect the precision stylus point from hitting a surface.
- Check surface finish instruments periodically against a master.

UNIT 5 REVIEW AND SELF-TEST

A. OPTICAL FLATS AND MICROSCOPES

1. a. Describe a design feature on a toolmaker's microscope that makes it possible to make measurements to accuracies of 0.001" (0.02mm) and finer.
   b. Give two applications of a toolmaker's microscope that is equipped with a vernier angle-measurement scale.

2. List three scale mounts that are interchangeable on an industrial measuring magnifier.

3. Interpret what each of the following patterns of interference bands means in relation to a measurement or the accuracy of a machined surface:
   a. Straight, parallel, uniform, and
   b. Straight, parallel, uniform bands that curve slightly at the ends

4. Give one reason for normalizing the temperature of optical measuring flats, other instruments, and the workpiece in a measurement, layout, or precision inspection setup.

5. List two precautions to observe when industrial microscopes and optical flats are used.

B. HIGH AMPLIFICATION COMPARATORS

1. Name four general types of high amplification comparison measuring instruments.

2. Compare the accuracy range of a standard, everyday shop dial indicator with the accuracy range of a high amplification dial indicator.

3. List the steps in measuring a 0.9375″ dimension on a flat gage to within +0.000020″/−0.000010″. Use a mechanical comparator.

4. Cite two design features of a reed comparator that differ from similar features of mechanical comparators.

5. Give two advantages of a contour measuring (optical) projector in comparison to a mechanical comparator.

6. State three advantages of using pneumatic gaging instruments over GO and NO-GO gages and other mechanical measuring methods.

7. Give two advantages of a meter readout on an electronic amplifier for precision measurements.

8. Cite three precautions a worker must follow when working with high amplification comparators.

C. QUALITY CONTROL

1. Identify four different kinds of testing that are necessary to quality control.

2. State the importance of *nominal size* to dimensions of mating features.

3. a. Refer to a table of classes of fits and gage maker's dimensional tolerances.
   b. Determine the gage maker's tolerance for parts (1) through (4) according to the class of fit given for each part: (1) 25.4mm, class XX; (2) 1.125″, class X; (3) 203.2mm, class Y; and (4) 10.375″, class Z.

4. Explain why interchangeable Unified and American National Form threads will also interchange with SI metric threads having corresponding pitches.

5. Give three corrective steps a machine operator takes when a quality control inspection shows a machined part falls outside the tolerance.

6. a. Use a sequential-sampling plan for the inspection of 900 parts at an acceptable quality level of 1.5.
   b. Indicate (1) the size of the next sample and (2) the cumulative size of sample if there are 4 or more rejects in the fourth sample.

7. a. Tell how information from a control chart may affect the part design and/ or the manufacturing processes.
   b. Indicate how control charts are valuable in time-motion studies.

D. SURFACE TEXTURE CHARACTERISTICS AND MEASUREMENT

1. a. Use a table of surface texture values for general machining processes.
   b. Give the *lay pattern* and the *roughness height in microinches* for the following five machining processes and roughness width values: (1) end milling, 0.005"; (2) straight turning, 0.002"; (3) shaping, 0.001"; (4) cylindrical grinding, 0.0005"; and (5) drilling, 0.005".

2. a. Describe the functions served by a surface comparator specimen and viewer.
   b. Identify the kind of information the worker obtains by visual and viewer inspection of surface texture measurement.

3. a. Use a table of surface roughness heights for general manufacturing processes.
   b. Give the lower surface finish range in microinches (where special precision manufacturing processes are used) for the following operations: (1) reaming, (2) finish grinding (cylindrical), (3) drilling, and (4) rough milling.

4. a. Cite two main differences between a surface finish analyzer and a comparator specimen.
   b. State one advantage of using a surface finish analyzer over the comparator specimen method.

5. Tell why work surface profiles are tape recorded on a profilometer.

6. State two advantages of using a surface finish microscope in preference to a profilometer for general surface texture applications.

7. Indicate one check that must regularly be made with surface finish measuring instruments.

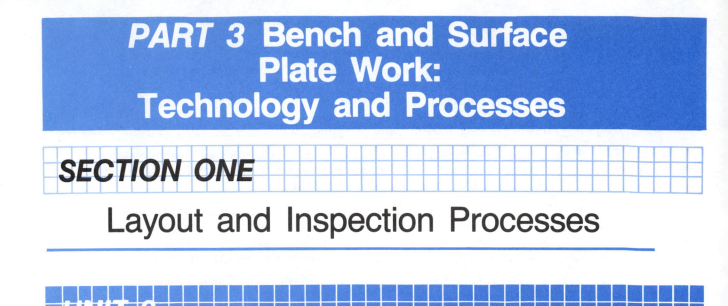

PART 3 Bench and Surface Plate Work: Technology and Processes

SECTION ONE

Layout and Inspection Processes

UNIT 6

Precision Accessories for Surface Plate Work

Generally, precision layout, measurement, and inspection accessories are made of metal. However, granite products are also used due to certain advantages over similar metal items. In addition to the qualities of granite stated earlier, consideration should be given to the following properties.

- *Hardness:* Black granite has an 80 Rockwell C scale hardness by test. It is harder than tool steel. Black granite is also tough enough to resist chipping and scratching.
- *Rigidity:* Black granite surface plates have extremely low deflection under load. They are, therefore, almost distortion free and retain their original accuracy.
- *Density:* The uniform density of black granite makes it possible to produce a fine-textured, lapped, and frosted surface. The surface finish prevents gage blocks and precision surfaces from wringing.
- *Low reflectivity:* A black granite surface provides a dark nonglare finish. Light reflected from illumination, the surface

plate, tools, instruments, and workpieces is limited. The dark surface is also an excellent background for layout lines that are scribed with a brass rod. Such lines may be removed with an ordinary eraser.
- *Porosity:* The uniform grain structure of black granite gives it a low water (moisture) absorption rate. A black granite surface is not affected by acids, chemical fumes, dust, or dirt. Surface plates and gages are, therefore, rustproof, and rusting and pitting of precision gages that are left on the plates are prevented.
- *Thermal (heat) stability:* Temperature changes have extremely limited effect on the accuracy of black granite gaging products.
- *Easy care and maintenance:* Black granite surface gages and accessories may be cleaned easily. Soap and water, detergent, or a manufacturer's solvent may be used. The surface will not stain or discolor.
- *Low cost:* The initial, maintenance, and relapping costs of a granite plate are lower than for a metal surface plate.

A. LINEAR AND ANGULAR LAYOUTS

GRANITE PRECISION ANGLE PLATES AND ACCESSORIES

Granite *precision angle plates*, such as the plates shown in Figure 6-1, are available for production or for inspection. The difference between production and inspection plates is in the degree of accuracy. Production plates are accurate to within ±0.00005″ in 6″ (0.001mm in 150mm). The accuracy of inspection, or master, plates is ±0.000025″ in 6″ (0.0006mm in 150mm). The accuracies relate to flatness and squareness. Angle plates are lapped on either two surfaces (base and face) or four surfaces (base, face, and two ends). These surfaces are finished square and flat.

Angle plates are available with *metal inserts*. Examples of these plates are shown in Figure 6-1. Some inserts are recessed into one side and the main gaging face. These inserts are used for magnetizing purposes. Threaded inserts are available for clamping applications.

The term *universal* indicates that all faces on a *universal right angle plate* may be used. All six faces of the universal right angle shown in Figure 6-2 are square and parallel.

PARALLELS AND OTHER FORMS OF GRANITE ACCESSORIES

Parallels are matched in pairs. Granite parallels serve the same purposes as ground steel parallels. They are machined and finished for production and inspection (master) standards of accuracy. Either the top and bottom or the top and bottom and two sides may be finished with parallel surfaces.

Black granite precision box parallels are matched pairs of parallels. The base area is smaller in comparison to the height than it is on regular parallels. Workpieces that are difficult to layout and inspect are usually elevated from a surface plate with box parallels. These parallels provide an accurate working surface that is parallel to the surface plate.

The granite *straight edge* has one precision finished plane face. If it is of inspection grade, the plane face is accurate to ±0.00005″ (1.0 μm) over 12″. The more precise laboratory-grade straight edge is finished to ±0.000025″ (0.6 μm) accuracy over 12″. Straight edges are used for such purposes as checking the straightness and parallelism of flat surfaces and the beds of machine tools.

Granite *precision step blocks* have flat, parallel, and square surfaces. These surfaces are held to close tolerances for flatness and parallelism.

A *cube* is really a combination of three sets of parallels. The opposite faces are parallel and square. The faces provide perpendicular, flat, and parallel reference surfaces in relation to the plane of a surface plate.

V-blocks support cylindrical workpieces during layout, manufacture, or inspection. The 90° V is centered and parallel to the bottom face. Some V-blocks are produced with the 90° V parallel to the bottom and two sides. The V is machined square (perpendicular) to the two ends.

Courtesy of DoALL COMPANY

Inserts for magnetic applications

Threaded inserts for clamping to the plate

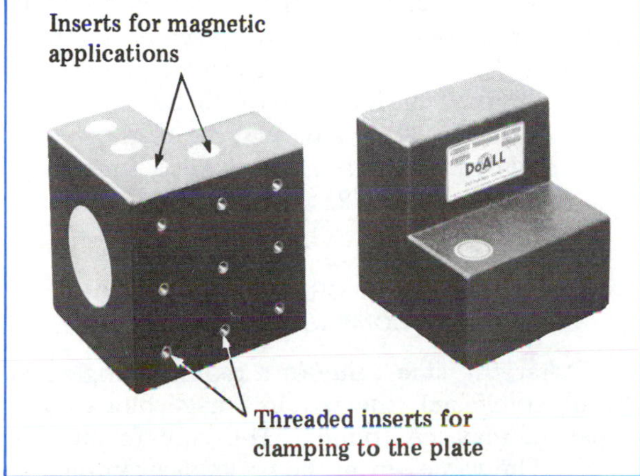

Figure 6-1 Angle Plates with Metal Inserts

Courtesy of DoALL COMPANY

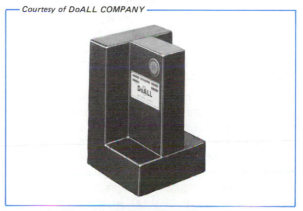

Figure 6-2 Universal Right Angle

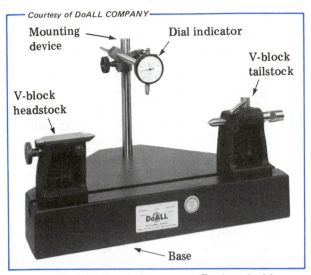

Courtesy of DoALL COMPANY

Mounting device

Dial indicator

V-block tailstock

V-block headstock

Base

Figure 6–3 Bench Comparator Equipped with V-Block Headstock, Tailstock, and Indicator

Courtesy of BROWN & SHARPE MANUFACTURING COMPANY

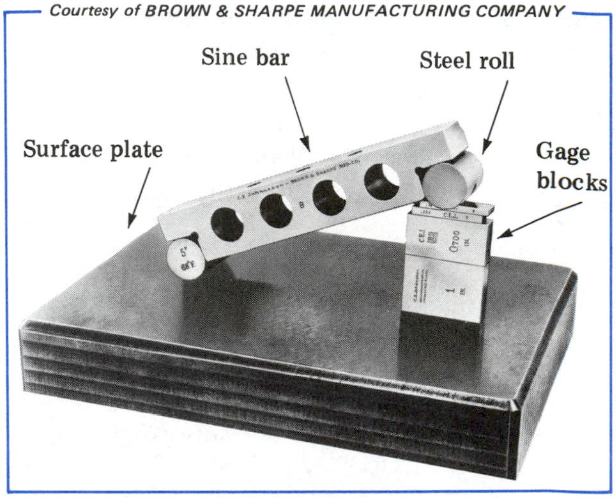

Sine bar

Steel roll

Surface plate

Gage blocks

Figure 6–4 Typical Angle Setting Using a 5″ Sine Bar, Gage Blocks, and Surface Plate

A *bench comparator* consists of a perfect plane surface (surface plate), a vertical post, a swivel arm, and a fine-feed adjustment with a dial indicator. The surface plate (plane reference surface base) may be produced from metal or black granite. The base may be slotted. Its shape may be modified to accommodate layout, inspection, and measurement accessories such as a headstock and an adjustable tailstock, V-blocks, or bench centers. An example of a V-block headstock and tailstock combination with a dial indicator is shown in Figure 6–3. The base, headstock, and tailstock are made of black granite.

Bench comparators are highly accurate for gaging roundness, flatness, parallelism, and concentricity. Measurements of length, width, height, steps, depths, and angles may be compared against linear standards.

PRECISION ANGLE MEASUREMENTS AND ACCESSORIES

Angular surfaces of workpieces or assemblies are often layed out, machined, or inspected to a degree of accuracy within *one minute of arc*. This is a finer measurement than can be taken with a vernier bevel protractor. A sine bar, sine plate, sine block, and compound (universal) sine block are four precision angular layout and measurement instruments. Each of these instruments may be set to a required angle by applying trigonometry.

COMPUTING THE SINE VALUE (GAGE BLOCK HEIGHT)

The name *sine bar* is derived from the use of triangles and trigonometric functions. The *sines* of required angles are used to calculate the height of a triangle. The triangle formed consists of the hypotenuse of a known length and the required angle.

Figure 6–4 shows a typical angle setting of a sine bar. Gage blocks and a surface plate are used. The center-to-center distance between the hardened steel rolls of equal diameter is a multiple of 5″. The basic center distances are 5″, 10″, or 20″.

Example: Determine the gage block measurement (height) required to set a workpiece at an angle of 20°28′. A 5″ sine bar and surface plate setup are required. The sine value of 20°28′ must first be established. This value may be read directly in a *table of natural trigonometric functions*. The sine value for 20°28′ is 0.34966″.

COMPUTING INCH- OR METRIC-STANDARD GAGE BLOCK COMBINATIONS

After the sine value of a required angle has been established, the required gage block combination may be found. The sine formula is used. The values from the previous example are used in the formula as follows:

$$\frac{\text{sine of angle}}{(20°28')} = \frac{\text{height (gage block combination)}}{\text{length of sine bar (5'')}}$$

height = sine of angle (0.34966) × length of sine bar (5'')

= 1.7483''

If metric-standard gage blocks are to be used, the problem may be solved according to the same trigonometric function. The 1.7483'' gage block height may be converted to the metric equivalent of 44.407mm. Metric gage blocks of this height will produce an angle of 20°28' when a 127mm (5'') sine bar is used.

ESTABLISHING A HEIGHT SETTING FOR DIFFERENT LENGTH SINE BARS

Sine bars with center distances of 5'' and 10'' are common. For larger layout, inspection, and measurement operations, a 20'' sine bar is used. The basic 5'', 10'', and 20'' lengths were especially selected in this unit to simplify the mathematical computations and setups. The height to which a 10'' sine instrument is set is established by simply moving the decimal point (in the natural trigonometric function value of the required angle) one place to the left. For example, with the 20°28' angle, by moving the decimal point in the trigonometric sine value of 0.34966, the height becomes 3.4966. The value represents the gage block combination height.

TABLES OF CONSTANTS FOR SINE BARS AND PLATES

The machine tool industry has further simplified layout, inspection, and machining practices with sine bars and sine plates. *Tables of constants* are available in handbooks or are furnished by prod-

uct manufacturers. Part of a table of constants is reproduced in Table 6–1. Note that values of degrees and minutes are given with the accompanying constant (inch value).

The values in the Table apply to 5'' sine bars or plates. Settings for 10'' instruments are found by multiplying by 2. For example, in the 20°28' setting, the constant value shown in a complete table is 1.7483'' for a 5'' sine bar. A 10'' sine bar is set at 1.7483'' × 2, or 3.4966''. For a 20'' sine bar, the constant value is multiplied by 4. By contrast, the constant value in the table is divided by 2 when a 2 1/2'' sine bar is used.

The same table of constants is used to determine the degrees and minutes in an angle. For a 5'' setup, the height (gage block measurement) is located in the inch values section of the table. The degrees and minutes corresponding to the inch value represent the angle of the setup.

SINE BAR, SINE BLOCK, AND SINE PLATES

Sine bars, sine blocks, and simple and compound sine plates and perma-sines are commonly used in layout, inspection, assembly processes, and measurement. They are essential in maintaining accuracies to one minute of arc.

SINE BAR

A *sine bar* (as displayed earlier in Figure 6–4) is a precise layout and measuring accessory. It consists of two rolls mounted on opposite ends of a notched steel or granite parallel. All metal parts are hardened, machined to precise limits, and accurately positioned. The center distance between the rolls is fixed at lengths that are a multiple of 5'' (127mm).

Table 6–1 Partial Table of Constants for Setting 5'' Sine Bars or Sine Plates (16°0' to 23°5')

Minutes	16°	17°	18°	19°	20°	21°	22°	23°
0	1.3782	1.4618	1.5451	1.6278	1.7101	1.7918	1.8730	1.9536
1	.3796	.4632	.5464	.6292	.7114	.7932	.8744	.9550
2	.3810	.4646	.5478	.6306	.7128	.7945	.8757	.9563
3	.3824	.4660	.5492	.6319	.7142	.7959	.8771	.9576
4	.3838	.4674	.5506	.6333	.7155	.7972	.8784	.9590
5	1.3852	1.4688	1.5520	1.6347	1.7169	1.7986	1.8797	1.9603

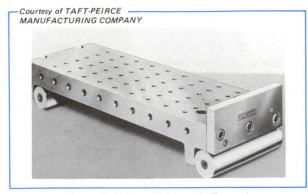

Figure 6–5 Sine Block with Tapped
Holes and End Plate (Stop)

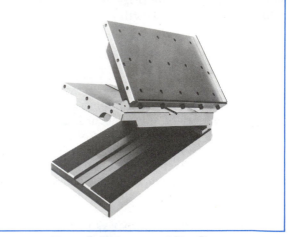

Figure 6–6 Compound Sine Plate

SINE BLOCK

A *sine block* is essentially a wide sine bar (Figure 6–5). A sine block (or modification) provides maximum efficiency for layout, inspection, and angle measurements. The exact angular degree that a 5″ or 10″ sine block makes with a plane reference surface is determined by a precise vertical height (combination of gage blocks). These blocks are placed under the elevated end of the sine block.

SIMPLE SINE PLATE

Single angles, which lie in one plane, are usually set by using a *simple sine plate*. This sine plate has a precision base, a movable sine plate, and two hardened precision rolls. The base and elevating sine plate are hinged.

The simple sine plate is set at the required angle by establishing the overall measurement of a gage block combination.

COMPOUND SINE PLATE

Compound angles are angles formed by edges of triangles that lie in different planes. Compound angles may be layed out, inspected, or measured by holding a workpiece on a *compound sine plate* (Figure 6–6). A series of tapped holes in the top, sides, and end permit the direct strapping and clamping of workpieces.

A typical compound sine plate has an intermediate hinged plate. Attached to it is the top

sine plate. The design of this plate includes the same features as the features on a simple sine plate.

The intermediate plate and the top sine plate are each set at a required angle in one of two planes. The two angles make up the compound angle. The gage block combination for each angle is established. These combinations represent the heights to which the intermediate plate and sine plate are set.

PERMANENT MAGNETIC SINE PLATE (CHUCK): PERMA-SINE

A *permanent magnetic sine plate* is a simple or compound angle sine plate (Figure 6–7) in

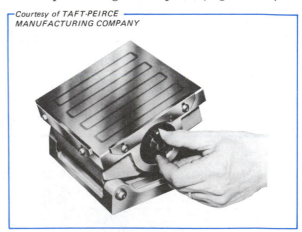

Figure 6–7 Simple Angle, Permanent Magnet
Sine Plate (Perma-Sine)

which a permanent magnetic chuck forms the working face. The name is shortened so that the combination is often referred to as a *perma-sine* or a *sine chuck*.

Perma-sines are especially adapted to surface grinder work. The workpiece may be layed out, inspected, or machined. Angles produced are within one minute of arc accuracy. The fine magnetic pole spacing on the compound angle plate permits holding small and large parts securely. The magnetic force is turned on for holding and off for releasing by moving a simple lever.

Courtesy of BROWN & SHARPE MANUFACTURING COMPANY

Dial Test Indicator with Long-Reach Swivel Point and Automatic Reversal (Graduations 0.0005″)

Figure 6–8 Dial Test Indicator with Long-Reach Swivel Point and Automatic Reversal (Graduations 0.0005″)

B. TECHNIQUES FOR CHECKING PARALLELISM, SQUARENESS, ROUNDNESS, AND CONCENTRICITY

Extremely precise checking for parallelism, squareness, roundness, and concentricity may be done with standard measuring instruments such as a vernier height gage and/or a dial test indicator. Greater sensitivity is achieved by using electronic and pneumatic gages and instruments.

CHECKING FOR PARALLELISM

To check for *parallelism*, the workpiece is placed with one of the parallel surfaces resting flat on the face of the surface plate. If the part has a projection on the bottom surface, precision parallels may be used. These parallels accurately elevate the surface to permit accurate measurement.

A test indicator is then selected with either a front- or top-mounted dial. Whether a swivel-point test indicator or a regular dial indicator (with plunger-type stem) is used depends on the location of the surface to be checked. A dial indicator with a magnet base is especially adapted for rigid mounting on any ferrous surface. An indicator with a regular base is adapted to either black granite or metal surface plates.

Like the dial indicator, the dial test indicator (Figure 6–8) is usually held in an attachment from a universal set. The dial test indicator is mounted on a column, magnetic base, or height gage. The contact point of the indicator is slid into position over one corner of the workpiece. The readings on the indicator dial face show any

deviation in parallelism. As in all precision measurement work, the readings are rechecked.

CHECKING FOR SQUARENESS AND FLATNESS

Usually the squareness of two surfaces is given in terms of *deviation per six inches of length*, particularly when stating the accuracy of measuring tools. One common practice in determining the amount that a workpiece or other part is out-of-square is to use a master cylindrical square. The cylindrical square is carefully brought against the surface to be checked. The cylindrical square is rotated until no light is visible. Any variation within the range of the cylindrical square may be read directly from the graduations.

CHECKING FOR ROUNDNESS

Roundness refers to the uniform distance of the periphery of a plug, ring, or hole in relation to the axis. Roundness is a measurement of a true diameter or radius. V-blocks, a surface plate, a dial indicator, and a holding device are generally used for checking roundness.

The workpiece is placed in the V-block. The contact point of a dial indicator (that is mounted securely) is moved onto the outside diameter. The indicator is positioned at the point where the highest reading is obtained—that is, at the centerline of the workpiece.

The dial face is turned until the indicator reading is zero. The part is then rotated carefully

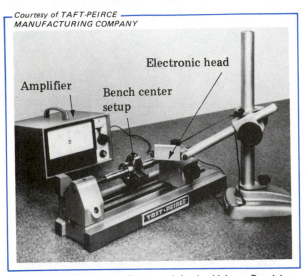

Figure 6–9 Measuring Concentricity by Using a Precision Bench Center and Electronic Height Gage and Amplifier

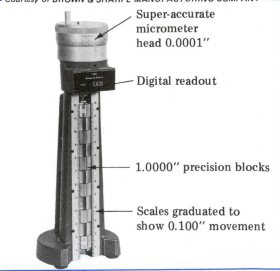

Figure 6–10 Precision Height Gage with Digital Readout and 1.0000″ Precision Blocks

so that the position of the V-block, part, and original indicator setting are not changed. Any out-of-roundness from the true diameter shows up on the direct dial readings of the indicator.

CHECKING FOR CONCENTRICITY

Concentricity relates to the periphery (outside surface) of a cylindrical part and an axis of rotation. A cylindrical surface is said to be concentric when every point is equidistant from the imaginary axis. The measurement of concentricity is often called *total indicated runout (TIR)*. Testing and measuring for runout are usually done on a metal or black granite surface plate. A bench center with two dead centers and a dial test indicator are generally used. Higher accuracy of measurement is obtained with an electronic height gage and amplifier unit (Figure 6–9).

CHECKING FOR AXIAL RUNOUT

Axial runout relates to the amount a shoulder or flange on a cylindrical part is out of alignment with the axis of rotation. A bench center, plane surface, and mounted dial test indicator are needed to test for axial runout.

The centered workpiece with a shoulder or flange is placed between centers on the bench center. The bench center is placed vertically on the surface plate. The part is turned one revolu-

tion. Any runout is indicated by the movement of the contact point. The amount of runout is read directly from the readings on the dial test indicator face.

GENERAL PRECISION LAYOUT AND MEASUREMENT PRACTICES

TAKING LINEAR MEASUREMENTS (LENGTH AND HEIGHT)

Linear measurements may also be checked, or measuring instruments may be set, with a combination gage block/height gage instrument. This instrument (Figure 6–10) consists of a base and post. The post supports a carrier with 1.0000″ (in the case of a metric instrument, 25mm) thick precision blocks. These blocks are arranged so that the top of one block is on the same plane as the bottom of an adjacent block. On some models, the lower gage block in the column is 0.090″ thick. This size permits measurements of 0.100″ height to be checked.

The 12″ instrument in Figure 6–10 has a super-accurate micrometer head. The thimble graduations may be read directly in 0.0001″. A metric micrometer head has readings from the spindle in increments of 0.002mm. The two scales on the sides of the precision blocks show each 0.100″ of movement. The direct readings are further simplified by a digital readout.

LAYING OUT AND MEASURING HOLE LOCATIONS

Measuring the exact locations of holes is another common surface plate process. Hole locations within ±0.0001″ (±0.002mm) may be layed out with a gage block accessory set. Gage blocks of the required height and the precision scriber point combination are applied.

INDIRECT MEASUREMENT OF A HOLE POSITION

Instruments for laying out and measuring hole locations are also used to measure the location of holes that are already formed. One major difference is that the hole location is not established directly. The measurement is made to the bottom of the hole or to the top of a plug inserted in the hole. Many parts in punch and die, jig and fixture, and job shop work require that the location of one or more holes be measured. The workpiece is positioned on a surface plate and strapped to a precision angle block with a clamp. A dial test indicator is secured to a vernier height gage.

The contact point of the indicator is set at the required height (the center line distance less one-half the diameter of the hole). The measurement is made with a gage block combination. The hole position is then checked against the setting of the dial test indicator.

MEASURING SMALL HOLE LOCATIONS

It is not always possible to position the indicator contact point in a hole. If the diameter of the hole is too small, a press or push fit pin or plug is inserted temporarily. The indicator contact point is set at zero at the required gage block height. This height equals the center distance plus the radius of the pin or plug. Any variation from the zero setting indicates the amount the hole position varies from the required dimension.

WORK ORDER, PRODUCTION PLAN, AND BLUEPRINT OR SKETCH

The *work order* and/or *production plan* provide the following types of information about a particular part or unit; name, function, quantity, work processes, schedule, and so on. The production plan provides technical information about tool requirements, the machining sequence, tolerances, sizes, machine setups, cutting speeds and feeds, and other production conditions. The work order and production plan are usually used with a mechanical drawing, freehand sketch, or blueprint.

The blueprint and the work order provide all essential information for the performance of a bench work process. This information includes the technical details of form, sizes, quantity, material, finish, and a part's relation to other fitted or mated parts. The same technical information is conveyed to the designer, draftsperson, engineer, mechanic, technician, and consumer.

PREPARATION OF THE WORKPIECE

One of the unique functions of all machine and metal trades is the production of accurately assembled parts of precise shape, size, and surface finish. The sizes, shapes, and other characteristics of materials drawn from supply sources should always be checked against the corresponding size, shape, and other physical specifications in the work order. The material should be sufficiently oversized to permit the worker to perform the hand and machine processes necessary to produce a particular part.

The most practical and universally used of any surface coloring agents are layout dyes. These dyes may be used with ferrous metals and with some nonferrous metals such as aluminum, brass, bronze, and copper. After the layout surface is burred and cleaned to remove any oil film, the dye solution is brushed or sprayed on.

Layout dye dries rapidly and provides a deep blue, lavender, or other-colored surface. Finely scribed lines stand out sharply and clearly.

The surfaces of rough castings and forgings may be painted with a thin white pigment that is mixed with alcohol. This white pigment spreads easily and dries rapidly. It produces an excellent coating for layout operations.

WORK HABITS AND SAFE PRACTICES

The skilled craftsperson works in an orderly manner. Attention is paid to the condition, placement, and handling of all layout, cutting, or other hand tools and instruments. The work place and the part or parts are all considered in terms of safety and accuracy.

A number of hand tools have cutting edges or are heavy. These tools must be placed where they do not rest on or scrape against finished surfaces or precision measuring tools. Rough, unmachined surfaces must also be positioned so that they are not in direct contact with finely finished surfaces or precision instruments. Accuracy and personal safety require that separate places on the work bench be planned for placing and working with different tools.

Scraping produces scratched or scored surfaces. The skilled worker, therefore, avoids scraping one surface against another. Instead, parts are gently picked *up* and carefully placed in position. Surfaces of different degrees of finish that are to be placed together should have a thin metal sheet or even paper placed between the surfaces to protect them.

A. SAFE PRACTICES WITH PRECISION LINEAR AND ANGULAR MEASURING ACCESSORIES AND INSTRUMENTS

- Remove dust and foreign particles from gage blocks and microfinished surfaces with a fine, soft camel's hair brush.
- Check along microfinished surfaces that are to be in contact in order to detect any nicks or burrs. Remove all imperfections properly.
- Allow time for measuring instruments and workpieces to normalize to the same temperature.
- Avoid working near heat sources that produce varying temperatures of a workpiece and measurement tool.
- Check gage blocks, measurement surfaces, and instruments against master standards to ensure that correct measurements can be established.

- Place granite angle blocks, parallels, gage blocks, and other precision measurement accessories and instruments in their appropriate cases.
- Apply the amount of force on straps and clamps that is needed to hold parts securely. Any excess force tends to spring or bend an assembly and thus produces inaccurate measurements.
- Clean measuring instruments and dry them thoroughly. Avoid using compressed air, which may drive dust and other foreign particles into an instrument.
- Lubricate moving parts. After an instrument is used and cleaned, apply a fine corrosive-resistant lubricant to it and then store the instrument in an appropriate container.
- Make sure that the dimensions, shape, and materials of a workpiece permit it to be drawn securely to a perma-sine. The magnetic force must be adequate to prevent any movement of a workpiece during a machining or other working process.

B. SAFE PRACTICES IN PRECISION LAYOUT, MEASUREMENT AND INSPECTION

- Protect the accuracy of a surface plate by not laying unnecessary tools, instruments, and workpieces on it. All items used on a surface plate should be checked for burrs, scratches, and nicks. Remove all imperfections before the items are used.
- Place only the workpiece and the measuring and layout tools and accessories required for a job on a surface plate.
- Elevate parts with projecting lugs, arms, and other surfaces. These surfaces should clear the face of a surface plate. Use protective materials between a rough-surfaced workpiece and a precision-finished surface.
- Recheck all measurements for repeatability of measurement.
- Lap work centers if any burrs or nicks exist on the angular surfaces.
- Follow the practice of noting measurements, particularly where a number of measurements are to be taken.

UNIT 6 REVIEW AND SELF-TEST

A. LINEAR AND ANGULAR LAYOUTS

1. List three advantages of using granite over similar cast iron and steel layout and measuring instruments and accessories.

2. a. Refer to a table of constants for 5″ sine bars (blocks).
   b. Determine the required gage block height for each of the following 5″ sine block angle settings: (1) 16°, (2) 18°5′, and (3) 22°57′.

3. a. State four main uses of compound sine plates.
   b. Indicate the design features of a compound sine plate that make compound angle settings possible.

4. State two precautions to take to safely store precision linear and angular measuring instruments and gages.

B. PARALLELISM, SQUARENESS, ROUNDNESS, AND CONCENTRICITY

1. Explain the importance of a perfectly flat plane surface to precision layout and inspection processes.

2. List the steps required to check the flat faces of a hardened and ground rectangular fixture plate for parallelism (within ±0.0001″).

3. Explain the difference between the *roundness* and the *concentricity* of a turned part.

4. List the steps to lay out three horizontal lines when a vernier height gage is used. The lines are located 25.4mm, 38.1mm, and 50.8mm from the same ground edge.

C. PREPARATION OF WORKPIECES

1. Give four reasons why the worker depends on a blueprint and/or work order (production plan).

2. Explain briefly what is meant by (a) the condition and (b) the preparation of a workpiece.

3. Set up a series of four conditions that constitute good work habits and attitudes.

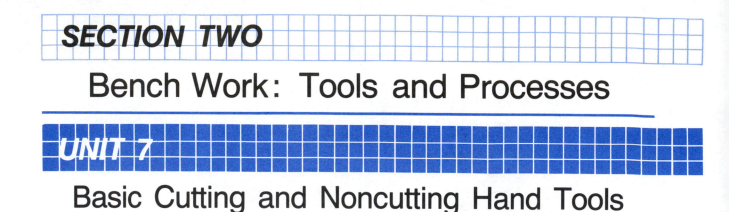

UNIT 7

Basic Cutting and Noncutting Hand Tools

Bench hand tools fall into two groups: (1) *noncutting tools* and (2) *cutting and shaping tools.* Each group of tools relates to a number of bench work processes as follows:

- Noncutting hand processes such as
  - Hammering,
  - Fastening,
  - Clamping,
  - Positioning;
- Cutting and shaping hand processes such as
  - Sawing,
  - Chiseling,
  - Punching and driving,
  - Filing,
  - Threading,
  - Reaming.

A. NONCUTTING HAND TOOLS

HAMMERING TOOLS AND PROCESSES

BALL-PEEN HAMMERS

One of the everyday tools used in the machine and metal trades is the *ball-peen* hammer. Two other tools that are not as common are the *straight-peen* and the *cross-peen* hammers. One end of each hammer head is shaped as a cylindrical solid. The face is ground slightly crowned and the edge is beveled. This face is used to deliver a flat blow (force) or to shape and form metal. The other specially shaped end of each head permits the operator to deliver a blow in a more confined area or to shape and form metal in a somewhat different manner than with the face. For example, a surface may be *peened* or shaped through a series of cupped indentations with a ball-peen hammer.

Hammers come in many sizes. The size used depends on the work size and required force. Hammer weights range up to three pounds. Ball-peen hammers weighing just an ounce are used in fine instrument work and for small, precise layout operations. The 12-ounce ball-peen hammer is one of the most practical sizes for general machine work. A large force may be delivered by a hammer at a desired point by firmly grasping the handle near the end and delivering a solid, sharp blow.

SOFT-FACE HAMMERS

There are many assembly and disassembly operations in which machined parts fit tightly together. These parts must be driven into place without damage to any part or surface. In these operations, therefore, the face of a hammer head must be softer than the workpiece. The pounding faces of a soft-face hammer may be made of rawhide, plastic, wood, hard rubber, lead, brass, or other soft metal or material. Some makes of soft-face hammers are designed so that the faces may be replaced when worn.

FASTENERS AND FASTENING TOOLS

Screws, nuts, and bolts are called *fasteners.* Fasteners are turned with two major groups of hand tools: screwdrivers and wrenches. The construction and design of these tools vary depending on the application. For instance, small jewelers' screwdrivers are used for fine precision work. By contrast, some screwdrivers for heavy-duty service have a square shank to which a wrench

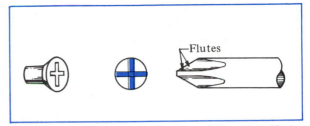

Figure 7-2 Features of a Phillips-Head Screwdriver Blade

and length of the screw slot. The tip of the blade should be flat so that it can be held firmly against the bottom of the slot.

The blade of the Phillips-head screwdriver has four *flutes* that cross at the center of the cone-shaped tip (figure 7-2). It fits screws and bolts that have two slots that cross in the center of the head. The crossed slots extend across the screw head but stop short of the outer edge.

OFFSET SCREWDRIVERS

The third type of screwdriver is the *offset* screwdriver. This screwdriver is made from one solid piece of metal and usually has a flat, wedge-shaped tip on each end. One tip is at a right angle to the "handle" (body of the screwdriver). The other tip is parallel to the handle. Two common offset screwdrivers are illustrated in Figure 7-3. Offset screwdrivers also come with Phillips and Reed and Prince type tips.

The offset screwdriver is used to hold and turn fasteners in spaces that are inaccessible to straight-shank screwdrivers.

HAND WRENCHES

Like screwdrivers, hand wrenches are also considered to be fastening tools. Therefore, hand wrenches are made of a high-quality alloy steel that combines strength, toughness, and durability.

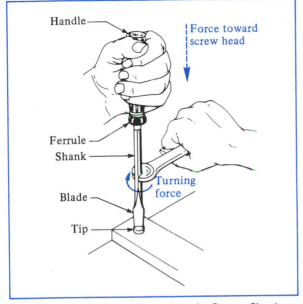

Figure 7-1 Wrench Attached to the Square Shank of a Heavy-Duty Flat-Head Screwdriver for Added Turning Force

can be attached to gain added leverage (force) as shown in Figure 7-1.

SCREWDRIVERS

There are many types of screwdrivers. The types most often used in the machine shop are the flat-head, Phillips-head, and offset screwdrivers.

FLAT- AND PHILLIPS-HEAD SCREWDRIVERS

A screwdriver usually combines four basic parts: handle, ferrule, shank, and blade (head).

A good screwdriver shank and blade are made of a high-quality steel that has been forged to shape, hardened and tempered. Screwdriver sizes are designated by the length of the shank; typical sizes are 6″, 8″, and 10″. The size of the tip is proportional to the blade length.

There are two basic shapes of tips: *flat head* and *Phillips head*. The flat-head and Phillips-head screwdrivers are similar, except for the shape of their blades and tips. The flat-head shape, ground to a slight wedge and flat across the tip, fits screw heads that are slotted completely across the head diameter. The worker should select a tip size that is almost the width

Figure 7-3 Offset Screwdrivers

Wrenches also are heat treated to prevent wear and to gain other desirable physical properties.

Wrenches derive their names from (1) their shape—for example, *open-end* wrench; (2) their construction—for example, *adjustable* wrench; or (3) the requirements of the job—for example, *pipe* wrench. Four common types of wrenches used in bench work are described in this unit. They are:

- Solid nonadjustable wrenches,
- Adjustable wrenches,
- Allen wrenches,
- Special spanner wrenches.

SOLID NONADJUSTABLE WRENCHES

The solid wrench derives its name from its construction. It is made of one piece. The size of the jaw opening is, therefore, permanent (nonadjustable). A solid wrench is designed so that the overall length is proportional to (1) the size of a bolt or nut and (2) the force normally required for tightening or loosening operations.

The three general types (shapes) of solid nonadjustable wrenches are: (1) *solid open-end*, (2) *box*, and (3) *socket*. Each of these wrench types is available in metric or inch sizes and singly or in sets.

Solid Open-End Wrenches. Some solid wrenches are open ended—that is, they are designed with either one open end or two open ends with different sizes. The open end (head) permits the wrench jaws to be slid onto the sides of a bolt or nut from either the top or side and for force to then be applied from whatever position is most advantageous.

Box Wrenches. Other solid nonadjustable wrenches, box wrenches, have closed heads (ends). Each head accommodates a particular size of hexagon bolt or nut. Box wrenches are used whenever space permits the heads to be fitted *over* bolts or nuts.

Socket Wrenches. Still other solid wrenches, socket wrenches, have hollow, fluted sockets. Hexagon sockets fit over hexagon-head nuts and bolts. Square sockets are designed to fit over square-head nuts and bolts.

Socket wrenches provide flexibility. They may be used with a ratchet handle, which permits turning nuts and bolts in confined spaces. They also may be used with flexible adapters and extension bars, which make it possible to turn bolts and nuts that are inaccessible with other wrenches. The parts of a typical socket wrench set are shown in Figure 7–4.

Socket and box wrenches should be used whenever possible in preference to open-end and adjustable wrenches. The proper size and shape of a socket or box wrench *nests*, or *houses*, the nut or bolt to prevent slippage. Thus, a greater force may be applied, and manipulating in confined places is easier.

Torque Wrenches. The proper functioning of many parts requires that the contact surfaces be secured together with a uniform (specified) force. The *torque* (tightening, or turning, force) must be controlled to prevent overstressing and stripping of the threads or shearing off of the bolt or nut.

Tension or torque wrenches are used to measure and control turning force (Figure 7–5). These wrenches have either a graduated dial or other type of scale that indicates the amount

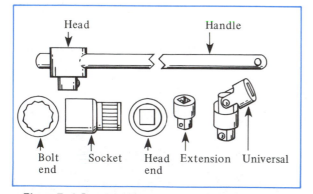

Figure 7–4 Parts and Features of a Socket Wrench Set

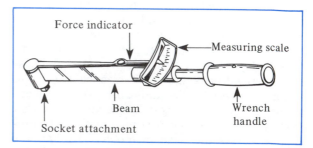

Figure 7–5 Torque Wrench

of applied turning force in foot- and/or inch-pounds.

ADJUSTABLE WRENCHES

Adjustable wrenches are adaptable to a wide range of nut, bolt, or workpiece sizes and shapes. Three common adjustable types are the *adjustable open-end* wrench, the *monkey* wrench, and the *pipe* wrench. One adjustable wrench can often replace many solid (fixed-size) wrenches. The sizes of monkey and pipe wrenches are designated by their overall length— for example: a 10″ monkey wrench and a 12″ pipe wrench. The size of an adjustable open-end wrench is indicated by the maximum dimension of the jaw opening.

Monkey, pipe, and adjustable open-end wrenches require that the direction of pull (turning force) be *toward* the adjustable jaw. For personal safety and to avoid damage to the workpiece or part, these wrenches must be checked and adjusted repeatedly during the operation to maintain a close fit between the jaws and the work. The pipe wrench is self-adjusting to a limited extent. The jaws move together when a force is applied.

The pipe wrench is used for turning objects that are round or irregularly shaped. It's two jaws have teeth. As pressure (force) is applied in the turning direction, the teeth grip the work. The turning force is transmitted to the part and causes it to turn. The teeth, however, leave indentations and burrs, which might cause the workpiece to be rejected.

ALLEN WRENCHES

Hollow or socket type setscrews and other bolts are turned by a hexagon-shaped offset wrench known as an *Allen* wrench. Figure 7–6

shows one of these wrenches. Allen wrenches are made of alloy steel and are heat treated to withstand great forces. These wrenches are available individually or in sets and in either millimeter or fractional-inch sizes.

SPANNER WRENCHES

Figure 7–7 shows two types of spanner wrenches. The *face spanner* has pins that fit into notches or holes. The *single-end (hook) spanner* is used to turn a special nut or bolt that is notched.

Spanner wrenches are widely used on machine tools. The wrenches turn large nuts that draw and hold a chuck or other device securely to a tapered machine spindle. A reverse turning process unlocks the chuck so that it may be removed.

CLAMPING AND GRIPPING DEVICES

BENCH VISES

Bench vises are devices that can be adjusted quickly to clamp work between their jaws. The vise is usually positioned on a bench at a height that permits the worker to comfortably perform bench work operations.

The bench vise top section can be turned horizontally on a simple swivel arrangement. A swivel lock secures or releases the vise top section from the base, which is attached to the bench.

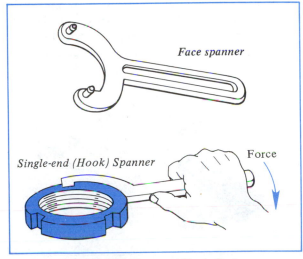

Figure 7–7 Face and Single-End (Hook) Spanner Wrenches

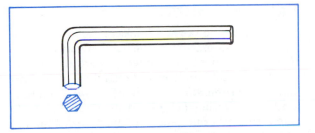

Figure 7–6 Allen Wrench

Regular vise jaws are *serrated* (grooved) to provide a good gripping surface. Some workpiece surfaces, however, must be protected against the scratches, burrs, and indentations caused by direct contact with the jaws. Therefore, *soft jaws*, which fit over the serrated vise jaws, are used to prevent scoring of such workpiece surfaces.

PLIERS

Pliers are used for gripping, turning, and drawing wire and small parts and for cutting metal wires and pins. The two most commonly used pliers are *slip-joint* pliers and *needle-nose* pliers.

CLAMPS

Clamps are used for holding work in place. The *C-clamp* is usually used for comparatively heavy-duty clamping. The C-clamp also permits the clamping of parts that are wider than can be accommodated with other (parallel) clamps.

Parallel clamps have two jaws that provide parallel clamping surfaces. The distance between the two jaws is controlled by two screws. When the jaws are parallel with the work faces, sufficient force may be exerted to securely hold the pieces. It is important that as much of the full surface of each parallel jaw be in contact with the work as is possible.

Figure 7–8 shows a simple toolmaker's parallel clamp. As holding devices, a set is generally used for clamping workpieces with finished

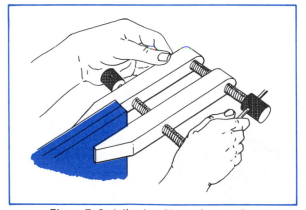

Figure 7–8 Adjusting Clamp Jaws to Be Parallel for Applying Force

surfaces during layout, inspection, and basic machining operations. Because the clamping faces of each jaw are ground smooth, they will not mar a finished surface.

INDUSTRIAL GOGGLES AND FACE SHIELDS

Although industrial goggles and face shields are not classified as tools, they are introduced here because their use is essential in preventing eye injuries. They are especially important in bench and floor work and in machining operations because of the ever-present danger of injuries from flying particles and other foreign matter.

The worker should obtain a personal pair of goggles or an appropriate shield. Goggles should be ventilated to prevent fogging but should still protect the eyes against grit and other small particles. Goggles that fit over vision-correcting eyeglasses are available.

Face shields are preferred as a lightweight and cooler eye protection device. Shields are practical in operations that do not involve dust, abrasives, and other fine particles.

B. HAND SAWING PROCESSES AND TECHNOLOGY

Two sets of related factors should be considered when selecting the correct hacksaw frame and blade:

- Shape, size, and kind (properties) of material to be cut;
- Features of a particular saw blade (pitch, set of the teeth, and length; the metal of which the blade is made; and its heat treatment).

HACKSAW FRAMES

Selection of the correct hacksaw frame is an important consideration in the hand sawing process. Hacksaw frames are named according to the shape of their handles—for example, *pistol-grip* and *straight-handle*. Each of these frames is available as either (1) a solid-steel nonadjustable frame that fits one length of blade or (2) an adjustable frame that accommodates various lengths

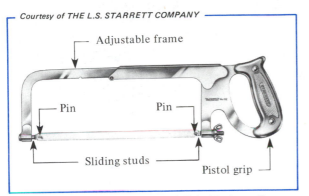

Courtesy of THE L.S. STARRETT COMPANY

Adjustable frame

Pin Pin

Sliding studs

Pistol grip

Figure 7–9 An adjustable Pistol-Grip Hacksaw Frame

of blades. An adjustable pistol-grip hacksaw frame is shown in Figure 7–9. This particular adjustable frame accommodates hacksaw blade lengths ranging from 8″ to 10″ (200mm to 250mm).

As shown, the front and handle ends of the frame are each equipped with a *sliding stud*. The studs may be turned to and locked in any one of four different positions. They permit turning the blade to 90°, 180°, 270°, and the normal vertical position.

There is a pin on each sliding stud. The frame is set to the blade length. With the teeth pointed forward, the blade is seated against the flat part of each stud. Tension is applied to hold the blade tightly in the frame.

HACKSAW BLADES

Another important consideration in the hand sawing process is the selection of the correct hacksaw blade. Hacksaw blades are made from high-grade tool steel, tungsten alloy steel, tungsten high-speed steel, molybdenum steel, and molybdenum high-speed steel. Tungsten adds *ductile* strength to steel and thus relieves some of its brittleness. Molybdenum adds to its wearing qualities.

All blades are hardened and tempered. There are three hardness classifications: all hard, semi-flexible, and flexible back. Each classification describes the heat treatment of a particular blade. *All hard* indicates that the entire blade is hardened. Although the all-hard blade retains a sharp cutting edge longer, it is relatively brittle and may be broken easily. *Flexible back* means

that only the teeth on the blade are hardened. Therefore, in contrast to the all-hard blade, the flexible-back blade can *flex* (bend slightly) during the cutting process without breaking.

Length is also a factor in selecting a hacksaw blade. *Length* refers to the center-to-center distance between the holes in the ends of the blade. Various lengths of blades are available. The most common lengths are 8″, 10″, and 12″ (200mm, 250mm, and 300mm). Blade width is standardized at approximately 1/2″ or 12mm. Hand hacksaw blades are 0.025″ (0.6mm) thick.

Two other important factors also enter into the selection of a blade. They are the *number* of teeth and the *set* of the teeth.

Hacksaw blades are made with a specified number of teeth per inch. This number is called the *pitch* of the blade. For example, a blade with 14 teeth per inch is a 14 pitch blade. Blades are made with pitches that vary from coarse (14 pitch) to medium (18 or 20) to fine (24 to 32).

The pitch to select depends on the hardness, the shape, and the thickness of the part to be cut. For large sections or mild materials, a 14 pitch provides good chip clearance. Harder materials like tool steel, high carbon and high speed steels require an 18 pitch blade. A 24 pitch is used on structural iron shapes, brass, copper, iron pipe, and electrical conduit. A 32 pitch is used for thin tubing and sheet metals.

In addition to being specified in number of teeth per inch (pitch), hacksaw teeth are *set* (positioned) to permit a *kerf* (groove) to be cut freely. The kerf is slightly wider than the thickness of the blade to prevent the blade from binding. A *standard* set means that one tooth is offset to the right and the next to the left, and so on. Some fine-pitch blades have a *double alternate* set, or a *wave* set (Figure 7–10).

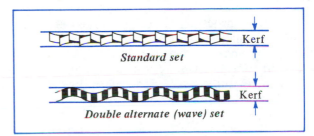

Kerf

Standard set

Kerf

Double alternate (wave) set

Figure 7–10 Standard and Double Alternate Set Patterns

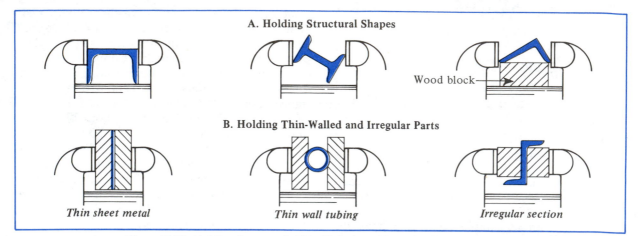

A. Holding Structural Shapes

Wood block

B. Holding Thin-Walled and Irregular Parts

Thin sheet metal   *Thin wall tubing*   *Irregular section*

Figure 7–11 Holding Structural, Thin-Walled, and Irregular Shapes in a Vise

CUTTING SPEEDS AND FORCE (PRESSURE)

For hand hacksawing, blade manufacturers recommend a cutting speed of 40 to 50 strokes per minute. At this speed the worker is able to control (relieve) the force on the teeth during the return stroke and saw without tiring.

The amount of force to apply when cutting depends, as with other hand tools, on the skill and judgment of the operator. Enough force must be applied on the *forward* cutting stroke to permit the teeth to cut the material. The forces are released and the blade is raised slightly from the work surface to prevent the teeth from scraping during the return stroke.

HOLDING IRREGULAR SHAPES AND THIN-WALLED MATERIALS

A general principle in hand hacksawing is to always engage at least two teeth with the work to prevent digging in or stripping of the teeth or bending of the part. Thin, flat materials should be held between blocks of wood. Tubing and other circular shapes should be nested in a simple form made of soft material. A wider area of clamping surface is thus provided.

On thin-gage tubing a wooden plug may be inserted in the center. Sometimes the hacksaw cut is made through both the wood support and the workpiece to permit a greater number of teeth to be engaged. Breaking through or damaging the part is thus avoided.

Common practices for holding structural shapes in a vise are illustrated in Figure 7–11A. Similarly, Figure 7–11B shows how thin-walled and irregularly shaped parts should be nested and held.

C. CHISELING, PUNCHING, AND DRIVING TOOLS AND PROCESSES

CHISELS

Many workpieces require the cutting of metal or other hard material by a hand process called *chiseling*. Chiseling usually refers to two operations: (1) *chipping*, which is the process of removing or cutting away material, and (2) *shearing*, which refers to the cutting apart of metals or other materials.

Cold chisels are used in both bench and floor work for chipping metals and other materials. Cold chisels are also used in shearing operations. The solid jaw of a vise may be used as one blade for shearing; the cold chisel is considered to be the other shearing blade. A scissor-like action of these blades shears the material.

The three main parts of a cold chisel are: (1) the cutting end (edge), (2) the body, and (3) the head. The body shape of the chisel may be hexagonal, octagonal, rectangular, square, or round. Cold chisels are usually forged to shape, hardened, and tempered. Cold chisels will cut materials that are softer than the hardness of the chisel itself. Cold chisels may be driven

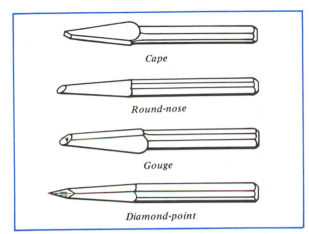

Figure 7–12 Common Types of Cold Chisels

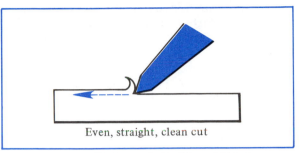

Figure 7–13 Correctly Positioned Chisel

(forced) with a hand hammer or a pneumatic (air) hammer.

COMMON TYPES OF COLD CHISELS

Cold chisels are named according to the shape of or the shape made by their cutting edge. The chisel that is most common is the *flat* chisel. This chisel is used to chip (remove) material from a surface. The flat chisel also shears (cuts through) rivets and other parts. The cutting faces of this chisel are ground, in general practice, to an included angle of 60° to 70°. The cutting edge is slightly crowned.

Other common types of cold chisels are shown in Figure 7–12. The *cape* chisel is the narrowest of the chisels in width. The cape chisel is, therefore, especially useful for cutting narrow grooves, slots, and square corners. The cape chisel is also used to cut shallow grooves in work surfaces that are wide. The flat chisel is then used to remove the *stock* (material) between the grooves.

The *round-nose* chisel has a round cutting edge (Figure 7–12). It is designed for chiseling round and semicircular grooves and inside, round *filleted* corners. The *gouge* chisel is used for chipping circular surfaces. The *square* or *diamond-point* chisel is used for cutting square corners and grooves.

HOLDING, POSITIONING, AND STRIKING CHISELS

The holding, positioning, and striking of all types of chisels is important. Skillful cutting

requires that the chisel be held steady in one hand, with the finger muscles relaxed. For a chipping operation the cutting edge is positioned for the depth of cut. The chisel is positioned for each successive rough cut and again for the final finish cut (Figure 7–13).

For a shearing operation the cutting edge of the chisel is set at the cutting line. On mild ferrous metals a 1/16″ or 2mm cut is considered a roughing cut; 1/32″ or 1 mm, a finish cut. The depth of the roughing cut may be increased to 3/32″ on softer nonferrous materials.

Quick, sharp hammer blows on the head of the chisel are most effective for cutting. For personal safety it is important to *follow the cutting edge* and not to watch the head of the chisel. The depth of cut depends on the angle at which the chisel is held. The sharper the angle between the chisel and the workpiece, the deeper the cut. After each cut the chisel is positioned for the next cut. The cutting process is then repeated. The chisel point may be lubricated with a light machine oil for easier driving and faster cutting.

Note: When chiseling cast parts like cast iron, brass, or aluminum, the corners may break away

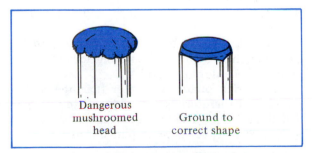

Figure 7–14 Corrected Shape of a Mushroomed Chisel Head

below the desired layout line. The chiseling of cast parts is started from the outside edges. The cut is then finished near the center of the workpiece.

Caution: Put on a pair of safety goggles or a protective shield.

Caution: A *mushroomed* head must be reground for safety. The corrected shape of a mushroomed head is illustrated in Figure 7–14.

HAND PUNCHES

Hand punches are used in bench work for four general purposes:

- To make indentations;
- To drive pins, bolts, or parts;
- To align parts;
- To punch holes.

Punches are known by the shape and diameter of their point, or shank end. Each type of punch is designed for a specific use. Solid, pin, and taper punches are made in a range of sizes. In practice, where parts are to be driven in or removed, the diameter of the punch should be as large as possible to permit the worker to deliver a solid, driving force with a hammer.

The *hollow* punch is used to cut holes in *shim* (very thin) stock or other thin materials. Hollow punches are made in sets from 1/4″ (6mm) to 4″ (100mm) in diameter.

PRICK AND CENTER PUNCHES

Two similar punches are used extensively in center layout work for making cone-shaped indentations: the *prick* punch and the *center* punch.

The center punch has a 90° included angle. This punch produces a centering location as a guide for positioning a drill.

The prick punch is smaller than the center punch and therefore is not subjected to as heavy a striking force. The included angle of the point is 30°. Primarily, the prick punch is used to:

- Form a small center hole from which circles may be layed out with dividers;

- Make small, easily identifiable, cone-shaped depressions along scribed lines. The prick-punched holes assist the worker during machining operations;
- Mark the location of parts for proper identification and assembly.

DRIFTS

Drifts are used to drive parts and are usually round, soft metal pieces. Drifts must be slightly smaller than the diameter of a shaft, pin, stud, or other part that is to be driven. Brass rods are often used as drifts. Brass is softer than steel and other alloys and will not score or mark the part that is being driven.

D. HAND FILE CHARACTERISTICS AND FILING PROCESSES

The *file* is a cutting tool. Files are used extensively in bench and other hand operations to:

- Alter the shape of a part,
- Reduce the size of a part,
- Remove burrs and edges,
- Finish surfaces that have tool marks,
- Touch up surfaces that require fitting in assembly.

CHARACTERISTICS OF FILES

PARTS OF A FILE

Terms that are commonly used to designate the parts of a file are shown in Figure 7–15.

The cutting edges of a file consist of a number of teeth. These teeth are cut in parallel rows diagonally across the face. Since files are made of high-grade tool steel and are hardened and tempered, the teeth are hard. A file, therefore,

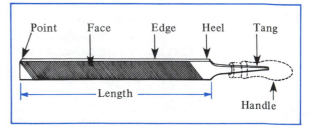

Figure 7–15 Parts of a File

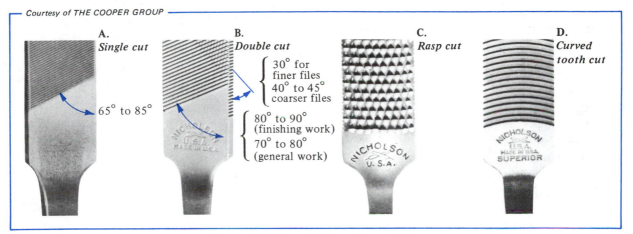

Courtesy of THE COOPER GROUP

A. *Single cut* — 65° to 85°

B. *Double cut*
- 30° for finer files
- 40° to 45° coarser files
- 80° to 90° (finishing work)
- 70° to 80° (general work)

C. *Rasp cut*

D. *Curved tooth cut*

Figure 7–16 Four Different Cuts of Files

can cut softer materials like unhardened steels, cast iron, brass, and nonmetallic products.

FILE CUTS

Files are produced with different *cuts* (parallel rows of teeth). A file with a single series of cuts is called a *single-cut* file. A *double-cut* file has two series of cuts. The two series of parallel rows of teeth cross each other at an angle with the edge. The angle of the cut on single-cut files varies from 65° to 85° with the edge (Figure 7–16A). Double-cut files for general metal work have one set of cuts at a 40° to 45° angle from the edge and a second crossing set at an angle from 70° to 80° (Figure 7–16B). The angles of the two cuts of finer double-cut files for finishing operations are 30° and 80° to 90°, respectively. Two other cuts of files are the *rasp cut* (Figure 7–16C) and the *curved tooth* (Figure 7–16D).

COARSENESS OF CUTS

The distance between each parallel row of cuts denotes the *coarseness* of a file. The coarsest cut on files larger than 10″ is known as *rough cut*. The finest cut is *dead smooth*. In between there are degrees of coarseness from a *coarse cut*, a *bastard cut*, a *second cut*, to a *smooth cut*.

Numbers are used to indicate the cut and relative coarseness of files that are smaller than 10″. There are ten different cuts: #00, #0, #1, #2, #3, #4, #5, #6, #7, and #8. The #00 cut is the coarsest, and #8 is the finest. Again, because

of the relation between coarseness of cut and size of file, a #1 cut 8″ file is coarser than a #1 cut 4″ file.

Each cut of file is used for a particular hand filing process. The most commonly used files in machine shops are: the bastard cut, for rough filing; the second cut, for reducing a part to size; and the smooth cut, for finish filing.

FILE NAMES

The name of a file is associated with either its cross section, particular use, or general shape. For example, a circular shape file is called a *round file*. A file with a rectangular cross section is generally designated as a *flat file*. However, if it is used for mill filing, it is identified as a *mill file* and if the edges are ground as safe filing edges the file is known as a *pillar file*. A cross section of a file having a curved and a flat face is called a *half-round* file.

PRECISION FILES

The files just described are generally classified as *machinists'* files. Three other groups of files, known as *precision* files, are next described from among the many designs of files. Precision files include the *Swiss pattern*, *needle-handle*, and *special-purpose* files.

SWISS PATTERN FILES

Swiss pattern files are used to accurately file and finish small precision parts. They are used commonly by tool and die makers, instrument makers, jewelers, and other highly skilled craftspersons. The files have cross sections similar to machinists' files.

The major differences between Swiss pattern and machinists' (American pattern) files are that:

- The points of Swiss pattern files are smaller,
- The tapers of Swiss pattern files toward their points and at their tangs are longer and sharper,
- The cuts of the Swiss pattern teeth are finer,
- The specifications of Swiss pattern files are more exacting.

Six of the most common Swiss pattern files —usually called just *Swiss* files—include half round, narrow pillar, knife, round, square, and three square (triangular) shapes. There are over 100 different shapes. The teeth are double cut on the faces. The teeth on the edges of the knife and three-square files are single cut. Some of these files are manufactured with safe edges (no teeth).

NEEDLE-HANDLE FILES

Needle-handle files are another group of small precision files. They are normally 4″ to 6″ long. They are used for fine, precision filing where a small surface area is to be formed or fitted with another part.

Needle-handle files also are designated by their shape, cut, and length. The characteristics and common shapes of some of these files are illustrated in Figure 7–17.

SPECIAL-PURPOSE FILES

Special-purpose files produce different kinds of cutting actions and surface finishes on various materials. They are designed for greater efficiency than general-purpose files. For example, although a mill file is widely used in lathe work, the special-purpose, long-angle, lathe file has a number of advantages:

- The teeth, cut at a longer angle, provide an efficient shearing action, in contrast to the tearing and scoring action of general-purpose files;
- Chatter marks are prevented;
- Clogging of the teeth is reduced;
- Materials are cut away faster;
- A finer surface finish is produced.

Soft metals, plastics, stainless steels, and other materials each require special-purpose files. These files permit maximum cutting efficiency, dimensional accuracy, and high-quality surface finishing. For very soft metals a special single-cut coarse file with a small angle provides an efficient shearing action (Figure 7–18).

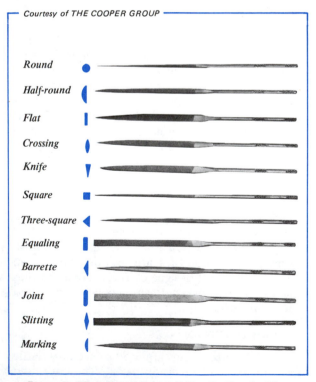

Courtesy of THE COOPER GROUP

Round
Half-round
Flat
Crossing
Knife
Square
Three-square
Equaling
Barrette
Joint
Slitting
Marking

Figure 7–17 A Set of Assorted Needle-Handle Files

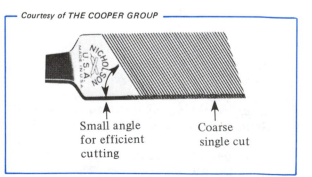

Courtesy of THE COOPER GROUP

Small angle for efficient cutting

Coarse single cut

Figure 7–18 A Small-Angle, Single-Cut Coarse File for Soft Metals

FEATURES OF FILES

There are three features of files that are especially important in filing. They are a *safe edge*, *convexity*, and *taper*.

SAFE EDGE

A safe edge is a file edge that has no teeth. The safe edge protects the surface against which the file is moved from any cutting action. The safe edge also permits filing into a sharper corner.

CONVEXITY

Convexity of a file means that the faces are slightly convex and bellied along their length. The faces curve so that they are smaller at the point end of the file. Convexity is important for five reasons:

- A perfectly flat face would require a tremendous downward force to make the teeth bite and a great forward force to produce a cutting action;
- Control of a perfectly flat file would be difficult;
- A flat surface could only be produced if every part of the file were perfectly straight;
- File warpage, resulting from the hardening process, would produce an undesirable concave face;
- Convexity makes it possible for the skilled worker to control the position and amount of cutting.

TAPER

Most files taper slightly in width. The taper on round, square, and three-square (triangular) files is more severe than the taper of other types. The cross-sectional area is reduced considerably from about the middle of the body to the point. A file is termed *blunt* if it has a uniform cross section along its entire length.

THE CARE OF FILES

One common problem in hand filing is *pinning*. Pinning occurs when a *pin* (small particle of the material being filed) gets wedged (pinned) on the lead edge of the file teeth. The pins, as they move on the file across the face of the work, tear into and scratch the filed surface.

Pinning is caused by:

- Bearing too hard (applying too much force) on a file, particularly on a new file;
- Failing to regularly clean the file teeth.

Pinning may be avoided by:

- Brushing the workpiece with a file brush to remove the filings,
- Brushing the file teeth with a wire file card,
- Pushing a metal scorer or a soft metal strip through the grooves of the teeth,
- Rubbing ordinary chalk across the file teeth to chalk them.

FILE HANDLES

Part of the skill in using a file depends on the size, shape, and feel of the handle. A file may be controlled better and hand fatigue may be avoided if the handle is easy to grasp and hold. The handle should be selected for shape and size. The handle (1) must feel comfortable and (2) must give balance to the file.

FILING PROCESSES

CROSS FILING

Cross filing, often referred to as just plain *filing*, is the process of diagonally crossing file strokes. First, the file face is forced against the work and pushed forward on the cutting stroke. The force is relieved on the return stroke. Then, after a few strokes, when a complete surface area has been filed, the position of the file and the filing operation are changed. The strokes are crossed. The procedure continues until the part has been *brought down* (filed) to a layout line or dimension. As much of the file face as is possible should be used during each cutting stroke. Checks for flatness and squareness of filed surfaces are shown in Figure 7–19A and B.

The surface of most materials may be filed directly. However, cast-iron parts require a special procedure. The *scale* (outer surface) of a cast-iron casting is often *file hard*. If a file is used directly on such a surface, the teeth will dull rapidly. The condition may be overcome either by rough grinding the surface or by chipping below the depth of the scale. Once the scale is removed, the cast iron is soft enough to be rough or finish filed.

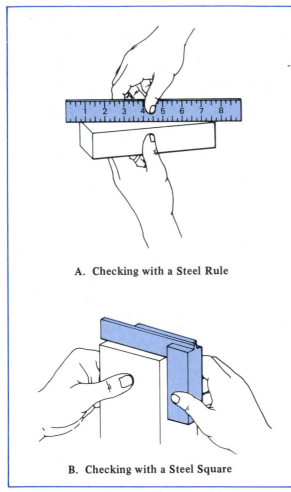

A. Checking with a Steel Rule

B. Checking with a Steel Square

Figure 7–19 Checking Flatness and Squareness

DRAWFILING

Drawfiling is a process of *graining* a work-piece—that is, producing a fine-grained surface finish. For drawfiling, a single-cut file is held with both hands, as illustrated in Figure 7–20. The file is then forced down onto the work-piece as it is moved over the surface. A cutting action is produced by *drawing* (pushing and pulling) the file alternately in each direction. Particular care must be taken to use a sharp file and to keep it clean.

Drawfiled surfaces are sometimes polished by holding a strip of abrasive cloth tightly against the bottom face of the file. The file is then drawn over the work to produce a polished surface.

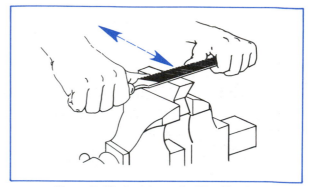

Figure 7–20 Positions of a File, Hands, and Work in Drawfiling

SOFT METAL FILING

The files and processes described thus far have dealt primarily with the filing of steels and other hard metals. These same files may not be suitable for soft materials because filings can clog and wedge between the teeth.

The curved-tooth file, mentioned earlier, is specially designed for efficiently filing soft metals like brass, copper, aluminum, and lead. This file has chisel-edged teeth that produce an excellent finished surface. Clogging of the teeth is reduced by the shape of the grooves between the rows of teeth. The curved-tooth file is also used for finish filing of soft metal parts turned on a lathe.

BURRS AND BURRING

A burr is the turned-up or razor-sharp project-ing edge of a material. Burrs generally are pro-duced by sawing, drilling, punching, shearing, and most cutting processes. Burrs are also raised through careless handling, hammering, dropping, and nicking. Damaged or spoiled work parts often result from failure to remove burrs before clamp-ing or performing hand or machine operations.

Burred edges, nicks, dents, and ragged or unwanted sharp corners are usually removed by hand filing. The sharp edges and burrs on hardened steel pieces may be removed easily with a small oilstone. The burrs on curved surfaces may be scraped with the three-cornered scraper. In summary, parts are burred to:

• Ensure dimensional accuracy,

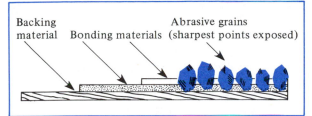

Figure 7–21 Prime Materials Used in the Manufacture of Coated Abrasives

- Provide for the exact fitting of mating parts,
- Permit safe handling,
- Avoid damage to tools and machines,
- Improve the finish and appearance of the part.

E. COATED ABRASIVES AND HAND FINISHING (POLISHING)

Abrasives and their applications to machine cutting and surface finishing processes are covered in detail in the units on grinders and grinding operations. Only basic information on coated abrasives that relates to bench and floor work is considered in this unit.

Coated abrasives are used in hand operations in machine shops. These abrasives produce a smooth or polished surface and remove very slight amounts of material. The edges of their abrasive grains, with their ability to fracture and form new sharp edges as they wear, make coated abrasives excellent for polishing.

STRUCTURE OF COATED ABRASIVES

Coated abrasives are produced by bonding a particular type and quantity of abrasive grains on a suitable flexible backing material. The relationship of the *coating*, the *bonding*, and the *backing* is shown in Figure 7–21.

The abrasive grains are electrostatically embedded in a bond in the backing material. The sharpest points of the grain are thus positioned outward and exposed.

The grains are spaced evenly as either an *open* or *closed* coating. The two basic open and closed coatings are shown in Figure 7–22. An open-coated abrasive cloth has only part of the backing surface covered with abrasive grains (Figure 7–22A). The open spaces between grains prevent the grains of the abrasive material from filling with chips or clogging.

By contrast, in closed coating, all of the backing surface is covered (Figure 7–22B). Closed-coated abrasive cloths, belts, and discs are used in heavy cutting operations that require a high rate of stock removal.

FORMS OF COATED ABRASIVES

Coated abrasives are produced in a number of forms. The most common forms are shown in Figure 7–23. These forms include *abrasive belts*, *grinding discs*, and *abrasive cloth* in sheet or roll form. Abrasive belts provide an efficient cutting material and technique for polishing sheet or strip steel. Grinding discs are used for smoothing welds and for removing surface

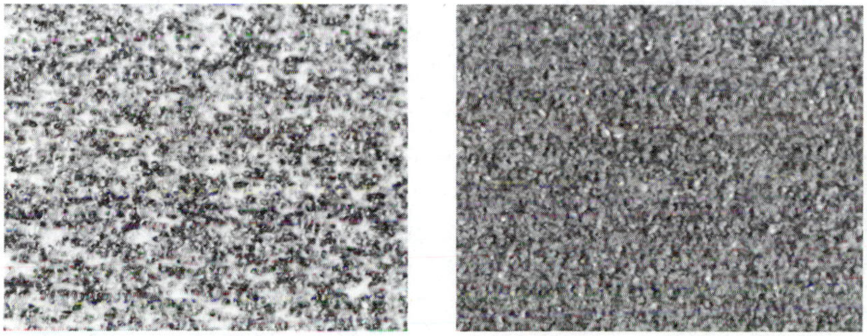

A. Open

B. Closed

Figure 7–22 Two Basic Abrasive Coatings

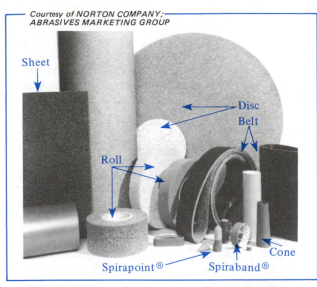

Figure 7–23 Common Forms of Coated Abrasives

irregularities, projections, and burrs. Abrasive cloth is used primarily for hand polishing and surface finishing.

When selecting a coated abrasive, consideration must be given to two sets of factors:

- The type of material in the part and the required surface finish,
- The qualities of the coated abrasive that are needed to cut efficiently and to produce the desired finish.

CHARACTERISTICS OF COATED ABRASIVES

The cutting efficiency of coated abrasives and surface finish depend on the:

- Abrasive material,
- Bonding and backing materials,
- Type of coating,
- Grain size.

ABRASIVE MATERIAL

An *abrasive* is a hard, sharp-cutting, crystalline grain. Abrasives are applied in machine, metal, and other industries to grind, polish, buff, and lap metal surfaces to specific sizes and surface finishes. Abrasives may be natural or artificial.

All abrasives possess at least the three following properties:

- *Penetration hardness*, which is the cutting ability due to the hardness of the grain;
- *Fracture resistance*, which is the ability of the grain to fracture when the cutting edge has started to dull and thereby produce a new sharp cutting edge;
- *Wear resistance*, which is the ability of the grain to resist rapid wear and to maintain its sharpness (wear resistance is affected by the chemical composition of the grains in relation to that of the work being ground).

Natural Abrasives. The natural abrasives that are still used in industry include *garnet*, *flint*, *crocus*, *diamond*, and *emery*. Coated garnet and flint abrasive papers and cloths are used in the woodworking industry. Crocus (red iron oxide) cloth and rouge (flour-sized powder) are used for polishing and producing a very fine finish. Extremely small amounts of metal are removed with rouge.

Artificial Abrasives. *Aluminum oxide*, *silicon carbide*, and *boron carbide* are synthetic (artificial) abrasives. The hardness of each of these abrasives is higher than the hardness of the natural abrasives (except diamond).

Silicon carbide is made by fusing silica sand and coke in an electric furnace at high temperatures. The sand provides the silica; the coke, the carbon. Silicon carbide is used for coated abrasives, grinding wheels, and abrasive sticks and stones. Silicon carbide has a high grain fracture. It is, therefore, desirable for grinding some of the softer metals like cast iron, bronze, and copper. Nonmetallic materials that are ground with silicon carbide include marble, pottery, and plastic.

Aluminum oxide is manufactured from bauxite ore. The ore is heated to extremely high temperatures, at which it is purified to crystalline form. The addition of titanium produces an abrasive that has added toughness qualities. With toughness, resistance to shock, and grain fracturing properties, aluminum oxide abrasives are suitable for grinding both soft and hard steels, wrought and malleable iron, and hard bronze.

Artificial abrasives with ever-higher hardness qualities are being developed. The important fact is that artificial abrasives are harder and have greater impact toughness than the natural abrasives. Artificial abrasives have the following advantages over natural abrasives:

- Faster cutting characteristics,
- Ability to retain their cutting properties longer,
- Uniformity of quality,
- Economical manufacture in a wider range of grain sizes.

BACKING MATERIALS

Paper, *cloth*, *fiber*, and combinations thereof provide the backing material for coated abrasives. Cloth and fiber backings are the most durable. They are used for hand and machine cutting, buffing, and polishing operations.

The most common forms of cloth-backed coated abrasives for hand use are sheets and rolls. The sheets are 9″ X 11″ (225mm X 275mm). The rolls range from 1/2″ to 3″ (12mm to 75mm) wide. Other forms of coated abrasives for high-powered grinding machines include belts, discs, spiral points, and cones.

BONDING MATERIALS AND COATINGS

Bonding material serves as an adhesive. This material holds the grain particles to the backing and indirectly contributes to their cutting properties. Since coated abrasives may be used on dry or wet processes, a combination of bonds is employed. *Glues* provide a bond for dry, light, hand grinding and polishing. *Synthetic resins* are practical for wet grinding with waterproof cloth. There are also modifications of glues, resins, and varnishes. The resin-bonded combina-

tions, in particular, provide moisture- and heat-resistant properties. These properties are required for rough grinding and polishing applications for toughness.

GRAIN SIZES

The abrasive grain sizes for coated abrasives range from extra coarse to flour. The extreme and the intermediate degrees of coarseness are indicated by numbers that range from 12 to 600. Coarseness classifications and their number designations are given in Table 7–1.

In trade language the terms *grain sizes* and *grits* are used interchangeably. Usually coarser grain sizes (grits) are used for heavy stock removal; finer grain sizes, for highly polished surface finishes. For hand polishing it is good practice to start with as coarse a grain as is possible. The grain size must be consistent with the required surface finish. Reducing the grain size from coarser to finer during a series of final finishing operations speeds the cutting action.

A. Safe Practices for Using Bench Tools

Hammering Tools

- Examine the hammer each time *before* it is used. The handle must be securely wedged in or bonded to the head. The flat (slightly crowned) face must not be chipped.
- Use a soft-face hammer if the work surface must not be dented by the hammer blow.

Fastening Tools

- Firmly press the screwdriver tip into the screw slot or cross-slots while applying turning force. Applying force helps to prevent the tip from slipping out of the slot(s), which can cause burring of the screw head slot(s), and personal injury.
- Turning force should be applied to an adjustable wrench only in the direction of the *movable* jaw.
- Apply turning force gradually, while carefully monitoring the position of the wrench and the work to detect signs of slippage. Excessive force can cause the jaws of any open-end wrench to spring.

Table 7–1 Coarseness Designations and Grain Sizes

Relative Coarseness	Grain Sizes
Extra coarse	12, 16, 20, 24, 30, 36
Coarse	40, 50
Medium	60, 80, 100
Fine	120, 150, 180
Extra Fine	220, 240, 280, 320, 360, 400
Flour	440, 500, 600

- Protect the surfaces of a workpiece from indentations produced by pipe wrenches.

Clamping Devices

- Use soft jaws or other soft material when holding finished work surfaces in a bench vise to prevent marring of the finished surface.
- Position the workpiece in a vise so that it does not overhang into an aisle or other area where a person might accidentally brush against it.

B. Safe Practices in Hand Hacksawing

- Position the workpiece area where the cut is to be made as close to the vise as is possible to prevent springing, saw breakage, and personal injury.
- Nest thin-walled tubing and other easily damaged parts in suitable forms. Otherwise, they may be collapsed or distorted by the clamping force.
- Select the blade pitch and set that is most suitable for the material and the nature of the cutting operation. Stripped teeth often result when the pitch is too coarse and fewer than two teeth are in contact with the workpiece at one time.
- Apply force only on the forward (cutting) stroke. Relieve the force on the return stroke.
- Reduce the speed and the force at the end of a cut. Slowing down prevents the teeth from digging into the small remaining section of the material. It also prevents the cut portion from breaking away.
- Start a new blade in another place when a blade breaks during a cut to prevent binding and blade breakage. Remember that the groove produced by a blade becomes narrower as the balde wears.
- Saw straight. If the cut *runs* (strays from the correct line), reposition the workpiece a quarter turn and start a new cut. Starting a new cut prevents cramping of the blade

and also helps guide the second cut in the correct line.
- Cut a small groove with a file in sharp corners where a saw cut is to be started. The groove permits accurate positioning of the saw and also prevents stripping of the teeth.

C. Safe Practices for Chiseling, Punching Holes, and Driving Parts

- Wear safety goggles or a protective shield when chipping and driving parts. Flying chips and other particles may cause eye injury.
- Place a chipping guard or frame between the work area where a part is being chipped and other workers.
- Grind off any overhanging metal (mushroom) that may form on the head of a chisel, punch, or drift. Fragments of a mushroomed head may break away and can fly off at high speed during hammering.
- Limit the depth of a roughing cut to 1/16″ (2mm). Allow 1/32″ (1mm) for a finish cut. Heavier roughing cuts require excess force, which may cause a chisel to break or the cutting edge to dull quicker.
- Avoid rubbing the edges of holes that have been punched in metal.

D. Safe Practices in Filing and Polishing

- Avoid hitting a file directly, scraping files together, or placing one file on another.
- Remove pins that cling to and filings that clog the file teeth.
- Use a file with a properly fitted, tight handle.
- Avoid handling parts with burrs or fine, sharp projections. Remove burrs as soon as is practical.
- Remove the outer scale on a cast-iron casting by grinding or chiseling before filing.
- Use an eye protective device for operations involving the use of abrasive materials.

UNIT 7 REVIEW AND SELF-TEST

A. NONCUTTING HAND TOOLS

1. List four noncutting and four cutting and shaping bench processes.

2. Identify (a) one type of hammer that is used to drive a pin and (b) another type that is used for assembling mating parts.

3. a. Describe a flat- and a Phillips-head screwdriver.
   b. State two advantages of the Phillips-head screwdriver over the flat-head screwdriver.

4. Recommend a hand wrench to use in each of the following applications:
   a. Tightening a hexagon nut in a confined recessed area;
   b. Tightening large-size, square-head lag screws;
   c. Assembling hexagon bolts in areas that are inaccessible to open-end wrenches and where each bolt must be uniformly tightened;
   d. Assembling and disassembling large diameter hexagon coupling nuts;
   e. Turning hollow or socket-type setscrews;
   f. Tightening (1) a shouldering collar that is fitted with a single slot on the periphery and (2) another collar that is fitted with two slots on the face.

5. Describe how holding workpieces with a C-clamp differs from gripping with a set of parallel clamps.

6. State two factors that should guide the worker in selecting industrial goggles or a protective face shield.

7. List four general safety precautions to observe when selecting and using hammers, screwdrivers, and wrenches.

B. HAND SAWING PROCESSES

1. Explain the design features of a hand hacksaw frame that permit the hacksaw blade to be positioned straight or at 90°, 180°, and 270°.

2. Identify four materials from which hacksaw blades are manufactured.

3. List the kinds of information that must be supplied in ordering hand hacksaw blades.

4. State one principle to follow for hand hacksawing thin-sectioned materials and structural shapes.

5. Define cutting speed as applied to hand hacksaw blades.

6. Tell how to start a cut at a sharp corner that may strip the saw teeth.

7. State a safe practice to follow in hand hacksawing when a blade breaks before the cut is completed.

C. CHISELING, PUNCHING, AND DRIVING TOOLS AND PROCESSES

1. a. Name three common hand chisels.
   b. Describe the main design features of the cutting end of each chisel named.
   c. List an application of each chisel named.

2. Tell what effect changing the angle of a chisel has on chiseling.

3. State three guidelines to follow when using a hand chisel for a shearing operation.

4. Distinguish between the design and use of a solid punch in comparison to a hollow punch.

5. Indicate two characteristics of drifts that differ from the characteristics of solid punches.

6. List the steps for aligning holes in two mating parts.

7. State three safety precautions for personal protection against injury during chiseling.

D. HAND FILE CHARACTERISTICS AND FILING PROCESSES

1. Identify (a) the types of cuts and (b) the angles of a general metal work file.

2. State how coarseness is designated for files under 10" long.

3. a. List three files that are used to produce a flat surface.
   b. Give one application of each file listed.
   c. Describe at least one distinguishing design feature of each file listed.

4. Give reasons why Swiss pattern files are preferred by many precision mechanics in place of American pattern files.

5. Enumerate three advantages of using a long-angle, lathe file to mill file lathe work.

6. Cite four reasons why files are produced slightly convex along their length.

7. State what corrective steps may be taken to avoid pinning.

8. Describe drawfiling.

9. List four safety precautions to follow in filing.

E. COATED ABRASIVES AND HAND FINISHING

1. Contrast open-coated with closed-coated abrasive cloths.

2. Name three properties of all abrasives.

3. State three advantages of using artificial abrasives in place of natural abrasives.

4. Give two guidelines the skilled mechanic uses to finish a machined surface by hand polishing.

Hand Reamers and Reaming Processes

HAND REAMING

Reaming is the process of shearing away small amounts of material from a previously prepared hole. The process produces a hole that is:

- Perfectly round,
- Held to a specified diameter within +0.001″/ –0.000″ (or to a closer tolerance),
- Smooth and has a high-quality surface finish,
- Straight.

Reaming is performed by a multiple-fluted cutting tool known as a *reamer*. Cutting action takes place by rotating and feeding the cutting edges on each flute of the reamer against the periphery (surface) of a hole. The cutting on many hand reamers is done along a tapered starting end. The amount of material removed in hand reaming ranges from 0.001″ to 0.003″ and occasionally up to 0.005″.

COMPARISON OF DRILLING AND REAMING PROCESSES

Many machined holes must be held to close tolerances for roundness, straightness, quality of surface finish, and dimensional accuracy. Drilling and other machining processes produce holes that have comparatively rough surfaces. These holes are larger in diameter than the drill size. They may not be perfectly straight and round. A drilled hole may vary up to +0.005″ in diameter for drill sizes up to 1/2″. The variation may be as high as 0.020″ for 1″ and larger sizes of drills. When a drill is guided by a drill bushing in a fixture, greater accuracy is possible.

Whereas a drill is used to produce a hole in a solid mass, the reamer is a *finishing* tool. The reamer reams a hole precisely to size. Parts may be hand reamed or machine reamed. Only a limited amount of material is removed by a finishing cut with a hand reamer. By contrast, a greater amount may be removed by machine reaming.

This unit is limited to hand reamers and hand reaming processes. *Solid* (fixed), *adjustable*, and *expansion* types of reamers are covered. Consideration is given to *straight*, *spiral*, and *taper* forms. The features of hand reamers are illustrated to help explain the shearing action. A few commonly used cutting fluids are considered. These fluids are related to cutting action and surface finish.

FEATURES OF HAND REAMERS

The main design features of the cutting end of a spiral-fluted hand reamer are identified in Figure 8–1. The hand reamer, whether straight or spiral (helical), has a number of *flutes*. These flutes are cut into and along the body. The flutes are machined so that the flute face is *radial* —that is, if the edge of a steel rule were placed on the front face of the reamer, it would pass through the center.

Each flute has a *land*. Each land has a small *margin*, or circular width behind the cutting edge of the reamer. The margin extends from the cutting face to a *relieved* area. Beyond the margin, the land is relieved to prevent the reamer from binding in the hole.

Shearing (cutting) action is produced by the *starting taper*. Beginning at the point of the reamer, the flutes are ground at a slight taper. This taper merges into the outside surface (periphery) of the reamer. The taper usually extends back for a distance slightly longer than the

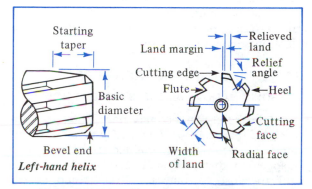

Figure 8–1 Main Features of the Cutting End of a Spiral-Fluted Hand Reamer

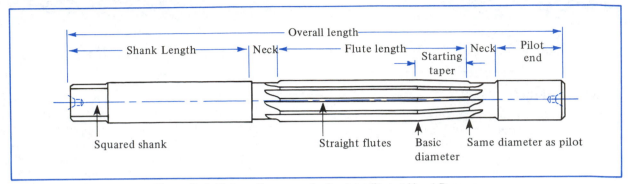

Figure 8-2 Unique Features of a Straight-Fluted Hand Reamer

diameter of the reamer. Additional features of a hand reamer are illustrated in Figure 8–2, which shows a straight-fluted hand reamer.

The *shank* of each reamer is smaller than its outside (basic) diameter. In reaming it is desirable to turn the reamer clear through the workpiece whenever possible to avoid scoring the inside finished surface. The end of the shank is square to permit turning a reamer with a tap wrench. Both ends of the reamer have center holes for two purposes:

• To provide a bearing surface for grinding and sharpening (the bearing surface is concentric with the periphery of the reamer);
• To position the reamer on the machine (where the hole may have been drilled or rough reamed), with a machine center to help to align the reamer with the line of measurement of the workpiece.

Hand reamers are usually made of carbon steel and high-speed steel. Reamers are manufactured in standard dimensional sizes. Special multiple diameters are designed for applications that require the simultaneous reaming of more than one diameter of a hole.

BASIC TYPES OF HAND REAMERS

SOLID HAND REAMERS (STRAIGHT HOLES)

Solid hand reamers may have flutes that are *straight* (parallel) or *helical* (spiral and at an angle). These reamers produce straight holes of a specified diameter, with good surface finish, and to close tolerances. Straight and helical reamers may be purchased in inch-standard and metric-standard sets. Reamer sizes are in increments of 1/64″ and 1/32″ and metric equivalents. Reamers are also furnished in special sizes. The outside diameter is usually checked with a micrometer.

The outside diameter of an old reamer may be ground to within 0.003″ to 0.005″ of a standard size. Such a reamer is a *roughing* reamer. A roughing reamer is used to correct inaccuracies caused by the cutting action of a drill. A *finishing* reamer is of the exact size as the finished diameter and is used as a *second* reamer to produce an accurately reamed hole.

Spiral-fluted reamers are machined with a *left-hand helix* design. This design permits the cutting edge to cut as the reamer is turned and fed into the work. If the helix were right-hand, the reamer would thread into the work, score the surface, and produce an inaccurate hole. Spiral-fluted reamers are also used for reaming holes with keyways.

SOLID HAND REAMERS (TAPERED HOLES)

The flutes of reamers may be tapered for reaming tapered holes. There are three basic forms of taper reamers: *pin*, *socket*, and *burring*.

Taper Socket Reamer. Taper socket reamers produce tapered holes. These reamers conform to Morse, Brown & Sharpe, or metric-standard tapered socket, tool shank, and other spindle standards. Since these reamers must remove a considerable amount of material, they are produced in a set of two reamers. One reamer is for roughing. A series of shallow grooves is cut

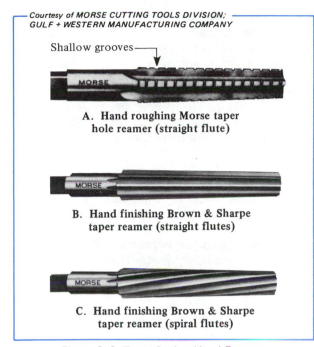

Shallow grooves

A. Hand roughing Morse taper hole reamer (straight flute)

B. Hand finishing Brown & Sharpe taper reamer (straight flutes)

C. Hand finishing Brown & Sharpe taper reamer (spiral flutes)

Figure 8–3 Taper Socket Hand Reamers

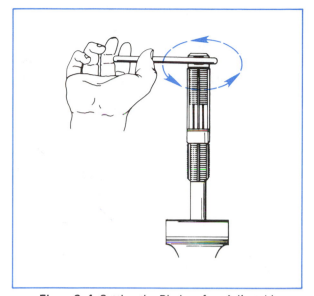

Figure 8–4 Setting the Blades of an Adjustable Hand Reamer to a Required Diameter

across each flute of a roughing reamer and in a staggered pattern (Figure 8–3A). The grooves relieve some of the force required to turn the reamer because they reduce the area over which the cutting takes place. A straight-flute finishing taper reamer is illustrated in Figure 8–3B. Figure 8–3C also illustrates a finishing reamer that has spiral flutes.

Burring Reamer. Burring reamers are not considered as finishing tools in terms of producing a precision surface finish or accurate diameter. A burring reamer has a steep taper. This reamer is used to burr tubing and pipes and to enlarge the diameter of a hole in thin metal. A burring reamer has straight or helical flutes.

EXPANSION HAND REAMERS

The body of an expansion reamer is bored with a tapered hole. This hole passes through a portion of its length. A section of each flute is slotted lengthwise into the tapered hole. The hole is threaded at the pilot end to receive a tapered plug. As this plug is turned into the tapered portion of the body, the flutes are expanded. The limit of expansion varies from

0.006″ for 1/4″ diameter reamers, to 0.012″ for 1″ to 1 1/2″ reamers, to 0.015″ for 1 9/16″ to 2″ reamers.

The pilot on the expansion reamer is undersized. It serves to align the cutting edges concentric with the hole.

The expansion reamer may be increased in size only to produce oversized holes. These holes require the removal of a small amount of material over a nominal size.

Expansion hand reamers may be straight fluted or left-hand helical fluted. The helical-fluted expansion reamer operates on the same principle of adjustment as the straight-fluted reamer. The helical flute is especially adapted to reaming holes that are grooved or in which the surfaces are interrupted (partly cut away).

ADJUSTABLE HAND REAMERS

Adjustable hand reamers (Figure 8–4) are practical for reaming operations that require a diameter that is larger or smaller than a basic size. A complete set of adjustable hand reamers makes it possible to produce holes over a range of 3″, starting at 1/4″ diameter. Each reamer may be expanded to produce intermediate diameters, which overlap the next size of adjustable reamer.

CHATTER MARKS
RESULTING FROM REAMING

Chatter marks are uneven, wave-like imperfections on reamed surfaces. These surfaces have undesirable high and low spots. Chatter marks that result from reaming may be caused by any one or combination of the following conditions:

- The speed at which a reamer is being turned is too fast;
- The distance a reamer is advanced each revolution (feed) is too little;
- A workpiece is not held securely, and there is too much spring to the tool or the work;
- A reamer is unevenly forced into a workpiece;
- Chips are not being removed and are clogging the flutes;
- The drilled or rough-reamed hole is incorrect in terms of size, surface condition, roundness, or straightness.

The remedy is to correct the particular condition and make whatever changes are necessary in speed, feed, proper preparation of the hole to be reamed, and so on. Chatter marks may also be avoided by using a left-hand spiral reamer that may be fed evenly. Some reamers are increment cut; their cutting edges are unevenly spaced. This spacing permits overlapping during the cutting process and thus prevents a ragged surface from being formed.

CUTTING FLUIDS

The use of a cutting fluid on certain materials makes the hand reaming process easier. Less force is required and a higher quality of surface finish is possible. Most metals except cast iron are reamed using a cutting fluid. Generally, for nonferrous metals like aluminum, copper, and brass, a soluble oil or lard oil compound is recommended. Cast iron is reamed dry. Soluble oils and sulphurized oils may be used on ordinary, hard, and stainless steels.

Recommended cutting fluids for various ferrous and nonferrous metals are listed in Appendix Table A–19. While recommended primarily for machine reaming, the cutting fluids may also be used for hand reaming.

Safe Practices in the Care and Use of Hand Reamers

- Examine the cutting edges and lands of reamers for nicks, burrs, or other surface irregularities. These irregularities produce surface scratches and imperfections. Burrs are removed by hand with an oilstone.
- Turn reamers in a right-hand (clockwise) direction only. Where possible, move the reamer completely through the workpiece while turning. Otherwise, keep turning the reamer clockwise while pulling outward on the tap wrench.
- Leave not more than 0.005″ or 0.1mm of material for hand reaming. Holes that must be held to tolerances of +0.0005″/–0.0000″ should be rough reamed to within 0.001″ to 0.003″ and then finish reamed.
- Examine each drilled hole before reaming it. The hole must not be tapered or bell-mouthed (out of round) or have a rough finished surface. Any of these conditions may cause tool breakage and an inaccurately reamed hole.
- Use a left-hand spiral-fluted reamer for reaming holes in which there is an intermittent cut.
- Rotate the reamer clockwise and slowly. Allow it to align itself in the hole. The feed should be continuous.

UNIT 8 REVIEW AND SELF-TEST

1. Compare the variation from the nominal hole size for 12.5mm and 25mm diameter holes that are produced by (a) drilling and (b) reaming.

2. Describe each of the following design features of a hand reamer: (a) radial flute face, (b) margin, and (c) starting taper.

3. State why it is common practice to turn a straight hand reamer through a workpiece.

4. Tell what function is served by the series of shallow grooves cut into each flute in a roughing taper socket reamer.

5. Indicate the limit of expansion on the following diameter ranges of expansion hand reamers: (a) up to 6mm, (b) 1″ to 1 1/2″, and (c) 39mm to 50mm.

6. Cite three advantages of adjustable blade reamers over expansion hand reamers.

7. Give three common corrective steps that are taken to eliminate chatter marks during hand reaming.

8. List four cautions that must be observed to avoid personal injury and/or damage to hand reamers.

Characteristics, Tools, and Hand Tapping Processes of Internal Threading

INTRODUCTION TO SCREW THREADS

BASIC THREAD PRODUCING METHODS

Eight basic methods are employed to produce threads. The threads meet high standards of form, fit, accuracy, and interchangeability within a thread system. The eight methods are:

- Hand cutting internal threads with taps and external threads with dies;
- Machine tapping using formed cutters;
- External and internal machine thread cutting with single-point cutting tools;
- Machine threading using multiple-point cutting tools;
- Machine milling with rotary cutters;
- Machine grinding with formed abrasive wheels;
- Machine rolling and forming;
- Casting, using sand, die, permanent mold, shell casting, and other techniques (one unique advantage of the casting method is that parts of some machines—for example, sewing and vending machines, typewriters, and toys—have internal threads cast in place).

THE NATURE OF SCREW THREADS

A *screw thread* is defined as a helical (spiral) ridge of uniform section (form and size). This ridge is formed on a cylindrical or cone-shaped surface. The basic features of a screw thread are illustrated in Figure 9–1. Screw threads that are produced on an outer surface, like the threads on bolts and threaded studs, are called *straight external* threads. Screw threads that are cut on cone-shaped surfaces, like pipe threads, are *tapered external* threads. Threads that are cut parallel on inner surfaces, as are the threads of an adjusting nut or a hexagon nut fastener, are known as *straight* or *parallel internal* threads. External and internal parallel threads are referred to in shop language as just *outside* threads or *inside* threads.

Some threads may be advanced (moved) or tightened into a part by turning them clockwise. These threads are classified as *right-hand* threads. Threads that advance when turned counterclockwise are *left-hand* threads. Right- and left-hand threads each have special applications. Unless otherwise indicated on a drawing, the thread is assumed to be right-hand.

AMERICAN NATIONAL AND UNIFIED SCREW THREADS

Each thread has a shape, or profile. This shape is called the *thread form*. Although the size of a thread form varies according to the dimensions of a threaded part, the shape remains constant. The symbol ∪ preceding a thread form on a drawing indicates that the shape conforms to Unified system standards and is accepted by Great Britain, Canada, and the United States.

There are four basic thread series and one special thread series in both the Unified thread system and the American National thread system. The four basic thread series are:

- *Coarse-thread series*, with threads designated as UNC in the Unified system and NC (National Coarse) in the American National system;
- *Fine-thread series*, with threads designated UNF and NF;

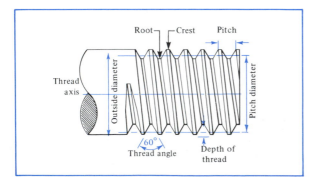

Figure 9–1 Basic Features of a Screw Thread

- *Extra-fine-thread series*, UNEF and NEF;
- *Constant-pitch series*, 8 UN, 12 UN, and 16 UN for the Unified system and 8 N, 12 N, and 16 N for the American National system.

There are a number of pitch series in the constant-pitch series. There is the 8-pitch, 12-pitch, and 16-pitch series, and others. *Constant-pitch* means there are the same number of threads per inch for all diameters that are included in the series. For example, the 12 N series always has 12 threads per inch regardless of the outside thread diameter. The constant-pitch series is used when the other three thread series fail to meet specific requirements.

There are also designations for special conditions that require a nonstandard or special thread series. Some drawings show the symbol UNS, which indicates that the threads are Unified Special, or NS, which indicates National Special threads.

ISO METRIC THREADS

The ISO metric threads for all general threading and assembling purposes are identified with the *ISO Metric Coarse series*. The *ISO Metric Fine series* is employed in fine precision work. Table 9–1 shows twelve common ISO metric thread sizes ranging from 2–24mm. These sizes cover about the same range as the fractional inch sizes in the National Coarse series up to 1″ outside diameter. There are additional standard sizes in

Table 9–1 Selected ISO Metric Thread Coarse Series (2–24mm)

Outside Diameter (mm)	Pitch	Tap Drill Diameter
2	0.4	1.6
2.5	0.45	2.1
3	0.5	2.5
4	0.7	3.3
5	0.8	4.2
6	1.0	5.0
8	1.25	6.8
10	1.5	8.5
12	1.75	10.3
16	2.0	14.0
20	2.5	17.5
24	3.0	21.0

the ISO Metric Coarse series beyond the 2–24mm range. The nine common thread sizes in the ISO Metric Fine series threads cover an equivalent range of fractional inch sizes in the National Fine series up to 1″.

In Table 9–1, for example, the coarse series shows that an ISO metric threaded part with an 8mm outside diameter has a pitch of 1.25mm. The information would appear on an industrial print or drawing as M8–1.25 for the tapped hole it represents. The tap drill diameter is 8mm - 1.25mm, or 6.75mm, which is rounded off to 6.8mm for the tap drill size.

More complete tables cover the whole range of sizes and series. These tables are contained in trade handbooks, Standards Association papers, manufacturers' technical manuals, and other printed manuals.

CUTTING INTERNAL THREADS WITH HAND TAPS

The process of cutting internal threads is referred to as *tapping*. Internal threads may be formed with a cutting tool called a *tap*. Taps are used to meet three general tapping requirements. Some holes are tapped *through* a workpiece. Other holes are tapped part of the way through and are called *blind* holes. Some blind holes are *bottomed* with a thread.

FEATURES OF HAND TAPS

A tap is a specially shaped, accurately threaded piece of tool steel or high-speed steel. After a tap has been machined almost to size and form, its threads are *relieved* (reduced in size) to produce cutting edges. The tap is then hardened and tempered. A fine tap is ground to size and shape.

The general features of a hand tap are represented in Figure 9–2. A number of *flutes* (grooves) are milled into the body for the length of the threaded portion. These flutes help to form the cutting edges. The flutes also provide channels so that chips can move out of the workpiece along the flutes during the cutting process. Regular taps have straight flutes. Some taps are *spiral fluted* to produce a different cutting action in which the shearing that takes place and the spiral shape push the chips ahead of the tap.

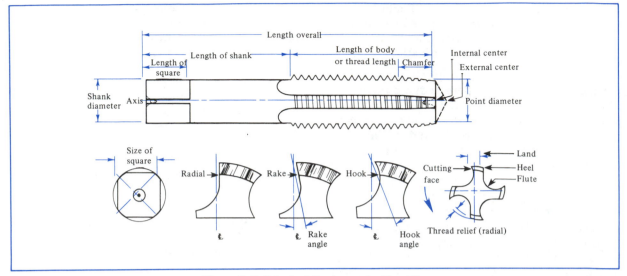

Figure 9–2 Principal Features of a Hand Tap

The *shank* on some taps is smaller in diameter than the tap drill size. This difference in diameter permits the last teeth on the tap to be employed in cutting a thread. Also, the tap may be turned until it completely moves through the workpiece. Besides being smaller in diameter, the shank end is square. The square shape provides a good bearing surface for a tap wrench and for turning the tap. Sometimes there is a *center hole* on the shank end. The center hole is used to align the tap at the line of measurement. Alignment is essential in starting the hand tap and in tapping on a machine.

TAP SETS

Standard Sets. Taps that are larger than 1/4″ usually come in standard size sets for each thread size and pitch. A set consists of three taps: a *taper* tap, *plug* tap, and *bottoming* tap.

The taper tap is sometimes called a *starting* tap. It may be used to start the hand tapping process, particularly when threading a blind hole, or to tap a hole that is drilled completely through a part. The taper tap has the first eight to ten threads ground at an angle (backed off). The taper begins with the tap drill diameter at the point end. It extends until the full outside diameter is reached. These angled threads start the tap and do the rough cutting. Because the taper tap has a greater number of angled teeth

engaged in cutting the thread than plug and bottoming taps, less force is required. Also, the taper tap is easier to align with the line of measurement, and there is less likelihood of breakage.

The plug tap has the first three to five threads tapered. The plug tap is usually the second tap that is used for tapping threads (the taper tap being the first). It may be used for tapping through a part or for tapping a given distance. Certain blind-hole tapping is done with a plug tap if the last few threads need not be cut to the total depth.

The bottoming tap is *backed off* (tapered) at the point end from one to one and one-half threads. This tap is the third tap that should be used when a hole is to be blind tapped and is to *bottom* (cut a full thread) at the end of the tap drill hole. The bottoming tap should be used after the plug tap. Extra care must be taken near the end of the thread to see that the tap is not forced (jammed) against the bottom of the hole.

Machine-Screw Taps. Tap sizes that are smaller than 1/4″ are designated by whole numbers rather than by fractions. For example, a designation of 10–32 means that the outside diameter (OD) of a machine-screw tap is equal to the number of the tap (#10) times 0.013 plus 0.060″. For the 10–32 tap the outside diameter,

calculated by using the formula, equals (10 ×
0.013) + 0.060″, or 0.190″. The "32" refers
to the number of threads per inch. Thus, a
10–32 tap is one that has an outside diameter of
0.190″ and cuts 32 threads per inch.

On smaller size taps the amount of material
to be removed in cutting the thread form is
limited. The use of two or three taps in a set is
therefore not required.

Serial Set Taps. There are times when deep
threads must be cut by hand in tough metals.
Taps that cut deep threads have a different design than standard taps. These taps are known
as *serial* taps. They differ in the amount of material each tap cuts away.

The #1 tap cuts a shallow thread. The #2
tap cuts the thread deeper. The #3 tap is the
final sizing tap. It cuts to the required depth to
correctly form the teeth.

Serial taps must be used in sets. A combination of these taps reduces the amount of force
exerted on any one tap and helps avoid tap
breakage.

OTHER TYPES OF HAND AND MACHINE TAPS

Gun Taps. Gun taps derive their name from
the action caused by the shape at the point end
of the tap. The cutting point is cut at an angle.
The design of the point causes chips to shoot
out ahead of the tap. The cutting action is
shown in Figure 9–3. Gun taps are applied primarily in tapping through holes in stringy metals
whose chips tend to bunch up and lodge in
straight-fluted taps. With these metals, tapping
requires that a greater force be exerted on a tap.
This force increases the probability of tap
breakage.

Gun taps are production taps. To withstand
the greater forces required in machine tapping,
there are fewer flutes on gun taps and the flutes
are cut to a shallower depth than standard taps.

Gun taps are furnished in three general forms.
Because each form performs a function that is
different from the function of a standard tap,
the names vary from the names of standard taps.
The three forms of gun taps are the *plug gun*,
bottoming gun, and *gun flute only* tap.

The plug gun tap is used for through tapping
where holes are open and chips may shoot out.

The bottoming gun tap is used for bottoming tapping operations.

The gun flute only tap is used on soft and
stringy metals for shallow through tapping. The
design features of this tap appear in Figure 9–4.
A large cross-sectional area gives added strength
and permits gun flute only taps to be powered
with higher-force turning devices than regular
machine taps.

Spiral-Fluted Taps. *Low-angle spiral-fluted*
taps are especially adapted for tapping soft and
stringy metals. Such metals include copper,
aluminum, die-cast metals, magnesium, brass,
and others. Instead of straight flutes, these taps
have spirals. The spiral flutes provide a pathway
along which chips may travel out of a workpiece.

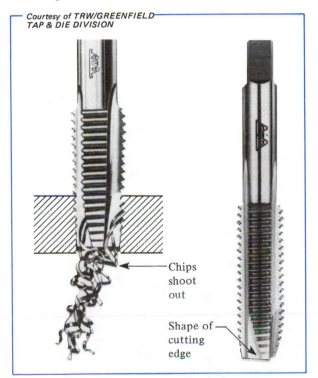

Courtesy of TRW/GREENFIELD TAP & DIE DIVISION

Chips shoot out

Shape of cutting edge

Figure 9–3 Cutting Action of a Plug Gun Tap

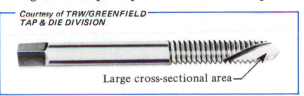

Courtesy of TRW/GREENFIELD TAP & DIE DIVISION

Large cross-sectional area

Figure 9–4 A Gun Flute Only Tap

High-angle spiral-fluted taps are excellent for tapping (1) tough alloy steels, (2) threads that are interrupted and then continue in a part, and (3) blind holes.

Pipe Taps. Three common forms of pipe taps are used to cut American National Form Pipe Threads. These threads include (1) standard taper pipe threads, (2) straight pipe threads, and (3) dryseal pipe threads. The processes of producing pipe threads are similar to processes used for tapping any other American National Form threads. The *taper pipe* tap forms NPT (National Pipe Taper) threads. The *straight pipe* tap cuts NPS (National Pipe Straight) threads. The *dryseal pipe* tap produces American National Standard Dryseal Pipe Threads.

Standard taper pipe threads and straight pipe threads are the same shape as the American National Form. Their thread angle is 60°. Their crests and roots are flattened. Their differences lie in diameter designation and pitch. Pipe threads taper 3/4″ per foot.

Parts cut with tapered pipe threads may be drawn together to produce a rigid joint. When a pipe compound seal is used on the threads, the line is made gas or liquid leakproof.

Straight pipe threads are used in couplings, low-pressure systems, and other piping applications where there is very limited vibration.

Dryseal pipe threads are applied in systems requiring pressure-tight seals. These seals are supplied by making the flats of the crests the same size or slightly smaller than the flats on the roots of the adjoining (mating) thread. Figure 9–5A shows the initial contact between crests and roots as a result of hand turning. A pressure-tight seal produced when parts are further tightened by wrench is illustrated in Figure 9–5B.

FLUTE DEPTH AND TAP STRENGTH

While referred to as hand taps, a number of taps also are used for machine tapping. The difference in some cases is the material from which a tap is made. Hand taps are usually manufactured from high-carbon tool steel; machine taps, from high-speed steel.

Tap sizes that are 1/2″ or larger normally have four flutes. Sizes smaller than 1/2″ may have two, three, or four flutes. For example, machine-screw taps usually have three flutes. The three flutes are cut deeper than flutes of large tap sizes and provide extra space for chips in blind-hole tapping or in tapping deep holes where chips are stringy.

TAP DRILL SIZES AND TABLES

A *tap drill* refers to a drill of a specific size. A tap drill produces a hole in which threads may be cut to a particular thread depth.

TAP DRILL SIZES

The *tap drill size* may vary from the root diameter required to cut a theoretical 100% full thread to a larger diameter. The larger diameter leaves only enough material to cut a fraction of a full-depth thread. According to laboratory tests, a 50% depth thread has greater holding power than the strength of the bolt. In other words, the bolt shears before the threads of the tapped hole strip.

The material in a part, the length of thread engagement, and the application of the thread itself are considerations for establishing the percent of the full-depth thread that is required. Finer threads for precision instruments may be cut close to their full depth. By contrast, bolts, nuts, and deep-tapped holes in the NC, UNC, or ISO Metric coarse series may require a smaller percentage of the full-depth thread. The possibility of tap breakage is reduced when holes are threaded to a fraction of the full depth.

When the tap drill size is to be computed, values may be substituted in the following formula:

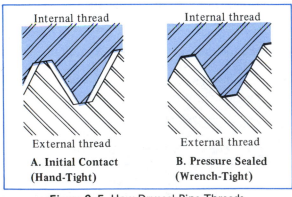

Figure 9–5 How Dryseal Pipe Threads Produce a Pressure-Tight Seal

Table 9–2 Tap Drill Sizes (75% thread) UNC/NC Series (Partial Table)

Thread Size and Threads per Inch	Major Outside Diameter (Inches)	Tap Drill Size	Decimal Equivalent of Tap Drill
1/4–20	0.2500	#7	0.2010
5/16–18	0.3125	F	0.2570
3/8–16	0.3750	5/16	0.3125
7/16–14	0.4375	U	0.3680
1/2–13	0.5000	27/64	0.4219
9/16–12	0.5625	31/64	0.4844

$$\text{Tap Drill Size} \atop (75\% \text{ of Total Depth}) = \text{Outside Diameter} \left(- 0.75 \times \frac{1.299}{\text{Number of Threads per Inch}} \right)$$

If a thread depth other than 75% is specified, substitute the required percent in place of 0.75. The nearest standard size drill is selected as the tap drill.

TAP DRILL TABLES

The Appendix contains a series of thread tables. One table is titled *Hole Sizes for Various Percents of Thread Height and Length of Engagement.* Two other tables relate to various thread dimension in the UNC/NC and UNF/NF series.

Table 9–2 provides just a few thread sizes to illustrate tap drill size tables. The sizes all relate to a thread depth of approximately 75%. Note, as an example, that a 1/2–13 UNC/NC thread requires a 27/64″ tap drill to produce a 75% thread.

TAP WRENCHES

Taps are turned by specially designed *tap wrenches.* Two general forms of wrenches are the *T-handle* and the *straight-handle* (Figure 9–6). Both of these wrenches have adjustable jaws that permit the use of a single tap wrench handle to accommodate a number of taps with different sizes of square heads. However, the range of tap sizes is so great that tap wrenches are made in a number of sizes.

The size of the square head of a tap is proportional to the tap size. Care must be taken to select the appropriate size of tap wrench. Many taps are broken because excessive force is exerted through the leverage of a wrench that is too large. While designed principally for use as tap wrenches, these wrenches are also used with other square-shanked cutting tools for hand turning operations.

How to Tap a Through Hole

STEP 1 Turn the tap (Figure 9–6). As the tap takes hold, the downward force may be released.

Note: Turn with a steady, gentle, even force. Straight-handle tap wrenches and taps are turned by grasping one end of the handle with one hand and the other end with the other hand.

STEP 2 Back off the tap after each series of two or three turns.

STEP 3 Continue to turn the tap, repeating the steps for cutting, lubricating, and backing off.

Note: It is good practice to follow the tap with a plug tap. The plug tap then

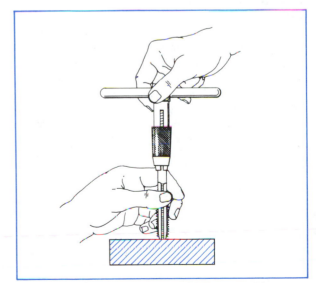

Figure 9–6 Practices in Hand Tapping with a T-Handle Tap Wrench

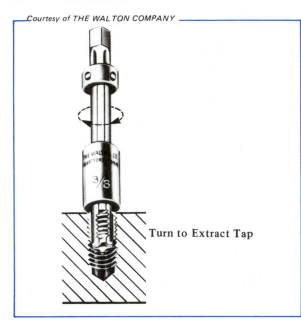

Courtesy of THE WALTON COMPANY

Turn to Extract Tap

Figure 9-7 Application of a Tap Extractor

serves as a finishing or sizing tap. A bottoming tap is used as a final tap for threading a blind hole.

TAP EXTRACTORS

The name *tap extractor* indicates that the tool is used to extract parts of taps that have broken off in a hole (Figure 9-7). The extractor has prongs. These prongs fit into the flute spaces of a broken tap. The prongs are pushed down into the flutes. The steel bushing on the extractor is moved down as far as possible to the work. The bushing thus provides support for the prongs. The extractor is turned counterclockwise with a tap wrench to remove a broken right-hand tap. After the broken tap is removed,

the tapped hole should be checked to see that no broken pieces remain in the hole.

Safe Practices with Threading Tools and Processes

- Reverse a standard hand tap during the cutting process. A reversing action breaks the chips, makes it easier to move them out of the tapped hole, and reduces the force exerted on the tap.
- Select a size and type of tap wrench that is in proportion to the tap size and related to the nature of the tapping process. Excessive force applied by using oversized tap wrenches is a common cause of tap breakage.
- Exercise care with bottoming taps, particularly in forming the last few threads before bottoming. Avoid excessive force or jamming the point of the tap against the bottom edge of the drilled hole.
- Back out a tap that is started incorrectly. Restart it in a vertical position, square with the work surface. Any uneven force applied to one side of a tap handle may be great enough to break the tap.
- Use caution when using an air blast. Chips should only be blown from a workpiece when (1) each operator in a work area uses an eye protection device and (2) the workpiece is shielded to prevent particles from flying within a work area.
- Handle broken taps carefully. Taps fracture into sharp, jagged pieces. Workpieces with parts of broken taps should be placed so that no worker may brush up against them. Use only a hand tool, such as a pair of pliers, to grasp or try to move a broken tap.

UNIT 9 REVIEW AND SELF-TEST

1. List four basic hand and machine (lathe cutting) methods of producing threads.

2. Differentiate between the pitch to diameter relationship of threads in the UNC series and the constant-pitch series (UN) for the Unified system.

3. Show how information about ISO metric threads with a 25mm outside diameter and a pitch of 4mm is represented on an industrial drawing.

4. Identify when each of the following taps is used: (a) taper, (b) plug, and (c) bottoming.

5. Tell when (a) low-angle spiral-fluted and (b) high-angle spiral-fluted taps are used.

6. Give one design feature of an American National standard dryseal pipe thread.

7. State why general-purpose threading to 75% of full depth is an accepted practice.

8. Explain how part of a tap that breaks in a hole may be removed safely.

9. List four safety practices to follow to avoid personal injury or tap breakage when threading.

Drawings, Dies, and External Threading

SCREW THREAD SPECIFICATIONS AND DRAWING REPRESENTATION

To provide all essential screw thread details, a drawing must include full specifications for cutting and measuring threads. Technical information about thread sizes, depth or length, class of threads (fits), and surface finish appear near the representation of the threaded portion. Sometimes the thread length is specified by a dimension on the part. Other instructions may be stated as notes.

Dimensions for internal and external threads are added to the part drawing. Simplified drawings of threads with dimensioning codes and dimensions are provided in Figure 10–1. Each drawing includes a series of encircled numerals. The type of dimension represented by a particular numeral is explained by a code (Figure 10–1C). Specifications of an external thread are given on a drawing as shown in Figure 10–1A.

Details of a blind threaded hole (internal thread) appear on a drawing as illustrated in Figure 10–1B.

MEASURING SCREW THREAD PITCH

Pitch is the distance at the same point on a thread form between two successive teeth. Two simple screw thread measurement techniques may be used to establish the pitch of a particular threaded part. The first technique requires the use of a screw pitch gage; the second technique, a steel rule.

SCREW PITCH GAGE TECHNIQUE

A *screw pitch gage* has a series of thin blades. Each blade has a number of teeth. The teeth match the form and size of a particular pitch. Each blade is marked for easy reading.

Pitch is checked by selecting one gage blade. The blade teeth are placed in the threaded grooves of an actual part. The blade is sighted to see

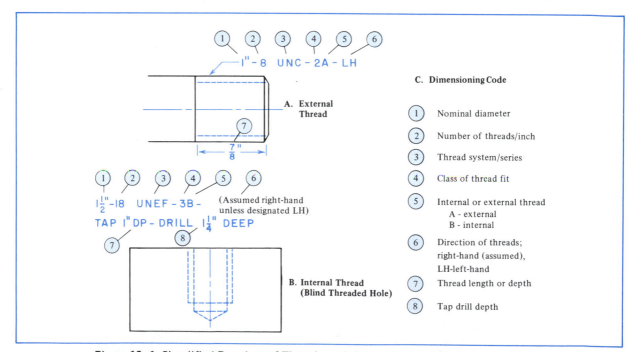

Figure 10–1 Simplified Drawings of Threads, with Dimensions and Dimensioning Codes

whether the teeth match the teeth profile of the workpiece. Different blades may need to be tried until a blade is found that conforms exactly to the workpiece teeth. The required pitch is thus established.

STEEL RULE TECHNIQUE

Screw pitch may also be measured by placing a steel rule lengthwise on a threaded part. An inch graduation is usually placed on the crest of the last thread. The number of crests are counted to the next inch graduation on the rule. This number of crests (threads) represents the pitch.

EXTERNAL THREAD CUTTING HAND TOOLS

External threads are hand cut with cutting tools known as *dies*. Dies are made in a variety of sizes, shapes, and types. Selection of dies depends on the thread form and size and the material to be cut. Dies are made of tool or high-speed steel and other alloys. Dies are hardened and tempered. Some dies are *solid* and have a fixed size. Other dies are *adjustable*.

SOLID DIES

The solid square-or hexagon-shaped die has a fixed size. Rarely used in the shop, it is applied principally to chase threads that have been poorly formed or have been damaged. Such threads require that a sizing die be run over them to remove any burrs or nicks or crossed threads. The damaged threads are reformed to permit a bolt or other threaded area to turn in a mating part. The hexagon shape may be turned with a socket, ratchet, or other adjustable wrench.

ADJUSTABLE SPLIT DIES

The adjustable split die is available in three common forms. The first form, shown in Figure 10-2, must be adjusted each time the die is changed in a *die holder*. In some cases the adjustment is made by turning the adjusting screws in the die holder (sometimes called a *die stock*). The split die is placed in the die holder and tried on a correctly sized threaded part. The screws

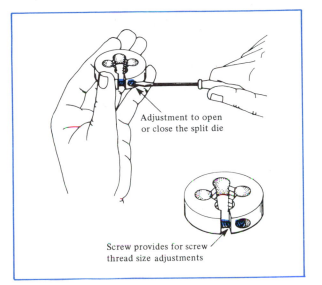

Figure 10–2 A Split Threading Die with a Screw Adjustment

are adjusted until there is a slight drag between the die and the threaded part. In other cases the die is adjusted by a cut-and-try method. The die is opened. A few full threads are cut. The workpiece is tried on the mating part. The die is adjusted until the correct size is reached. The die is then locked at the setting for this size.

The second form of adjustable split die has a screw within the die head. This *screw adjustment* opens or closes the die thread teeth. Thus a range of adjustments for cutting oversized or undersized threads is provided.

The third basic form of adjustable split die has three essential parts: a *cap*, a *guide*, and the *die halves* (which are the threading die).

The guide serves two purposes:

- It forces the tapered sides of the die halves against similarly tapered sides in the cap. The force holds the two halves tightly in position so that the size cannot change.
- It pilots the die (halves) so that it is centered over a workpiece. Once the die halves are set to the required size, they are locked in position with the cap. The assembled die is then locked in the die holder.

PIPE THREAD DIES

Pipe threads are tapered in order to make tight joints in air and liquid lines. Pipes are

Table 10–1 National Standard Taper (NPT) Pipe Threads

Nominal Pipe Size	Threads per Inch	Outside Diameter	Tap Drill Size
1/8	27	0.405	11/32
1/4	18	0.504	7/16
3/8	18	0.675	37/64
1/2	14	0.840	23/32
3/4	14	1.050	59/64
1	11 1/2	1.315	1 5/32
1 1/4	11 1/2	1.660	1 1/2
1 1/2	11 1/2	1.990	1 47/64
2	11 1/2	2.375	2 7/32

Figure 10–3 Right-Hand Thread Screw Extractor

measured (sized) according to their inside diameter. Because of this inside measurement, pipe threads are larger in diameter than regular screw threads. Pipe threading taps and dies are, therefore, larger for a specific diameter than regular taps and dies.

A 1/2" standard taper pipe thread (NPT) requires a tap drill of 23/32". This drill size permits a hole to be tapered with a thread taper of 3/4" per foot. The actual outside diameter of the 1/2" threaded pipe is 0.840". Other common National taper pipe thread sizes are given in Table 10–1.

Dies for pipe threads may be solid or adjustable. The adjustable die halves are held in an *adjustable stock* that serves as the die holder. There are indicating marks on both the die and the stock. When the die halves are aligned with these marks, the die is set to cut a standard pipe thread. The adjustable stock is provided with a positioning cap. This cap centers the work with the die halves.

SCREW EXTRACTOR

Bolts, screws, studs, and other threaded parts may be sheared off, leaving a portion of the broken part in the tapped hole. This portion may be removed by using a tool called a *screw extractor*. This tool is a tapered metal spiral. A right-hand thread screw extractor is illustrated in Figure 10–3. The spiral is formed in the reverse direction from the threaded part that is to be removed. In other words, a left-hand spiral is needed for a right-hand thread.

A pilot hole is first drilled into the broken portion. The diameter of the hole must be smaller than the root diameter of the thread. The drill size to use is generally given on the shank of the screw extractor.

Then, as the point of the spiral taper is inserted and turned counterclockwise (for right-hand threads), the screw extractor feeds into the hole. The spiral surfaces are forced against the sides of the hole in the broken part. The extractor and part become bound as one piece. The threaded portion then may be removed by turning the extractor.

Safe Practices in Cutting External Threads by Hand

- Analyze and correct the causes of torn threads. Torn threads indicate any one or all of the following conditions:
  - The jaws may be set too deep,
  - The threads are being cut at an angle,
  - The diameter of the workpiece is too large,
  - The cutting teeth are dull,
  - A proper lubricant (if required) is not being used.
- Remove burrs produced by threading.
- Exercise care in threading long sections. Turn a die holder from a position in which the hands do not pass across the top of a threaded part.
- Use an eye protective device when removing chips from the die or workpiece.

1. Explain the statement: A drawing provides full specifications for cutting and measuring a required thread.

2. Give three reasons why adjustable split dies are preferred over solid dies.

3. State three major differences between pipe thread dies and regular screw thread dies.

4. List the steps required to thread a workpiece on the bench using an appropriate size adjustable threading die.

5. Identify how a broken threaded part may be removed.

6. State three conditions that result in cutting torn external threads when threading by hand.

PART 4 Basic Cut-off and Grinding Machines: Technology and Processes

SECTION ONE

Metal-Cutting Saws

This section describes the types, features, and operation of power hacksaws and horizontal band machines. Basic cutoff operations performed with these machines are considered in terms of everyday applications. Cutting speed, feeds, cutting fluids, machine adjustments, and cutting blades or saw bands are described in depth because the craftsperson must make correct selections. Also examined are step-by-step procedures for mounting workpieces, setting up the machines, and performing other processes essential to accurate and efficient cutting off.

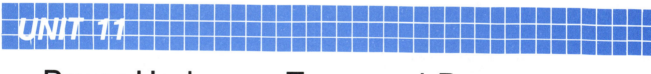

UNIT 11

Power Hacksaws: Types and Processes

Power hacksaws are used in metal sawing to cut sections of materials such as rods, bars, tubing, and pipes to length. They can also be used to cut castings, forgings, and other parts. The hardness range of these parts and materials is from comparatively soft nonferrous metals (like aluminum, brass, and bronze) to mild steels, tool steels, and harder alloys. The sections of stock may be cut square (straight) with the work axis or at an angle up to 45°.

The power hacksawing process is similar to the hand hacksawing process. In both processes the cutting action takes place by applying force on the teeth of a saw blade as it moves across a work surface. A series of chips are cut by the shearing action of the teeth. Cutting action

continues until a section is cut off. Materials that are smaller than 1/2″ are usually cut by hand. Larger sizes are cut by machine.

POWER DRIVES AND MACHINE OPERATION

Most utility power hacksaws, whether wet or dry cutting, are equipped with an oil *hydraulic system*. This system provides smooth uniform control for the speed and cutting force. The hydraulic system actuates the saw frame. The system controls the downward feed (force) during the cutting stroke. It also automatically raises the frame and blade on the return stroke. At the end of the cut, the blade and frame are brought up to the highest position to clear the work.

Usually, dry-cutting power hacksaws have two cutting speeds: 70 and 100 strokes per minute. Where a greater range of materials and sizes is to be cut, wet-cutting machines are made with a four-speed drive motor. Four-speed machines may cut at 45, 70, 100, or 140 strokes per minute. The general range for six-speed machines is 45, 60, 80, 85, 110, and 150 strokes per minute. These machines are provided with a speed/feed chart for different materials. The cutting speeds are obtained by positioning the speed-change lever at the required speed setting.

MAJOR MACHINE PARTS

The power hacksaw is a comparatively simple machine. A utility-type, wet-cutting power saw is shown in Figure 11–1 to illustrate the major parts of a power hacksaw.

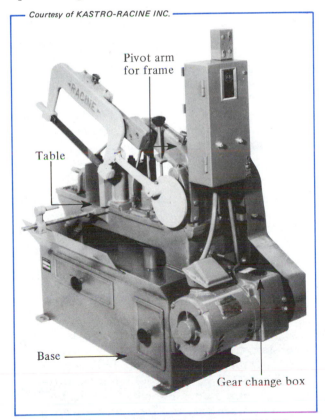

Courtesy of KASTRO-RACINE INC.

Figure 11–1 Utility-Type, Four-Speed (6″ × 6″), Wet-Cutting Power Hacksaw

The workpiece is held in a vise. The vise has a fixed jaw and a movable jaw (Figure 11–2). Some vises are designed for making square cuts. Other vises have elongated slots in the base. These slots permit the fixed (solid) jaw to be positioned at right angles to the blade. The fixed jaw may also be positioned at any angle up to 45°. The movable jaw may be swiveled. Force is applied by the movable jaw to hold the workpiece securely against the face of the fixed jaw.

An adjustable work stop is attached to the table. The stop is set at a particular linear dimension and locked in place. The workpiece then is positioned against the stop. Additional pieces of the same linear dimension may be cut without further measurement. Long, overhanging bars of stock are usually supported by a floor stand.

HEAVY-DUTY AND PRODUCTION HACKSAWS

Heavy-duty and production power hacksaws are built heavier than the utility type. These hacksaws have additional accessories.

Production power hacksaws have hydraulically operated work tables (carriage). When the

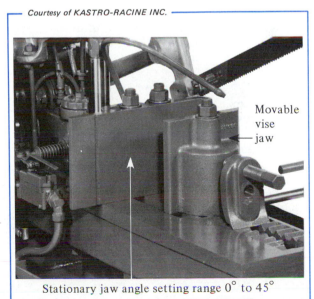

Courtesy of KASTRO-RACINE INC.

Stationary jaw angle setting range 0° to 45°

Figure 11–2 Vise Jaws Positioned to Permit Angle Settings to 45°

Table 11-1 Recommended Pitches, Cutting Speeds, and Feeds for Power Hacksawing Ferrous and Nonferrous Metals (Partial Table)

Material	Pitch (teeth per inch)	Cutting Speed (feet per minute)	Feed (force in pounds)
Iron			
cast	6 to 10	120	125
malleable	6 to 10	90	125
Steel			
carbon tool	6 to 10	75	125
machine	6 to 10	120	125
Aluminum			
alloy	4 to 6	150	60
Brass			
free machining	6 to 10	150	60

movable carriage is fully loaded and the cutoff length is set, the cutoff process is a continuous one.

Square and round bars and structural and other shapes may be stacked for production cutting. *Stacking* refers to the grouping of many pieces together. They are clamped at one time, positioned, and cut to the required length. The operation of the vise jaws, the resetting of the stacked bars to the required length, the raising of the frame, the application of the cutting force, and the circulating of the cutting fluid are all controlled hydraulically.

POWER HACKSAW BLADES

Tables of machinability ratings of metals indicate the range of ease or difficulty in cutting a given metal. The tables provide information for establishing whether a cutting process should be carried on wet or dry and what kind of cutting fluid should be applied.

Other tables are used to determine the correct pitch, cutting speeds, and feeds for power hacksawing. The particular material to be cut and the size and shape of its sections must be considered. Table A-5 in the Appendix lists the recommended pitches, cutting speeds, and feeds for selected ferrous and nonferrous metals. A portion of this table is illustrated in Table 11-1. The pitch recommended represents an average for material sizes of 2″ or smaller. Coarser pitches

may be used for thicker or larger sizes. Finer pitches are used for thinner sections.

SELECTION OF POWER HACKSAW BLADES

Some of the principles governing the selection of hand hacksaw blades apply equally as well to power hacksaw blades:

- There must be at least two teeth in contact with the work surface at all times;
- The greater the cross-sectional area is, the coarser the required pitch (the coarser pitch provides greater chip clearance);
- Easily machined and soft materials require a coarse pitch and large chip clearance;
- Hard materials and small cross-sectional areas require finer pitches.

While some power hacksaw blades are made of carbon alloy steel, high-speed and other alloy steel blades are more durable. High-speed tungsten, high-speed molybdenum, and molybdenum steels are excellent cutting-blade metals.

Standard blade lengths range from 12″ to 14″, 16″, 18″, and 24″. Length depends on machine size and the nature and size of the material to be cut. The general range of pitch is from 4 to 14 teeth per inch. General metric size blade lengths are 300mm, 350mm, 400mm, 450mm, and 600mm. The common metric pitch range is from 6mm to 2mm.

SET PATTERNS FOR SAW BLADE TEETH

A common set pattern for the teeth on power saw blades is called *raker set*. The raker set pattern consists of a repeat design for every three teeth. Figure 11-3 shows the raker set. One tooth is unset. This tooth is followed by two teeth, one of which is offset to the right. The other tooth is offset to the left. This pattern continues for the length of the blade. The raker set blade is recommended for heavy work on bar stock, forgings, die blocks, and parts that have a constant cross section.

A second common set pattern is called *wave set* (Figure 11-3). Wave-set teeth are offset in groups. These groups alternate from right to left to form a wave pattern. Wave-set teeth are used where there is a considerable range of material sizes to be cut. Blades with wave-set teeth are

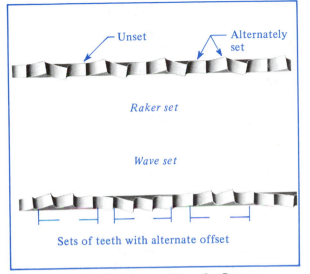

Unset

Alternately set

Raker set

Wave set

Sets of teeth with alternate offset

Figure 11-3 Raker and Wave Set Patterns for Saw Blade Teeth

used for cutting materials with cross-sectional areas that change. Structural forms and pipes are examples of parts that have changing area sizes.

A third set pattern, which has limited use in metal working, is called *straight set*. In this older pattern the teeth are offset alternately to the right and left of the blade.

Safe Practices in Power Hacksawing

- Position the cutting teeth and the blade to cut on the cutting stroke.
- Tighten the blade tension until it is adequate to hold the blade taut during the cutting operation.
- Check the blade pins regularly to see that they are not being sheared.
- Check the workpiece to be sure that it is tightened securely before starting the cut.
- Make sure the blade is moved away from the work before starting the power hacksaw.
- Start a new blade (after a cut has been started) in a new location. Otherwise the teeth may bind in the old kerf and break the blade.
- Direct the flow of the cutting fluid (when fluid is required) over the cutting area. The flow must be as close to the cutting saw teeth as possible.
- Support the ends of long pieces that project from the power hacksaw by using a roller stand.
- Place a protecting screen or a danger flag at the end of parts that extend any distance from the saw frame.
- Cool the cutoff section before handling. Cooling helps to avoid burns and cuts resulting from hot, burred pieces.
- Remove cutting fluids and clean the reservoir regularly.

UNIT 11 REVIEW AND SELF-TEST

1. Describe one cutting cycle of a common 6″ × 6″ (150mm × 150mm) utility power hacksaw.

2. Cite three guidelines that especially apply to selecting power hacksaw blades.

3. Give a general application of the following set patterns for power hacksaw blade teeth: (a) raker set, (b) wave set, and (c) straight set.

4. Explain two design features of a six-speed power hacksaw that make this machine more versatile than a two-speed machine.

5. Describe how stacked bars are automatically cut on a production power hacksaw.

6. State three safe practices that relate particularly to power hacksawing.

Horizontal Band Machines and Cutoff Sawing Processes

The horizontal (cutoff sawing) band machine is a second type of power saw. This machine is often referred to as a *metal-cutting band saw*, a *cutoff band saw*, or a *cutoff band machine*. The term *band* indicates that the cutoff sawing process requires a closed saw blade.

There are also vertical band machines with saw bands and file bands. These machines are used for intricate and precise sawing and filing processes.

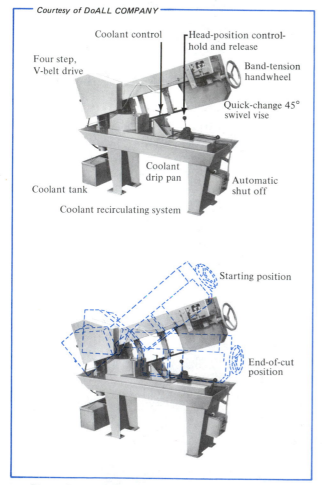

Courtesy of DoALL COMPANY

Coolant control
Head-position control-hold and release
Four step, V-belt drive
Band-tension handwheel
Quick-change 45° swivel vise
Coolant drip pan
Coolant tank
Automatic shut off
Coolant recirculating system

Starting position
End-of-cut position

Figure 12–1 Utility-Type Horizontal Band Machine

BASIC TYPES AND FEATURES OF HORIZONTAL BAND MACHINES

Metal-cutting horizontal band machines for cutoff sawing may be either *dry cutting* or *wet cutting*. They may also be of a *utility* or *heavy-duty* design. The utility design is illustrated in Figure 12–1.

The saw frame of the band machine has two wheels. These wheels hold and drive a continuous (closed) *saw blade* (band). A tension control adjusts the blade to track properly at the correct tension. The teeth are thus forced through the work. Blade guide inserts are provided for positioning the blade vertically at the cutting area. The inserts guide the blade to cut squarely. The saw frame is hinged to permit raising the saw band (blade) to clear the work and lowering it to take a cut. A *pneumatic system* controls the circulation of the cutting fluid and many of the machine mechanisms.

OPERATING PRINCIPLES

The continuous saw band revolves around the *driver* and *idler wheels*. Attached to the frame are two adjustable *(blade) supports* with *guide inserts*. These supports and guides serve two functions:

- Guide the blade in a vertical position so that it does not bend from the work.
- Support the blade so that a cutting force can be applied.

DRY CUTTING AND WET CUTTING

Materials are dry cut in both power hacksawing and band machine sawing when:

- A slow cutting speed is used and the frictional heat that is generated during the cutting process is minimal;
- A metal is comparatively soft and easy to cut;

- A material, such as gray and malleable cast iron, produces loose graphite that acts as a lubricant;
- Wet cutting hard materials has a tendency to produce a work-hardened surface.

Wet cutting is recommended to:

- Dissipate the heat generated over the small surface area of each saw tooth,
- Remove heat to prevent softening the cutting edges of the saw teeth,
- Reduce the friction between the chips and the saw teeth,
- Prevent the depositing of metal (caused by the cutting action) at or near the edges of the teeth,
- Clear chips away from the workpiece,
- Keep the sides of the saw blade from being scored,
- Increase productivity and tool wear life.

CUTTING ACTION AND CUTTING FLUIDS

The action of the saw teeth cutting through a workpiece usually produces a tremendous amount of heat. Heat is generated in three main places (Figure 12–2):

- At the point of contact and along the cutting face,
- On the cutting edge,
- At the shear plane in the forming of a chip.

The heat must be carried away, and the temperature must be held within a specific range. Excess heat can cause the teeth to soften and dull.

The chart in Figure 12–3 shows the effect of heat on the hardness of three major kinds of saw blades. Note that the teeth of the carbon alloy saw blade soften at a comparatively lower temperature than do teeth of the high-speed steel saw blade. As the saw teeth dull and wear, their cutting efficiency is reduced drastically. Over the same temperature range, tungsten carbide saw blade teeth maintain their cutting qualities.

Cutting fluids are used to dissipate the heat produced by sawing and cutting. The fluids also wash chips away from the cut and increase the saw life.

There are three basic groups of cutting fluids that are widely applied in wet cutting: straight oils, soluble oils, and synthetic (chemical /water) cutting fluids.

STRAIGHT OILS

Straight oils are mineral oils. They are used on very tough materials that must be cut at slow speeds. Straight oil has *lubricity*—that is, high lubricating properties in contrast with its heat removing capability. Straight oils may have sulphur or other chemical compounds added (additives).

SOLUBLE OILS

Soluble oils are used in the cutting off of a wide variety of materials. Since band machine

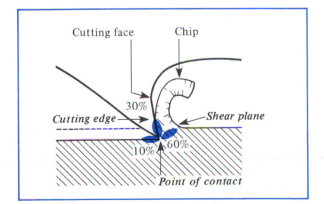

Figure 12-2 Three Main Places Where Heat Is Generated by the Cutting Action of Saw Teeth

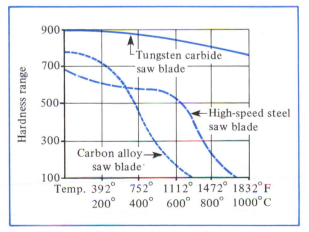

Figure 12-3 Effect of Heat on the Hardness of Saw Band Teeth

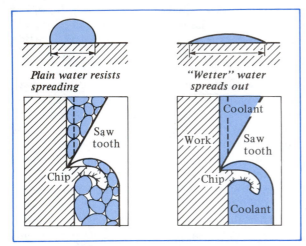

Figure 12–4 Heat Dissipation Properties of Cutting Fluids Improved by Chemical Additives

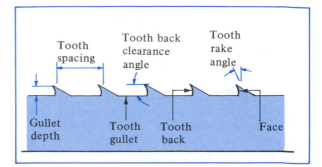

Figure 12–5 General Features of Saw Tooth Forms

sawing is done at high speeds, a great deal of heat is generated. It is necessary to remove the heat rapidly to maintain efficient cutting and to protect the life of the saw teeth and saw band.

Soluble oils are mineral oils that are readily *emulsified* (suspended) into fine particles, or globules, when mixed with water. The resulting mixture combines the properties of straight oil for lubricity with the high cooling, or heat dissipating, rate of water.

The more concentrated the mixture- for example, one part of soluble oil to three parts of water (1:3)—the greater the lubricity. The more dilute the mixture—for example, 1:7—the greater the heat removal capability.

SYNTHETIC (CHEMICAL/WATER) CUTTING FLUIDS

Synthetic cutting fluids do not contain mineral oils. They are a mixture of chemicals added to water to produce "wetter" water. Water tends to resist spreading because of its surface tension. This surface tension may be reduced by chemical additives, as shown in Figure 12–4. Synthetic cutting fluids then have better properties of spreading and flowing easily and quickly. These properties provide for a high rate of heat removal. Another desirable property is that many synthetic cutting fluids are transparent. This transparency makes the cutting area and action more visible.

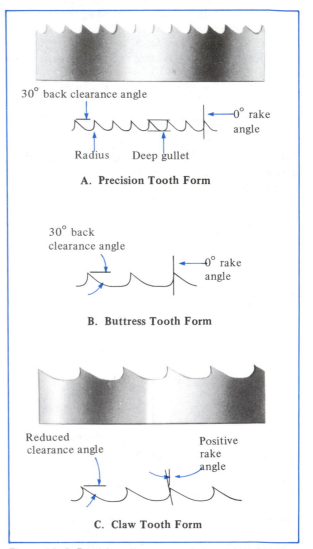

A. Precision Tooth Form

B. Buttress Tooth Form

C. Claw Tooth Form

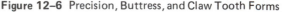

Figure 12–6 Precision, Buttress, and Claw Tooth Forms

KINDS AND FORMS OF SAW BLADES (BANDS)

There are three general kinds of cutoff saw blades: *carbon alloy*, *high-speed steel*, and *tungsten carbide*.

Carbon alloy saw blades are generally used in the toolroom and maintenance shop and in light manufacturing where accuracy is required. High-speed steel blades are used in heavy-duty and full-time production work. Tungsten carbide blades are suited for heavy production and for rough cutting through tough materials.

Carbon alloy and high-speed steel saw bands have three forms of saw blade teeth: *precision*, *buttress*, and *claw*. Tungsten carbide saw bands have a special tooth form. The general design features of saw tooth forms are illustrated in Figure 12–5.

PRECISION TOOTH FORM

The precision form is the most widely used tooth form. The tooth has the following features: a rake angle of 0°, a back clearance angle of 30°, a deep gullet, and a radius at the bottom. Each of these features is shown in Figure 12–6A. The precision form produces accurate cuts and a fine finished surface. The clearance angle and gullet provide ample chip capacity for most cut-off sawing operations.

BUTTRESS TOOTH FORM

The buttress form is also known as a *skip tooth* (Figure 12–6B). The 0° rake angle and the 30° back clearance angle are similar to the precision form. However, the teeth are spaced wider apart. Wider spacing provides greater chip clearance. Buttress-form teeth cut smoothly and accurately. The buttress tooth form is recommended for thick work sections, deep cuts, and soft material.

CLAW TOOTH FORM

The claw tooth is also called a *hook tooth* (Figure 12–6C). The tooth has a positive rake angle. The clearance angle is smaller than the clearance angle of the precision and buttress forms. The gullet is specially designed to be stressproof. The claw tooth form makes it possible to cut at a faster rate and at reduced feed pressures—features that provide longer tool life.

TUNGSTEN CARBIDE TOOTH FORM

The tungsten carbide tooth form has a positive rake angle and a smaller clearance angle than any of the other tooth forms (Figure 12–7). The tungsten carbide teeth are fused into a fatigue-resistant blade that is necessary for heavy, tough cutoff sawing operations.

CUTOFF SAWING REQUIREMENTS AND RECOMMENDATIONS

More cutoff sawing operations are required in the manufacture of parts than any other single machining process. The craftsperson must make a number of decisions that center around the saw band alone. Some of the major considerations are as follows:

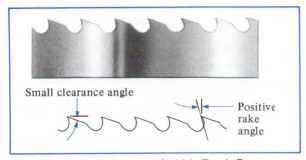

Figure 12–7 Tungsten Carbide Tooth Form

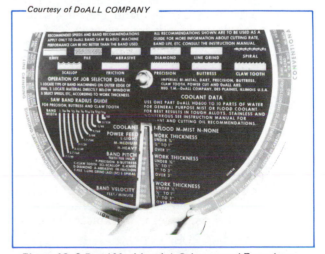

Courtesy of DoALL COMPANY

Figure 12–8 Band Machine Job Selector and Functions

- Kind of saw blade,
- Form of tooth,
- Cutting fluid (if required) and its rate of flow,
- Pitch of the teeth,
- Velocity of the saw band in feet per minute.

The decisions must be based on certain information:

- Size and shape of the workpiece,
- Properties of the material to be cut,
- Required quality of the finish of the sawed surface,
- Quantity to be cut off,
- Overall cross-sectional area if multiple pieces are cut at one setting.

Band machine and saw band manufacturers provide tables to aid the worker in selecting the correct saw band for maximum cutting efficiency. Cutoff sawing requirements and recommendations are further simplified and combined in a *job selector*. A section of job selector is illustrated in Figure 12–8.

How to Select the Saw Blade and Determine Cutoff Sawing Requirements

Sample Job Requirement: Select the correct saw blade for cutting off a quantity of workpieces. The required lengths are to be cut from a 2″ diameter bar of low carbon steel in the 1015 to 1030 range. A band machine with a welded and dressed blade is to be used.

STEP 1 Turn the job selector to the kind of material to be cut. Figure 12–9 at ①.

> Note: The band machine settings and operating conditions are read in the window above the material setting. The feed, band pitch, kind of saw blade, and band velocity are given in the column to the left of this window. The thickness of the workpiece and the rate of coolant flow are given in the column to its right. In this example, a workpiece thickness of 1 1/2″ is assumed.

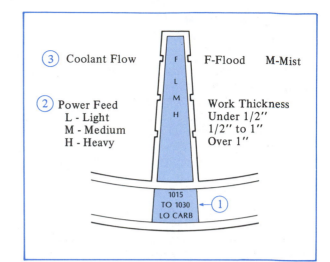

③ Coolant Flow   F-Flood   M-Mist

② Power Feed
L - Light
M - Medium
H - Heavy

Work Thickness
Under 1/2″
1/2″ to 1″
Over 1″

1015 TO 1030 LO CARB ①

Figure 12–9 Reading Coolant and Feed Requirements in the Job Selector

STEP 2 Read the setting for the power feed (H, or heavy, in Figure 12–9). Set the machine for a heavy power feed (H), as shown at ②.

STEP 3 Read the rate of coolant flow. (The selector shows F [flood] at ③.)

STEP 4 Refer to a machinability table on cutting fluids for ferrous metals. (Appendix A–19 is an example.) Select the recommended fluid in the low-carbon steel column (machinability group II) under sawing processes. Either a sulphurized oil (Sul), a mineral-lard oil (ML), or a soluble or emulsified oil or compound (Em Sul ML) is recommended.

> Note: Some job selectors provide information on cutting fluids.

STEP 5 Read on the job selector the recommended kind of saw blade and pitch. (The 6P at ⑤, Figure 12–10 indicates 6 pitch, precision form, carbon alloy blade.)

STEP 6 Select and mount the appropriate saw band.

STEP 7 Read on the job selector the recommended band velocity. (The setting recommended at ④ is 155 feet per minute.) Set the band velocity.

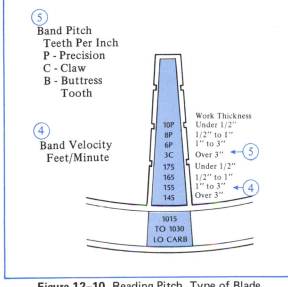

Figure 12-10 Reading Pitch, Type of Blade, and Velocity on the Job Selector

Safe Practices in Band Machining (Cutoff Sawing)

- Check the blade tension. It must be adequate to permit proper tracking on the driver and idler wheels and to transmit sufficient force to cut off the workpiece.
- Determine the speed, feed, kind, and form of saw blade teeth, and the cutting fluid. Cutoff sawing specifications are obtained from machine and blade manufacturers' tables.
- Lock the frame guards in the locked position before any control switches are turned on.
- Direct the cutting fluid to provide maximum lubrication between the blade, the guides, and the workpiece. The cutting fluid must also provide for dissipating the heat generated at the cutting edges.
- Position the power controls in the OFF position when making blade or work adjustments.

UNIT 12 REVIEW AND SELF-TEST

1. List three distinct design features of the band machine and blade.

2. Explain why the horizontal band machine cuts faster than a reciprocating power hacksaw.

3. State three conditions under which dry band machine sawing is recommended over wet cutting.

4. Identify one saw blade material of which the cutting quality of the saw blade teeth is maintained longer (over the same temperature range) than the cutting quality of a high-speed steel blade.

5. State why synthetic cutting fluids are used in cutoff sawing processes that require a high heat-removal rate.

6. Distinguish between (a) the precision form and (b) the tungsten carbide form of saw band teeth.

7. Indicate the kind of information that is provided by a job selector on a horizontal cutoff band machine.

8. List the preliminary steps that must be taken before mounting and cutting off parts on a horizontal cutoff band machine.

9. Identify three safe practices to observe when band machining (cutoff sawing).

SECTION TWO

Bench and Floor Grinders

This section describes the principles of and procedures for correctly using general-purpose bench and pedestal (floor) grinders to shape and size small tools and workpieces made of hardened steel. Included are abrasives and grinding wheel technologies related to bench and floor grinding; procedures for accurately and safely mounting, truing, and dressing grinding wheels; and step-by-step procedures for grinding.

UNIT 13

Hand Grinding Machines, Accessories, and Basic Processes

GRINDING AND STONING PROCESSES

Grinding is the process of cutting away particles of a material. Grinding reduces a part or tool to a particular size and shape. The cutting is done by countless numbers of abrasive grains that are bonded together. As the cutting edge of each grain is forced into the surface of a workpiece, it cuts away a small chip. By rotating a grinding wheel, a continuous series of cutting edges are placed in contact with the work or tool. These abrasive edges grind away the surface.

Stoning is a hand grinding process. In this instance, an abrasive *oilstone* is used. One purpose of *stoning* is similar to burring with a file. However, with hardened materials an abrasive stone is required to do the cutting. Stoning also is used to produce a fine cutting edge that results in a smoother cutting action, a finer surface finish, and longer tool life.

BENCH AND FLOOR (PEDESTAL) GRINDERS

The grinding of the cutting edges of tool bits, drill points, and chisels; the forming of blades and other tools; and various other metal-removing processes all may be performed on a simple grinding machine. This machine is called a *bench grinder* if it is mounted on a bench as shown in

Figure 13–1. The terms *floor* or *pedestal grinder* are used interchangeably when the grinding head is mounted on a pedestal for floor work. Pedestal grinders range in size from grinders used for small tool grinding operations to large grinders used for heavy-duty, rough snagging processes on large forgings and castings.

CONSTRUCTION AND FEATURES OF GRINDERS

Grinding operations on the bench or floor grinder are referred to as *offhand grinding*. In offhand

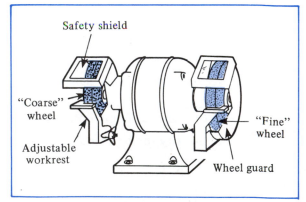

Figure 13–1 Bench Grinder

grinding the object is held and guided across the revolving face of the grinding wheel by hand.

Regardless of the type, bench and pedestal (floor) grinders have a direct or belt power driven *spindle*. The spindle is flanged and threaded at both ends to receive two *grinding wheels*. The grinding wheel at one end is coarse, for rough cutting operations. The wheel on the opposite end is fine, for finish grinding.

Each grinding wheel is held between two *flanged collars*. There is a simple *adjustable tool rest* for each wheel. Each rest provides a solid surface upon which the part may be placed and steadied while being ground. The tool rest may be positioned horizontally or at an angle to the wheel face.

Wheel guards (Figure 13–1) are an important part of the grinder. Safety glass shields permit the work surface to be observed and provide additional protection against flying abrasives and other particles. However, because of the hazards of machine grinding it is also necessary for the worker to use personal eye protection devices.

GRINDING WHEELS

MOUNTING ON A SPINDLE

Grinding wheels are mounted directly on the spindles of bench and floor grinders. Figure 13–2 shows a mounted grinding wheel assembly. One end of the spindle has a right-hand thread; the other end, a left-hand thread. The thread direction safeguards the wheel from loosening during starting and stopping and during cutting action.

Grinding wheels have a soft metal core that is bored accurately to the diameter of the spindle. The wheel slides on the shaft with a snug fit. The wheel should never be forced on the spindle or tightened with too great a force against the flanged collars. A soft compressible material (washer) is placed between the sides of the wheel and the flanged collars. The washer, which is sometimes called a *blotter*, is often used as the label for the manufacturer's specifications. This washer helps to avoid setting up strains in the wheel. The clamping nut is drawn against the flanged collar only tight enough to prevent the grinding wheel from turning on the spindle during cutting.

TRUING AND BALANCING GRINDING WHEELS

A wheel is tested for trueness and balance before grinding. *Trueness* refers to the concentricity of the outside face and sides of a grinding wheel as it revolves at operating speed. *Balance* relates to whether the mass of a grinding wheel is evenly distributed about its axis of rotation. Excessive force may cause a wheel to fracture and fly apart. A fractured wheel is a safety hazard and should be replaced. Grinding wheel manufacturers true and balance wheels so that they are ready to be mounted and used directly. Wheels over 12″ (304.8mm) diameter should always be balanced before mounting.

Truing is a process of shaping a grinding wheel. A wheel is *trued* to restore its concentricity. The

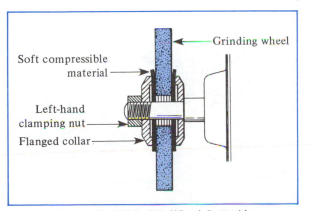

Figure 13–2 Grinding Wheel Assembly

Grinding wheel

Soft compressible material

Left-hand clamping nut

Flanged collar

Courtesy of DESMOND-STEPHEN MANUFACTURING COMPANY

A. Glazed (Loaded) Wheel Before Dressing

B. Open, Dressed Wheel

Figure 13–3 Glazed and Dressed Grinding Wheels

wheel may be dressed so that it has a straight, angular, or special contoured face. However, (after truing), the grinding face (periphery of the grinding wheel) must run concentric with the spindle axis.

DRESSING AND HAND DRESSERS

DRESSING

During the grinding process some of the sharp-edged abrasive grains become dull (glazed). The grains may not fracture as they should to provide new, sharp cutting edges. Also, metal particles from the workpiece or tool may become imbedded in the wheel. Such a wheel is said to be *loaded* and must be *dressed*. A loaded, glazed wheel is shown in Figure 13–3A.

Dressing is the process of reconditioning a grinding wheel. Reconditioning a wheel is necessary for efficient grinding and to produce a good quality of surface finish. Figure 13–3B illustrates an open, dressed wheel. The grains are sharp. Fast cutting takes place.

Hand dressers are used for both dressing and truing grinding wheels. The two common hand types to be considered next are the *abrasive stick* and the *mechanical dresser*.

DRESSERS

Abrasive Stick. The abrasive dressing stick is used to remove particles that produce a glazed surface and to sharpen the cutting face. The abrasive stick shown in Figure 13–4 is held by hand. It is brought carefully into contact with the grinding wheel. Usually a few passes across the face are sufficient to correct the glazed condition and produce sharp-cutting abrasive grains.

Mechanical Dressers. Mechanical dressers are usually used for heavier dressing and truing operations than the abrasive stick. Mechanical dressers are designed with many different forms of metal wheel discs (Figure 13–5). These discs are mounted on a shaft at one end of a hand holder. The discs have star or wavy patterns. They are made of hardened steel and are replaceable. There is a circular disc separator between each pair of star discs that allows for a great deal of side play between the discs.

As the disc faces are brought into contact with the face of the grinding wheel, they turn at considerable velocity in comparison to the speed of the grinding wheel. The action between the dresser discs and the grinding wheel causes the abrasive grains to fracture. As new, sharp cutting edges are exposed, the imbedded particles are torn out. The wheel is also trued during the process. A diamond dresser, guided under more controlled conditions, is used to true a grinding wheel to a finer degree of accuracy.

Caution: Extreme care must be taken when hand dressing. The grinding wheel guard and shield must be in place, the tool rest must be correctly positioned to support and serve as a guide for the dresser, and personal goggles must be worn.

Mechanical dressers are practical where speed in reconditioning a grinding wheel is more essential than accuracy. The heavy-duty dresser

Courtesy of DoALL COMPANY

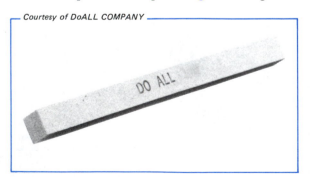

Figure 13–4 Abrasive Dressing Stick

Courtesy of DESMOND-STEPHEN MANUFACTURING COMPANY

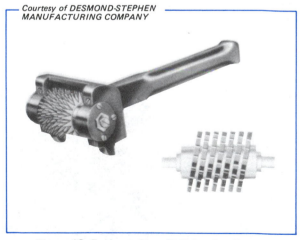

Figure 13–5 Heavy-Duty Ball Bearing Dresser

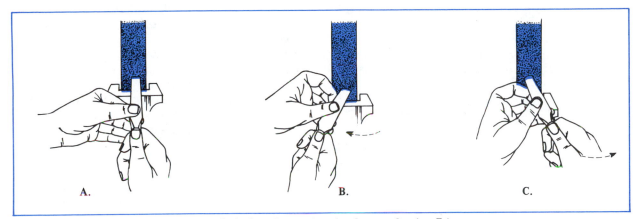

Figure 13–6 Grinding an Angular Convex Cutting Edge

in Figure 13–5 has dust-protected ball bearings for the cutter spindle. This dresser is practical on large, coarse grinding wheels, particularly wheels that are bakelite and rubber bonded.

WHEEL CONDITIONS REQUIRING DRESSING

The craftsperson must make decisions as to when a grinding wheel must be dressed. Dressing is needed under the following conditions:

- An uneven chatter surface is produced,
- The workpiece is heated excessively during grinding,
- Colored burn marks appear on the work surface,
- There are load lines on the workpiece that indicate the wheel is loaded with metal particles,
- There is considerable vibration when the wheel is in contact with the workpiece.

How to Grind an Angular Surface

Grinding an Angular Surface

STEP 1 Set the work rest at or near the required angle. Move the work rest to within 1/8″ of the grinding wheel.

STEP 2 Hold the tool or workpiece firmly on the work rest with one hand.

STEP 3 Bring the tool into contact with the revolving grinding wheel face and guide the tool across the entire wheel face with the other hand.

STEP 4 Continue grinding until the required width of the angular surface is reached.

Note: A slightly different procedure may be followed in grinding the convex angular face on a cold chisel. An angular and convex surface is produced by holding the flat surface of a chisel against the work rest at an angle (Figure 13–6A) while the chisel point is swung through a slight arc (Figures 13–6B and C). This cutting procedure produces the convex cutting face.

Caution: Quench the tool point frequently in water so that the tool retains its hardness. Nonhardened workpieces should also be quenched in water to keep the temperature within the range in which the part may be handled safely.

Safe Practices in Hand Grinding

- Secure the collars against the sides of a grinding wheel by applying a slight tightening force to the spindle nuts. Apply only enough force to prevent the wheels from turning on the spindle during the grinding process. Excessive force can strain and fracture the wheels and thus create an unsafe condition.
- Store grinding wheels carefully in a special rack. Damage may result if the wheel is laid flat or placed where it can be hit.

- Dress glazed and dull grinding wheels that are inefficient and produce surface imperfections in the workpiece. After a grinding wheel is trued or dressed, the tool rest must be re-adjusted to within no more than 1/8" of the wheel face.
- Provide a small space between the tool rest, the dresser, and the grinding wheel. The tool rest supports and guides the dresser and permits it to be positioned at a cutting angle.

- Keep machine guards and shields in position to provide good visibility. Put on safety glasses before the grinder is started. Keep them on throughout the grinding process.
- Maintain a grinding wheel speed below the maximum revolutions per minute specified by the manufacturer. The manufacturer's recommended speed is marked on the grinding wheel.
- Perform grinding operations from a position that is out of direct line with the revolving wheel.

UNIT 13 REVIEW AND SELF-TEST

1. Compare the stoning of burrs with the removal of burrs by filing.

2. Describe the mounting of grinding wheels on bench and floor grinders.

3. State how a heavy-duty grinding wheel may be hand dressed.

4. Indicate three unsafe conditions that require the worker to dress a grinding wheel.

5. List the main steps for offhand grinding the flat face on a tool.

6. Give three safe practices to follow before starting a bench or floor grinder.

SECTION ONE

Drilling Machines and Accessories

This section identifies the design features and functioning of conventional and unconventional drilling machines. "Conventional" hole-making processes involve tools such as drills and reamers. Unconventional machines employ "unconventional" hole-making processes that involve newer technologies such as ultrasonics, laser beam, and electrochemical cutting "tools" are covered later in Part 15. Also described in this section are the work-holding devices and accessories commonly used in hole-making processes.

UNIT 14

Drilling Machines and Processes

CONVENTIONAL DRILLING MACHINES

Hole making refers to many different processes. Some of these processes are drilling, reaming, counterboring, countersinking, spotfacing, and tapping (Figure 14–1). These processes are performed on a group of machines that are identified as *drilling machines*. Some drilling machines are light and simple. Operations are controlled by the operator, who feeds a cutting tool into the work. Simple drilling machines are usually of the bench or floor type. Other drilling machines are large, for heavy-duty operations. In mass manufacturing there are multiple-spindle drilling machines that are used for the rapid production of many holes.

Some work may be brought conveniently to a machine and positioned for drilling and other operations. Large, heavy, cumbersome parts that are cast, forged, or produced by other methods are secured on a machine table. Hole-making operations are then performed by positioning the movable tool head and the cutting tool at a desired location. The radial drill press is used for such operations.

Most holes are machined straight, at a right angle to a surface. Some holes are drilled at various angles. Other holes require deep-hole drilling, in which the workpiece is rotated instead of the drill. Drilling machine tables may be stationary or adjustable for angle operations. Whether holes are machined straight, angled, or deep, conventional hole-making machines are used.

Six different groups of conventional machines are described in this unit:

- Sensitive and heavy-duty drilling machines (bench and floor types);

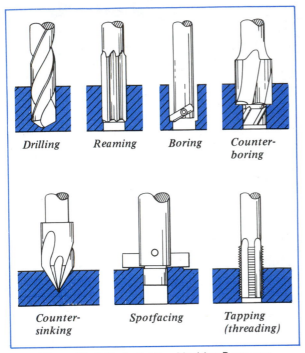

Figure 14–1 Basic Drilling Machine Processes

- Multiple-spindle types, for production;
- The deep-hole drilling machine;
- The radial drill press, for operations requiring a long overhanging arm and a machine head that may be swung through a large radius;
- The multi-operation turret drilling machine;
- A numerically controlled unit, the operations of which are programmed.

SENSITIVE DRILL PRESS (BENCH AND FLOOR TYPES)

Of all conventional drilling machines, the simplest drilling machine design and the easiest machine to operate is the *sensitive drill press* (Figure 14–2). When the length of the drill press column permits the machine to be mounted on a bench, it is referred to as a sensitive *bench* drill press. The same machine is also produced in a floor model. This model is referred to either as a sensitive *pedestal* or sensitive *floor* drill press.

The machine size indicates the maximum work diameter that can be accommodated. For instance, a 24″ drill press will accommodate a

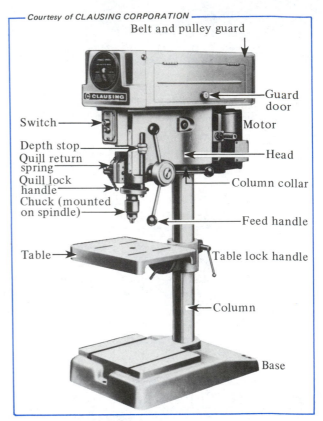

Figure 14–2 Features of a Sensitive Drill Press (Bench Model)

workpiece with up to a 12″ radius (24″ diameter). The speed range of the sensitive drill press permits drilling holes from 1/2″ down through the smaller fractional, number, letter, and metric drill sizes. The smaller the drill size, the higher the speed required. Bench and floor drill presses are widely used for drilling, reaming, boring, countersinking, counterboring, spotfacing, and other operations.

MAJOR COMPONENTS OF THE SENSITIVE DRILL PRESS

Regardless of the manufacturer, there are four major components of the sensitive drill press: base, work table, head, and drive mechanism.

Machine Base. The base of most drilling machines has a stationary upright column. The axis of the column is at a right angle to the base. The upright column and base are rigid. The workpiece may rest on the machined surface of the base. Long parts may be accurately positioned and held on the machined base.

Work Table. Mounted on the upright column is a work table. The work table may be round, square, or rectangular. The table may be moved vertically and then locked at a required height to accommodate different work lengths. The work table may be held square in relation to the spindle, or it may be adjustable for angle work. Tables have either elongated slots or grooved slots. Grooved slots provide a surface against which T-head bolts or nuts may be tightened. The slots permit locating the work in relation to the spindle.

Drill Press Head. The head includes a sleeve through which a spindle moves vertically. Vertical movement of the spindle is controlled by a feed lever that is hand operated. Where there is a power feed, the spindle may be moved up and down both by hand and by power.

The end of the spindle has a taper bore. Some drilling attachments, chucks, and cutting tools have a corresponding taper shank and tang. There are adapters to fit the taper bore. The adapters accommodate tool shanks or other parts that have a taper smaller or larger than the taper bore.

Drill chucks are a practical and widely used device for holding straight-shank tools and pieces of round stock. The chuck jaws are adjustable over a wide range of sizes. For example, one drill chuck may hold all diameters from a #60 drill to 3/8″ (or 10mm).

Drive Mechanism. The spindle may be turned at different speeds (revolutions per minute) by a V-belt drive mechanism. The required operating speed depends on the material, nature and size of the operation, type of cutting tool, and desired surface finish. Speed adjustments are made by changing the step-cone pulley over which the V-belt rides. The range of bench and floor drill press speeds is from 30 to 5,000 revolutions per minute.

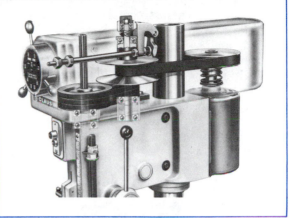

Figure 14–3 A Quick-Setting Variable Speed Drive

THE STANDARD AND HEAVY-DUTY UPRIGHT DRILL PRESS

The *standard upright drill press* is of heavier construction than the sensitive drill press. In the shop the standard upright drill press (drilling machine) is usually called just a *drill press*. The column may be round, cast, or fabricated in a rectangular form. Speed variations on the lighter upright drill presses are obtained by a variable speed drive like the speed drive illustrated in Figure 14–3. The operator sets the variable speed control at the desired speed.

The *heavy-duty upright drilling machine* uses a gear drive system. The selected speed is obtained by shifting gear selector handles while the machine is stopped. Different combinations of gears are thus engaged. The range of speeds is increased further by using a two-speed motor. This motor provides speed versatility for a low-speed range and a high-speed range.

The floor-type standard drill press has both a manual (hand) feed and an automatic mechanical feed. A variable feed range from 0.002″ to 0.025″ is standard. The feed control lever is set at the required feed per revolution. The cutting tool automatically advances the feed distance at each complete turn.

THE GANG DRILL

The *gang drill* is a type of multiple-spindle drilling machine. A number (gang) of drilling heads are lined up in a row. They may be driven individually or in the gang.

The number of spindles is determined by the number of operations to be performed. Holes in a workpiece may be drilled, reamed, bored, tapered, and so on. Each hole may be of a different diameter.

Each cutting tool is usually positioned in relation to a specific dimension by using drill jigs or special work-holding fixtures. On especially large drilling machines, the table on which the work is held may be positioned by hydraulic feed mechanisms that are numerically controlled.

THE MULTIPLE-SPINDLE DRILLING MACHINE

The *multiple-spindle*, or *multispindle drilling machine* is a mass-production hole-making machine (Figure 14–4). It consists of a number of

Figure 14–4 A Multiple-Spindle Drilling Machine

drilling heads that are positioned at desired locations. Each spindle has at least two universal joints. The drivers may be expanded or closed to bring a spindle to a new location.

It is important that a guide plate (mounted with the drill head above the work) or a drill jig be used in multiple-spindle drilling operations. Also, the drill lengths must be staggered to distribute the starting load.

THE WAY-DRILLING MACHINE

The *way-drilling machine* is an adaptation of the multiple-spindle drilling machine. The difference is that the spindles of the way-drilling machine operate from more than one direction. The production of automotive engine blocks is an application of way-drilling. A series of holes is produced simultaneously on several sides of these blocks.

THE DEEP-HOLE DRILLING MACHINE

One problem in drilling a deep hole is producing a hole that continues to follow a required axis. The longer the hole, the greater the tendency for the drill to be deflected. A true, straight hole may be drilled by holding the tool stationary and rotating the work. The cutting tool then follows the axis of rotation. A straighter and more precisely drilled hole is produced.

In deep-hole drilling, the drill is supported and guided by a guide bushing. It is always necessary to keep a coolant on the drill point. The chips must be removed continuously.

Special drills are available that permit a cutting fluid to flow through the drill body to the cutting edges. The fluid cools the cutting edges, drill, and workpiece while chips are forced out of the drilled hole.

THE RADIAL DRILL

The *radial drill* derives its name from its design. A radial drill, with its major components identified, appears in Figure 14–5.

A heavy workpiece is positioned and clamped to the base. The tool head may be moved with a handwheel. It may also be motor driven rapidly close to the desired hole location. The cutting tool is usually brought by hand to the exact point where an operation is to be performed.

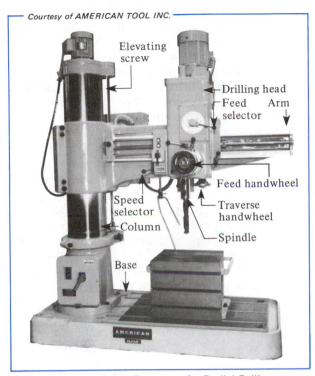

Figure 14–5 Features of a Radial Drill

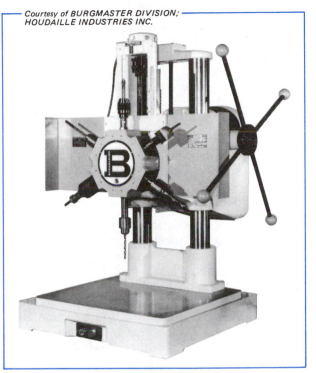

Figure 14–6 A Six-Station, Bench Model Turret Drilling Machine

The spindle is moved to the work until the cutting tool is almost in contact with the work. The automatic feed is then engaged. Some machines have automatic trips to regulate the depth. A predetermined depth may be set. As the spindle and cutting tool reach the required depth, the feeding stops.

The radial drill may produce straight holes or angular holes. The holes may be located from a direct layout on the workpiece or with a template or a drill jig.

THE TURRET DRILLING MACHINE

The *turret drilling machine* has a number of drill head units. The turret of the drilling machine may have six, eight, or ten separate heads (stations). Figure 14–6 provides an example of a six-station turret drilling machine. A different cutting tool or size may be inserted in each head. For example, the hexagonal turret in Figure 14–6 may be set up with a combination of center drill, pilot drill, large-sized drill, spotfacer, counterbore, and reamer. This combination represents the sequence in which the hole is started with the center drill, drilled with the pilot drill at a high speed, followed with the reamer drill, spotfaced, counterbored to a particular depth, and finally reamed.

The turret drilling machine eliminates tool changing time. Also, a single machine may be used for all processes. Small-sized machines may be hand fed. Some large machines, on which operations like face milling are performed, may also be manually operated. Other turret drilling machines may be tape operated by a numerical control unit.

SPECIAL HOLE-PRODUCING MACHINES

The drilling head is versatile. A single operation like drilling may be performed. Other operations may be combined to include drilling, reaming, boring, and so on. One head may be used, or any number of heads may be combined. They may be positioned vertically, horizontally, or at a simple or compound angle.

A workpiece may be positioned and held in a drill fixture. A part may also be held in a drill jig with a drill bushing to guide the drill. The table may be round, rectangular, or rotary. The combination of machine and work-holding devices and cutting tools depends on the special quantity production requirements of the job. *Special hole-making machines* incorporate many of these requirements and design features.

WORK-HOLDING DEVICES

The accessory in which the work is positioned and held may be a *work-holding,* or *clamping, device.* Some of the common methods of positioning and holding a workpiece include:

- Clamping directly in a vise.
- Mounting on parallels or V-blocks. The part is clamped in position with hold downs, straps, clamps, and T-bolts.
- Nesting and holding in a specially designed form. The drill or cutting tool may be guided for location by a drill bushing. The drill bushing also helps to produce a concentric hole. Another holding device is called a drill jig or fixture.

WORK-HOLDING VISES

Although there are other vise designs, the standard, angle, and universal vises are the three basic vise types. Workpieces of regular shape (square, rectangular, and round) may be held in a *standard drill press vise.* After the hole location is layed out, the work is placed on parallels. The work then is securely tightened in the vise and positioned on the drill press table for the hole-making operation. The drill press vise should be clamped to the work table for heavy drilling operations.

An *angle vise* is used for some angular machine operations. This vise holds the work at the required angle and eliminates the need for tilting the work table. Compound angles require that the work be positioned and held in a *universal machine vise* like the vise shown in Figure 14–7. This vise permits three separate angular adjustments: 360° in a horizontal plane, within 90° from a horizontal to a vertical plane, and through 90° in the third plane.

The *safety work holder* (Figure 14–8) is another holding device. It fits around the machine column and has two easily adjusted arms. These arms hold the work in position on the work table.

V-BLOCKS

V-blocks are widely used to position and hold cylindrical parts for layout, machining, and assembly operations. A V-block is a rectangular metal solid. It has V-shaped angular surfaces on at least two opposite sides.

Courtesy of UNIVERSAL VISE & TOOL COMPANY

Figure 14–7 A Universal (Three-Way) Vise

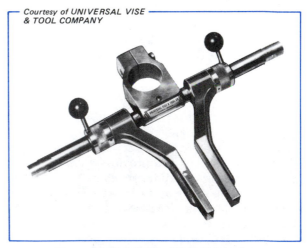

Courtesy of UNIVERSAL VISE & TOOL COMPANY

Figure 14–8 A Safety Work Holder for Column-Mounting on a Drill Press

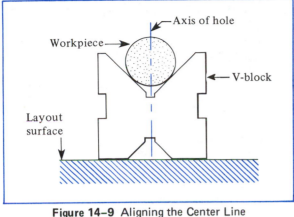

Figure 14–9 Aligning the Center Line of a Hole to Be Drilled

Round workpieces of different diameters may be centered and nested in V-blocks. A U-shaped clamp is provided to hold the workpiece in position.

When a hole is to be drilled through the axis of a cylindrical workpiece, a center line is usually scribed on the end. The workpiece is mounted in V-blocks. The center line (axis) is then aligned with a steel rule or other square head. The square is held against the surface of the machine table or other layout plate. The alignment for the workpiece is shown in Figure 14–9.

STRAP CLAMPS AND T-BOLTS

Irregular-shaped workpieces are sometimes held directly on machine tables with a number of differently shaped *strap clamps*. (The terms *strap*, *clamp*, and *strap clamp* are used interchangeably.) Each clamp has an elongated slot. A *T-bolt* fits in the slot or in a machined groove in a table and passes through the clamp. The opposite end of the clamp is raised by *blocking*. It is then positioned to apply a maximum force on the workpiece.

The flat surface of the clamp should be parallel to the table to permit the force applied with the T-bolt to hold the workpiece securely and in the position required for the machining operation. Examples of square head, cutaway head, and tapped T-head bolts are shown in Figure 14–10.

The judgment of the craftsperson is important. The force exerted in the clamping process

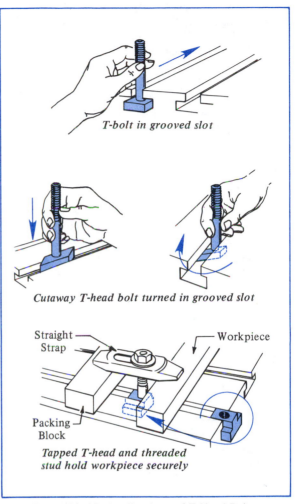

Figure 14–10 Applications of T-Head Bolts, Straight Strap, and a Packing Block

should not cause conditions in which the part is bent or fractured. Applying excess force causes such conditions.

Blocking to raise the clamp to a correct height may be done in several ways. A step block with a number of steps of different heights may be used. Sometimes sets of blocks are used. These blocks may be added together to reach a particular height. Use of an adjustable jack permits even greater variation in height and makes adjustment easier.

The four common shapes of strap clamps, each of which has a number of variations, are the *straight, finger, bent,* and *U*-strap clamps.

Straight Strap Clamp. The straight strap clamp is a rectangular bar with an elongated slot running lengthwise. The two ends are flat. One end rests on the workpiece; the other, on the blocking. The necessary force is produced by tightening the T-bolt, washer, and nut assembly (Figure 14–10). Some straight strap clamps have an adjusting screw on one end.

Other examples of clamps include the *finger strap*, *bent-tail strap (gooseneck clamp)*, and *U-strap* with a finger.

THE ANGLE PLATE

The *angle plate* that is used on a drilling machine usually has a number of holes or elongated slots. These holes or slots permit strapping the angle plate to the table and securing a workpiece to the right-angle face. Figure 14–11 shows a typical angle plate setup.

DRILL JIGS AND FIXTURES

Mentioned earlier as devices for positioning and holding workpieces, drill jigs and fixtures are widely used with workpieces of irregular shape when a quantity must be machined accurately.

A *drill fixture* is a specially designed holding device. The fixture contains a nesting area into which the workpiece fits. This area positions and holds the workpiece. The relationship is fixed between the workpiece, the spindle of the machine, and the cutting tool.

A *drill jig* is an extension of the drill fixture. It serves the same functions and, in addition, has guide bushings. These bushings guide cutting tools such as drills and reamers to a precise dimensional location. The guide bushings help to produce accurate concentric holes. Figure 14–12 shows a simple drill jig for accurately drilling a hole in a cylindrical part.

MACHINE ACCESSORIES

The term *accessories* refers to the tools and the mechanical parts and machines that are used in conjunction with a machine tool. These accessories aid in setting up and holding a workpiece so that a particular process may be performed. Five accessories that are widely used in hole-producing operations are: *parallels*, *jack screws*, *step blocks* and *packing blocks*, *shims*, and *universal swivel blocks*. Clamps (straps), bolts, and V-blocks, which are also accessories, were described earlier.

JACK SCREW

The jack screw, or jack, is a simple, easily adjusted support device. The jack is designed to provide for a wide range of adjustment. With a jack it is possible to make up whatever length is required between blocks, a machine table, and a work surface. All jacks consist of a base, an elevating screw, and a swivel head. Jacks are placed under a workpiece and brought into contact with the surface that is to be supported. A jack may also be used under a clamp to replace a number of blocks.

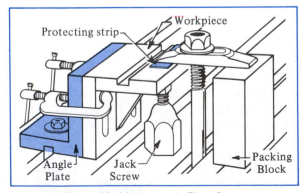

Figure 14–11 An Angle Plate Setup

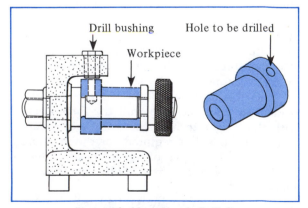

Figure 14–12 A Simple Drill Jig

STEP BLOCKS AND PACKING BLOCKS

The step block derives its name from the fact that each block consists of a series of steps. These steps permit the end of a clamp to be located at whatever height (step) is needed for a job. Each step also provides a surface against which a clamping force may be applied.

Packing blocks are individual blocks of different sizes. The blocks may be used alone or in any combination. When a desired height is reached, a force may be applied correctly on a workpiece.

SHIMS AND PROTECTING STRIPS

Shims are thin metal strips. They are used to compensate for surface variations in measurement. Since it is necessary to have a good bearing surface so that a workpiece will not move during a cutting process, shims may be inserted between the workpiece and a parallel. Shims level an uneven surface.

Protecting strips are thin metal pieces. These strips are placed under any metal part, such as a clamp. Protecting strips are used to prevent damage to a finished work surface or machine table when a force is applied in a work-holding setup.

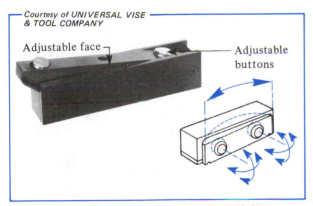

Courtesy of UNIVERSAL VISE & TOOL COMPANY

Adjustable face

Adjustable buttons

Figure 14–13 Universal Movement of the Adjustable Face and Buttons of a Universal Swivel Block

UNIVERSAL SWIVEL BLOCK

Many surfaces that are tapered or irregularly shaped must be held securely. A universal swivel block is used with a vise for such purposes (Figure 14–13). The universal swivel block consists of two major faces. One face on the body is flat. This face rests against the stationary jaw of the vise. The other face has two buttons. This face is adjustable. It may be positioned against an irregular or tapered surface. The universal motion of the swivel block permits it to accommodate the shape of a workpiece.

Safe Practices in Drilling Machine Operation and Maintenance

- Check the table surface, work-holding device, and the workpiece. Each must be free of burrs, chips and foreign particles. Cleaning should be done with a brush or a wiping cloth with the machine turned off.
- Fasten each workpiece securely to withstand all cutting tool forces, to ensure personal safety, and to avoid damage to the machine, work, cutting tool, and accessories.
- Stop the machine when a feed change requires the shifting of a gear lever.
- Check to see that covers and guards are in place and secured. All extra tools and parts should be removed before starting the machine.
- Check personal clothing. Loose clothing may be caught by a revolving spindle, chuck, or cutting tool.
- Use eye protection goggles or a protective shield.
- Direct carefully any air blast used for cleaning the work, machine, or tools. The blast should be downward and away from any other operators. Also, cover the work and the area around it to restrict chips and other particles from flying.

UNIT 14 REVIEW AND SELF-TEST

1. State three main differences between the operation of (a) a sensitive bench drill press and (b) a multiple-spindle drilling machine.

2. Distinguish between gang drilling and turret drilling machines.

3. Indicate the kinds of specifications the craftsperson must code in a series of letters and numbers to transfer to a tape for a numerically controlled drilling machine.

4. Identify (a) three unconventional drilling machines, (b) the form of energy used by each machine, and (c) the type of hole-producing tool or process required for each machine.

5. Give three examples of the unique hole-producing functions of unconventional drilling machines.

6. Cite the advantages of using a safety work-holder attachment on a drill press to position and hold a workpiece.

7. Describe briefly the function served by each of the following work-holding devices or accessories as related to drilling machines: (a) step blocks, (b) adjustable jack, (c) gooseneck clamp, (d) drill fixture, and (e) universal swivel blocks.

8. List three safety precautions to follow in setting up for drilling to avoid personal injury.

SECTION TWO

Cutting Tools and Drilling Machine Operations

This section presents the technologies involved in and procedures for hole-making processes performed on drilling machines. These processes include drilling, reaming, threading, countersinking, counterboring, spotfacing, and boring.

UNIT 15

Drilling Technology and Processes

DESIGN FEATURES OF TWIST DRILLS

MATERIALS

There are four common materials used in the manufacture of drills: *carbon steel*, *high-speed steel*, *cobalt high-speed steel*, and *tungsten carbide*. Although some drill design features are similar, each of these materials is adapted to special drilling requirements.

Carbon Steel Drills. Drills of carbon steel are used in comparatively slow-speed drilling operations. The limited amount of heat that is generated by the cutting action must be dissipated quickly, because the hardness and cutting ability of a carbon steel drill is lost once the drill point is overheated and the temper is drawn. *Temper* refers to the degree of hardness and the relieving of internal stresses within a tool. Overheating causes the drill to soften and lose its cutting edges. A carbon steel drill may be identified by white sparks in a spangled pattern that are evident during drill grinding.

High-Speed Steel Drills. High-speed drills have largely replaced carbon steel drills. As its name implies, a high-speed steel drill retains its cutting edges at high cutting speeds and at temperatures up to 1,000°F. The cutting speeds are twice as fast as for carbon steel drills.

High-speed steel drills are used in production. Such drills are versatile cutting tools for ferrous and nonferrous metals and other hard materials. The shank end of a high-speed steel drill is stamped HS or HSS. The mark identifies the metal used in the drill. When a high-speed steel drill is ground, it gives off a dull red spark.

Cobalt High-Speed Steel Drills. The addition of cobalt to high-speed steel produces properties that are necessary for tough drilling operations. Cobalt high-speed drills are used for drilling tough castings and forgings, armor plate, work-hardened stainless steel, and other hard steels. Cobalt high-speed steel drills are run at high cutting speeds. A steady, uninterrupted feed is used. Such drills may be operated at higher temperatures than high-speed steel drills. Figure 15–1 illustrates a heavy-duty, cobalt high-speed steel drill.

Tungsten Carbide Drills. Carbide-tipped and solid carbide drills are used for drilling abrasive

Courtesy of DoALL COMPANY

Figure 15–1 A Heavy-Duty, Cobalt High-Speed Steel Drill

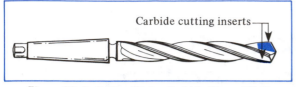

Figure 15–2 A Two-Flute, Carbide-Tipped Drill

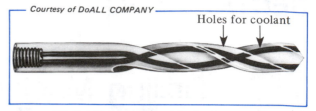

Figure 15–3 A Phantom View of a Coolant-Feeding Drill

materials such as glass-bonded plastics; brass, aluminum, bronze, and other nonferrous metals; and hard-scaled castings.

A carbide-tipped drill is produced by altering the spiral flute of a high-speed steel body. A tungsten carbide cutting tip is inserted and the clearance angle is changed. A common two-flute, carbide-tipped drill is shown in Figure 15–2. Carbide-tipped drills must be sharpened on special grinding wheels. A silicon carbide or diamond-impregnated grinding wheel is used.

TYPES OF TWIST DRILLS

There are several types of twist drills. These types include *two-flute*, *three-* and *four-flute*, *sub-land*, and *coolant-feeding* drills.

Two-Flute Twist Drill. The most commonly used twist drill for all-purpose and production drilling is the two-flute, high-speed steel drill. As the name indicates, there are two flutes. These flutes may be either straight or spiral. The depth of the flutes varies. The depth depends on the degree of strength and rigidity required. In addition to the standard form, heavy-duty two-flute drills are manufactured.

Three- and Four-Flute Drills. Three- and four-flute drills are used where holes are to be enlarged and close tolerance control is necessary. These drills are referred to as *core drills*. The enlarging is done on holes that may be formed by casting, drilling, punching, and other production methods.

Sub-Land Drills. A sub-land, multidiameter drill is an adaptation of a two-flute drill. The cutting end is reduced in diameter for a distance along the body. The cutting edges between the first and second diameters may be ground to produce a square, round, or angular connecting area.

The advantage of using a sub-land drill is that it is possible to drill two or more concentric holes of different diameters in one operation. Only one drill and setup are required to drill two holes, drill and counterbore, or drill and recess holes.

Oil-Hole (Coolant-Feeding) Drills. Many machining processes require a coolant and a lubricant. The use of a cutting fluid increases cutting tool life, production, dimensional accuracy, and quality of surface finish. An adequate supply of cutting fluid may be provided from a coolant nozzle. The coolant is flushed into a drilled hole under a slight pressure.

However, for deep-hole and heavy-duty drilling, it may be necessary to use a *coolant-feeding drill*. A phantom view of this drill is shown in Figure 15–3. A coolant-feeding drill has a hole drilled in the core or through the clearance area. The hole extends to the point of the drill. The cutting fluid is forced under pressure through the drill. The fluid thus reaches the area where the cutting action takes place. The cutting fluid keeps the cutting lips flooded and cooled. The cutting fluid also helps to remove the drill chips from the workpiece.

FUNCTIONS OF TWIST DRILL PARTS

There are two main parts of a twist drill: the *shank* and the *body*. The shank provides a gripping surface. A force may be exerted through the shank to turn and feed the drill. The body is designed to provide efficient cutting action, chip removal, and heat control. Other parts of a twist drill are the *flutes*, the *web*, the *land*, and the *point*. The principal design features and part names of a twist drill are given in Figure 15–4.

The Land. The outer surface between the flutes is the land of the drill. If the land were

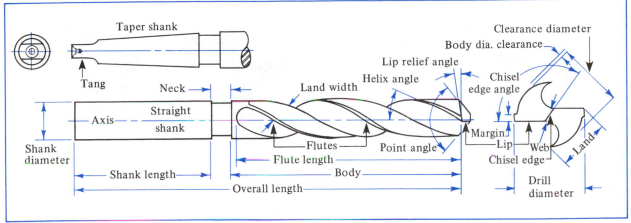

Figure 15–4 Main Design Features of a Twist Drill

the same size as the outside diameter, there would be a great amount of friction in drilling. Friction is reduced by leaving only a small margin of the land equal to the outside diameter of the drill. The remaining portion of the land is relieved (the diameter is reduced).

The Point. The cutting action of a twist drill results from the design of the point. The point includes the cone-shaped area formed with the flutes (Figure 15–5A). The angle of the cone depends on the material to be drilled, the drill size, and the nature of the operation. Cutting edges, or *lips*, are formed by grinding a lip clearance on the angular surface of each flute

(Figure 15–5B). In other words, a lip is ground at a clearance angle. The area behind the cutting edge slopes away to produce a chisel-edged shape that provides the shearing, cutting action. Since the clearance angles are on opposite sides of a two-flute drill, a flat surface is formed across the drill axis (dead center). A chisel point angle of 120° to 135° is used for general drilling.

DRILL POINT SHAPES

Cutting conditions vary for different materials. To compensate for these conditions, it is necessary to change the shape of the cutting edge and the cutting angle. The *standard drill angle* for general-purpose drilling operations is

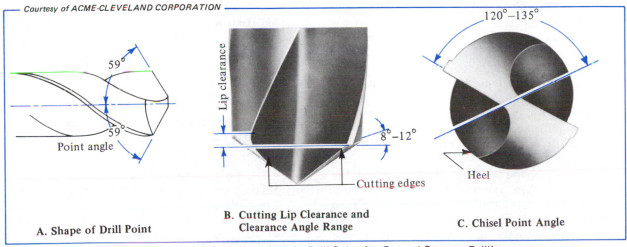

A. Shape of Drill Point

B. Cutting Lip Clearance and Clearance Angle Range

C. Chisel Point Angle

Figure 15–5 Shape and Angles of a Drill Point for General-Purpose Drilling

Table 15–1 Recommended Lip Clearance Angles
(Inch-Standard Drill Ranges)

Drill Sizes	Drill Diameters	Lip Clearance Angle
#80 to #61	0.0135 to 0.0390	24°
#60 to #41	0.0400 to 0.0960	21°
#40 to #31	0.0980 to 0.1200	18°
1/8 to 1/4	0.1250 to 0.2500	16°
F to 11/32	0.2570 to 0.3438	14°
S to 1/2	0.3480 to 0.5000	12°
33/64 to 3/4	0.5156 to 0.7500	10°
49/64 and up	0.7656 to	8°

Courtesy of ACME-CLEVELAND CORPORATION

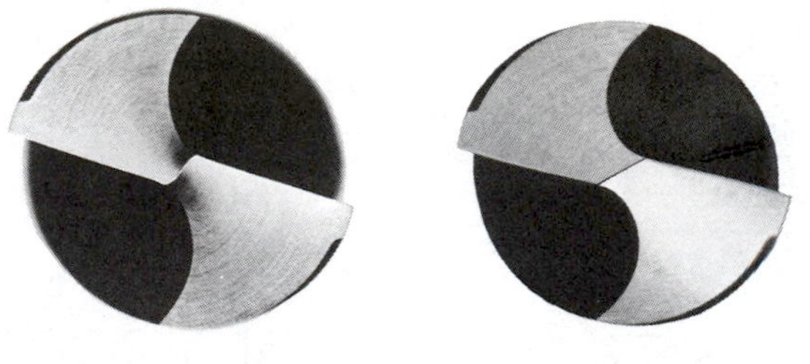

Spiral point   *Chisel point*

Figure 15–6 Features of Drills Ground
to Spiral and Chisel Points

118°. There is a 59° angle on each side of the drill axis. The angles are shown in Figure 15–5A, B, and C.

The cutting edge is formed by relieving the flute. A general *clearance angle* of from 8° to 12° is illustrated. Too small a clearance angle causes the drill to rub instead of cut. At too steep an angle, the drill tears into the work. Thus, the cutting edges may dull rapidly or parts of the drill may break away. After extensive research drill manufacturers have prepared tables of recommended lip clearance angles. Table 15–1 contains recommended lip clearance angles for drill sizes #80 to 1″ and larger. The same lip clearance angles apply to equivalent metric drill sizes.

The precision to which a drill produces an accurate size hole depends on the grinding of the cutting lips. They must be equal in length and at the same angle.

Thus far, only the *chisel point* of a general-purpose drill has been described. The flat shape of this drill point helps to produce a hole that meets most work requirements. These requirements relate to conditions of roundness and size within rough tolerance limits. Roundness may be more accurately produced when the chisel point is guided by a drill bushing in a drill jig.

Another shape of drill point is called the *spiral point*. The spiral-point drill, in contrast to the chisel-point drill, has a sharply pointed end. With this sharp end, the drill may be started accurately without center punching, center drilling, or using a drill jig. The hole produced with the spiral point is concentric with the axis and is round. The photos in Figure 15–6 provide a comparison of a spiral- and a chisel-point drill.

Spiral points may be ground on standard drills. A special spiral-point drill-grinding machine is used for the grinding.

The self-centering feature of the spiral-point drill eliminates the need for a drill-positioning device. The spiral-point drill is particularly adaptable to numerically controlled drilling operations. Spiral-point drills have longer tool life than drills ground with a chisel point. Test records of spiral-point drills show that a greater number of workpieces may be produced between each spiral-point grinding in contrast to the number of workpieces produced between drill-point conventional grindings.

SPLIT-POINT DRILL

Another point, the *split point*, reduces the web to a minimum shape. This point forms two additional cutting edges. Figure 15–7 shows that the clearance angle of the added cutting edges is formed by grinding. Part of the web section is ground parallel with the drill axis. In contrast to the chisel and spiral points, the design of the split point permits:

- Easier and faster cutting,
- Easier access for the chips to clear the cutting edges and workpiece,
- Less force to be used in the drilling process,
- Cutting with less power,
- Less heat to be generated from cutting.

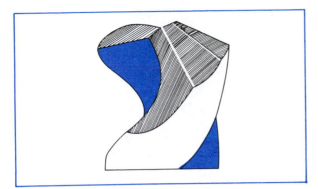

Figure 15–7 A Split-Point Drill Grind

The split point reduces tool wear. The grinding does not disturb the cutting lips. There is also less tool breakage when drilling into hard outer surfaces (scale) of castings and forgings.

THE DRILL POINT GAGE

The accuracy of the drill point angle and the dead center may be measured with a *drill point gage*. A common drill point gage has a graduated blade shaped to the required drill angle. This blade is attached to a steel rule.

The body of the drill is held against the edge of the steel rule. The point end is brought up to the blade. Any variation between the cutting edge and the correct angle may be sighted. The location of dead center is read on the graduated blade. The drill is then turned 180°. A fractional dimension reading is again taken. The drill point angle is correctly ground when there is no light between the cutting edge and the angular blade. The lips are correct when the dimensional reading to the chisel point of each lip is the same.

TAPER DRILL SHANK ADAPTERS

Adapters are used to accommodate the difference between a taper-shank drill and the drill press spindle taper. Two common types of adapters are called a *drill sleeve* and a *taper socket*.

DRILL SLEEVE

When the taper shank of a drill is smaller than the spindle taper, the shank may be brought up to size with a drill sleeve (Figure 15–8). The

Figure 15–8 A Drill Sleeve

standard taper that is used on taper-shank drills is known as a *Morse taper*. The same standard is used on many other cutting tools like reamers, boring tools, and counterbores. Morse tapers are numbered from 0 through 7. The number 2, 3, and 4 tapers are used on drills that range from 3/8″ to 1 1/2″ diameter (10mm to 38mm). For example, a drill sleeve is required for a 1/2″ drill that has a #2 Morse-taper shank when it is to fit a spindle with a #3 or a #4 Morse taper.

FITTED (TAPER) SOCKET

The taper socket is also known as a *fitted socket* (Figure 15–9). It has an external Morse-taper shank. The body has an internal taper. There is an elongated slot at the small end of the taper. This slot is where the tang of the drill or cutting tool fits. It also serves as a way of removing a tool.

A fitted (taper) socket often provides a method of extending the length of a cutting tool. Different combinations of taper sizes are available. For example, a fitted (taper) socket with a #3 Morse-taper shank may have a #2 taper in the body. The #3 taper fits a #3 taper spindle. The #2 taper accommodates a drill or reamer that has the same size taper shank.

THE DRILL DRIFT

The *drill drift* is a flat, metal, tapered bar. It has a round edge and a flat edge. The round edge fits the round portion (end) of the elongated

Figure 15–9 A Fitted (Taper) Socket

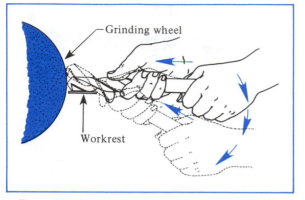

Figure 15–10 Correct Hand Action in Drill Grinding

slot of a drill sleeve or a fitted socket. The flat edge rests and drives against the tang end of a cutting tool. The joined tapered surfaces are freed by a sharp, slight tap on the drift.

The *safety drill drift* is a type of drift. A floating handle is attached to the drift. The body of a cutting tool may be held in one hand. The other hand is used to quickly slide the handle so that it delivers a sharp tap. The force disengages the tapers. It is good practice to protect the surface under a drill so that it will not be hit by a cutting tool when it is freed.

DRILL POINT GRINDING

Drill points may be ground by several methods. They may be ground off hand (Figure 15–10) or on a *drill point grinding machine.* A method in which a drill point attachment is fitted to a bench or floor grinder may also be used (Figure 15–11). Regardless of the method, the drill must be positioned, held, and fed into the grinding wheel to produce a cutting edge and clearance in correct relation to the flute.

There are two main differences between off-hand grinding and grinding with a drill point machine or attachment. First, the skill and judgment of the operator controls the accuracy of the drill point when grinding by hand. Second, the face of a straight grinding wheel is used in hand grinding. The cutting is done in machine grinding on the side face of a *cup wheel.*

In drill point grinding, particular attention must be paid to the following factors: clearance angle, heat generated, lips, and web.

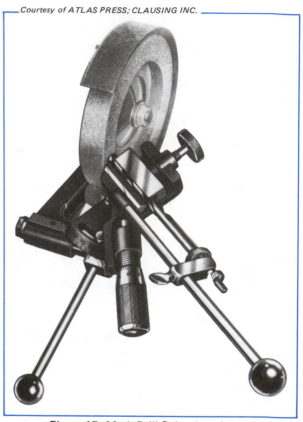

Figure 15–11 A Drill Point Attachment for a Bench Grinder

DESIGNATION OF DRILL SIZES

There are three standard designations of drill sizes in the inch-standard system and one designation in the metric system. In the inch-standard system, a drill may be given a *wire number size, letter size,* or *fractional inch size.* Each drill size has its equivalent decimal value. Metric sizes are given in terms of millimeters. The common sizes of straight-shank drills in the metric system, for example, range from 0.20mm to 16.00mm.

NUMBER SIZE DRILLS (WIRE GAGE DRILLS)

The sizes of the number size series of drills range from #80 (0.0135″) to #1 (0.228″). The most widely used set of drills ranges from #1 to #60 (0.040″).

LETTER SIZE DRILLS

The sizes of the letter size series of drills are designated by letters from A (0.234″) to Z (0.413″). Letter size drills are larger than the largest diameter in the number size drill series, which is #1 (0.228″).

FRACTIONAL SIZE DRILLS

The sizes of the fractional size series of drills begin at 1/64″ diameter. Each successive drill is 1/64″ larger. The general range of diameters extends to 3″. The term *set* is used for a series of fractional size drills in a range, such as the 1/64″ to 1/2″ range. When a series of fractional size drills is part of a set—for example, from 1/64″ to 1/2″—it is sometimes called a *jobber's set*.

Part of a drill size table with selected number, letter, fractional, and metric sizes is represented in Table 15-2. Note that there are variations in diameter among all four drill size designations. These variations provide an adequate range for most hole drilling from a #80 drill (0.0135″) to 3/4″ or 19.0mm (0.748″).

DRILL SIZE MEASUREMENT

Drill sizes may be measured most accurately with a micrometer. A micrometer measurement is taken across the margins of a two-flute drill.

A quick, common practice in the shop is to use a *drill gage*. A drill gage is a flat plate with

Table 15-2 Selected Drill Sizes and Decimal Equivalents

Number (wire)	Letter	Fractional (inch)	Metric (mm)	Decimal Equivalent
1				0.2280
			5.8	0.2283
			5.9	0.2323
	A			0.2340
		15/64		0.2344
			6.0	0.2362
	B			0.2380
			6.1	0.2402
	C			0.2420

the same number of holes as there are drills in the series. There are number size, letter size, fractional size, and metric size drill gages. The decimal equivalent appears for each drill size.

CAUSES AND CORRECTION OF DRILLING PROBLEMS

A number of problems can arise in drilling if correct practices are not followed. These problems affect dimensional accuracy, surface finish, and drilling efficiency. Tool life is increased and the accuracy and quality of drilling are improved when correct practices are followed.

Table 15-3 lists a number of common problems that relate to drills and drilling procedures. Possible causes of each problem are given. These causes are followed by a series of check points for correcting each problem. The craftsperson must be able to recognize a problem, establish its cause, and take corrective steps.

CUTTING SPEEDS FOR DRILLING

Every cutting tool performs most efficiently and with greater tool life when it is cutting at the correct cutting speed. *Cutting speed* is expressed as the *surface feet the circumference of a drill travels in one minute*. This speed is stated as surface feet per minute (sfpm).

Manufacturers' tables contain recommended cutting speeds for different types of cutting tools and materials. Table 15-4 shows part of a table that lists the high-speed steel drill speeds for various steels. These speeds are reduced by 40–50% when a carbon steel drill is used instead of a high-speed drill.

The craftsperson must determine the spindle *revolutions per minute* (RPM) from the information provided in a cutting speed table. The RPM must produce the required cutting speed for a particular drilling operation.

CALCULATING RPM FOR A CUTTING SPEED

The rpm required for a particular surface feet per minute cutting speed (or circumferential speed) may be calculated with the simple formula:

$$RPM = \frac{Cutting\ Speed \times 12}{Drill\ Circumference}$$

Table 15–3 Drilling Problems: Causes and Corrective Action

Problem	Possible Cause	Corrective Action
Drill breaking	–Dull drill point	–Regrind drill point
	–Insufficient lip clearance	–Check angle of lip clearance
	–Spring in the workpiece	–Reclamp workpiece securely
	–Feed too great in relation to cutting speed	–Decrease feed or step up the cutting speed
Drill breaking (nonferrous and other soft materials)	–Flutes clogged with chips	–Use appropriate drill for the material
	–Cutting point digging into the material	–Remove the drill a number of times so the chips clear the hole
		–Increase the cutting speed
		–Grind a zero rake angle
Rapid wearing of cutting edge corners	–Cutting speed too fast	–Reduce cutting speed
	–Work hardness; hard spots, casting scale, or sand particles	–Sand blast or tumble castings, or heat treat
		–Change method of flowing the coolant and check for correct cutting fluid.
Tang breaking	–Taper shank improperly fitted in sleeve, socket, or spindle	–Check condition of adapter or spindle
		–Clean taper hole and remove nicks or burrs by reaming carefully
Margin chipping (using drill jig) Cutting edge and lip chipping	–Jig bushing too large	–Replace drill bushing with same size as drill
	–Lip clearance too great	–Regrind to proper lip clearance
	–Rate of feet too high	–Reduce the feed rate
Cracking or chipping of high-speed steel drill	–Excessive feed	–Reduce the feed rate
	–Improper quenching during grinding or drilling	–Quench gradually with warmer coolant when grinding
Change in cutting action	–Drill dulling; cutting edge damaged	–Regrind the drill point correctly
	–Variations in the hardness of the workpiece	–Anneal the workpiece
	–Change in the amount of cutting fluid reaching the drill point	–Increase the direction or rate of flow of the cutting fluid
Oversize drilled hole	–Improper grinding of point with equal angles and length of cutting edges	–Check reground cutting angle and length of cutting edges
	–Excessive *play* in spindle	–Adjust or correct amount of free movement in spindle
	–Slightly bent drill	–Straighten drill carefully
Cutting on one lip	–Angle of cutting lips varies –Unequal length of cutting lips	–Regrind and check accuracy of cutting angle and length of cutting edges
Poor quality of surface finish (rough drilled hole)	–Improperly ground drill point	–Regrind and check accuracy of drill point
	–Excessive drill feed	–Decrease drill feed
	–Wrong cutting fluid or limited rate of flow	–Change cutting fluid for the kind of material
	–Workpiece not correctly secured	–Reposition and secure the workpiece securely
Drill splitting (along the center)	–Excessive feed	–Decrease drill feed
	–Insufficient lip clearance	–Increase the lip clearance to the correct angle

Table 15–4 Partial Table of Cutting Speeds for High-Speed Steel Drills

Material	Cutting Speed Range (surface feet per minute)
Steel	
Low-carbon	80 to 150
Medium-carbon	60 to 100
High-carbon	50 to 60
Tool and die	40 to 80
Alloy	50 to 70

Table 15–5 Feeds for Selected Drill Sizes

Drill Size (inches)	Feed (inches per revolution)
Small # sizes to 1/8	0.001 to 0.002
1/8 to 1/4	0.002 to 0.004
1/4 to 1/2	0.004 to 0.007
1/2 to 1	0.007 to 0.015
Larger than one inch	0.015 to 0.025

Values are rounded off to produce a more simplified formula that is used in the shop:

$$RPM = \frac{CS \times 4}{d}$$

For example, assume that a series of holes is to be drilled in a low-carbon steel. The cutting speed of 150 sfpm is to be used with a 1/2″ drill. The known values are inserted in the simplified RPM formula as follows:

$$RPM = \frac{150 \times 4}{1/2} = 1,200$$

The drill must turn 1,200 RPM to produce a cutting speed of 150 sfpm.

FACTORS AFFECTING CUTTING SPEED

The cutting speed that the worker finally selects is influenced by seven different factors:

- Material of the workpiece,
- Material of which the cutting tool is made,
- Size of the hole,
- Quality of surface finish that is required,
- Nature and flow rate of the cutting fluid,
- Type and condition of the drilling machine,
- Manner in which the workpiece is mounted and held.

CUTTING FEEDS FOR DRILLING OPERATIONS

Cutting feed indicates the distance a tool cuts into the workpiece for each revolution of a cutting tool. Tables of cutting feeds are provided by cutting tool manufacturers. The tables provide a recommended range of feeds according to the size of the tool. Table 15–5 gives feeds for selected drill sizes. These sizes range from the smallest number size drill to drills of 1″ diameter and larger.

The *feed indicator* on a machine that is equipped with an automatic feed device is set within the recommended feed range.

How to Drill Holes

Flat Surfaces Held in a Drill Vise

STEP 1 Select the correct size drill. Check the shank to see that it is free of burrs. Check the drill point for the accuracy of the cutting and clearance angles.

STEP 2 Determine the required speed and the feed if a power feed is to be used. Set the speed selector and the rate of feed.

STEP 3 Clamp the work-holding device securely to the table. On small-sized drills the vise may be held by hand.

Caution: The vise should be held by hand only when there is no possibility that the vise may turn. A safety precaution is to place a stop in a table slot.

STEP 4 Chuck a straight-shank drill or center drill.

STEP 5 Move the workpiece so that the center of the required hole aligns with the axis of the spindle and drill.

STEP 6 Start the machine. Bring the drill point to the work. Apply a slight force to start the drill.

STEP 7 Set the coolant nozzle if considerable heat might be generated. The cutting fluid should flow freely to the point of the drill.

> Note: In general practice, where there is limited heat, a cutting fluid is brushed on.

STEP 8 Continue to apply a force to feed the drill uniformly through the workpiece.

> Note: On deep holes (depth in excess of five to six times the diameter of the drill), feed the drill until it is the equivalent of one diameter. Then remove the drill to help clear the chips and permit the cutting fluid to flow freely into the holes. The power feed is set so that it trips off after the drill clears the workpiece on a through hole. The depth of a blind hole may be read directly on a graduated scale on the drill head.

STEP 9 Stop the machine. Remove the chips with a brush. The burrs may be cut away with a triangular scraper or a countersink.

Safe Practices in Drilling

- Check the straight and taper shanks and tangs of drills, sleeves, and drill sockets. They must be free of burrs and nicks. The tapers and tangs are checked for correct taper fit.
- Set the tapers on drills, adapters, chucks, and so on, by a firm tap with a soft-face hammer.
- Grind the cutting edges of a drill to the same length and angle. Otherwise oversized, roughly drilled, and uneven holes will be produced.
- Move (flow) chips out of the drilled hole to prevent clogging, the generation of excess heat, and drill breakage.
- Quench carbon steel drills regularly during grinding to prevent softening (drawing the temper). A burned drill must be reground beyond the softened portion.
- Avoid excessive speed, feed, heat, and force, all of which can cause rapid wearing of a drill point and drill margin or drill breakage.
- Check the lip clearance angle. Too great a lip clearance angle causes chipping of the cutting edge. Too small a clearance angle and excessive speed causes drill splitting.
- Wear a protective shield to prevent eye injury and to keep hair away from a revolving spindle and cutting tools.
- Avoid wearing loose clothing near rotating parts. Remove all wiping cloths from the drilling area before starting a machine.

UNIT 15 REVIEW AND SELF-TEST

1. a. Cite the effect the addition of cobalt has on the cutting ability of high-speed drills.
    b. Identify the kinds of materials that may be efficiently drilled with cobalt high-speed steel drills.

2. Explain the principal design feature of a coolant-feeding drill.

3. Tell what purpose a drill bushing in a drill jig serves in drilling.

4. Give three advantages of using a spiral-point drill over a chisel-point drill.

5. Differentiate between two common types of adapters for drills.

6. Indicate three checkpoints to observe in grinding drill points.

7. Name four common drill size gages.

8. List five factors a drilling machine operator must consider when selecting the cutting speed for a particular drilling operation.

9. Give two cautions to observe in grinding a drill offhand.

10. List the steps to follow in setting up a part and drilling a hole at a right angle to the axis.

11. Provide a list of four safety precautions to follow in drilling, in addition to the cautions listed in test item 9.

Machine Reaming: Technology and Processes

Reaming serves four main purposes:

- To enlarge a hole to a precise dimension,
- To form a straight (concentric) hole when the hole first is bored concentric and then reamed to size,
- To produce a smooth hole with a high-quality surface finish,
- To correct minute out-of-round conditions of a drilled hole.

UNDERSIZED HOLE ALLOWANCE FOR MACHINE REAMING

A sufficient amount of material must be left in a predrilled or machined hole to permit accurate reaming. The amounts and conditions for hand reaming to close tolerances were covered as a bench work process. The amount stated for hand reaming ranges from 0.002″ to 0.005″ (0.05mm to 0.12mm).

By contrast, larger amounts of material must be left for machine reaming. A common practice is to drill a hole 1/64″ smaller than required, or *undersized*. This practice is followed in a *one-step reaming process* with a fluted reamer. The reamer is able to correct minor surface and shape irregularities of the hole.

Table 16–1 gives maximum material allowances for five hole sizes. The sizes range from 1/4″ to 3″, and metric equivalents from 6mm to 72mm. Holes that require machining to finer dimensional tolerances and a higher quality surface finish are often produced by a *two-step reaming process*. An undersized hole is first rough reamed to within 0.002″ to 0.005″. A second fluted machine reamer is used for finish reaming.

FEATURES OF MACHINE REAMERS

The shank of a machine reamer is either straight or tapered. The taper shank has a tang. The flutes may be straight or helical (spiral) with either a right-hand or left-hand helix. The body and shank may be one piece, or the fluted body may be a separate cutting tool fitted to an arbor.

The cutting action is performed by the cutting edges. The cutting end features of a machine reamer are labeled in Figure 16–1. Each cutting edge is formed by a radial rake angle, a chamfer angle from 40° to 50°, and a chamfer relief angle. A machine reamer cuts to the outside diameter.

Table 16–1 Material Allowances for Reaming (General)

Inch Standard		Metric Standard	
Reamer Diameter	Maximum Allowance	Reamer Diameter	Maximum Allowance
1/4″	0.010″	6mm	0.2mm
1/2″	0.016″	12mm	0.4mm
1″	0.020″	24mm	0.5mm
2″	0.032″	48mm	0.8mm
3″	0.047″	72mm	1.2mm

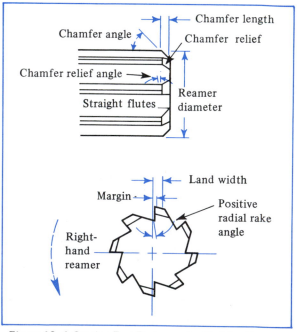

Figure 16–1 Cutting End Features of a Machine Reamer

TYPES OF MACHINE REAMERS

The following eight common types of machine reamers are described.

- General jobber's reamer,
- Shell reamer,
- Fluted chucking (machine) reamer,
- Rose chucking reamer,
- Expansion chucking reamer,
- Adjustable chucking reamer,
- Step reamer,
- Taper (machine) reamer.

Each of these types is available with a straight or taper shank and straight or spiral flutes.

JOBBER'S REAMER

The *jobber's (chucking) reamer* is a one-piece machine reamer. Jobber's reamers are general-purpose reamers.

SHELL REAMER

The *shell reamer* consists of two parts. An arbor and shank form one part. The second part is a reamer shell that includes the flutes and cutting teeth. The bore of the shell has a slight taper that fits the arbor taper. A pin (lug) in the arbor fits the two slots in the shell. Figure 16–2 illustrates straight- and spiral-fluted shell reamers and an arbor.

A shell reamer is used as a finishing reamer for dimensional accuracy. The shell is replaceable, making this reamer an economical type.

Also, the same arbor may be used with a number of differently sized shell reamers.

FLUTED CHUCKING REAMER

The *fluted chucking reamer* (Figure 16–3) is designed for cutting on the end and along the teeth. The teeth have a slight chamfer ground on the ends, for end cutting. A clearance angle is also ground along the entire length of each tooth. The fluted reamer is used for finish and difficult reaming operations.

A circular margin (outside diameter) of 0.005″ to 0.020″ (0.12mm to 0.5mm) runs the length of the flute. The lands are backed off to provide body clearance. The cutting end has a 45° bevel. Helical flutes are especially adapted to produce smooth, accurate holes and a free-cutting action. A helical fluted reamer must be used when there is an interruption in the roundness of a hole, such as a keyway or a cutaway section.

ROSE CHUCKING REAMER

A *rose chucking reamer* is a coarse-toothed reamer in comparison to a fluted machine reamer. The teeth are ground at a 45° end-cutting angle. The beveled teeth do the cutting. The rose chucking reamer is used when a considerable amount of material is to be removed. Under these conditions the rose reamer produces a rough reamed hole.

These reamers are used to rough ream undersized holes to within 0.003″ to 0.010″ (0.08mm to 0.3mm). The teeth have a back taper of 0.001″ per inch running the entire length of

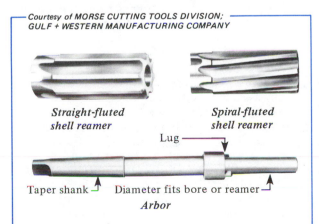

Straight-fluted
shell reamer

Spiral-fluted
shell reamer

Lug

Taper shank — Diameter fits bore or reamer
Arbor

Figure 16–2 Shell Reamers and an Arbor

Straight-shank

Taper-shank

Figure 16–3 Straight- and Taper-Shank
Fluted Chucking Reamers

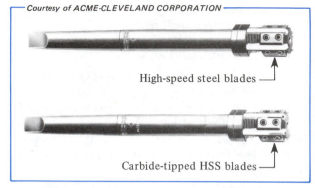
Courtesy of ACME-CLEVELAND CORPORATION

High-speed steel blades

Carbide-tipped HSS blades

Figure 16–4 Adjustable Chucking Reamers

Table 16–2 Selected Cutting Speeds for High-Speed Steel (HSS) Drills

Material	Cutting Speed Range	
	Surface Feet per Minute	Meters per Minute

Steel
    Low carbon
    (.05–30%)
	80 to 100	24 to 30

Cast iron Soft gray	100 to 150	30 to 45

Nonferrous
    Aluminum and
    alloys
	200 to 300	60 to 90

Plastics Bakelite, etc.	100 to 150	30 to 45

each flute. The back taper provides hole clearance so that the reamer does not bind in deep-hole reaming. The outside diameter is ground concentric. There is no land.

EXPANSION CHUCKING REAMER

As its name suggests, an *expansion chucking reamer* can be increased slightly in diameter. An adjustment feature on the end permits enlarging the reamer to produce a hole that is larger than a standard size. Compensation can also be made for wear. This feature increases the tool life.

ADJUSTABLE CHUCKING REAMER

Adjustable chucking reamers are produced with high-speed steel (HSS) blades or carbide-tipped, high-speed steel blades (Figure 16–4). The blades can be adjusted to increase the range of the reamer up to 1/32″ (0.8mm) in diameter. The blades are replaceable.

The carbide-tipped chucking reamer is especially adaptable to the reaming of ferrous and nonferrous castings and other materials. These materials require the cutting tool to have high abrasion-resistant qualities. The cutting tool must withstand high cutting speeds and temperatures.

STEP REAMERS

Step reamers produce two or more concentric diameters in one operation. A common step reamer has one portion of the fluted area ground at a small diameter for rough reaming, followed

by the standard diameter for final precision reaming. Other examples include the cutting of two concentric diameters which have angular or square shoulders.

TAPER REAMERS

Machine taper reamers are of similar design to hand taper reamers. The tapers conform to standard Morse, Brown & Sharpe, SI metric system, and other special tapers.

MACHINE REAMING SPEEDS AND FEEDS

MACHINE REAMING SPEEDS

The speed at which a hole is reamed depends on the following conditions:

- Required dimensional accuracy,
- Material to be reamed,
- Kind and flow of the cutting fluid on the cutting edges and workpiece,
- Desired quality of surface finish,
- Type and condition of the machine spindle,
- Size of the hole to be reamed.

The cutting speed for general machine reaming of a material is approximately 60% of the cutting speed required for drilling the same material. A table may be used to determine the correct cutting speed. Table 16–2 gives some examples of cutting speeds for HSS drills in surface feet per minute (sfpm) and meters per minute (m/min). The values are converted to cutting speeds for reaming by multiplying by 0.6 (60%).

MACHINE REAMER FEEDS

Like hand reamers, machine reamers should be fed two or three times deeper per revolution than in drilling. The amount of feed depends on the material and required quality of surface finish. As a general rule the highest rate of feed should be used to produce the desired finish. Operator judgment is important. A starting feed of from 0.0015″ to 0.004″ (0.04mm to 0.1mm) per flute per revolution may be increased by observing the cutting action, dimensional accuracy, and surface finish.

REAMER ALIGNMENT AND CHATTER

Reaming accuracy depends on alignment. The axis of the hole to be reamed, the reamer, and the machine spindle must be aligned. When a reamer bushing is used on a jig, this, too, must be aligned with the reamer axis. Any misalignment produces excessive wear on the bushing and reamer. The hole is also reamed inaccurately. A *floating reamer holder* reduces misalignment.

Chatter always affects dimensional accuracy and the quality of the machined hole. As mentioned earlier, chatter is caused by:

- Excessive cutting speed;
- Too light a feed;
- Improper grinding with excessive clearance;
- An incorrect setup in which the workpiece is not held securely;
- Play, or looseness, in the spindle or floating holder.

When chatter is produced, the operation must be stopped and the cause of the chatter corrected.

CUTTING FLUIDS FOR REAMING

The reaming of most materials requires the use of a cutting fluid. As in other machining operations, the cutting fluid is a coolant. Where necessary, the cutting fluid also serves as a lubricant. A cutting fluid helps to produce a finer cutting action and surface finish and prolongs tool life.

Single workpieces of gray cast iron are reamed dry. A stream of compressed air is used as a coolant in production reaming of cast iron. Mineral-lard oils and sulphurized oils are recommended in general reaming practices. Tables of cutting fluids and manufacturers' data are referred to in selecting an appropriate cutting fluid.

COMMON CAUSES OF REAMER BREAKAGE OR EXCESSIVE WEAR

Excessive wear and breakage of reamers may be caused by any one or a combination of the following incorrect practices or conditions:

- Burrs or other foreign particles on the surfaces of the taper socket or machine spindle;
- Misalignment of one or more parts of the setup that causes excessive wear on the lands of the reamer and produces a bell-mouthed hole;
- Cutting speeds that are too fast or too slow;
- Inappropriate composition of the cutting fluid used on the material being reamed or lack of lubrication between a guide bushing and the reamer;
- Too much or insufficient stock allowance for reaming;
- A rough, tapered, or bell-mouthed drill hole that causes a wedging action;
- Faulty grinding.

How to Machine Ream Holes

Reaming Through Holes

STEP 1 Align the hole axis and spindle axis. Center drill. Drill the hole with the reamer drill.

STEP 2 Measure the reamer diameter. Check the taper shank, spindle, and adapter (if used) for burrs or nicks.

STEP 3 Determine the reamer speed and feed. Set the spindle speed at the required sfpm and feed per revolution.

STEP 4 Guide the reamer carefully to start in the drilled hole. Apply a cutting fluid with a brush or spout container so that it reaches the cutting lips.

Note: If the reamer starts to chatter, withdraw it. Reduce the surface speed.

STEP 5 Ream through. Withdraw the reamer while it is still revolving.

STEP 6 Clean the workpiece. Check the dimensional accuracy. Reream if necessary.

STEP 7 Remove burrs from the workpiece.

Safe Practices in Machine Reaming

• Examine the shank, tang, and margins of the reamer for nicks and burrs. Nicks and burrs impair accuracy and produce scratches in the work surface. Remove burrs with an oilstone.

• Store reamers in separate storage sections to prevent the reamers from rubbing against or touching any tool, instrument, or hardened part.

• Set a taper-shank reamer in a sleeve, socket, or taper spindle by tapping firmly and gently with a soft-face hammer.

• Turn the reamer in a clockwise direction only to prevent damage to the margins, cutting edges, and work surface.

• Align the axis of the reamer, workpiece, and spindle to avoid drifting of the reamer and assure reaming to specifications.

• Strap workpieces securely. Use adequate blocking and protecting strips.

• Stop the operation whenever chatter occurs. Diagnose the cause. Take steps to remedy the condition.

• Withdraw the machine reamer while it is still turning.

UNIT 16 REVIEW AND SELF-TEST

1. Tell how the cutting edges of a machine reamer are formed.

2. Distinguish between general applications of fluted chucking reamers and rose chucking reamers.

3. Make a rule-of-thumb statement about the cutting speeds for reaming compared to the cutting speeds for drilling the same material and hole size.

4. State three considerations and/or checks by the machine operator before increasing the cutting feed for a machine reaming process.

5. List five incorrect practices requiring corrective operator action to avoid excessive reamer wear and breakage.

6. Explain the function of a floating reamer holder.

7. Set up the series of steps to be followed in producing reamed holes.

8. Tell how to safely withdraw a machine reamer to prevent scoring a finish reamed hole.

Machine Threading on a Drill Press

This unit deals with the thread-cutting tools and accessories that are generally used on drilling machines. These thread-cutting tools either tap internal threads or cut external threads. The American Unified system and the SI metric system of dimensioning screw threads are illustrated. Common tapping troubles and their causes are diagnosed. Recommended corrective measures are presented.

MACHINE TAPPING ATTACHMENTS

The skilled craftsperson often uses power to cut a thread with a tap on a drill press. The machine must be equipped with a spindle-reversing mechanism.

For such an application the tap is secured in the drill chuck and a hole is tapped. The spindle is then reversed. The tap is backed out and removed. Care must be taken to apply only a limited force and to see that the tap starts to cut immediately. Further, the tap must thread completely out of the workpiece without rubbing on the first threads.

Tapping is also done with other machine attachments and accessories that provide for quality production and reduce tap breakage. There are a number of different types of machine tapping attachments:

- All-purpose drill press tapping attachment,
- Reversing-spindle tap driver,
- Nonreversing-spindle tap driver,
- Torque-driven tapping attachment,
- Micro-tapping attachment,
- Heavy-duty tapping attachment.

An all-purpose drill press tapping attachment is shown in Figure 17–1. This tapping attachment works equally well with mild steels, tool steels, and other nonferrous metals. Lubrication systems are provided to concentrate the cutting fluid within the cutting area.

REVERSING-SPINDLE TAP DRIVER

A *reversing-spindle tap driver* is used on a drilling machine that has a reversing spindle.

The tap driver holds the tap securely while it is turned to cut the thread and while the tap is removed. Spindle direction is reversed to remove the tap.

NONREVERSING-SPINDLE TAP DRIVER

A *nonreversing-spindle tap driver* is adapted to a drilling machine where the spindle rotates in one direction only. The tapping attachment may be mounted in the drill chuck or in the tapered spindle. The tap is chucked in the attachment and fed to the workpiece. The downward force causes a *forward clutch* to engage and drive the tap. Raising the spindle lever releases the torque on the tap. A *reversing clutch* in the attachment is then engaged. The tap is backed out.

Courtesy of BUCK SUPREME INC.

Figure 17–1 An All-Purpose Drill Press Tapping Attachment

TORQUE-DRIVEN TAPPING ATTACHMENT

Conditions such as accumulation of chips in tap flutes, forcing of a tap at the bottom of a blind-hole tapping operation, and improper use of cutting fluids often require that excessive force be applied to a tap. These conditions and the resultant force may fracture the tap. Tap breakage is prevented by using a *torque-driven tapping attachment*. A *friction clutch mechanism* is provided for closely controlling the amount of torque (force) that may be applied. A *torque-setting device* is set according to the size and strength of a tap. When the torque setting is exceeded, the mechanism releases the tap to prevent tap breakage.

MICRO- AND HEAVY-DUTY TAPPING ATTACHMENTS

Microtapping attachments are designed for small taps in the #00 to #10 range. By contrast *heavy-duty self-contained tapping attachments* have speed reducers and are used for tapping operations that normally require a heavy-duty drill press. The speed-reduction feature increases the torque-driving capability of the attachment. A floating-driver feature compensates for any slight misalignment.

FACTORS AFFECTING MACHINE THREADING EFFICIENCY

CUTTING SPEED CONSIDERATIONS

There are many factors that affect the cutting speed required for machine tapping and other thread-cutting processes. Each of the following factors must be considered before the craftsperson makes a judgment about cutting speed:

- Kind of material to be tapped,
- Hardness and heat treatment of the workpiece,
- Pitch of the thread,
- Length of the threaded section,
- Chamfer in the hole or the beveled edge of the workpiece,
- Required percent of full thread depth,
- Quality and fit (class) of the thread,
- Quantity and type of cutting fluid,

- Design of the drilling machine and the machine tapping attachment.

The recommended cutting speeds for general machine tapping are given in Appendix Table A–12. These cutting speeds must be adjusted to compensate for the preceding factors. In addition the following conditions must be considered:

- As the length of a tapped hole increases, the cutting speed must be *decreased* (the chips tend to accumulate and thus prevent the full flow of the cutting fluid and produce additional heat on the cutting edges);
- Taps with long chamfers (taper taps) may be run faster than plug taps in holes that are tapped for a short distance (plug taps have just a few chamfered threads);
- Plug taps may be run faster than taper taps in holes that are to be tapped to a great depth;
- A 75% depth thread may be cut at a faster speed than a full-depth thread;
- The cutting speed for coarse-thread series taps that are larger than 1/2″ is slower than fine-thread series taps of the same diameter;
- The cutting speed for taper thread (pipe) taps is from one-half to three-quarters of the speed used for straight thread tapping;
- Greater tapping speeds are employed when tapping is automatically controlled (versus manual operation).

GENERAL CUTTING FLUID RECOMMENDATIONS

Cutting fluids serve the same functions in machine tapping as they do in reaming, countersinking, counterboring, and other machine processes. Friction is reduced, tap life is increased, surface finish is improved, and thread dimensions may be better controlled when the correct cutting fluid is used. Table 17–1 lists a number of ferrous and nonferrous metals and their recommended cutting fluids.

Plastic and cast-iron parts are machine tapped dry or with compressed air. The air jet removes the chips and cools the tap.

Table 17–1 Recommended Cutting Fluids for
Tapping Selected Metals (Partial Table)

Metal	Recommended Cutting Fluid for Machine Tapping
Ferrous	
Plain carbon and alloy steels	Sulfur-base oil with active sulfur
Malleable iron	
Monel metal	
Tool steel	Chlorinated sulfur-base oil
High-speed steel	
Stainless and alloy steels (heat treated to a higher degree of hardness)	
Nonferrous	
Aluminum, brass, copper, manganese bronze, naval brass, phosphor bronze, Tobin bronze	Mineral oil with a lard base
Aluminum and zinc die-casting metal	Lard oil diluted with up to 50% kerosene

ANALYSIS OF COMMON TAPPING PROBLEMS

Table 17–2 analyzes the causes of eight common tapping problems and suggests corrective measures that may be taken for each problem.

FEATURES OF MACHINE TAPS

Machine taps are similar to hand taps except for two major differences. First, the shank end of a machine tap is shaped to fit the tapping attachment holder. Instead of the square head that provides a gripping surface for hand taps, the whole shank of a machine tap is round. The shank with a groove cut lengthwise fits a particular size and shape of chuck.

Second, machine tapping requires the use of a high-speed steel (HSS) tap. The HSS tap withstands higher operating temperatures and speeds than a carbon steel hand tap.

CHECKING AND MEASURING A THREADED PART

Internal and external threads may be measured in various ways. When a thread is machine tapped or cut with a die on a drilling machine, its size may be checked with a bolt, nut, or the mating part. Threads that are held to closer tolerances may be gaged with *thread plug* or *ring gages*. The pitch diameter of a thread may be measured with a thread micrometer or by using three wires and a standard micrometer.

How to Machine Tap (Drill Press)

STEP 1 Select the size and type of machine tap to use.

STEP 2 Check the drawing specifications for the percent of full depth of thread required. Check the size of the tap drill hole.

STEP 3 Select a tapping attachment that accommodates the tap size and is appropriate to the design of the drill press.

STEP 4 Mount the tapping attachment in a chuck or in the tapered spindle. Adjust the torque.

STEP 5 Set the spindle speed. Position the cutting fluid nozzle.

STEP 6 Align the axis of the tapping attachment and tap with the center line of the workpiece. Check to see that the work is held securely.

STEP 7 Start the drill press. Bring the tap carefully to the chamfered edge of the drilled hole. Apply a slight force so that

Table 17–2 General Tapping Problems: Causes and Corrective Action

Problem	Probable Cause	Corrective Action
Tap breaking	—Tap drill size too small	—Use correct size tap drill
	—Dull tap	—Sharpen cutting edges
	—Misalignment	—Check holder and tap; align concentric with axis of tapped hole.
	—Excessive force in bottoming	—Drill tap hole deeper if possible —Reverse the tap direction earlier —Correct the torque (driving force) or replace the machine tapper
	—Tapping too deep	—Use spiral point or serial taps
Teeth chipping	—Chips loading on cutting edges	—Back the tap to break the chips —Check the cutting fluid
	—Jamming the tap at the bottom in blind hole tapping	—Correct the reversing stop or reverse sooner —Drill a deeper hole
	—Chips packed in the blind hole	—Remove and clean the tap during tapping
	—Work hardening	—Check prior hole-producing processes
Excessive tap wear	—Sand and abrasive particles	—Machine tumble or wire brush to remove foreign matter
	—Incorrect or inadequate quantity of cutting fluid	—Consult chart for appropriate cutting fluid —Position the lubricant nozzle so that the cutting fluid reaches the cutting edges.
	—Worn and dull tap	—Sharpen a dull tap —Replace a worn tap
Undersized threads	—Enlargement and shrinking of workpiece during tapping	—Use an oversized tap —Sharpen the cutting faces so that the tap cuts freely
Oversized threads	—Loading	—Check both the quality of the cutting fluid and the quantity that reaches the cutting edges.
	—Misalignment	—Align the tap and work axes before starting to tap
	—Worn tapping attachment or worn machine spindle	—Adjust the play in the spindle or tapping attachment
Bell-mouthed hole	—Excessive floating of machine spindle or tapping attachment	—Position the machine spindle and tapping attachment accurately
	—Misalignment	—Align the axes of the tap and workpiece
Torn, rough threads	—Incorrect chamfer or cutting angle	—Grind the starting teeth and cutting angle properly on a tool and cutter grinder
	—Dull tap teeth	—Grind the cutting faces of the tap
	—Loading	—Back the tap and remove the chips
	—Inadequate cutting fluid	—Consult the manufacturer's recommendations for the appropriate cutting fluid
Wavy threads	—Misalignment	—Align the axes of the spindle, tap, tap holder, and workpiece
	—Incorrect thread relief or chamfer of teeth	—Grind the cutting edges and chamfered teeth concentrically

the tapered threads of the tap begin to engage.

STEP 8 Continue to cut the thread. Continuously apply a cutting fluid.

STEP 9 Back the tap out of the hole. Brush away the chips in deep-hole or blind-hole threading. Chips tend to accumulate.

> **Note:** Avoid using any force when withdrawing a tap, particularly when the last thread is reached.

Safe Practices in Machine Threading and Tapping

- Align the axes of a thread-cutting attachment, tap or die, and workpiece. Alignment helps to prevent damage to a tap or die and produces an accurately threaded part.

- Direct an adequate supply of cutting fluid to reach the cutting edges of a threading tool. A cutting fluid improves tool life, increases cutting efficiency, and helps to produce a quality surface finish.

- Avoid applying excessive force or jamming of a tap. Excessive force and jamming cause a tap to fracture.

- Free a drill press spindle so that a tap, when once started, works freely into or out of a workpiece.

- Observe general safety precautions related to rotating machines and the safe handling of cutting tools.

UNIT 17 REVIEW AND SELF TEST

1. Describe how a revolving die head works.

2. Compare the operation of torque-driven and heavy-duty tapping attachments.

3. a. List three factors the craftsperson must consider for machine tapping.
   b. List three conditions that govern the selection of cutting speeds for machine tapping.

4. State what effect the use of a correct cutting fluid has on machine threading.

5. Suggest three general safety precautions that must be followed to cut quality external threads and to prevent damage to a tap, die, or workpiece

Countersinking, Counterboring, Spotfacing, and Boring

The outer surface of a hole may be cut away at an angle, recessed with a shoulder, or machined with a flat circular area. The general machining processes that produce such shapes on the outer surface are called countersinking, counterboring, and spotfacing. A hole may also be enlarged by boring. These processes and the cutting tools and accessories that are used in each process are treated in this unit.

COUNTERSINKS AND COUNTERSINKING

Countersinking refers to the machining of a cone-shaped opening or a recess in the outer surface of a hole. The angle of the recessed indentation corresponds with the angle of a screw, a rivet head, or other tapered object that seats against the angular surface. A hole that is machined at an angle is called a *countersunk hole.* The tool that produces this cone shape is a *countersink* (Figure 18–1).

Standard countersinks are manufactured with an included angle of either 60° or 82°. Countersinks may be made of carbon tool steel, high-speed steel, or carbides. Holes may be countersunk by chucking and revolving a counter-

sink. The depth of cut is controlled with a stop gage or by measuring across the outside (large) diameter. The mating part may also be tried in the countersunk hole. Countersinking should be performed at about one-half the speed used in drilling the hole.

Chatter is produced when the speed is too great, the workpiece is insecurely held, or the countersink is forced.

COMBINATION DRILL AND COUNTERSINK

One widely used shop practice for accurately guiding a large drill is to drill a combination pilot and countersunk hole. The pilot and countersunk hole provides a bearing surface for 60° angle machine centers. This practice is known as *center drilling.*

A *combination drill and countersink* is used for center drilling. The point of the cutting tool is positioned in a center-punched hole. The small-diameter point end produces a pilot hole. The angular surface of the countersink portion enlarges the hole and forms a 60° cone-shaped surface. The countersunk hole is drilled to the depth at which the outside diameter is smaller than the required finished hole size.

Combination drill and countersinks are available either with a single cutting end or with double cutting ends. Center drill tips are also available. These drill tips fit into adapters from which they may be easily removed and replaced. A single-point center drill tip and an adapter are illustrated in Figure 18–2. Center drills are made of carbon tool steel, high-speed steel, or carbide. The small-diameter point ends are fragile. Care must be taken not to exert excessive force on them during the drilling process. The feed per revolution is less on the countersink portion than for normal drilling.

Combination drill and countersinks are produced with many sizes of points and outside diameters.

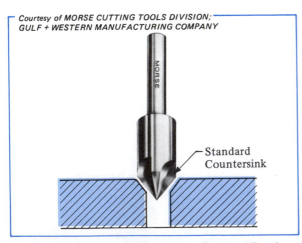

Courtesy of MORSE CUTTING TOOLS DIVISION; GULF + WESTERN MANUFACTURING COMPANY

Standard Countersink

Figure 18–1 Countersinking to a Required Depth

Courtesy of DoALL COMPANY

— Adapter

— Single-point center drill tip

Figure 18–2 A Center Drill Tip and Adapter

COUNTERBORES AND COUNTERBORING

Counterboring refers to the process of enlarging and recessing a hole with square shoulders. An enlarged and recessed hole that has square shoulders is known as a *counterbored hole*. Holes are counterbored to permit pins with square shoulders, screw heads, or other mating parts to fit *below* the surface of a workpiece. The cutting tool that is used for enlarging a hole is called a *counterbore*. A counterbore consists of a shank, body, and pilot.

The shank may be ground straight or to a standard Morse taper. A taper shank standard Morse taper counterbore with pilot are illustrated in Figure 18–3.

Different sizes of pilots may be used with one counterbore. Other counterbores that are used for only one size of screw head are made solid. Counterbores are usually three or four fluted. Counterbores are available in outside diameter sizes that vary, by increment of 1/16",

from 1/4" to 2". Similar metric sizes range from 6mm to 50mm.

The cutting action of the counterbore is produced by the cutting lips. These lips are formed on the end (face) of each land. The cutting lips are radial. The lips are relieved to form the cutting angle. The flutes are spiral and provide for the flow of chips out of the workpiece.

Counterbores made of high-speed steel are widely used. Carbide-tipped counterbores are practical where tough materials are to be counterbored. Rough operations, where the outer surface of a workpiece is uneven or hard (as in the case of scale on castings), require a carbide-tipped counterbore.

General counterboring operations are performed at a slower speed than the speed used for drilling. A cutting fluid must be used on materials that require a lubricant.

DRILL/COUNTERBORE COMBINATION

Drilling and counterboring for socket-head and other standard fasteners may be performed in one operation. A short-length *drill/counterbore combination* is designed for such an operation. The drill/counterbore may be a solid type with a body of regular length or it may be a tip type.

The tip type fits into an adapter body. The body has a slightly tapered hole that holds and secures the tip. The adapter has a hole drilled through the body. A knockout rod is inserted into the hole and tapped against the tip to remove it. A typical set of combination drill/counterbore tips accommodates socket-head cap screw sizes #5, 6, 8, and 10, and 1/4", 5/16", 3/8", 7/16", and 1/2". Holes may be drilled and counterbored to diameter tolerances of +0.005".

SPOTFACERS AND SPOTFACING

Spotfacing refers to the machining of a flat surface that is at right angles (90°) to a hole. Spotfacing produces an area against which a square-shouldered part may fit accurately. The operation may be performed on a drilling machine. The end-cutting tool is called a *spotfacer*. The surface is said to be *spotfaced*.

A counterbore may be used as a spotfacer when its diameter is large enough to produce a spotfaced area of the required dimension.

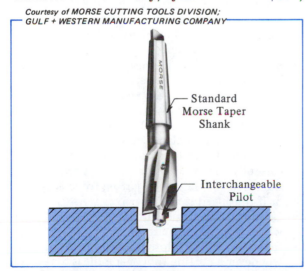

Courtesy of MORSE CUTTING TOOLS DIVISION; GULF + WESTERN MANUFACTURING COMPANY

— Standard Morse Taper Shank

— Interchangeable Pilot

Figure 18–3 Required Dimensions Met by the Pilot and Outside Diameter of a Carbide-Tipped Counterbore

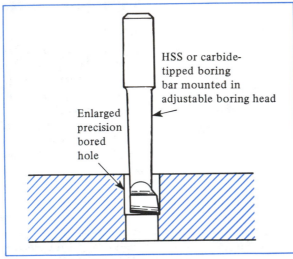

HSS or carbide-tipped boring bar mounted in adjustable boring head

Enlarged precision bored hole

Figure 18–4 Boring to Enlarge a Hole

BORING AND BORING TOOLS

Boring describes the process of enlarging a hole by using a *single-point cutting tool* (Figure 18–4). The cutting edge may be adjusted to accommodate different work sizes. Holes that have been formed by casting, punching, drilling, or any other process may be bored.

The cutting tool may be held in a *boring bar* or a *boring head*. The amount the tool is adjusted is indicated on a graduated scale. The boring process permits the taking of heavier finish cuts than is possible with reamers. The single-point cutting tool is adjustable to bore any diameter within the range of the boring head.

The taper shank of a boring head may be inserted directly in the taper bore of the drill press spindle. The cutter is secured in the offset section of the boring head. Small holes are usually bored on drilling machines. The same tables of recommended cutting speeds that are used for similar boring operations on lathes and boring mills are used.

How to Counterbore a Hole

STEP 1 Select a counterbore with a pilot that is a few thousandths of an inch smaller than the drilled or reamed hole.

STEP 2 Determine the cutting speed, spindle RPM, and feed. Set the machine speed and feed.

Note: Reamer speeds for the same diameter may be used.

STEP 3 Position the pilot in the hole. Bring the cutting edges carefully to the work surface. Set the spindle stop to the required depth.

STEP 4 Apply cutting fluid to the cutting area. Start to counterbore. Continue until the hole is counterbored to the required depth.

Note: It is good practice during the process to periodically withdraw the counterbore and check the cutting action and quality of surface finish.

STEP 5 Clean the workpiece. Burr the edge of the counterbored hole.

Safe Practices in Countersinking, Counterboring, Boring, and Spotfacing

- Fit the spindle taper and cutting tool shank and tang properly. Securely seat them. Use a soft-face hammer to carefully drive the tapers together.
- Rotate all right-hand cutting tools in a clockwise cutting direction.
- Examine the cutting edges of the cutting tool and stone any burrs.
- Stop the machine if there is chattering. Correct the causes of chatter.
- Use goggles. Cover the workpiece and tools if compressed air is used. Direct the air blast downward and away from other workers.
- Feed the small-diameter cutting point of a combination drill and countersink slowly and carefully into a workpiece. The point may be fractured if a heavy or jarring force is applied or if the speed is too fast.
- Use a brush to apply cutting fluid to the point where the cutting action is taking place.
- Remove wiping cloths from the work area whenever a cutting tool or spindle is to revolve.
- Secure loose clothing before operating any rotating machinery.

UNIT 18 REVIEW AND SELF-TEST

1. State two main functions served in the machine shop by holes that are countersunk to either 60° or 82°.

2. Differentiate between a combination drill and countersink and a center drill tip.

3. Indicate two functions that are served by an interchangeable pilot on a counterbore.

4. Tell what the difference is between spotfacing and boring with respect to the cutting tools that are used for each process.

5. List the steps for boring a hole when using a drill press.

6. List three safety precautions to observe when working with countersinks, counterbores, spotfacers, and boring tools.

PART 6 Turning Machines: Technology and Processes

Engine Lathes: Design Features and Operations

The lathe is used for *external* and *internal cylindrical machining processes*. Basic external operations include facing, turning, knurling, cutting off, threading, and polishing. Basic internal operations relate to drilling, boring, turning, undercutting, countersinking, counterboring, reaming, tapping, and threading.

All of these internal and external operations are covered in detail. Principles and applications of cutting tools, work-holding and driving devices, and machine setups are described. Cutting speeds, feeds, cutting fluids, and other machining practices are also included.

Lathe work skills, the principles of machining, and related technology may also be applied to other machine tools. Turret lathes, hand production and automatic screw machines, boring machines, and vertical turret lathes (some of which are numerically controlled) are adaptations of the lathe. Common production turning machines are covered in Section Two.

UNIT 19

Engine Lathes: Functions, Controls, and Maintenance

A. FUNCTIONS, TYPES, AND FEATURES

The engine lathe remains a basic machine tool. A modern, conventional lathe with major features identified is shown in Figure 19–1. The parts and mechanisms of a lathe serve one of the following three major and interrelated functions:

- To provide power to all moving parts,
- To hold and rotate a workpiece,
- To position, hold, and move a cutting tool.

THE LATHE BED

The body to which other components of a lathe are fitted is called the *lathe bed*. It consists

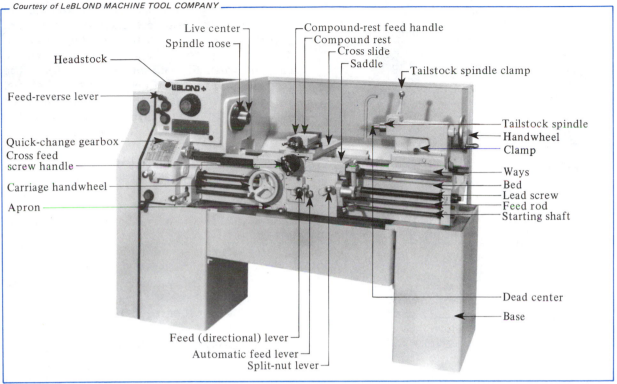

Courtesy of LeBLOND MACHINE TOOL COMPANY

Live center
Spindle nose
Compound-rest feed handle
Compound rest
Cross slide
Saddle
Tailstock spindle clamp
Headstock
Feed-reverse lever
Tailstock spindle
Handwheel
Clamp
Quick-change gearbox
Cross feed screw handle
Ways
Bed
Lead screw
Feed rod
Starting shaft
Carriage handwheel
Apron
Dead center
Base
Feed (directional) lever
Automatic feed lever
Split-nut lever

Figure 19–1 Major Features of a Conventional Lathe

of a ruggedly designed casting or weldment. *V-ways* and other *ways* are machined on the top surface of the bed. A headstock, gearbox, lead screw and feed rod drive, and coolant system are usually attached to the lathe bed. The carriage and tailstock mechanisms are aligned on the ways of the bed and may be moved longitudinally.

MOTION-PRODUCING MECHANISMS

Motion for all moving parts of a lathe is transmitted from a power source such as a *motor*. The motor is connected by belts or gears to a *headstock* through which a spindle is driven. The speed and feed of each moving part of the lathe is directly related to the spindle movement.

End gears are used to establish the precise RPM of the spindle in relation to the *gear train* in the *quick-change gearbox*. The gear ratios, in turn, control the RPM of the *lead screw* and *feed screw* according to the spindle RPM.

The rotary motion and speed and feed of the lead screw are converted in the *apron*. Linear

movements are produced for the *carriage* (longitudinal feed) and *cross slide* (transverse feed). These same movements of the carriage and cross slide may also be performed by hand feeding. On some lathes the lead screw serves the function of controlling the movement for both threading and feeding.

LATHE COMPONENTS FOR PRODUCING WORK AND TOOL MOVEMENTS

The headstock and the spindle serve a work-rotating function. The end gears, quick-change gears, and carriage mechanism are all related to tool movement functions. The tailstock serves a work- or tool-supporting function as well as a rotating function.

THE HEADSTOCK

The headstock is located at the head, or left-hand end, of the lathe bed. The headstock contains the spindle drive mechanism and the lathe spindle.

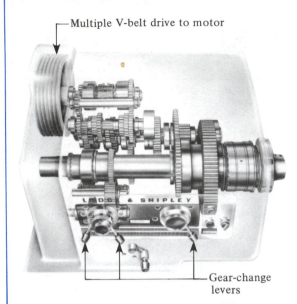

—Multiple V-belt drive to motor

Gear-change levers

Figure 19–2 A Phantom View of a Gear-Drive Headstock

Power and motion from a motor are transmitted through the headstock drive mechanism to the lathe spindle. The spindle may be driven by *multiple V-belts*, a *variable-speed drive*, or by *gear transmission*. Figure 19–2 is a phantom view of a gear-drive headstock.

The spindle speed of small cone-pulley belt-driven lathes is controlled by the position of the belt on one of the step-cone pulleys. The drive may be direct. Speed changes on belt drives are made by shifting the belt to a different step-cone pulley combination.

Some variable-speed drives provide for a wide range of speeds by engaging a set of *back gears*. Still other variable speed drives may be controlled electrically or mechanically while the lathe is in motion.

Heavy powered lathes have completely geared headstocks. The gears are engaged in a manner similar to shifting an automobile transmission. The gear ratio on a geared-head lathe is changed by *speed-change levers*. Spindle speed changes must be made when the gears are *not* in motion to prevent stripping the gear teeth. *Gear* combinations *(trains)* provide double, triple, quadruple, and other spindle speed ranges.

On the end of the headstock there is an *end gear train*. The gears are enclosed by modern guards that are provided with a *safety electrical disconnect switch*. The switch shuts down the machine when the end guard is removed. The gears within the geared headstock are of special alloy steel. The gears are induction hardened and precision ground. All parts are lubricated with an oil bath system. The oil level is sighted on an oil level indicator.

The required spindle speed is obtained by setting the speed-selector levers. The spindle is actuated by a *forward/stop/reverse lever*. This lever is located on the apron to provide on-the-spot control by the operator from any operating position.

New machines are being designed for increases in spindle speeds. Fluid and electro-magnetic devices, variable-speed motors, and other mechanisms are being used to meet increased machine tool requirements.

THE HEADSTOCK SPINDLE

The dominant feature of the lathe is the spindle. The condition, accuracy, and rigidity of the spindle affect the precision of most machining operations and the quality of surface finish. The headstock spindle is mounted in opposed, preloaded, precision antifriction bearings. *Spindle runout*, or *eccentricity*, is reduced to a minimum. The spindle is hollow to permit small parts, bar stock, and lathe attachments to pass through it. The face end, or nose, of the spindle is accurately ground with an internal taper. Adapters are fitted to reduce the size of the spindle hole to permit accessories, like a sleeve for the live center, to be inserted. Spindles are hardened and ground.

SPINDLE NOSE TYPES

The spindle nose is accurately ground to receive nose mounting plates. These plates align chucks and faceplates in relation to the spindle. The three basic designs for the outside of a spindle nose include the

- Older threaded spindle nose with a squared shoulder,
- Long taper spindle nose, and the
- Cam-lock spindle nose.

The *threaded spindle nose* receives a threaded mounting. The shoulder helps to align the mounted device. The threads hold and secure the device in place and transfer the spindle driving force.

The *long taper spindle nose* provides a tapered surface. This surface receives the corresponding taper of a spindle nose mounting plate. A key transfers the spindle driving force to the nose mounting. The threaded ring on the spindle screws onto a matching thread. The threaded ring draws and securely holds the mounting on the taper. A spanner wrench is used to tighten or loosen the threaded ring nut.

The *cam-lock spindle nose* has a short taper. This taper accurately positions the spindle mounting. The spindle mounting has cam studs that fit into a ring of holes on the face. The spindle mount is held securely on the taper and against the face of the spindle lock nose by the cam studs. These studs are secured by turning a chuck key. The spindle mounting is driven by the cam studs.

END GEARS AND QUICK-CHANGE GEARBOX

The gears on the end of the headstock between the spindle gear and the drive gear on the gearbox are called *end gears*. End gears serve two functions:

- To transmit motion from the spindle to the lead screw and feed screw through the gearbox,
- To change the direction of the lead and feed screws.

The relationship between the spindle speed and the lead and feed screws is established through a series of gears. The gears in a *quick-change gearbox* are engaged by levers.

A gear-change mechanism for thread pitches (Unified and SI metric) is pictured on the lathe headstock in Figure 19–3. Gear shifting is done by positioning one or more of the three knobs *when the spindle is stopped*. The range of Unified threads that may be cut on this lathe is from 11 to 100 threads per inch. The metric thread combinations may be set from 0.275mm to 2.7mm pitches.

An *index plate* on a quick-change gearbox indicates the various lever positions for setting

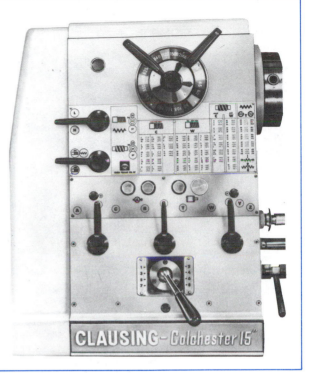

CLAUSING - *Colchester 15"*

Figure 19–3 Feed and Thread Pitch Change Knobs (Unified and SI Metric Systems)

the gearbox for either feed or thread cutting. Pitches and feeds are given on the quick-change gearbox plate in terms of both inch- and metric-standard measurements.

THE CARRIAGE

The lathe cutting tool is secured, moved, and controlled by different attachments and mechanisms on the lathe carriage. The carriage consists of a *saddle*, *cross slide*, and *apron* (Figure 19–4). With this combination, a cutting tool may be fed into or away from a workpiece. The rotary motion of the lead screw and the feed screw is converted to a horizontal motion. The movement may be across the length of a workpiece and/or at a right angle to it. A *compound rest* is attached to the cross slide to produce small tapers and other angle surfaces.

The Apron. The controls and mechanisms for all movements of the carriage are housed in the apron. A selector lever positions the proper

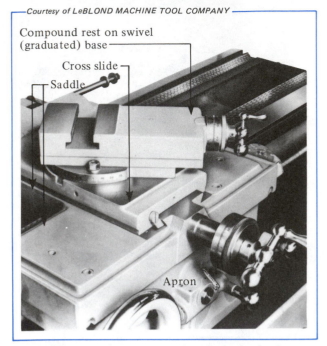

Compound rest on swivel (graduated) base

Cross slide

Saddle

Apron

Figure 19–4 Major Design Units of a Lathe Carriage

gears for longitudinal and cross feeds. A friction clutch engages the feed. A *half-nut lever* permits engaging the lead screw for thread cutting.

The Saddle. The saddle supports the apron and the cross slide, on which a compound rest is mounted. The cross slide mechanism provides tool movement along the work axis or at right angles to it. The crosswise motion is produced whenever the cross slide feed screw is turned either by hand or power.

The Cross Slide and Compound Rest. The compound rest is mounted on the cross slide. The compound rest may be swiveled through 360° to any angular position. The base is graduated in degrees to permit angular adjustments. The compound rest may also be swung to conveniently position a cutting tool. The compound rest is fed by hand.

The basic functions of the compound rest include:

- Cutting angles and short tapers,
- Moving the cutting tool to face to close tolerances (when set parallel to the ways of the lathe),
- Feeding the tool in thread cutting.

Cross Slide (Traverse) and Compound Feed Dials. The amount the cross slide is moved may be measured on a cross slide feed dial. Similar measurements for compound rest movements are read on a compound feed dial.

The feed dials are graduated according to the precision built into the lathe and the general job requirements for which it was designed. Some feed dials have readings in thousandths of an inch (0.001″) or two-hundredths of a millimeter (0.02mm). Higher precision lathes have dials (collars) that read to one ten-thousandth of an inch (0.0001″) and one one-hundredth of a millimeter (0.01mm).

Many new lathes have dual inch-standard and SI metric-standard dials. These dials are located on the cross slide and compound rest feed screws and on other lathe micrometer measuring devices.

THE TAILSTOCK

There are two units to a tailstock: a *base* and a *head.* The base may be positioned quickly and secured *longitudinally* along the bed of the lathe. The head may be moved *transversely* at right angles to the lathe central axis. These movements position the tailstock either *on center* (the headstock and tailstock axes aligned) or *off center* (for turning tapers). A *zero index line* and other graduations are machined on the base and head sections. These graduations are used to center or offset the tailstock.

The head includes the *tailstock spindle.* The spindle has a standard Morse internal taper. A *dead center* is secured in the tapered spindle.

LATHE ACCESSORIES FOR HOLDING WORK OR TOOLS

TOOLHOLDERS (TOOL POSTS)

Cutting tools are held and positioned in holders by three general types of cutting-tool attachments. These attachments are called *tool posts.*

Type 1 is designed for a single operation at one setting. The cutting tool is held in a holder. The holder is positioned in a *tool block* (rectangular) or *tool post* (round) holding device.

Type 2 provides for multiple operations (Figure 19–5). Each tool is adjusted in the holding device. Up to four successive operations are performed by a quick-release and positioning lever. This type of cutting-tool holder and positioning device is often called a *four-way turret*.

Type 3 is a *precision* block and toolholder that permits rapid tool changes. The cutting tools are *preset* in each holder. The cutting tool and holder are accurately positioned on a locating pin and in an accurately slotted tool block (Figure 19–5).

Each of the three basic types of toolholders has a T-shaped base. The base fits the T-slot of the cross slide. The base is tightened and held securely in the T-slot.

LATHE CENTERS

Lathe centers are available with tapered bodies. The *live center* requires a tapered adapter or sleeve. The center and sleeve are held directly in the headstock spindle. The *dead center* is mounted in the tailstock spindle. A workpiece revolves on the dead center.

Both live and dead centers may be of the solid, hardened-steel type. Carbide-tipped dead centers are available. These centers have the ability to withstand high temperatures caused by friction. However, it is often desirable to use a revolving dead center to reduce friction.

THE FACEPLATE

A *faceplate* is a common work-holding device (Figure 19–6). The elongated slots permit unusually shaped workpieces to be positioned and strapped securely against the vertical (90° to the lathe axis) plane surface of the face plate.

Care must be taken in faceplate work to use balancing weights. These weights offset any

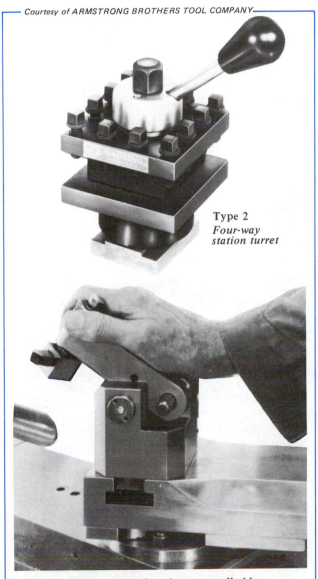

Courtesy of ARMSTRONG BROTHERS TOOL COMPANY

Type 2
Four-way station turret

Type 3 *Precision block and preset toolholders*

Figure 19–5 Basic Types of Tool-Post Holders

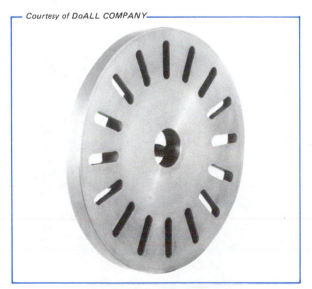

Courtesy of DoALL COMPANY

Figure 19–6 A Face Plate

centrifugal forces that may be created by the unequal distribution of a mass.

THE DRIVER PLATE

A great deal of work is machined on a lathe between centers. A centered workpiece is driven by a *lathe dog* and *driver (dog) plate*. The flat, flanged driver plate fits on and is secured to the spindle nose. The tail of a lathe dog rides in one of the two radial slots to drive a centered workpiece.

LATHE ATTACHMENTS

There are five general lathe attachments. Their names indicate their prime functions: taper attachment, threading attachment, micrometer stop, center rest (steady rest) and follower rest, and rapid traverse mechanism.

TAPER ATTACHMENT

The *taper attachment* is one of the most accurate, fastest, and practical methods of cutting a taper on a lathe (Figure 19–7). An adjustable angle attachment is clamped against one of the back ways. The mechanism consists of an adjustable block that clamps to the cross slide.

The block moves at a fixed rate according to the angular (tapered) setting of the attachment. This produces the same movement in the cross slide. As the longitudinal feed is engaged and the work turns, a taper is produced. There are two sets of graduations on the tapered attachment. These graduations read in *degrees of taper* and *inches of taper per foot* (or metric equivalents). On the model illustrated, graduations at the headstock end provides an accurate vernier setting.

THREADING ATTACHMENTS

The *thread-chasing dial* is one type of threading attachment. It is fitted on the side of the apron. The chasing dial rotates freely when the split nut is not engaged for threading. The dial stops rotating when the split nut is engaged.

Another common threading attachment is the *high-speed threading attachment*. The attachment shown in Figure 19–8 fits on the side of the apron. It contains its own split nut and a tripping mechanism. An adjustable stop is preset for the length of thread that is to be cut. The split nut automatically disengages each time the preset length is reached. The feed-reverse lever is used to change the direction of the lead screw to right- or left-hand threads.

Courtesy of DoALL COMPANY

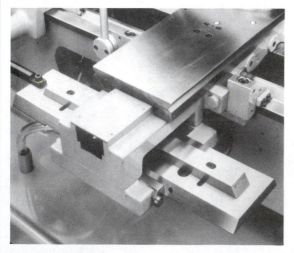

Figure 19–7 A Taper Turning Attachment

Courtesy of DoALL COMPANY

Figure 19–8 Automatic, Adjustable Stop, High-Speed Threading Attachment

MICROMETER STOP

The *micrometer stop* is a convenient device consisting of a single or multiple stop micrometer head and an attachment for securely clamping the stop on one of the front ways. A micrometer setting provides an accurate stop. The carriage is brought to the micrometer stop when duplicating multiple pieces or when finishing parts or surfaces to exact lengths.

STEADY (CENTER) REST AND FOLLOWER REST

A long workpiece must be supported at regular intervals and continuously while a cut is being taken. The *steady rest (center rest)* performs this function. It helps to prevent springing of a workpiece under the cutting action of a cutting tool. The steady rest is secured to the ways at a point along a workpiece that provides maximum support.

The *follower rest* performs a similar function except that it follows the cutting action. The follower rest is attached to the saddle and therefore moves along the work (Figure 19–9). The work-supporting shoes of the follower rest support a revolving workpiece at two places. The shoes prevent the forces of the cutting tool against a workpiece from springing the workpiece.

RAPID POWER TRAVERSE

Considerable time is saved, particularly on heavy models of lathes, by the addition of a *rapid traverse attachment* (Figure 19–10). The attachment relates to increased speed in making forward and reverse longitudinal movements of the carriage and the cross slide.

Rapid power traverse is controlled through longitudinal and cross feed friction levers. A safety clutch is incorporated to automatically disengage one of the feeds and thus prevent overloading.

LATHE SIZES AND SPECIFICATIONS

Lathe models are usually designated by specifications such as *swing over bed* and the *center-to-center distance between the headstock and tailstock* (Figure 19–11). For example, a 13″ (330mm) by 25″ (635mm) general-purpose lathe will swing a 13″ (330mm) diameter workpiece over the bed. However, only an 8 1/4″ diameter (210mm) part will clear the cross slide. The 25″ (635mm) center distance will accommodate a maximum length of 25″ between the centers of this lathe.

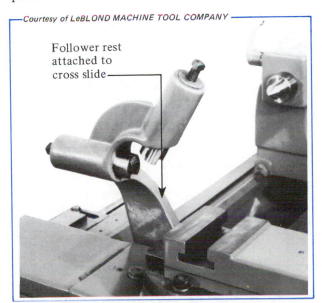

Courtesy of LeBLOND MACHINE TOOL COMPANY

Follower rest attached to cross slide

Figure 19–9 A Follower Rest

Courtesy of DoALL COMPANY

Figure 19–10 A Rapid Traverse Attachment

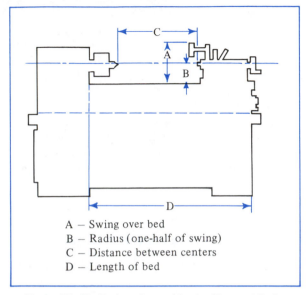

A — Swing over bed
B — Radius (one-half of swing)
C — Distance between centers
D — Length of bed

Figure 19-11 Designations of Lathe Size and Swing

CATEGORIES OF LATHES

Lathe models may be grouped in three broad categories: all-purpose lathes, production lathes, and special production lathes.

ALL-PURPOSE LATHES

Small sizes of *all-purpose* lathes are mounted on benches and are designated as *bench* lathes. Lathes of still smaller size that are used by instrument makers are called *instrument* lathes. Larger-sized (diameter and length) lathes like 10″ to 48″ (250mm to 1200mm) models are known as *engine* lathes.

A high-precision model may be specified as a *toolroom*, or *toolmaker's*, lathe. The toolroom lathe has a wide range of speeds, especially high speeds, and is usually equipped with a complete set of accessories. The tolerances of all parts of the toolroom lathe are held to closer limits than the limits of a standard lathe. The toolroom lathe makes it possible to machine to the high degree of accuracy required in toolmaking, instrument making, and other precision work.

A lathe that has a swing of 20″ or larger and is used for roughing (hogging) and finishing cuts is often referred to as a *heavy-duty* lathe. The swing (capacity) of a lathe may be increased. A

section of the lathe bed adjacent to the headstock may be cut away to permit large diameter workpieces to be turned. In other words, there is a gap in the bed next to the headstock. This lathe is referred to as a *gap* lathe.

PRODUCTION LATHES

Large quantities of duplicated parts that require lathe operations to be performed at high-production rates are produced on adaptations of the lathe. *Turret*, *automatic*, and *vertical turret* lathes are three common types of *production* lathes. The design of these machine lathes, multiple tool setups, simultaneous cutting with several tools, and other processes and advanced technology are treated in detail in Units 32 and 33.

The wide range of turret lathe operations has been further extended by programming using numerical control and multiple tool heads.

SPECIAL LATHE EXAMPLES

Two examples of *special production* lathes are the *duplicating* lathe and the *floturn* lathe. These lathes perform processes (contour turning and roll flowing) that increase the range of common lathe work.

Duplicating Lathe. The duplicating, or contour, lathe is used in machining irregular contours and blending sections. The cutting tool movement is guided by air, hydraulic, mechanical, or electrical devices. The shape that is produced is the result of a tracing device. This device follows the contour of a *template* and controls the movement of the cross slide tool.

Floturn Lathe. The floturn lathe is used for *roll flowing* as a method of cold-forming metal. A heavy pressure is applied spirally with two hardened rollers against a metal block. The lathe and the floturn process may be used with flat plates and forgings and castings and for extruded and other shapes.

The metals may range in hardness from semi-hard steels to softer ferrous metals. In operation, the mandrel and workpiece turn. A roll carriage forces the metal ahead of it. The squeezing and stretching of the metal cause the workpiece to take the contour of the mandrel. The wall thicknesses can be held accurately to within ±0.002″ (0.05mm).

B. LATHE CONTROLS AND MAINTENANCE

HEADSTOCK CONTROLS FOR SPINDLE SPEEDS

STEP-CONE PULLEY DRIVES

Spindle speed (RPM) changes on lathes with step-cone pulley drives are made by stopping the lathe. The belt tension lever is then moved to loosen the belt. The manufacturer's spindle speed table is usually mounted on the headstock cover. The table (index plate) gives the full range of spindle speeds.

BACK-GEAR DRIVES

On some designs of lathes there is a back-gear drive for slower than normal speeds. The back gears increase the force the spindle may exert for heavier cuts. The speed is reduced to provide a slower range of RPM. Back gears should be changed only when the power is off.

The back gear mechanism is engaged or disengaged by moving the lever or control knob.

VARIABLE-SPEED DRIVE CONTROLS

Spindle speed changes are made on variable-speed drives while the motor is running. The speeds are changed hydraulically by positioning the speed-control dial at a desired speed. The spindle RPM is read directly from the dial at the

Courtesy of CLAUSING CORPORATION

Figure 19–12 A Variable-Speed Control Dial for High and Low Spindle Speed Ranges

index line. Figure 19–12 shows a variable-speed control dial. The speed range for the back-gear drive is from 52 to 280 RPM. The direct drive range (open belt) is from 360 to 2,000 RPM.

Back-gear (low spindle) speed combinations on variable-speed drives are engaged by a second lever on the headstock while the lathe spindle is stopped.

GEARED-HEAD CONTROLS

The geared-head lathe incorporates all the features of direct spindle drive and back-gear reduction. Spindle RPM changes are made when the spindle is stopped and different sets of gears are engaged.

The spindle speed changes appear on an index plate that is attached to the headstock. The positions of the gear-change levers are indicated for each speed.

MACHINE CONTROLS FOR FEEDS AND THREADS

QUICK-CHANGE GEARBOX CONTROLS

The *quick-change gearbox* serves the twofold function of turning the feed screw and the lead screw. An index plate on the gearbox indicates the range of feeds per revolution and the pitch of inch-standard or metric threads.

The positions and combinations of the quick-change gear levers for a particular feed or screw thread are given on the index plate. Two levers control the feed and thread pitch ranges. Another lever known as the *feed-reverse lever* changes the direction of rotation of the lead and feed screws.

APRON CONTROLS

The actual automatic feeding of the cutting tool on the carriage is controlled by a *feed-change lever* or automatic feed knob. Usually, when the lever is in the up position, the cutting action is toward the headstock. The lever is moved to a neutral position to stop the feed. The downward position of the feed lever reverses the direction of feed. Similarly, a second feed lever or knob engages the automatic cross feed. The direction of the cross feed may be changed by moving the feed-reverse lever.

LOST MOTION

All machine tools and instruments have *lost motion*. Lost motion is the distance a male or female screw thread or a gear tooth turns before the mating slide, gear, or other mechanism moves. Lost motion is also known as *backlash*, *slack*, *play*, or *end play*.

Lost motion may be corrected by withdrawing or turning (backing off) the screw thread or gear away from a required point. The direction is then reversed, and the cutting tool is advanced by feeding it in one direction only.

OVERLOAD SAFETY DEVICES

There is always the possibility of overloading gear trains, lead and feed screws, and other machine parts. Manufacturers use *shear pins* and *slip clutches* to prevent damage.

A shear pin is designed to be of a particular diameter and shape. The shear pin withstands a specific force. The pin shears and the machine motion stops when the design force is exceeded. The shear pin in the end gear train is used to prevent overload and the stripping of gear teeth. A shear pin makes it almost impossible to apply an excessive force on a shaft, rod, or lever.

A slip clutch protects the feed rod and connecting mechanisms. The slip clutch is designed to release the feed rod when a specific force is exceeded. The slip clutch mechanism has an added feature that permits the feed rod to be automatically reengaged when the force is reduced. Figure 19–13 provides an interior view of a spring-ball slip clutch applied to a feed rod.

TAILSTOCK ALIGNMENT

How to Align the Tailstock

Checking Alignment with a Test Bar

STEP 1 Mount the live and dead centers.

STEP 2 Place and hold a centered, cylindrical, parallel test bar between centers (Figure 19–14).

> Note: Clean the center holes of the test bar before mounting between centers.

STEP 3 Mount a dial indicator in the tool post.

STEP 4 Bring the point end of the indicator into contact with the test bar. Note the reading.

STEP 5 Move the dial indicator carefully toward the second end (Figure 19–14).

STEP 6 Note any variation in the readings.

> Note: The distance to move the tailstock is equal (approximately) to the difference in the two dial readings.

STEP 7 Return the dial indicator to the tailstock portion of the test bar. Loosen the binding nut and adjust the tailstock until the dial reading is the same as the reading at the headstock end.

STEP 8 Secure the head and base sections of the tailstock. Recheck the alignment.

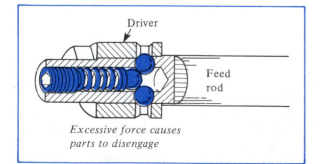

Figure 19–13 Application of a Spring-Ball Slip Clutch to a Feed Rod

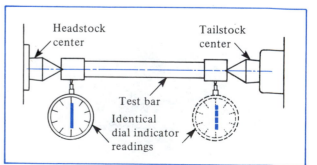

Figure 19–14 Precision Alignment of the Centers with a Test Bar and Dial Indicator

LATHE MAINTENANCE

CLEANING THE LATHE

Lathe cleaning means the safe removal of cutting chips and foreign particles from the machine, accessories, cutting tools, and workpiece. It also involves cleaning the cutting fluid reservoir and system. The purpose of cleaning, as discussed before, is to protect the worker, the machine, and the workpiece. *Chips should be removed when the machine is stopped.* A chip rake should be used when continuous chips are formed into a large mass.

When chip breakers are applied, the broken chip fragments may be brushed away from the machine areas. Comparatively large quantities of chips are shoveled out of the chip pan or tray. A chip disposal container should be readily available so that none of the cutting fluid is spilled on the floor.

After the chips are removed, the headstock, tailstock, and carriage are first brushed and then wiped with a clean cloth. The painted surfaces are wiped first; the machine surfaces, last. All accessories and cutting tools must be brushed and wiped clean.

The lead screw must also be cleaned. When the lead screw is cleaned, the machine may be started with all feeds disengaged. A hard, short-fiber brush may be moved in the threads along the entire length of the lead screw as it turns slowly.

With continuous use, gum and oil stains form on the exposed machine surfaces of the lathe. After these surfaces are wiped with a clean cloth, they should be wiped again with a cloth saturated with a small amount of kerosene or other gum or nontoxic stain-cutting solvent.

LUBRICANTS AND LUBRICATION

Lubricants serve four main functions:

- Lubricating,
- Cooling,
- Cleansing,
- Protecting against corrosion.

Lubricating Function. A lubricant provides a *hydrostatic fluid shim* between all mating parts when the machine is in motion. In other words, the lubricant fills all clearance spaces. The metal surfaces are cushioned and slide with the least amount of friction. Less force is required to move mating parts, and friction-generated heat is decreased. Finer surface movement conditions produce smoother operation. There is less power consumption, and wear is reduced. A fluid shim also serves the important function of helping to maintain the precision of a machine tool.

Cooling Function. The heat that normally is generated in the machine tool itself is carried away by the flow of a lubricant. A cooling effect also results from the reduction of friction. Cooling also helps to prevent seizing due to expansion.

Cleansing Function. A lubricant washes away minute foreign particles. These particles are produced by the abrasion and wearing of surfaces that move against each other.

Protection against Corrosion. Exposure of unprotected ferrous metals to air, use of certain water-soluble cutting fluids, and turning of corroded surfaces may produce rust. Rust on a machined part results in a pitted and rough surface that has abrasive properties.

Rust-preventive mists are available to protect unfinished parts or machine surfaces. The rust preventive is sprayed as a mist. The rust-preventive mist has a water-displacing property. It is effective on wet or dry surfaces. Workpieces and machine parts are thus protected during idle periods or when operating without a cutting fluid.

FACTORS AFFECTING MACHINING CONDITIONS

A number of common lathe problems are listed in Table 19–1. The most probable causes of each problem and the corrective actions the worker may take are also given.

Table 19–1 Factors Affecting Machining Conditions and Corrective Action

Problem	Probable Cause	Corrective Action
Machine vibration	—Work-holding device or setup is out of balance	—Counterbalance the unequal force —Reduce the spindle speed
	—High cutting speed	—Reduce the spindle speed
	—Floor vibration	—Secure the base firmly to a solid footing
	—Overhang of cutting tool or compound rest	—Shorten the length the toolholder extends beyond the tool post —Position the toolholder so that the compound rest has maximum support from the bottom slide surfaces
Vibration of the cross slide and compound rest	—Compound rest slide extends too far from the bottom slide support surfaces	—Reset the cutting tool so that there is greater area in contact between the top and bottom slide surfaces
	—Loose gib	—Adjust the gib to compensate for wear and lock it in the adjusted position
	—Overhang of cutting tool or compound rest	—Shorten the length of the cutting tool in the holder and the tool-holder in the tool post
Play in feed screw	—Wear on the feed screw and nut	—Back the cutting tool away from the work. Take measurements by maintaining continuous contact, turning in one direction only
Chatter	—High rate of cutting speed or feed	—Reduce the spindle speed and rate of feed
	—Incorrect rake or relief angles, or form or cutting angles	—Regrind the cutting bit to the recommended rake, relief, and cutting angles and form
	—Overhang of cutting tool or toolholder	—Shorten the length the cutting tool extends from the holder and the tool post
Misalignment of headstock and tailstock axes	—Tailstock offset	—Move the tailstock head in its base until the spindle axis aligns with the headstock spindle axis

Safe Practices in the Care and Maintenance of the Lathe

- Secure all machine guards in place before starting the lathe.
- Make sure the lathe spindle is stopped when making speed changes on step-cone pulley, geared-head, and back-gear drives.
- Change the spindle speeds of variable-speed motor drives when the spindle is in motion.
- Check for excessive force if there is continuous slipping of a slip clutch on a feed rod. Damage may result if the condition is not corrected.
- Provide compensation for lost motion when setting a cutting tool.
- Remove chip fragments only when the lathe is shut down. Use a brush or a chip rake.
- Place chips in a chip container. Avoid splashing cutting fluid on the shop floor.
- Check all machine lubricating levels before the lathe is started.
- Use machine lubricants in forced-feed systems that meet the lathe manufacturer's specifi-

cations. Lubricants must be maintained at the level indicated on the sight indicator.

- Coat exposed machine surfaces with a rust-preventive film to protect against possible corrosion, particularly during a machine shutdown period.

- Avoid wearing loose, hanging clothing near moving parts. Also, all wiping cloths must be removed from the machine.

- Wear safety goggles when cleaning the lathe and when performing lathe operations.

UNIT 19 REVIEW AND SELF-TEST

A. LATHE FUNCTIONS, TYPES, AND FEATURES

1. Indicate the relationship of end gears to the spindle RPM of a lathe.

2. Identify the design feature of a tailstock that permits offsetting for taper turning.

3. Indicate a modern safety feature that is incorporated in the design of guards that cover end gear trains.

4. State the functions served by external and internal tapers on a lathe spindle nose.

5. Give two functions that are served by the index plate on a quick-change gearbox.

6. List two advantages of the four-way turret type 2 cutting toolholder and positioning device over a type 1 rectangular tool block for producing a number of identical parts.

7. Describe briefly how a multiple-head micrometer stop works.

B. LATHE CONTROLS AND MAINTENANCE

1. State the prime function that is served by leveling.

2. Describe briefly the operation of (a) variable-speed drive controls and (b) geared-head controls on an engine lathe.

3. Explain how lost motion in mating threads or gears on a lathe is corrected.

4. Name two general protective overload devices on a lathe.

5. a. Explain the effect of corrosion on machined surfaces that slide on each other.
   b. Give a preventive step that may be taken to prevent corrosion during shut-down periods.

6. Select one method and list the steps for checking the alignment of the head-stock and tailstock spindle axes.

7. Name five terms that are associated with the cleaning of the coolant system on a lathe.

8. State three safety precautions to observe in the care and maintenance of a lathe.

SECTION TWO

Lathe Cutting Tools and Toolholders

Lathe work processes begin with the selection of the most suitable cutting tool and toolholder. This section presents:

- Design characteristics and specifications of basic lathe cutting tool bits and inserts and the coding systems used to identify them,
- Design characteristics and general applications of toolholders and chip breakers
- Procedures for grinding high-speed steel tool bits.

UNIT 20

Cutting Tools, Specifications, and Holders

A. SINGLE-POINT CUTTING TOOLS: TECHNOLOGY AND PROCESSES

SINGLE-POINT CUTTING TOOL DESIGN FEATURES

The particular shape and the angles to which a tool bit are ground are determined by the process to be performed and the material to be cut. Conditions that relate to speed, feed, and finish also are considered. General terms used in the shop in relation to a tool bit are illustrated in Figure 20–1.

Cutting tools are classified as *right-* or *left-hand* and *side-cutting* or *end-cutting*. Typical side- and end-cutting operations are illustrated in Figure 20–2. The cutting edge on a right-hand tool bit is on the left side. The left-hand tool bit has the side-cutting edge ground on the right side. End-cutting edge tools cut with the end-cutting edges and are not provided with side-cutting edges. Side-cutting tools cut as the carriage moves longitudinally. End-cutting tools cut by the cross (transverse) movement of the cross slide.

In addition to the *nose radius* there are six principle tool angles. These angles include *end-*

and *side-relief angles*, *back-* and *side-rake angles*, and *end-* and *side-cutting-edge angles*.

RELIEF ANGLES

A relief angle, whether for the side-cutting or end-cutting edge or the nose, permits a cutting tool to penetrate into a workpiece. The relief angle of the nose radius is a combination and blending of the end- and side-relief angles.

Tool life and tool performance depend on the relief angles. If the relief angle is too large, the cutting edge is weakened and may wear quickly, chip, or break away. Too small a relief angle may limit proper feeding. It may also limit cutting action by causing rubbing that generates considerable heat.

High-speed steel cutting tools are ground with relief angles of 8° to 16°. The relief angle for general lathe processes is 10°. Smaller relief angles are used with harder materials. The harder and more brittle cemented carbide cutting tools require greater support under the cutting edge. The relief angle range for cemented carbides is from 5° to 12°.

BACK- AND SIDE-RAKE ANGLES

The side-rake angle indicates the number of degrees the face slants from the side-cutting edge.

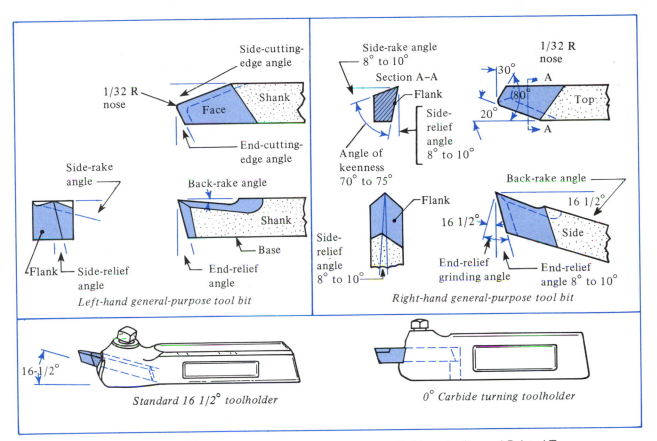

Figure 20–1 General-Purpose Single-Point Turning Tools, Toolholders, Angles, and Related Terms

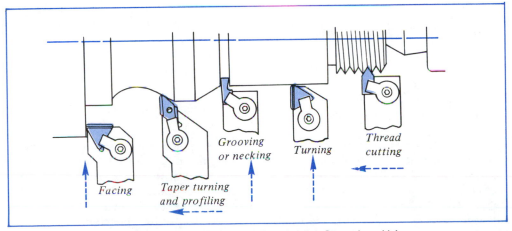

Figure 20–2 Typical Side and End-Cutting Operations Using Carbide Inserts and Toolholders

Table 20–1 Recommended Rake Angles for Selected Materials
(High-Speed Steel and Carbide Cutting Tools)

Material	Hardness Range (Bhn)	HSS Rake Angle (°)		Carbide Rake Angle (°)	
		Back	Side	Back	Side
Carbon Steel (plain)	100 to 200	5 to 10	10 to 20	0 to 5	7 to 15
. . .	400 to 500	−5 to 0	−5 to 0	−8 to 0	−6 to 0
Cast iron (gray)	160 to 200	5 to 10	10 to 15	0 to 5	6 to 15
Brass (free cutting) . . .		−5 to 5	0 to 10		

The side-rake angle is shown in Figure 20–1. It may be positive or negative. Similarly, the back-rake angle represents the number of degrees the face slants, as shown in Figure 20–1. The angle may slant downward, *positive rake*, or upward, *negative rake*, in moving along the cutting edges.

Negative side-rake and back-rake angles strengthen the cutting edge. The added strength is important when cutting hard and tough materials under severe machining conditions. Negative rake angles are used on throwaway carbide cutting tool inserts.

Manufacturers' recommendations and trade handbooks are available for determining correct rake angles for high-speed steel and carbide cutters and inserts for use with different materials of varying degrees of hardness. Sample listings in Table 20–1 are taken from the more complete Appendix Table A–17.

SIDE- AND END-CUTTING-EDGE ANGLES

The *side-cutting edge* is ground at an angle to the side of a tool bit or tool shank. The range is from 0° to 30°, depending on the material to be cut and the depth, feed, and so on. A 20° angle is common for general lathe turning processes.

The *end-cutting angle* is formed by a perpendicular plane to the side of a tool bit and the end-cutting surface. This angle should be as small as possible and still permit the position of the side-cutting edge to be changed.

In addition to clearing the workpiece, the end-cutting angle provides a backup area for the nose and adjoining end-cutting area of the cut-ting tool. A 15° end-cutting-edge angle may be used for rough turning. An angle of 25° to 30° may be used for general turning processes.

LEAD ANGLE, CHIP THICKNESS, AND CUTTING FORCE

The positioning of the side-cutting edge in relation to a workpiece influences such factors as the direction the chip flows, the thickness of the chip, and the required cutting force. The *lead angle* also affects these factors. The lead angle is the angle the side-cutting edge forms with a plane that is perpendicular to the axis of the workpiece. A large lead angle has the following disadvantages:

- A greater force is required to feed the cutting tool into a workpiece,
- There is a tendency for the workpiece to spring away from the cutting tool and produce chatter,
- The vibration of the workpiece may cause the cutting edge of the tool to break down.

THE NOSE RADIUS

Cutting tools are rounded at the nose to machine a smooth, continuous, straight surface. Cutting tools with a sharp point may turn a fine helical groove in the surface. The size of the radius depends on the rate of feed and the required quality of surface finish. There is a tendency to produce chatter when too large a radius or too fast a speed is used. A nose radius of from 1/64″ to 1/8″ is common.

B. CUTTING TOOL MATERIALS, FORMS, AND HOLDERS

Cutting tools should possess three properties: *hardenability*, *abrasion resistance*, and *crater resistance*. *Hardenability* relates to the property of a steel to harden deeply. A cutting tool must retain hardness at the high temperature developed at its cutting edge. *Crater resistance* is the ability to resist the forming of small indentations, or craters, that form on the side or top face of a cutting tool. A crater tends to undercut in back of a cutting edge. A crater weakens a cutting edge and in some cases causes it to break away under heavy cutting forces.

CUTTING TOOL MATERIALS

General cutting tools used for lathe work fall into the following groups: *high-speed steel, cast nonferrous alloy, cemented carbide, ceramic*, and *diamond*. The earlier carbon steel tool bits are seldom used for general lathe work.

HIGH-SPEED STEEL

High-speed steel is the most extensively used material for cutting tools. Its usage is based on factors such as tool and machining cost, tool wear, work finish, availability, and ease of grinding. High-speed steel single-point cutting tools are capable of withstanding machining shock and heavy cutting operations. A good cutting edge is maintained during red heat conditions.

Grades of High-Speed Steel. There are many grades of high-speed steel. Each grade depends upon the chemical composition of the steel. The grades are identified according to American Iron and Steel Institute (AISI) designations. Some of the commonly used T and M grades are indicated in Table 20–2. The addition of cobalt increases the ability of a cutting tool to retain its hardness during heavy roughing cuts and at high temperatures.

CAST ALLOYS

Cast alloy tool bits are used for machining processes at extra-high speeds. These speeds range from almost one-quarter to two times more than the speeds used for high-speed steel. Such tool bits have better wear-resistant qualities than high-speed steel. Cast alloys, like Stellite, make it possible to efficiently cut hard-to-machine cast iron and steel. However, cast alloy tool bits are brittle. They are not as tough as high-speed steel bits.

Cast alloys are manufactured as solid tool bits—for example, tips that may be brazed on tool shanks—and as cutting tool inserts. These tool bits have the advantage of maintaining a good cutting edge at high cutting speeds.

Cast alloy tool bits are popularly known by such trade names as Stellite, Rexalloy, Armaloy, and Tantung. Each trade name tool bit relates to a specific composition of chromium, tungsten, carbon, and cobalt that produces a tool bit with a high degree of hardness and high wear resistance. Cast alloy tool bits also have the ability to maintain a cutting edge while operating at red heat cutting temperatures below 1400°F.

CEMENTED CARBIDES

Cemented carbides are powdered metals. They are compressed and sintered into a solid mass at temperatures around 2400°F. The composition and applications of a few selected

Table 20–2 Common Grades of High-Speed Steel for Cutting Tools

Industrial Designation		
AISI	Trade Name	Application
T–1	High-speed steel	Single-point cutting tools; general purpose
T–4	Cobalt high-speed steel	Machining cast iron
T–6	Cobalt high-speed steel	Machining hard-scaled surfaces and abrasive materials
T–15*	Cobalt high-speed steel (super HSS)	Machining difficult-to-machine materials, stainless steels, high-temperature alloys, and refractory materials
M–2	General high-speed steel	Single-point and multiple-point cutting tools: milling cutters, taps, drills, and counterbores

* Extremely hard; requires slow grinding using a light grinding force

Table 20–3 Applications of Selected Cemented Carbide Cutting Tools

Composition of Tool	Application
Tungsten carbides	Cuts on gray cast iron, malleable iron, stainless steel, nonferrous metals, die-cast alloys, and plastics
Tungsten carbide and cobalt (with titanium and tantalum carbide added)	Medium and light cuts on carbon and alloy steels, alloy cast iron, monel metal, and tool steels
Titanium carbide and nickel or molybdenum	Light precision cuts and precision boring cuts requiring a high-quality surface finish

carbides appear in Table 20–3. Cemented carbide cutting tools possess the following qualities:

- Excellent wear resistance,
- Greater hardness than high-speed steel tools,
- Ability to maintain a cutting edge under high temperature conditions,
- Tips and inserts that may be fused on steel shanks and cutter bodies for economical manufacture of the cutting tools,
- Cutting speeds that may be increased three to four times the speeds of high-speed steel cutters.

CERAMICS

The sintering of fine grains of aluminum oxide produces a ceramic material that may be formed into *cutting tool inserts*. These inserts are particularly adapted for machining highly abrasive castings and hard steels. Ceramic cutting tools are harder than cemented carbide tools and may be used at extremely high cutting speeds on special high-speed lathes. Ceramic cutting tools may be used in turning plain carbon and alloy steels, malleable iron (pearlitic), gray cast iron, and stainless steel.

INDUSTRIAL DIAMONDS

Industrial diamonds may be designed as single-point cutting tools. Diamonds are used on workpieces that must be machined to extremely close dimensional and surface finish tolerances. Diamond-tip cutting tools are used for light finish-cut turning and boring operations. The depth of cut may range between 0.003″ to 0.030″ (0.08mm to 0.8mm) using a feed of from 0.0008″ to 0.005″ (0.02mm to 0.1mm) per revolution.

Industrial diamonds may be operated at extremely high speeds for machining nonferrous and nonmetallic materials. Single-point diamond cutting tools are especially adapted for finish machining hard-to-finish metals like soft aluminum. The factors of additional cost and brittleness of industrial diamonds must be considered against the advantages.

CUTTING SPEEDS FOR SINGLE-POINT TOOLS

Cutting speed is influenced by eight major factors:

- Material in the cutting tool;
- Material in the workpiece (its heat treatment, hardness, abrasiveness, internal hard spots, and other surface conditions);
- Rake, relief, and clearance angles and shape of the cutting areas;
- Use of a chip breaker to form and to clear away chips;
- Desired economical, efficient tool life;
- Use of a cutting fluid where required;
- Type of cut (roughing or finishing) and the depth;
- Rigidity of the work setup, the capability and construction of the lathe, and the nature of the toolholding device.

Cutting speed for lathe work refers to the rate at which a workpiece revolves past a fixed point of a cutting tool. In lathe work the cutting point is at the turned diameter. Cutting speed is measured in surface feet per minute (fpm, or sfpm) or meters per minute (m/min).

Table 20–4 Cutting Speed (sfpm and m/min) Using
High-Speed Steel Cutting Tools (Partial Table)

Material to Be Machined		Cutting Speed (sfpm and m/min) for Lathe Operations*		
		Turning and Boring		Screw Thread Cutting
		Roughing	Finishing	
Low-carbon (C) steel 0.05 to 0.30%C ...	sfpm	90	100	35–40
	m/min	27	30	10–12
High-carbon (C) tool steel 0.06 to 1.7%C ...	sfpm	50	70	20–25
	m/min	15	21	6–8
Brass ...	sfpm	150	300	50–60
	m/min	45	90	15–18
Aluminum ...	sfpm	200	350	50–70
	m/min	60	105	15–21

* Without using a cutting fluid

$$CS \text{ in sfpm} = \frac{(\text{Diameter in Inches} \times \pi) \times RPM}{12 \text{ Inches}}$$

When the metric system is used, the cutting speed (CS) is expressed in meters per minute (m/min). The metric equivalent of the formula is:

$$CS = \frac{(\text{Diameter in mm} \times \pi) \times RPM}{1000}$$

In actual practice the operator usually determines the RPM at which to set the spindle speed. The recommended cutting speed is found in a handbook table (Table 20–4).

$$RPM = \frac{CS \times 12}{\text{Diameter} \times \pi} \quad \text{or} \quad \frac{CS \text{ (m/sec)} \times 1000)}{\text{Diameter} \times \pi}$$

(Inch Standard)   (Metric System)

SYSTEM FOR DESIGNATING BRAZED-TIP CARBIDE TOOL BITS

Brazed-tip, single-point carbide tool tips are designated by a system of two letters and a numeral. As shown in Table 20–5, the first letter indicates the tool shape and the general process(es) for which the tool may be used. The second letter identifies the side-cutting edge and the direction of cut. The numeral gives the size of the square shank in sixteenths of an inch.

The designation (BR–6) of the carbide tool bit used in the example in Table 20–5 indicates that it has a lead angle of 15° (B). The tool bit may be used for any general purpose. The brazed tip is ground as a right-hand cutting tool (R). The shank (6) is 6/16″, or 3/8″, square.

SPECIFICATIONS OF THROWAWAY CARBIDE INSERTS

Carbide cutting tool inserts are preformed. A system for identifying eight design features of throwaway inserts provides complete information. The system consists of letters and numerals. For example, a throwaway carbide insert may be specified as SNEF–334E. Standard industrial designations for each of these sample letters and numerals are provided in handbook tables. The SNEF–334E sample designation indicates the insert *shape* is square (S), has a 0° *clearance angle* (N), has a *cutting point* and *thickness class* of 0.001″ (E), is a clamp-on *type* with a chip breaker (F), is 3/8″ *square* (3), is 3/16″ *thick* (3), has a cutting point of 1/16″ *radius* (4), and is *unground* and *honed* (E).

Table 20–5 System for Designating Brazed-Tip Carbide Tool Bits

SAMPLE DESIGNATION: B   R   6

Tool Characteristics		Hand	Shank Size
Shape	**Applications**	R right hand	*Shank Size* (Expressed in 1/16″)
A 0° lead	Shouldering, turning		
B 15° lead	Forming to lead angle	L left hand	
C Square nose	Grooving and facing		
D 30° lead	Undercutting and chamfering		
E 45° lead	Threading		

CHIP FORMATION AND CONTROL

CUTTING FORCES

There are three prime forces that act during the cutting process: *longitudinal, radial,* and *tangential.* As a cutting tool feeds into a workpiece, it produces the *longitudinal force.* The force of the workpiece against the front face of the cutting tool is a *radial force.* A *tangential force* results from the downward action of the workpiece against the top of the tool.

CHIP FORMATION

The cutting action of a cutting tool produces three major types of metal chips: *continuous chip, discontinuous chip,* and *segmental chip.* A continuous chip represents the flow of a ductile metal (like steel) over the cutter face. The chip is in the form of a continuous spiral ribbon.

A discontinuous chip is produced when cutting brittle metals (like cast iron). Since these metals do not flow, they fracture into many discontinuous particles.

A segmental chip has a file tooth appearance on the outer side. The inner side that flows over the cutting tool face has a smooth, burnished surface. A segmental chip is produced when taking a heavy feed on ductile materials. The internal metal fractures spread to the outer surface of the chip. The particles are welded together by the heat of the cutting force.

CHIP CONTROL

Chips that form continuously or segmentally at high speeds must be controlled. *Chip control* means the disposing of chips in a safe manner. The accumulation of chips near a revolving workpiece; along the bed, carriage, and cross slide; or on cutting tools is dangerous. Chips may become wrapped around a work or chucking device and thus often cause damage to the workpiece, cutting tool, or lathe. More important is the hazard of personal injury. Revolving chips or hot metal particles may fly in a work area. Operational dangers are increased with carbide cutting tools. A great quantity of chips form and accumulate rapidly when heavy cuts are made at high speeds.

Continuous chips are broken into small, safe sizes with a *chip breaker.* The chip breaker feature may be ground as an area on the top of a tool bit, preformed on an insert, or added as a chip groove. An example of a parallel-type chip breaker for brazed carbide cutting tools is shown in Figure 20–3. The preformed chip grooves for carbide inserts are also illustrated.

Provision is made to accommodate chip breakers on toolholders. Chip breakers for throwaway insert toolholders are commercially available. Chip breakers are designed for positive-, neutral-, and negative-rake toolholders and for general, light, medium, and heavy feeds.

Three general forms of chip breakers, which are ground on the top face of high-speed steel tool bits, are shown in Figure 20–4.

CUTTING-TOOL HOLDERS

A lathe cutting tool bit may be supported in a heat-treated toolholder. The cutting tool and holder are then secured in a tool post. Cemented

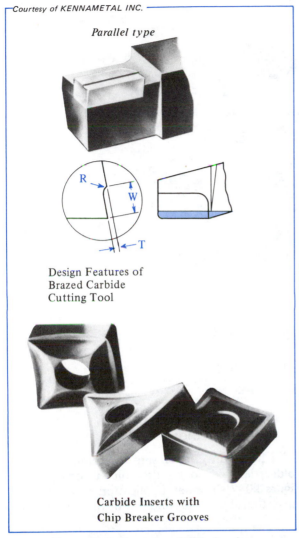

Courtesy of KENNAMETAL INC.

Parallel type

Design Features of
Brazed Carbide
Cutting Tool

Carbide Inserts with
Chip Breaker Grooves

Figure 20–3 Examples of Chip Breakers

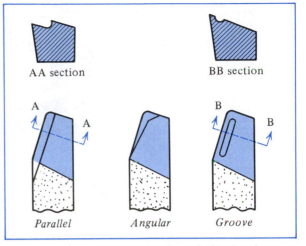

AA section   BB section

Parallel   *Angular*   *Groove*

Figure 20–4 Three Basic Types of Chip Breakers
Ground on High-Speed Steel Tool Bits

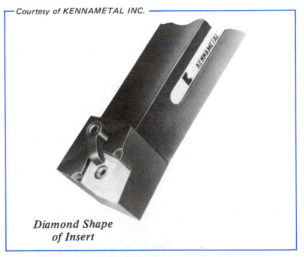

Courtesy of KENNAMETAL INC.

*Diamond Shape
of Insert*

Figure 20–5 Toolholder for Carbide Inserts

carbide inserts are usually brazed on a steel
shank. The shank is held in a tool post or tool
rest. Disposable carbide inserts are held and secured in a special type of carbide toolholder
(Figure 20–5).

CARBIDE AND CERAMIC INSERTS

A square, diamond-point, or parallelogram-
type insert has a total of eight cutting edges.
When the four top cutting edges become dull,
the insert is reversed. Each new cutting edge
is then used. The insert is thrown away when
the eight cutting edges are dull.

A toolholder for disposable carbide or
ceramic inserts is designed as a negative- or
positive-rake toolholder. The inserts for the
positive-rake toolholder must be ground with a
side-relief angle. Thus, only the top face edges
may be used. The general shapes of inserts are
square, rectangular, triangular, round, and
diamond.

QUICK-CHANGE TOOL SYSTEM

Figure 20–6 illustrates a *quick-change tool
system*. The system consists of a quick-change tool
post and a number of toolholders for different

Courtesy of ARMSTRONG BROTHERS TOOL COMPANY

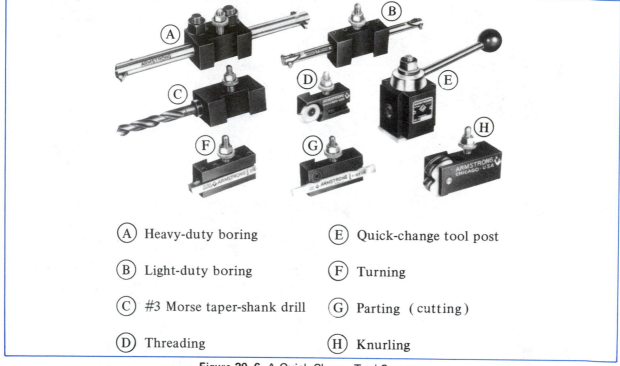

(A) Heavy-duty boring   (E) Quick-change tool post

(B) Light-duty boring   (F) Turning

(C) #3 Morse taper-shank drill   (G) Parting (cutting)

(D) Threading   (H) Knurling

Figure 20–6 A Quick-Change Tool System

cutting tools. Each cutting tool is mounted quickly, aligned, and secured in position. The combination illustrated is used for performing general external and internal lathe operations.

Courtesy of CINCINNATI MILACRON INC.

Figure 20–7 A Heavy-Duty, C-Shaped Block Secured on a Compound Rest

HEAVY-DUTY, OPEN-SIDE TOOL BLOCK (HOLDER)

A heavy-duty, open-side tool block, or toolholder, is used to rigidly hold a cutting tool. Figure 20–7 shows an application of the heavy-duty tool block mounted on the compound rest.

C. OFF-HAND GRINDING HSS TOOL BITS

How to Grind High-Speed Steel Tool Bits

Note: The steps in this example apply to a right-hand turning tool (roughing cut).

Grinding Side-Cutting-Edge and Side-Relief Angles

STEP 1 Position the tool rest. Hold, position, and guide the cutting tool with the fingers of both hands (Figure 20–8).

The bottom (heel) is tilted toward the grinding wheel.

STEP 2 Rough grind the side-cutting-edge and side-relief angles.

STEP 3 Check the side-relief angle. Use a bevel protractor or a side-relief angle gage.

> **Note:** Rough grind to within approximately 1/16″ from the top of the tool bit.

Grinding End-Cutting Edge and End-Relief Angles

STEP 1 Raise the front end of the tool bit to increase the angle of contact with the face of the grinding wheel.

STEP 2 Rough grind the end-cutting-edge and end-relief angles to within 1/16″ of the tool bit point.

> **Note:** The formed surface must blend the end-cutting and side-cutting angles, the flanks, and the nose radius.

Grinding Side- and Back-Rake Angles

STEP 1 Hold the tool bit against the face of the grinding wheel so that the side-cutting edge is horizontal.

STEP 2 Tilt the tool bit surface adjacent to the cutting edge inward toward the grinding wheel to grind the back-rake angle.

STEP 3 Rough grind the face to within 1/16″ of a sharp edge.

Finish Grinding

STEP 1 Repeat all of the previous grinding steps. Use the fine-grained wheel. Take light finishing cuts.

Safe Practices in Using Cutting Tools and Toolholders

• Grind high-speed tool bits and cemented carbide and ceramic inserts to the specified clearance, rake, and cutting-edge angles.

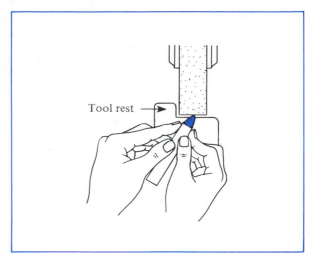

Figure 20–8 Holding and Guiding a Tool Bit While Grinding

These angles affect cutting efficiency, tool life, quality of surface finish, and dimensional accuracy.

• Grind a chip breaker into a cutting tool bit or place a separate chip breaker adjacent to the cutting edges. The continuous breaking away of small chips provides safe working conditions and prevents damage to the lathe and workpiece.

• Select lathe speeds and feeds that permit turning processes to be performed economically and safely.

• Pay attention during rough turning (taking hogging cuts) to the effect of the longitudinal, radial, and tangential cutting forces and the heat generated on a workpiece and cutting tool.

• Quench the high-speed tool bit regularly when dry grinding.

• Position the machine safety shields and use safety goggles.

• Stop the lathe to remove chips. Use a chip rake.

• Secure all loose clothing. Remove wiping cloths from the lathe before starting the machine.

UNIT 20 REVIEW AND SELF-TEST

A. SINGLE-POINT CUTTING TOOLS: TECHNOLOGY AND PROCESSES

1. State what effect crater resistance of a lathe cutting tool has on its cutting ability.

2. Indicate what relationship exists between the RPM of a lathe spindle and the cutting speed (sfpm).

3. Identify (a) the two cutting edges of a single-point cutting tool and (b) the purpose of relieving the flank.

4. Describe the effect of negative side- and back-rake angles on the cutting edge of a single-point cutting tool.

5. Indicate what safety precaution must be taken to control the accumulation of continuous or segmental chips at high speeds.

B. CUTTING TOOL MATERIALS, FORMS, AND HOLDERS

1. a. Give an application of a cast alloy tool bit.
   b. Tell why it is preferred in this application to a standard high-speed steel tool bit.

2. Distinguish between the manufacture of cemented carbide and ceramic inserts.

3. a. List five common shapes of cemented carbide and ceramic inserts.
   b. Give an advantage of disposable cemented carbide or ceramic inserts.

4. Cite two cutter precautions to take when selecting diamond, ceramic, or cemented carbide cutting tools.

5. List four of the eight design features that provide specifications for throw-away carbide inserts.

6. Explain the meaning of tangential force during a cutting process on a lathe.

7. a. Describe a quick-change tool system.
   b. List the advantages of such a system over a multiple-station turret.

C. OFF-HAND GRINDING HSS TOOL BITS

1. Tell how to grind the side- and end-cutting edges, the relief angles, and the face of a single-point cutting tool.

2. State two safe practices to observe in selecting, grinding, or cutting with a single-point cutting tool for lathe work.

SECTION THREE
Lathe Work between Centers

Many lathe work processes are performed with the workpiece mounted between the lathe centers. This section describes the principles of and procedures for performing the following basic between-centers processes:

- Centering and lathe setups;
- Facing and straight turning;
- Shoulder turning, chamfering, and rounding ends;
- Grooving, form turning, and cutting off;
- Filing, polishing, and knurling;
- Taper and angle turning.

UNIT 21

Center Work and Lathe Setups

CENTER WORK AND CENTER DRILLING

FUNCTIONS OF CENTER HOLES

A *center hole* is a cone-shaped or bell-mouthed bearing surface for the live and/or dead centers. Each center hole must be of a particular size and depth, depending on the size of the part to be machined and the nature of the operation. An improperly drilled center hole limits the bearing surface, wears away rapidly, and may result in inaccurate turning. The center drilled hole should have a fine surface finish to reduce friction on the cone- or bell-shaped bearing area as shown in Figure 21–1B.

TYPES AND SIZES OF CENTER DRILLS

Table 21–1 gives recommended sizes of *plain* center drills for different diameters of workpieces.

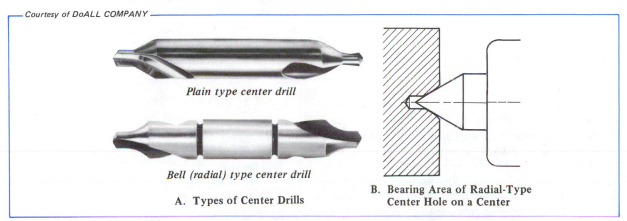

Courtesy of DoALL COMPANY

Plain type center drill

Bell (radial) type center drill

A. Types of Center Drills

B. Bearing Area of Radial-Type Center Hole on a Center

Figure 21–1 Two Types of Center Drills and a Radial Center Hole

Table 21–1 Recommended Center Drill Sizes (Plain Type)
for Different Diameter Workpieces

Size No.	Diameter (inches)			
	Workpiece	Countersink	Drill Point	Body
00	3/32 to 7/64	3/64	0.025	5/64
0	1/8 to 11/64	5/64	1/32	1/8
1	3/16 to 5/16	3/32	3/64	1/8
2	3/8 to 1/2	9/64	5/64	3/16
3	5/8 to 3/4	3/16	7/64	1/4
4	1 to 1 1/2	15/64	1/8	5/16
5	2 to 3	21/64	3/16	7/16
6	3 to 4	3/8	7/32	1/2
7	4 to 5	15/32	1/4	5/8
8	6 and over	9/16	5/16	3/4

Note in each case that the diameter to which the hole is countersunk is always less than the body diameter of the combined drill and countersink (center drill).

A *radial* type of center drill has an 82° angle above the 60° bearing band. This type of center drill automatically produces a *safety center*. The radial center hole is lubricated easily. The radial type center drill is available in sizes ranging from #00 (0.025″ drill and 5/64″ body diameters) to #18 (1/4″ drill and 3/4″ body diameters). Similar metric sizes and styles of center drills are available. The sizes ranging from #00 to #8 are adapted for precision tool, gage, and instrument work. The heavy-duty type ranges from #11 to #18. These sizes are used for heavy production parts, especially forgings and castings. The features of a plain- and a radial-type center drill are shown in Figure 21–1A.

PREPARING THE LATHE FOR TURNING BETWEEN CENTERS

THE LIVE CENTER

The accuracy to which a part may be machined on centers depends on the *trueness* of the headstock (live) center and the alignment of the headstock and tailstock centers.

The trueness of the live center may be checked using a dial indicator. The indicator point is brought into contact with the surface of the live center. The spindle is revolved by hand. If the center runs out-of-true, it either may be ground

in place or another live center must be inserted and tested.

TURNING A LIVE CENTER

When a chuck is already mounted, it is common practice to chuck a piece of steel and turn a center. The compound rest is swung to a 30° angle with the spindle axis. A center is roughed out and then finish turned. The side of the chuck jaw serves to drive the lathe dog and workpiece.

ALIGNING THE DEAD CENTER

The three basic methods of checking the alignment of centers were described earlier. Another fast, accurate method of aligning centers is to use a *micro-set adjustable live center* in the tailstock (Figure 21–2). The micro-set adjustable live center is used on light work.

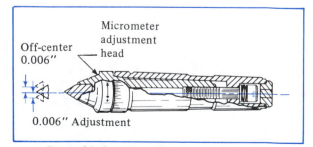

Figure 21–2 A Micro-Set Adjustable Center

Combination
center drill and
countersink

Workpiece

Figure 21–3 Drilling a Center Hole on a Drill Press

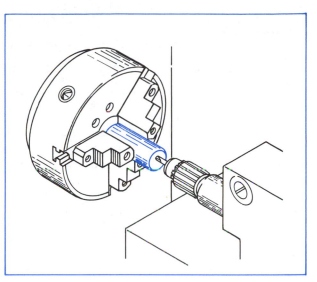

Figure 21–4 Center Drilling Work Held in a Chuck

COMMON CENTER DRILLING TECHNIQUES

How to Drill Center Holes

Drilling on a Drill Press

STEP Position the workpiece using a V-block
1 or the V-groove of a drill press vise.

STEP Select the number of center drill that
2 will produce a large enough center hole
for the job requirements.

STEP Chuck the center drill. Set the spindle
3 speed. Align the drill point with the
center-punched mark.

STEP Brush on a thin film of cutting fluid.
4 Bring the drill carefully into contact
with the workpiece (Figure 21–3). Drill
to depth.

Note: Use a fine feed and a cutting
fluid to finish the center drilling.

Drilling a Center Hole on a Lathe

Center Drilling Work Mounted in a Chuck

STEP Chuck the workpiece so that the outside
1 surface runs concentric. Face the end.

STEP Mount the drill chuck and center drill in
2 the tailstock spindle.

STEP Bring the tailstock up toward the work-
3 piece and tighten it. Put a light pressure

on the tailstock spindle by partially
clamping it. Feed the center drill to
start the process (Figure 21–4).

STEP Continue to feed carefully to the re-
4 quired depth. Use a cutting fluid.

Center Drilling Long Workpieces

Long workpieces are center drilled by chucking one end. The second end is supported in a
steady rest (Figure 21–5). The steady rest jaws
are adjusted until the center of the workpiece is
aligned with the tailstock center.

Figure 21–5 Using a Steady Rest to Center Drill
a Long Workpiece

After the end is faced, the center drill is fed carefully into the revolving workpiece until the required outside diameter of the countersunk portion is reached.

Safe Practices in Center Work and Lathe Setups

- Check the depth of a center-drilled hole. It must provide an angular or radial bearing surface adequate to support a workpiece and withstand cutting action.
- Use a rotating dead center when turning workpieces that revolve at high speeds.
- Use lathe dogs equipped with safety screws or safe locking devices.
- Check the bent tail of a lathe dog to see that it clears the bottom or sides of the driver plate slot. The dog position must permit a workpiece to ride on the center.
- Lubricate the angular or radial bearing surface of a workpiece that revolves on a solid dead center.
- Rotate a new workpiece and tool setup by hand to ensure that all parts clear without touching.
- Cover the lathe bed and other machined surfaces whenever a tool post grinding attachment is used.

UNIT 21 REVIEW AND SELF-TEST

1. Identify (a) the functions and (b) the characteristics of a precision, radial center-drilled hole.

2. State three practices the skilled mechanic follows when using a live center.

3. Cite two advantages of the micro-set adjustable center for light turning processes as compared to using a solid dead center.

4. List five general practices for setting up a lathe and taking basic cuts.

5. State the method by which long workpieces are supported for center drilling.

6. Tell why a center is often turned from a mild-steel chucked piece rather than by using a standard, hardened live center.

7. Give four precautionary steps to take before starting to turn a workpiece between centers.

Facing and Straight Turning

The process of machining a flat plane surface on the end of a workpiece on a lathe is referred to as *facing*. The process of turning one or more diameters is known by such terms as *straight turning*, *parallel turning*, *cylindrical turning*, or *taking a straight cut*.

Facing and turning processes require differently formed cutting tools and different workholding devices and tool setups. Some of the common shapes of high-speed steel turning tool bits are illustrated in Figure 22–1.

SETTING UP FOR FACING OR STRAIGHT TURNING

POSITIONING THE COMPOUND REST

Compound Rest Set at 30°. The compound rest is graduated on the base and may be rotated through 360°. Angular settings such as 30°, 45°, and 60° are common. The 30° angle provides a convenient method of producing a fixed cutting tool movement. With the compound rest set at 30°, the amount of side (longitudinal) movement of a cutting tool is equal to one-half the distance the compound rest is moved (Figure 22–2). When the compound rest is set at 30° and the handwheel is turned 0.010″ (or 0.2mm on a metric-graduated dial), it produces a side motion of 0.005″ (0.1mm).

Compound Rest Set at 90°. A 90° angular setting (parallel to the work axis) of the compound rest is practical for facing and shoulder turning. The amount the cutting tool is fed is read directly from the compound rest graduated collar setting. If the compound rest is moved 0.5mm (0.020″), the cutting tool advances the same distance (0.5mm or 0.020″).

Compound Rest Set at 84°16′. Although not a common practice, precise tool settings to approximately 0.0001″ (0.002mm) may be made with the compound rest set at an angle of 84°16′. The cutting tool feeds into the workpiece 0.0001″ (0.002mm) for each 0.001″ (0.02mm) movement of the compound rest feed screw. Because the angle graduations on compound rests are not graduated in minutes, the 16′ must be estimated.

PRACTICES FOR SETTING THE DEPTH OF CUT

The graduated micrometer collar on feed screws is a practical device for economically producing accurately machined parts. There are, however, certain procedures that must be followed to compensate for lost motion (backlash) and possible collar and cutting tool movement.

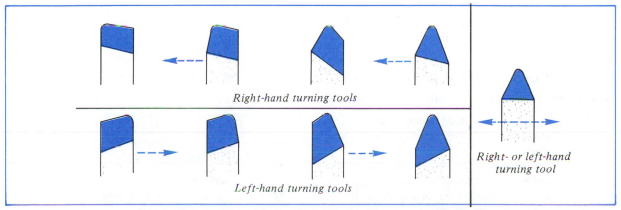

Right-hand turning tools

Left-hand turning tools

Right- or left-hand turning tool

Figure 22–1 Common Shapes of High-Speed Steel Turning Tool Bits

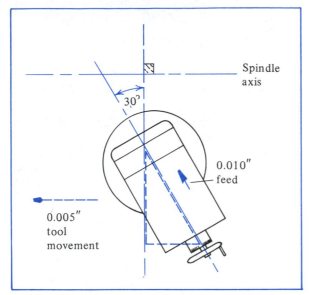

Figure 22–2 Longitudinal Tool Movement Equal to One-Half the Feed of the Compound Rest Set at 30°

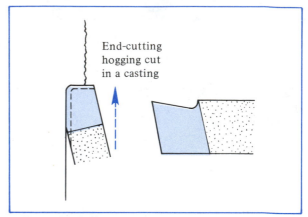

Figure 22–3 End-Cutting High-Speed Steel Tool Bit Used for a Rough Facing Operation

Some of the steps that must be taken are:

- Turn the compound rest feed screw at least one-half turn in the direction in which the depth of cut is to be set;
- Set the locking device on the graduated feed collar before taking any readings;
- Feed the cutting tool into (not away from) the workpiece when setting the depth of cut;
- Back out the cutting tool at least one-half revolution whenever the feed hand-wheel has been moved too far. The cutting tool is then moved clockwise to remove the backlash and permit feeding the cutting tool to the correct setting.

FACING TOOLS, SETUP, AND PROCESSES

A workpiece is *faced* for three main reasons:

- To produce a flat plane to provide a reference surface for other measurements and operations,
- To machine a flat-finished area that is at a right-angle to the axis of the workpiece,
- To turn a surface to a required linear dimension.

THE CUTTING TOOL

Facing may be done with a side-cutting or end-cutting tool. The cutting point has a radius. The cut may be fed either toward or away from the center. The cut may be a roughing or a finishing cut. The workpiece may be held in a chuck, between centers, on a faceplate, or in a holding fixture.

Although the high-speed steel tool bit is commonly used, all factors and conditions for selecting a cutting tool must be considered. The tool bit that is selected must be the most productive for the job. Cemented-tip carbides with solid shank and carbide and ceramic inserts cut more efficiently at high speeds and on harder and tougher materials than do high-speed steel tool bits.

Two shapes of cutting tool high-speed steel bits are generally used for facing. The end-cutting tool is adapted to hogging cuts (Figure 22–3).

The facing tool bit with a 55° cutting angle and a slight radius nose is used when facing the end of a workpiece that is mounted between centers.

How to Face a Workpiece

Chuck and Faceplate Work

Rough Facing

STEP Lay out the required length.
1

STEP 2 Chuck or mount the workpiece. Select and set the spindle RPM. Spindle speed is based on the largest diameter to be faced.

STEP 3 Use an end-cutting or plain side-cutting rough turning tool. Set the depth of cut. Use a coarse feed. Feed the cutting tool across the face almost to the layout line.

> **Note:** The end-cutting tool is fed directly into the workpiece (Figure 22–3).
>
> **Note:** When a considerable amount of material is to be faced off a workpiece, a series of longitudinal cuts are first taken. The diameter is reduced for each successive cut. The cutting tool is fed to almost the required length.

STEP 4 Feed the cutting tool across the face to even the steps that are formed.

Finish Facing

STEP 1 Replace the end- or side-cutting tool with a facing tool. Position the cutting point on the center line.

STEP 2 Increase the speed. Decrease the feed. Take a light finishing cut. Start at the center and feed to the outside diameter.

Facing Work between Centers

STEP 1 Select a facing tool with a 55° cutting angle. Mount the tool bit so that the point is at center height.

STEP 2 Advance the cutting tool into the workpiece to the required depth.

STEP 3 Feed the tool from the center outward across the face of the work.

GENERAL RULES FOR STRAIGHT TURNING

Straight, or *parallel turning* means that a workpiece is machined to produce a perfectly round cylinder of a required size. The general rules for all lathe processes apply.

- The cutting tool should be fed toward the headstock wherever possible, especially on work between centers. The cutting forces should be directed *toward* the live center.
- The number of cuts should be as few as possible. Sufficient stock should be left for a final finish cut if a fine surface finish is required.
- The size **and** condition of the lathe, the shape and material of the workpiece and cutting tool, and the type of cut all affect the depth of cut.
- The roughing cut must be deep enough on castings, forgings, and other materials to cut below the hard outer surface (scale).
- Deep roughing cuts should be taken using the coarsest possible feed.
- The cutting tool and holder must be set in relation to the workpiece. Any movement of the holder produced by the cutting forces should cause the cutting tool to swing *away* from the work.
- Trial cuts are taken for a short distance. Each trial cut permits the workpiece to be measured before a final cut is taken. Any undersized positioning of the cutting tool may be corrected before the part is machined.

How to Do Straight (Parallel) Turning

Taking Roughing and Finishing Cuts

STEP 1 Select the kind and shape of tool bit that is appropriate for rough turning the particular part.

STEP 2 Use a straight or left-hand toolholder that will clear the workpiece and the machine setup.

STEP 3 Position and solidly secure the tool bit and toolholder to the compound rest.

STEP 4 Determine and set the spindle speed at the required RPM.

STEP 5 Set the feed lever for a roughing cut.

STEP 6 Move the cutting tool to the end of the workpiece. Adjust the depth of cut until at least half of a true diameter is turned.

STEP 7 Take the trial cut for a distance of from 1/8″ to 1/4″. Leave the tool set. Return it to the starting end. Stop the lathe.

STEP 8 Measure the turned-diameter portion only.

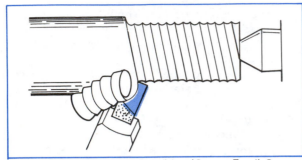

Figure 22–4 Taking a Roughing (Coarse Feed) Cut

STEP 9 Note the reading on the graduated collar. It may also be set at zero. This reading is the reference point. Subtract the measured diameter from the required diameter. An oversized allowance is made for the finish cut,

STEP 10 Move the cutting tool to depth. Start the lathe. Engage the power feed. Take the required number of roughing cuts (Figure 22–4).

STEP 11 Replace the roughing tool with a finish turning tool.

STEP 12 Increase the spindle speed and decrease the cutting feed for finish turning

STEP 13 Take the trial cut. Check the accuracy of the turned diameter. Adjust further for depth. Engage the power feed. Turn the required diameter to the specified length.

Safe Practices in Facing and Straight Turning

- Check the edges of the compound rest for clearance between it and the revolving workpiece or work-holding device. Special care must be taken when the compound rest is positioned at an angle to the spindle axis.
- Take up all lost motion in the cross feed screw. Then lock the graduated collar at the starting point of a measurement.
- Turn the cross feed handwheel clockwise to apply a force to set the depth of cut. The inward force causes the tool to remain at the depth of cut even when cutting forces are present.
- Position the cutting tool and holder so that any movement during the cutting process moves the cutting edge away from the workpiece.
- Wear a protective eye device and observe general machine safety practices.

UNIT 22 REVIEW AND SELF-TEST

1. State why the compound rest is generally set at a 30° angle for straight turning processes.

2. List the steps to take to turn a steel part from an outside diameter of 100mm down to 80 mm.

3. Identify two general forms of facing tools.

4. State four practices the lathe operator follows that apply to all straight turning processes.

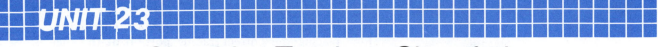

Shoulder Turning, Chamfering, and Rounding Ends

A *shoulder* is the area that connects two diameters. *Shoulder turning* is the process of machining the connecting area to conform to a particular size and shape. There are three basic shoulder shapes:

- The *square* shoulder,
- The *beveled* (angular) shoulder,
- The *filleted* (round) shoulder.

A *chamfer* is an outside edge or end that is beveled or cut away at an angle. A *rounded end* or corner differs from a chamfer in that the round shape is a radius.

COMMON MEASUREMENT PROCEDURES FOR SHOULDER TURNING

The length of a shoulder is usually marked by cutting a light groove around the workpiece. The cutting tool is positioned by measuring the required shoulder length with a steel rule.

Shoulders that are to be machined to a precision linear dimension may be measured with a gage or depth micrometer. In precision shoulder turning, the final cut is positioned using a micrometer stop for the carriage. The compound rest feed screw may also be moved the required distance. The depth is read on the graduated collar.

The diameter and shoulder length are generally turned to within 1/64″ of the finished size to leave sufficient material at the shoulder to permit turning to the required shape and dimension. One shoulder-turning technique locates the length of a square shoulder by cutting a groove in the workpiece with a necking tool. After grooving, the body diameter is turned to size. One or more cuts is taken until the required dimension is reached.

Another shoulder-turning technique is to *block out the shoulder* and leave enough material at the shoulder to finish turn it to the required form. The adjacent cylindrical surface is then turned to size. Figure 23–1 shows how a shoulder is turned by in-feeding an end-cutting tool to the shoulder diameter. Excess shoulder material is removed by feeding to the right in this case.

APPLICATION OF THE RADIUS GAGE

The *radius gage* provides a visual check on the radius of the cutting tool or the filleted shoulder.

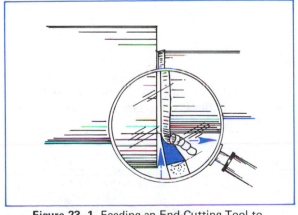

Figure 23–1 Feeding an End-Cutting Tool to Cut a Square Shoulder

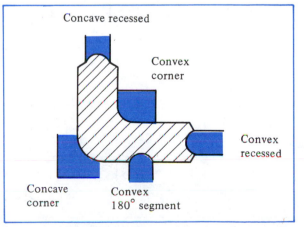

Figure 23–2 Applications of the Five Gaging Areas of a Radius Gage

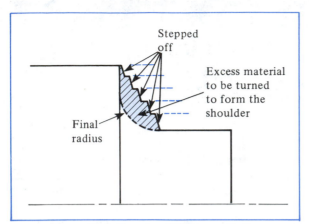

Figure 23–3 Stepping Off a Large Radius

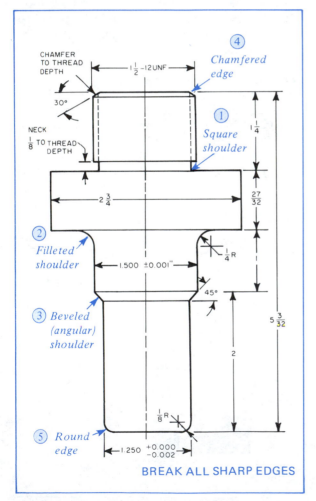

BREAK ALL SHARP EDGES

Figure 23–4 Representation and Dimensioning of General Types of Shoulders and Edges

The accuracy of the cutting tool is checked by placing the correct radius portion of the gage over the ground radius. A radius gage and five general applications of it are pictured in Figure 23–2.

Low spots are sighted by any background light that shines between the gage and the cutting tool. The dark, high-spot areas are ground or honed if there are only a few thousandths of an inch to be removed. When no light is seen along the radius, the cutting point is formed to the correct shape and depth.

One common practice that is repeated in turning a shoulder to a *large radius* is to *step off* the shoulder. Successive facing or straight turning cuts are taken. These cuts end in a series of steps. The steps, or excess corner material, are rough turned. A stepped-off section is shown in Figure 23–3.

The lathe operator often *contour forms* a large radius by moving the longitudinal and cross feed screws simultaneously by hand. This action sweeps the cutting tool through the desired radius. Rounded corners are generally formed with a tool ground to the correct radius.

Commercial flat, metal radius gages (as previously illustrated) are available. The general sizes range in radius from 1/32″ R to 1/2″ R. There are usually five different gaging areas on each blade. These areas permit gaging the radius of a corner, a shoulder, or a recess in a circular workpiece.

REPRESENTING AND DIMENSIONING SHOULDERS, CHAMFERS, AND ROUNDED ENDS

The techniques of representing and dimensioning the three general types of shoulders and rounded and beveled ends are illustrated in Figure 23–4. The note BREAK ALL SHARP EDGES indicates that the corners be broken by filing a slight bevel or radius as the workpiece is turned.

How to Turn a Beveled (Angular) Shoulder

Method 1: Setting a Side-Cutting Tool

STEP 1 Position the side-cutting edge at the required angle. (The turning of an angular shoulder is shown in Figure 23–5.)

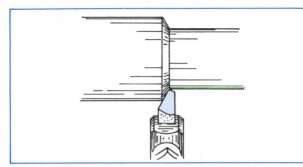

Figure 23–5 Turning an Angular Shoulder

STEP 2   Remove excess stock from the angular shoulder.

STEP 3   Take a trial cut on the body diameter. Measure the diameter; then move the cutting tool to the required depth. Set the cross feed graduated collar at zero.

STEP 4   Use a power feed and take the cut almost to the shoulder. Disengage the power feed.

STEP 5   Decrease the spindle speed for a large deep-cut shoulder. Reducing speed tends to reduce chatter. Continue to slowly feed the cutting tool by hand into the angular shoulder face.

STEP 6   Feed to the required length. Use a cutting fluid.

Method 2: Setting the Compound Rest

STEP 1   Position the compound rest at the required angle.

STEP 2   Set up a side-cutting tool.

STEP 3   Take cuts from the center outward. Feed the cutting tool by hand by turning the compound rest feed screw.

STEP 4   Continue to machine the bevel. The final beveled surface starts at the required length of the small diameter. It continues at the required angle to the larger diameter.

How to Turn a Rounded End

A rounded end with a small radius is generally formed by using a tool bit which is ground to the required radius. A larger radius is turned by stepping-off the excess stock. A formed cutter is then used to finish the rounded end. Slow speeds, fine feeds, and a cutting compound are used to produce a high-quality rounded end.

Safe Practices in Turning Shoulders, Chamfers, and Rounded Ends

• Blend the stepped sections of a filleted or square shoulder to the desired form before a final round-nose or blunt angular-nose formed facing tool is used for the finishing cut.

• Align the cutting point of a facing tool on center.

• Reduce the spindle speed and use an appropriate cutting fluid whenever a large radius, angular shoulder, or a rounded or beveled end is to be machined. The speed reduction helps prevent chatter.

• Check the accuracy of a turned concave or convex radius with a radius gage.

• Use the left-hand method of filing a workpiece on the lathe. This method is a safety precaution to prevent hitting the revolving workpiece, lathe jaw, or chuck.

• See that the file handle fits solidly on the tang.

UNIT 23 REVIEW AND SELF-TEST

1. Indicate what the worker is expected to do when a drawing of a part has the note: BREAK ALL SHARP EDGES.

2. Explain what it means in shoulder turning to block out the shoulder.

3. List three lathe work applications of a radius gage.

4. Describe the meaning of a note on a shop sketch that reads: CHAMFER TO THREAD DEPTH.

5. State two common methods of (a) turning a square shoulder, (b) turning a beveled or angular shoulder, and (c) rounding a corner.

6. Give two safety precautions to take in turning a large radius or an angular (beveled) shoulder.

Grooving, Form Turning, and Cutting Off

Turning processes that produce a specially shaped cut-away section around a workpiece are referred to as *grooving*. A groove that is cut into an outside diameter is called a *neck*. A groove that is formed inside a hole is known as a *recess* or an *undercut*.

Form turning is used to produce a regular or irregular shape around a cylindrical surface. *Cutting off (parting)* is the separating of material by cutting through a revolving workpiece. The purposes, general cutting tools, and setups for grooving, form turning, and cutting off, are covered in this unit.

COMMON SHAPES OF GROOVES

There are three basic shapes of grooves: square, round, and angular (V-shaped). The cutting tools that produce these shapes are named: square-, round-, and V-nosed grooving tools.

SQUARE GROOVE

The square groove may be cut at a shoulder or at any point along the workpiece. As stated earlier, the square groove provides a channel. This channel permits a threading tool to ride into the groove or a mating part to shoulder squarely. Parts that are to be cylindrically ground are often undercut at a shoulder. The undercut makes it possible for the face of the grinding wheel to grind the diameter to size along the entire length.

The square groove may be turned with a cutoff (parting) tool. A standard high-speed steel tool bit or a carbide-tipped bit ground to the required width are commonly used.

ROUND GROOVE

The round groove eliminates the sharp corners of the square groove. A round groove is particularly important on parts that are hardened or that are subjected to severe stresses. On these parts a sharp corner may crack and fracture easily.

The round groove also provides a pleasing appearance for ending such work as knurled areas. Round grooves may be produced by regular cutting tools that are formed to the required shape and size.

V-GROOVE

Deep V-grooves are usually turned by offsetting the compound rest at the required angle. One angular side of a V-groove is cut at a time. Shallow V-grooves may be cut by using a formed V-cutting tool like the tool shown in Figure 24–1.

Special care must be taken when cutting a deep groove in a long, small-diameter workpiece or in work that has been turned between centers. A follower rest should be positioned to support the workpiece behind the cutting edge.

Grooving tools are end-cutting tools. They are ground with relief for the cutting end and sides. Cutting speeds are decreased in grooving because the cutting area is greater than in straight turning. Cutting fluids and fine feeds are required to produce a fine surface finish.

FORM TURNING

PREFORMED CUTTING TOOLS

Form turning is a practical way of producing regular shapes as well as intricate contours.

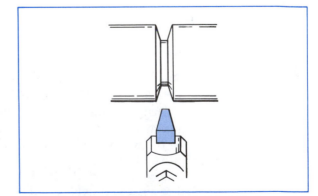

Figure 24–1 A Form-Ground Tool Bit for Turning a V-Groove

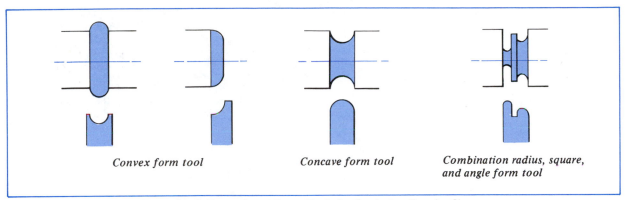

Convex form tool　　　　*Concave form tool*　　　　*Combination radius, square, and angle form tool*

Figure 24–2 Examples of Form Tools for Producing Regular Shapes

Preformed tools—that is, tools that are precision ground to a required shape and size—are used in one method of form turning. Such tools are sharpened on the top cutting face to maintain the accuracy of the form. One advantage of using an accurately ground form tool is that multiple pieces may be reproduced to the same shape and size. Gaging and measuring each part thus may be eliminated.

The contours produced are usually a combination of a round, angular, or square groove or a raised section. The radius may be concave or convex. A convex radius requires the workpiece to be turned first with a raised collar section. The collar is then reduced to shape and size with a convex-radius form tool. Figure 24–2 provides examples of forms that are turned with preformed cutting tools having concave and convex radii and a contour consisting of concave radii, square, and angular sections.

The cutting action with formed turning tools takes place over a large surface area. Accordingly, the spindle speed must be reduced to about one-half the RPM for regular turning. On very large and deep contours a gooseneck toolholder is used. This holder helps to eliminate chatter and produces a fine surface finish.

PRODUCING A GROOVED FORM BY HAND

Another common practice in form turning is to form the groove by *hand*. This method is used when one part or just a few parts are to be machined and no template is available. The form is produced by the operator. A partially formed cutting tool is fed into the revolving workpiece. The longitudinal and transverse feeds are applied manually and simultaneously. The amount the carriage moves in relation to the movement of the cross slide governs the shape that is turned.

CUTTING-OFF (PARTING) PROCESSES

Cutting-off processes are usually performed with a specially formed thin blade that is rectangular in shape. The sides of the cutting tool are shaped so that there is a relief angle. The blade is held solidly against a corresponding angular surface of a straight or offset toolholder. Cutoff blades are made of high-speed steel, carbides, and cast alloy. Four common shapes of blades are shown in Figure 24–3. Blade holders are commercially available in straight and right- and left-hand offset styles.

Small workpieces are often cut off with a standard tool bit that is ground as an end-cutting parting tool. The depth to which the narrow blade or point enters the workpiece, the extent the blade must extend, and the fragile nature of this cutting-off tool requires that special precautions be taken.

PROBLEMS ENCOUNTERED IN CUTTING OFF STOCK

The lathe operator must recognize four major problems in cutting off (parting) stock. A simplified statement of each problem, the probable cause, and the recommended correction of each problem is given in Table 24–1.

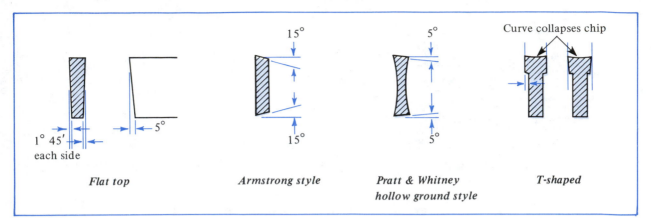

15°

5°

Curve collapses chip

1° 45'
each side

5°

15°

5°

Flat top

Armstrong style

*Pratt & Whitney
hollow ground style*

T-shaped

Figure 24–3 Types of HSS, Carbide, and Alloy Cutoff Blades

How to Cut Off (Part) Stock

A common cutting-off setup is shown in Figure 24–4. After positioning and setting the spindle speed at about one-half the rpm for straight turning, the parting tool is fed slowly and continuously through the workpiece. Cutting fluids are most always used for cutting-off processes.

Table 24–1 Cutting-Off Problems: Causes and Corrective Action

Cutting-Off Problem	*Probable Cause*	*Corrective Action*
Chatter	Speed too great	Decrease the cutting speed
	Tool extended too far	Clamp the cutting tool short
	Play in cross slide and compound rest	Adjust the gibs on cross slide and compound rest
	Backlash in compound rest feed screw	Back the compound rest away from the work; turn the feed screw to the left at least one-half turn
	Play in lathe spindle	Adjust the end play in the spindle bearings
Rapid dulling of cutting tool	Cutting speed too high	Decrease the cutting speed
	Too much front clearance	Decrease the front clearance angle
	Lack of cutting lubricant	Apply lard oil or required cutting compound for part being turned
	Overheating and drawing the tool temper due to insufficient clearance	Grind front, side, or back clearance
	Overheating due to excessive cutting speed	Decrease spindle speed
Cutting tool digging into work and work climbing on cutting tool	Tool set too low	Set tool at height of lathe center
	Play in compound and cross slide	Adjust gibs on cross slide and compound rest
	Backlash in compound rest feed screw	Back the compound rest away from the work; turn the feed screw to the left at least one-half turn
	Too much front clearance	Decrease the front clearance angle
	Too much top rake	Decrease the top rake
Cutting tool riding on work	Tool too high	Set tool at the height of the lathe center
	Insufficient front clearance	Increase the amount of the front clearance angle

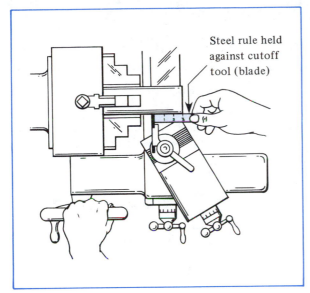

Steel rule held against cutoff tool (blade)

Figure 24-4 Locating the Cutoff Tool at a Required Length

Safe Practices in Grooving, Form Turning, and Cutting Off

- Set the spindle speed at about one-half the RPM that is used for straight turning with high-speed steel tool bits or cutoff blades. Apply this general rule to grooving, form turning with a preformed cutting tool, and cutting off.
- Back up a workpiece with a follower rest when turning deep grooves that may weaken the workpiece.
- Use a straight or right-hand offset toolholder. Hold the stock so that the least amount extends beyond the holding device. Cut off as close to the holding device and spindle nose as possible.
- Use a protective shield or goggles. The hot chips tend to spatter. Avoid handling hot pieces, particularly at the final stage of parting.

UNIT 24 REVIEW AND SELF-TEST

1. List four reasons for grooving.

2. Identify two methods of form turning on a lathe.

3. Explain why a gooseneck toolholder is used for form turning large, deep contours.

4. State four common cutting-off problems in lathe work.

5. Explain what the drawing notation CUT TO THREAD DEPTH means in relation to a groove.

6. Indicate the main work processes to turn a groove.

7. List the steps to follow in cutting off a workpiece to a specified length.

8. List four safe practices that must be observed in grooving, form turning, and cutting off.

Lathe Filing, Polishing, and Knurling

Filing and polishing are two lathe processes that produce a smooth, polished surface finish. A minimum amount of 0.002″ to 0.003″ is left on a workpiece outside diameter for filing and polishing.

FILING TECHNOLOGY AND PROCESSES

TYPES OF FILES

Two common types of lathe files are used on ferrous metals. The *single-cut bastard mill file* is a general-purpose file for lathe work. A 10″ to 12″ *long-angle lathe file* is also commonly used (Figure 25–1). This file has sides (width and thickness) that are parallel. It has two uncut (safe) edges that permit finish filing to a shoulder. The long-angle lathe file has features that partially eliminate chatter, provide for rapid cleaning of chips, and reduce the probability of scoring a workpiece. Bastard, second-cut, and smooth-cut mill files produce a fine-grained surface.

Three other types of files are used on non-ferrous metals. They are the *aluminum file*, the *brass file*, and the *super-shear file*. The first two files are shown in Figure 25–2.

The teeth of the aluminum file are designed to quickly clear chips and to eliminate clogging. The combination of deep undercut and fine overcut features produces a tooth form that tends to break up the file particles. The chips are cleared during the filing process. The tooth form also helps to overcome chatter.

Brass, bronze, and other tough, ductile metals are finish filed with a brass file. The teeth on this file have a deep, short upcut angle and a fine, long-angle overcut. The design breaks up the chips so that the file clears itself.

Aluminum, copper, and most other non-ferrous metals may be filed on the lathe with a super-shear file. The teeth are cut in an arc (Figure 25–3). The arc is off center in relation to the file axis. The milled, double-purpose tooth permits free cutting and easy removal of chips from the teeth and produces a smooth surface finish.

FILING A TURNED SURFACE

Cutting is always done on the forward stroke when filing by hand. Strokes are overlapped to about one-half the width of the file, as is done

Courtesy of NICHOLSON DIVISION; THE COOPER GROUP

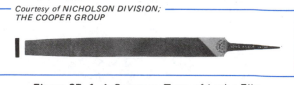

Figure 25–1 A Common Type of Lathe File

Courtesy of NICHOLSON DIVISION; THE COOPER GROUP

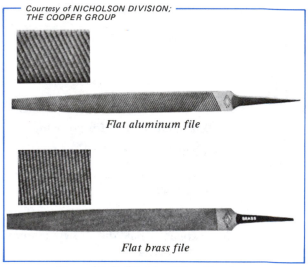

Flat aluminum file

Flat brass file

Figure 25–2 Files for Filing Aluminum and Brass on a Lathe

Courtesy of NICHOLSON DIVISION; THE COOPER GROUP

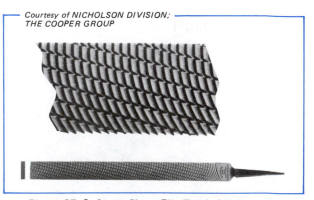

Figure 25–3 Super-Sheer File Teeth Cut in an Arc

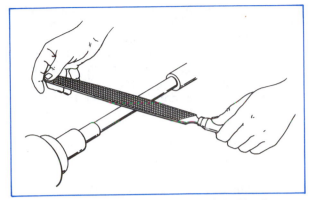

Figure 25–4 Correct Positions of the Hands and File for Filing Left-Handed on a Lathe

for bench filing. The spindle speed used for filing should be about twice as fast as the speed used for turning. A revolving dead center is preferred for filing and polishing work that is held between centers. A speed of 30 to 40 strokes per minute provides good control of a file in removing the limited amount of material that is left for filing. Filing is also intended to remove light tool marks.

Only a slight force should be applied to a file. Excessive force produces an out-of-round surface and tends to clog the teeth and score a workpiece. A file must be cleaned frequently with a file card and brush. The teeth are sometimes chalked to prevent clogging and to simplify cleaning. The extreme sharpness of the fine cutting edges of a new file is removed by filing the flat surface of a cast-iron block. Filing lathe work left-handed is the preferred, safer method (Figure 25–4).

POLISHING ON THE LATHE

After completion of the filing process, a smoother polished surface is obtained by using an abrasive cloth. Polishing is done by pressing and moving the abrasive cloth across the revolving workpiece. Aluminum oxide abrasive cloths are used for polishing steels and most ferrous metals. Silicon carbide is widely used for non-ferrous metals. Grit sizes of 80 to 100 are generally used. Finer grained and more highly polished surfaces are produced with grit sizes of 200 or finer.

High spindle speeds are required for polishing on the lathe. During the initial polishing, greater force may be applied by holding an abrasive strip against a file. Long, regular strokes are used against the revolving workpiece. A still higher gloss finish may then be produced by using a worn abrasive cloth and applying a few drops of oil.

Polishing straps are common when polishing large diameters. The abrasive cloth is held by two half sections of a form (polishing strap). Force is applied on the polishing strap to press it against the surface that is turning.

KNURLS AND KNURLING

A *knurl* is a raised area impressed upon the surface of a cylindrical part. *Knurling* is the process of forcing a pair of hardened rolls to form either a *diamond-shaped* or *straight-line pattern*. Knurling is a *material displacement* process. It requires great force for the knurl rolls to penetrate a workpiece surface and impress a pattern.

Courtesy of J.H. WILLIAMS DIVISION; TRW INC.

Helical-ridge knurl rolls form diamond pattern (three pitch sizes)

Three pitches of straight-line pattern knurl rolls

Figure 25–5 Knurl Rolls for Diamond and Straight-Line Pattern Knurls

The *diamond pattern* is formed by overlapping the shape produced by two hardened rolls. Each roll has teeth or ridges. The set of teeth on one roll has a right-hand helix (lead). The teeth on the other roll are cut with a left-hand helix. The general patterns and pitches of knurl rolls are shown in Figure 25–5. One roll forms a series of right-hand ridges; the other roll forms left-hand ridges. These two series of ridges cross to form the diamond-shaped pattern. The *straight-line pattern* is formed by using two hardened steel rolls that have grooves cut parallel to the axis of each roll.

There are three general purposes for knurling:

- To provide a positive gripping surface on tools, instruments, or work parts (this surface permits ease of handling and precise adjustment);
- To raise the surface and increase the diameter to provide a press fit or an irregular surface (the knurl produces a gripping surface that prevents two mating parts from turning, as in the case of a plastic handle on the shank of a screwdriver);
- To add to the appearance of the work.

KNURL ROLL SIZES AND TYPES OF HOLDERS

Knurl rolls are available in three basic sizes that are identified by *pitch*. The three sizes and their pitches are *coarse* (14 pitch), *medium* (21 pitch), and *fine* (33 pitch). Pitch refers to the

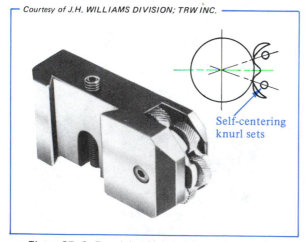

Self-centering knurl sets

Figure 25–6 Revolving Head with Six Knurl Rolls Mounted in Dovetail Slide for Quick-Change Tool Post

number of teeth per linear inch. A knurl roll is heat treated and rides on a hardened steel pin.

A single set of knurl rolls is usually held on a self-centering-head toolholder. Some holders are made for a single set and pitch. A multiple-head holder contains three sets of coarse, medium, and fine knurl. The knurl rolls are mounted in a revolving self-centering head that pivots on a hardened pin. The multiple heads are also designed to fit a dovetailed quick-change tool post (Figure 25–6).

FORMING A KNURLED PATTERN

The following practices must be followed to avoid a few problems that are common in knurling:

- The knurl rolls must be kept in continuous contact with the workpiece until the process is completed.
- The automatic feed must be engaged continuously to traverse the full length of the knurled surface. Otherwise a ring is formed that produces a knurled surface with a varying pattern that is impractical to correct.
- A double-impression knurled pattern is produced when uneven pressure is applied on both knurl rolls. Double impressions may be corrected by
  - Positioning the knurl rolls on center,
  - Tracking the teeth carefully,
  - Infeeding the knurl rolls again so that the teeth penetrate evenly.
- The knurl teeth must be cleaned before the operation and *while the machine is stopped.*
- The flow of cutting fluid must be adequate to wash any particles away from the work surface and the rolls.
- A follower rest and/or a steady rest may be required to offset the force applied during the forming of the knurl.

How to Knurl

STEP 1 Check the condition of the knurl rolls. Clean the teeth. Mount the knurl rolls on center with the faces parallel to the workpiece (Figure 25–7).

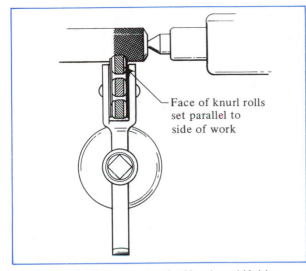

Face of knurl rolls
set parallel to
side of work

Figure 25-7 Positioning the Knurls and Holder

STEP 2 Set the spindle speed at about one-half the RPM used for turning. Select a carriage feed of from 0.020" to 0.030" (0.5mm to 0.8mm). The amount of feed depends on the pitch of the knurl. Start the lathe.

STEP 3 Move the knurling tool until about one-half of the face bears on the workpiece. Feed in the tool to a depth of 0.020" to 0.025" (0.5mm to 0.6mm).

STEP 4 Stop the lathe. Check the correctness of the pattern.

STEP 5 Position the cutting fluid nozzle to flow a small quantity of cutting fluid on the knurls.

Note: Use a follower rest or steady rest when support is required.

STEP 6 Engage the automatic feed. Start the lathe and feed for the full length of the knurled section.

STEP 7 Stop the lathe. Use a stiff brush to remove particles of the workpiece material from the knurls.

STEP 8 Check the knurled pattern. Continue to feed the rolls in to depth. Reverse the automatic feed. Take successive passes across the workpiece until a clean, smooth crest pattern is produced.

Safe Practices in Filing, Polishing, and Knurling

- Be sure the file handle fits correctly before beginning the lathe filing. A revolving workpiece tends to force the file toward the operator.
- File left-handed.
- Chalk the file teeth of a new file to remove excessive sharpness from the tooth edges. Chalking the file and using a light force prevent scoring and produce a fine, smooth finish.
- Keep the fingers, wiping cloth, or brush away from a revolving workpiece and the knurl rolls.
- Check the support of a workpiece. A follower and/or steady rest may be needed to prevent bending or distortion of the workpiece during knurling.
- Disengage the lead screw and feed rod when filing and polishing to prevent them from being accidentally engaged at a high spindle speed.

UNIT 25 REVIEW AND SELF-TEST

1. State the functions served by the deep undercut and the fine overcut features of an aluminum file.

2. Indicate the characteristics of a brass file.

3. Describe the shape and functions served by super-shear file teeth.

4. State what purpose is served by filing a workpiece on a lathe.

5. Indicate the kind of abrasive cloth to use (a) to polish steels and other ferrous metals and (b) to polish nonferrous metals.

6. State three guidelines to follow in polishing a workpiece on a lathe.

7. State two reasons for knurling.

8. Specify the basic knurl sizes and patterns.

9. Identify three common knurling problems and their causes.

10. List three safety precautions to observe when filing on a lathe.

11. Explain why the lead screw and feed rod on a lathe must be disengaged when filing or polishing.

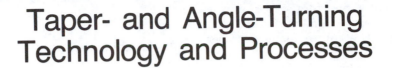

Taper- and Angle-Turning Technology and Processes

PURPOSES AND TYPES OF TAPER AND TAPER SYSTEMS

The six standard taper series that are widely used in the machine and metal industries are:

- American Standard Taper Pin,
- Jacobs Taper,
- American Standard (Morse) Taper,
- Brown & Sharpe Taper,
- Jarno Taper,
- American Standard Self-Releasing Steep Taper.

AMERICAN STANDARD TAPER PINS

American Standard Taper Pins are widely used when two mating parts must be accurately positioned and held in a fixed position. The taper pin permits easy assembling and disassembling. Taper pins are furnished in standard lengths in relation to the taper size. Taper pins have a standard *taper per foot* of either 0.2500″ or 0.0208″ per inch.

The general sizes are numbered from #7/0 to #2/0 and #0 to #11. Taper pin holes are usually reamed. The reamer sizes are designed so that one size overlaps the next succeeding size. The diameters at the small end of the taper range from approximately 1/16″ (for the #7/0 size) to 41/64″ (for the #10 size). American Standard Taper Pin tables give the large and small diameter, length of each pin, and the drill size.

JACOBS TAPER

The *Jacobs Taper* is often used with a standard Morse or Brown & Sharpe taper that fits standard machine spindles. The Jacobs is a short, self-holding taper that corresponds with a short taper bore. The bored holes are used on drill chucks or external tapers of a shaft, shank, or spindle. The Jacobs taper is widely applied on portable power-driven tools.

AMERICAN STANDARD (MORSE) TAPER

The *American Standard (Morse) Taper* has a small taper angle. This angle produces a wedging action and makes the Morse a self-holding taper. The Morse is the most extensively used taper in the machine industry. It is used on common cutting tool features such as the shanks of twist drills, reamers, counterbores, and countersinks and on machine tool features such as drill press spindles and lathe spindles.

The Morse taper series is numbered from #0 to #7. The small-end diameters of this series

Table 26–1 Sample Dimensions: American Standard (Morse) Taper

Number of Taper	Diameter at Small End (d)	Standard Plug Length (L)	Diameter at Gage Point (D)	Taper per Foot	Taper per Inch
#1	0.369″	2″	0.475″	0.5986″	0.0499″
#3	0.778	3 3/16	0.938	0.6024	0.0502
#7	2.750	10	3.270	0.6240	0.0520

range from 0.252″ to 2.750″. The lengths vary from 2″ to 10″. Unlike the other taper series, the amount of taper per foot varies in the Morse series with each number except #0, #4 1/2, and #7. The approximate taper per foot is 5/8″.

Table 26–1 lists the kind of information that is required for designing and machining an American Standard (Morse) taper.

BROWN & SHARPE TAPER

The *Brown & Sharpe (B & S) Taper* is another self-holding taper. The B & S taper series is numbered from #1 to #18. Within this series the diameters at the small end of the taper range from 0.200″ to 3.000″. The B & S taper is 0.502″ per foot (0.0418″ per inch) for #1, #2, #3, and #13. The taper varies from 0.5024″ to 0.4997″ for numbered tapers #4 through #14. The taper for #14 through #18 is 0.5000″.

JARNO TAPER

The *Jarno Taper* series (#1 to #20) uses a uniform taper per foot of 0.600″. The number of a Jarno taper is directly related to the small diameter, the large diameter, and the length. The small-end diameters of the series range from 0.100″ to 2.000″. The following simple formulas may be used to calculate the sizes:

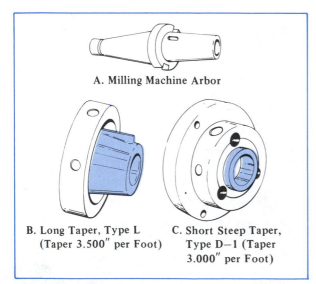

A. Milling Machine Arbor

B. Long Taper, Type L (Taper 3.500″ per Foot)

C. Short Steep Taper, Type D–1 (Taper 3.000″ per Foot)

Figure 26–1 Examples of Steep, Self-Releasing Tapers

$$\text{Diameter at Small End} = \frac{\text{\# of Jarno Taper}}{10}$$

$$\text{Diameter at Large End} = \frac{\text{\# of Jarno Taper}}{8}$$

$$\text{Length of Jarno Taper} = \frac{\text{\# of Jarno Taper}}{2}$$

For example, a #6 Jarno taper is 6/10″ (0.600″) at the small end of the taper and 6/8″ (0.750″) at the large end and is 6/2″ (or 3″) long.

AMERICAN STANDARD SELF-RELEASING STEEP TAPER

The *American Standard Self-Releasing Steep Taper* series is similar to the earlier Milling Machine Taper series. The taper of 3.500″ per foot ensures easier release of arbors, adapters, and similar accessories from the spindles of milling and other machines. Because the steep tapers are not self-holding, they require slots or keys to drive an accessory (Figure 26–1A). The mating parts are usually drawn and held together with a draw bolt, cam-locking device, or nut. The lathe type-L spindle nose has a steep taper, a key drive, and a threaded ring nut (Figure 26–1B). The features of this nose accurately position, hold, and drive a mounted faceplate, driver plate, or chuck.

The short steep taper (3.000″ per foot) of the type D–1 lathe spindle nose centers the arbor or accessory (Figure 26–1C). The cam-lock device holds the arbor or accessory accurately and firmly on the spindle nose.

TAPER DEFINITIONS AND CALCULATIONS

TAPER FORMULA TERMS

The following terms are used in the formulas for calculating taper features:

- T designates *taper* (the difference in size between the large diameter and the small diameter),
- T_{pi} is the *taper per inch* (the amount the workpiece diameter changes over a 1″ length),
- T_{pf} is the *taper per foot* (the change in diameter in a 1′ length),
- D refers to the *large-end diameter* of the taper,

- d denotes the *small-end diameter* of the taper,
- L_t is the *length* of the taper in inches,
- L_o is the *overall length* of the workpiece.

CALCULATING TAPER VALUES

The *taper* (T) equals the difference in the large (D) and small (d) diameters:

$$T = D - d$$

The *taper per foot* (T_{pf}) equals the difference between the large (D) and small (d) diameters in inches multiplied by 12 and divided by the required length of taper (L_t). Expressed as a formula,

$$T_{pf} = \frac{(D - d) \times 12}{L_t} \quad \text{Also,} \quad T_{pf} = T_{pi} \times 12$$

The *taper per inch* (T_{pi}) equals the difference between the large diameter (D) and small diameters (d) divided by the length of the taper in inches (L_t):

$$T_{pi} = \frac{D - d}{L_t} \quad \text{or} \quad T_{pi} = \frac{T_{pf}}{12}$$

When the taper per inch (T_{pi}) is known, the taper per foot (T_{pf}) is found by simply multiplying by 12:

$$T_{pf} = T_{pi} \times 12$$

CALCULATING THE DIAMETERS

The *large diameter* $D = (T_{pi} \times L_t) + d$ If the taper per foot (T_{pf}) is given,

$$D = \frac{(T_{pf} \times L_t)}{12} + d$$

The *small diameter (d)* may be calculated by using the formula: $d = D - (L_t \times T_{pi})$. If the taper per foot (T_{pf}) is given,

$$d = D - \frac{(L_t \times T_{pf})}{12}$$

CALCULATING THE TAILSTOCK OFFSET

The offset tailstock method is used to cut external shallow tapers. The amount to offset a tailstock depends on the overall length of the workpiece (L_o) and the amount of taper (T_{pi} or T_{pf}). Simple formulas are used to compute the required tailstock offset (T_o).

When the taper is given as *taper per inch* (T_{pi}),

$$T_o = \frac{T_{pi} \times L_o}{2}$$

When the taper is expressed as *taper per foot* (T_{pf}),

$$T_o = \frac{T_{pf} \times L_o}{24}$$

When a taper is dimensioned with a *small diameter* (d), *large diameter* (D), *length of taper* (L_t), and the *overall length* (L_o),

$$T_o = \frac{L_o \times (D - d)}{L_t \times 2}$$

DIMENSIONING TAPERS ON CYLINDRICAL PARTS

TAPERS ON CYLINDRICAL PARTS

The taper on a round workpiece is represented by two symmetrical tapering lines. The amount of taper is usually specified in terms of taper per foot or taper per inch. For example, a standard taper pin that has a taper of one-quarter inch per foot is indicated by the note:

0.250″ (or 1/4″) TAPER PER FOOT

MEASURING AND GAGING TAPERS

Tapered surfaces must be precision machined for two mating parts to fit and align accurately and to provide maximum holding ability for small-angle tapers. There are four common methods of measuring the accuracy of the taper angle and/or the size of the large and small diameters:

- Gaging with a taper plug or ring gage,
- Measuring with a standard micrometer,
- Measuring with a taper micrometer,
- Testing with a sine bar setup.

TAPER PLUG GAGE METHOD

The *taper plug gage* is hardened and precision ground (Figure 26–2A). The face and a ground area are identified as steps (1) and (2). When the small end of a correctly turned taper

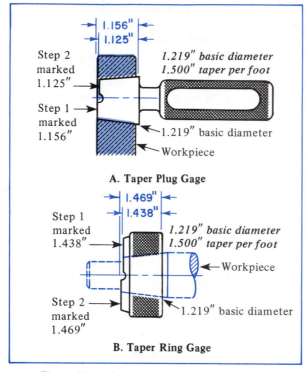

Figure 26-2 Features of Taper Plug and Ring Types of Go-Not-Go Gages (Step Style)

falls between these steps, the part is turned to size within the allowable taper limits.

TAPER RING GAGE METHOD

A hardened *taper ring gage* has a precision-ground tapered hole (Figure 26-2B). Part of the outside of the body is cut away (1). One or more index lines are marked on the flat section (2). When the taper is correct and the end of the tapered part cuts an index line, the outside diameters are accurately machined to size.

The taper must first be checked for accuracy. Usually, a fine coating of Prussian blue is applied to the machined part. The gage is then placed carefully on the taper. If the machined part is turned slightly and then removed, the high areas on the machined tapered surface are identified easily. Once the taper fits accurately, the diameters may then be checked by gage or micrometer measurement.

A few simple precautions must be taken. The mating part must be clean and free of burrs and nicks. Since a small-angle taper is self-locking when a small force is applied, the workpiece should be turned clockwise carefully. A counterclockwise withdrawing motion makes it easy to turn and remove the tapered plug or ring gage. Excessive turning causes undue wear on the gage.

TAPER MICROMETER

The *taper micrometer* provides a more reliable and accurate measuring instrument and method than does the standard micrometer. The taper micrometer includes an adjustable anvil and a 1″ sine bar that is attached to the frame. The sine bar is adjusted by the movement of the spindle.

The accuracy of the taper is measured by placing the taper micrometer over the workpiece. The thimble is adjusted until the jaws just touch the tapered surface. The reading on the spindle indicates the taper per inch. This reading may need to be converted to taper per foot or angle of taper.

TAPER TURNING PROCESSES

TAPER ATTACHMENT

The taper attachment provides a quick, economical, practical way of accurately turning internal and external tapers. In principle the taper attachment guides the position of the cutting tool in an angular relation to the lathe center axis. A sliding block moves along the guide bar of the taper attachment. The angular setting of the guide bar is transmitted by the sliding block to produce a similar movement of the cutting tool. There are two basic types of taper attachments. These types are the *plain taper attachment* and the *telescopic taper attachment*.

The plain taper attachment requires the removal of the binding screw that connects the cross feed screw to the cross slide. The cross slide is then moved by securing the sliding block to the extension arm on the cross slide. The compound rest feed handwheel is turned to set the depth of cut with the plain taper attachment.

The cross slide and cross feed screw and nut are not disengaged with the telescopic taper

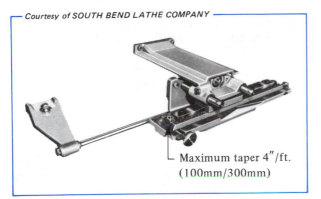

Courtesy of SOUTH BEND LATHE COMPANY

Maximum taper 4″/ft.
(100mm/300mm)

Figure 26–3 A Permanently Mounted
Telescopic Taper Attachment

attachment (Figure 26–3). The depth of cut may be set directly with the cross feed hand-wheel.

ADVANTAGES OF TURNING A TAPER WITH A TAPER ATTACHMENT

One advantage of using a taper attachment is that tapers may be turned accurately, economically, and with less danger of work (center) spoilage. There are also many other advantages of turning a taper with a taper attachment:

- Alignment of the live and dead centers, required by other methods, is eliminated.
- The workpiece is supported by the full angular bearing surface at the center holes.
- Tapers may be cut on work held in a chuck, between centers, or on a faceplate, fixture, or other work-holding setup.
- A greater range of inch- and metric-standard taper sizes may be turned.

- Internal and external tapers may be cut with the same setting of the taper attachment. This one setup permits machining the matching parts accurately.
- The taper attachment may be positioned directly at the required taper or in degrees of taper. The base of the taper attachment has a scale graduated in taper per foot, millimeters of taper, and degrees of taper. Calculation of these dimensions is thus eliminated.
- The angle of taper is not affected by any variation in the length of the workpiece. The one taper setting may be used to produce multiple parts with the same taper, regardless of any variation in the length of the workpiece.

ANGLE TURNING WITH THE COMPOUND REST

GRADUATIONS ON THE COMPOUND REST SWIVEL SLIDE

While all compound rest swivel slides are graduated in degrees, the arrangement is not standardized. For example, a graduated swivel slide may be marked from 90° to 0° to 90° or 0° to 90° to 0°. Two zero (0) index lines on the cross slide provide reference points.

Angles are usually stated as an *included angle* or as an *angle with a centerline* (one-half the amount of the included angle). The major parts of a compound rest with a zero reference and graduated slide are shown in Figure 26–4.

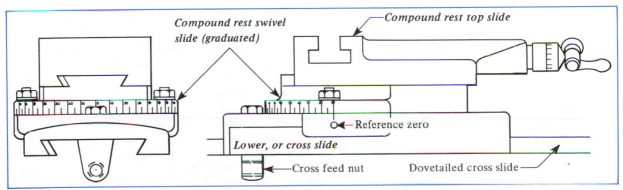

Compound rest swivel slide (graduated)

Compound rest top slide

Reference zero

Lower, or cross slide

Cross feed nut

Dovetailed cross slide

Figure 26–4 Major Parts of a Lathe Compound Rest

COMPOUND REST SETUP
FOR ANGLE TURNING

The angle at which the compound rest is set usually corresponds to the angle with a center line. Thus the angle produced is the included angle. The degree reading on the swivel slide depends on the axis from which the compound rest is set and the arrangements of the numbered graduations.

Short, steep angles are usually turned by swiveling the compound rest. Three common setups are described. If an included angle is shown on a drawing, the compound rest is positioned at one-half of this included (required) angle. For example, to turn the 60° center, the compound rest is swiveled to 30°. A 45° angle may be produced by swiveling the compound rest 45° to the transverse axis. The two sides of a 70° included angle are turned by swiveling the compound rest to a 35° angle and turning one side. The compound rest is then swiveled 35° in the opposite direction. The turned included angle is then 70°.

The position of the compound rest for feeding in relation to the lathe center or cross slide axis is influenced by safety and convenience considerations. Feeding with the compound rest handwheel from a position in front of the workpiece is the safer and most convenient method.

COMPUTING COMPOUND
REST ANGLE SETTING

The compound rest is used occasionally to turn a short taper. The dimension of the taper is usually stated in terms of taper per foot (T_{pf}). This dimension must be converted to degrees and minutes to correspond with the graduations on the swivel slide. The angle may be computed by one of the following two methods.

CONVERTING T_{pf} TO ANGLE
WITH CENTERLINE SETTING

The angle to which the compound rest must be set in relation to the centerline is calculated by using the following formula:

Compound Rest Angle = T_{pf} × 2.383

Example: Calculate the compound rest angle in relation to the centerline for turning a short taper of 1″ T_{pf}.

$$\text{Compound Rest Angle} = T_{pf} \times 2.383$$
$$= 1'' \times 2.383$$
$$= 2.383°$$

Note: The graduations on the compound rest are in divisions of one degree. In this example the worker must estimate 3/8 of one degree (0.383°). After the compound rest is set, a trial cut is taken. The T_{pf} is measured. Further adjustment of the compound rest then is made if needed.

TANGENT OF THE ANGLE
WITH THE CENTER LINE

The following formula is used to calculate the tangent (tan) of the angle to which the compound rest must be set in relation to the center line:

$$\tan = \frac{T_{pf}}{24}$$

Example: Using a T_{pf} of 1″, the tangent is equal to 1/24, or 0.04167. A table of natural tangent values shows that 0.04167 represents an angle of 2°23′.

Safe Practices in Turning
Tapers and Angles

- Check the size of the bearing surface of the center holes. They must be large enough to permit the workpiece to be moved out of alignment when using the offset tailstock method.
- Lubricate the dead center hole. Carefully adjust the workpiece so that it may turn freely.
- Test the lathe dog to see that the tail is free to ride in the driver plate slot.
- Move the cutting tool away from the workpiece and then position it for the next cut. This movement takes up the lost motion of the taper attachment and setup.

- Turn the tapered workpiece clockwise carefully when testing for accuracy. To release the tapered surfaces, reverse the turning direction (counterclockwise) and gently pull outward to release the tapered surfaces without damaging them.

UNIT 26 REVIEW AND SELF-TEST

1. Identify four main purposes that are served by tapers.

2. a. Give the general range of numbered sizes of taper pins.
   b. State two major applications of American Standard Taper Pins.

3. Distinguish between a Jacobs taper and an American Standard (Morse) taper.

4. State the taper per inch for (a) Brown & Sharpe tapers and (b) Jarno tapers.

5. Differentiate between the holding devices for an American Standard self-releasing steep taper and a type D–1 short steep taper (3.000″ per foot).

6. Define each of the following taper terms or give the formula for calculating each value: (a) taper, (b) taper per foot, (c) large diameter, and (d) small diameter.

7. a. Identify two common methods of turning shallow tapers.
   b. Tell how a taper is represented and dimensioned on a drawing.

8. List three general methods of measuring the accuracy of the taper angle and/or the large and small diameters.

9. State four advantages of turning a taper with a taper attachment compared to the offset tailstock method.

10. List the steps required to turn an included angle of 60° on a workpiece between centers and using a compound rest.

11. Tell how to turn two tapered mating parts when testing the accuracy of the tapers.

12. State two safe practices the lathe worker must observe when turning tapers.

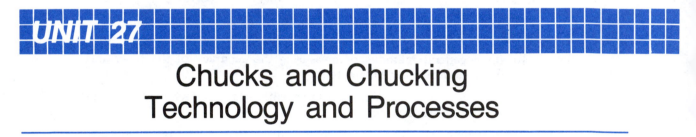

SECTION FOUR
Lathe Work Held in a Chuck

Boring, reaming, drilling, internal and external thread cutting, and other lathe processes generally require that the workpiece be mounted in a chuck or a collet. This section opens with descriptions of the types of chucks typically used for these processes and the methods of mounting and truing work and centering or off-centering for eccentric turning. Safety practices are stressed to prevent personal injury or damage to the machine, accessories, or tools. Subsequent units deal with the principles of and procedures for performing the following lathe processes:

- Centering, drilling, countersinking, and reaming;
- Straight hole boring, counterboring, recessing (undercutting), taper boring, and mandrel work;
- Cutting and measuring 60° form internal and external threads;
- Advanced thread cutting.

UNIT 27

Chucks and Chucking
Technology and Processes

THREADED AND STEEP-TAPER SPINDLE NOSES AND CHUCK ADAPTER PLATES

TYPE-L SPINDLE NOSE

As stated earlier, there are three main types of spindle noses. The *threaded spindle nose* is the oldest type. The chuck flange and the chuck itself are centered by the accuracy of the threads, a straight-turned section, and the spindle nose shoulder.

The American Standard *type-L spindle nose* has a taper of 3.500″ per foot (Figure 27–1). The taper bore of the chuck flange centers on the corresponding taper of the spindle nose. Note also that the chuck flange has a keyway. This keyway fits the key in the tapered nose of the spindle. The key and keyway prevent the chuck from turning on the tapered surface of the spindle nose.

The chuck is held securely on the taper and concentric with the spindle by a lock ring. A spanner wrench is used for both tightening and loosening the lock ring.

TYPE D–1 CAM-LOCK SPINDLE NOSE

There are five main design features of the *D–1 cam-lock spindle nose*. These features are

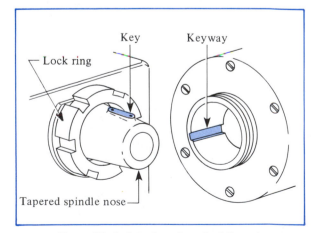

Figure 27–1 American Standard Type-L Spindle Nose with a Key

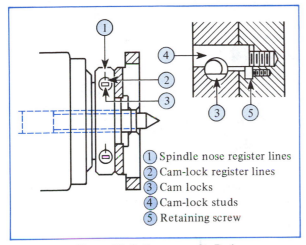

Figure 27–2 Features of a D–1 Cam-Lock Spindle Nose

1 Spindle nose register lines
2 Cam-lock register lines
3 Cam locks
4 Cam-lock studs
5 Retaining screw

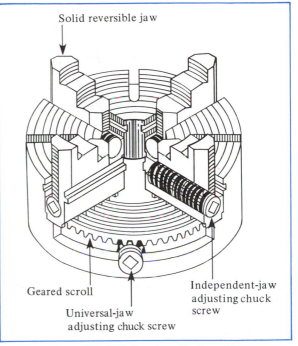

Figure 27–3 A Phantom Section of a Four-Jaw Combination Chuck

identified in Figure 27–2. The short, steep taper of the D–1 nose is 3.000″ per foot. There are from three to six cam-lock studs ④ depending on the chuck size. The studs extend from the chuck or other accessory. The studs fit into corresponding holes in the face of the spindle nose. There are also eccentric cam locks ③ that match. As the eccentric cam lock turns against the stud, the accessory is drawn firmly onto the taper and against the face of the spindle flange.

There are register lines ① for each of the cam locks. The index line of each cam lock ② must match with the register position to mount or remove the accessory. The cam locks are given a partial turn clockwise to mount a chuck; counterclockwise, to remove it.

UNIVERSAL THREE-JAW CHUCK

The name of the *universal three-jaw chuck* indicates that there are three accurate, self-centering jaws. These jaws are controlled by a bevel gear-driven scroll. All three jaws may be actuated (moved) by turning any one of the adjusting sockets. Since the universal chuck is self-centering, the jaws do not require individual setting as do independent-jaw chucks.

Due to the shape of the screw thread on their back sides, the jaws are not reversible. Other jaw sets are available to accommodate

large-diameter work and work that requires inside chucking.

FOUR-JAW INDEPENDENT CHUCK

Each jaw of the *four-jaw independent chuck* is adjusted independently (Figure 27–3).

The chuck jaws are stepped to take work of small and large diameters. They may be used to hold workpieces on either an inside or an outside diameter. The jaws are also reversible. The four-jaw independent chuck can accommodate a wider range of shapes and sizes than the universal chuck.

The four-jaw chuck may be used to grip square, round, or irregular-shaped pieces. The jaws may be positioned concentric or off center, depending on the job requirements. The work surfaces may be finished or rough as are the outer surfaces of castings and forgings or surfaces produced by flame cutting or other processes.

The face of the chuck has a series of concentric rings cut into it at regular intervals. These rings, or grooves, aid the operator in centering a

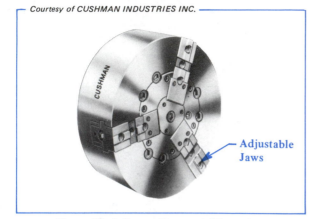

Courtesy of CUSHMAN INDUSTRIES INC.

Adjustable
Jaws

Figure 27–4 Power-Operated, Three-Jaw,
Self-Centering Chuck

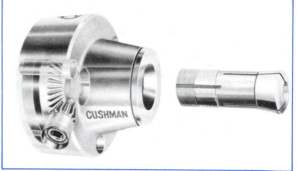

Courtesy of CUSHMAN INDUSTRIES INC.

Figure 27–5 A Spring Collet Chuck and
a Standard Collet

workpiece. Adjustments are made by loosening one jaw and tightening the opposite jaw. The rough setting is then checked by other approximate or precise methods of truing a workpiece.

The *combination chuck* combines the distinguishing features of both the universal and the independent chucks. The general-purpose, four-jaw combination chuck permits the jaws to be adjusted either universally or independently.

A chuck cradle of hardwood is usually used to install or to remove any type of heavy chuck. The cradle is grooved to fit and slide on the ways and under the chuck. The cradle protects the operator from injury and prevents damage to the lathe ways.

POWER-OPERATED, SELF-CENTERING CHUCKS

Power-operated, two- and three-jaw chucks with adjustable and nonadjustable jaws are widely used in production. These chucks may be *self-centering power chucks.*

The nonadjustable-jaw power chuck is recommended for general manufacturing service. It is particularly suited for repetitive operations, especially on heavy workpieces. This chuck is balanced to eliminate chatter and vibration at high spindle speeds (RPM).

The adjustable-jaw power chuck provides independent jaw action (Figure 27–4). The jaws are designed to hold irregular-shaped workpieces. Once the jaws are adjusted, the initial accuracy of positioning is maintained with successive workpieces.

COLLET CHUCKS

Collet chucks are simple, accurate, practical work-holding devices. They are used principally for holding regular-shaped bars of stock and workpieces that are finished on the outside. The work is generally of round, square, or hexagonal shape. Two common types of collets are the *spring collet* and the *rubber-flex collet.*

SPINDLE NOSE COLLET CHUCK

The *spindle nose collet chuck* takes a wider range of work sizes than does the draw-in bar type of collet chuck. The phantom view of a spring collet chuck in Figure 27–5 shows a bevel gear threaded disc. This disc is mounted inside the chuck body. The bevel teeth are moved by turning the bevel socket on the chuck body with a chuck wrench. As the spring collet is drawn into the chuck body, the taper on the split jaws tightens against the workpiece. Spring collet chucks are designed to fit standard types of lathe spindle noses. Some spindle nose collet chucks use a spur gear designed to tighten or loosen the collet.

JACOBS SPINDLE NOSE COLLET CHUCK

The *Jacobs spindle nose collet chuck* takes rubber-flex collets. This chuck is also made to fit different types of spindle noses. The usual range of each collet is 1/8″ (1/16″ over and under a nominal size). A single rubber-flex collet and Jacobs type of spindle nose chuck serves a wide range of sizes (Figure 27–6).

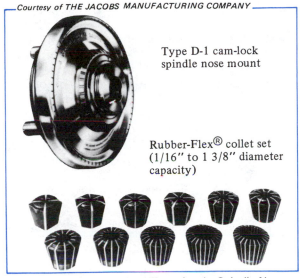

Type D-1 cam-lock spindle nose mount

Rubber-Flex® collet set (1/16″ to 1 3/8″ diameter capacity)

Figure 27–6 Handwheel-Type Jacobs Spindle Nose Collet Chuck with Rubber-Flex Collets

Figure 27–7 A Standard Drill Chuck Fitted with a Taper Shank

The work is first inserted in the collet. The chuck handwheel is turned clockwise to tighten the collet jaws against the workpiece. If the workpiece is not long enough to extend into the rubber-flex collet for at least 3/4″, then a plug of the same diameter is needed. This plug is placed in the back of the collet. The plug helps to ensure that the collet grips the work securely. It prevents the work from springing away if a heavy force is applied during the machine operation.

DRILL CHUCK

The standard *drill press chuck* is commonly used for drilling, reaming, tapping, and other operations on the lathe (Figure 27–7). This chuck is fitted with a taper shank. The shank fits the spindle bore of the tailstock. A sleeve or taper socket may be used to accommodate the chuck in the spindle headstock.

Another type of drill chuck is designed with a hollow core. This core permits the holding of long bars of small cross section that extend through the spindle.

WORK TRUING METHODS

APPROXIMATE METHODS

There are a number of approximate and precise methods of truing chuck work. The con-

centric rings on the chuck face provide one approximate method.

Once chucked, the work may be further tested for trueness by revolving the spindle slowly. A piece of chalk is held in one hand. The hand is steadied and the chalk is brought to the workpiece so that any high spot produces a chalk mark. The jaw (or jaws) opposite the high spot is moved out. The opposing jaw (or jaws) is moved in. The process is repeated until the work is concentric (no high spots).

Trueness may also be roughly checked by using a toolholder. Reduction of the light (space) showing between the back end of the toolholder and the slowly turning workpiece indicates the high spot. A piece of white paper may be placed on the cross slide under the toolholder to show the light (space) and thus identify which jaws must be adjusted.

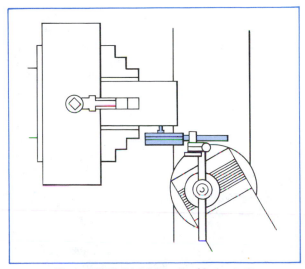

Figure 27–8 Dial Indicator Method of Precision Truing

PRECISION TRUING BY THE DIAL INDICATOR METHOD

Out-of-trueness may be corrected by a more accurate method. A dial indicator and holder are positioned and held in the tool post (Figure 27–8). The indicator point is carefully brought into contact with the workpiece. The work is revolved slowly by hand. The amount that the work runs out-of-true is indicated by movement of the dial indicator pointer. The difference between the high and low spots is noted. The jaw opposite the high spot is moved out one-half the distance. The opposing jaw is moved in the same distance. The workpiece is again revolved and further adjustments are made. Each chuck jaw is then checked for tightness. The work is concentric when there is no movement of the dial indicator pointer.

STEP 2 Slide the cleaned and burr-free chuck on a cradle. Position it on the nose spindle.

STEP 3 Turn the lathe spindle slowly by hand. Align the cam-lock studs with the clearance holes in the spindle nose.

STEP 4 Slide the chuck onto the tapered portion of the spindle and up to the shoulder.

STEP 5 Turn each cam lock in a counterclockwise direction. Apply equal force to each cam lock. The chuck should be drawn tightly against the spindle shoulder. The short tapered surfaces ensure that the chuck is correctly aligned and runs concentrically.

How to Mount Lathe Spindle Accessories

Assembling a Steep-Taper Spindle Nose Collet Chuck

STEP 1 Select a spring collet chuck with an adapter plate. The chuck and plate must fit either the type-L or D–1 spindle nose on the lathe.

STEP 2 Proceed to assemble the collet chuck on the tapered spindle nose. Secure the collet chuck to the spindle nose. Tighten either the lock ring or the cam locks.

STEP 3 Insert the collet and align the keyway and key. Engage a few threads.

STEP 4 Place the burr-free workpiece in the collet.

STEP 5 Turn the chuck handwheel. Draw the collet against the taper portion until the work is held securely.

Mounting Universal and Independent-Jaw Chucks

Cam-Lock Spindle Nose

STEP 1 Align the index line on each cam-lock stud with the register line on the spindle.

Safe Practices with Chucks and Chuck Work

- Shut off the power to the lathe when mounting or removing a chuck.
- Obtain assistance when handling a heavy chuck or other machine accessory.
- Use a wooden cradle to slide a heavy chuck onto or off a spindle nose.
- Keep the fingers out from under the chuck and the cradle.
- Stone away any burrs on the spindle nose, chuck adapter plate, or other machined part.
- Wipe all mating surfaces with a clean wiping cloth. Carefully move the palm and fingers over the parts to feel if there are any foreign particles left. Apply a drop of oil on the taper or thread of the spindle nose and the shoulder.
- Revolve the spindle by hand before starting any operation. Check to see that the chuck jaws and the workpiece clear the carriage. Also check the tool setup.

UNIT 27 REVIEW AND SELF-TEST

1. Explain how a D–1 cam lock spindle nose operates.

2. List the basic differences in the operation of a universal three-jaw chuck and the operation of an independent four-jaw chuck.

3. Tell why rubber-flex collets mounted in a Jacobs or Sjogren spindle nose collet chuck are more flexible than spring collet chucks.

4. Identify (a) two rough methods and (b) one precision method of truing work held in a chuck.

5. List the steps in removing a chuck or plate from a steep-taper (type-L) spindle nose.

6. State two precautions to take when changing the jaws on a universal chuck.

7. List four safe practices to observe when handling chucks and doing chuck work.

Centering, Drilling, Countersinking, and Reaming on the Lathe

CENTERING AND CENTER DRILLING WORK HELD IN A CHUCK

Work to be centered is usually held in a universal chuck or collet chuck. Irregular and rough surfaces are centered using an independent-jaw chuck. The center drill is held in a standard drill chuck.

After the workpiece end has been faced, the center drill is fed into the revolving workpiece. Although the cutting speed remains the same, the diameter of the angular body increases. Care must be taken to feed the center drill slowly, particularly when feeding the small-diameter pilot drill portion at the start. Excessive force may cause the center drill to fracture.

While not common practice, a center starting hole may also be spotted by grinding a tool bit to a steep-angle point (Figure 28–1). The area in back of the cutting face is ground away sharply to permit the cutting edge to cut without interference from rubbing against the angle hole.

A special 60° countersink may also be used when a center hole needs to be trued quickly and fairly accurately. Only one lip of the countersink is ground to do the cutting. The countersink is held in a drill chuck and is mounted in the tailstock. A high spindle speed is used. The tailstock spindle is brought back as far as possible to cut down on the overhang. The spindle clamp screw is tightened lightly. The countersink is then fed slowly into the revolving workpiece. These steps help prevent play in the tailstock spindle. Play may be caused by the interrupted, uneven cut that exists until a centered hole is produced.

DRILLING PRACTICES ON LATHE WORK

A drill is held on a lathe in a number of different holding devices. Straight-shank drills may be held in a drill chuck, especially if they are smaller than one-half inch in diameter. The drill chuck with taper arbor may fit directly into the matching tapered tailstock spindle (Figure 28–2). With an adapter the chuck may also fit the spindle nose. Large-diameter, straight-shank drills are usually held from turning by a drill holder. Sometimes a drill is gripped in a regular lathe chuck. Long workpieces, which extend a considerable distance from the chuck, are supported by a steady rest (Figure 28–3).

REAMING PRACTICES ON LATHE WORK

Jobber's shell, fluted-chuck, expansion, and adjustable reamers are all used in lathe work. Two or more step reamers and taper reamers are also used. All of these were described earlier under machine reamers and reaming processes.

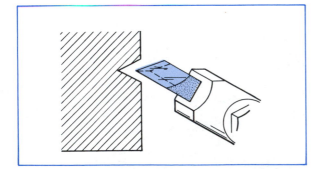

Figure 28–1 Spotting, Truing, or Centering with a Specially Ground Tool Bit

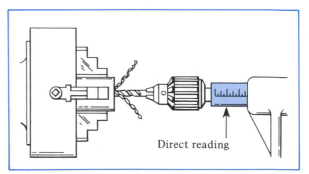

Direct reading

Figure 28–2 Measuring the Drill Depth on the Graduated Tailstock Spindle

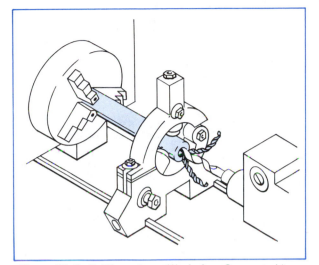

Figure 28-3 Drilling a Long Workpiece Supported by a Steady Rest, with the Drill Mounted Directly in the Tailstock Spindle

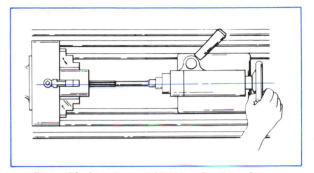

Figure 28-4 A Typical Machine Reaming Setup

In reaming, the cutting angles, speeds, feeds, and fluids are similar to those used for drilling, boring, and other machine tool processes. Tables of recommended operating conditions for reaming different materials are included in the Appendix.

The selection of a high-speed steel, cobalt high-speed steel, carbide-tipped, or other machine reamer depends on:

- The material of the workpiece,
- Production requirements,
- Whether the hole is interrupted,
- Other factors similar to the factors that affect hand reamers.

The workpiece requirements and machine conditions determine whether a straight-fluted or right-hand or left-hand spiral-fluted reamer is the most practical. The reamer may be of solid, expansion, or adjustable design. The machine reamer may have a straight shank that permits direct chucking. Or, it may have a taper shank that is adaptable for holding in a taper socket or sleeve or directly in a tailstock or headstock spindle (Figure 28-4). The same machine and tool safety precautions described in earlier units must be followed to prevent breakage and excessive reamer wear.

Extremely concentric holes require that the hole first be bored, leaving a minimum amount for reaming. Precise dimensional accuracy also may require both boring and hand reaming to meet the specified tolerances.

Another type of carbide-tipped reamer has flutes that are carbide tipped for the full length. These carbide tips provide a good bearing surface in the reamed hole. An inserted plug may be driven into the body to permit fine adjustments of the reamer to within 0.0001''. The reamer may also be reground to size numerous times.

How to Drill Holes on the Lathe

Drill Held in a Quick-Change Toolholder

STEP 1 Select the correct type and size of twist drill for the hole to be drilled. Stone any burrs from the shank.

STEP 2 Select an adjustable drill holder (Figure 28-5) that will accommodate the drill taper shank. Insert and tighten the drill in the holder.

STEP 3 Slide the drill holder on the quick-change toolholder. Lock it in position with the drill point at center height.

STEP 4 Determine the required spindle speed (RPM). Set the spindle speed. Start the lathe.

STEP 5 Set the feed rod at the recommended cutting feed. Start the flow of cutting fluid.

STEP 6 Move the drill into the center spot or center drilled hole. Feed the drill

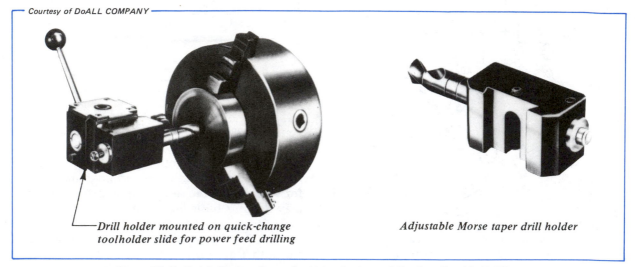

Drill holder mounted on quick-change
toolholder slide for power feed drilling

Adjustable Morse taper drill holder

Figure 28–5 Quick-Change Setup for Using Lathe and Carriage Feed in Drilling

by hand until its outside diameter contacts the workpiece.

STEP 7 Stop the lathe if a blind hole is to be drilled. Locate and adjust a carriage stop at the required depth.

STEP 8 Start the lathe. Engage the power feed. Feed a through hole automatically. Discontinue the feed on a blind hole when the drill has almost reached full depth.

STEP 9 Feed to final depth by hand on a blind hole. Clear the chips. Test for depth with a depth gage.

Note: Deep-hole drilling requires the same precautions for chip removal as in drilling by other methods.

STEP 10 Remove burrs caused by the drilling operation.

Safe Practices in Center Drilling, Drilling, Countersinking, and Reaming on the Lathe

- Remove any raised point that may be left from facing at the center before attempting to center drill.
- Reduce the speed and feed when drilling large-diameter holes to a depth of more than twice the drill size.
- Remove a tool bit or other cutting tool when it is not in use to prevent injury caused by brushing against the cutting point and edges.
- Grind the area steeply beyond the cutting point of a center-hole-spotting tool bit. The angular cutting edge must be able to cut to center hole depth without rubbing.
- Stop the lathe to clean out a drilled or reamed hole. Use a wiping cloth around a rod if the hole is deep.
- Burr the edges of each drilled or reamed hole if the job permits.

UNIT 28 REVIEW AND SELF-TEST

1. Tell what functions are served by (a) the pilot drill and (b) the angle cutting faces of a center drill.

2. State two differences between drilling and reaming on a lathe.

3. Identify three different (a) reamer types, (b) materials of which reamers are made, and (c) factors influencing the selection of a reamer.

4. Tell how an out-of-true center spot may be corrected before a hole is drilled.

5. Indicate the advantages of a quick-change toolholder for holding cutting tools.

6. List two cautions to observe with reamed holes.

7. List three functions that are served by a cutting fluid for deep-hole drilling.

8. State two safety precautions to follow in through drilling or reaming.

Boring Processes and Mandrel Work

PURPOSES OF BORING PROCESSES

Boring serves four main functions:

- To enlarge the diameter of a hole, particularly very large holes;
- To true up a hole. The surface is bored concentric and straight in relation to the axis of the workpiece;
- To produce an accurate, high-quality surface finish in an odd-sized hole;
- To machine a true hole as a pilot for subsequent cutting tools.

Holes that are to be reamed concentric and to size with a hand or machine reamer should be bored whenever possible. A hole is usually bored to within a tolerance (allowance) of 0.005″ (0.1mm) to 0.007″ (0.2mm) for machine reaming. This tolerance may be increased to 0.010″ (0.3mm) for 1/2″ (12.5mm) diameter reamed holes; 0.016″ (0.4mm) for 1″ (25mm) diameter; 0.030″ (0.8mm) for 2″ (50mm) diameter; and up to 0.045″ (1.1mm) for a 3″ (76mm) diameter. Tolerances (allowances) for hand reaming on diameters up to 1″ (following a boring operation) range from 0.002″ (0.5mm) to 0.005″ (0.1mm).

BORING TOOLS, BARS, AND HOLDERS

Four groups of cutting tools are in common use for boring processes. Forged boring bits, standard-size square high-speed steel cutting tool bits, carbide-tipped bits, and ceramic inserts are widely used. Each may be ground or formed for internal boring, grooving, threading, or form turning.

FORGED BORING TOOL AND HOLDER

The *forged single-point boring tool* has an offset end (Figure 29–1). It is generally used for light boring operations. The cutting tool may be made of high-speed steel, or the cutting end may be tipped with a carbide insert. The forged tool is secured in an offset V-grooved toolholder.

The toolholder is reversible to permit the holder to be used as both a right- and a left-hand toolholder. A single toolholder can accommodate a number of different cutting tool diameters. The forged cutting tool is used in small-diameter holes, particularly for sizes from 1/8″ (3mm) to 1/2″ (12mm).

The cutting end of a forged boring bit is ground similar to the end of a left-hand turning tool. The cutting end must have front clearance, side clearance, and side rake like a turning tool (Figure 29–1). The amount of front clearance increases sharply for smaller diameters. The front clearance angle must be adequate to permit the cutting edges to cut freely without rubbing at the heel. However, the amount of front clearance should not be excessive. Otherwise the cutting edge will not be properly supported and it will break away.

The shape of the cutting edges depends on the operations to be performed. For example, a forged boring bar ground for shoulder turning is similar to the shape of a left-hand facing tool. The cutting edge of a cutting tool for boring a straight hole is shaped like a regular left-hand turning tool.

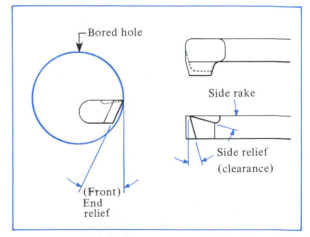

Figure 29–1 Rake and Relief Angles of a Solid, Forged Boring Tool

BORING BARS AND HOLDERS

A *boring bar* is a round steel bar. The bar positions and holds a tool bit. Boring bars are positioned for maximum rigidity. One type has a broached square hole in one end. This hole accommodates a regular-sized, square, high-speed steel or carbide-tipped tool bit. Boring bits may be held in a boring bar at 90°, 45°, and 30° angles. The tool bit may be secured in either end by means of a setscrew.

Another type is called an *end-cap* boring bar. The cutting tool is held in position by the wedging action of a hardened plug.

WEB-BAR BORING TOOLHOLDER

The *web-bar* boring toolholder is slotted on one end. A cutting tool is slipped into the slot. The nut on the opposite end is tightened to draw a tapered drawbar that transmits a tremendous clamping force on the boring bar slots. These slots grip the cutting tool so securely that the cutting tool will break before it moves in the bar.

Correct adjustment of a boring tool is particularly important. The boring bar and/or cutting tool must be clamped as short as possible to minimize the overhang. Greater tool rigidity is thus provided for the cutting process.

HEAVY-DUTY BORING BAR SET

The *heavy-duty boring bar set* consists of a combination toolholder/tool post and three sizes of boring bars. The availability of three different sizes permits the operator to use the largest (and strongest) size possible for a particular job. With greater strength in the boring bar, it is possible to take longer cuts at increased speeds.

Figure 29–2 shows a heavy-duty boring bar set. Note that the cutter may be positioned at 90°, 45°, and 30° in the three bars. The boring bar and cutting tool are positioned and secured by turning the nut. This nut clamps the boring bar in the holder body and secures the body in the compound rest T-slot.

BORING BAR AND QUICK-CHANGE TOOLHOLDER

A boring bar may easily be set in a quick-change toolholder. One design of bar has three angular ends. These ends permit boring with the tool bit held at an angle or at 90° to the work axis. The quick-change toolholder may be slid

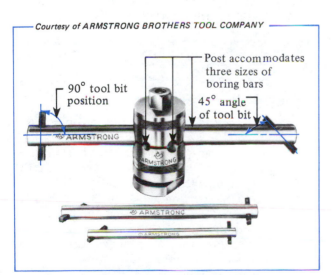

Courtesy of ARMSTRONG BROTHERS TOOL COMPANY

90° tool bit position
Post accommodates three sizes of boring bars
45° angle of tool bit

Figure 29–2 A Three-Bar, Heavy-Duty Boring Toolholder and Boring Bars

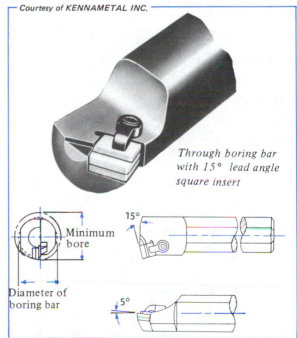

Courtesy of KENNAMETAL INC.

Through boring bar with 15° lead angle square insert

Minimum bore
Diameter of boring bar
15°
5°

Figure 29–3 Fixed-Head, Steel-Shank Boring Bar for Carbide Inserts

quickly in the dovetailed slots. The cutting tool may be adjusted vertically and locked in position.

BORING BARS FOR PRODUCTION WORK

Different types of boring bars are used with turret lathes and other semiautomatic turning machines. Such boring bars require throwaway carbide or ceramic inserts.

A fixed-head boring bar is illustrated in Figure 29–3. The boring bar uses square inserts for through boring. Another boring bar may be used for both threading and grooving by just changing the precision insert.

Other types of boring bars are also available. Holes may be bored to a square shoulder by using boring bars with triangular inserts. The shanks of these boring bars are made of steel or tungsten carbide. Tungsten carbide provides greater rigidity than standard steel boring bars.

CARBIDE INSERT SETUPS FOR BORING

The 90° slot position for the cutting tool is used for straight through boring. The cutting tool is set on center. Feeding is (right-hand) toward the spindle. A side rake of 5° to 7° provides a good cutting-edge angle. The amount of end relief depends on the inside diameter. The side-relief angle should be between 12° and 15°.

COUNTERBORING, RECESSING, AND BORING TAPERS

Three other common boring operations, in addition to internal threading, are counterboring, recessing (undercutting), and taper boring.

For most workpieces *counterbored holes* are formed with a counterbore. On larger diameters the counterbored hole is formed by boring. The shoulder of a counterbored hole is machined either square or with a small corner radius.

An *internal recess* or *groove* requires a formed cutting tool. Usually a standard tool bit is ground to shape and used with a boring bar. Small-diameter recesses may be formed with solid, forged cutting tools.

Tapered holes are generally bored using a taper attachment. The process is similar to regular taper turning but with two major exceptions: (1) Although the tool is fed toward

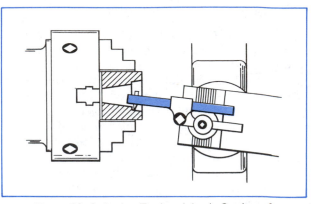

Figure 29–4 Boring Tool and Angle Setting of Compound Rest for Boring Short, Steep Tapers

the spindle, it is ground for turning left-hand; (2) the end-relief angle on smaller diameters must be increased to prevent the end of the cutting tool from rubbing.

The boring of *short steep tapers* combines angular turning with boring. The compound rest is positioned at the taper angle (Figure 29–4). The carriage is set in position. Angular feeding is done with the compound rest handwheel. The length of taper that can be bored is limited to the movement of the compound rest slide.

Large steep tapers are often cut using a standard right-hand offset toolholder. A standard toolholder provides greater rigidity than a boring bar. It is preferred wherever the work size and operation permit boring with it.

How to Bore a Hole

Straight-Hole Boring

STEP 1 Mount the boring bar in the toolholder, with the cutting tool as close to the toolholder as possible. The boring bar length should permit clearing the depth of the hole to be bored.

STEP 2 Set the cutting edge at center height (Figure 29–5). Check to see that the heel of the cutting tool clears the diameter.

STEP 3 Start the lathe. Move the cutting tool to take a trial cut for a short distance.

STEP 4 Stop the lathe.

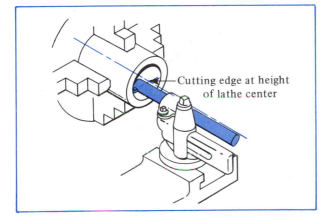

Figure 29–5 Tool and Holder Setup for Straight Boring

STEP 5 Position the boring tool to take the deepest possible cut with the coarsest feed.

> **Note:** If chatter marks are produced, check the correctness of the cutting and relief angles of the boring tool. Reduce the speed, feed, and depth of cut if necessary.

STEP 6 Take a trial finish cut with a fine feed for about 1/8''. Measure the diameter. Adjust the depth of cut if required.

STEP 7 Apply cutting fluid and take the finish cut.

> **Note:** The diameter of the bored hole should be checked at several places along its length to be sure the hole is not bell mouthed. A bell-mouthed hole indicates the tool is springing away from the work. Take an additional cut or two at the same setting to produce a parallel bored hole.

STEP 8 Break the sharp edge of the bored hole with a triangular hand scraper.

MANDRELS AND MANDREL WORK

Many workpieces that are bored or reamed require further machining on the outside surface. Such workpieces may be positioned accurately in relation to the work axis by using a *mandrel*.

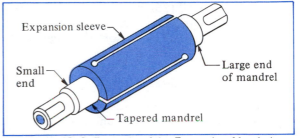

Figure 29–6 Example of An Expansion Mandrel

A mandrel is a hardened, cylindrical steel bar that is pressed into the finished hole of a workpiece.

Two common types of mandrels are the *solid* and the *expansion* mandrel. Other types include the *gang*, *threaded*, and *taper-shank* mandrel.

THE SOLID MANDREL

A mandrel is a hardened, cylindrical steel bar. The solid mandrel has a recessed center hole in each end. The ends are turned smaller than the body size. A flat is machined to provide a positive clamping surface. The setscrew of a lathe dog may be tightened against this surface. The large end has the mandrel size stamped on it. This marking also indicates the end on which the lathe dog should be clamped.

The body is ground with a slight taper of 0.0005'' per inch of length. The small end of mandrels under 1/2'' diameter (12mm) is usually a half-thousandth of an inch under the standard diameter. On large mandrels the small end is ground up to 0.001'' undersize. Due to the taper the large end is a few thousandths of an inch larger than the normal diameter. The accuracy of a mandrel depends on the condition and accuracy of the center holes.

THE EXPANSION MANDREL

The *expansion mandrel* accommodates a wider variation in hole sizes than the solid mandrel (Figure 29–6). The expansion mandrel consists of a taper mandrel and a slotted sleeve. The taper bore of the sleeve corresponds to the taper of the mandrel. The expansion mandrel may be expanded from 0.005'' to 0.008'' (0.1mm to 0.2mm) over the nominal size for diameters up to 1″ (25mm). The slotted sleeves come in different diameters. The same taper mandrel may be used with more than one diameter sleeve.

THE GANG MANDREL

The *gang mandrel* provides for the machining of multiple pieces. The gang mandrel has a flanged, parallel-ground body and a threaded end. Workpieces are placed side by side, and a collar and nut are tightened to hold them in place for machining.

The gang mandrel is mounted between centers. A lathe dog, tightened against the flat surface on one end, provides the drive force.

THREADED AND TAPER-SHANK MANDRELS

The *threaded mandrel* is used for mounting threaded parts that are to be turned. The workpiece is screwed onto the threaded end of the mandrel. The recessed thread and square flange permit the workpiece to be mounted squarely.

The threaded mandrel may be designed for chuck work or for machining work between centers. The *taper-shank mandrel* may be fitted to the headstock spindle by using an adapter. An adaptation of the taper-shank mandrel has the end turned straight. This end is slotted. A special flat-head screw applies a force against the slotted segments. These segments, in turn, hold the workpiece.

MOUNTING AND REMOVING WORK ON A MANDREL

THE ARBOR PRESS

A mandrel is pressed into the finished hole of a workpiece. The force is sufficient to permit the machining of the outside surfaces of the workpiece. Sometimes the mandrel is driven into position by a soft-faced hammer. The workpiece is placed on a mandrel block and the mandrel is driven in.

When work is turned on a mandrel, the cutting tool is set so that the cutting force is directed toward the large end of the mandrel. Light cuts are taken on large-diameter work to prevent the work from turning on the mandrel.

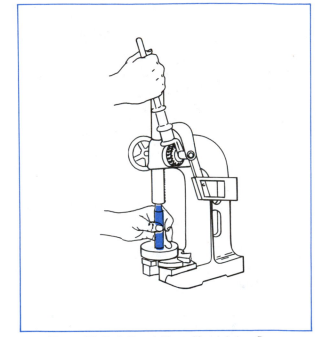

Figure 29–7 A Bench Type Hand Arbor Press

An *arbor press* is designed for mounting work on mandrels and arbors. Figure 29–7 shows one common type of arbor press. The workpiece is mounted on the table plate. Force is applied to the lever. This force is multiplied through a pinion gear to the rack teeth on the ram. The force is applied by the ram to move the mandrel into the bored or reamed hole. The arbor press is also used widely in the assembling of shafts, pins, bushings, and other parts that require a force fit.

Safe Practices in Internal Boring

- Clamp the boring bar so that the cutting tool is as close to the holder as possible. Check to see that the head end clears without rubbing into the chuck or the end of the workpiece.
- Grind the cutting point (front) relief angle steep enough so that the heel will not rub on the workpiece.
- Stop the lathe before taking a measurement.
- Position the nozzle so that the lubricant flows inside a bored hole.

- Shorten the holder length if chatter is produced. Also decrease any or all of the following: speed, feed, and depth of cut.
- Burr the edge of a bored hole or recessed area with a triangular scraper.

- Stop the lathe to remove chips and clean a bored hole.
- Use a protective shield or safety goggles and observe all machine safety practices.

UNIT 29 REVIEW AND SELF-TEST

1. Determine and state the amount of material to be allowed for machine reaming the following bored holes: (a) 25mm diameter, (b) 2'', and (c) 90mm.

2. Name four groups of cutting tools that are used for boring processes.

3. Give two advantages of a heavy-duty boring bar set compared to a plain boring bar.

4. a. Identify three different boring operations performed with throwaway carbide or ceramic inserts.
   b. Name the shape of the insert that is used for each operation.

5. State two differences between grinding a cutting tool for taper boring and for regular turning.

6. List the steps for boring a counterbored hole.

7. Describe undercutting (internal recessing).

8. a. List five common types of lathe mandrels.
   b. Cite the advantage of a multiple-piece work-holding mandrel compared to a mandrel used for an individual part.

9. Indicate two different applications of arbor presses.

10. State two safe practices that must be followed to prevent damage to work that is mounted on a mandrel.

11. Indicate corrective steps to take if chatter is produced during boring.

Cutting and Measuring 60° Form External and Internal Screw Threads

Thread cutting was related in earlier units to the hand tapping of internal threads and the cutting of external threads with a die. Many threads are machined on a lathe with a single-point thread-cutting tool. This unit covers the gages needed to measure the different cutting-tool angles or the thread pitch. The lathe set-ups and tools; the cutting of single-pitch, right- and left-hand (internal and external) threads; and new terminology are also included. Thread measurements are applied. Formulas are used to compute required thread dimensions.

GAGES, CUTTING TOOLS, AND HOLDERS

CENTER GAGE

A small, flat gage called a *center gage* is used to check the accuracy of the sides of a thread-cutting tool. This gage has a series of 60° angles. The parallel edges have a number of fractional graduations. The graduations are used to measure thread pitches. The center gage is also used to position a thread-cutting tool in relation to the axis of a workpiece.

SCREW THREAD GAGE

Another type of *screw thread gage* is a circular disc with a series of V-shaped (thread form) openings around the circumference (Figure 30–1). The thread sizes on this gage conform to the standards for American National or Unified Threads. It is used for checking Acme threading tools. A similar flat plate gage is available for checking SI metric thread-cutting tools.

THREADING (THREAD-CUTTING) TOOL

A *threading tool* is a single-point cutting tool that has an included angle and a point shape that meet a specific thread form standard. The thread-cutting tool requires that the flank of the side-cutting edge be ground with additional clearance. The cutting edges are therefore ground at an angle equal to the *thread helix angle* plus the *regular relief angle*. The clearance angle prevents the flank of the cutting tool from rubbing against the sides of the threads as the tool advances along a workpiece.

Figure 30–2 shows a correctly ground right-hand thread-cutting tool. The point is rounded slightly for a Unified thread. The front clearance angle permits the cutting tool to clear the

Courtesy of BROWN AND SHARPE MANUFACTURING COMPANY

Figure 30–1 Circular (60°) V-Thread Form Gage

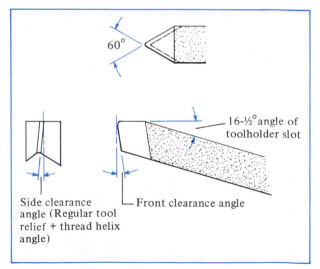

Figure 30–2 Correct Clearance Angles for a Right-Hand Thread-Cutting Tool

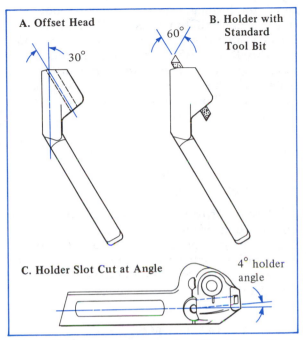

Figure 30–3 Design Features of an Offset Spring-Head Holder for a Thread-Cutting Tool

Table 30–1 Positions of the Thread-Chasing Dial to Engage the Split Nut for Cutting Threads

Nature of Threads per Inch to Be Cut	Graduation for Engaging the Thread-Chasing Dial
Even number of threads	Any graduation 1/2 1 1 1/2 2 2 1/2 3 3 1/2 4
Odd number of threads	Any numbered graduation 1 2 3 4
Fractional number of threads	Every other odd or even graduation 1/3 or 2/4
Threads that are a multiple of the lead screw pitch	Any point where the split nut meshes

diameter of a revolving workpiece. The top face is ground at the angle of the cutting-tool holder. When secured in the toolholder at center height for threading, the top face is horizontal. The clearance angle for cutting a left-hand thread is formed on a left (side) cutting flank.

FORMED THREADING TOOL

A circular blade ground to a particular thread form is called a *formed threading tool*. The blade has appropriate relief and clearance angles. The circular-form cutter is held securely against the side face of the toolholder. A hardened stop screw is used to adjust the cutter at center height.

SPRING-HEAD THREAD-CUTTING TOOLHOLDER

A *spring-head* thread toolholder (Figure 30–3) is generally used with a standard high-speed steel square tool bit. The spring head may be tightened with a locking nut to permit the taking of heavy and roughing cuts. When loosened, the holder has a spring feature. This feature is especially desirable for finishing threads.

THREAD-CHASING ATTACHMENT

A *thread-chasing attachment* is a threading device that is attached to the lathe carriage. Its function is to locate a position at which a lead screw may be engaged or disengaged. The exact point of engagement permits the threading tool to follow in the helix of the previously cut groove.

The device has a dial with lines and numbers. Table 30–1 gives the even and odd lines on the chasing dial. The correct engagement of the split nut on the lead screw is indicated for even, odd, and half threads.

SETTING THE LEAD SCREW

The amount a lead screw moves in relation to each revolution of the lathe spindle is controlled by gear combinations in the quick-change gearbox. The correct lead screw setting may be made by positioning the gears as indicated on the index plate. There are usually two levers to be positioned. A third lever controls the direction the lead screws turns. One position of this lever is for cutting right-hand threads. The second position disengages the lead screw. The third position reverses the direction of rotation to cut left-hand threads.

Caution: The feed rod must be disengaged when cutting threads. Otherwise the feed may be accidentally engaged, thus producing excessive forces on the feed screw. Damage may result.

SPINDLE SPEEDS FOR THREAD CUTTING

The spindle speeds required for threading are slower than the speeds used for turning. The spindle speed for cutting coarse threads on 3/4″ (18mm) and larger diameters is one-fourth the speed for turning. A faster speed, about one-third to one-half the speed for turning, is used for fine pitches and smaller diameters. The cutting speed may be increased still further when machining brass, aluminum, and other soft materials. Steels that are tougher and harder than low-carbon (soft-cutting) steels require slower speeds.

DESIGN FEATURES AND THREAD FORM CALCULATIONS

THREAD CALCULATIONS

A series of formulas is used to compute the different dimensions of screw threads. While the letters designating a design feature may be different, the same values are included in the formulas. Table 30–2 shows four standard basic thread forms and a few of the formulas for each

Table 30–2 Partial Set of Formulas for Common Thread Forms

American National Standard Threads	*British Standard Whitworth Threads*	*Unified Threads*
$D = 0.6495 \times P$ or $\dfrac{0.6495}{N}$ $C = 0.125 \times P$ or $\dfrac{0.125}{N}$	$D = 0.6403 \times P$ or $\dfrac{0.6403}{N}$ $r = 0.1373 \times P$ or $\dfrac{0.1373}{N}$	D for external thread $= 0.6134 \times P$ or $\dfrac{0.6134}{N}$ D for internal thread $= 0.5413 \times P$ or $\dfrac{0.5413}{N}$ C for external thread $= 0.125 \times P$ or $\dfrac{0.125}{N}$ C for internal thread $= 0.250 \times P$ or $\dfrac{0.250}{N}$

SI Metric Threads		
$D = 0.7035\,P$ (maximum) $0.6855\,P$ (minimum) $C = 0.125\,P$ $R = 0.0633\,P$ (maximum) $0.054\,P$ (minimum)		

Symbols

P = Screw thread pitch	C = Width of flat at crest
N = Number of threads per inch	R = Width of flat at root
D = Single depth of thread	r = Radius at crest or root
W = Width of groove	

form. There are five important dimensions not shown in Table 30–2: major, minor, and pitch diameters and normal and actual size.

The *pitch diameter* is important in design and measurement. It represents the diameter at that point of the thread where the groove and the thread widths are equal. The pitch diameter is equal to the *major diameter* minus a single thread depth. Thread tolerances and allowances are given at the pitch diameter.

The *minor diameter*, formerly called *root diameter*, is the smallest thread diameter. Minor diameter applies to both external and internal threads.

ALLOWANCE, TOLERANCE, LIMITS, AND SIZE

Screw threads are machined to various specifications. These specifications depend on the application, the material from which the part is made, the method of generating the thread form, and other factors. The size and fit depend on *allowance*, *tolerance*, and *limits*.

Allowance refers to the difference allowed between the largest external thread and the smallest internal thread. Allowance is an intentional difference between mating parts. The allowance may be positive (clearance) so that parts will fit freely. A negative allowance is specified for parts that must be assembled with force (force fit). Thus, a given allowance produces the tightest acceptable fit. Reference tables are used to establish allowances. Maximum and minimum pitch diameters are given for any classification of fit.

Example: The allowance for a 1″–8UNC class 2A (outside) and 2B (inside) fit is the difference between the minimum pitch diameter of the inside thread and the maximum pitch diameter of the outside thread.

Tolerance, as defined earlier, is the acceptable amount a dimension may vary in one direction or in both directions. The allowable tolerance for threads in the American National and the Unified thread systems is plus (+) on outside threads and minus (–) on inside threads.

Example: The tolerance for the 1″–8UNC class 2A (external) thread equals the:

maximum pitch diameter =	0.9168″
minimum pitch diameter =	0.9100″
tolerance (or acceptable variation) =	0.0068″

Limits represent maximum and minimum dimensions. In the previous example the upper limit of the pitch diameter for a class 2A fit is 0.9168″. The lower limit is 0.9100″.

The *basic size* is the theoretical exact size of the designated thread. It is from the basic size that size limitations are made. The basic size of a 1 3/8″–6NC thread is 1.375″. The thread notation on a part drawing indicates the 1 3/8″ size.

The *actual size* is the measured size.

THREAD MEASUREMENT AND INSPECTION

SCREW THREAD MICROMETER

The *screw thread micrometer* measures the pitch diameter (Figure 30–4A). These micrometers are designed to measure 60°-angle threads in the inch-standard or SI metric-standard systems. Screw thread micrometers have a 1″ (25mm) range.

The screw thread micrometer spindle has a 60° conical point. The swivel anvil has a corresponding cone shape. The line drawing in Figure 30–4B shows the shapes and the point of

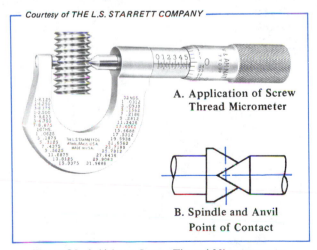

Courtesy of THE L.S. STARRETT COMPANY

A. Application of Screw Thread Micrometer

B. Spindle and Anvil Point of Contact

Figure 30–4 Using a Screw Thread Micrometer to Measure the Pitch Diameter of a Screw Thread

contact. The micrometer measurement is taken at the pitch diameter of the thread, along the helix plane. A slightly inaccurate measurement is produced. For extremely precise threads the helix angle must be considered. For these threads the micrometer is usually set with a master *thread plug gage*. The threaded part may then be measured precisely.

THREAD COMPARATOR MICROMETER

The *thread comparator micrometer* compares a thread measurement against a thread standard. The micrometer has a 60° cone-shaped spindle and anvil. The micrometer reading is established by first gaging a standard thread plug gage.

OPTICAL COMPARATOR

The common *optical comparator* magnifies a part from 5 to 250 times and projects it upon a screen. Design features of a thread are measured in one step. The measurements include: form; major, minor, and pitch diameters; pitch; and lead error. The surface illuminator of the comparator permits examining the surface finish of the thread.

Measurements are established by using a micrometer or end measuring rods on the comparator. Angles are measured by rotating the hairlines on the viewing screen. A vernier scale permits the making of angular measurements to an accuracy of one minute of one degree. Sometimes a chart is substituted for the viewing screen.

TOOLMAKER's MICROSCOPE

The *toolmaker's microscope* makes it possible to measure the same thread features as can be measured with the optical comparator. With this microscope the table is moved by precision lead screws. A micrometer thimble is used to turn each lead screw. Measurements, read directly from the graduations on the thimble, are within 0.0001″ (0.002mm).

THREE-WIRE METHOD

The *three-wire method* is used to check the pitch diameter of 60°-angle screw threads in the American National, Unified, or SI Metric systems.

The three-wire method is recommended by the National Bureau of Standards and the National Screw Thread Commission. The method provides an excellent way of checking the pitch diameter. Any error in the included thread angle has a very limited effect on the pitch diameter.

Three wires of the same diameter are required. Two of these wires are placed in thread grooves on one side. The third wire is placed in a thread groove on the opposite side. The wires and threads are positioned between the anvil and spindle of a standard micrometer (Figure 30–5).

Best Wire Size. The wire size to use depends on the thread pitch. There are three possible wire sizes that may be used: *largest*, *smallest*, and *best*. The best wire size is recommended.

Table 30–3 Formulas for Wire Size and Three-Wire Measurement

Dimension	Formula
Measurement over wires (M)	$M = D + 3W_b - \dfrac{1.5155}{N}$
Largest wire size (W_l)	$W_l = \dfrac{0.900}{N}$ or 0.900P
Best wire size (W_b)	$W_b = \dfrac{0.5155}{N}$ or 0.5155P
Smallest size wire (W_s)	$W_s = \dfrac{0.540}{N}$ or 0.540P

Symbols

D = Major thread diameter
W_l = Largest diameter of wire
W_b = Best wire diameter
W_s = Smallest wire diameter
N = Number of threads per inch
M = Three wire measurement

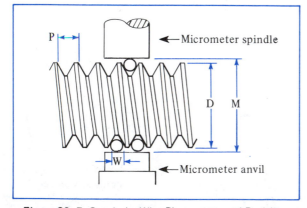

Figure 30–5 Symbols, Wire Placement, and Position of the Micrometer Spindle for Three-Wire Measurement

Calculating Wire Size and Thread Measurement. Four formulas for measuring American National and Unified threads are given in Table 30–3.

CALCULATING THE LEADING AND FOLLOWING SIDE ANGLES FOR CUTTING TOOLS

Tables are usually used to establish the clearance angles for the leading and following side-cutting edges for all forms of threads. There are times, however, when these angles must be calculated.

The helix angle of the *leading side* is represented by the

$$\text{tan of leading-side angle} = \frac{\text{lead of thread}}{\text{circumference of } \textbf{minor} \text{ diameter}}$$

The tangent of the *following side* is equal to the lead divided by the circumference of the major diameter:

$$\text{tan of following-side angle} = \frac{\text{lead of thread}}{\text{circumference of } \textbf{major} \text{ diameter}}$$

The helix angle is found by using a table of natural trigonometric functions to convert the numerical value of the tangent to the angle equivalent, which is usually given in degrees (°) and minutes (′).

Usually a relief angle of 1° is *added to* the helix angle for the leading side. One degree is *subtracted from* the helix angle for the following side.

DEPTH SETTINGS FOR AMERICAN NATIONAL FORM (60°) THREADS

The feeding of a thread-cutting tool is usually done by turning the compound rest handwheel. Some threads are cut by directly infeeding the cutting tool with the compound rest set at 0″. Where a precision form thread is to be produced, the thread-cutting tool may be fed with the compound rest set at 30°. The tool is fed at the 30° angle to thread depth. However, common shop practice is to set the compound rest at 29° to provide a slight angular clearance. The right side of the cutting tool just shaves the thread to produce a fine finish.

Table 30–4 gives examples of the depth-of-feed for each thread for compound rest settings of 0°, 29°, and 30°.

Note: The thread depths are based on the correct width (0.125P) of the flat cutting-tool point. Unless this width is accurate, the depth readings will not be correct.

How to Set Up a Lathe for Thread Cutting
American National and Unified (60°-angle) Thread Forms STEP 1 Set the compound rest at 0°, 29°, or 30°. The angle depends on the selected method of cutting.
Note: Set the compound rest counterclockwise to cut a right-hand thread.

Table 30–4 Examples of Depth Settings for American National Form Threads

Threads per Inch	Compound Rest Angle Setting and Depth of Feed* (in thousandths of an inch)		
	0°	29°	30°
4	0.1625	0.1858	0.1876
6	0.1080	0.1235	0.1247
56	0.0116	0.0133	0.0134
64	0.0101	0.0115	0.0117

STEP 2 Determine the cutting speed and spindle RPM. Set the spindle at this speed.

STEP 3 Engage the lead screw gears.

STEP 4 Position the lead screw lever to turn in the correct direction for a right- or left-hand thread.

STEP 5 Disengage the feed rod by moving the feed-change lever to the neutral position.

How to Cut an Outside Right- or Left-Hand Thread

STEP 1 Back the thread-cutting tool away from the workpiece.

STEP 2 Start the lathe. Bring the cutting tool in until it just touches the workpiece.

Caution: A beginner should use a slow spindle speed until all steps are coordinated.

STEP 3 Set the micrometer collars on the cross feed and compound rest feed screws at the zero setting. Lock the collars at this setting.

STEP 4 Move the carriage until the cutting tool clears the workpiece.

STEP 5 Turn the compound rest handwheel to feed the tool to a depth of 0.002″ to 0.003″.

STEP 6 Engage the half nut when the correct line on the chasing dial reaches the index line.

STEP 7 Turn the cross feed handwheel counterclockwise as quickly as possible at the end of the cut to clear the tool of the thread groove.

STEP 8 Disengage the split nut at the same time as the cross feed handwheel is turned counterclockwise.

STEP 9 Stop the lathe. Return the carriage to the starting position. Check the pitch.

STEP 10 Start the lathe. Feed the threading tool for a roughing cut. Set the depth of cut with the compound rest.

Note: Engage the half nut when the chasing dial reaches the index line.

Apply a cutting fluid over the work surface with a brush.

STEP 11 Continue to take roughing cuts. The depth of each successive cut should be decreased.

Note: Determine in advance the final depth reading on the compound rest micrometer collar.

STEP 12 Take the last two cuts as finish cuts. The depth may be set for 0.003″ (0.08mm), then 0.002″ (0.05mm).

Note: In some instances the finish thread-cutting tool is fed in the last 0.001″ (0.025mm) by infeeding at 30° or 90°.

STEP 13 Check or measure the thread size.

How to Clean Up the Back Side of the Thread

The cutting tool is fed to depth for each successive cut by turning the compound rest handwheel. Since the compound rest is set at a 29° or 30° angle, one side-cutting edge and the nose of the tool do the cutting. One chip is formed and flows freely over the top of the tool. If both side-cutting edges and the point of a

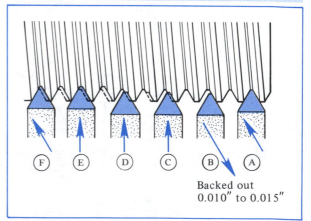

Figure 30–6 Successive Infeeding Cuts Taken to Clean Up the Back Side of a Thread

cross slide infeed were used, the cutting tool would tend to tear the thread and produce a rough surface finish.

Often, the side opposite the thread flank that is being cut (the back side) is rough. The second side-cutting edge of the threading tool is then used to clean up the back side. The process of cleaning up the back side of a right-hand thread is shown in Figure 30–6.

LATHE SETUP FOR CUTTING SI METRIC THREADS

MODERN INCH-STANDARD AND SI METRIC LATHES

The ratios in the gear trains of late-model lathes permit thread cutting for the inch-standard and SI metric systems. The spindle and lead screw gears in the end gear train do not need to be replaced to establish the 50/127 ratio between the inch-standard and metric systems of measurement (1″ to 2.54 cm).

The levers used to obtain different spindle speeds, feeds, and lead screw movements and to reverse the direction are mounted on the headstock and quick-change gearbox. The lever for starting, stopping, and reversing the lathe is located on the apron.

INTERNAL THREADING

Two common methods of cutting internal threads include *tapping* and *machine threading*. Larger and coarser thread sizes are sometimes roughed out with a single-point thread-cutting tool. A tap is then used to cut the thread to size.

THREADING USING A SINGLE-POINT CUTTING TOOL

A center gage or Acme thread gage, boring bar or other form of toolholder, and a forged or standard cutting tool are needed for internal threading. A bored hole is preferred because it is concentric.

A groove (recess) should be cut a few thousandths of an inch below the maximum thread diameter for blind holes. The groove

width should be about 1 1/2 times the thread pitch. If the part permits, a shallow recess may be cut to the same depth on the end of the workpiece. The outside of the bored or recessed hole should be chamfered 14 1/2° to accommodate the Acme thread form and 30° for 60°-angle thread forms. The chamfer reduces burrs. Also, the mating threads fit easily.

MACHINE SETUP FOR TAPER THREAD CUTTING

The recommended method of cutting an accurate taper thread is to use a taper attachment. Taper threads that are cut by offsetting the tailstock center are not as accurate as taper threads cut by the taper attachment method.

The thread-cutting tool is set perpendicular to the axis of the workpiece. If there is no cylindrical, solid surface on the workpiece from which the cutting tool can be set squarely, it may be necessary to hold the edge of a center gage against the tailstock spindle. Once the thread-cutting tool is positioned and secured, the center gage is replaced with the workpiece.

Safe Practices in Cutting Internal and External Threads

- Disengage the feed rod before engaging the feed screw. Otherwise, the feed and lead forces opposing each other may damage apron parts and the accuracy of the lead screw.
- Use slower speeds for thread cutting than speeds used for other turning operations.
- Avoid excessive overhang of the cutting tool or boring bar.
- Grind the thread-cutting tool with clearance adequate to compensate for the thread helix angles.
- Grind the flat or round cutting point accurately. Thread depth measurements depend on the width of the point.
- Keep fingers and wiping cloths away from the revolving workpiece and cutting tool.

UNIT 30 REVIEW AND SELF-TEST

1. Cite the different functions served by a center gage and a circular screw pitch gage.

2. Tell how the cutting-edge clearance angles of a thread-cutting tool are established.

3. Explain the function of a thread-cutting attachment.

4. State why the pitch diameter is important in thread calculations and measurements.

5. Tell what the differences are between the allowance, tolerance, and limits of screw threads.

6. Identify two thread (a) measuring tools and (b) inspection instruments.

7. Explain why a relief angle of 1° is added to the helix angle for the leading side and is subtracted from the helix angle of the following side in thread cutting.

8. List four settings or dimensions that the lathe operator must check before cutting a thread.

9. Explain the meaning of cleaning up the back side of a thread (that is rough).

10. Explain how the cross feed and compound rest handwheels are used to reset a threadcutting tool.

11. Explain the relationship between the 50-tooth spindle gear and the 127-tooth gear on the lead screw.

12. State two characteristics of internal threads cut with a single-point threading tool that make such threads superior to die-cut threads.

13. Indicate why a taper thread may be cut more accurately by using a taper attachment than by offsetting the tailstock.

14. List three safety precautions to observe when thread cutting on a lathe.

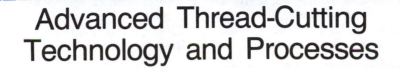

Advanced Thread-Cutting Technology and Processes

The square and the 29° thread form are also important thread forms. Three series in the 29° included angle forms are considered: the general-purpose American Standard Acme, the Stub Acme threads, and the worm thread (Brown & Sharpe). Characteristics, design features, and the cutting of these threads on an engine lathe are covered in the unit. Production thread-making processes and the technology and cutting of multiple-start threads are also considered.

PRODUCTION METHODS OF MAKING THREADS

The seven basic methods of producing screw threads are as follows:

- *Casting,*
- *Rolling,*
- *Chasing,*
- *Die and tap cutting,*
- *Milling,*
- *Grinding,*
- *Broaching.*

CASTING METHODS

Die Casting and Permanent Mold Casting. Threads produced by die casting and permanent mold casting have a high degree of accuracy and a good surface finish. The threads (internal) are cast in parts that are usually fastened together and are not disassembled.

The disadvantage of die cast and permanent-mold cast threads is in the low melting point alloy that is used. The parts are comparatively soft and have limited durability if reused. Many die cast parts are designed with steel and other inserts. These inserts are either cast in place or threaded into a hole. The inserts overcome the problem of rapid wear that results when a soft, die cast metal is used.

Plastic Mold Casting. Plastic mold casting is used on plastic materials only. Metal inserts (aluminum, brass, and steel) may be cast in place. Where a plastic material is strong enough to hold a fastener without stripping, it is economical to tap the threads.

ROLLING METHOD

Rolled threads are formed by displacing metal using flat or round dies. These dies are shaped in the exact form of the finished thread. A sliding motion of the dies burnishes and work hardens the threads. No burns are left. The accuracy of the thread lead, pitch diameter, and thread angle is maintained over longer runs with cold rolling than with any multiple-point cutting tool process.

MILLING METHOD

A milled thread is formed by a revolving milling cutter. The shape of the cutter conforms to the required thread form. Both internal and external threads may be milled. This method produces a more accurate thread than a thread produced by using taps and dies. Several advantages of the milling process are as follows:

- The formation of a coarse-pitch or a long thread is particularly suited to this process;
- Lead screws may be milled to close tolerances under fast production conditions;
- The full thread depth may be cut in one pass;
- A simple cutter may be used to mill more than one thread size.

GRINDING METHOD

Hardened parts are threaded by grinding. Grinding is the most precise method of generating a screw thread. Pitch diameters may be held to an accuracy of ±0.0001″ per 1″ (±0.002mm per 25.4mm). The accuracy of lead may be ground within 0.0003″ in 20″ (0.01mm in 508mm).

Both internal and external threads may be ground. Single or multiple-form grinding wheels are used. The threads may be completely ground to depth from solid stock or finish ground. Some advantages of grinding are as follows:

• Distortion resulting from heat treating may be eliminated. The threads may be ground from solid stock after hardening;
• Parts that may be distorted by milling can have the threads ground to depth without distortion;
• Hardening and stress cracks from preformed threads are eliminated;
• Grinding threads to close dimensional and form accuracies is practical.

The work speed for general thread grinding is from 3 to 10 inches per minute (ipm) or 75mm to 250mm per minute (mm/min). During each revolution of the work, the work is moved past the grinding wheel a distance equal to the thread pitch.

Thread Grinding Wheels. Resinoid-bond wheels are used when a fine edge must be maintained and where a limited degree of accuracy is required. These wheels operate at 9,000 to 10,000 sfpm (2,750 to 3,050 m/min). Vitrified-bond wheels are more rigid than resinoid-bond wheels. Vitrified-bond wheels are used for extreme accuracy. The recommended speed range is from 7,500 to 9,500 sfpm (2,300 to 2,900 m/min).

Threads are ground in carbide and hard alloys by using diamond wheels. These wheels are of rubber or plastic bond. Diamond chips are set in the bond. Threads may also be produced by centerless grinding. Product examples of such grinding include taps, worms, lead screws, thread gages, and hobs.

BROACHING (SCRU-BROACHING)

Presently, broaching has limited applications to internal threads. The method, often referred to as *scru-broaching*, is used principally in the automotive field. Workpieces such as steering-gear ball nuts, lead screws, and rotating ball-type automobile assemblies may be scru-broached.

An opened positioning fixture and tooling setup for a broaching machine are shown in

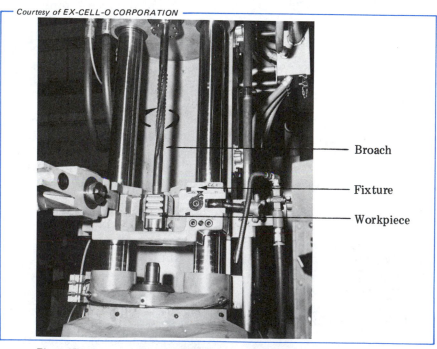

Courtesy of EX-CELL-O CORPORATION

Broach

Fixture

Workpiece

Figure 31–1 Opened Fixture, Broach, and Scru-Broach Machine Setup for Broaching an Internal Thread

Figure 31–1. The broach has a pilot end and spiral-formed teeth on the body. The broach is guided by a lead screw. As the broach turns, the workpiece and fixture are drawn up to cut the thread.

AMERICAN STANDARD UNIFIED MINIATURE SCREW THREADS

SIZES IN THE UNM SERIES

The *American Standard Unified Miniature Screw Thread Series* (UNM) is an addition to the standard coarse and fine thread series of the American National and the Unified thread series. The UNM thread series provides general-purpose fastening screws for applications on instruments and microminiature (exceedingly small, precise) mechanisms. UNM threads are also known as *Unified Miniature Screw Threads*. The 60° UNM basic thread form and series range from 0.0118″ to 0.0551″ (0.30mm to 1.40mm) in diameter.

Fourteen sizes and pitches have been endorsed as the foundation for the Unified standard. These sizes and pitches coincide with a corresponding range endorsed by the International Organization for Standardization (ISO). The UNM screw thread form is compatible with the basic profiles of both the Unified (inch-standard) and ISO (metric-standard) systems. Threads in either series are interchangeable.

UNM Thread Form Tables. Tables are available that give the *limiting diameters* according to size and tolerances for internal and external threads. The limiting diameters correspond to the major, pitch, and minor screw thread diameters. Table 31–1 provides an example.

The minimum flat on a thread-cutting tool equals 0.136 × pitch. The thread height at the minimum flat is 0.64 × pitch. The thread flat and depth may be either computed or read from UNM tables of design data.

TECHNOLOGY AND CUTTING PROCESSES FOR SQUARE THREADS

The depth and the width of a square thread at the root and crest is equal to 0.5 pitch (P). A clearance of 0.002″ is generally added to the depth measurement.

How to Cut a Square Thread
STEP 1   Calculate the width of the tool bit.

Note: For roughing out coarse-pitch threads, grind the tool point width from 0.010″ to 0.015″ (0.2mm to 0.4mm) smaller than the thread groove width. The square cutting tool point is ground 0.002″

Table 31–1 Sample of Size and Tolerance Limits of Unified Miniature Screw Threads

Size Designation*	Metric Pitch (mm)	Metric-Standard External Threads (mm)						Lead Angle at Basic Pitch Diameter	
		Major Diameter		Pitch Diameter		Minor Diameter			
		Maximum[a]	Minimum	Maximum[a]	Minimum	Maximum[b]	Minimum[c]	Deg.	Min.
0.30 UNM	0.080	0.300	0.284	0.248	0.234	0.204	0.183	5	52

Size Designation*	Threads per Inch	Inch-Standard Internal Threads (0.001″)						Lead Angle	
		Minor Diameter		Pitch Diameter		Major Diameter			
		Minimum[a]	Maximum	Minimum[a]	Maximum	Minimum[d]	Maximum[c]	Deg.	Min.
0.30 UNM	318	0.0085	0.0100	0.0098	0.0104	0.0120	0.0129	5	52

*Boldface denotes preferred series
[a]Basic dimension
[b]Use minimum minor diameter of internal thread for mechanical gaging.
[c]Reference only. The thread tool form is relied on for this limit.
[d]Reference only. Use the maximum major diameter of the external thread for gaging.

to 0.003" (0.05mm to 0.08mm) larger for finish threading.

STEP 2 Position the quick-change gears for the required threads per inch or metric pitch in millimeters.

STEP 3 Set the compound rest at 0°.

STEP 4 Set up the workpiece in a chuck, between centers, or in a fixture or other work-holding device.

STEP 5 Turn a groove to the minor diameter at the end of the threaded section, if possible.

STEP 6 Set the threading tool square with the work axis.

STEP 7 Start the lathe. Set the cross feed graduated collar at zero.

Note: Some workers feed the cutting tool by using the compound rest.

STEP 8 Proceed to cut the thread to depth. Feed the tool from 0.005" to 0.010" (0.1mm to 0.2mm) for each roughing cut.

Note: Use feeds of approximately 0.002" to 0.003" (0.05mm to 0.08mm) for each finishing cut.

TECHNOLOGY AND MACHINING PROCESSES FOR ACME THREADS

All Acme screw threads have a 29° included angle. Although not as strong, the Acme thread is replacing the square thread. The American Standards for Acme threads provide for two categories of applications: *general-purpose* and *centralizing*.

GENERAL-PURPOSE (NATIONAL) ACME THREAD FORM

The *general-purpose Acme screw thread* (National Acme Thread Form) consists of a series of diameters and accompanying pitches. Table 31–2 shows some recommended sizes, pitches, and basic minor diameters for all thread classes (of fits).

Basic Thread Classes. Three classes of general-purpose Acme threads are: 2G, 3G, and 4G.

Table 31–2 Recommended Threads per Inch and Basic Minor Diameters for General-Purpose (National) Acme Thread Series (1/4" to 1 1/2")

Size	Number of Threads per Inch	Basic Minor Diameter
1/4	16	0.1875
5/16	14	0.2411
3/8	12	0.2917
1/2	10	0.4000
5/8	8	0.5000
3/4	6	0.5833
7/8	6	0.7083
1	5	0.8000
1 1/4	5	1.0500

Each class has clearance on all diameters to provide free movement. The class of mating external and internal threads should be the same for general-purpose screw thread applications. Class 2G is used for such assemblies. Classes 3G and 4G provide for finer fits with less backlash or end play.

Basic Dimensions. A number of general-purpose Acme screw thread tables are found in trade handbooks. These tables provide basic dimensions; formulas; limiting diameters (minimum and maximum major, minor, and pitch diameters for 2G, 3G, and 4G thread classes); and pitch diameter allowances and tolerances. The pitch diameter allowance on external threads may be calculated by multiplying the square root of the outside diameter by 0.008 for class 2G, 0.006 for class 3G, and 0.004 for class 4G.

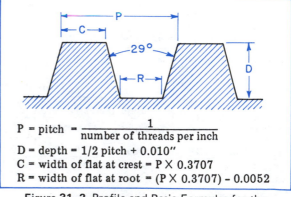

$P = pitch = \dfrac{1}{\text{number of threads per inch}}$

$D = depth = 1/2\ pitch + 0.010"$

$C = \text{width of flat at crest} = P \times 0.3707$

$R = \text{width of flat at root} = (P \times 0.3707) - 0.0052$

Figure 31–2 Profile and Basic Formulas for the General-Purpose Acme Thread Form

Table 31–3 Basic Formulas and Dimensions of Two Selected American Standard Stub Acme Screw Threads

Design Features ➡	Threads per Inch	Pitch	Height of Thread (Basic)	Total Height of Thread	Thread Thickness (Basic)	Width of Flat	
						Crest of Internal Thread (Basic)	Root of Internal Thread
Formulas ➡	N	P = 1/N	0.3P	0.3P + 1/2 allowance*	P/2	0.4224P	0.4224P – (0.259 × allowance)*
Basic dimensions (rounded to four decimal places) ➡	16	.0625	.0188	.0238	.0313	.0264	.0238
	14	.0714	.0214	.0264	.0357	.0302	.0276

*The allowance for 10 or more threads per inch is 0.010″; for less than 10 threads per inch, 0.020″.

The thread profile drawing in Figure 31–2 shows the formulas used by the worker to calculate the basic dimensions of the general-purpose Acme thread form.

General Thread Clearance. A clearance of 0.010″ (0.2mm) is generally provided at the crest and root of the mating threads for pitches that are finer than 10 threads per inch (or 2.5mm pitch). The hole diameter on an internal Acme thread, for coarser pitches than 10 threads per inch, is 0.020″ (0.5mm) larger than the minor diameter of the screw. The outside diameter of an Acme tap is correspondingly larger than the major (outside) diameter of the screw.

AMERICAN STANDARD STUB ACME SCREW THREADS

The *Stub Acme screw thread* has a 29° included angle with a flat crest and root. The thread is used for unusual applications where a coarse-pitch, shallow-depth thread is needed. While the formula for pitch is the same as for general-purpose Acme threads, the formulas for thread height, tooth thickness, basic width of flat at crest and root, and the limiting diameters are different. Table 31–3 provides formulas for the basic dimensions of two selected Stub Acme threads with sizes of 16 and 14 threads per inch.

29° WORM THREAD (BROWN & SHARPE)

The *worm thread* (Brown & Sharpe) is also a 29° form thread. Worm threads are generally combined with worm gears to provide mechanical movement for transmitting uniform angular motion rather than power. Such applications permit the design of a deeper thread form than on general-purpose Acme threads. The tooth thickness at the crest and root are correspondingly smaller. The 29° worm thread form and simple formulas for the basic dimensions are shown in Figure 31–3.

How to Cut a Right-Hand Acme (29° Form) Thread on a Lathe

STEP 1 Grind the tool bit to fit the 29° Acme thread gage. The slot used should correspond with the thread pitch.

STEP 2 Set the quick-change gears to cut the required number of threads per inch or metric thread pitch.

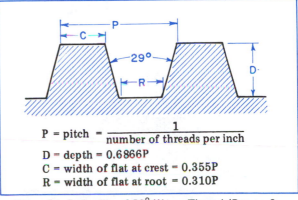

$$P = \text{pitch} = \frac{1}{\text{number of threads per inch}}$$

D = depth = 0.6866P
C = width of flat at crest = 0.355P
R = width of flat at root = 0.310P

Figure 31–3 Profile of 29° Worm Thread (Brown & Sharpe) and Formulas for Basic Dimensions

STEP 3   Position the compound rest at 14 1/2° to the right.

STEP 4   Locate the thread-cutting tool at center height and position it with a 29° Acme thread gage at an angle of 90° with the workpiece axis.

STEP 5   Feed the cutting tool to thread depth (minor diameter). Turn the end of the workpiece to this minor diameter for about 1/16″ (1.5mm) long.

STEP 6   Back the cutting tool out to position it at thread cutting depth for the first roughing cut.

STEP 7   Feed the tool in at the 14 1/2° angle. Take a series of roughing and finishing cuts.

> **Note:** Coarse-pitch Acme threads may be roughed out with a square-point cutting tool.

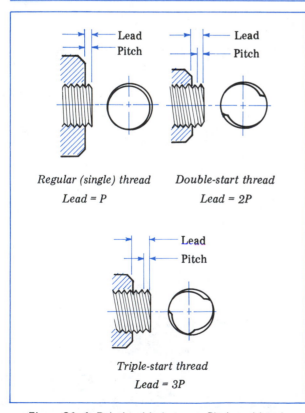

Figure 31–4 Relationship between Pitch and Lead for Single-, Double-, and Triple-Start Threads

MULTIPLE-START THREADS

Multiple-start threads are used to increase the rate and distance of travel along a screw thread per revolution. The lead of a multiple-start thread is increased without increasing the depth to which a thread is cut. The three common multiple-start threads are known as *double*, *triple*, and *quadruple threads*. The relationship between the pitch and leads of double and triple multiple-start threads and a regular single-pitch screw thread are shown in Figure 31–4.

DRAWING REPRESENTATION OF ACME THREADS

Acme screw threads are designated on drawings by the major (outside) diameter, number of threads per inch (or pitch in millimeters for metric threads), lead, thread class, and Acme (thread form). If the threads are left-hand, the letters LH follow the thread class.

METHODS OF CUTTING MULTIPLE-START THREADS

The positioning of the single-point chasing (cutting) tool, the setting of the quick-change gears, cutting speeds, depths of cuts and the rough and finish cutting of multiple-start threads are the same as for single-pitch screw threads. However, there are two major differences:

- The quick-change gears must be set to cut the thread to the required *lead*;
- The workpiece must be advanced a fractional part of one revolution for a multiple thread. For example, a double thread has two starting places located exactly 180° from each other.

If a thread-chasing dial is used, the cutting tool may be positioned at the next entry position after each cut. Thus, on a double thread, each thread is cut alternately to the same depth. Another method of positioning for the next start is to advance the cutting tool by using the compound rest to advance the tool a distance equal to the pitch.

Safe Practices in Advanced Thread-Cutting Applications

- Operate vitrified-bond thread-grinding wheels within a 7,500 to 9,500 sfpm (2,300 to 2,900 m/min) range. Resinoid-bond wheels may be operated between 9,000 to 10,000 sfpm (2,750 to 3,050 m/min).
- Check the leading and following side angles on single-point threading tools to permit the cutting edges to cut without interference from rubbing against the sides of the threads.
- Remove the thread-cutting tool at the end of each cut. Reset it at the next depth to start each new cut. Check the cutting-tool depth before taking each cut.
- Back the cutting tool out and away from the workpiece before positioning it for the next successive multiple-start thread.
- Remove the fine, wedge-shaped burr at the starting and end threads, particularly if they are square or Acme threads.

UNIT 31 REVIEW AND SELF-TEST

1. a. Describe two production methods for manufacturing screw threads, other than chasing on an engine lathe.
   b. Cite one advantage of each production method in comparison to the chasing of threads.

2. a. Identify the general purpose for which the American Standard Unified Miniature Screw Thread (UNM) Series was designed.
   b. Describe the basic thread form and size range of the UNM series.
   c. Tell why a UNM screw thread is interchangeable with a same pitch ISO miniature screw thread.

3. Interpret the meaning of the following dimension found on a drawing of a fast-operating clamping screw for a drill jig:

   1.500 – 0.25 PITCH, 0.75 LEAD
   ACME, CLASS 2G, LH

4. a. Refer to appropriate Acme screw thread tables.
   b. Give the dimensions needed for both machining and inspecting three parts that are threaded to conform to general-purpose (National) screw thread series standards. The outside diameters of the three parts are 0.500″, 1.000″, and 2.000″.

5. Explain the difference in setting up an engine lathe to chase a multiple-start thread as compared to cutting a single-lead regular screw thread.

6. State what precautions must be taken to avoid the following screw-thread cutting problems:
   a. Thread groove rubs against the following side of a square-thread cutting tool.
   b. Burrs project beyond the threaded-end face of the workpiece.

Production Turning Machines: Technology and Processes

Turret lathes vary in design and size from hand-feed to sophisticated numerically controlled models. This section deals with general-purpose turret lathe classifications, descriptions of major parts and components, and design features. Tooling setups are related to basic processes. Principles of tooling, using single- and multiple-point cutting tools, work and cutter-holding devices, and the planning of machine setups, cutting speeds, and feeds are considered.

Plain and automatic single- and multiple-spindle screw machines, their controls, operation, and maintenance (troubleshooting) are covered in detail, including cam functions and design.

UNIT 32

Turret Lathes and Screw Machines

A. TURRET LATHES: COMPONENTS AND ACCESSORIES

AUTOMATED BAR AND CHUCKING MACHINES

A modern *fully automatic chucker and bar machine* is shown in Figure 32–1. This machine may be programmed directly. Trip blocks operate microswitches, which control the functions of the carriage, cross slide, vertical slide, and threading head, as well as speed and feed changes. The reverse feed is also independent of the forward feed so that rapid return may be programmed. The machine is capable of being programmed for the following operations:

- Straight turning,
- Taper turning,
- Facing,
- Forming,
- Chamfering (beveling),
- Single-point threading,
- Die-head threading,
- Cutting off,
- Drilling,
- Straight or taper boring,
- Reaming,
- Recessing,
- Tapping.

There are 34 spindle speeds on this machine. The range is from 125 to 3,000 RPM. Four speeds within this range are available throughout the machining cycle. Tolerances within 0.0001″ (0.002mm) are obtainable on the cross slide by using a "tenth" (0.002mm) dial indicator. Similarly, a carriage dial indicator reading in increments of 0.001″ (0.02mm) is used for dimensions of length. This chucker and bar machine is a small model, having one 2 HP spindle drive motor and another 2 HP motor for the hydraulic pump.

TURNING CENTERS (UNIVERSAL TURNING SYSTEMS)

Another modern adaptation of the turret lathe is a group of machine tools identified as *turning centers* or *universal turning systems*. These machine tools are usually numerically controlled. Part of a universal numerically controlled (NC) turning system is illustrated in Figure 32–2. A multi-tool single turret design

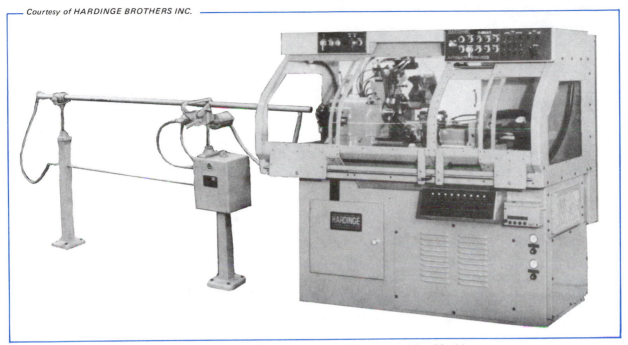

Figure 32–1 Modern Fully Automatic Chucker and Bar Machine

and chucking setups are shown. The turret slants 35°. This feature places the tooling for outside diameter operations at a plane perpendicular to the plane of operation. Thus, maximum turning force and interference-free turning are provided.

Note that the inner row of eight tools on the turret are for inside diameter operations, as shown in Figure 32–2A. The boring and inside diameter turning tools are positioned in this inner row of the turret. This single turret accommodates 16 tools. There are three spindle speed ranges: (1) 65 to 2,000 RPM, (2) 80 to 2,500 RPM, and (3) 90 to 3,000 RPM. This model heavy-duty turning machine can accommodate bar stock up to 2 1/2″ (64mm) diameter. It has a chucking capacity up to 12″ (305mm). Boring bars up to 2″ diameter, with a maximum length of 9 1/2″ for bar and holder, may be mounted in the turret. This particular model requires a 40 HP motor to drive the spindle.

CLASSIFICATION OF BAR AND CHUCKING MACHINES (TURRET LATHES)

Multiple tooling is provided on a turret lathe by replacing the usual lathe tailstock spindle and as-sembly with a hexagonal turret. Also, the regular tool post on a compound rest is replaced by a square turret to hold four separate cutting tools. A square turret may also be positioned at the back of the cross slide. Thus, 16 different cutting tools may be set up at one time.

Horizontal turret lathes may be classified into *bar machines* and *chucking machines*. Bar machines are designed for feeding bar stock through a spindle or for machining castings or forgings that may be held in collets. Chucking machines are used for workpieces that are held and driven by a chuck or work-holding fixture. Chucking machines are used for machining regular and irregular-shaped castings, forgings, and large-size cut bar stock.

RAM AND SADDLE TYPES OF TURRET LATHES

Bar and chucking machines may be of the *ram type* or the *saddle type*. On the ram type, the base on which a ram moves is clamped securely on the bed of the turret lathe. The turret is mounted on the slide or ram. The ram movement toward or away from the spindle is controlled by hand or power feed. The ram type is adapted for bar work and chuck work where it

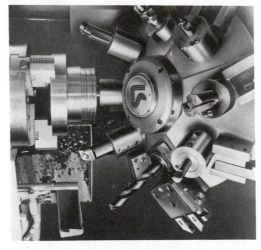

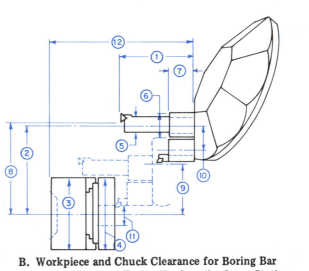

A. Boring and Inside Diameter Turning Tools on Inner Turret Row

B. Workpiece and Chuck Clearance for Boring Bar and Right Angle Facing Tool on the Same Station

Figure 32–2 Multi-Tool 16-Station Slant Turret and Chucking Setups on Universal (NC) Turning System

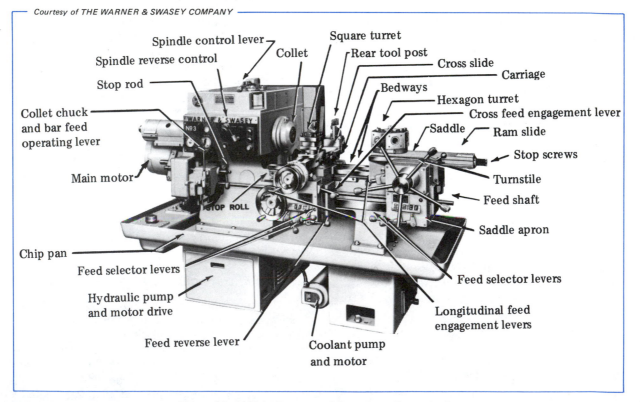

Figure 32–3 Main Features of Ram-Type Turret Lathe

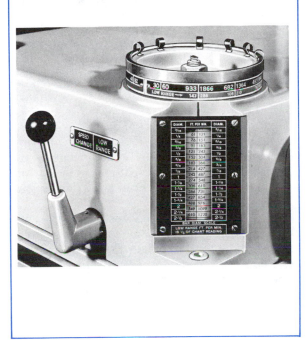

Figure 32–4 Direct Reading Spindle Speed Preselector for Dial Settings

is possible to control the overhang of the ram. The main features of a bar machine with a ram slide are identified in Figure 32–3.

On the saddle type of turret lathe, the turret is mounted on a saddle. The saddle with its apron and gearbox move back and forth on the machine ways. The saddle type has a longer stroke than the ram type and a more rigid turret mounting. The tool overhang remains constant as the whole carriage and turret setup advances as a unit. The side-hung carriage makes it suitable for heavier chuck work and long turning and boring cuts.

Preselector and Cross Slide Features of the Ram-Type Turret Lathe. There is a direct-reading speed preselector on many geared-head models. It permits shifting to a preselected speed by setting the preselector dial. The preselector shown in Figure 32–4 reads directly in surface speed, RPM, and work diameter.

The cross slide may be *plain* or *universal*. The plain cross slide is adjusted along the bed and locked at a required position. The universal cross slide is mounted on a carriage. This cross

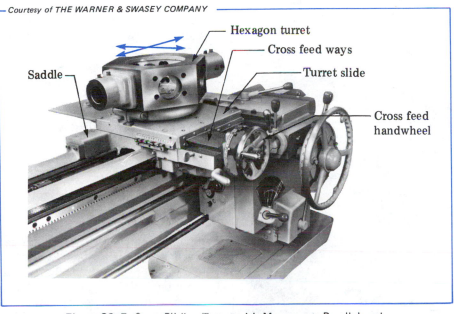

Figure 32–5 Cross-Sliding Turret with Movements Parallel and at Right Angles to Spindle Axis

slide permits movement in two directions to cut parallel (longitudinally) and at right angles, cross-wise to the bed ways (Figure 32–5).

Classification of Saddle-Type Turret Lathes. Saddle-type turret lathes may be of the *fixed-center turret, cross-sliding turret,* or *compound cross slide* type.

The *fixed-center turret* is fixed to align with the spindle axis. The turret moves parallel to this axis. The *cross-sliding turret* (Figure 32–5) may be fed by power or hand.

The *compound cross slide* (Figure 32–6) has a compound rest that may be set directly at any angle. This feature permits cutting bevels and steeper tapers than it is possible to machine with a taper attachment.

MAJOR COMPONENTS OF THE TURRET LATHE

HEADSTOCK

Automatic turret lathes have an infinite number of variable spindle speeds and may be programmed from 150 to 3,000 RPM. The spindle on a turret lathe is designed to run clockwise or counterclockwise. Spindle brakes are available as a machine and personal safety precaution.

A quickly adjusted *stop rod* is located under the headstock. Individual feed-stop screws

Figure 32–6 Compound Cross Slide

control each tool position to within 0.001″ (0.02mm).

CARRIAGE OR CROSS SLIDE

The two basic types of cross slides are: (1) *side-hung* and (2) *reach-over.* The *side-hung cross slide* provides maximum swing capacity over the cross slide.

The *reach-over cross slide* is also called a *bridge cross slide.* This cross slide is supported on both bed ways and has the advantage over the side-hung type because it is possible to add a tool post or cross slide turret in the rear position.

Cross Slide Position Stops. The cross slide may be positioned for different depths of cut by adjustable stops. The set of stops on the front end controls the depth of cut of the front tools. The rear stops control the depth of the rear turret tools. The cross slide travel is controlled by dogs, which engage the stops.

Carriage Feeds. The carriage may be fed longitudinally or transversely, by hand or power. The power feed may be rapid traverse or regular. Trip dogs are used to move the feed lever to the off position. The feed may be reversed to provide tool movement in either direction. Hand-wheels are provided for cross slide and longitudinal movement of the main turret.

A second set of feed stops, at the left side of the carriage, controls the longitudinal movement of all carriage tools. The feed stops are part of a *stop roll.* This stop roll has a center stop that permits adjusting all of the stops the same amount, without disturbing the individual stop screws.

MAIN AND SECONDARY TURRETS

Main turrets are generally hexagonal or octagonal. Some main turrets are designed with an inside and an outside row. Such turrets accommodate two sets of tools: one set of eight tools for internal operations; another set of eight for external operations. Some main turrets are mounted with the ram or saddle in a horizontal plane. Other turrets are mounted vertically on the ram or are designed with the bed and turrets inclined for easier access and machine tool setups.

Secondary turrets on the cross slide may be mounted horizontally or vertically.

BASIC MACHINE ATTACHMENTS

THREAD-CHASING ATTACHMENTS

The two basic kinds of threading attachments are: (1) *leader and follower* and (2) *independent lead screw*. These attachments are used for chasing threads with single-point thread-cutting tools or for accurately leading-on taps and die heads from the main turret. The *leading-on attachment* attached to the main turret on a ram-type turret lathe feeds the turret at the correct pitch for the thread being cut. An *automatic knock-off mechanism*, operated by the turret stop screws, provides a depth control for threading cuts to shoulders or in blind holes.

Lead Screw Attachment for Ram-Type Turret Lathe. The cross slide is actuated for thread cutting by opening and closing half nuts with a lever located on the side of the cross slide apron. The length of travel is controlled by the automatic knock-off mechanism. This mechanism may be used for right- and left-hand threads.

Lead Screw Attachment for Saddle-Type Turret Lathe. The main units required in thread cutting on a saddle-type turret lathe include the *selective quick-change gear box* and the *lead screw gear box*. Movement for thread cutting is set by the *independent lead screw selector lever*. Threads are cut using either the *carriage* or the *saddle half-nut control lever*.

TAPER TURNING METHODS AND ATTACHMENTS

Tapers are turned on turret lathes by three general methods. Tapers may be turned by using (1) a formed tool, (2) a roller rest taper turner, or (3) a taper attachment.

A *roller rest taper turner* (Figure 32–7) may be used to produce long, accurate tapers on bar jobs. The taper turner can be set quickly for size by adjusting a graduated dial setting. The angle of taper is controlled by the *taper guide bar*. As the cutting tool advances toward the headstock, the cutter and rolls in the roller rest taper turner recede to produce the required taper.

A *taper attachment* (Figure 32–8) provides a third general method of turning an internal or external taper or chasing threads. When turning a taper, the attachment is clamped to the bed. The cross slide or the cross sliding main turret moves along the bed. The *pivoted guide plate* of the taper attachment is set at the angle of the

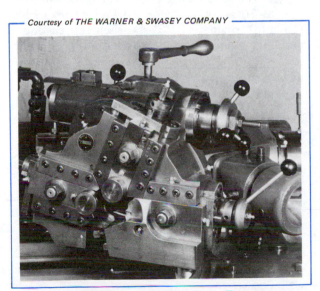

Figure 32–7 Roller Rest Taper Turner

Figure 32–8 Taper Attachment for Ram-Type Turret Lathe

required taper. The guide plate guides the cross slide and the mounted cutter along the taper angle. This movement produces the required taper.

KINDS OF MACHINE CUTS

The four kinds of cuts that may be taken on a turret lathe are: (1) *single*, (2) *multiple*, (3) *combined*, and (4) *successive*. A *single cut* involves one cutting tool that performs one operation at one time. *Multiple cuts* are two or more cuts taken at one time from one turret station. *Combined cuts* are cuts taken by tools mounted on both the cross slide and hexagon turret at the same time. *Successive cuts* relate to cuts that are made from successive faces of the hexagon turret in consecutive order.

HEADSTOCK SPINDLE TOOLING

COLLET CHUCKS

Standard special alloy steel collets have a spring-tempered body. These collets are manufactured in round and other regular shapes in fractional, metric, decimal, letter, and number sizes. Special-shaped collets are available for holding odd-shaped parts and extruded stock.

Extended-nose collets have a soft face and pilot hole. These collets may be machined to accommodate a special size or odd shape. The collet may be drilled, bored, or stepped out to the exact required size. The extended nose permits deep counterbore and tool clearance for extended work processes.

Positive-stop collets (Figure 32–9) have all the design features of the standard collet plus a precision-threaded section at the back end of the collet bore. The positive-stop collet may be fitted with a *solid positive* stop, an *ejector* stop, or a *long* stop. All stops are threaded into a positive-stop collet. Each stop pin is adjustable to the desired part length.

Dead-length collets and *dead-length step chucks* permit shoulders and faces to be machined to exact length regardless of variations in the outside diameter. An adjustable solid stop is threaded into an inner collet. The inner collet is spring-loaded against the spindle face so that no lateral movement occurs.

Precision expanding collets are used for chucking on internal surfaces. The expanding collet assembly holds the workpiece in a previously machined bore. This assembly permits the part to be machined with concentric and square shoulders, faces, and diameters in relation to the bore.

STEP CHUCKS AND CLOSERS

Step chucks and *step chuck closers* are designed for accurately holding large-diameter work. The step chuck and closer shown in Figure 32–10 accommodates workpieces up to 6″ (152mm) diameter. Parts such as castings, molding, stampings, and tubing may be held

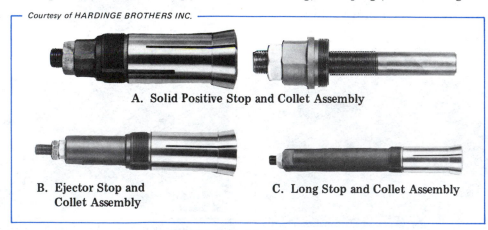

Courtesy of HARDINGE BROTHERS INC.

A. Solid Positive Stop and Collet Assembly

B. Ejector Stop and Collet Assembly

C. Long Stop and Collet Assembly

Figure 32–9 Positive-Stop Collets

rigidly and accurately, without crushing or distortion.

CHUCKS

Standard chucks include independent-jaw chucks, universal (scroll) chucks, and combination independent-jaw and universal chucks. The step design of the jaws permits their use over a wide range of diameters.

One main difference between turret lathe and engine lathe chucks is in the use of standard reversible-top soft jaws. These jaws are used for inside and outside chucking operations on certain production jobs.

Chuck Jaw Serrations and Jaw Types. Standard chuck jaws are *serrated*. Grooves are cut at 60° to a medium depth of 0.078" (2mm) and pitch of 1/8" (3mm). A limited-depth (fine) serrated jaw is used for finer gripping requirements.

Rocking jaws are used on castings, forgings, weldments, and other rough surfaces. Surface roughness and irregularities in the gripping area require the use of rocking jaws. The rocking jaw is undercut so that the clamping forces grip the part at the ends only.

Wide or wrap-around jaws are often used on second operation work. Thin-wall, fragile parts require wide or wrap-around jaws. These jaws distribute the gripping forces and minimize distortion.

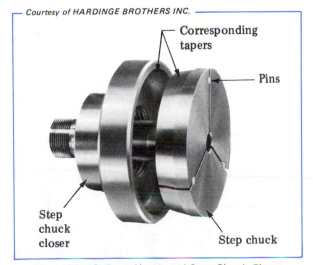

Corresponding tapers

Pins

Step chuck closer

Step chuck

Figure 32–10 Step Chuck and Step Chuck Closer

It is important that the radius of the chuck jaws be ground to within ±0.001" (±0.02mm) of the workpiece diameter to ensure that the gripping forces around a fragile part are distributed evenly over the entire chucking area.

SPECIAL FIXTURES AND FACEPLATES

Special fixtures are often used under the following conditions:

- A difficult-to-machine job must be held very rigidly to withstand heavy or extremely accurate cuts;
- The use of a fixture is more economical on a high-production job;
- The part may be loaded and unloaded faster and easier than by conventional chucking.

Fixture designs vary considerably. Some fixtures are mounted on standard chucks and complement the chuck jaws. Other fixtures are used for second and third operations to nest workpieces on a previously machined surface. Indexing is used on some fixtures for the multiple positioning of a workpiece.

Faceplate Applications. The use of a T-slotted faceplate for holding a fixture is a common shop practice. Angle plates that fasten directly to the T-slots of the faceplate are used.

Angle fixtures are used when a surface to be machined is in an exact relation to a previously machined flat surface.

Indexing fixtures permit the movement of workpieces that require similar machining operations in two or more locations. A common holding problem in turret lathe work is encountered with workpieces that require identical turning, boring, facing, chamfering, and threading operations on two or more bosses.

Fixture plates are available commercially. They are round plates with a flange for direct application to the headstock spindle. Fixture plates may be machined to become a fixture and for mounting special-purpose chucks or other fixtures.

UNIVERSAL BAR EQUIPMENT FOR PERMANENT SETUPS ON TURRET LATHES

A permanent setup of universal bar equipment for a six-station turret generally includes two

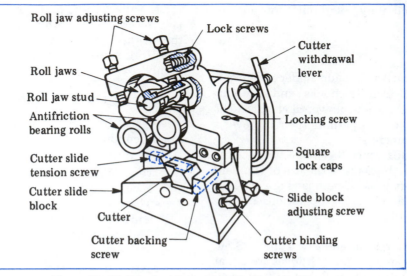

Figure 32–11 Principal Parts of Roller-Type Bar Turner

short, flanged tool holders (one of which holds a revolving stock stop), a large, flanged tool holder, a combination end face and turner, and two single cutter turners. Other cutting tools and accessories are: a starting drill and drill chuck, adjustable knee tool, clutch tap and die holder, floating tool holder, die head, combination stock stop and starting drill, center drilling tool, and a revolving center. With this combination of holders and cutting tools in the hexagon turret, it is possible to position a workpiece, face, turn, chamfer, undercut (groove), center, drill, bore, ream, and thread. Other tooling may be provided on the square turret and rear tool post to provide for additional single, multiple, or combined cuts. External straight and taper turning, facing, shoulder turning, thread chasing, and cutting off processes are added.

BAR TURNER FUNCTIONS AND OPERATIONS

The *bar turner* (Figure 32–11) is the most widely used cutting-tool holder and work-support device for bar work. Each bar turner consists of a rigid cutter holder and self-contained rolls or flat carbide shoes that support the cutting action and serve as a steady rest. The rolls or shoes also support the workpiece to produce a concentric, turned surface.

The rolls are adjustable to accommodate a wide range of finished diameters. During the cutting action, the cutting forces produce a constant vibrationless relationship between the workpiece and the cutting edge. The rolls are usually set just slightly beyond the turned radius (nose of the cutter). In this position, the cutting forces that keep the turned surface against the rolls burnish the diameter. The effect is a finished diameter that is free of tool marks.

A shank-type cutter of high-speed steel, brazed-tip carbide, or insert-tip carbide is used. The cutter is set on end and is ground to cut in this position.

STARTING BAR TURNERS PROPERLY

Supporting the Workpiece. Bar turners are used when a bar extends far enough beyond a collet chuck or chuck to require support. Support may be provided by chamfering or center drilling the end and using a center support.

After a starting cut is made by the tool on the square turret, the bar turner is positioned to turn the entire outside diameter for the required length.

Adjusting the Cutter Slide Block. The bar holder is designed to accommodate variations in cutter thickness. A cutter slot, set at an angle in the block, permits gripping the cutter and automatically establishing front and side clearances.

A pictorial view of a roller-type bar turner (Figure 32–11) shows the adjustments of the

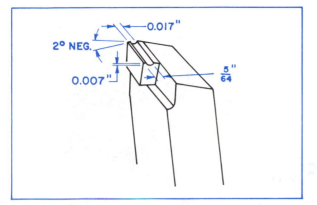

Figure 32–12 Specifications of Ground Chip Groove in Brazed-Tip Carbide Bar Turner Cutter

Selecting Cutters and Materials. The three recommended bar turner cutter materials are cobalt high-speed steel, brazed-tip carbides, and insert-tip carbides. The insert-tip carbides are recommended for long production jobs where cutters must be changed during the run. Chip breakers are used on insert-tip cutting tools.

Maintaining Chip Control. The shape, angles, and dimensions given in Figure 32–12 are recommended for brazed-tip carbide bar turner cutters. In the case of insert-tip carbide cutters, separate chip breakers are used. The chip breaker is clamped on top of the insert tip.

SLIDE TOOLS FOR TURRETS

A *box tool* (Figure 32–13) is designed for right- or left-hand general turning work on soft metals such as brass and aluminum. It is adaptable for light finishing cuts where a fine-quality surface finish and close tolerances are to be maintained. Carbide back rests are used to extend wear life. These V-rests may be single or double. Box tools are made for right- or left-hand turning with one or two cutters. Some box tools are arranged to hold a centering tool or center drill in the shank.

A *balance turning tool* (Figure 32–14) consists of blade blocks, a blade block retainer, and fine adjusting screws. These parts are interchangeable with equivalent-sized *roller back box tools*. One-piece setting gages are available for adjusting each blade. The blades may be set for equal depth or for roughing and finish turning

slide block. The slide block adjusting screw permits adjustment of the cutter for size. The spring, screw, and shoe provide tension against the adjusting screw. The screw is turned clockwise to feed the cutter toward the work.

Compensation must be made for variations in cutter thickness. An adjustable backing screw repositions the cutter (after it is reground or changed) in relation to the rolls.

At the end of a cut, the cutter withdrawal lever is turned to retract the cutter.

The cutter slide block is adjusted to increase the tension when turning large diameters. The adjustment is made by tightening the cutter slide tension screw. Similarly, the slide block is adjusted to the extreme forward position for small work where less tension is required.

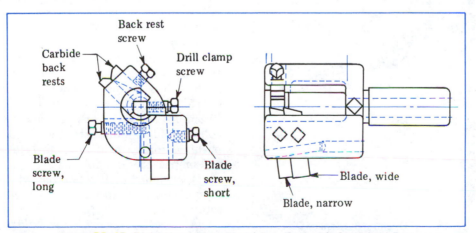

Figure 32–13 Right-Hand (Two-Blade) Box Tool with Carbide V-Rest

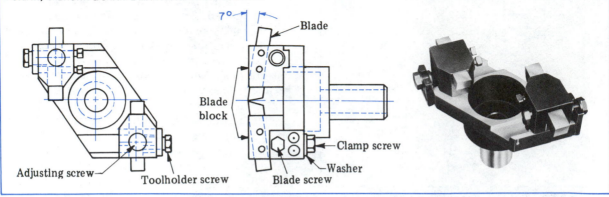

7°

Blade

Blade block

Clamp screw

Washer

Adjusting screw

Toolholder screw

Blade screw

Figure 32–14 Principle Parts of a Balance Turning Tool

cuts. The equal-depth setting permits machining fine finishes to close tolerances with increased tool life. Setup time is decreased and repeatability of settings is ensured.

Slide tools provide a rugged, movable slide that permits the turning of tapers, turning behind shoulders, turning irregular shapes or contours, necking, forming, or cutting off. The cutting tool remains on center during the full length of slide travel. The three common interchangeable heads for a slide tool body include

Figure 32–15 Self-Opening Stationary Die Head (Outside Trip Type)

a turning head, a recessing tool head, and a knurling head.

An *adjustable toolholder* accommodates drilling, reaming, counterboring, chamfering, and similar tools. The arm extending from the side of an adjustable holder holds a square-face chamfering tool bit at an angle of 45°. Round chamfers (rounded corners) may be produced by using a chamfering tool bit ground to the required radius. Drills may be held by a bushing in an adjustable head. The adjustable toolholder permits setting center drills, drills, reamers, and other end-cutting tools at the exact center of the workpiece.

A *self-aligning floating reamer holder* floats freely on antifriction bearings. These bearings provide alignment when the reamer enters a hole. After being adjusted to an approximate alignment with the work, the reamer aligns itself with the hole that is being reamed. This alignment eliminates bell-mouthed and egg-shaped holes.

An *adjustable stub collet holder* permits tool and collet changes without realigning the holder. Stub collets are used. They provide a range of 0.015″ (0.4mm): +0.005″ (0.12mm) to –0.010″ (0.25mm) for variations in tool diameters. Stub collet holders eliminate the need for special bushings for fractional, decimal, letter, number, and metric sizes of tool shanks.

A *standard drill chuck* has wide application in universal bar tooling and other turret lathe setups.

A *releasing "acorn" die holder* is used for cutting standard external threads. A *releasing*

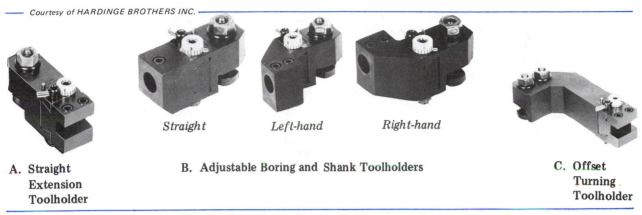

Straight   *Left-hand*   *Right-hand*

A. Straight Extension Toolholder

B. Adjustable Boring and Shank Toolholders

C. Offset Turning Toolholder

Figure 32–16 Cutting-Tool Holders for Turret Tooling

tap holder is used for tapping internal threads.

A *self-opening stationary die head* (Figure 32–15) is used to cut threads on machines where the die head does not rotate. The chasers (thread form cutters) are adjustable to a required pitch diameter. Die heads are available with an inside or outside trip. The length of thread on an inside trip model is accurately controlled.

CUTTING-TOOL HOLDERS

While the designs of toolholders used on hexagon and octagon turrets differ, the functions are

Figure 32–17 Planned Production Tooling and Setups That Fall within a Tool Circle

similar. The general design features of one manufacturer's cutting-tool holders are shown in Figure 32–16. A *straight extension* toolholder (Figure 32–16A), *boring* and *shank* toolholders (Figure 32–16B), and an *offset turning* toolholder (Figure 32–16C) for turning large-diameter work are displayed. These holders are used for boring tools, square tool bits, and shank-type tooling. One unique feature is the fine adjustment provided by the graduated collar. The graduations are in increments of 0.0002″ (0.005mm). Each fifth graduation reads in 0.001″ (0.025mm). Each graduation indicates the amount of change on the diameter of the workpiece. A movement between two graduated lines shows the diameter is changed by 0.0002″ (0.005mm). The cutter is locked in position when the required diameter is reached.

A *slide tool* is used for turret turning and boring operations. The fine pitch adjusting screw permits dial settings of 0.0002″ (0.005mm). Each fifth dial graduation is marked to represent 0.001″ (0.025mm).

CUTTING-OFF TOOLS

Cutting-off processes are generally performed by mounting the cutoff tool in a holder on the square turret or rear tool post. On some turret lathes, the cutting-off process is done from a vertical cutoff slide. In all instances, the cutter must be on center.

One grinding technique for increasing tool life on insert carbide cutoff tools that are used for parting solid stock is to grind the face to a

3/4″ (19mm) radius. The edges are ground at a 45° angle chamfer. The radius form stabilizes the cutter. A smoother, flatter surface results. When tubing is to be parted, a slight lead angle of 3° to 5° is ground on carbide inserts. The lead angle reduces the fine burr normally produced on the workpiece.

THE PLANNED TOOL CIRCLE

Production efficiency requires that operator effort and operating time be reduced, where practical. Efficient machining may be accomplished by positioning the cutting tools and setups so that they fall within a *tool circle*. The saddle should also be positioned so that each cutting tool travels the shortest possible distance before machining begins.

In a tool circle, all positioning, support, and cutting tools extend an equal distance from the hexagon turret. Figure 32–17 shows a production setup where all of the tools and holders fall within a tool circle. As each position of the turret is indexed, the longitudinal travel to the end of the workpiece is approximately the same. The operator is able to set up uniform operation timing motions to index and position each tool to the cut. The longitudinal turret movement stroke and the turnstile movement are kept short. Operator fatigue is reduced when the saddle is locked in a position that permits the operator to grip and move the turnstile comfortably.

Note: The distance between the saddle and the end of the workpiece must be great enough for the binder clamp to lock the hexagon turret securely.

Note: Each tool on the turret must be checked for interference. When indexed, each tool must clear the square turret and rear tool post tools.

PERMANENT SETUPS WITH CHUCKING TOOLING

Castings, forgings, large-diameter rolls, collars, gear blanks, and flanged parts may be machined with chucking tooling. Machining usually involves combined and multiple cuts. As many processes as possible are completed in chucking in order to ensure concentricity and squareness between internal and external surfaces. Rough castings and forgings require three distinct cuts:

- A first cut for heavy stock removal,
- A second cut for truing up,
- A third cut for dimensional accuracy.

REPRESENTATIVE CHUCKING TOOLING (FLANGES)

Cast or forged flanges are considered as a representative sample of a group of chucking jobs. Each part may be chucked on a turret lathe and machined on the inside and outside.

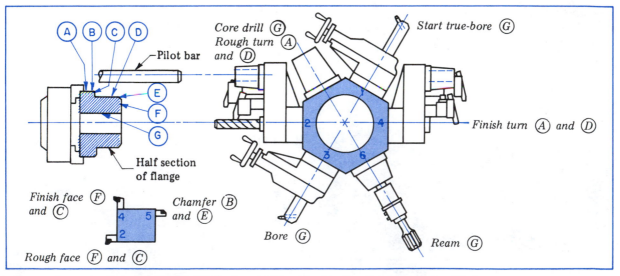

Figure 32–18 Permanent Chucking Tooling Setup for Flange Work

A permanent setup for machining flanges with chucking tooling is shown in Figure 32–18. The representative flange is shown as a half section. The flange is held in a chuck. The surfaces to be machined are lettered Ⓐ through Ⓖ. The square turret and hexagon turret tooling corresponding with each process are also lettered Ⓐ through Ⓖ. The sequence of each cut is identified by a numeral.

TOOL CHATTER PROBLEMS

Tool chatter produces a high-pitched, shrieking, whistling sound. Most tool chatter problems are caused by excessive tool overhang. The tool set-up is not rigid enough to withstand the cutting force and speed. Corrective steps for tool chatter problems are as follows:

- Keep the overhang of the toolholder and cutter to a minimum;
- Use the largest possible size of cutting tool, boring bar, or shank-type cutter holder;
- Grip workpieces close to the chuck or spindle nose;
- Use a revolving center to support heavy cuts;
- Tighten and recheck all clamping screws in cutting-tool holders;
- Position cutters on center;
- Replace a dull cutter with one especially sharpened for a specific job and keep the cutter sharp;
- Reduce the feed rate;
- Eliminate *dwell* so that a cutting tool does not remain in contact at the same machined surface position after the process is completed.

A. Safe Practices in the Care and Maintenance of Turret Lathes

- Shut down the turret lathe before checking all automatic oiling systems. Make sure that there is an adequate supply of oil and that the forced feed systems are working.
- Go through a complete cycle at reduced speeds to establish that all tools and stops are positioned correctly. Check the machined part for dimensional accuracy and quality of surface finish.

- Check the aligning hole on the turret at each station before inserting or sliding a positioning or cutting tool into the turret.
- Check the cutting fluid systems for the main turret, secondary turret, square tool post, and back position toolholder. The nozzles must be capable of delivering a required flow of coolant to the place where it properly serves to cool and aid in the cutting process.
- Test the cutting fluid with a manufacturer's gage to establish the strength and condition of the solution.

B. Safe Practices in Setting Up Spindle, Cross Slide, and Turret Tooling

- Shut down the turret lathe. Clean the spindle bore, collets, and other spindle accessories before mounting. Use a wire hand brush to remove foreign particles from threads. An air hose should not be used. The air pressure forces dirt and grit into bearing surfaces.
- Check all seating areas on holders to be sure each cutting tool rests solidly.
- Select cutters, holders, and square, hexagon, octagon, etc. turret accessories that provide maximum rigidity for holding, driving, and machining.
- Select chuck jaws so that two jaws grip an area that is less than one-half the diameter of the workpiece. The two jaws thus tend to center the workpiece against the third jaw.
- Regrind the chip groove on a bar turner cutter so that chips are formed and removed properly.
- Check for possible interference between the workpiece, tooling setups, and the indexing of each turret face.
- Tighten the cutter and binding screws on a bar turner so that the cutting edge does not move below center.
- Make sure each stop roll is adjusted to trip the power longitudinal and transverse feeds at the place where the workpiece is machined to the specified dimension.
- Use rocking jaws on rough surfaces (castings, forgings, weldments, and so on). Such jaws compensate for surface inaccuracies and imperfections.
- Withdraw the cutters and stop the turret lathe if machine vibration or tool chatter occurs.

A. TURRET LATHES: COMPONENTS AND ACCESSORIES

1. a. Describe a bar machine and a chucking machine in terms of holding and feeding workpieces.
   b. Distinguish between a ram-type and a saddle-type bar and chucking machine.
   c. Indicate the difference between a fixed-center turret and a cross-sliding turret on a saddle-type turret lathe.

2. Cite three functions of a secondary turret on a cross slide.

3. Describe briefly the location and operation of a leading-on thread-chasing attachment.

4. Describe how a taper is produced using a roller rest taper turner.

5. Cite three possible problems that may develop with improper care and maintenance of an indexing turret.

6. State three important safety checks the turret lathe operator makes before starting a production run.

B. SPINDLE, CROSS SLIDE, AND TURRET TOOLING AND PERMANENT SETUPS

1. Describe the difference between combined cuts and successive cuts as re-lated to turret lathe work.

2. Cite two advantages of an extended-nose collet as compared to a conven-tional spring collet.

3. a. State the purpose of positive-stop collets.
   b. Name three general types of stops for positive-stop collets.
   c. Explain a common design feature for positive-stop collets.

4. a. Give applications for the use of rocking jaws on a standard independent-jaw chuck on a turret lathe.
   b. Tell what function wrap-around jaws serve.

5. State two conditions under which special fixtures and other faceplate set-ups are more practical for turret lathe operations than are standard chucks or step chucks and closers.

6. a. Describe how cutter wear life is increased by features that are incorporated in the design of a bar turner.
   b. Name two cutter design features for controlling chips in carbide-tip or brazed-tip insert applications.

7. a. Name two slide tools for turret lathe work.
   b. Describe the function of each slide tool.

8. State why air pressure should be avoided in cleaning the spindle bore, work-holding accessories, and cutting tools on a turret lathe.

9. Explain the functions of a stop roll in relation to the cross slide carriage.

10. Explain three characteristics of a planned tool circle in production machining on a turret lathe.

11. Identify three possible causes of turning a finished diameter undersize when a bar turner is used.

12. List two practices to consider when taking external and internal cuts at the same time.

13. State three corrective steps to take to avoid tool chatter problems.

14. List three safety checks the turret lathe operator makes with respect to a tooling circle setup.

Basic External and Internal Machining Processes

Basic external cuts refer to straight, taper, and bevel turning; facing; grooving; thread cutting; cutting off; and knurling, which, technically, is not a cutting process.

Basic internal cuts relate to center drilling and drilling; straight, angle, and taper boring; counterboring; countersinking; reaming; undercutting (recessing); and threading either chased or tapped threads. Typical bar machined parts are shown in Figure 33–1.

CUTTING SPEEDS (sfpm) AND FEEDS (ipr) FOR BAR TURNERS

Table 33–1 gives one manufacturer's recommended starting cutting speeds and feeds for bar turner processes on ram-type turret lathes. The use of these cutting speeds and feeds provides for maximum machining production rates, quality of surface finish, and dimensional accuracy (concentricity and diametral setting).

All cutting speed and feed tables provide a baseline of values that serve to guide the turret lathe operator. Compensation may need to be made for factors such as chip control, rate of tool wear, machine horsepower, specifications, and other conditions governing all machine tools and setups.

BASIC EXTERNAL CUTS

TURNING CUTS

Straight turning is generally performed by one or a combination of three different methods: (1) side turning from the square turret; (2) overhead turning from the hexagon or octagon turret; and (3) long-bar turning using a bar turner.

Side Turning from the Square Turret. In side turning, the cutting tool is mounted in the square turret. Turning cuts may be taken simultaneously with processes that are carried on from turret tooling stations. Roughing and finish turning cuts may be taken by positioning the cutting tools at successive square turret stations.

Straight Turning with a Bar Turner. Bar turners are capable of machining a fine surface to tolerances within 0.001" (0.02mm). Bar turners, equipped with either brazed-tip or insert-type carbide cutters, are used to perform most of the heavy metal turning processes.

Overhead Turning from the Hexagon Turret. The multiple turning head on the hexagon turret provides rigid tool support that is essential in

Figure 33–1 Typical Bar Work Machined on a Turret Lathe

Table 33–1 Starting Points for Cutting Speeds and Feeds for Bar Turner Processes on Ram-Type Turret Lathes

	Metals to Be Machined				
	C-1070	A-4340	C-1045	C-1020	B-1112
	Cutting Speeds (sfpm)				
High-speed steel cutter	50	70	90	120	150
Premium Carbide cutter	220	245	320	420	490
	Feeds				
Depth of cut	1/8″	1/4″	3/8″	1/2″	5/8″
Feeds (ipr) (0.001″ per revolution)	0.030	0.018	0.012	0.0075	0.0045

heavy metal-removing processes. Cutter holders are mounted in the multiple turning head (Figure 33–2A). A reversible plain cutter holder (Figure 33–2B) is used for rough turning cuts. A reversible adjustable cutter holder (Figure 33–2C) has a built-in micrometer feature that permits a cutter to be set to size accurately and quickly.

Taper Turning. Internal or external tapers may be machined by using a taper attachment.

The cutting tool is mounted in the square turret on ram-type and fixed-center saddle-type turret lathes. The taper attachment is set at the required taper angle and locked to the bed. The cutter travel is controlled by the taper attachment guide plate (slide block).

FACING CUTS

Square Turret Facing. The square turret is used for most facing cuts. A single-point facing

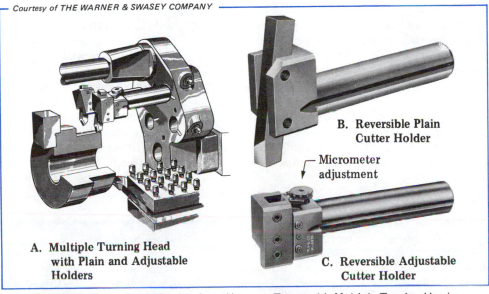

Courtesy of THE WARNER & SWASEY COMPANY

B. Reversible Plain Cutter Holder

Micrometer adjustment

A. Multiple Turning Head with Plain and Adjustable Holders

C. Reversible Adjustable Cutter Holder

Figure 33–2 Overhead Turning from Hexagon Turret with Multiple Turning Head

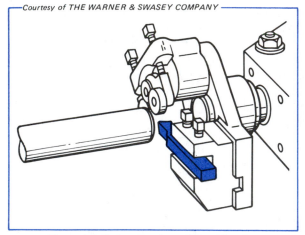

Courtesy of THE WARNER & SWASEY COMPANY

Figure 33-3 Application of Combination Turner and End Former to Facing

cutter is generally mounted in one station of the square turret. The cross slide carriage is positioned along the bed ways at the required length. It is clamped in position for facing cuts. Workpieces may also be faced by combining the facing cut with drilling, boring, or other processes.

Hexagon Turret Facing. Short facing cuts are often taken with tools mounted on the turret in a *quick-acting slide tool.* The slide tool is actuated by a hand-operated lever. This lever feeds the mounted cutter at right angles to the spindle axis.

Facing across large-diameter workpieces is performed on cross-sliding turret machines. The cutter is held on the turret in a through-slot cutter holder, boring bar, or tool-base cutter block.

End Facing. An *end former* or *combination turner and end former* (Figure 33-3) is used to support the end of a workpiece that would otherwise spring under the cutting action. The rolls are set ahead of the cutter to support the workpiece.

FORMING AND CUTOFF CUTS

Forming Cuts. A fast method of producing a finished shape and diameter is by *forming.* The four basic forming cuts are: (1) necking (cutting a recess close to a shoulder), (2) chamfering a beveled surface, (3) radius forming, and (4) grooving. Single forming cuts are usually taken

with the cutter mounted in the square turret. A *necking cutter block* mounted on the rear of the cross slide is used when several necking or grooving cuts are required.

The hexagon turret may also be used for forming cuts. A bar and cutter mounted in a quick-acting slide tool may be positioned and fed to cut a groove or chamfer. An internal facing cutter is usually held in a boring bar or tool-base cutter block on cross slide turret machines.

Cutoff Cuts. Cutting-off operations are performed by mounting the cutoff-tool holder in the square turret or in the rear cutter block. High-speed steel and carbide cutoff blades, mounted in a holder, permit the rapid changing of cutters as they wear. Maximum rigidity in the setup ensures fast, smooth cutoff.

THREADING

Threading with Die Heads. A continuous, even force must be applied when tapping so that the turret moves with the die head. For most threads, the die head may be led on by hand—that is, the turnstile is turned manually in starting and following the thread.

Long, accurate threading requires that a positive lead be provided for uniformly feeding the die head across the workpiece. A leading-on attachment or a thread-chasing attachment set at the required thread pitch, is generally used.

Precision Threading with Single-Point Cutters. Single-point thread cutting is used to produce threads that must be concentric with other diameters and meet precise tolerances.

Some ram- and saddle-type turret lathes are equipped with a thread-chasing attachment. This attachment is fitted to the cross slide and feeds the thread cutter in the square turret. Threads may also be cut from the hexagon turret when it is fitted with a thread-chasing attachment.

How to Sequence External Cuts

Example: Set up a ram-type turret lathe to mass-produce the threaded shaft shown in Figure 33-4.

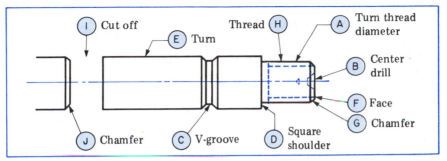

Figure 33–4 Sequence of External Cuts for a Threaded Shaft

> **Note:** The thread is to be die cut. The die head is to be led on by hand.

STEP 1 Feed the stock to length against a revolving bar stop.

> **Note:** It is assumed that the bar end is already chamfered.

STEP 2 Position (index) the bar turner that is mounted on the hexagon turret.

STEP 3 Turn the thread diameter Ⓐ.

STEP 4 Index the center drilling tool on the turret. Center drill the end of the bar Ⓑ.

STEP 5 Index to position the revolving center. Support the workpiece with the center.

STEP 6 Form the V-groove with a cutter mounted in the square turret Ⓒ.

STEP 7 Position the shoulder-turning cutter on the square turret. Square the shoulder at the end of the thread Ⓓ.

STEP 8 Turn the outside diameter by using a turning tool mounted in the square turret Ⓔ.

STEP 9 Remove the center. Position the end former or combination turner and end former on the turret.

> **Note:** The tools have rolls that hold the workpiece securely against the cutter.

STEP 10 Face Ⓕ and chamfer Ⓖ the end of the threaded shaft.

STEP 11 Turn the hexagon turret to the self-opening die head position.

STEP 12 Check the thread length trip stop. It should be adjusted so that the die head opens for the chaser insert to clear the workpiece just before the die head reaches the shoulder. Cut the thread Ⓗ.

STEP 13 Feed the cutoff tool mounted on the rear tool post to cut the workpiece to length Ⓘ.

STEP 14 Position the chamfering tool in the square turret. Chamfer the end of the next workpiece Ⓙ.

BASIC INTERNAL PROCESSES

DRILLING

Parts are drilled to produce holes to specifications, to center stock, and to remove material from existing holes. Drilling cuts require different drills. Parts may be center drilled or spotted with a *start drill*. Holes may be produced from the solid by using a *twist drill*. Holes that are larger than 2″ (50mm) diameter are drilled with a *spade drill*. Previously drilled, cored, or pierced holes may be enlarged with a *core drill*. Deep-hole drilling sometimes involves a *coolant-feeding drill*.

Core drilling requires a three- or four-fluted cutter. The core drill is used to enlarge previously drilled, cast, or pierced holes. Such holes must first be chamfered. Long holes are often started by boring to the core drill diameter for about 1/2″ (12mm).

Spade drilling is a practical method of drilling holes that are 2″ (50mm) or larger in diameter. The face of the cutter is ground to a cutting angle similar to the angle of a twist drill face. A spade drill has two cutting edges.

Deep-hole drilling relates to holes that are four or more times deeper than the drill diameter. Most deep-hole drilling is done with drills designed so that a cutting fluid may be pumped to the cutting edges.

BORING CUTS

Boring cuts are used to turn concentric, straight holes to precise dimensional limits. Boring bars are designed for internal rough and finish turning, facing, counterboring, undercutting, chamfering, bevel and taper turning, and thread chasing. Boring bars hold square high-speed steel tool bits, brazed-carbide or carbide-tip cutters, ceramic inserts, and chip breakers.

Straight Boring. *Turret boring* on a ram-type or fixed-center saddle-type turret lathe requires the cutter to be mounted in a slide tool. The boring bar and cutter are thus permitted to be adjusted to the bore size. The slide tool set-up shown in Figure 33–5 permits micrometer adjustment of the cutter to establish the depth of cut and bore diameter. Once the cutter is set to depth, the slide is clamped securely and provides the rigidity necessary to take roughing cuts or to finish bore a hole.

Cross-sliding hexagon turret boring on a saddle-type machine does not require the use of a slide tool. Adjustments for bore diameter are made by moving the cross slide directly.

Square turret boring is used for limited operations. These include the boring of short holes, taking lighter cuts, and using slower feeds. The solid, forged cutter designed for square turret use is generally less rigid than the heavier, solid boring bars.

Taper Boring. The square turret is also adapted for *taper boring*. The taper attachment is set at the required taper and is secured to the bed. The bore size is established by moving the cross slide handwheel. For accurate taper boring the cutter is fed *away* from the spindle. The cut is in the direction of the largest taper diameter.

Large taper boring operations are usually performed on a cross-sliding hexagon turret machine which permits the use of heavier, rigid boring tools adapted to heavy cuts.

REAMING CUTS

Reaming usually follows boring. Boring permits machining to a specific size and leaving the proper amount of material to remove in the finish reaming process.

Solid and *expandable reamers* are used. These reamers are held in a *floating reamer holder* which is mounted on the turret and is designed to float. Thus, the reamer is able to align itself with a previously drilled or bored hole.

Taper reamers are used to finish a taper bore or to produce an internal taper directly. A pair of roughing and finish cutting taper reamers is used, especially when steep tapers are to be cut.

Adjustable floating-blade reamers are designed for reaming large-diameter holes. There are two adjustable cutting blades in a floating blade reamer. This reamer is mounted in a flanged toolholder. The two reamer blades are aligned in a previously formed hole by the floating action of the cutter blades.

INTERNAL RECESSING (UNDERCUTTING), FACING, AND BACK-FACING CUTS

Relief areas for threading, grinding, shouldering, and taper surfaces must often be produced as internal cuts. A combination of longitudinal and cross slide movements is required. The cutting tools are brought to position inside the workpiece. The tools are then fed by cross feeding to undercut (internally recess) or produce an internally faced surface. Turret lathe setups for internal recessing and right- and left-facing processes are illustrated in Figure 33–5.

TAPPING AND SINGLE-POINT THREAD CUTTING

Sufficient force must be exerted to start the tap in a tap hole. Once started, the tap must be self-feeding. A tap that is forced during cutting, or *dragged* while it is being withdrawn, will produce defective threads. The turret and holder must be fed at a rate equal to the thread pitch.

Tapping with Solid and Collapsing Taps. *Solid taps* are generally used for thread sizes up to 1 1/2″ (38mm). These taps are held on the

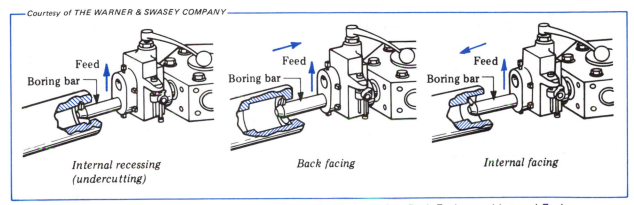

Courtesy of THE WARNER & SWASEY COMPANY

Feed

Boring bar

Internal recessing (undercutting)

Feed

Boring bar

Back facing

Feed

Boring bar

Internal facing

Figure 33–5 Slide Tool Setup on Hexagon Turret for Undercutting, Back Facing, and Internal Facing

hexagon turret in a releasing tap holder. The forward motion of the turret and tap is stopped by the setting on the turret stop. The tap is withdrawn by reversing the direction of spindle rotation.

Collapsing taps are used for thread sizes larger than 1 1/2″ (38mm). These taps are designed to collapse. The cutting edges pull away from the workpiece at the end of the cutting stroke. Collapsing taps may be mounted directly on a hexagon turret or in a flanged toolholder. One advantage of a collapsing tap over a solid tap is that the cutter is withdrawn without reversing the spindle direction.

Single-Point Thread Cutting. The single-point threading tool is held in a boring bar that is mounted in a hexagon turret slide tool. A flanged holder is used for mounting the threading tool on cross-sliding turrets. The cutter may also be held in a thread-chasing cutter holder that is mounted on the square turret of the cross slide. Multiple passes are required until the cutter is fed to thread depth.

How to Sequence Internal Cuts

The tooling up and sequence of processes for a production run of a typical part are identified

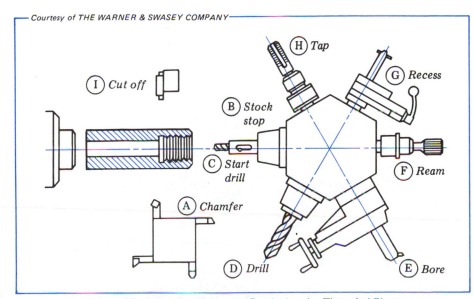

Courtesy of THE WARNER & SWASEY COMPANY

(I) Cut off

(H) Tap

(G) Recess

(B) Stock stop

(C) Start drill

(F) Ream

(A) Chamfer

(D) Drill

(E) Bore

Figure 33–6 Tooling Setup for Producing the Threaded Sleeve

in Figure 33–6 by the letters (A) through (I). Note that the chamfered end, outside diameter, and cutting-off processes are external cuts from the square and rear tool positions on the cross slide.

Safe Practices for Basic External and Internal Machining Processes

- Use manufacturers' tables of recommended cutting speeds and feeds. Adjust the recommended values based on the specifications of the piece part, the accuracy and quality of surface texture, rate of tool wear, chip control, and machine capability.
- Check the working clearance between the cross slide and turret tooling.
- Mount bar turner rolls on a finished diameter ahead of the cutter to produce a precise, concentric diameter.
- Mount bar turner rolls behind the cutter to support the cutting force and burnish the turned surface.
- Use a multiple turning head on the hexagon turret to provide rigid tool support for heavy cuts.
- Use a leading-on attachment to cut fine, precise threads.
- Chamfer or bore the end of a cored or pierced hole for a short distance to center guide the lips of a drill.
- Use a spade drill for drilling holes over 2" diameter to minimize the overhang of the drill.
- Use the heaviest possible boring bar with the shortest overhang for heavy, long boring cuts.
- Bore an internal taper on the finish cut with the direction of feed toward the large diameter.
- Set the turret stop about 1/8" (3mm) less than thread depth to control threading to a shoulder or other fixed point of a workpiece.
- Check the composition of the cutting fluid and the rate of coolant flow at each station. Position splash guards to contain the coolant and chips.
- Keep the work area around the turret lathe dry and free of chips.

1. Explain how a burnished, quality surface finish is produced by using a bar turner.

2. Describe how large-diameter workpieces may be faced on machines equipped with a cross-sliding turret.

3. a. List four basic forming cuts for turret lathe work.
   b. Indicate two general locations for cutting-off tools. State why such locations are functional and desirable.

4. Use machinability and cutting speeds and feeds tables for turret lathe operations. Determine the starting cutting speeds in sfpm for the following three jobs:
   a. Rough facing class A machinability parts with a carbide-tipped cutter.
   b. Cutting off class B machinability parts with a high-speed steel cutter.
   c. Rough cutting class B machinability parts with a high-speed steel cutter by using a bar turner.

5. Determine the starting feed rates for machining on a ram-type turret lathe for each of the following processes:
   a. Turning a workpiece, using a roller turner setup, for a 3mm depth of cut;
   b. Forming steel parts to a 1/4″ depth of cut;
   c. Cutting off a 1″ diameter workpiece with a high-speed steel cutter.

6. Tell what function is served by each of the following drills: (a) start drill, (b) spade drill, and (c) core drill.

7. Describe two turret lathe setups that permit a boring bar and cutter to be adjusted for each successive cut when straight boring.

8. Explain the difference between the cutting of an internal, square recess and back facing a square shoulder on a turret lathe.

9. Give two advantages of using collapsing taps in preference to solid taps for certain applications on a turret lathe.

10. State three safety precautions the turret lathe operator should take to prevent personal injury or damage to the machine, setup, or workpiece.

Single- and Multiple-Spindle Automatic Screw Machines

CLASSIFICATIONS OF SCREW MACHINES

PLAIN SCREW MACHINE

Screw machines are designed to perform a single machining operation or several operations at one time or in a rapid sequence. The main slide, turret, and other slides of the *plain screw machine* are operated manually. By contrast, these components and the feeding of bar stock on the semiautomatic (wire feed) screw machine are operator controlled.

AUTOMATIC SCREW MACHINE

The *automatic screw machine* is fully automatic in operation. Once the cutting tools are set up on the cross slide and turret and the controlling cams are positioned, the stock is automatically advanced and clamped in the chucking device. The turret and tools are automatically indexed in sequence.

Single-Spindle Automatic Screw Machine. The *single-spindle* automatic screw machine is a general-purpose machine in which a single piece of stock is fed through the spindle at one time. The workpiece is machined in the sequence in which the tools in the indexing turret and on the cross slide are set up.

Usually, round, hexagon, square, and regular-shaped bar stock is processed on the single-spindle automatic screw machine. The bar stock is securely held in a spring collet chuck. The chuck is actuated by a friction clutch inserted between two pulleys on the spindle. The spindle drive allows the direction of rotation to be changed (reversed).

Multiple-Spindle Automatic Screw Machines. The *multiple-spindle* automatic screw machine has four, five, six, eight, or more work-rotating spindles. Additional production capacity is possible due to the increased number of operations that can be carried on at the same time on one machine.

MAJOR COMPONENTS OF AUTOMATIC SCREW MACHINES

There are many designs for screw machines. A Brown & Sharpe ultramatic screw machine is shown in Figure 34–1 as an example. However, all designs include cams, levers, clutches, trip dogs, and collet or chucking devices. Bar stock in 10, 12, and 20 foot (250mm, 300mm, and 500mm) lengths is automatically advanced after each workpiece is cut off.

SPINDLE

The spindle is mounted in antifriction bearings. The capacity for the Brown & Sharpe ultramatic screw machine (Figure 34–1) is from 3/4″ (19mm) to 1 5/8″ (41mm) diameter. Large-capacity #3 machines have either a 2″ (50mm) or 2 3/4″ (70mm) capacity. Two-speed and four-speed screw machines are available. There are 18 spindle speeds in the two-speed range from 20 RPM to 5,018 RPM for the 3/4″ (19mm) capacity machine. The #2 machine, 1 5/8″ (41mm) size, has 17 speeds in the two-speed range from 14 RPM to 2,906 RPM. High speeds and low speeds are produced in a clockwise and counterclockwise direction.

FEED CYCLE CHANGE GEARS

Feed cycle change gears (Figure 34–2) control the rate of production. The gears are installed at the right end of the machine. The safety cover provides for the automatic disengagement of the drive shaft.

The gears that are used are determined by the time (in seconds) required to machine one piece. The feed change gear plate on the machine shows the position of the driver and driven gears on the stud. Gear combinations are given for a range of time (in seconds) to machine one piece from 1.6 to 1,000. The right column on the gear plate shows the gross production per hour from 2,250 (1.6 seconds per piece) to 3.6 parts (1,000 seconds).

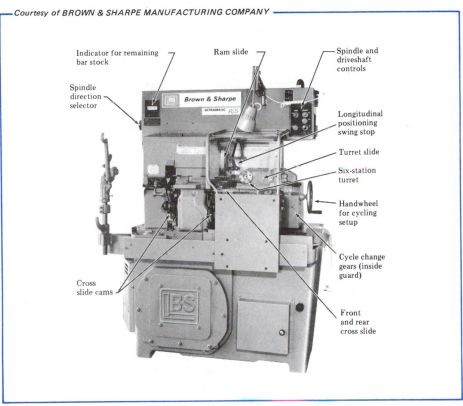

Figure 34-1 Major Features of an Ultramatic Screw Machine

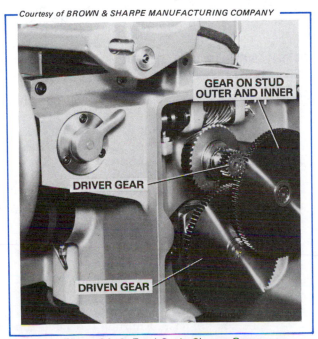

Figure 34-2 Feed Cycle Change Gears

TRIP LEVERS, CAMSHAFTS, DOGS, AND DOG CARRIERS

The chuck clutch trip lever controls the chuck clutch. The chuck clutch operates mechanisms through gearing to the chuck camshaft. A chuck and feed cam on this shaft operates a chuck fork (Figure 34-3). The fork causes the collet to open and then to close. The stock is automatically advanced and positioned by an adjustable feed slide. Adjustable dogs on dog carriers control the opening of the chuck before the stock is advanced during the stock feeding cycle.

An automatic operating mechanism controls such operational movements as indexing the turret and changing the spindle speeds. One camshaft carries the control drum, upper slide cams, chuck dog carrier, and the cross slide cams. A second camshaft carries the lead cam for the turret slide and the rapid pull-out cam.

Dog shaft carriers control the turret dog carrier (for timing the turret indexes) and the reversing dog carrier (for spindle clutch and spindle

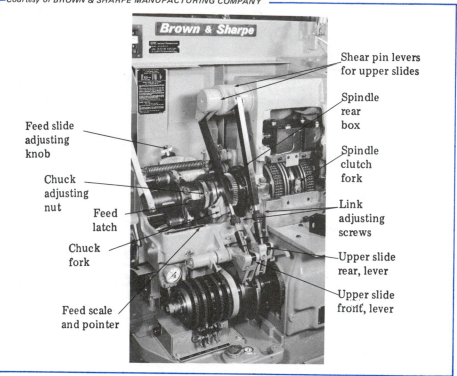

Figure 34–3 Spindle and Feed and Upper Slide Adjustments

speed changes). Adjustable trip dogs on the turret dog carrier are set to move the trip lever for single or double indexing. Single indexing is used to index each position of the hexagon turret. Double indexing is used to index every other tool position.

UPPER FRONT AND REAR TOOL SLIDES

Upper slides provide additional capacity for turning and cutoff processes. Figure 34–4 illustrates an upper front slide equipped with a forming tool post. This slide accommodates right- and left-hand turning tools. The upper rear slide has two cutoff toolholders. These are adapted for right- and left-hand "T" type cutting-off blades.

TURRET SLIDE

The turret slide may be moved toward or away from the spindle. Movement is controlled by the relationship between the lead lever, withdrawal cam, and the rack of the turret slide. The turret slide is provided with a positive stop

screw. This screw is used to maintain extreme accuracy of depth to close dimensional tolerances.

TOOLING FOR AUTOMATIC SCREW MACHINES

Standard tool bits are used with many holders. The cutters may be altered slightly to be converted for direct use. Tools for machining external surfaces are mounted on the turret. These tools include balance turning tools, box tools, knee tools, plain and adjustable hollow mills, and swing tools. End-of-work machining tools include centering and facing tools, pointing tools, and pointing toolholders for circular tools. Internal surfaces are cut with tools such as drills, reamers, counterbores, and recess cutters. These tools may be held in rigid or floating holders.

Internal threads are cut with solid and adjustable taps. External threads are cut with self-opening die heads or thread rolls mounted in cross slide toolholders. Knurling tools and top and side knurl holders are held on the cross slide.

Courtesy of BROWN & SHARPE MANUFACTURING COMPANY

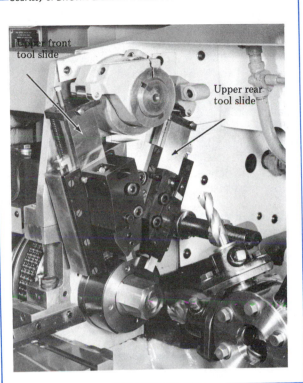

Figure 34–4 Upper Front and Rear Tool Slides

Courtesy of BROWN & SHARPE MANUFACTURING COMPANY

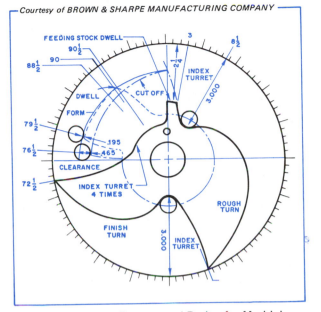

Figure 34–5 Cam Features and Design for Machining a Workpiece on an Automatic Screw Machine

Other knurling tools and adjustable knurl holders are mounted on the turret. Circular cutting-off and forming tools are used in the regular cross slide tool post or an adjustable tool post. Thin, straight-blade cutting-off tools and square tools are supported on the cross slide.

The *turret-mounted tools* include: angular cutting-off tools, support and auxiliary tools, back rest for chuck, fixed and adjustable guides for operating spring tools, and spindle brakes.

OPERATION OF FULLY AUTOMATIC SCREW MACHINES

Bars are loaded in each of the hollow spindles on a fully automatic screw machine. Operations at each spindle are carried on simultaneously and in sequence. The end-of-work machining tools (except for the die head) are all fed as a unit by movement of the tool slide. The die head is fed independently. Similarly, the tools on the cross slide are fed separately by the drum cams.

Each spindle rotates as a single-spindle machine. Each spindle has a spring collet chuck. Cutting tools such as drills, reamers, counterbores, and taps are mounted in front of each spindle. Turning, forming, and cutting-off tools are secured on the cross slides. When an operation is finished, each spindle turns to the next position. The spindles are rotated until the cycle is completed.

FUNCTIONS AND DESIGN OF CAMS

Cams control the timing and position of all cutting tools, and the action of work-holding devices. Cams impart motion to the turret and the front, rear, and vertical cross slides. This motion controls the cuts. Cams are designed to *rise, fall,* or *dwell*—that is, they cause the cutting tools to advance, be withdrawn, or remain stationary.

An example of three superimposed cams is shown in Figure 34–5. The cams are used for rough turning, finish turning, and cutting-off operations. This particular cam is used to automatically machine a part on a Brown & Sharpe automatic screw machine. Note that the cam perimeter is divided into 100 parts. The speed of the camshaft determines the number of seconds represented by each division.

CUTTING FLUIDS

As with all other machining processes and machine tools, a cutting fluid serves as a coolant and a lubricant. In addition, the cutting fluid must contain properties that permit it to separate rapidly from the chips.

Straight mineral oils are usually used as a blending medium for general-purpose screw machine operations. The mineral oil is mixed with a base cutting oil that possesses the necessary lubricating qualities.

Mineral lard oils are used on parts that require a higher quality surface finish than can be obtained by using straight mineral oils. Mineral lard oils have noncorrosive properties. These properties make mineral lard oils especially suitable for machining copper and copper alloys.

Sulphur cutting oils and *sulphurized oils* are recommended to increase tool life. Sulphur increases the cooling and lubricating characteristics of oils. Sulphurized oils wet surfaces faster than regular cutting oils—that is, sulphurized oils penetrate into remote places. Such oils have excellent film strength and help prevent chips from welding on the tool point.

Mineral oils are adapted for light machining operations on some steels and brass. Mineral oils are used for tapping and threading some nonferrous metals and for other difficult work.

RECENTLY DEVELOPED CUTTING OILS

The major claims for new, light-colored, transparent, and odorless cutting oils are as follows:

- The transparency of recently developed oils makes the machined part clearly visible to the machine operator during machining;
- The cutting oils have maximum lubricity, which results in reduced heat and tool wear;
- Antiweld qualities minimize buildup at the cutting edge of the tool for a considerable distance from the cutting edge;
- Increased extreme-pressure properties reduce the frictional heat developed by the rubbing chips on the extreme-pressure cutting areas;
- Exceptional cooling characteristics permit heat that is generated by the rubbing

of the chips and workpiece to dissipate rapidly;
- The oils are corrosive and are recommended for all kinds of machining operations on all grades of steel. Noncorrosive oils are used for nonferrous metals.

Major oil producers have also developed special sulphurized and other cutting oils. These oils are all compounded to be noncorrosive and nonstaining. In addition to the previously listed claims, such cutting oils are claimed to produce equally satisfactory results on ferrous and nonferrous metals.

AUTOMATIC SCREW MACHINE ACCESSORIES

Second machine operations are often eliminated by adding special attachments to the machine being used. Attachments are available for slotting

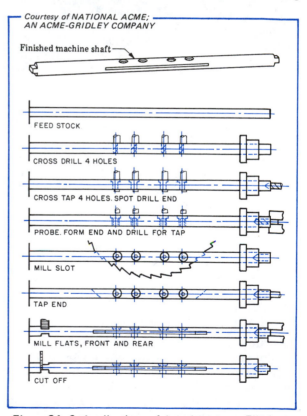

Figure 34–6 Applications of Attachments to Eliminate Second Machine Operations on a Shaft (Produced on an Eight-Spindle Automatic Screw Machine)

Table 34-1 Examples of Automatic Screw Machine Troubleshooting

Problem	Possible Cause	Corrective Action
Stock Stop Adjustment Variation in length	—Projection left because the cutting-off tool does not remove all of the teat.	—Reset the cutting-off tool on center.
	—Feed-out period on the lead cam is too short.	—Replace with the correct cam.
	—Turret cutting tools push the stock back into the collet.	—Sharpen each cutting tool. Check for correct feed.
	—Worn roll and pin on lead cam lever.	—Remove and replace the worn parts.
	—Chuck or feed finger tension too loose (or worn).	—Adjust the chuck to the required tension or replace a worn feed finger.
Turret Tooling *Knee Tools* Cutting with a taper	—Bar too small to use a knee tool; length of cut too long.	—Replace with a box tool, balance turning tool, or hollow mill.
	—Too much force required for the cut.	—Grind the correct clearance on the tool bit.
	—Tool not on center; too much feed; spindle speed too fast.	—Set tool on center or a few thousandths above; correct the feed and/or speed.
Balanced Turner Cutting rough	—Front clearance too high; rake too deep.	—Regrind the tool bit to the correct rake and clearance angles.
Turret Clamp bolts do not hold the turret tools	—Turret tool clamp bushing too short; radius worn; or bolt too long.	—Replace the turret tool clamp bushing.

screws, burring, tapping nuts, rear-end threading, cross drilling (at right angles to the longitudinal machine axis), slotting, and machining flats. An example of a machine shaft produced on an eight-spindle automatic screw machine is shown in Figure 34-6. The drilling and tapping of the four holes and the milling of the center slot and the two slots on the end are performed on the same screw machine. Attachments are used for these operations. The production rate for this particular drilled, slotted, and threaded shaft is 200 parts per hour.

AUTOMATIC SCREW MACHINE TROUBLESHOOTING

Examples of automatic screw machine troubleshooting problems, causes, and corrective action are provided in Table 34-1. Some screw machine manufacturers furnish complete troubleshooting information for rod feeding and stocking, hole forming, turning, knurling, and other processes, turret adjustments, and accessories.

Safe Practices in Setting Up and Operating Screw Machines

- Use noncorrosive oils on machine tools where there is a possibility of leakage into the bearing system.
- Place enclosing guards around chucking and machining areas before operating a screw machine to prevent personal injury that can be caused by tool breakage, flying chips, or the workpiece being thrown out of a chuck.
- Check the proper placement of each tool and how securely each is held. In addition to personal injury, work damage may result from a tool working loose or being forced out of a holder.

- Shut down the machine if any tooling changes or setup adjustments are to be made.
- Safeguard bar stock that projects beyond the lathe spindle. Guard rails are placed to cover the extended length in order to shield workers from a revolving workpiece. Note that sections of piping through which bar stock may be fed should be installed to prevent long bars from whipping around.
- Use short sheet metal or plastic screens to intercept and direct the cutting fluid to the chip pan, screens, and oil reservoir.
- Install a fine mesh, nonskid floor mat for worker protection against any oil that may be thrown from the screw machine.
- Change the sawdust or oil-absorbent compound around a screw machine as common practice. (Many departments have adopted nonslip, nonflammable floors for areas surrounding screw machines.)
- Wash hands and face thoroughly at regular intervals. Avoid contact with the cutting fluid as much as possible to prevent skin disorders that are caused by certain cutting fluids.
- Change work clothes frequently. Otherwise, they may become saturated with the cutting fluid.

UNIT 34 REVIEW AND SELF-TEST

1. List three attachments that permit additional second machine operations to be performed on a screw machine.

2. Describe briefly the function of (a) feed cycle change gears, (b) the chuck clutch trip lever, and (c) the dog shaft carriers of an automatic screw machine.

3. a. Name five standard tools for automatic screw machines.
   b. Give the major process for which each tool is designed.

4. Indicate five workpiece specifications (in addition to dimensions) that are required for the designer to lay out screw machine cams.

5. State six conditions that control screw machine speeds and feeds.

6. a. Identify four properties of new screw machine cutting oils that make them different from standard mineral oils.
   b. Describe the importance of each property in relation to either tool wear life, quality of surface finish, heat removed, or machining.

7. Record the corrective steps to follow in troubleshooting the following screw machine problems:
   a. Stock stop adjustment: Scored end of workpiece; stock stop center hole too large.
   b. Turret tooling: (1) Knee tool cutting with a taper; tool not on center. (2) Second cutter of balanced turner back of center; shoulder steps produced.

8. Describe three safety precautions an operator should observe before starting a screw machine.

PART 7 Milling Machines: Technology and Processes

SECTION ONE

Horizontal Milling Machines, Cutters, and Accessories

This section presents introductory information about milling machines and milling processes, as follows.

- Designs and functions of principal features and accessories;
- Maintenance procedures;
- Design features and applications of standard milling cutters;
- Speeds, feeds, and cutting fluids used for milling processes; and
- Cutter-holding and work-holding devices and accessories.

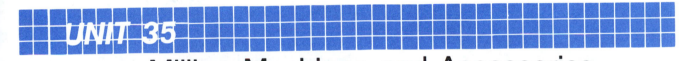

UNIT 35

Milling Machines and Accessories

A. FUNCTIONS AND MAINTENANCE OF MILLING MACHINES AND ACCESSORIES

Milling is the process of cutting away material by feeding a workpiece past a rotating multiple-tooth cutter. The cutting action of the many teeth around the *milling cutter* provides a fast method of machining. The machined surface may be flat, angular, or curved. The surface may also be milled to any combination of shapes. The machine for holding the workpiece, rotating the cutter, and feeding it is known as a *milling machine.*

INDUSTRIAL TYPES OF MILLING MACHINES

The major types of general-purpose and common manufacturing machines are:

- Knee-and-column,
- Fixed-bed,
- Rotary-table,
- Planetary,
- Tracer-controlled.

These types and the design features of numerically controlled machines are described briefly in this unit.

There are other types of milling machines (millers) for particular machining processes. For example, the *thread miller* is used for milling lead screws. The *cam miller* is designed for machining disc cams. The *skin miller* is adapted to special milling processes in the aircraft industry where, for example, the outer covering (skin) of wings is tapered to accommodate the stresses within each wing section.

287

The range of milling machine sizes is also extensive. Table lengths and movements extend from the small bed lengths of bench millers to bed lengths of 100 feet to 150 feet. Some millers have stationary spindles. On other millers the spindle carriers may be swiveled about a vertical or horizontal plane. Together with the cross feed (transverse), longitudinal, and vertical movements, these millers each have five axes of control. Many milling machines are programmed and are numerically controlled.

KNEE-AND-COLUMN TYPES OF MILLING MACHINES

There are five basic types of *knee-and-column* milling machines:

- Hand,
- Plain,
- Universal,
- Omniversal,
- Vertical-spindle.

Figure 35–1 A Hand Milling Machine

Hand Milling Machine. As the name implies, the hand milling machine is small. It is entirely hand operated. The hand miller pictured in Figure 35–1 may be mounted on a bench or on a floor base. This miller is particularly useful in small milling machine manufacturing operations. Slotting, single- and multiple-cutter milling, direct indexing, and other simple milling operations may be performed economically. The hand feed permits a rapid feed approach and withdrawal after a cut.

Plain Horizontal Milling Machine. The plain miller has three principal straight-line movements. The table moves in a longitudinal path. The movement is parallel to the face of the machine column. The table may be moved (cross-fed) in or out and/or up and down (vertically). The spindle is mounted in a horizontal position. The cutter is rotated by the spindle movement. The feed is in a straight line. The plain miller is a general-purpose, standard machine tool.

Universal Milling Machine. In addition to straight-line movements, the universal milling machine has the added movement of a swivel table. A standard universal miller with the main features identified is shown in Figure 35–2. The table is mounted on a saddle. The base of the table is graduated and designed for setting the table either straight (0°) or at an angle. The angle setting permits the table to travel at an angle to the column face. The universal miller is especially adapted to produce straight and spiral cuts. Common examples are the spiral flutes on reamers, drills, and other cutters and the spiral teeth on helical-form gears.

Omniversal Milling Machine. The omniversal milling machine is primarily a precision tool-room and laboratory miller. The rotation of the knee about the column face is in a perpendicular axis to the knee. This rotation makes it possible to machine tapered spirals and tapered holes.

Vertical-Spindle Milling Machine. The spindle on the vertical-spindle milling machine is mounted vertically. The two basic types of vertical-spindle millers are classified as: (1) fixed-bed and (2) knee-and-column. A knee-

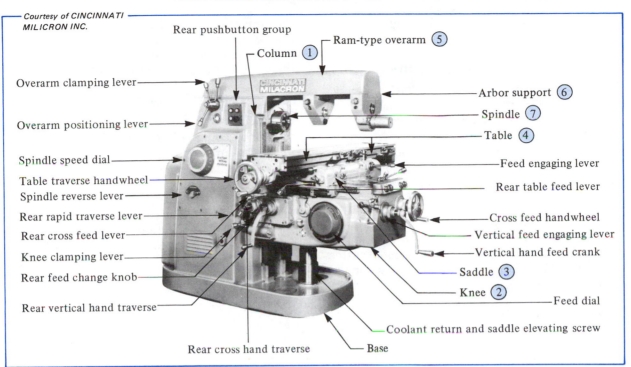

Rear pushbutton group

Column ①

Ram-type overarm ⑤

Overarm clamping lever

Arbor support ⑥

Overarm positioning lever

Spindle ⑦

Table ④

Spindle speed dial

Feed engaging lever

Table traverse handwheel

Rear table feed lever

Spindle reverse lever

Rear rapid traverse lever

Cross feed handwheel

Rear cross feed lever

Vertical feed engaging lever

Knee clamping lever

Vertical hand feed crank

Rear feed change knob

Saddle ③

Knee ②

Rear vertical hand traverse

Feed dial

Coolant return and saddle elevating screw

Rear cross hand traverse

Base

Figure 35–2 Main Features of a Knee-and-Column Type of Milling Machine

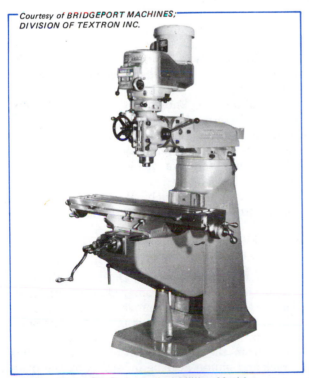

Figure 35–3 A Vertical Milling Machine

and-column type of universal toolroom vertical-spindle machine is shown in Figure 35–3. The machine accessories, attachments, cutters, set-ups, processes, and maintenance are covered in detail in successive units.

FIXED-BED MILLING MACHINE

The *fixed-bed miller* is usually a heavier, more rigidly constructed machine tool than the knee-and-column type of miller. A fixed-bed milling machine has an adjustable spiral head. This head permits adjustment of the cutter in relation to the fixed position of a workpiece and table. Only the table travels longitudinally along the fixed ways of the bed.

The fixed-bed milling machine is primarily an automatically-cycled manufacturing type of machine. Milling cutters such as face mills, shell end mills, and other cutters mounted on arbors are used. Face milling, slotting, and straddle milling are common operations performed on this machine.

Some fixed-bed machines are equipped for automatic rise and fall of the spindle head to permit the milling of surfaces at different levels. These machines are also called *rise and fall millers*.

ROTARY-TABLE MILLING MACHINE

The design features of the *rotary-table milling machine* differ considerably from the features of the flatbed, plain, and universal millers. The table revolves. The worker loads and unloads workpieces in fixtures attached to the table. The table rotates under a face-type cutter. Usually there are two spindle heads.

PLANETARY MILLING MACHINE

A workpiece is held stationary on the *planetary milling machine*. One or more cutters may be held in vertical and/or horizontal spindles. Production units are also designed with magazine loading. This machine performs many operations that normally are done on the lathe. The planetary milling machine is used for workpieces that are heavy, difficult to machine, or are so unbalanced or delicate that they cannot be rotated,

TRACER-CONTROLLED MILLING MACHINE (DUPLICATOR OR DIE SINKER)

There are two basic types of *tracer controlled* milling machines. One type is called a *profile miller* (profiler). The second type has a number of names, including *duplicator*, *die sinker*, and *Keller machine*.

Profile Miller. The profile miller is similar to the vertical-spindle milling machine. Machining is usually done by small-diameter end mills. The cutting path is controlled by a *tracer (stylus)* with the same diameter and form as the end mill. The tracer follows a template of the required size and shape. The movement of the cutter is controlled by hand or automatic feed.

Die Sinking Machines (Profiler, Automatic Tracer-Controlled Miller). Many form dies and formed parts require three-dimensional machining. Die sinkers are designed for operation manually or in combination with electronic and hydraulic controls. A template is used. The template is a replica of the work to be produced.

The tracer provides three-dimensional information as it is moved over the entire surface. This information is duplicated by the movement of the corresponding profiler head and form cutter.

FUNCTIONS AND CONSTRUCTION OF MAJOR UNITS

The terms commonly used to identify the major units of all milling machines are given in Figure 35-2.

The major units of a horizontal milling machine include the knee, column, saddle (universal or standard), table, overarm, table controls, spindle, arbor support, overarm braces, drives, speed and feed controls, and the coolant system.

TABLE AND KNEE CONTROLS AND GRADUATED DIALS

Table and knee movements are made with feed screws and specially designed nuts. These screws and nuts are incorporated in the knee, saddle, and table design.

The knee is moved by an *elevating screw*. This screw extends from the inside of the knee to the base. The motion of the screw is controlled by the *vertical hand feed crank*.

A second *feed screw* is provided to move the saddle crosswise, toward or away from the column. The *cross feed handwheel* is used to turn this screw.

A third screw, known as the *table lead screw*, runs the length of the table. It is turned by a handwheel located at the end of the table.

There are micrometer collars (dials) on each of the three screw-movement controls (Figure 35-4). The collars are graduated to read in thousandths of an inch (0.001″) or in two-hundredths of a millimeter (0.02mm).

OVERARM (5), ARBOR SUPPORT (6), AND BRACES

The dovetailed ways on the top of the milling machine column accommodate the *overarm*. The overarm may be moved in a horizontal plane. The overarm axis is parallel to the spindle axis. The overarm slides toward (into) or away from the column. On heavy-duty millers a rack-and-pinion gear permits moving the overarm to position.

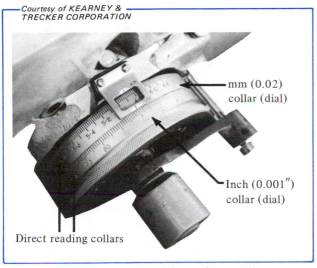

Courtesy of KEARNEY & TRECKER CORPORATION

mm (0.02) collar (dial)

Inch (0.001″) collar (dial)

Direct reading collars

Figure 35–4 Precision-Graduated Inch/Metric Micrometer Collars for Table, Knee, and Saddle Movements

Arbor supports are designed to slide on the overarm. Two common designs of arbor supports are available. The *outer arbor support* extends down below the milling machine arbor. The lower section of the support contains a bearing into which the pilot end of the arbor fits. The arbor support and overarm are used together to keep the arbor aligned and to prevent it from springing. Otherwise chatter marks, poor surface finish, and an inaccurate workpiece may be produced.

The *intermediate support* has a large center hole. A bearing sleeve rides in this center hole. The arbor fits through and is supported by the bearing sleeve. The intermediate support and bearing sleeve are used when roughing cuts and multiple cuts are to be taken.

A self-contained oiling system is incorporated in many arbor supports. The system consists of an oil reservoir, a sight gage that indicates the supply of oil, and a plunger. A small quantity of oil is forced directly into the bearing sleeve by depressing the plunger.

On heavier models the overarm is rigidly supported for heavy cuts by using *overarm braces.*

THE SPINDLE ⑦

The *spindle* is housed in the upper section of the column. Four main functions are served by the spindle:

- Aligning and holding cutting tools and arbors;
- Transmitting motion and force from the power source for cutting;
- Ensuring that the machining processes are dimensionally accurate;
- Driving a slotting, vertical-spindle, or other attachment.

The spindle itself is hollow bored. The front end has a tapered hole. Standard arbors, adapters, and cutting-tool shanks fit the tapered hole. A draw-in bar is usually used to secure the tools against the taper. The front end of the spindle is fitted with two set-in lugs that extend beyond the face. Arbors, adapters, and cutting tools have identical slots. These slots match the positions of the lugs. The lugs fit into the slots and provide a positive drive. Four other holes are threaded in the spindle face. These holes are used for mounting face mills.

The direction of the spindle rotation is reversed by one of two common methods. Some models have two motor switches, one switch for clockwise rotation and a second switch for counterclockwise rotation. Other machines are equipped with a spindle-direction change lever.

MILLING MACHINE DRIVES

Spindle Speeds. Speed changes are made on a constant-speed drive by a sliding gear transmission housed in the column. The various spindle speeds (revolutions per minute) are obtained by positioning the speed-control levers. The index chart on the column gives the lever settings for a required speed.

Feeds and Feed Levers. Feed is the rate at which a workpiece is fed to a cutter. The rate is specified as so many thousandths of an inch or millimeters per revolution of the spindle. The feed rate may also be given in inches or millimeters per minute when the milling machine is equipped with a constant-speed drive.

A feed index chart on the knee or column gives the range of feeds. The position of the feed-change levers and a dial design are illustrated in Figure 35–5. The lever or dial settings regulate the lengthwise movement of the table, the cross feed movement of the saddle, and the

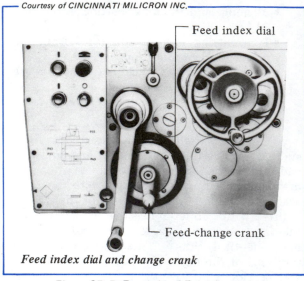

Feed index dial

Feed-change crank

Feed index dial and change crank

Figure 35–5 Example of Feed Controls

vertical travel of the knee when the automatic feed is applied.

Other levers are used to manually engage or disengage the automatic feed mechanism. These levers are located on the knee and saddle. Some machines are also designed for rapid traverse.

Trip Dogs. Trip dogs are simple adjustable guides and a safety device. They control the length of movement. The trip dog moves against a pin and causes it to move and disengage the feed. Trip dogs may be placed on the front of the table, under the saddle, and behind the knee.

MAINTENANCE OF THE LUBRICATING SYSTEMS

PRESERVING ACCURACY AND SMOOTHNESS OF OPERATION

Oiling systems are incorporated into the design of plain, universal, and other types of milling machines. The systems maintain a continuous film of oil between the mating surfaces of moving parts. The oil provides good, regular lubrication and also maintains the precision of the mating parts.

Pressure, *gravity*, and *splash systems* are widely used. The oil is forced through tubing to the surfaces, threads, and bearings that require

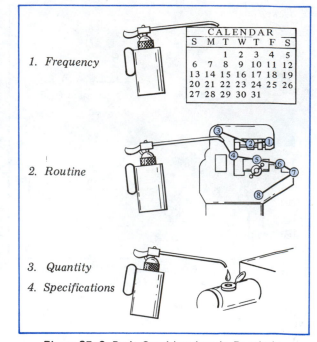

1. *Frequency*

2. *Routine*

3. *Quantity*
4. *Specifications*

Figure 35–6 Basic Considerations in Regulating Machine Lubrication

lubrication. Glass-faced *sight gages*, indicate whether the oil is circulating, the amount of oil in the reservoir, and the level at which it should be maintained.

REGULAR PROCEDURES FOR MACHINE LUBRICATION

Technical manuals identify the lubricating systems and points and specify the procedures to follow. The machine tool builder recommendations center around four basic considerations as illustrated in Figure 35–6.

LUBRICANT SPECIFICATIONS AND QUANTITY

The specifications (type) and quantity of lubricant depend on the application. Geared spindles are in constant rotation and produce the forces necessary for cutting. The spindle lubricant must withstand variations in temperature and must have the required viscosity to flow, cool, and lubricate under varying machining conditions. The lubricant on the column ways requires different properties than the oil used to meet the heat-removing needs of the spindle.

The quantity of lubricant to use is established by the manufacturer according to the conditions of the system. The indicated range between *full* and *add oil* on a sight indicator, for example, provides the operator with the designer recommendations.

B. MILLING CUTTER - AND WORK-HOLDING DEVICES

CUTTER-HOLDING DEVICES

The spindle of the milling machine provides a bearing surface that aligns the arbor, adapter, or tool shank. The power to drive the cutter is transmitted from the spindle to the cutter.

Modern milling machine spindle noses are ground to a standard steep machine taper size of 3 1/2″ per foot, or an included angle of 16°36′. The angle of taper makes the taper *self-releasing*. A locking device such as a quick-releasing collar or draw-in bar is used to secure an arbor or adapter to the spindle taper.

A milling machine cutter is generally held by one of four methods:

- Mounting the cutter on the nose of the spindle,
- Inserting a shank directly in the spindle bore,
- Using adapters and collets,
- Mounting the cutter on an arbor.

Cutter-holding devices thus include the spindle nose; arbor; keyseat and key; spacing collars, bearing sleeve, and arbor nut; draw-in bar; adapter; sleeve and collet; and cutting-tool holder.

SPINDLE NOSE

The four common steep taper sizes of spindle noses are #30, #40, #50, and #60. The

Table 35–1 National Standard Steep Machine Tapers

Taper #	Large Diameter
30	1 1/4″
40	1 3/4″
50	2 3/4″
60	4 1/4″

range of the large diameters of the common taper sizes is given in Table 35–1.

ARBOR STYLES

There are three styles of arbors: A, B, and C. The *style A arbor* has a pilot (Figure 35–7A). The pilot fits into a corresponding bearing in the arbor support. Style A permits a workpiece to be brought up close to the arbor. The advantage is that smaller-diameter cutters may be used with style A arbors that can be used with style B. Where needed, a bearing sleeve and an intermediate arbor support are added to provide additional bearing support and prevent a long arbor from vibrating.

The *style B arbor* is supported by one or more bearing sleeves and the same number of intermediate supports (Figure 35–7B).

Style B arbors are used where a minimum of clearance is needed between the cutters, workpiece, and holding device.

Multiple milling operations often require the use of two arbor supports for heavy-duty milling. This setup provides maximum rigidity and

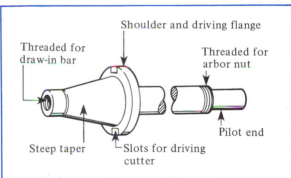

A. Style A Arbor with Pilot for End Support

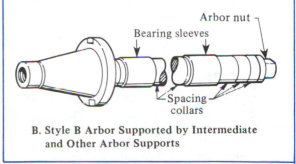

B. Style B Arbor Supported by Intermediate and Other Arbor Supports

Figure 35–7 Style A and B Milling Machine Arbors

eliminates chatter. The bushing and bearing sleeves have a free-running fit.

The *style C arbor* is a short arbor. It is sometimes called a *shell end mill arbor.* The cutter is mounted on a shouldered end. It is held against the face of the arbor by a screw. A special wrench is used to reach into the counterbored face of the shell end mill. The cutter is driven by two lugs on the arbor. These lugs mate with corresponding slots in a shell end mill.

SPACING COLLARS, BEARING SLEEVE, AND ARBOR NUT

Spacing collars are used for spacing or locating purposes. They also hold one or more cutters on an arbor. Spacing collars have parallel faces ground to accuracies as close as ±0.0005″ (±0.01mm). A *bearing sleeve* also fits on the arbor and is keyed to fit the arbor support bushing (Figure 35–7B).

An *arbor nut* is a round nut with two flat surfaces machined along part of the length. The milling machine arbor wrench fits this nut. The thread direction is opposite the direction of the milling operation.

THE DRAW-IN BAR

A *draw-in bar* has two threaded ends, an adjustment nut, and a threaded collar. The draw-in bar extends almost the length of the spindle. One threaded end screws into the tapered end of the arbor or adapter. The draw-in nut on the other end is adjusted against the left (outside) face of the spindle. The threaded collar is tightened to draw the tapered arbor and spindle surfaces together.

The process is reversed to release the surfaces and remove the cutter-holding device. The end of the draw-in bar is tapped gently but firmly with a soft-face hammer to release the tapers.

ADAPTERS

Adapters are available to accommodate older styles of arbors and cutting tools that have self-holding tapers for Brown & Sharpe or Morse taper-shank cutting tools.

Another adapter is made for interchanging face milling cutters from one type of spindle to a standard spindle. With this adapter shell end mills that have an internal taper designed for the old short taper spindle nose can be mated to the standard steep taper spindle nose. A special long draw-in bar is used to pull the tapered surfaces together and to secure the face milling cutter.

The *cam-lock adapter* is designed to hold cutting tools that have steep taper shanks (Figure 35–8). As the cam-lock is turned against a corresponding surface on the tool shank, there is a positive lock that prevents the cutter or holder from turning.

SLEEVES AND COLLETS

The milling machine *sleeve* serves the same function as a standard sleeve used on a drill press, lathe, or other machine tool (Figure 37–9E). A sleeve has an internal and external taper. The external taper surface fits into an adapter. The taper on the cutting tool corresponds with the internal taper of the sleeve. The sleeve makes it possible to use small-taper-diameter cutters in a large spindle bore.

A *collet* serves a function similar to the function of a sleeve. A collet is a precision-ground, cylindrical sleeve (Figure 35–9F and H).

Examples of common cutter-holding devices like adapters, arbors, collets, and bushings are illustrated in Figure 35–9.

WORK-HOLDING DEVICES

PLAIN MILLING MACHINE VISE

One of the most widely used and practical holding devices is the *milling machine vise.* Usually this vise is heavier than the all-purpose vise

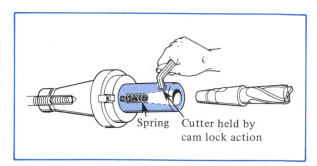

Spring   Cutter held by cam lock action

Figure 35–8 A Cam-Lock Adapter

Courtesy of CINCINNATI MILICRON INC.

(A) Collet adapter	(D) Arbor adapter	(G) Bushing
(B) Shell end mill Style C arbor	(E) Reducing collet (sleeve)	(H) Split collet
(C) Style A arbor (steep taper shank)	(F) Solid collet	

Figure 35–9 Standard Arbors, Adapters, and Collets

used with drill presses. The milling machine vise base is slotted (grooved) in two directions. Keys fit the vise grooves and the table slot. These keys align the vise jaws either parallel or at right angles to the column face (Figure 35–10).

SWIVEL VISE

The *swivel vise* has the added feature of a movable, graduated base. The workpiece may

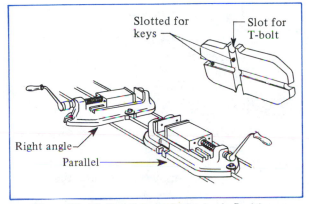

Figure 35–10 Parallel and Right-Angle Positions of a Plain Milling Vise on the Table

be machined parallel to the column face or at a required angle. A *universal vise* is used when workpieces must be held at a compound angle.

Another milling machine vise is the *cam-action vise*. It is especially adapted to machining multiple pieces.

VISE JAWS AND FIXTURES

Most milling machine operations require that the workpieces be held with conventional (straight) vise jaws. However, special jaws may be added to a vise to serve as a fixture. Irregularly shaped parts may be positioned, secured, and machined accurately with a special set of jaws. For production of a large quantity of irregularly shaped parts that require precise locating, a fixture is usually used.

OTHER WORK-HOLDING DEVICES

Accessories and work-holding devices are strapped to the table with standard straps and clamps. A number of shapes and sizes are available to accommodate different workpieces and milling processes. Flat, finger, adjustable, and V-strap clamps are examples.

Standard milling machine bolts, like the bolts used on drilling machines, are of three common types. The T-heads that fit the T-slot of the table may be designed with a square-head, cutaway T-head, or a tongue block with a stud.

Flat and step blocks support the ends of clamps. Adjustable jacks are used to accommodate different work heights. Wedges and shims are used to compensate for unevenness.

Round work is generally held in *V-blocks.* The underside of the base is fitted with a tongue and may be mounted over a table slot. The V-block faces are parallel to the table. Milling machine *angle plates* provide a surface that is at 90° to the horizontal plane of the table. The base of the angle plate has two grooves. These grooves are cut at right angles to each other. A tongue is fitted to the groove. The angle plate may be positioned parallel with or at a right angle to the column face.

Parallels serve auxiliary functions in layout and work positioning. Solid and adjustable parallels are used in milling machine work.

The types, sizes, and applications of C-clamps and parallel clamps are similar to the clamps used in bench and drill press work.

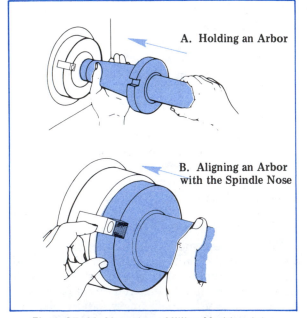

A. Holding an Arbor

B. Aligning an Arbor with the Spindle Nose

Figure 35–11 Mounting a Milling Machine Arbor

How to Mount and Remove a Milling Machine Arbor

STEP 1   Hold the tapered-shank end of the arbor in one hand and the shaft portion in the other (Figure 35–11A).

STEP 2   Insert and guide the tapered shank carefully into the spindle nose. Align the slots (Figure 35–11B). Move the arbor in as far as possible.

STEP 3   Move in the draw-in bar and turn it to engage the threads of the arbor.

STEP 4   Turn the draw-in bar nut by hand as far as possible.

STEP 5   Use an open-end wrench to turn the draw-in bar nut. Tighten the nut until the arbor is held securely in the taper and against the spindle face.

STEP 6   Unlock the spindle. Start the machine. Check the end of the arbor to see that it runs perfectly true.

How to Mount and Remove a Cam-Lock Adapter

Mounting a Cam-Lock Adapter

STEP 1   Select the adapter that fits the spindle nose taper and the shank of the cam-lock toolholder (or cutting tool).

STEP 2   Lock the spindle. Be sure all surfaces are clean and free of burrs.

STEP 3   Mount the adapter. Follow the same steps as used for mounting an arbor.

STEP 4   Recheck the taper hole of the adapter and the taper shank of the arbor to be sure that they are clean and free of burrs.

STEP 5   Insert the cam-lock arbor in the adapter. Match the cam lock and slot.

STEP 6   Turn the setscrew clockwise with a setscrew wrench.

STEP 7   Continue to turn the setscrew until the arbor is securely seated and the tapers are locked together.

STEP 8   Unlock the spindle. Start the machine and test the arbor for trueness.

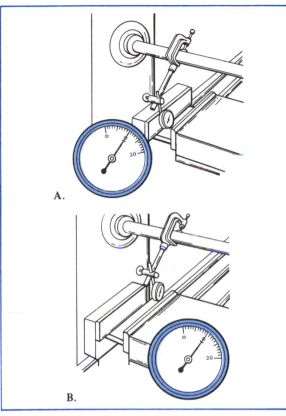

A.

B.

Figure 35–12 Checking the Accuracy of the Alignment of a Solid Vise Jaw (Parallel to Column)

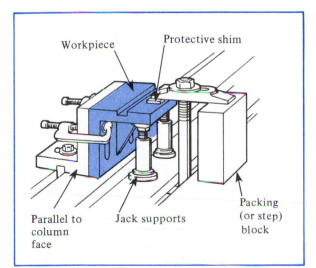

Workpiece

Protective shim

Parallel to column face

Jack supports

Packing (or step) block

Figure 35–13 Clamping Setup for Machining a Workpiece Mounted on an Angle Plate

How to Position Swivel Vise Jaws

Using a Dial Indicator

A dial indicator is used to accurately align vise jaws either parallel or at right angles to the table movement. The accuracy of the setting is limited by the dimensional accuracy of the dial indicator. The fixed jaw can thus be set to within 0.0005″ (0.01mm) or less, depending on the indicator.

The indicator setup for checking the vise jaws for parallelism with the column face is shown in Figure 35–12. When the vise jaws are to be positioned at a right angle to the column face, the dial indicator is turned 90°. The stationary jaw is then moved (transversely) past the indicator pointer by turning the cross slide handwheel. If there is a difference in the dial readings, the vise is tapped gently in a direction away from the higher reading until there is no change in the indicator readings (Positions A and B).

How to Use Work-Holding Accessories

Using an Angle Plate, Parallel or C-Clamp, and Jack

An example of a typical clamping setup which requires the use of an angle plate, T-bolt, strap clamp, step block, and C-clamps is illustrated in Figure 35–13.

A. Safe Practices in Lubricating and the Maintenance of Milling Machines

- Shut down the milling machine when examining, cleaning, and lubricating moving parts or before opening the motor compartment.
- Avoid leaning or rubbing against the machine. Any slip against a lever or a handwheel may cause the machine to start or a cut to be taken accidentally.

- Treat the table top and flat top surface of the knee as a precision surface plate. Layout tools, wrenches, hammers, excess straps, and so on should be placed on an adjacent stand.
- Use a T-slot metal cleaner to remove the extremely sharp chips from the T-slots. A stiff brush and small chip pan should be used to clear the chips from the table, saddle, knee, and base.
- Keep the coolant strainer on the end of the table clear to permit recirculating the cutting fluid.
- Check all sight gages. Maintain the proper level of lubricant in each reservoir.
- Clean the area around an oiler to remove foreign matter before lubricating. Wipe up any excess oil after lubricating.
- Wipe up any lubricant that has been spilled or has overflowed onto any machine surface or the floor.
- Place oily and dirty wiping rags or cloths in a metal container.
- Observe the standard personal safety precautions against loose clothing and the use of wiping cloths around moving parts.

B. Safe Practices with Milling Cutter- and Work-Holding Devices

- Remove burrs or nicks from the spindle face and nose and the arbor adapter taper, shaft, bearing, and spacing collars; and other work-holding devices.
- Tighten the draw-in bar with enough force to pull together the tapered and face surfaces of the spindle and arbor.
- Locate milling cutters on an arbor as close to the column as possible to provide maximum support for the cutting tool during machining.
- Disengage steep tapers by striking the draw-in bar a sharp blow. The bar must not be loosened more than one thread to prevent damage to the threads.
- Mount and remove a style A arbor by holding it with two hands.
- Store arbors vertically in a rack to prevent them from hitting together or being bent.
- Use a properly fitted protective shield or safety goggles during all milling operations.

UNIT 35 REVIEW AND SELF-TEST

A. FUNCTIONS AND MAINTENANCE OF MILLING MACHINES

1. Distinguish between the principal straight-line table movements of plain and universal horizontal millers and plain vertical milling machines.
2. Classify three nonstandard milling machine types.
3. Identify the major construction parts and mechanisms of a milling machine.
4. Describe the functions of (a) a dividing head and (b) a rotary table.
5. Explain the functions of the automatic lubrication systems of modern milling machines.
6. Indicate four precautionary practices to follow when lubricating and maintaining a milling machine.

B. MILLING CUTTER- AND WORK-HOLDING DEVICES

1. Describe style A, B, and C milling machine arbors.
2. Explain briefly the function served by the draw-in bar.
3. List five work-holding devices for general milling processes.
4. Give the steps to follow in removing a style A, B, or C milling machine arbor.
5. Describe three practices for positioning vise jaws on a milling machine.
6. Explain how V-blocks and other work-holding accessories are used to nest a round workpiece and hold it securely on a milling machine table.
7. State three machine or tool safety precautions the milling machine operator must observe.

Technology and Applications of Standard Milling Cutters

A *milling cutter* is a cutting tool that is used on the milling machine. Milling cutters are available in many standard and special types, forms, diameters, and widths. The teeth may be straight (parallel to the axis of rotation) or at a helix angle. The cutter may be right-hand (to turn clockwise) or left-hand (to turn counter-clockwise).

Dimensional and other design standards for milling cutters have been approved by the American National Standards Institute (ANSI).

This unit covers general features of milling cutters, forms of cutter teeth, materials used in milling cutters, and standard milling cutter types. Speeds, feeds, and cutting fluids are included. The actual milling processes are covered in later units.

A. FEATURES AND APPLICATIONS OF MILLING CUTTERS

Milling cutters may be broadly grouped in two categories: *standard* and *special*.

Some of the terms used to identify the major features of milling cutters appear in Figure 36–1.

There are a number of other terms that are used to describe a milling cutter or to order a cutter from a toolroom. The following features are described by these terms:

- Outside diameter,
- Cutter width (measurement across the face),
- Hole size (usually +0.001″ of nominal arbor size),
- Keyway (to match arbor keyseat),
- Shank (straight or tapered: B & S or Morse self-holding taper),
- Cutter teeth.

BASIC FORMS OF CUTTER TEETH

Circular milling cutters are designed with three basic forms of teeth. These forms are the *standard tooth*, *formed tooth*, and *inserted tooth*.

STANDARD-TOOTH FORM

Some of the common design features of the standard tooth form are illustrated in Figure 36–2. Teeth are said to be *radial* when each tooth face lies along a line that cuts through the center of the cutter. Teeth are cut either with a radial tooth face or at a 0° *(zero), positive*, or *negative rake* angle to the radial line.

The tool life and performance of carbide-tipped teeth, which operate at exceedingly high speeds and coarse feeds, are improved when the teeth have negative rake.

The purpose of zero or negative rake is to protect the cutting edges of carbide tools. Their edges are brittle in comparison to the edges of

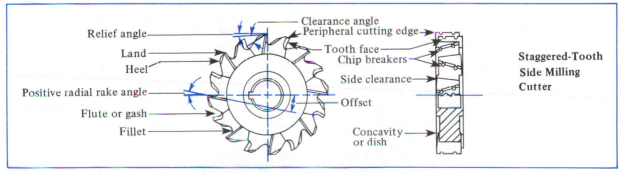

Figure 36–1 General Features of Milling Cutters

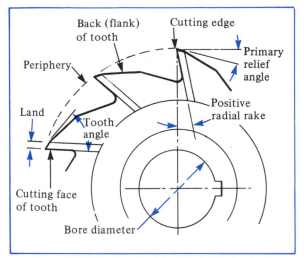

Figure 36–2 Design Features of and Terms Used with a Standard Milling Cutter

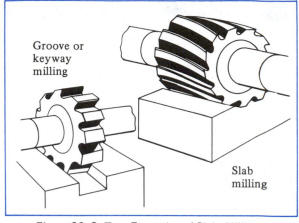

Figure 36–3 Two Examples of Plain Milling

high-speed steel tools. With a negative rake the cutting forces fall within the cutter body. Surprisingly, carbide cutters are capable of producing a high-quality surface finish at high machining speeds.

There may be two or three clearances on the standard-tooth form. The teeth are ground at a slight *clearance angle (primary clearance)* around the periphery. This clearance extends from the tooth face and produces a cutting edge. The narrow, flat surface directly behind the cutting edge is called the *land*. The teeth on the saw-tooth form are sharpened by grinding the land. The angle formed by the cutting face and the land is the *tooth angle*. The body of the tooth is cut at a secondary angle and at a third angle. The design provides maximum support for the cutting edge. It also provides a *chip space* for the fast removal of chips.

FORMED-TOOTH MILLING CUTTER

As the name suggests, a formed-tooth cutter has a contour, or tooth outline, of a particular shape. A concave milling cutter with a specified diameter produces a round shape on a workpiece. A gear cutter machines a gear tooth that conforms to specific design requirements. The right- or left-hand radius cutter mills a round corner. The flute cutter cuts flutes on drills, reamers, taps, and other cutting tools. A single

formed cutter is used for some applications. In applications that involve the milling of a wide contour, formed cutters may be set up in combination.

The clearance, or *eccentric relief* of a formed cutter, follows the same contour as the cutting edge. Each semicircular cutting face on a radial-tooth formed cutter retains the original shape when radially ground at a zero rake angle.

STANDARD TYPES OF MILLING CUTTERS

PLAIN MILLING CUTTERS

Two examples of *plain milling cutters* are shown in Figure 36–3. The diameter and the width of a cutter depend on whether a part is to be *slab milled* (milling a wide, flat surface) or requires a narrow-width slot. There are three broad classes of plain milling cutters: *light-duty*, *heavy-duty*, and *helical*.

Light-Duty Plain Milling Cutters. Light-duty plain milling cutters up to 3/4″ (18mm) wide generally have straight teeth. These teeth are formed around the periphery only and are parallel to the cutter axis. High-speed cutters have from four to five teeth for each inch of diameter.

Cutters over 3/4″ (18mm) wide usually have helical teeth. The helix angles range from 18° to 25°. These angles produce a shearing, cutting action. Less force is required, vibration and chatter are reduced, and a better quality of surface finish is produced with a helical-tooth cutter than with a straight-tooth cutter.

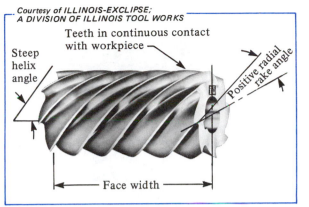

Figure 36–4 A Helical-Type Plain Milling Cutter

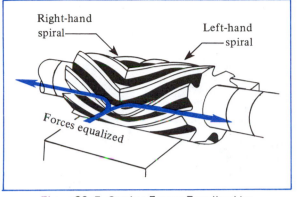

Figure 36–5 Cutting Forces Equalized by Right- and Left-Hand Helical Cutters

Heavy-Duty Plain Milling Cutters. Heavy-duty plain milling cutters are also called *coarse-tooth cutters*. These cutters average two to three teeth for each inch of diameter. For example, a 3″ heavy-duty plain milling cutter usually has eight teeth. The helix angles of the teeth of heavy-duty cutters are steeper than the angles of light-duty cutters. Different cutter manufacturers vary the helix angles from 25° to 45°.

Wide, flat surfaces are usually slab milled with a particular type of heavy-duty cutter. This cutter is called a *slab mill, slabbing cutter*, or *roughing cutter*. This type of cutter has interrupted teeth. Part of each tooth is relieved at a different place along the length to break up the chip. A slabbing cutter is able to produce a good surface finish under heavy cutting conditions.

Helical Plain Milling Cutters. Helical plain milling cutter teeth are formed at a steep helix angle (Figure 36–4). This angle ranges from 45° to 60°, or steeper. The teeth are designed so that they engage the work at a steep right- or left-hand helix angle. The cutting force is absorbed in end thrust. The teeth are continuously engaged in comparison with the intermittent cutting action of a straight-tooth cutter. The helical plain milling cutter therefore eliminates chatter.

Wide surfaces are often milled by interlocking a right- and a left-hand helical cutter. The forces exerted by each cutter are thereby offset (canceled) (Figure 36–5).

Helical plain milling cutters are efficient for wide, shallow cuts. They are not as practical as the heavy-duty cutters for deep cuts, slab milling, or coarse feeds.

SIDE MILLING CUTTERS

Side milling cutters are similar to plain milling cutters. However, in addition to teeth around the periphery, other teeth are formed on one or both sides. Most of the cutting is done by the teeth around the periphery. The side cutting teeth cut the side of a workpiece. These cutters are mounted on and are keyed on the arbor.

Side mills are not recommended for milling slots. There is a tendency for the side cutting teeth to mill wider than the specified cutter width.

Four types of side milling cutters are in general use. They are the *plain, half, staggered-tooth*, and *interlocking* side milling cutters.

Plain Side Milling Cutters (Figure 36–6). The plain side milling cutter has teeth on the periphery and on both sides. The teeth on the sides taper slightly toward the center of the cutter. The concavity of the teeth provides clearance, or side relief. The plain side milling cutter is adaptable for general-purpose side milling, slotting, and straddle milling.

Half Side Milling Cutters. One side of the half side milling cutter has teeth. The other side is flat and resembles a plain milling cutter. The teeth around the circumference are helical.

Left-hand half side
milling cutter

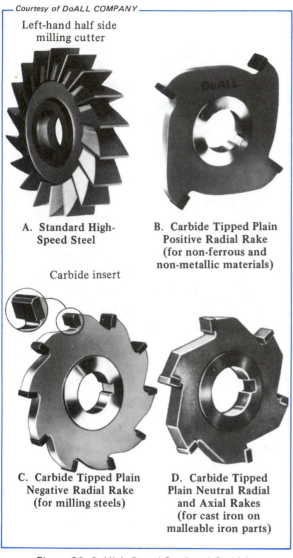

A. Standard High-
Speed Steel

B. Carbide Tipped Plain
Positive Radial Rake
(for non-ferrous and
non-metallic materials)

Carbide insert

C. Carbide Tipped Plain
Negative Radial Rake
(for milling steels)

D. Carbide Tipped
Plain Neutral Radial
and Axial Rakes
(for cast iron on
malleable iron parts)

Figure 36-6 High-Speed Steel and Carbide-
Tipped Plain Side Milling Cutters

Figure 36-7 Interlocking Side Milling Cutters

One side of each tooth forms a side tooth. Each land is ground from 1/64″ (0.4mm) to 1/32″ (0.8mm) wide. There is a minimum back clearance of 0.001″ per inch. This clearance prevents the cutter from binding in the cut.

The staggered-tooth side milling cutter is a heavy-duty cutter. It is designed to remove large amounts of metal, to mill with a minimum of vibration and chatter, and to produce a high-quality surface finish in deep cuts.

Interlocking Side Milling Cutters. The interlocking side milling cutter is made in two halves. These halves are placed side by side and interlock (Figure 36-7). The teeth around the circumference of the interlocking cutters are alternately long and short. The alternate teeth interlock. The cutters may be separated for a specific width by using spacing washers. The amount of separation is controlled by the overlap range of the interlocking teeth.

SLITTING SAWS

Slitting saws are a type of plain or side milling cutter. They are designed for cutoff work on the milling machine and for cutting narrow slots. The three general-purpose types are the *plain* metal-slitting saw, metal-slitting saw *with side teeth*, and *staggered-tooth* (Figure 36-8) metal-slitting saw. The *screw-slotting cutter* is a modification of a slitting saw. Details relating to features and applications of slitting saws and screw slotting cutters follow in Unit 40.

These teeth produce a shearing, cutting action that minimizes chatter. The top and side teeth are deep cut to serve two purposes: (1) to permit the free flow of chips and (2) to extend tool life. The teeth may be reground a greater number of times than can the teeth of a standard side milling cutter.

Staggered-Tooth Side Milling Cutters. The staggered-tooth side milling cutter has cutting teeth with alternate right- and left-hand helixes.

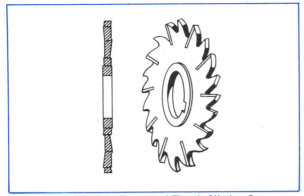

Figure 36–8 A Staggered-Tooth Slitting Saw

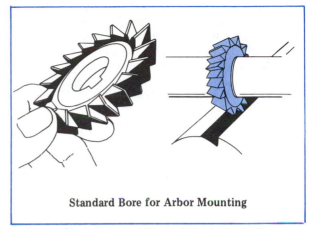

Standard Bore for Arbor Mounting

Figure 36–9 Standard Type of Single-Angle Cutter

ANGLE MILLING CUTTERS

There are two basic types of *angle milling cutters: single-angle* and *double-angle*. Some single-angle cutters are of the hole type (Figure 36–9). These cutters are mounted on a regular arbor. Other single-angle cutters are threaded for mounting on a threaded adapter.

Double-Angle Milling Cutters. The teeth of double-angle cutters are V-shaped. They are cut in the two angular faces. Double-angle milling cutters are available with standard included angles of 45°, 60°, and 90°. Double-angle milling cutters with unequal angles are also available.

T-SLOT AND WOODRUFF KEYSEAT CUTTERS

T-Slot Cutters. The T-slot cutter is a single-process cutter (Figure 36–10). It is used to mill a T-slot after a groove has been milled with a side or end milling cutter. It is an unusual cutter. It mills five sides of a T-slot at the same time: the bottom, the two interrupted faces, and the right and left sides.

Woodruff Keyseat Cutters. Woodruff keys are widely used between two mating parts that are to be keyed together. The Woodruff keyseat cutter is the cutting tool used for milling a semi-circular keyseat. Cutter details and keyseating processes are covered in Unit 40.

END MILLING CUTTERS

End milling is the process of machining horizontal, vertical, angular, and irregular-shaped surfaces. The cutting tool is called an *end mill*. End mills are used to mill grooves, slots, keyways, and large surfaces. They are also widely used for profile milling in die making. End mills are coarse-tooth cutters. The teeth are cut on the periphery as well as on the face.

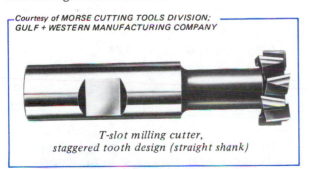

T-slot milling cutter, staggered tooth design (straight shank)

Figure 36–10 Straight Shank T-Slot Cutter

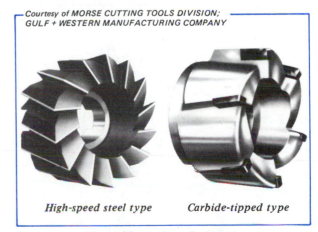

High-speed steel type   *Carbide-tipped type*

Figure 36–11 Shell End Mills

Just two examples of the larger diameter shell end mills are shown in Figure 36–11. Full descriptions, features, specifications, and processes are included later in Unit 38. Small diameter solid end mills are covered in Unit 46.

FACE MILLING CUTTERS

The *face milling cutter* is esentially a special form of end mill. The size is usually 6″ or larger in diameter. The teeth are beveled or rounded at the periphery of the cutter.

The body of a face milling cutter, like the cutter shown in Figure 36–12 is usually made of an alloy steel. High-speed steel, cast alloy, carbide, or carbide-tipped blades (inserted teeth) are available. The blades are adjustable. They may be either reground and resharpened or replaced.

Face milling cutters are heavy-duty cutters. Heavy cuts, coarse feeds, and high cutting speeds are essential. Hogging cuts also require cutting below the scale on castings, rough edges on parts that are cut out by burning processes, and forgings.

FLY CUTTERS

A *fly cutter* consists of an arbor or other holding device, one or more single-point cutters, and setscrews or other fasteners. The cutting tool is usually a tool bit ground to a desired shape. A fly cutter may be used for internal boring or for external operations. Plain, angular,

and form milling are a few examples. Fly cutters are made of high-speed steels, cast alloys, and cemented carbides.

ROTARY FILES AND BURRS

Although they are circular in shape like a milling machine cutter, *rotary files* and *burrs* are not classified as milling cutters. The cutting edges consist of closely spaced, shallow grooves cut around the circumference of the cutter.

Rotary files and burrs are adapted to metal-removing applications using a flexible or other portable hand unit. Trimming a weld, preparing parts to be welded, and removing small amounts of metal in form dies are examples of the type of work performed with these cutting tools. Rotary files and burrs are made of high-speed steel or cemented carbide.

B. SPEEDS, FEEDS, AND CUTTING FLUIDS FOR MILLING

FACTORS THAT AFFECT CUTTING FEEDS

INCREASING AND DECREASING THE RATE OF FEED

Feeds per tooth for general milling processes are suggested in Table 36–1. The range extends from feeds for rough milling operations on hard, tough ferrous metals to light cuts on nonferrous metals. Feeds are given in 0.000″ and 0.00mm. An average starting feed for each process, in both fractional inch and millimeter measurements, appears under the "Average Starting Feed" column in the table.

Feeds may be increased for certain operations like slab milling; heavy roughing cuts; abrasive, scaled surface conditions; and easily machinable materials. On light cuts it is possible to increase both the feed and the speed.

COMPUTING CUTTING FEEDS

Feed is a combination of distance and time. The time interval is expressed as *per minute*. The distance represents how far a workpiece moves during a milling process. Feed may be given in terms of inches per minute (ipm) or millimeters per minute (mm/min). Feed may also be stated as the thickness of the chip that is cut away by each tooth.

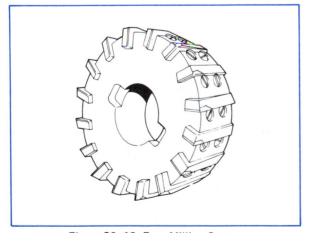

Figure 36–12 Face Milling Cutter

Table 36–1 Suggested Feeds per Tooth for General Milling Processes

Process	Average Starting Feed		Range of Feed	
	0.000"	0.00mm	0.000"	0.00mm
Face milling	0.008	0.20	0.005–0.030	0.10–0.80
Straddle milling	0.008	0.20	0.005–0.030	0.10–0.80
Channeling or slotting	0.008	0.20	0.005–0.020	0.10–0.50
Slab milling	0.007	0.20	0.005–0.020	0.10–0.50
End milling or profiling	0.004	0.10	0.002–0.010	0.05–0.25
Sawing	0.003	0.10	0.002–0.010	0.05–0.25
Thread milling	0.002	0.05	0.001–0.005	0.02–0.10
Boring	0.007	0.20	0.005–0.020	0.10–0.50

Table 36–2 Recommended Feed per Tooth (0.000") for Cemented Carbide Milling Cutters (Partial Table)

Material	Face Mills	Spiral Mills	Side and Slotting Mills	End Mills	Form Relieved Cutters	Circular Saws
Malleable iron and cast iron (medium hard)	0.016	0.013	0.010	0.008	0.005	0.004

Table 36–3 Formulas for Computing Feeds and Speeds for Milling Processes

Required Value	Formula
Feed per Tooth (F_t)	$F_t = \dfrac{F''}{n \times RPM}$
Feed per Revolution of Cutter (F_r)	$F_r = F_t \times n$ $F_r = \dfrac{F''}{RPM}$
Feed in Inches per Minute (F'')	$F'' = F \times n \times RPM$ $F'' = \dfrac{i^3pm}{D \times W}$
Revolutions of Cutter per Minute (RPM)	$RPM = \dfrac{sfpm \times 12}{C\ in\ Inches}$
Simplified formula:	$RPM = \dfrac{sfpm \times 4}{d}$
Cutting speed (sfpm)	$sfpm\ (CS) = \dfrac{C \times RPM}{12}$
Simplified formula:	$sfpm = \dfrac{d \times RPM}{4}$

Symbols

F''	=	Feed in inches per minute	sfpm	=	Surface feet per minute
F_t	=	Feed per tooth	CS	=	Cutting speed
n	=	Number of teeth in cutter	i^3pm	=	Cubic inches of material removed per minute
C	=	Circumference of cutter	D	=	Depth of cut
d	=	Diameter of cutter in inches	W	=	Width of cut
RPM	=	Revolutions of cutter per minute			

Some feed tables list recommended feeds (0.000" or 0.00mm) per tooth for selected materials and milling processes. Table 36–2 shows one line from the feed table in a handbook. Six milling machine processes are recorded. The recommended feed per tooth is given for using cemented carbides to mill malleable and medium-hard cast-iron parts.

FEED AND SPEED FORMULAS (MILLING PROCESSES)

Table 36–3 gives the formulas for computing different speed and feed values for milling machine processes. Table 36–4 provides a single example from a larger table of cutting speeds. The cutting speeds in sfpm are given for milling soft cast iron using four different cutter materials.

FEED MECHANISMS AND CONTROLS

The feed and traverse controls for the table, knee, and saddle of a heavy-duty plain milling machine are identified in Figure 36–13.

Table 36–4 Example of Cutting Speed (sfpm)

Material to Be Milled	Milling Cutter Material				
	High-Speed Steel	Super High-Speed Steel	Stellite	Tantalum Carbide	Cemented Carbides
Cast Iron Soft	50 to 80	60 to 115	90 to 130		250 to 325

How to Set the Feed (Table, Saddle, and Knee)

STEP 1 Determine the correct feed.

STEP 2 Move the power feed lever(s) to the required feed position(s). The lever position(s) is usually shown on the feed index plate or dial.

STEP 3 Move either the longitudinal (table), transverse (saddle), or vertical (knee) feed-control lever. The lever to use depends on whether the table, saddle, or knee is to be power fed. The direction in which the lever is moved depends on the required direction of power feed.

STEP 4 Turn the appropriate clamp lever to permit the table, knee, or saddle to slide. The other clamp levers are tightened to securely hold the corresponding milling machine units.

Note: If necessary, the trip dogs should be set. The dogs will trip (stop) the power feed at set positions and thus prevent damage to the workpiece, setup, and machine.

Courtesy of KEARNEY & TRECKER CORPORATION

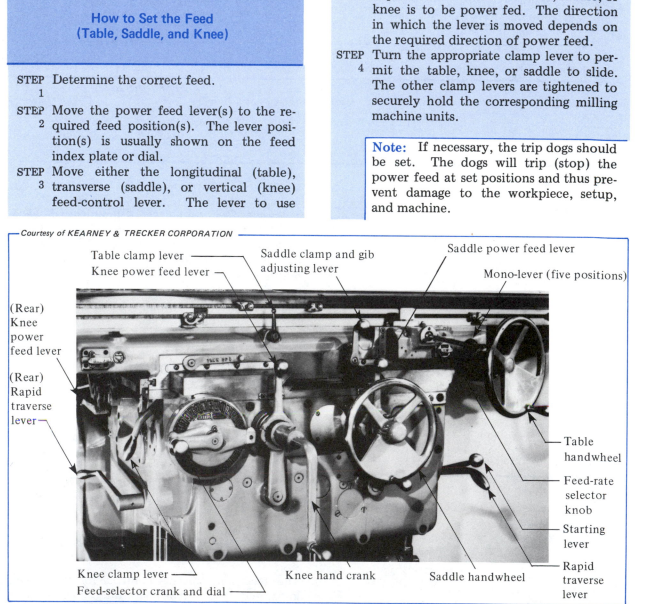

Figure 36–13 Feed and Traverse Controls of a Heavy-Duty Plain Milling Machine

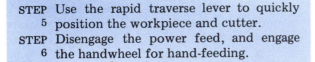

STEP 5 Use the rapid traverse lever to quickly position the workpiece and cutter.

STEP 6 Disengage the power feed, and engage the handwheel for hand-feeding.

> **Note:** Hand-feeding is used to start a cut. When the operator is assured that the milling process can be carried on safely, power feed is then engaged.

CUTTING FLUIDS AND MILLING MACHINE SYSTEMS

COOLING PROPERTIES OF CUTTING FLUIDS

A cutting fluid must have at least the following important properties:

- A high specific heat factor, which relates to the ability to absorb heat (the greater the capacity to absorb heat, the faster the cooling action);
- Sufficiently low viscosity, which regulates the property of the cutting fluid to cling and to flow.
- Maximum fluidity to penetrate rapidly to the cutting edges.
- Resist deterioration resulting from excessive heat.

LUBRICATING PROPERTIES OF CUTTING FLUIDS

Figure 36–14 illustrates the considerable force required to shear material. The friction of the cutting action generates heat. A good cutting fluid reduces friction in the following ways:

- By providing a film of lubricant between the contact surfaces and work.
- By lubricating the chip, the contact surfaces of the workpiece, and the milling cutter.
- By maintaining a film of lubricant between the cutting edges and faces of the cutter and the chip. The frictional heat generated by chips sliding over a cutting face may become so intense that the chips fuse to the cutter teeth.
- By quickly flowing chips away from the cutting area.

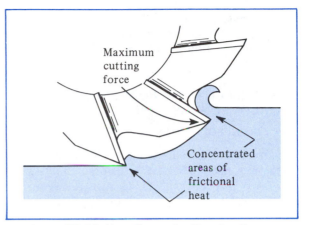

Figure 36–14 Heat-Generating Shearing Forces of Milling Cutter Teeth

Not all metals require a lubricant. Cast iron should be cut without a lubricant because it contains *free graphite*. The free graphite lubricates the cutting tool. Any additional oily lubricant produces a gummy substance that retards chip movement. The chips tend to clog the chip spaces, and the cast iron becomes glazed.

INCREASING TOOL LIFE

A good cutting fluid prevents the development of frictional heat and resists the adhesion of chip particles to the cutting edges and faces. The cutting edge is also protected from wearing away.

A great deal of milling with carbide cutters is done dry. However, when a cutting fluid is used, there must be a large, constant flow. Interrupted flow of the cutting fluid on a carbide tool causes thermal cracking, which, in turn, produces chipping of the tool.

IMPROVING AND PRESERVING SURFACE FINISHES

A good surface finish may be produced by dry machining certain metals like cast iron and brass. The surface finish of other materials can be improved by using a cutting fluid to remove the chips.

Preserving the surface finish means selecting a cutting fluid that will not discolor or rust a finished surface. Discoloration and corrosion of nonferrous metals are produced by additives in the cutting fluid. Rusting of ferrous parts may be avoided by rapidly evaporating the water in a cutting fluid solution. The remaining oil coats and protects the surface.

KINDS OF CUTTING LUBRICANTS AND COMPOUNDS

WATER SOLUTIONS

Water has excellent properties to rapidly absorb and carry off heat. Unfortunately, water has two drawbacks. First, rust forms on the workpiece and machine surfaces of the miller. Second, water thins the lubricating oils needed between mating surfaces.

Some soluble oils are mixed with water to form an *emulsion*. The emulsion is milky in appearance. The soluble oils are usually a mineral oil, a vegetable oil, or an animal oil. An *emulsifying agent*, such as soap, is added. The consistency of the emulsion depends on the ratio of water to soluble oil. The emulsion consists of fine droplets of oil that are suspended in the water, soap, and oil mixture. Other soluble oils are available in a clear, transparent mixture.

CUTTING OILS

Generally, mineral, animal, and vegetable cutting oils are *compounded* (blended). The blending is done for economical reasons as well as for improving certain cutting qualities. The straight cutting oils include lard oil and untreated mineral oils. Lard oil is mixed with mineral oils to prevent it from becoming rancid and clogging feed lines. Lard oil is, however, an excellent tool lubricant under severe cutting conditions.

When lard oil is mixed with cheaper mineral oils, the percentage of lard oil is determined by the nature of the cutting action and the hardness of the material to be cut. Mineral oils generally have better lubricating qualities than soluble oils. Lard oil is added when the cutting forces are too severe for straight mineral oils.

SULPHURIZED OILS

The addition of sulphur to a cutting fluid permits the cutting speed to be increased significantly. The cutting fluid is also able to withstand the greater forces that accompany heavy cuts on tough materials. Sulphurized oils are dark in color. They continue to darken as more sulphur is added, until all transparency is lost.

A pale yellow, transparent cutting fluid is produced by mixing mineral oil with a *base oil* (such as lard oil) to which sulphur is added. The amount of sulphur depends on the nature of the operation.

KEROSENE

Kerosene is used primarily on nonferrous metals like aluminum, brasses, bronzes, magnesium, and zinc. The machinability ratings on these metals are above 100 percent.

AIR AS A COOLANT

Although air is not a cutting fluid, it performs two cutting-fluid functions. First, a flow of air under pressure helps to remove chips from the cutter and workpiece. Second, the air cools the cutting tool, the chips, and the part. Air is used on cast iron and other metals and materials where a cutting fluid cannot be applied. The air stream may be produced by suction or as a blast.

APPLICATION OF CUTTING FLUIDS

The three most common cutting fluid distributors are designed for (1) a flow directed across a wide area, (2) a flow directed over multiple cutters, and (3) a concentrated flow for narrow-width cutters.

A. Safe Practices in the Care and Use of Milling Cutters

- Stop the milling machine before setting up or removing a workpiece, cutter, or accessory.
- Loosen the arbor support arm screw slightly. The arm should move but should still provide support while the arbor nut is loosened.
- Make sure that all chips are removed from the cutter and workpiece and that the cutter is not hot. A wiping cloth should be placed on the cutter to protect the hands. Use two hands when sliding the cutter off the arbor.

- Determine the hand of the cutter, the required direction of spindle rotation, and whether the arbor nut is right- or left-hand.
- Protect the table surface with a wiping cloth or protective tray if setup and measuring tools are to be placed on it.

B. Safe Practices in Operating Speed, Feed, and Coolant System Controls

- Reduce the spindle RPM if there is excessive wear on the lands.
- Shift the speed-selector lever to change the spindle speed in a geared head when the gears are stopped.
- Start a cut by hand-feeding. Once it is established that the work and cutter are correctly positioned and there is good cutting action,

either the speed or feed, or both, may be increased.
- Set the trip dogs before engaging a power feed.
- Check clearances between the cutter, arbor support, work-holding device, and workpiece before engaging the rapid traverse or regular power feed.
- Sample the cutting fluid by testing for composition, viscosity, and sediment. Bring the cutting fluid back to strength before using it on a cut.
- Wipe up any cutting fluid that spills on the floor. Dispose of oily waste and wiping cloths in a metal container.
- Avoid skin contact with cutting fluids. Direct contact with cutting fluids may cause skin infections.
- Wear a protective shield or safety goggles. Secure all loose clothing when working with or around moving machinery.

UNIT 36 REVIEW AND SELF-TEST

A. FEATURES AND APPLICATIONS OF MILLING CUTTERS

1. State the general specifications that are used to order milling cutters for style A and B arbors.

2. Define (a) rake angle and the effect on cutting using milling cutters with (b) positive rake or (c) negative rake.

3. Explain the function that is served by providing eccentric relief on formed cutters.

4. Describe how milling cutter teeth that are formed at a steep helix angle affect: (a) the direction of force and (b) the cutting forces.

5. Differentiate between a half side milling cutter and an interlocking side milling cutter.

6. a. Name three different types of slitting saws.
   b. Give an application of each slitting saw named.

7. Indicate differences between a solid end mill and a shell end mill.

8. Describe (a) the function of a face milling cutter and (b) its major design features.

9. List three safe practices to follow in the care and use of milling cutters.

B. SPEEDS, FEEDS, AND CUTTING FLUIDS FOR MILLING

1. Express how cutting speed and feed rates are specified for milling machine work in inch- and metric-standard systems.

2. Calculate the cutting speed of a 4″ diameter cutter traveling at a spindle speed of 100 RPM.

3. Explain how each of the following conditions affects cutting speeds for general milling machine processes: (a) increasing the width and depth of cut and the chip thickness, (b) changing from a standard tooth to a steep helix angle where a minimum amount of heat is generated, (c) using fine feeds on thin-sectioned workpieces, and (d) improving the efficiency of a cutting fluid.

4. Indicate how each one of the following conditions affects cutting feeds for general milling processes: (a) cutting through scaled surfaces, (b) roughing cuts, (c) deep grooving cuts, and (d) chipping of the cutter teeth at the cutting edges.

5. State three ways in which a quality cutting fluid (where applicable) reduces friction between a cutting tool, the workpiece, and the chips.

6. List three machine safety practices to follow in setting speed, feed, and coolant system controls.

SECTION TWO

Typical Milling Setups and Processes

This section describes the principles of and procedures for performing three of the most common milling processes carried out on the horizontal milling machine:

- Plain milling, face milling, and side milling.

UNIT 37

Plain Milling on the Horizontal Milling Machine

In plain milling the competent operator must be able to apply the technology and procedures that relate to each of the following processes:

- Preparing and laying out a workpiece;
- Determining the safest and most practical method of holding a workpiece;
- Setting up a workpiece accurately for each successive operation;
- Selecting and mounting a milling cutter appropriate to the job requirements;
- Determining the correct spindle speed, cutting feed, and cutting fluid;
- Setting up the machine;
- Taking the necessary milling cuts and judging the cutting action of a cutter;
- Measuring a workpiece for dimensional accuracy;
- Cleaning and maintaining the machine and accessories.

CONVENTIONAL AND CLIMB MILLING

CONVENTIONAL (UP) MILLING

The two common methods of removing metal are called *conventional* and *climb milling*. Conventional milling is also referred to as *up milling*. Figure 37–1 shows the direction a cutter turns in relation to the table feed. The cutter rotates clockwise. The workpiece is fed to the right. If the cutter rotates counterclockwise,

the table feed is to the left (into the cutter). The cutting action takes place from the bottom of the cut to the face of the workpiece.

The major forces in conventional milling are upward. These forces tend to lift a workpiece. In all milling it is important to use a *work stop*. This stop prevents a part and/or work-holding device from moving on the table. A stop is bolted to the table immediately ahead of a workpiece.

CLIMB (DOWN) MILLING

Climb milling is also known as *down milling*. The work is fed in the same direction as the rotation of the cutter teeth. In Figure 37–2, the cutter rotates in a clockwise direction. The

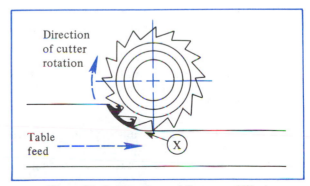

Figure 37–1 Directions of Cutter and Work for Conventional Milling

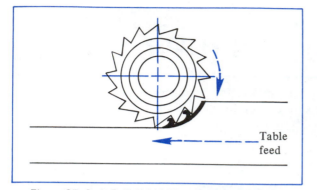

Figure 37–2 A Full Chip Milled from the Top of a Work Surface in Climb Milling

workpiece is fed into the cutter from right to left. If the cutter rotates in a counterclockwise direction, the cutting feed is from left to right.

The cutting forces in climb milling are mostly downward. A stop is usually placed ahead of a workpiece to prevent any movement during a heavy milling operation. There is also a tendency to pull the work into the revolving cutter. Therefore, climb milling is avoided unless the milling machine is rigidly constructed and/or equipped with an *automatic backlash eliminator*. Once the device is engaged, it is automatically activated during climb milling.

The chip formation in climb milling is opposite the chip formation in conventional milling. Figure 37–2 shows that the cutter tooth is almost parallel with the top surface of the workpiece. The cutter tooth begins to mill the full chip thickness. Then, the chip thickness gradually diminishes.

Climb milling has several advantages:

- Climb milling tends to eliminate surface burrs that are normally produced during conventional (up) milling.
- Because revolution and feed marks are minimized, climb milling produces a smoother cut than conventional milling.
- Climb milling is more practical than conventional milling for machining deep, narrow slots. Narrow cutters and saws have a tendency to flex and crowd sideways under the force of a cut. The cutting action in climb milling permits these cutters to cut without springing under a heavy force.

- Climb milling forces the work against the table, fixture, or surface to which it is clamped. Thus, climb milling is desirable for machining thin or hard-to-hold workpieces and for cutting off stock.
- Laboratory tests indicate that less power is required for climb milling than for conventional milling.
- Consistently parallel surfaces and dimensional accuracy may be maintained on thin-sectioned parts.
- Cutting efficiency is increased with climb milling because more efficient cutter rake angles may be used than with conventional milling.

However, climb milling has two significant disadvantages:

- Climb milling is *dangerous*. The milling machine should be equipped with a backlash eliminator. All play must be removed between the lead screw and nut. The ways and sliding surfaces must also be free of lost motion. Backlash, play, and lost motion cause the cutter to pull the work into the teeth.
- Climb milling is not recommended for castings, forgings, hot-rolled steels, or other materials that have an abrasive outer scale or surface. The continuous contact of the cutting teeth on a rough, hard surface causes the teeth to dull rapidly.

MILLING FLAT SURFACES

SETTING UP THE MACHINE, WORK-HOLDING DEVICE, AND WORKPIECE

Small, regularly shaped workpieces are usually mounted in a vise. Large pieces may be fastened directly to the table. Irregularly shaped parts are nested and secured in fixtures.

Most jobs require a roughing cut and a finish cut. The roughing cut should be as deep as possible. The depth is determined by the rigidity of the setup, the capacity of the machine, and the surface condition. Usually 1/64″ (0.015″ to 0.020″, or 0.4mm to 0.5mm) is left for a finish cut. A finish cut is necessary for dimensional accuracy and to produce a high-quality surface finish.

Most workpieces may be conveniently held in a milling machine vise. The vise jaws are positioned accurately. The work is generally set on parallels. The height of the jaw must permit a workpiece to be held securely. The correct setup provides maximum seating and holding power. It is good practice to seat a workpiece on two narrow parallels so that the operator may determine when the part is properly seated.

A protecting strip is placed between any rough surface, the ground jaws of the vise, and the ground faces of the parallels. Often when seating the work, the movable jaw tends to lift the part slightly off the parallels. The condition is overcome by applying a slight force with the movable jaw. The workpiece is then gently but firmly tapped. Too hard a blow tends to make the work rebound. The vise is tightened further. The work is rechecked and reseated if necessary.

THE MILLING CUTTER AND SETUP

A light-duty, heavy-duty, or helical plain milling cutter should be selected. The selection of a correct type of cutter depends on the material, the nature of the operation, and the time within which the part is to be milled.

The cutter or combination of cutters should be wide enough to mill across the width of the workpiece. If two cutters are required, it is good practice to use interlocking cutters. Interlocking cutters permit the cutting forces to be equalized. Usually cutters with the smallest possible diameter are used. The larger the diameter, the longer the feed time required to travel across the length of a workpiece.

When heavy slab cuts are to be taken, consideration must be given to selecting and mounting a cutter so that the cutting forces are directed toward the column. The arbor support must also be positioned to give the greatest possible rigidity.

How to Mill Surfaces Parallel and/or at Right Angles

The surface that is milled first becomes a reference plane. If the opposite surface is to be milled parallel, the first surface is seated on parallels and secured. The usual procedure for

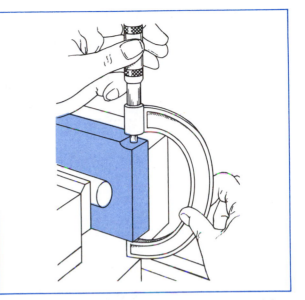

Figure 37–3 Checking Workpiece Size after a Trial Cut

removing burrs is followed prior to positioning. A trial cut is usually taken across the face of the second surface. The ends are measured for parallelism and thickness.

When four sides are to be milled square, the first milled side is held against the solid vise jaw. The adjacent side, if it is straight, may be placed on parallels. Otherwise the face to be milled may need to be set level by using a surface gage or other indicator. The third side is milled by seating the second side on parallels. A trial cut is taken for about 1/4″. The part is measured with a steel rule, caliper, or micrometer (Figure 37–3). The knee is adjusted to produce the required dimension.

If the fourth side is rough, a round bar may be placed above center between the movable jaw and the workpiece. When the vise is tightened, the force is toward the solid jaw and parallels. This force tends to seat the workpiece. The workpiece is then measured for size (Figure 37–3). The knee is adjusted if required. The part is then milled to size.

PROBABLE CAUSES OF AND CORRECTIVE STEPS FOR COMMON MILLING PROBLEMS

Vibration of the machine, cutter, or work (or of all three) causes unsafe machining conditions.

Table 37–1 Common Milling Problems: Probable Causes and Corrective Action

Milling Problem	Probable Cause	Corrective Action
Chatter	—Lack of rigidity in the machine, work-holding device, arbor, or workpiece.	—Increase rigidity; machine support and secure workpiece, cutter, and arbor more effectively
	—Excessive cutting load	—Use cutter with smaller number of teeth
	—Dull cutter	—Resharpen
	—Poor lubrication or wrong lubricant	—Improve lubrication and check lubricant specifications
	—Straight-tooth cutter	—Use helical-tooth cutter
	—Peripheral relief angle too great	—Decrease relief angle
Poor quality of surface finish	—Feed too high	—Decrease feed or increase speed
	—Dull cutter	—Resharpen accurately
	—Cutting speed too low	—Increase sfpm (spindle speed) or decrease feed
	—Cutter has insufficient number of teeth	—Use finer tooth cutter
Cutter digs (hogs) in	—Peripheral relief angle too great	—Decrease relief angle; use recommended angles
	—Rake angle too large	—Decrease rake angle
	—Improper speed	—Check recommended speed; adjust accordingly
	—Failure to tighten saddle or knee clamping levers	—Tighten saddle and knee clamping levers
Vibration	—Cutter rubs; insufficient clearance	—Use staggered-tooth cutter or cutter with side-relief teeth
	—Arbor size and support	—Use larger arbor and adjust support arm

Vibration also often results in damaged work. Some milling machine models have a *vibration damping unit* located inside the overarm. This unit is shown in the cutaway section of the overarm in Figure 37–4. Its function is illustrated by the vibration amplitude line drawing. The vibration (chatter) damping capability permits greater depths of cuts and higher feeds and speeds than are ordinarily possible. Cutting is more efficient and smoother. The vibration damping overarm may be used with both climb and conventional milling.

Chatter, vibration, and other common milling problems, probable causes, and corrective steps are suggested in Table 37–1.

APPLICATIONS OF THE SURFACE GAGE

Many jobs require plain milling on a surface that is irregular, warped, or varied in thickness. In milling such a surface, it is impractical to seat the part directly on parallels. Surfaces of castings, forgings, and uneven parts may be positioned (leveled) for machining by using a surface gage. The part may be held in a vise or it may be strapped to the table.

The surface gage may also be used for layouts (Figure 37–5). Guide lines may be scribed on the sides and ends of the workpiece. These lines are particularly helpful for determining whether the rough surface will clean up when the final finish cut is taken.

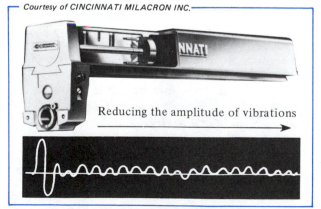

Courtesy of CINCINNATI MILACRON INC.

Reducing the amplitude of vibrations

Figure 37–4 Vibration Damping Unit Inside the Overarm

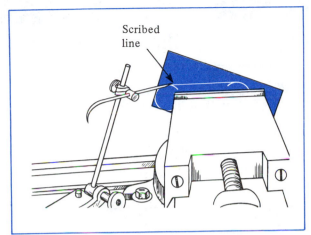

Figure 37–5 Positioning a Layed-Out Workpiece with a Surface Gage

Safe Practices in Plain Milling Setups and Work Processes

- Lock the spindle in the off (stopped) position so that it cannot be tripped acidentally during a machine setup or the taking of a measurement.
- Feed the work in climb milling in the same direction in which the cutter rotates.

- Avoid climb milling with a standard miller unless it is rigidly constructed or is equipped with an automatic backlash eliminator.
- Take a deep first cut below the hard outer scale on castings, forgings, and other irregular or abrasive, rough surfaces.
- Feed a milling cutter into a workpiece by hand. When it is evident that the part will not shift under the cutting force, engage the power feed.
- Mill as close as practical to the work-holding device and arbor support arm. The cutting forces should also be directed toward the spindle.
- Mill toward the solid vise jaw or the vertical leg of an angle plate.
- Lock knee and saddle before engaging the table feed.
- Stop the machine if any one of the common milling problems develops. Check the most probable cause(s). Take the suggested corrective steps.
- Store each tool and machine accessory in its correct rack, bin, or other protected location.
- Place rags and oily cloths in a metal container. The area around the machine should be dry. After wiping up oil on the floor, use a nonskid compound on the oil spots.

UNIT 37 REVIEW AND SELF-TEST

1. Explain why climb milling is preferred over conventional milling under the following conditions: (a) machining deep, narrow slots; (b) cutting off or machining thin, hard-to-hold workpieces; and (c) eliminating surface burrs and feed marks.

2. Indicate the five major functions the operator performs in setting up a milling machine.

3. List three considerations that guide the operator in selecting the appropriate milling cutter.

4. State what advantages a vibration damping unit has over the standard overarm in controlling chatter.

5. Identify four practical methods of milling ends square.

6. Tell how the surface gage is used to position a rough-surfaced part that is held in a milling machine vise.

7. Indicate why a deep first cut is taken below the hard outer scale on castings, forgings, and weldments.

8. State two safe practices to observe in climb milling.

Face Milling on the Horizontal Milling Machine

Three common methods of *face milling* on the horizontal milling machine are covered in this unit. These methods include applications of inserted-blade face mills, shell end mills, and solid-shank end mills. Procedures are examined for milling shoulders which are produced with shell end mills and solid end mills. Face milling as applied to these mills is referred to as *end milling*.

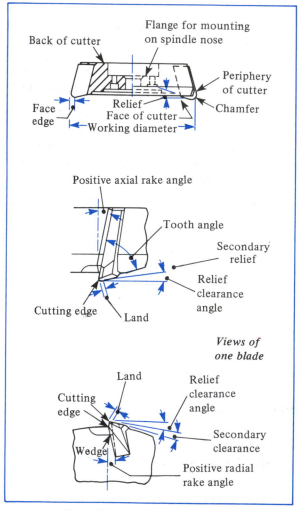

Figure 38–1 General Features of a Face Milling Cutter

APPLICATIONS OF FACE MILLING CUTTERS AND END MILLS

Solid end mills are generally used to mill flat surfaces that are smaller than 1 1/2″ (38mm) wide. The range of commercial solid end mills is from 1/8″ (3mm) to 2″ (50mm) in diameter. Shell end mills are normally employed to face mill surfaces from approximately 1 1/4″ (32mm) to 5″ (130mm) wide. Face milling cutters are used in machining surface areas that are wider than 5″ (130mm).

FACE MILLING CUTTER DESIGN FEATURES
RAKE ANGLES

Figure 38–1 shows some of the design features of a face milling cutter. A face milling cutter is selected in terms of the material to be machined, the nature of the work processes, and the machine setup. Reference is made to manufacturers' tables of rake angles. From these tables the operator establishes whether to use *positive*, *zero*, or *negative radial* and *axial rake angles*.

EFFECT OF LEAD ANGLE AND FEED

LEAD ANGLE

The same principle of *lead angle* that is applied in lathe work for a single-point cutter applies to face milling. Small lead angles of from 0° to 3° are often used to machine close to a square shoulder. Figure 38–2A shows that with a limited lead angle, a chip thickness of 0.015″ (0.4mm) is practically equal to the feed of 0.015″ (0.4mm).

Using the same depth of cut but increasing the lead angle to 45° changes the chip thickness (Figure 38–2B) to 0.010″ (0.25mm). Note that as the lead angle increases, the chip becomes wider. A steep lead angle limits the depth of cut. For practical purposes the maximum lead angle on face milling cutters is 30°.

There are a number of advantages to using a sizable lead angle (Figure 38–2C):

- The cutter contacts the workpiece along the blade rather than at the tip of the cutting edge,
- Cutting forces are applied where the strength of the blade is greater than at the edge,
- Only a partial chip is formed on initial impact and at the end of the chip,
- Cutting a thinner chip thickness adds to effective cutting edge life.

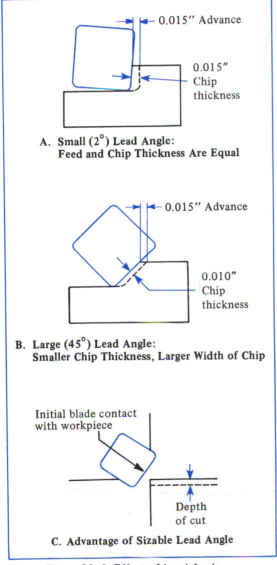

A. Small (2°) Lead Angle:
   Feed and Chip Thickness Are Equal

B. Large (45°) Lead Angle:
   Smaller Chip Thickness, Larger Width of Chip

C. Advantage of Sizable Lead Angle

Figure 38–2 Effect of Lead Angle on Chip Thickness

FEED AND CUTTER LINES

A circular pattern of *feed* and *cutter lines* may be produced in face milling. These lines result from the:

- Size of the nose radius of the cutter blades,
- Coarseness of the feed,
- Limited rigidity in the setup,
- Nature of the machining process,
- Cutter size,
- Sharpness of the cutting edges.

The ridges produced by the face mill may be reduced by grinding the cutting edge of each blade flat. The flat should be wider than the feed lines.

FACTORS AFFECTING FACE MILLING PROCESSES

The following major factors affect dimensional accuracy and quality of surface finish:

- Free play in the table movements;
- Eccentricity, or cutter runout. The trailing edges of the cutter produce light cuts that follow the main cut;
- Lack of support of the workpiece at the point of cutting action;
- Need for more rigid stops. Additional stops may be needed to prevent any movement of the workpiece during cutting;
- Position of the table. The table must be positioned as close as possible to the spindle;
- Dull cutting edges. These edges require more power and produce excessive heat;
- Cutter revolution at too slow a speed. A slow speed causes a buildup on the cutting edge;
- Cutter revolution at too high a speed. A high speed produces excessive cutter wear;
- Too great a cutter feed. This feed causes chipping and breaking of the cutting edges;
- Positive angle of entry. Contact is made at the cutting tip, or weakest area, of the blade.
- Negative angle of entry. The initial force is applied along the cutting face, or strongest area, of the blade and away from the cutting edge;
- Failure of the cutting fluid to flood the cutting area. The intermittent cutting

Table 38–1 General Feeds for End Mills (0.000″ feed per tooth, HSS): 1/8″ to 2″.

Diameter of End Mill	Steel				Cast Iron	Nonferrous Metals		
	Low Carbon	High Carbon	Medium Hard Alloy	Stainless		Aluminum	Brass	Bronze
1/8	.0005	.0005	.0005	.0005	.0005	.002	.001	.0005
1/4	.001	.001	.0005	.001	.001	.002	.002	.001
3/8	.002	.002	.001	.002	.002	.003	.003	.002
1/2	.003	.002	.001	.002	.0025	.005	.003	.003
3/4	.004	.003	.002	.003	.003	.006	.004	.003
1	.005	.003	.003	.004	.0035	.007	.005	.004
1 1/2	.006	.004	.003	.004	.004	.008	.005	.005
2	.007	.004	.003	.005	.005	.009	.006	.005

process and the speed of carbide face mills produce a *fanning action*. A noncontinuous cooling action produces thermal, or heat, cracks in the blades. The problem is overcome by forcing the cutting fluid through a fine spray, or mist, to continuously reach the cutting edges.

FACE AND SHOULDER MILLING WITH A SOLID END MILL

Two basic processes may be performed with *solid end mills*. A plane surface may be milled, or two right-angled surfaces may be produced at the same time. These surfaces are referred to as a *step*, or *shoulder*. Solid end mills with two or more flutes may be used for face milling. These same end mills produce a shoulder when the face and peripheral teeth are set to cut the two surfaces at one time.

RATE OF FEED

Considerable judgment must be exercised by the operator in terms of cutting speeds, cutting feeds, and depth of cut. Table 38–1 provides general recommendations of feed per tooth for different end mill diameters and materials.

The small sizes of end mills are comparatively fragile. Excessive feeds cause tool breakage and chipping at the edges of the teeth. The rate of feed for solid end mills is calculated by substituting the number of flutes for the number of teeth in the formula. Slower feeds are used for deeper cuts.

CUTTING SPEEDS

Tables of cutting speeds, mentioned previously for other machining processes, apply equally to end mills. The cutting speeds given in such

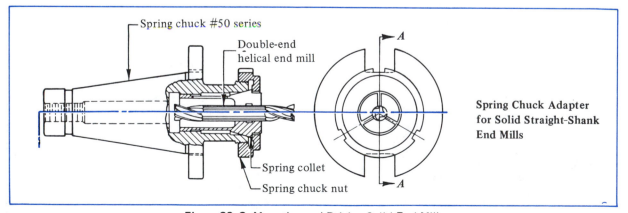

Figure 38–3 Mounting and Driving Solid End Mills

Courtesy of CINCINNATI MILACRON INC.

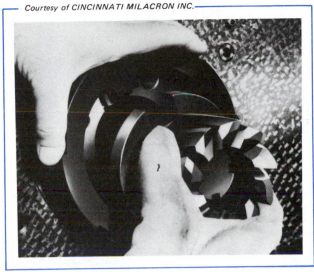

Figure 38–4 Mounting and Locking a Shell End Mill and Adapter into a Quick-Change Spindle Nose Adapter

tables are used to establish the spindle and cutter RPM.

HOLDING SOLID AND SHELL END MILLS

Taper-shank solid end mills that have a tang are held in a *tang-drive collet* (often called a *sleeve*). A threaded-end shank is held in a plain collet adapter. Straight-shank single and double end mills are usually mounted in a spring chuck adapter. The adapters are then secured in a type C arbor (Figure 38–3).

Figure 38–4 shows how an end mill adapter, reducing collet, and a shell end mill are held for mounting. The setup is locked into the quick-change adapter on the spindle nose by a partial turn of the clamp ring.

How to Face Mill with a Face Milling Cutter

STEP Select a face milling cutter (Figure
1   38–5).
STEP Hold the cutter with both hands and
2   mount it on the spindle nose (Figure
      38–6A). Start the fastening screws and
      securely tighten them (Figure 38–6B).

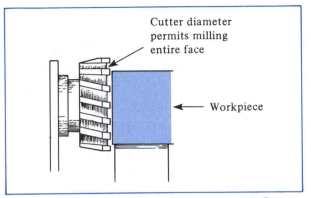

Figure 38–5 Relationship of an Inserted-Tooth Face Milling Cutter to the Width to Be Machined

STEP Calculate the spindle RPM. Set the
3   speed dial or gear-change levers at the
      closest lower RPM. Check the direction
      of spindle rotation.
STEP Calculate the rate of feed. Set the feed
4   dial or feed-change lever to the required
      feed.
STEP Set the trip dogs.
5
STEP Check the cutting fluid. Position the dis-
6   tributor to flow (or spray on a cemented
      carbide cutter) the cutting fluid.

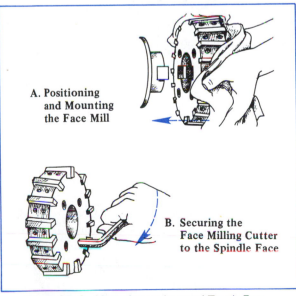

A. Positioning and Mounting the Face Mill

B. Securing the Face Milling Cutter to the Spindle Face

Figure 38–6 Mounting an Inserted-Tooth Face Milling Cutter

STEP 7 Center the cutter at the height of the centerline of the workpiece. Lock the knee.

Note: If the workpiece is wider than the cutter diameter, the distance of the first cut should be not more than three-fourths of the cutter diameter. About one-fourth of the cutter should extend above the top edge of the workpiece. The second cut to clean up the face must overlap the first cut.

STEP 8 Start the spindle. Feed the saddle inward by hand until the cutter just grazes the workpiece. Set the cross feed micrometer collar at zero. Lock the collar.

STEP 9 Move the table to clear the cutter. Turn the saddle handwheel to the required depth of cut. Lock the saddle.

STEP 10 Start the flow of cutting fluid. Feed the face mill by hand. Take a trial cut for a short distance.

STEP 11 Stop the spindle and the coolant flow. Check the workpiece for size. Make whatever adjustment is needed.

STEP 12 Start the machine and the coolant flow. Engage the power feed. Take the cut across the face of the workpiece.

STEP 13 Stop the feed, cutting fluid, and spindle. Return the cutter to the starting position.

Note: Unlock the knee and move the workpiece up to overlap any second cut that may be required.

STEP 14 Increase the speed and decrease the feed for a finish cut.

Safe Practices in Face, End, and Shoulder Milling

- Lock the knee and saddle before face milling.
- Handle the cutter with a cloth and only when the cutter is cool enough.
- Use work stops and heavy-duty strap clamps to rigidly hold a workpiece for deep cuts and coarse feeds.
- Remove all wiping cloths from the machine before it is placed in operation.
- Place a tray or wooden cradle under large face mills when mounting or dismounting them to prevent damage to the cutter or table.
- Engage the power feed only after it is established that the workpiece is supported rigidly enough to withstand the cutting forces.
- Stop the spindle and lock it during setup, measurement, and disassembling steps.
- Follow standard personal, tool, and machine safety precautions.

UNIT 38 REVIEW AND SELF-TEST

1. Give three advantages to using a steep lead angle (to a maximum of 30°) on face milling cutters.

2. State five factors that affect dimensional accuracy and quality of surface finish when face milling.

3. Identify three cutter-holding adapters for straight- and/or taper-shank solid end mills.

4. Give the steps to follow in safely mounting a face milling cutter.

5. Tell how to position a face milling cutter to mill an area that is wider than the cutter diameter.

6. Set up the series of steps to follow to end mill a shoulder with a solid or shell end mill.

7. List three safety precautions to take when disassembling a face milling setup.

8. State two safe practices to follow in end milling with small-diameter end mills.

Side Milling Cutter Applications

This unit deals with applications of single and multiple side milling cutters, straddle milling, and gang milling.

SETUPS AND APPLICATIONS OF SIDE MILLING CUTTERS

Side milling cutters are of the following three basic designs:

- *Half* side milling cutters with teeth on one side and on the periphery (one, two, or three surfaces may be cut at one time);
- *Full* side milling cutters with staggered teeth (narrow, deep grooves may be milled);
- *Interlocking* cutters ground with cutting teeth on one side and on the periphery (the combination of two interlocking cutters permits the simultaneous milling of three surfaces, such as the surfaces of wide grooves).

HALF SIDE MILLING CUTTER APPLICATIONS

Side Milling an End Surface or Step. One of the common applications for a half side milling cutter is milling an end square. The workpiece is usually held in a vise. The vise is accurately

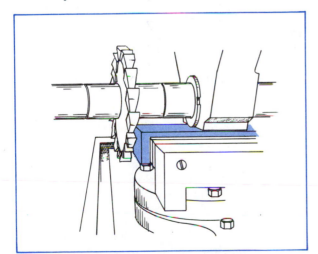

Figure 39–1 Width of a Workpiece within the Cutting Range of a Side Milling Cutter

aligned parallel to the arbor axis. The half side milling cutter is used with a type A, short arbor. The diameter of the cutter must permit milling the face in one cut—that is, the cutter must clear the work step (Figure 39–1).

When the cutter is mounted close to the spindle or if the workpiece is shorter than the width of the vise, the part must be turned 180° to face the second end. There are many applications in which both ends of a workpiece are milled parallel at one setting using a full side milling cutter.

Milling Square Steps, or Shoulders. The half side milling cutter is also used to mill square steps, or shoulders. The side teeth mill the vertical face. The peripheral teeth mill the bottom surface. A roughing cut is usually taken, followed by a finish cut. Cutter RPM, cutting speeds, and feeds are determined in the same manner as for plain and face milling.

FULL SIDE MILLING CUTTER APPLICATIONS

Milling Two End Surfaces. The full side milling cutter is more versatile than the half side mill. With the full side mill, the setup for milling one end of a workpiece can also be used to mill the second end at the one setting. The cutter in this setup is located farther along the arbor support to permit machining with each cutting face. After the first cut the cutter is moved to the other end of the workpiece.

The cutter may be set with a steel rule. The part may be machined to a more precise linear measurement by positioning the cutter with the aid of the cross slide micrometer collar. The distance the work is moved is equal to the required dimension plus the width of the cutter.

Backlash is taken out of the cross feed screw. The saddle is then moved back to within 0.004″ to 0.006″ (0.1mm to 0.2mm) of the required linear dimension. The saddle is locked. A trial cut is taken. The machine is stopped. The length is measured by micrometer and the cut is

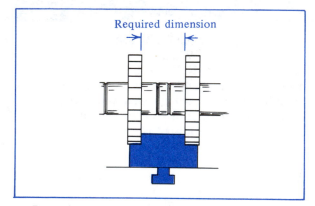

Figure 39–2 Required Width between Side Milling Cutters Produced by Spacing Collars

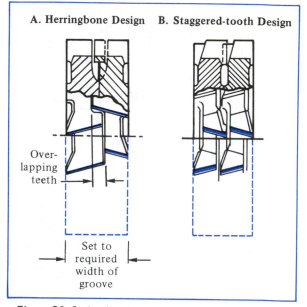

Figure 39–3 Application of Interlocking Side Milling Cutters to Machine a Groove to an Accurate Width

adjusted if required. This procedure helps to correct any machining error that may be produced by the cutter.

Milling a Step. A half side or full side milling cutter may be used to mill a step. The cutter is positioned to cut the vertical face to a required size. The saddle is locked in position. The knee is then raised until the cutter is set at depth. The knee is also locked. The settings may be made by using the micrometer collars or according to layout lines on the workpiece.

Milling a Groove. The sides and depth of a groove are usually layed out. The workpiece is then secured in a vise or strapped directly on the table. A full side milling cutter is used because groove milling requires cutting on the two sides and face of the cutter. A staggered-tooth side milling cutter is used for machining a deep groove.

If the groove is wider than the cutter, it is roughed out. From 0.010″ to 0.015″ (0.2mm to 0.4mm) is left on all sides for a finish cut. When the cutter groove and cutter size are the same, climb milling is recommended because it is easier to hold to a dimensionally accurate groove.

STRADDLE MILLING WITH SIDE MILLING CUTTERS

A straddle milling setup requires two side milling cutters and spacing collars. Spacing collars that correspond in length to the required dimension of a workpiece feature are placed between the inside hub faces of the cutters

(Figure 39–2). If necessary, a spacing collar may be ground to a needed size.

The accuracy to which the width is milled depends on how true the cutters run, how well the teeth cut, the removal of chips between the side face of the cutter and the workpiece, the nature of the cut, the cutting fluid (if required), and the work setup.

Many square, hexagonal, octagonal, and rectangular parts are machined by straddle milling. Straddle milling is particularly useful for round parts. Round parts are usually held in a chuck on a dividing head or on a rotary table. After each cut the workpiece is accurately positioned by indexing.

INTERLOCKING SIDE MILLING CUTTERS

Interlocking half side milling cutters are used in pairs for milling a wide groove or channel in one operation. The interlocking cutters may be spaced within the limits that the teeth overlap. Interlocking cutters are used to mill a groove or channel to an accurate width.

Figure 39–3A and 3B shows two sets of interlocking side milling cutters. The width is adjusted by adding thin, metal collars (shims) between the individual cutters. The left-hand

Figure 39–4 A Rigid Cutter, Workpiece, and Table Setup for Gang Milling

and right-hand helical flutes help counteract the cutting forces.

GANG MILLING

Figure 39–4 shows a gang milling setup and process. Note how close to the spindle the operation is being performed. The table is near the column face. A type B overarm and large end bearing provide solid support for the cutters. This is a rigid setup adapted for gang milling.

A gang milling operation may include any combination of different types of cutters. The illustration shows a number of different diameters and widths of plain milling cutters and a pair of side milling cutters. Oftentimes other form milling cutters are included.

MACHINING PARALLEL STEPS

Parallel steps are milled with cutter, workpiece, and machine setups similar to the setups used to straddle mill parallel sides. However, if the parallel sides are to be milled to two different depths, the diameters of the cutters will differ. The smaller cutter diameter is equal to the larger diameter minus one-half the difference of the depth dimensions. For example, if cutters with 6″ and 4″ diameters are used, there will

be a variation of 1″ between the steps that are milled.

The way in which the workpiece is held also usually differs. When deep steps are to be milled, the part is either held in a fixture or clamped on the ends directly on the table to provide rigid support. Clamping straps also make it possible for the cutter to mill to the required depth without interference.

Safe Practices for Side Milling Setups and Processes

- Check for clearance between the arbor collars or overarm and any projecting surfaces on a workpiece or work-holding device.

- Use care in positioning a side milling cutter so that the teeth are not brought against the vise or other holding device.

- Position the cutters and cutting action as close to the spindle as work conditions permit.

- Tighten the arbor nut only after the overarm is supporting the arbor.

- Use a staggered-tooth, full side mill for cutting deep grooves.

- Flood the flutes and area of cutting action to flow the chips away from a cutter and workpiece.

- Stop the spindle and lock it when taking measurements, adjusting the workpiece or cutter, or disassembling the setup.

- Provide added cutter and work support for wide, deep cutting with interlocking side milling cutters and for multiple cutter setups for gang milling.

- Observe standard safety practices related to protective goggles, setup, assembly and disassembly procedures, proper clothing, and chip removal and cleanup.

UNIT 39 REVIEW AND SELF-TEST

1. List the major applications of (a) half side milling cutters and (b) full side milling cutters.

2. Differentiate between milling with staggered-tooth interlocking side milling cutters and gang milling.

3. Indicate the advantage of straddle milling parallel sides compared to machining each side with a single side milling cutter.

4. Tell how to machine parallel steps at different heights with a straddle milling setup.

5. State why the full side milling cutter flutes and cutting areas are flooded with cutting fluid during deep-groove milling.

6. Name two precautions to take in straddle milling.

7. List three safe practices to follow in mounting side milling cutters and cutter setups for gang milling.

Sawing, Slotting, and Keyseat and Dovetail Milling

METAL-SLITTING SAWS AND SLOTTING CUTTERS

Sawing refers to ordinary slitting, slotting, and cutting-off processes. *Slitting* relates to the use of a saw to cut through a tube or ring or to separate it into a number of parts. *Slotting* involves the machining of a narrow-width groove.

Manufacturers' tables usually provide guidelines for the number of teeth to use for slitting and sawing steel. Saws having two-thirds the number of teeth as saws used for steel are used on brass and deep slots. Copper is one of the more difficult metals to cut on the milling machine. Cutters for copper are specially shaped and have a small number of teeth compared to cutters for steel.

The cutters are thin and are called *slitting saws*. The *plain metal-slitting saw* is ground concave on both sides for clearance. The saw is made of high-speed steel with standard keyways. The cutter hub has parallel sides. The hub is the same width as the cutter teeth.

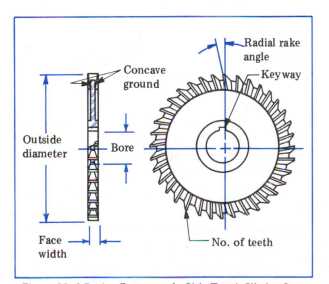

Plain slitting saws are available to fit standard arbor diameters. The general widths range from 1/32" (0.8mm) to 3/16" (5mm). The outside diameters vary from 2 1/2" (64mm) in narrow widths to 8" (200mm) for wide cutters.

The *metal-slitting saw with side teeth* (Figure 40–1) provides side chip clearance. The additional side clearance spaces between teeth and the recessed hub permit chips to flow freely away from the cutter teeth. Thus, binding and scoring of the work are prevented and the amount of heat generated is reduced. This saw is suitable for regular and deep slotting and sinking-in cuts.

Metal slitting saws with side teeth are commercially available in face widths that range from 1/16" (1.58mm) to 3/16" (4.76mm). The number of teeth on a cutter in this range is from 28 to 48. The outside diameters are from 2 1/2" (64mm) to 8" (200mm).

The *metal-slitting saw with staggered peripheral teeth* has an axial rake angle. The side teeth on this high-speed steel saw have a positive radial rake angle. The alternate helical (side) teeth help to eliminate chatter. The staggered teeth and side clearance provide the necessary chip space for deep cuts where heavy feeds are required. The side teeth are not designed for cutting. Staggered-tooth slitting saws have coarser teeth than do regular slitting saws. The staggered-tooth slitting saw is best adapted for cuts 3/16" and wider. Standard and heavy feeds are recommended.

SCREW-SLOTTING CUTTERS

The *screw-slotting cutter* is used principally to slot screw heads and to slit tubing, thin-gage materials, piston rings, and similar work. The sides of the cutter are ground concave. Screw-slotting cutters are available in three groups of standard sizes. These sizes and other specifications are listed in Table 40–1. Additional special

Figure 40–1 Design Features of a Side-Tooth Slitting Saw

Table 40–1 Specifications of Standard Screw-Slotting Cutters

Cutter Diameter	Bore Size	Teeth in Cutter	Ranges		
			Face Width	Wire Gage	Screw Head Diameter
2 3/4″ (69.85mm)	1″ (25.4mm)	72	0.020″ to 0.144″ (0.5mm to 3.7mm)	7–24	1/8″ and smaller through 1″ (3mm and smaller through 25mm)
2 1/4″ (57.15mm)	5/8″ (15.9mm)	60	0.020″ to 0.064″ (0.5mm to 1.5mm)	14–24	1/8″ and smaller through 3/8″
1 3/4″ (44.45mm)	5/8″ (15.9mm)	90	0.020″ to 0.064″ (0.5mm to 1.5mm)	14–24	(3mm and smaller through 9mm)

screw-slotting cutters are produced to meet unusual specifications. Technical information relating to cutter diameter; bore size; number of teeth in cutter; and ranges of face width, wire gage, and screw head diameter is then furnished by the manufacturer.

WORK-HOLDING SETUPS

Rigidity is an important consideration in cutting-off, slitting, and slotting operations. Usually when screw-slotting cutters are used, the parts to be slotted are nested in a fixture.

Sections of a workpiece are often cut off on the milling machine. A solid part, where no deflection would be produced by cutting action, is often mounted directly in a vise. The area to be cut off extends beyond the vise jaw. A plain cutter may be used, or a cutter with straight peripheral and side teeth or staggered peripheral teeth and alternate side teeth may be selected. The depth of the cutting-off operation, the nature and rigidity of the setup, the cutting fluid,

and the machinability of the material are factors that guide the craftsperson in cutter selection.

Another common setup is to strap the workpiece directly to the milling machine table. The position of the cut and cutter must be over one of the table T-slots. This setup is illustrated in Figure 40–2. Climb milling is then used. Added rigidity is provided in climb milling as the cutter forces the work against the table or other supporting surface.

Cast iron and other materials that have a hard outer scale may require conventional milling. Conventional milling minimizes the cutting effect that otherwise would require the teeth to be continuously cutting into the scale.

A common practice followed in the shop to overcome the tendency of the work to spring during conventional sawing is to place a strap across the face of the workpiece.

Slotting saws are sometimes used to mill narrow and other slots, where a side mill is impractical. If the slot is wider than the cutter face width, two or more cuts may be required. The saw is moved to machine the wider slot so that each cut overlaps the preceding one.

Angle pieces requiring a slot to be machined are often mounted on a right-angle plate (for surfaces at 90° to each other). The sine plate is used for cutting precise slots at any angle.

KEYSEAT MILLING

In general practice, *keyseat milling* refers to the machining of a groove in a shaft (keyseat). Three basic forms of keyseats are shown in Figure 40–3. The *plain (open) keyseat* may be milled with a single side milling cutter. When the cutter is the

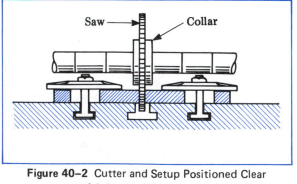

Figure 40–2 Cutter and Setup Positioned Clear of Arbor and Table T-Slot

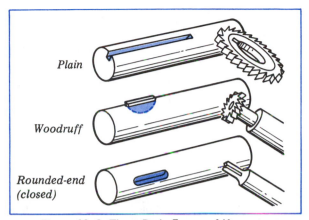

Figure 40–3 Three Basic Forms of Keyseats

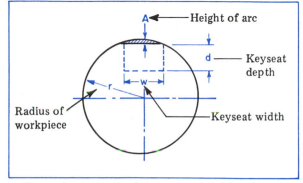

Figure 40–4 Depth of Keyseat and Height of Arc

correct width, the keyseat is produced by taking a single cut. Other widths may be milled by taking second cuts. The *Woodruff keyseat* is produced by a fixed-sized Woodruff keyseat cutter. The *rounded-end (closed or sunk) keyseat* may be cut at any position along an arbor or drive shaft. This keyseat is produced with an end mill.

COMPUTING THE DEPTH OF CUT FOR A KEYSEAT

Shop drawings for keyseats are dimensioned from the depth of cut to a theoretically sharp edge. This edge is formed where the vertical side of the keyseat and the outside diameter of the workpiece meet. In keyseat milling, the workpiece must be raised to the keyseat depth dimension plus the vertical height of the arc. The arc represents the additional material to be removed. The height of the arc varies according to the radius of the workpiece and the width of the keyseat. The following formula (illustrated in Figure 40–4) is used to compute the height of the arc:

$$A = r - \sqrt{r^2 - (1/2w)^2}$$

where A = height of arc,
  r = radius of workpiece,
  w = keyseat width.

WOODRUFF KEYSEAT CUTTER SPECIFICATIONS

NOMINAL SIZES OF WIDTH AND DIAMETER

Comercially available Woodruff cutters conform to American National Standard (ANSI)

specifications. The number of the cutter and the American Standard key numbers are identified. A notation such as #506 AMERICAN STANDARD WOODRUFF CUTTER indicates both the cutter size and the nominal key dimensions of width and diameter.

Stock cutter sizes range from 1/16″ (1.5mm) to 3/8″ (9.5mm) wide and 1/4″ (6.35mm) to 1 1/2″ (38.1mm) diameter. These sizes are nominal dimensions. The 1/2″ (12.7mm) cutter shanks are of uniform diameter.

NOMINAL HEIGHT ABOVE SHAFT

Handbook tables provide dimensional data for standard and special Woodruff keys. Key numbers are recommended for different shaft diameters. Woodruff keys project above the shaft diameter a distance equal to one-half the cutter width ±0.005″ (±0.12mm).

The depth of the keyseat in a shaft must be held to a tolerance of +0.005″/–0.000″ (+0.12mm/–0.00mm). Table 40–2 is a partial table of ANSI dimensions for Woodruff keys and keyseats. An important dimension for the machine operator is the depth (B in Table 40–2) of the keyseat from the top of the shaft. This depth represents the distance the workpiece/knee is raised from the initial zero cutter setting.

How to Mill a Keyseat with a Woodruff Cutter

STEP 1 Select a Woodruff keyseat cutter with a width and a diameter that meet the dimensional requirements of the job.

Table 40–2 ANSI Dimensions (in Inches) for Sample Woodruff Keys and Keyseats (ANSI B 17.2)

Woodruff		Keyseat Shaft					Key above Shaft	Keyseat Hub	
		Width A*†		Depth B†	Diameter F		Height C†	Width D†	Depth E
Key Number	Nominal Size Key	Min.	Max.	+0.005 -0.000	Min.	Max.	+0.005 -0.005	+0.002 -0.000	+0.005 -0.000
202	1/16 × 1/4	0.0615	0.0630	0.0728	0.250	0.268	0.0312	0.0635	0.0372
606	3/16 × 3/4	0.1863	0.1880	0.2143	0.750	0.768	0.0937	0.1885	0.0997
806	1/4 × 3/4	0.2487	0.2505	0.1830	0.750	0.768	0.1250	0.2510	0.1310

*Values for width A represent the *maximum* keyseat (shaft) width that will assure the key will stick in the keyseat (shaft). The *minimum* keyseat width permits the largest shaft distortion acceptable when assembling a maximum key in a minimum keyseat.

†Dimensions A, B, C, and D are taken at the side intersection.

STEP 2 Mount the cutter in the machine spindle. Use a spring collet or adapter for a standard Woodruff keyseat cutter. Use an arbor for hole-type cutters.

STEP 3 Align and secure the workpiece on V-blocks in a vise or other suitable work-holding device.

STEP 4 Center the cutter at the required distance from the end of the workpiece and on its vertical axis. Lock the table and saddle.

STEP 5 Start the cutter and coolant flow. Set the micrometer collar at zero at the starting position for the cut.

STEP 6 Feed the workpiece upward by hand feed to the required depth. Note this depth reading.

> **Caution:** The force on a Woodruff keyseat cutter increases from zero to a maximum at the end of the cut. The part is hand-fed to permit the depth of cut and force on the cutter to be decreased as the cutting action increases.

STEP 7 Stop the machine. Lower the knee. Clean the cutter, workpiece, and workpiece, and working area. Remove burrs from the milled edges of the keyseat.

STEP 8 Insert a Woodruff key into the keyseat. Use a soft-face hammer to tap the key slightly to seat it.

> **Note:** The key is removed if, after measurement, further machining is required.

CHARACTERISTICS OF DOVETAILS

Two common methods are used to machine *dovetail slides* in jobbing shops. The corresponding angular sides and the common flat base of mating parts may be formed by feeding a single-point cutting tool at the required angle. This method is the shaper method. The second method is to machine the dovetail on a vertical or horizontal miller. A formed single-angle *dovetail milling cutter* is used. Figure 40–5 shows a vertical milling machine setup for machining a dovetail slide.

Courtesy of CINCINNATI MILACRON INC.

Figure 40–5 Milling a Dovetail Slide with a Single-Angle Cutter on a Vertical Milling Machine

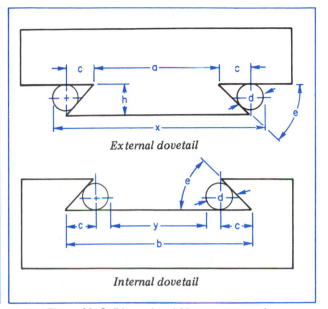

External dovetail

Internal dovetail

Figure 40–6 Dimensional Measurements of a Dovetail Slide

On cast parts, dovetail slides are usually preformed. Sufficient material is provided (left on) to permit machining below (under) the scale surface. Machined dovetail slides are produced within precision limits and with low microfinish surface textures.

Other dovetail slides require milling from a solid piece. In such milling, a rectangular area is rough machined. The angle is produced by roughing out with a formed angle cutter. The sides and base are then finish machined with a dovetail cutter. The same cutter is used to machine internal and external dovetails.

COMPUTING AND MEASURING INTERNAL AND EXTERNAL DOVETAILS

The dimensional accuracy of dovetail slides may be gaged or measured. An optical comparator may be used to establish the accuracy of the angles and other linear dimensions. Dovetail slides may also be measured with a gage or a micrometer.

Figure 40–6 illustrates the dimensional measurements of a dovetail slide. Dimensions X on the external dovetail and Y on the internal dovetail may be measured by micrometer or vernier caliper. Dimension X for the measure-ment of an external dovetail may be calculated by the following formula:

$$X = d \times (1 + \cot 1/2 \text{ angle } e) + a$$

Dimension Y for the measurement of an internal dovetail is also found by formula:

$$Y = b - d \times (1 + \cot 1/2 \text{ angle } e) \text{ and}$$
$$c = h \times \cot e$$

Safe Practices for Sawing, Slotting, and Milling Keyseats and Dovetails

- Stop the spindle and lock it. The cutter must be at rest before the width and depth to which a slot is being milled are measured.
- Stand to one side and out of line of travel of a revolving saw or slotting cutter.
- Use climb milling, where practical, for sawing, slotting, and Woodruff keyseat cutting operations.
- Make sure there is sufficient operating space between the cutter, arbor and support arm, and the workpiece and work-holding setup.
- Use a hand feed for milling a Woodruff keyseat. Hand-feeding permits the feed to be decreased as cutting forces increase.

UNIT 40 REVIEW AND SELF-TEST

1. State the difference (a) between regular slotting and slitting on a milling machine and (b) between the cutters used for each process.

2. Name one common work-holding setup for (a) sawing off pieces from a rectangular plate on a miller and (b) cutting narrow, shallow grooves that are equally spaced around the periphery of a cylindrical part.

3. a. Secure a milling machine cutter manufacturer's catalog.
   b. Give the specifications of an appropriate slitting saw to use for each of the following jobs: (1) cutting off pieces from 1/8″ × 3″ (3mm × 76mm) flat, cold-drawn steel plates; (2) sawing sections from cast iron castings where the cuts are 30mm (1 3/16″) deep; (3) heavy-duty sawing cuts wider than 3/16″ (4.7mm) and 1″ to 1 1/2″ (25.4mm to 38.1mm) deep.

4. Identify the design features of screw-slotting cutters that make them practical for milling narrow, shallow slots.

5. State two special design features of a Woodruff keyseat that make it more functional for certain applications than a square keyseat and key.

6. List three cutter and machine safety precautions to observe in slitting and in keyseat and dovetail cutting on a milling machine.

Indexing Devices:
Principles and Applications

UNIT 41

Direct and Simple Indexing

The *rotary* (or *circular milling) table* and the *dividing head* are two basic *indexing devices*. The function of each device is to accurately locate (index) workpiece features in relation to one another around a common axis. Machining operations are performed after the workpiece is rotated a required angular distance (degrees).

The functions, operation, and setting up of indexing equipment and accessories for direct and simple (plain) indexing are described in this unit. Advanced applications of the dividing head to compound (differential) indexing and to the

milling of helical forms, gears, and cams are covered in the next Section.

RANGES AND TYPES OF INDEXING DEVICES

DIRECT INDEXING DEVICES

A very simple form of indexing is often carried on with standard and compound vises. Many workpieces are indexed for a number of different cuts that are at an angle to one another by simply swiveling the vise jaw section. The angle setting is read directly from the graduations on the vise.

Workpieces held in collets are indexed directly. A *collet index fixture*, like the fixture shown in Figure 41–1, is practical for direct indexing. The two base faces are machined at a right angle. They serve as bases for mounting the fixture horizontally or vertically.

Special milling fixtures are designed with ratchet positioning devices. Once a workpiece is machined on one face, the position is unlocked. The nest in which the workpiece is held is then moved to the next angular setting.

GENERAL PRECISION INDEXING DEVICES

Rotary Tables and Attachments. The *rotary (circular milling) table* is a precision indexing device. Different models range from the line-graduated rotary table for direct indexing to extremely precise inspection devices. Power

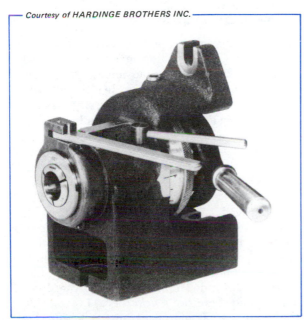

Figure 41–1 A Direct-Indexing Collet Index Fixture

feed, footstock, and other attachments are available.

Rotary tables permit a workpiece to be turned in a horizontal plane to precise positions in degrees, minutes, and seconds. With the addition of a right-angle bracket, the rotary table is adapted for work in a vertical position as shown in Figure 41–2.

Some rotary tables are graduated in degrees. An accuracy within 30 seconds of arc is maintained through a complete rotation of the table. There is an adjustable graduated collar on the handwheel. Readings may be taken directly to each minute. Increased accuracy is provided by a vernier plate. Direct readings may be taken with a vernier plate to within 5 seconds of arc.

Ultraprecise and *optical rotary tables* are applied primarily in high-precision jig boring and inspection processes and for the *calibration* (standards setting) of master tools. Precision indexing attachments are available to measure within *one-tenth (1/10) second of arc*, or 1/36,000 part of one degree.

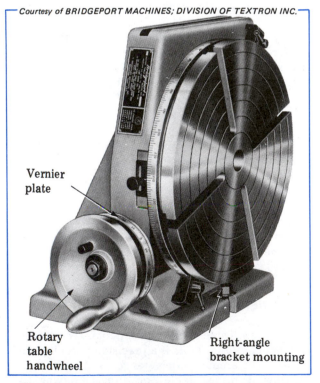

Courtesy of BRIDGEPORT MACHINES; DIVISION OF TEXTRON INC.

Vernier plate

Rotary table handwheel

Right-angle bracket mounting

Figure 41–2 Rotary Table with Vernier Plate Mounted on Right-Angle Bracket for Working in Vertical Position

The Standard Dividing Head. The most practical, everyday dividing device that is used in shops and laboratories is the *standard dividing head*. It is used to divide a circle, to mill one surface in relation to another, or to machine a number of surfaces around a workpiece.

Some workpieces require the machining of equal spaces—for example, four, six, or eight spaces—around the periphery. Such combinations may be positioned easily by using a *simple index head*. A workpiece is mounted in a chucking device. The chuck is connected directly to a spindle. The spindle has a direct-mounted dial plate. The work is turned (positioned) to a required location according to the divisions on the dial plate.

THE UNIVERSAL DIVIDING HEAD

The regular *universal dividing head* is a versatile, practical, and widely used dividing device. By means of a gear train, it is possible to divide a circle into thousands of combinations of equal and unequal parts. Precision linear and circular dimensions are involved.

ADVANCED MOVEMENTS AND PRECISION DIVIDERS

When cuts require a helix angle, it is necessary to swivel the milling machine table to an angle. A cutter is then advanced (fed) into a workpiece according to a fixed lead. The *universal spiral dividing head* is generally used for helical milling and is covered in detail in Unit 44.

The *wide-range divider adapter* in Figure 41–3 permits indexing from 2 to 400,000 divisions. This divider has an index plate, an index crank, and sector arms. These parts are mounted in front of the large index plate of the universal dividing head and add a ratio of 100:1 to the original 40:1 index ratio. The combination of the 40:1 ratio (40 turns of the index crank to turn a workpiece 1 complete revolution) and the 100:1 ratio increases the ratio to 4,000:1 (4,000 turns of the index crank to rotate the divider spindle, and therefore a workpiece, 1 revolution). The accuracy, as stated earlier, is within 1/10 second of arc.

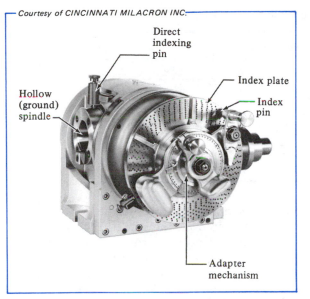

Figure 41–3 A Wide-Range (2 to 400,000 divisions) Divider Adapter on a Universal Dividing Head

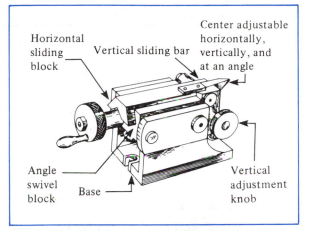

Figure 41–4 A Footstock with Vertical, Horizontal, and Angular Positioning Parts

DESIGN AND FUNCTIONS OF THE UNIVERSAL DIVIDING HEAD

A universal dividing head unit consists of three major components: *dividing head*, *footstock*, and *center rest*. The subassemblies of the dividing head include the *housing*, *swivel block*, and *spindle mechanism*.

MAJOR PARTS OF THE DIVIDING HEAD

The *housing* serves two main functions: (1) As a base the entire head may be bolted to the table and (2) as a bearing, the clamping straps permit the swivel block to be moved to an angle. The *swivel block* may be positioned at any angle from 5° below a horizontal position to 50° beyond a vertical position. The spindle axis of some heads may be moved through a 145° arc.

The *dividing head spindle* is hollow ground. Generally the nose has an internal Brown & Sharpe standard, self-locking taper. Centers and other mating tapered tools and parts may be mounted in the spindle. The nose is machined on the outside to position and hold chucks and other accessories.

The precise positioning of a workpiece is done by turning the index crank. The partial movement is indicated by the index plate and the index pin.

DIVIDING HEAD ACCESSORIES

Many parts are milled between centers and require mounting the dividing head, (spindle) center, and the footstock center. A driver is secured to the spindle center. Other workpieces are conveniently positioned and held in a *dividing head chuck*. Some dividing head chucks are of the single-step type. Other chucks include conventional or reversible three-step jaws.

The chucks are usually self-centering. The chuck itself is fitted with an adapter plate on the back side. The plate permits the chuck to be centered accurately and to be secured on the dividing head spindle.

Common work setups are made with the chuck and dividing head positioned horizontally, vertically, or at an angle. Some workpieces are held by the chuck alone. Other workpieces are also supported by the footstock.

THE FOOTSTOCK (TAILSTOCK)

The terms *footstock* and *tailstock* are used interchangeably. The footstock (Figure 41–4) serves two basic functions: (1) to support one end of a workpiece and (2) to position the workpiece in a horizontal plane or at an angle.

Table 41–1 Holes in Circles on Standard Index Plates

Brown & Sharpe Index Plates		
#1	*#2*	*#3*
15–16–17–18–19–20	21–23–27–29–31–33	37–39–41–43–47–49

Standard Cincinnati Index Plate	
First Side	*Reverse Side*
24–25–28–30–34–37 38–39–41–42–43	46–47–49–51–53–54 57–58–59–62–66

THE CENTER REST

The milling machine center rest serves the same functions as a lathe steady rest. It is used principally for milling operations on long workpieces or machine setups where the cutting force would otherwise cause the workpiece to bend away from the cut. The center rest consists of a base and a V-center that may be adjusted vertically.

DIRECT INDEXING

Direct indexing is also called *rapid indexing* or *quick indexing*. When direct indexing is to be used on a universal dividing head, the worm is disengaged from the worm and wheel by an eccentric device (collar). Disengagement of the worm makes it possible for the spindle and

direct indexing plate to be turned freely. This plate is mounted on the spindle. It usually has three circles of holes on the back side with 24, 30, and 36 holes in each circle. The range of direct divisions is limited to any number that is divisible into 24, 30, or 36. There are additional holes around the rim of the plate. These holes act as guides to rapidly locate some of the small divisions.

SIMPLE (PLAIN) INDEXING

In *simple indexing* a workpiece is positioned by using the crank, index plate, and sector arms. The dividing head spindle is connected by engaging the worm and worm wheel.

INDEX PLATES

The index plates for simple indexing, like the plates used for direct indexing, have a number

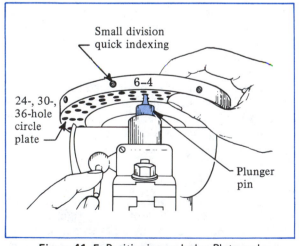

Figure 41–5 Positioning an Index Plate and Pin in Direct Indexing

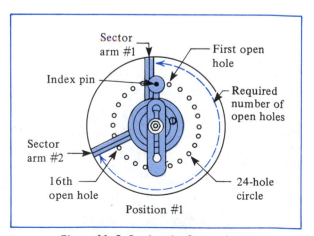

Figure 41–6 Setting the Sector Arms

of circles that are each divided into a different number of equally spaced holes (Figure 41–5). Two common series of index plate hole circles are listed in Table 41–1. There are three plates in the Brown & Sharpe (B & S) series. The standard plate for Cincinnati heads has a combination of eleven different hole circles on each side.

SIMPLE INDEXING CALCULATIONS

The standard gear ratio on the universal dividing head is 40:1. When a fractional part of a revolution is needed, the number of turns of the crank is equal to 40 divided by the number of required divisions. For example, each tooth in a 20-tooth gear is indexed by turning the crank 40 divided by 20, or 2 complete turns.

Sometimes angular dimensions are given on a drawing in place of a number of required divisions. With angular divisions the number of degrees in a circle is divided by the required angular measurement.

The setting of the sector arms to obtain a fractional part of a revolution (or a required angular dimension) is illustrated in Figure 41–6.

Safe Practices in Setting Up and Using Dividing Heads and Accessories

- Clean the table and base of the dividing head and footstock. Remove any burrs. Check to see that the tongues fit accurately into the table slots.
- Position the dividing head as near the center of the table as is practical.
- Tighten the milling machine dog securely to the driver plate.
- Check the alignment of the dividing head and footstock centers for milling parallel and taper surfaces and for proper seating in the center holes.
- Lock the index head spindle after each indexing and before the milling process is started.
- Remove any backlash if the index crank is moved beyond the correct hole.
- Observe personal, machine, tool, and work safety precautions in carrying out the milling processes.

UNIT 41 REVIEW AND SELF-TEST

1. Distinguish between a general and a universal spiral dividing head.

2. Tell how an indexing range up to 400,000 divisions and an accuracy of 1/10 second of arc are possible with a wide-range divider.

3. Indicate what movements are possible with the horizontal sliding block and the vertical sliding bar of the footstock.

4. Differentiate between direct indexing and plain indexing.

5. Explain how to eliminate backlash caused by turning the index crank of a dividing head too far.

6. List the steps to follow in (a) changing the index plate and (b) setting the sector arms on a dividing head for simple indexing.

7. Indicate three safety precautions to take in setting up and using a dividing head and accessories.

SECTION FOUR

Gear and Cam Milling

The actual machining of cams and gears involves a knowledge of design features, the interpretation of drawings and specifications, and the use of formulas to establish machining dimensions. This section provides information on methods of forming and representing spur, helical, bevel, and worm gears, and cams. Actual jobbing shop setups for the milling machine, dividing head, and other accessories; cutter selection; and step-by-step milling procedures and measurement practices are covered in detail.

UNIT 42

Gear Production, Design Features and Computations

FORM MILLING

The first *gear-cutting machines*, dating back to 1800, required a formed milling cutter. The milled form of the tooth and the space between teeth permitted mating teeth to fit.

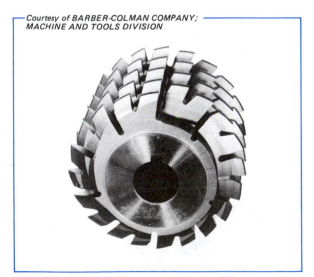

Figure 42–1 Gear Hobbing Cutter Features

GEAR HOBBING

By 1900, *hobbing machines* were produced for highly specialized production gear machining. Hobbing machines are still used today, particularly for extremely large gears of 15 or more feet (4.51m) in diameter. They are constructed with horizontal and vertical spindles.

The hobbing cutter (hob) resembles a worm thread (Figure 42–1). Cutting edges are formed on the cutter by gashes made parallel to the axis of the bore. The tooth form is the same as a gear rack. A gear tooth is generated when the hobbing (worm) cutter turns one revolution while the gear moves one space. The tooth is formed by feeding the cutter to tooth depth.

Spur, helical, and herringbone gears; splines; gear sockets; and other symmetrical shapes may be hobbed.

GEAR SHAPING

Gear shapers were first produced around the turn of the twentieth century. They use a cutter formed like a mating gear but relieved to produce cutting edges. The cutter requires a reciprocating movement. As the cutting progresses and the tool is fed to the required depth, the cutter and gear blank slowly rotate together. A gear is thus produced in one rotation of the blank.

Gear shaping is practical for cutting symmetrical spur, helical, and herringbone gears; internal and cluster gears; racks; elliptical gears; and other special shapes.

OTHER METHODS OF FORMING GEARS

BROACHING AND GEAR SHAVING

Broaching is a fast, economical method of producing individual and nonclustered gears. Broaching is a practical method for machining an accurate straight-tooth or helical-tooth gear form having a pressure angle of less than 21°.

Gear shaving is a technique for producing gear teeth on soft metals, mild steels, and other materials up to a 30 Rockwell C hardness. Surface finishes in the $32R_C$ to $16R_C$ range are possible. Gear shaving requires the use of a rotating cutter that may be positioned at an angle to the axis of a workpiece for diagonal shaving. The teeth are formed as the shaving cutter is fed to depth into a free-rotating blank (Figure 42–2).

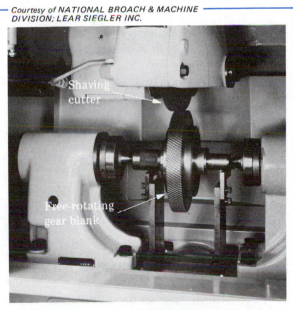

Courtesy of NATIONAL BROACH & MACHINE DIVISION; LEAR SIEGLER INC.

Shaving cutter

Free-rotating gear blank

Figure 42–2 Gear Shaving a Helical Gear

CASTING

The *sand casting* method is still used for large gears where the teeth are later machined. Other casting methods include permanent-mold, shell-mold, and plastic-mold casting. Many gears are *die cast* or *lost-wax cast* and require limited, if any, machining.

HOT AND COLD ROLLING

Hot rolling refers to the production of gear teeth from a solid, heated bar of steel. Gear-shaped rollers are impressed into a revolving heated bar. The force between the rollers and heated material causes the metal to flow and conform to the gear tooth shape and size.

Cold rolling is becoming widely used for the high-volume production of spur and helical gears. Material is saved in producing each gear. Second and third machining operations are eliminated from most gear requirements. Cold rolling produces a smooth, mirror-finish surface texture.

Hardened and precision-ground rolling dies are used. The gear teeth are formed by applying a tremendous force (as in hot rolling) for the rolling dies to impress the tooth form in the blank.

EXTRUDING

Extruding is limited to brass, aluminum, and other soft materials. In some cases, bars are formed to gear tooth shape by forcing (extruding) the material through forming dies. Each gear is then cut off to width. Second and other operations are required on extruded parts. For example, the hole must be pierced. The teeth are often finish machined to precise limits.

STAMPING

Gears for watches, small-gear mechanisms, and many business and household devices are mass-produced by *blanking (stamping)*. Fine blanking permits the stamping of gear parts up to 1/2″ (12.7mm) thickness within close tolerances and to high-quality surface finish requirements.

POWDER METALLURGY

Powder metallurgy is a method that produces gears in three major steps:

- *Mixing*, in which selected metals and alloys in powdered form are mixed in specific proportions;
- *Compressing*, in which the mixture in a particular size of gear-forming die is subjected to extreme force;
- *Sintering*, in which the shaped gear is brought to high temperatures and the metal particles are sintered (fused) together to form a gear of a required form and size.

Large gears produced by this method require the *preform* (gear as it comes from a powder metallurgy press) to be machined to a precise size and surface finish.

GENERATING MACHINES FOR BEVEL GEARS

GENERATING STRAIGHT-TOOTH BEVEL GEARS

Gear tooth generators are built to machine straight-tooth bevel gears that range in size from 0.025″ to 36″ (0.64mm to 914mm) in diameter. Two reciprocating cutting tools are mounted on the generating machine cradle. The cutting tools form the top and bottom sides of a tooth respectively. The cradle and gear blank roll upward together.

At the top of the upward movement, a tooth is completely generated and the cutting tools are automatically withdrawn. The machine indexes while the cradle moves down to the starting position for the next tooth.

GENERATING SPIRAL-TOOTH BEVEL GEARS

Spiral bevel gear generators produce spiral teeth. Tooth profiles for spiral bevel gears (with intersecting axes) and *hypoid gears* (with *nonintersecting axes*) may be generated on the same machine. The tooth profile and spiral shape are produced by combining straight-line and rotary motions between the gear blank and the cutter.

GEAR FINISHING MACHINES AND PROCESSES

More accurate gear tooth shapes and precise surface finishes are produced by such second operations as shearing, shaving, and burnishing. Heat-treated gears are finished by grinding, lapping, and honing.

SHEARING

Fini-shear is a new process for accurately producing and finishing gears. It uses a carbide cutter resembling a master gear. The cutter is rotated with and reciprocates against a gear blank. The cutting tool is brought into contact with the work at an angle to the work axis. The shearing, cutting action produces a high-quality surface finish.

GEAR BURNISHING

Gear burnishing is a cold-working process. The machine rolls a gear in contact with three hardened burnishing gears. Burnishing produces slightly work-hardened, smooth gear teeth.

FINISHING GEAR TEETH ON HARDENED GEARS

Two common gear tooth finishing processes for hardened gears include *grinding* and *lapping*. Gear teeth may be form ground or generated. Form grinding is similar to form milling. The grinding wheel is formed by diamond dressing.

Gear teeth on hardened gears may also be finish ground on a generating-type grinder. The tooth flanks are ground by the flat face of the wheel. The gear tooth is rolled past the revolving grinding wheel.

GEAR TOOTH MEASUREMENT AND INSPECTION

Gear tooth measurements and tests relate to:

- Concentricity (runout),
- Tooth shape,
- Tooth size,
- Surface texture,
- Lead (on helical gears),
- Tooth angles (on bevel gears).

TOOTH INSPECTION WITH AN OPTICAL COMPARATOR

In gear tooth inspection, as in other uses of an optical comparator, a transparent drawing of the tooth is placed on the screen. The drawing

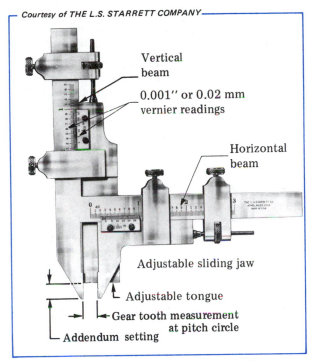

Vertical beam

0.001″ or 0.02 mm vernier readings

Horizontal beam

Adjustable sliding jaw

Adjustable tongue

Gear tooth measurement at pitch circle

Addendum setting

Figure 42–3 Gear Tooth Vernier Caliper Features and Application

Table 42–1 Selected Van Keuren Wire Diameters for External and Internal Gears

External Gears		Internal Gears
Wire Diameter = $\dfrac{1.728}{\text{Diametral Pitch}}$		Wire Diameter = $\dfrac{1.440}{\text{Diametral Pitch}}$
Diametral Pitch	Wire Diameter	Wire Diameter
2	0.8640	0.7200
2 1/2	0.6912	0.5760
3	0.5760	0.4800
4	0.4320	0.3600
64	0.0270	0.0225
72	0.0240	0.0200
80	0.0216	0.0180

scale conforms to the magnification power of the lens. The enlarged shadow of the gear tooth outline that is reflected on the screen is compared directly with the drawing. The nature and extent of error in tooth shape, size, and pitch are visually established.

SPECIAL TESTING FIXTURES

Special fixtures employing a master gear are used for testing concentricity (runout). The testing relates to out-of-trueness on spur gears (parallelism of the teeth and gear axes), angular runout on bevel gears, and the lead of helical and worm gears.

GEAR TOOTH VERNIER CALIPER MEASUREMENT

The *gear tooth vernier caliper*, illustrated in Figure 42–3, is designed to measure (1) the chordal thickness of a gear tooth at the pitch circle, and (2) the addendum distance from the top of the tooth to the pitch circle.

The vernier caliper is graduated in thousandths of an inch (0.001″). It is available in two sizes for measuring gear teeth from 20 to 1 diametral pitch. Metric gear tooth vernier calipers are graduated to fiftieths of a millimeter (0.02mm). The comparable gear tooth range in the English diametral pitch unit system is from 1.25 to 25 millimeter modules.

TWO-WIRE METHOD OF MEASURING GEAR TEETH

A micrometer measurement over the outside diameter of two wires is another technique for checking the accuracy of gear teeth. Two cylindrical wires (pins) of a specified diameter are placed in diametrically opposite tooth spaces for an even number of teeth or for an uneven number of teeth.

Standard wire (pin) diameters are specified in handbook tables according to the diametral pitch of the gear. The *diametral pitch* of a gear designed according to ANSI standards equals the number of teeth in each inch of pitch diameter. *Pitch diameter* represents the diameter of the pitch circle at which the tooth thickness (chordal distance) is measured.

Van Keuren Wire Diameters. Selected Van Keuren wire diameters for external and internal gears are listed in Table 42–1.

Table 42-2 Sample Dimensions for Checking External Measurement over Wires

Even Number of Teeth	Standard Pressure Angle			
	14 1/2°	20°	25°	30°
	Dimension over Wires*			
14	16.3746	16.3683	16.3846	16.4169
16	18.3877	18.3768	18.3908	18.4217
18	20.3989	20.3840	20.3959	20.4256

*The table dimensions are for 1 diametral pitch gear teeth using Van Keuren standard wire sizes.

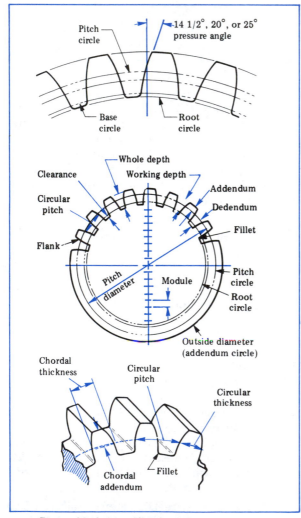

Figure 42-4 Identification of Common Spur Gear Terms and Measurements

Tables of Measurements over Wires. Once the standard Van Keuren wire size is determined, information on the measurement over wires is found in other tables. Handbook tables provide measurements for 1 diametral pitch, external and internal teeth, even and odd numbers of teeth in a gear, and a range from 6 through 500 teeth (Table 42-2). Values are given for four different pressure angles: 14 1/2°, 20°, 25°, and 30°.

$$\frac{\text{measurement}}{\text{over wires (M)}} = \frac{\text{table value for pressure angle}}{\text{diametral pitch of the gear}}$$

Allowance for Play between Teeth. The measurement over wires indicates the overall dimension when the pitch diameter is correct and no allowance is made for play (backlash) between mating teeth.

Handbook tables are available on standard backlash allowances for external and internal spur gears.

SPUR GEAR DRAWINGS (REPRESENTATION, DIMENSIONING, AND TOOTH DATA)

Drawings of gears conform to ANSI standards. The drawings contain such information as material and heat treatment (where required), quality of surface texture, gear tooth data, and manufacturer's identifying markings, and finished dimensions. Common spur gear terms and measurements are illustrated in Figure 42-4.

The view or views used depend on design features. Gears, pinions, or worms that are machined on a shaft are represented by a view that is parallel with the axis. This view is the preferred representation.

Gears with holes and hubs are generally drawn as a section view. The section cuts through the axis. The section may be partial or full. Usually just one view is used. When a drawing needs to be clarified, a section (front) view and a few teeth may be drawn, as shown in Figure 42-5.

Other dimensions relating to the gear teeth are included as *gear tooth data*. Sometimes, *reference dimensions* are included for purposes of computing other needed values.

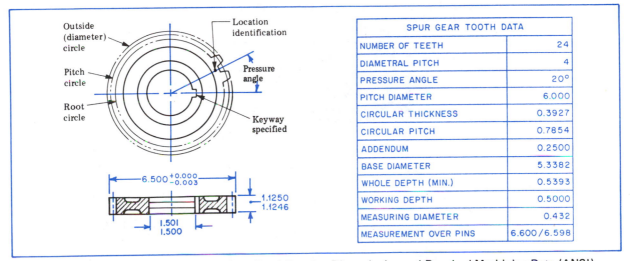

SPUR GEAR TOOTH DATA	
NUMBER OF TEETH	24
DIAMETRAL PITCH	4
PRESSURE ANGLE	20°
PITCH DIAMETER	6.000
CIRCULAR THICKNESS	0.3927
CIRCULAR PITCH	0.7854
ADDENDUM	0.2500
BASE DIAMETER	5.3382
WHOLE DEPTH (MIN.)	0.5393
WORKING DEPTH	0.5000
MEASURING DIAMETER	0.432
MEASUREMENT OVER PINS	6.600/6.598

Figure 42–5 Representation of Cast Spur Gear, Showing Dimensioning and Required Machining Data (ANSI)

MODULE SYSTEMS OF GEARING

The *English module* represents the *ratio* between pitch diameter in inches and the number of teeth. That is, diametral pitch equals the number of teeth per inch of pitch diameter. A 10 diametral pitch gear, for example, has 10 teeth for each inch of pitch diameter.

The *SI Metric module* is an *actual dimension:*

$$\text{SI Metric module} = \frac{\text{pitch diameter (mm)}}{\text{number of teeth in gear}}$$

For example, a 25-tooth gear with a pitch diameter of 75mm has a module of 3—that is, there are 3mm of pitch diameter for each tooth.

Most of the gearing produced in the United States is designed around the diametral pitch system. This system provides a series of standard gear tooth sizes for each diametral pitch. For example, in a 10 diametral pitch series of gears, a 40-tooth gear has a pitch diameter of 4″; a 42-tooth gear, 4.2″; a 44-tooth gear, 4.4″; and so on.

METRIC MODULE EQUIVALENT

A metric module equivalent of a diametral pitch may be found by dividing 25.4 by the diametral pitch. Where necessary, the answer is rounded off to the closest module in the standard

Table 42–3 SI Metric Spur Gear Rules and Formulas for Required Features of 20° Full-Depth Involute Tooth Form (Partial Table)

Required Feature	Rule	Formula*
Module (M)	Divide the pitch diameter by the number of teeth	$M = \dfrac{D}{N}$
	Divide the outside diameter by the number of teeth plus 2	$M = \dfrac{O}{N + 2}$
Addendum (A)	Multiply the module by 1.000	$A = M \times 1.000$
Clearance (S)	Multiply the module by 0.250	$S = M \times 0.250$
Whole depth (W^2)	Multiply the module by 2.250	$W^2 = M \times 2.250$
Tooth thickness (T)	Multiply the module by 1.5708	$T = M \times 1.5708$

*Dimensions are in millimeters.

Table 42–4 ANSI Spur Gear Rules and Formulas for Required Features of 20° and 25° Full-Depth Involute Tooth Forms (Partial Table)

Required Feature	Rule	Formula*
Diametral pitch (P_d)	Divide the number of teeth by the pitch diameter.	$P_d = \dfrac{N}{D}$
	Add 2 to the number of teeth and divide by the outside diameter.	$P_d = \dfrac{N + 2}{O}$
	Divide 3.1416 by the circular pitch.	$P_d = \dfrac{3.1416}{P_c}$
Whole depth of tooth (W^2)	Divide 2.250 by the diametral pitch.	$W^2 = \dfrac{2.250}{P_d}$
Thickness of tooth (T)	Divide 1.5708 by the diametral pitch.	$T = \dfrac{1.5708}{P_d}$
Center distance (C)	Add the pitch diameters and divide the sum by 2.	$C = \dfrac{D^1 + D^2}{2}$
	Divide one-half the sum of the number of teeth in both gears by the diametral pitch.	$C = \dfrac{1/2\,(N^1 + N^2)}{P_d}$

*Dimensions are in inches.

series. For example, if the metric module equivalent of a 10 diametral pitch is required, the nearest standard module to 2.54 is 2.5.

Table 42–3 is a partial listing of the SI Metric spur gear rules and formulas for required features of a 20° full-depth involute tooth form. Table 42–4 is a partial listing of the ANSI spur gear rules and formulas for required features of 20° and 25° full-depth involute tooth forms. The complete table is included in the Appendix.

CHARACTERISTICS OF THE INVOLUTE CURVE

The term *involute curve* identifies the almost exclusive profile used for gear teeth. An involute curve is generated as the path of a point. The point, theoretically held taut on a straight line, scribes an involute curve as it unwinds from the periphery of the base circle.

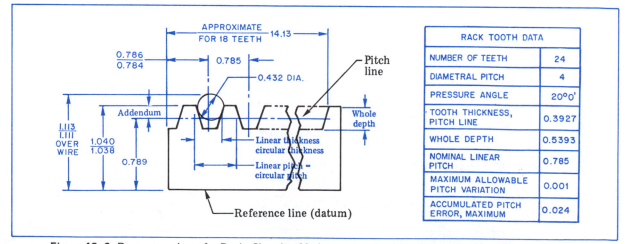

Figure 42–6 Representation of a Rack, Showing Methods of Dimensioning and Required Machining Data

RACK AND PINION GEARS

A *rack* is a straight bar with gear teeth cut on one side. The profile and design features of the teeth on a rack are identical with the mating spur gear *pinion* teeth (Figure 42–6).

Some different terms are used with rack and pinion gears because all circular dimensions on the rack become linear. The *lineal pitch*, as shown in Figure 42–6, corresponds with circular pitch:

$$\text{lineal pitch} = \frac{3.1416}{\text{diametral pitch}}$$

The *lineal thickness* is equal to the circular thickness of a spur gear.

REPRESENTING AND MACHINING A RACK

A rack is usually represented and dimensioned on a drawing as shown in Figure 42–6. Tooth data provide additional information for machining and inspecting the gear teeth.

Racks are machined in jobbing, maintenance, and small shops on a milling machine. Short rack lengths may be held on parallels in a vise or strapped to the table.

The teeth on a helical rack require the workpiece to be set at the gear helix angle in relation to the cutter axis.

HELICAL GEARING

Helical gears transmit rotary motion between two shafts. The shafts may have parallel axes that lie in the same plane or the axes may be at any angle up to 90° to each other.

Helical gear teeth permit continuous, smooth, quiet operation as compared to spur gear teeth. More than one helical gear tooth is engaged at one time. Thus, the gear strength is increased for teeth of equal size and pitch.

The *herringbone (double-helical) gear* used on parallel shaft movements combines the features of a right- and left-hand helical gear in one gear. Since the end thrusts are equalized, no end thrust bearings are required.

CALCULATING HELICAL GEAR DIMENSIONS

The basic features, symbols, rules, and formulas for helical gear calculations and measurements are found in handbook tables and manufacturer's technical data sheets.

NORMAL CIRCULAR AND DIAMETRAL PITCHES

The terms for helical gears are similar to the terms used for spur gears. Two terms, *normal circular pitch* and *normal diametral pitch*, require further discussion. The normal diametral pitch represents the diametral pitch of the helical gear tooth cutter.

The normal circular pitch of a helical gear is the distance between a reference point on one gear tooth and the corresponding reference point on the next tooth. The measurement is taken at the pitch circle on a reference plane that is at a right angle to the tooth helix angle (Figure 42–7).

BEVEL GEARING

DESIGN FEATURES

Bevel gears are used to transmit motion and power between two shafts whose axes intersect. Most bevel gear axes intersect at 90°. The axes may also be at any angle. The design features of a bevel gear and the accompanying terms appear in Figure 42–8.

The teeth on bevel gears conform to standard 14 1/2°, 20°, and 25° pressure angles. The tooth form follows the same involute form as a spur gear. However, the teeth taper toward the apex of the cone. Two right-angle (90° axes) bevel gears of the same module, pressure angle, and

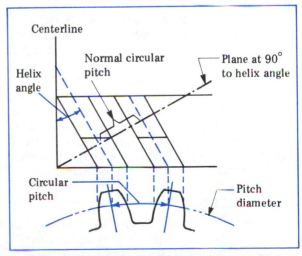

Figure 42–7 Relationship of Normal Circular Pitch of a Helical Gear to Circular Pitch

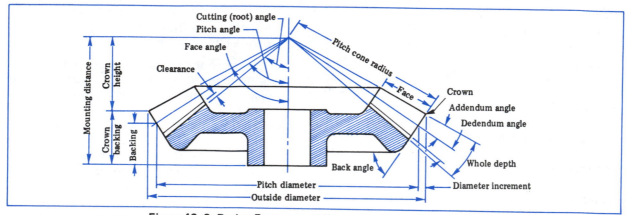

Figure 42–8 Design Features and Terms Applied to Bevel Gears

number of teeth are called *miter gears*. The small gear in a bevel gear combination is identified as a *pinion*.

COMPUTING BEVEL GEAR DIMENSIONS

The formulas for computing spur gear dimensions are also used for the following bevel gear features:

- Addendum,
- Dedendum,
- Whole tooth depth,
- Module,
- Diametral pitch,
- Circular pitch,
- Chordal thickness,
- Circular thickness.

Dimensions for such bevel gear features as: *pitch cone distance*, *diameter increment*; *addendum*, *dedendum*, *face*, and *cutting angles*; *face width*; and the *whole depth* and *tooth thickness* at the *small end of the tooth* are calculated. Formulas and other technical data are provided by gear tooth manufacturers and are found in handbooks. Formulas are also given for determining the precise and the approximate amount a *gear blank* is *offset* from the center line when bevel gear teeth are cut on a milling machine.

WORM GEARING

Worm gearing is used to transmit rotary motion and power between two nonintersecting shafts. The shafts generally are at right angles to each other. There are two gears in a worm gearing setup. The *worm* serves as the driver gear; the *worm gear*, as the driven gear. The teeth in the worm are similar to a coarse-pitch Acme form screw thread with a single-, double-, or triple-lead thread.

CLASSES OF WORM GEARING

The two general classes of worm gearing are the *fine-pitch* and the *coarse-pitch* class. Fine-pitch worm gears are largely used to transmit uniform, accurate angular motion rather than power. Gear terms and design features conform to American National Standard (ANSI) specifications (Figure 42–9). There are eight standard axial pitches (0.030″ to 0.160″). Fifteen standard lead angles (30′ to 30°) have been established. A standard 20° pressure angle cutter or grinding wheel angle is used.

SINGLE- AND MULTIPLE-THREAD WORMS AND VELOCITY RATIO

A worm and worm gear combination is capable of carrying greater loads than helical gears and at a *high velocity ratio*. This ratio depends on the thread lead on the worm and the number of teeth on the gear. For example, the velocity ratio of a single thread (single pitch lead) to a 36-tooth gear is 36:1; of a double thread, 18:1; a triple thread, 12:1. The worm gear and worm permit high speed reductions in confined areas.

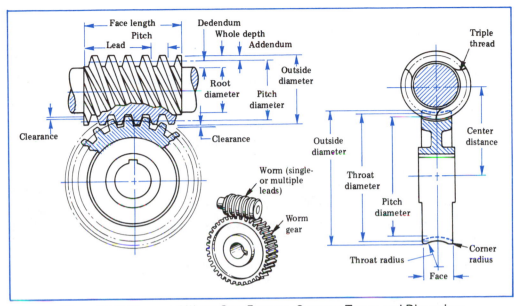

Figure 42-9 Worm and Worm Gear Features, Common Terms, and Dimensions

WORM GEARING TERMINOLOGY

The following terms and formulas are specific to worm gearing:

The *axial pitch* is the distance that extends between corresponding reference points on adjacent threads in a worm. The axial pitch of a worm is equal to the circular pitch of the mating worm gear.

The *axial advance (lead)* is the distance the worm thread advances in one revolution. A single thread advances a distance equal to the thread pitch; a double thread, twice the pitch; and so on.

The *lead angle* is formed by the tangent to the helix of the thread at the pitch diameter and a plane perpendicular to the worm axis. By formula,

$$\text{Lead Angle (tangent to Thread Helix Angle)} = \frac{\text{lead}}{\pi \times \text{Pitch Diameter of Worm}}$$

The *throat diameter of a worm gear* is the overall measurement of the worm at the bottom of the tooth arc. The throat diameter equals the pitch diameter plus twice the addendum.

GEAR MATERIALS

The three categories of materials used on gears are *ferrous*, *nonferrous*, and *nonmetallic*. Cast iron and steel are widely used. Low-carbon steels are used if gears are left soft or are to be case hardened. Higher-carbon steels are used for gears that are to be hardened, tempered, and ground.

The nonferrous materials are generally applied to light-duty operations or where corrosion resistance is required. Bronze; die cast metals; and aluminum, copper, and magnesium base metals are the most common.

Nonmetallic gears that are corrosion resistant to air and chemicals are used for quiet, high-speed mechanisms. Such gears are often meshed with metallic gears. Gears are made of plastics (where temperature is controlled), thermoplastics (for example, nylon), and resin-impregnated canvas (for example, formica).

UNIT 42 REVIEW AND SELF-TEST

1. a. Describe gear hobbing and gear shaping.
   b. List four different gear forms that may be cut on a gear shaper.

2. List five basic categories of gear manufacturing methods.

3. Give two distinguishing design features of a bevel gear generating machine as contrasted to a conventional milling machine.

4. List three important measurements or tests a craftsperson makes in milling a spur gear.

5. Describe briefly the shop practice of measuring the thickness of standard spur gear teeth with a gear tooth vernier caliper.

6. Calculate the pitch diameter, outside diameter, and whole depth of tooth for the following 20° full-depth spur gear tooth forms: (a) 4 diametral pitch, 48 teeth; (b) 16 diametral pitch, 89 teeth; (c) 6 module SI Metric, 24 teeth; and (d) 1 module SI Metric, 127 teeth.

7. State the difference between normal circular pitch and normal diametral pitch as applied to helical gearing.

8. a. Use a table for cutter selection for helical gear cutting.
   b. Determine the cutter number to use for cutting the following helical gears: (1) 20 teeth, 22 1/2° helix angle; (2) 25 teeth, 30° helix angle; (3) 36 teeth, 45° helix angle; and (4) 50 teeth, 30° helix angle.

9. Determine the following dimensions for a pair of 36-teeth, 10 diametral pitch, miter bevel gears: (a) pitch diameter, (b) pitch cone angle, (c) cutting (root) angle, (d) tooth thickness at large end, and (e) tooth thickness at small end.

10. State two design features of worm gearing that differ from bevel gearing features.

Machine Setups and
Measurement Practices for Gear Milling

INVOLUTE GEAR
CUTTER CHARACTERISTICS

Standard involute gear cutter sizes range from 1 to 48 diametral pitch. Figure 43–1 shows examples of gear tooth sizes from 4 to 16 diametral pitch (P_d). Special cutters are available for teeth that are smaller in size than 48 diametral pitch.

A *set* of gear cutters is available for each diametral pitch. Each set consists of eight cutters (#1 through #8). Sometimes, these eight *full-size* cutters are combined with seven *half-size* cutters (#1 1/2 through #7 1/2). The gear tooth range of full- and half-size standard involute form gear cutters in given Table 43–1.

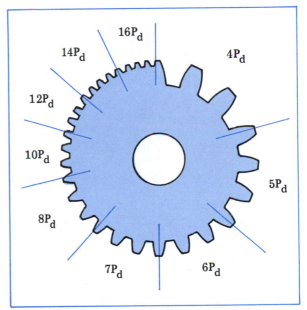

Figure 43–1 Examples of Comparative Gear Tooth Sizes from 4 to 16 Diametral Pitch (P_d)

How to Mill a Spur Gear

STEP 1 Mount the dividing head so that the cutting action takes place with the table positioned centrally.

Table 43–1 Range of Teeth Cut by Full- and Half-Size Standard Involute Form Gear Cutters

Courtesy of NATIONAL TWIST DRILL AND TOOL DIVISION; LEAR SIEGLER INC.

Cutter Number		Range of Teeth
Full Size	Half Size	
1		135 to rack
	1 1/2	80 to 134
2		55 to 134
	2 1/2	42 to 54
3		35 to 54
	3 1/2	30 to 34
4		26 to 34
	4 1/2	23 to 25
5		21 to 25
	5 1/2	19 and 20
6		17 to 20
	6 1/2	15 and 16
7		14 to 16
	7 1/2	13
8		12 to 13

STEP 2 Position the involute form gear cutter on the arbor as close to the column as the setup will permit.

STEP 3 Mount the gear blank and mandrel between centers. Tighten the footstock center and lock.

STEP 4 Align the center line of the cutter teeth with the vertical center line of the gear blank.

STEP 5 Lock the cross slide.

STEP 6 Check to see that the cutting teeth clear the circumference and face of the gear blank.

STEP 7 Select and mount the required index plate. Set the index pin.

STEP 8 Position the spindle speed and cutting feed dials and the coolant nozzle.

STEP 9 Set the table trip dogs.

STEP 10 Start the coolant flow and the spindle. Raise the table until the cutter teeth just graze the gear circumference of the gear blank.

STEP 11 Set the graduated knee feed collar at zero.

STEP 12 Move the cutter clear of the gear blank. Then, raise the table to three-fourths of the tooth depth.

STEP 13 Rough out the first tooth.

STEP 14 Repeat the indexing and cutting steps for the remaining teeth.

STEP 15 Loosen the knee clamp. Raise the table to the working gear tooth depth. Lock the knee clamp.

STEP 16 Reduce the cutting feed and/or increase the cutting speed to produce the desired surface finish.

STEP 17 Finish cut the first tooth. Stop the spindle. Return the cutter to starting position.

STEP 18 Continue to index and finish cut the remaining teeth.

Note: Sometimes, the table is set to machine within 0.002″ (0.05mm) of finish size. Adjustments may then be made, after measurement with a gear tooth vernier micrometer caliper, to machine the gear teeth within specified limits of accuracy.

RACK MILLING

Teeth on short racks may be indexed by moving the cross slide the distance equal to the circular pitch of the gear. The teeth on racks longer than the movement of the cross slide are positioned by moving the table longitudinally. The circular pitch distance is read directly on the graduated collar of the table lead screw.

RACK MILLING AND INDEXING ATTACHMENTS

Milling cutters are often held, positioned, and driven by a rack milling and indexing attachment.

The *rack milling attachment* is driven by the spindle. The attachment permits setups for milling straight teeth or teeth at an angle. The cutter is held at 90° to the position used in conventional milling (Figure 43–2).

The *rack indexing attachment* ensures accuracy in positioning the workpiece for successive teeth to be milled. A standard rack indexing attachment may be used for indexing as follows:

- From 4 to 32 diametral pitches;
- From 1/8″ to 3/4″ in increments of 1/16″;
- For table movements of 0.1429″, 0.1667″, 0.2000″, 0.2857″, 0.1333″, and 0.4000″.

HELICAL GEARING

How to Mill Helical Gear Teeth

STEP 1 Swivel the table to the helix angle and for the direction of the helix.

STEP 2 Calculate or use handbook tables for the change gears that will produce the required lead. Use the formula:

$$\frac{\text{lead of helix to be cut}}{\text{lead of table feed screw}} = \frac{\text{product of driven gears}}{\text{product of driver gears}}$$

STEP 3 Mount the required change gears between the table feed screw and the worm gear shaft on the dividing head. Use an idler to change direction, if needed.

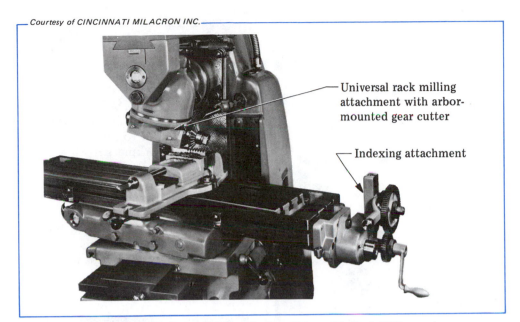

Courtesy of CINCINNATI MILACRON INC.

Universal rack milling attachment with arbor-mounted gear cutter

Indexing attachment

Figure 43–2 Horizontal Milling Machine Setup with Rack Milling and Indexing Attachments for Cutting Teeth on a Rack

STEP 4 Mount the workpiece between centers or hold it securely in a chuck or other work-holding attachment on the dividing head.

> **Note:** The workpiece should be positioned so that the large-diameter end of the mandrel is centered on the dividing head center.

STEP 5 Center the cutter approximately over the groove, flute, or tooth layout.

STEP 6 Set the spindle speed, cutting feed, and table trip dogs according to job requirements. Adjust the cutting fluid nozzle.

STEP 7 Raise the table to the work depth of the tooth. Lock the knee.

STEP 8 Hand-feed until the complete setup is checked for secureness and correct cutting action. Then, engage the power feed.

STEP 9 Return the cutter to starting position.

STEP 10 Index for the next tooth. Reposition the cutter at the work depth. Machine the tooth.

STEP 11 Stop the spindle. Clean the workpiece. Remove burrs from the machined surface.

STEP 12 Measure the normal circular pitch with a gear tooth vernier caliper. Make a working depth adjustment if needed.

STEP 13 Repeat step 10 until each tooth is milled.

How to Mill Teeth on Miter Bevel Gears

STEP 1 Center the cutter (Figure 43–3) over the center line of the gear blank. Set the knee graduated collar at zero. Clear the cutter and gear blank.

STEP 2 Raise the table to the whole tooth depth at the large end of the teeth. Lock the knee.

STEP 3 Mill the first tooth. Return the cutter to starting position.

> **Caution:** Lock the dividing head spindle before each cut is taken.

STEP 4 Index and cut the next tooth. Stop the spindle.

> **Note:** Many milling machine operators rough machine all teeth. The sides are then trimmed to size.

Figure 43–3 Setup for Machining Bevel Gear Teeth with a Gear Blank Mounted on a Dividing Head Spindle

STEP 5 Measure the tooth thickness at the large and small ends to determine the additional amount of material that may need to be trimmed from both sides of the tooth form.

STEP 6 Move the gear blank *off center* the distance calculated for the table offset.

STEP 7 Index and cut each successive tooth until all teeth have been milled on the one side. Stop the spindle.

STEP 8 Unlock the cross slide. Return the table and indexing plate to the original setting.

STEP 9 Move the table the calculated offset distance in the opposite direction.

STEP 10 Turn the index crank in the opposite direction the same amount as was required for trimming during the first cut.

STEP 11 Take a trial cut just far enough to permit measuring the large end of the tooth. Make whatever adjustments are needed.

STEP 12 Proceed to finish cut the first tooth. Recheck the measurement of the small end. Adjust the setup if required.

STEP 13 Index and trim the second tooth side until all teeth are milled.

WORM THREAD MILLING

Worm threads may be milled on a plain or universal milling machine. A dividing head is used for indexing and to serve as the mechanism for turning the workpiece past a revolving thread milling cutter. The required helix movement is produced by the gear train mounted between the table feed screw and the worm shaft on the dividing head.

When the helix angle to be milled exceeds the swivel angle of the machine table, a universal *spiral milling attachment* is used.

WORM GEAR MILLING

How to Mill a Worm Gear

STEP 1 Set the gear blank assembly between the dividing head and footstock centers.

STEP 2 Swivel the table to the helix angle and thread direction of the worm.

STEP 3 Align the center line of the cutter with the vertical axis of the mounted gear blank. Also, center the cutter in relation to the concave radius of the gear.

STEP 4 Start the spindle and flow of cutting fluid.

STEP 5 Feed the cutter to within 0.015″ (0.4mm) of the whole tooth depth.

STEP 6 Lower the table. Index for the next tooth. Gash the tooth.

STEP 7 Continue to index and gash until all teeth are cut. Stop the spindle.

STEP 8 Replace the gear cutter with the required hob. Remove the dog from the mandrel. Adjust the centers so that the mandrel turns easily on the centers, but without end play (Figure 43–4).

STEP 9 Center the hob in relation to the throat radius and the outside diameter of the gashed gear blank.

STEP 10 Start the hob revolving while carefully engaging the hob teeth in the tooth spaces.

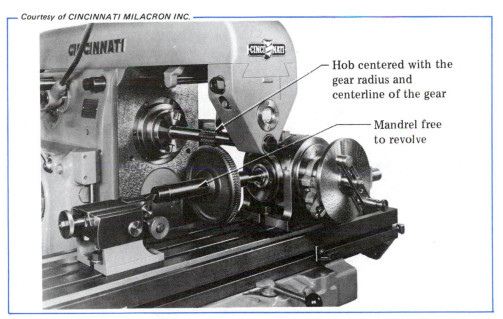

Courtesy of CINCINNATI MILACRON INC.

Hob centered with the gear radius and centerline of the gear

Mandrel free to revolve

Figure 43–4 Milling Machine Setup for Hobbing the Teeth on a Worm Gear

STEP 11 Feed the table up until the hob cuts the teeth to the required depth.

Safe Practices for Setting Up, Milling, and Measuring Gear Teeth

- Use a dividing head dog with the proper fit in the driving fork. Secure the mandrel assembly by tightening the setscrews in the driving fork.
- Align the center line of a formed spur gear cutter with the center line of the gear blank. The teeth sides will then be milled axially for correct mesh.

- Mill bevel gear teeth so that the cutting action and forces are toward the dividing head.
- Clear the formed gear tooth cutter or stop the spindle rotation when the cutter is returned to the starting position for the next tooth.
- Stop the spindle, clean the workpiece, and remove any machining burrs before a precise gear tooth measurement is taken.
- Standard machine safety precautions must be observed. Trip dogs are required to control table movement. Backlash must be compensated for in helical gear cutting and when reversing the direction for settings. Setup and adequate machining clearance are essential.

UNIT 43 REVIEW AND SELF-TEST

1. Give one reason why a set of full- and/or half-size gear cutters is required for each diametral pitch or metric module.

2. Set up a work processing plan to mill a spur gear.

3. State two milling problems that may be overcome by using a rack milling attachment instead of a conventional horizontal spindle setup.

4. Cite the principal advantage for using a rack indexing attachment instead of indexing teeth with a dividing head.

5. Identify two milling machine setups for milling helical gears that differ from the milling of spur gears.

6. Assume that all teeth on a miter bevel gear are rough cut to depth. List the work processes for finish trimming the gear teeth.

7. Give one advantage to the milling of a coarse-pitch, double-lead worm thread in comparison to cutting the same thread on a lathe.

8. Explain the difference between gashing and hobbing the teeth on a worm gear.

9. State three special safety precautions to follow when setting up and milling gears.

Helical and Cam Milling

This unit deals with the principles, terminology, and processes related to the milling of helical forms and common cams and followers.

LEAD, ANGLE, AND HAND OF A HELIX

The *lead of a helix* is the distance the helix advances longitudinally on a part as it makes one complete revolution about its axis. The lead of a helix generated by a single-point cutting tool is illustrated in Figure 44–1.

Each helix has an angle. The *helix angle* is the angle formed by the groove (helix) and the axis of the workpiece.

RIGHT- AND LEFT-HAND HELIX

The direction of a helix is called the *hand of the helix*. A *right-hand helix* (when it is viewed from the end) wraps around a workpiece in a clockwise direction. A *left-hand helix* (when seen with the axis of the workpiece running in a horizontal direction) advances in a counterclockwise direction and slopes down to the left (Figure 44–1).

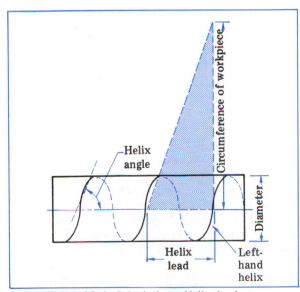

Figure 44–1 Calculating a Helix Angle to Establish Correct Table Setting

DIRECTION FOR SWIVELING THE TABLE

A left-hand helix requires swiveling the milling machine table in a clockwise direction when viewed from the front of the machine. Similarly, a right-hand helix requires swiveling the table counterclockwise to the required helix angle.

Calculating the Helix Angle. The helix angle to which the table is set is calculated in relation to the helix lead and workpiece circumference (Figure 44–1).

$$\text{tangent of helix angle} = \frac{\pi \times \text{diameter}}{\text{lead}}$$

GEARING FOR A REQUIRED LEAD

A helix is produced by advancing a workpiece past a revolving cutter according to the helix angle and lead. The universal dividing head is generally used to turn the workpiece. The workpiece is fed into the cutter by advancing the table, which has been positioned at the helix angle.

The ratio between the rotation of the dividing head spindle and the longitudinal movement of the table requires the use of *change gears*. The change gears provide the required gearing between the worm shaft of the dividing head and the table feed (lead) screw.

Gearing for the dividing head worm shaft and the table feed screw is illustrated in Figure 44–2. An idler gear is inserted in the gearing setup for a right-hand helix. The idler is not a driver or driven gear and is not included in the calculations for change gears. By using an additional idler gear, the direction of rotation of the workpiece is reversed for a left-hand helix.

CALCULATING CHANGE GEARS

A series of change gears is used in a gear train to accommodate different ratios of gears on the worm shaft and the feed screw. A set of change gears is provided as standard equipment with universal dividing heads and universal milling

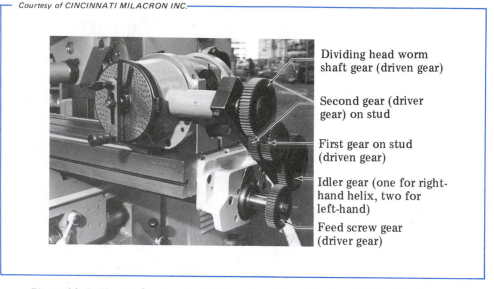

Dividing head worm shaft gear (driven gear)

Second gear (driver gear) on stud

First gear on stud (driven gear)

Idler gear (one for right-hand helix, two for left-hand)

Feed screw gear (driver gear)

Figure 44–2 Change Gearing for Dividing Head Worm Shaft and Table Feed Screw

machines. The gears and the number of teeth on each gear are as follows: 24, 26, 28, 32, 40, 44, 48, 56, 64, 72, 86, and 100. The combination of gears that are available to produce a required ratio may be determined by using the following formula:

$$\frac{\text{Lead of Helix}}{\text{Lead of Machine}} = \frac{\text{Product of Driven Gears}}{\text{Product of Driver Gears}}$$

Producing a Helix by Simple Gearing. To produce a helix by simple gearing, the gear ratio is found by dividing the required lead by the machine lead.

Example: A helix lead of 12″ is to be milled on a workpiece. Standard change gears are available. Calculate the gear ratio for the driven and driver gears. Use the formula,

$$\text{Gear Ratio} = \frac{\begin{array}{c}\text{Lead of Required Helix}\\ \text{(Driven Gears)}\end{array}}{\begin{array}{c}\text{Lead (Feed) of Machine}\\ \text{(Driver Gears)}\end{array}}$$

The lead of a standard milling machine is 10″ for each revolution of the workpiece (lead of helix). By inserting the lead values in the formula,

$$\text{Gear Ratio} = \frac{12}{10}$$

There are no 10- or 12-tooth gears to provide this ratio. However, if each value is multiplied by the same number, appropriate gears may be found. The ratio remains the same. By using 4 as a multiplier,

$$\frac{12 \times 4}{10 \times 4} = \frac{48 \text{ (Driven Gear)}}{40 \text{ (Driver Gear)}}$$

Thus, a 40-tooth gear on the feed screw driving a 48-tooth gear on the worm shaft will produce a 12″ helix lead. The gearing setup is called *simple gearing*.

Producing a Helix by Compound Gearing. Many helixes require more complicated setups that involve *compound gearing* in which a number of gears are formed into a *train*.

Instead of computing change gears for different leads, the craftsperson refers to handbook tables. Change gear combinations are given for leads ranging from 0.670″ (1.7mm) to 60.00″ (1,524mm). Table 44–1 provides examples of change gears for different helix leads. For instance, note the gears recommend for a 36″ lead. A 40-tooth driver gear is secured at the end of the feed screw. This gear drives a 64-tooth gear in the second position on the stud. The 32-tooth gear in the first position on the stud serves as a driver for the 72-tooth driven gear on the worm shaft.

GEARING FOR SHORT-LEAD HELIXES

Short-lead helixes, having leads smaller than 0.670″, require a different gearing setup. Short leads are produced by disengaging the dividing

Table 44–1 Examples of Change Gears for Different Helix Leads (Selected Leads from 32.14″ to 60.00″)

Helix Lead (in ″)	Driven Worm Shaft	Driver 1st Gear on Stud	Driven 2nd Gear on Stud	Driver Gear on Feed Screw
		Position		
35.83	86	32	64	48
36.00	72	32	64	40
36.36	100	44	64	40
36.46	100	48	56	32
36.67	48	24	44	24
36.86	86	28	48	40

head worm and worm wheel. Gearing is direct from the table feed screw to the dividing head spindle. The gears in a direct setup produce a lead that is 1/40 of the lead given in a handbook table. That is, the dividing head spindle turns the workpiece at an increased number of revolutions in relation to the table lead.

In the case of a standard table feed screw having four thread per inch,

$$\text{Gear Ratio} = \frac{\text{Lead to be Milled}}{0.250} = \frac{\text{Driven Gears}}{\text{Driver Gears}}$$

MILLING A HELIX

Helical grooves may be machined on a miller by using a double-angle, concave, fluting, or other formed cutter or end mill.

How to Mill a Helix on a Universal Milling Machine

The setup and milling of a plain helical-tooth milling cutter are used to demonstrate how design features and part specifications (Figure 44–3) are applied to the processes involved in milling the helix.

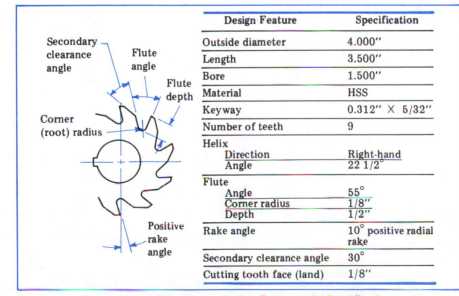

Design Feature	Specification
Outside diameter	4.000″
Length	3.500″
Bore	1.500″
Material	HSS
Keyway	0.312″ × 5/32″
Number of teeth	9
Helix	
Direction	Right-hand
Angle	22 1/2°
Flute	
Angle	55°
Corner radius	1/8″
Depth	1/2″
Rake angle	10° positive radial rake
Secondary clearance angle	30°
Cutting tooth face (land)	1/8″

Figure 44–3 Milling Cutter Design Features and Specifications

STEP 1 Lay out the helix angle of 22 1/2° on the circumference and the 10° positive radial rake angle (Figure 44–4) on the workpiece.

STEP 2 Select an index plate and set the sector arms to permit indexing the nine required teeth.

Note: The locking device for the index plate must be disengaged.

STEP 3 Calculate the helix lead (31.10″) by formula.

STEP 4 Refer to a handbook table of change gears for helical milling different leads. Select the closest lead to 31.10″. In this example, the lead is 31.11″.

STEP 5 Note the number of teeth and position, mount, and secure each gear in the gear train.

gear on feed (lead) screw: 48 teeth (driver)
second gear on stud: 56 teeth (driven)
first gear on stud: 24 teeth (driver)
gear on worm shaft: 64 teeth (driven)

STEP 6 Swivel the table in a counterclockwise direction to the helix angle of 22 1/2°.

STEP 7 Mount the flute milling cutter (55° angle with 1/8″ radius).

STEP 8 Adjust the angle position of the workpiece in relation to the edge of the fluting cutter.

Note: The cutter blank is rotated until the 10° radial layout line of the flute is aligned (same angle) with the angle face of the fluting cutter (Figure 44–4).

STEP 9 Move the table transversely until the center line of the cutter intersects the center of the layout (Figure 44–4).

STEP 10 Return the table (counterclockwise) to the 22 1/2° helix angle. Secure this angle.

STEP 11 Move the cutter to starting position to mill the first tooth. Raise the table to the 0.500″ depth.

STEP 12 Take a trial cut for a short distance. Stop the machine. Check the helix, tooth angle, and depth of helix. Then, continue to mill the first flute.

STEP 13 Index, reset the cutter, and machine each successive flute.

How to Machine the Secondary Clearance Angle

The secondary clearance angle (30°) in the example in Figure 44–3 extends from the point of intersection with one angular side of the flute to the land. The clearance angle is machined to provide a 1/8″ wide land.

The workpiece must be indexed to rotate it so that the 30° angle is in a correct relation to the side of the flute:

$$\text{Indexing} = 90° - 30° + \frac{\text{Flute Angle of } 55°}{2}$$
$$= 32\ 1/2° \text{ or } 32°30'$$

The 32°30′ indexing requires 3 full turns and an additional 11/18 of a revolution.

STEP 1 Replace the fluting cutter with a plain helical milling cutter.

STEP 2 Index the 3 and 11/18 turns.

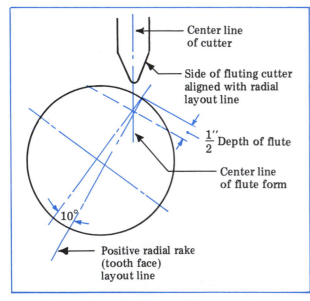

Figure 44–4 Cutter Axis Aligned at Center of Tooth Layout

Center line of cutter

Side of fluting cutter aligned with radial layout line

$\frac{1}{2}''$ Depth of flute

Center line of flute form

10°

Positive radial rake (tooth face) layout line

STEP | Adjust the cutter so that it clears the
3 | cutting face of the flute form.

STEP | Raise the table to cut the 30° angle face
4 | to depth.

> **Caution:** Stop the cutter rotation to check the width of the land.

STEP | Continue to machine. Index to mill the
5 | secondary clearance angle on each tooth.

MILLING A HELIX BY END MILLING

Cams and other parts—for example, actuators for transmitting motion—are often designed with helical grooves that have vertical sides. End mills are commonly used to mill such grooves on either plain or universal milling machines. Since the width of the groove produced by the cutter does not vary with the helix angle, it is not necessary to swivel the table.

The helical groove shape is determined by the form of the end mill. Standard square-face, angular-face, or ball-end end mills may be used. The helix may be milled with the end mill mounted in the spindle of a horizontal or a vertical milling machine.

FUNCTIONS OF CAMS IN CAM MILLING

The three most common cam motions deal with:

- Uniform (constant velocity) motion,
- Parabolic motion,
- Harmonic motion.

DESCRIPTIONS OF CAM MOTIONS

UNIFORM (CONSTANT VELOCITY) MOTION

A mechanism, cutting tool, or device that must rise and fall at a constant rate of speed requires a cam that produces a *uniform motion*. The rate of movement of the follower is the same from the beginning to the end of the stroke.

PARABOLIC MOTION

Parabolic motion is identified as uniformly accelerated or decelerated motion in 180°. A curve similar to the curve for accelerated motion follows for the retarded (decelerated) motion in the remaining 180° part of one cycle or revolution.

HARMONIC MOTION

Harmonic motion relates to a cam and follower that produce a true eccentric motion. The cam moves a follower in a continuous motion at a constant velocity.

COMMON TYPES OF CAMS AND FOLLOWERS

POSITIVE AND NONPOSITIVE TYPES OF CAMS

Plate or bar cams transform linear motion into the reciprocating motion of the follower. Templates that are used to control motion on a tracer miller; profiler; or contour, surface-forming lathe are another form of cam. Cams that impart motion are referred to as *positive* or *nonpositive*.

On a positive design cam, the follower is positively engaged at all times. The cylindrical (drum) cam (Figure 44–5) and grooved recessed

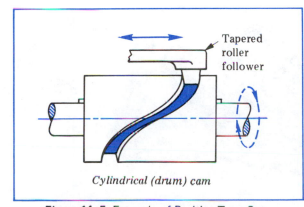

Cylindrical (drum) cam

Figure 44–5 Example of Positive-Type Cam

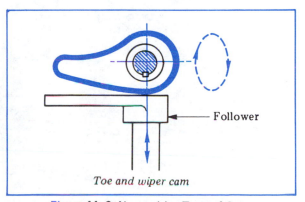

Toe and wiper cam

Figure 44–6 Nonpositive Type of Cam

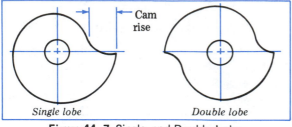

Figure 44–7 Single- and Double-Lobe
Uniform Rise Cams

plate cam are examples of positive-type cams. Engagement between the cam and follower is accomplished by a pin or roller riding in the groove.

The plate cam and knife-edge follower and the toe and wiper cam and follower (Figure 44–6) are common examples of nonpositive types. The cam produces the direction of motion and acceleration. The follower is pushed against the cam by gravity, a spring, or some other outside force.

BASIC FOLLOWER FORMS

There are four basic cam follower designs. The *roller (flat) type* has a free, rolling action that reduces frictional drag to a minimum. The *tapered roller type* is tapered to form a cone shape. The angle sides ride in a corresponding groove form. The *flat or plunge type* is widely used to transmit heavy forces in actuating a mechanism through a particular motion, and time sequence. The *knife-edge and pointed type* is used for intricate and precise movements that require a limited contact surface.

GENERAL CAM TERMS

While the field of cam and follower designs and applications is extensive, only six terms used daily in the shop are described here:

- *Lobe* refers to a projecting area of a cam; the lobe imparts a reciprocal motion to the follower; and single- and double-lobe uniform rise cams (Figure 44–7) are common examples.
- *Rise* is the distance one lobe on a revolving cam raises or lowers the follower.
- *Uniform rise* is the distance a follower moves as generated at a uniform rate around the cam.

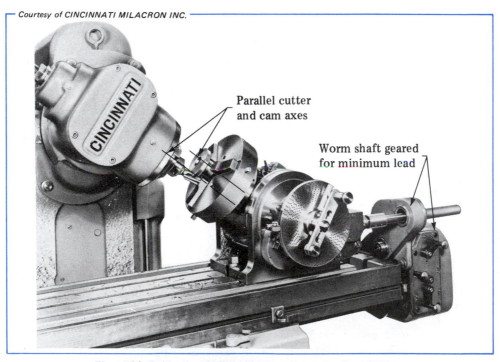

Parallel cutter
and cam axes

Worm shaft geared
for minimum lead

Figure 44–8 Vertical Milling Machine Setup for Cam Milling

- *Lead* is the total distance a uniform rise cam (with only one lobe in 360°) moves a follower in one revolution.
- *Cam profile* is the actual contour of the working surface of a cam.
- *Base circle* is the smallest circle within the cam profile.

BASIC CAM MACHINING PROCESSES

MACHINING IRREGULAR RISE PLATE CAMS

A number of plate cam applications require an uneven, irregular motion. For these applications, the cams are generally layed out. The cam form is produced by *incremental cuts*. The cam blank is rotated a number of degrees called an *angular increment*. A series of cuts is taken so that each splits the layout line. Each cut is continuously adjusted as necessary to produce the desired form in the angular increment. The ridge irregularities between the milling cuts are then filed and polished.

How to Mill a Uniform Rise Cam

Example: Mill a double-lobe uniform rise cam. Each lobe occupies 180°. Each lobe has a rise of 0.250″.

STEP 1 Calculate the cam lead. Use the formula,

$$\text{Lead} = \frac{\text{Lobe Rise} \times 360°}{\text{Lobe Space (° of Circumference)}}$$

$$= \frac{0.250 \times 360}{180} = 0.500″$$

STEP 2 Refer to a handbook table of change gears for cam milling. Select the gearing for the smallest lead.

Note: A table lead of 0.670″ is produced by positioning an 86-tooth gear on the feed screw (driver), a 24-tooth gear in the first position on the stud (driven), a 100-tooth gear in the second position (driver), and a 24-tooth gear on the worm shaft (driven).

Note: The locking device for the index plate must be disengaged.

STEP 3 Color the cam face with dye. Lay out a center line to indicate the starting point of each lobe.

STEP 4 Set up the workpiece in the dividing head and an end mill in the vertical head spindle.

Note: The end mill must be long enough to provide adequate clearance at the start and finish of each cut.

STEP 5 Calculate the required angle for the dividing head and the vertical head spindle. Use the formula,

$$\text{Sine of Angle} = \frac{\text{Required Lead}}{\text{Shortest Machine Lead}}$$

$$= \frac{0.500}{0.670}$$

$$= 0.7463 \ (48°16′)$$

STEP 6 Set the dividing head at 48°16′ (Figure 44–8). Set the vertical head spindle at the same angle.

STEP 7 Align the center line of the cutter with the scribed center line of the workpiece. Adjust the table until the cutter grazes the underside of the cam blank.

Note: A more rigid setup is produced when the cut is taken from this position. Also, chips flow away faster and layout lines remain visible.

STEP 8 Set the vertical feed collar at zero. Start the flow of cutting fluid. Feed the cutter through 180° by rotating the cam blank with the index crank.

STEP 9 Lower the table at the end of the cut so that the cutter and workpiece clear. Disengage the gear train or the dividing head worm.

STEP 10 Move the table back to the starting position.

STEP 11 Position the work so that the center line on the circumference is aligned with the axis of the cutter.

STEP 12 Reengage the gear train or the dividing head worm.

STEP 13 Mill the second lobe.

STEP Shut down the machine. Check the di-
14 mensional accuracy of each lobe.

Safe Practices in Setting Up and Milling Helixes and Cams

- Reduce the force against the sharp-angle cutting edges of a formed cutter at the start of a helical milling process. Avoid forcing or bringing the cutter teeth into sharp contact with the workpiece.
- Use cutting fluids to flow away chips and to aid in producing a quality surface finish.
- Hand-feed small end mills and formed cutters until all cutting conditions have been checked and the processes may be carried on by power feed.
- Stop the spindle and cutter before any measurement is taken.
- Check the clearance between the work-holding, cutter, and machining setups and the helix angle setting of the universal table before a cut is started.
- Set trip dogs to ensure that the table will not be moved accidentally beyond the place where the cutter clears the workpiece.
- Disengage the index plate locking device in short-lead milling when the index crank is used to rotate the workpiece.

UNIT 44 REVIEW AND SELF-TEST

1. State three basic functions served by helical forms that are milled into a cylindrical surface.

2. a. Describe the lead of a helix.
   b. Tell how the hand of a helix affects the table setting of a milling machine when an arbor (hole) type of milling cutter is used.
   c. Calculate the helix angle setting for a 50.8mm diameter part that has a lead of 254mm.

3. a. Determine the gear ratio for milling a machine part with an 18″ lead helical groove.
   b. Select a pair of gears for a direct-drive (simple) gearing setup to cut the 18″ lead.

4. a. Refer to a handbook table of change gears for milling different helix leads.
   b. Determine the number of teeth and position of each gear in the change gears that will produce a lead of 190.5mm.

5. Set up a vertical milling machine or prepare a work production plan to mill a three-lobe (120° apart) uniform rise cam. The rise on each lobe is 0.188″.

6. State two personal safety precautions to observe when milling helical flutes or cams.

SECTION FIVE

Vertical Milling Machines

This section deals with major design features of modern vertical milling machines. These features are related to combination vertical/horizontal machines, the two-axis tracer, the bed-type, and numerically controlled milling machines.

The versatility of the vertical milling machine is extended by attachments such as all-angle, right-angle, multiple-angle, and high-speed milling heads; horizontal milling machine attachment systems; and the cross slide milling head. These attachments as well as several tooling accessories, cutter-holding devices, and a digital readout system, are also described.

UNIT 45

Machine and Accessory Design Features

The *vertical milling machine* is used to accurately produce flat, angular, rounded, and multishaped surfaces. These surfaces may be machined in one, two, or three planes (X, Y, and Z axes). The milling processes may be vertical, horizontal, angular (single or compound angle), or spiral.

The longitudinal and cross (transverse) feeds may be by either hand or power or both. A single part may be produced or the machine may be set up for the machining of multiple, duplicate parts. A machine with numerical control may be tooled for mass production.

DESIGN FEATURES OF MODERN VERTICAL MILLING MACHINES

Modern vertical milling machines are versatile in design. Spindle speeds are usually controlled by variable-speed drives to permit a wide range of cutting speeds and feeds. Rapid traverse power feeds are available to reduce setup time in bringing the work to the cutter. The turret assembly is designed for making precision angular cuts. Handwheels and feed control levers provide for safe and convenient operator control of feeds.

Figure 45–1 shows the design features of a general-purpose vertical milling machine. The specifications for this machine give an idea of

the longitudinal feed (37 1/2″ or 950mm), cross feed (15 3/4″ or 400mm), vertical feed (17 5/8″ or 448mm). This particular quill may be fed 5 1/8″ or 130mm.

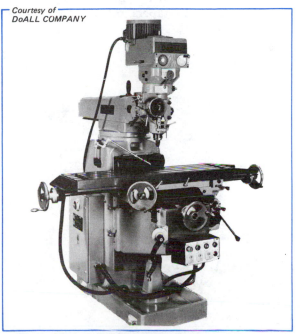

Courtesy of
DoALL COMPANY

Figure 45–1 Design Features of a General-Purpose Vertical Milling Machine

SPINDLE SPEEDS AND FEEDS

The specifications also indicate that the variable-speed drive provides a low spindle speed range of 60–490 RPM and a high range of 545–4,350 RPM. There are three power feeds for the quill. These feeds range from 0.0013″ to 0.004″ (0.03mm to 0.10mm) per revolution. The twelve longitudinal and cross feeds (table and saddle power feeds) range from 1/2″ to 20 7/8″ per minute (ipm) or 12mm to 530mm per minute (mm/min). The spindle nose has a #40 National Standard taper (NS). The table size is 12″ × 49″ (305mm × 1,240mm). Rapid longitudinal and cross feed traverses are available at the rate of 84 5/8 ipm or 2,150 mm/min. The rapid vertical advance is half the rate of the other feeds (42 5/16 ipm or 1,075 mm/min).

There are variations among machine tool manufacturers of power feed rates for the quill. Generally, the feeds per revolution are 0.0015″ 0.003″, or 0.006″ (0.04mm, 0.08mm, and 0.015mm). Spindle speeds may be *infinitely variable* or back gear and direct drive. Low and high speed ranges extend from 60 RPM to 5,400 RPM with a standard head. Spindles are also designed for a #7 Brown & Sharpe taper, quick-change tapers, and tapers other than National Standard (NS).

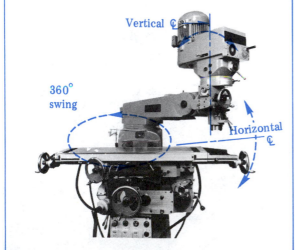

Courtesy of DoALL COMPANY

Vertical ℄

360° swing

Horizontal ℄

Figure 45–2 Positioning the Rigid Turret Assembly and the Swivel Head along the Major Axes

RIGID TURRET AND SWIVEL HEAD ASSEMBLY

The rigid turret assembly may be swung on its base through 360° (Figure 45–2). The head that is secured to the ram is moved by rack and pinion. The head is counterbalanced so that the operator can easily position it for angular cuts. Compound angles or single angles in the front-to-back plane may also be set (Figure 45–2). The head is easily swiveled through a worm gear arrangement. The head may be positioned to right or left of the 0° vertical axis (Figure 45–2).

HANDWHEELS AND FEED CONTROL LEVERS

Three handwheels control the longitudinal feed. Two of these handwheels are located on the sides of the table. The third longitudinal handwheel is located on the saddle. A fourth handwheel on the saddle controls the cross feed movement. The vertical movement of the knee is controlled by the crank. For safety, the handwheels are spring loaded—that is, they are *free wheeling* and remain stationary when the rapid traverse is engaged. The direction of table feed is established by the position of the feed lever. This lever operates both the rapid traverse and feed.

COMBINATION VERTICAL/HORIZONTAL MILLING MACHINE

A *combination vertical/horizontal milling machine* may be quickly and easily converted from a vertical to a horizontal mode. The term *mode* indicates an adaptation of a machine tool from one series of major processes to another. For example, by retracting (bringing back) the forearm, loosening the head bolts, and rotating the head, a vertical milling machine may be adapted to a horizontal mode for horizontal milling processes. Stops at 0° and 90° provide for vertical and horizontal alignment, respectively.

TRACER MILLING MACHINE

Tracer milling relates to the use of a template or sample part. The template serves as a guide for simultaneously controlling the movement of a cutter to reproduce a desired contour and/or form. A *tracer milling machine* has the versatility of straight-line milling (by locking the cut-

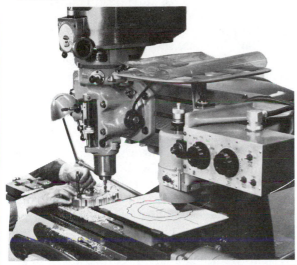

Figure 45–3 Tracer Milling Machines

Figure 45–4 Bed-Type Vertical Milling Machine

ter at any of the axes) or profiling (by operating the cutter in two or three planes). Feeding may be manually or automatically controlled.

The manufacturer of the tracer milling machine illustrated in Figure 45–3 claims a part-to-template accuracy of ±0.002" at 5 ipm (0.05mm at 127 mm/min). The accuracy at a higher feed rate of 30 ipm is ±0.0025" (0.052mm at 762 mm/min).

The part-to-part accuracy is ±0.002" at both the 5 and 30 ipm (0.05mm at 127 and 762 mm/min) feed rate. The part-to-pattern vertical axis tracing accuracy is within ±0.004" (0.1mm).

BED-TYPE VERTICAL MILLING MACHINE

A *bed-type vertical milling machine* provides for longitudinal and cross slide (transverse) movement of a workpiece under a heavy-duty head and quill. The machine pictured in Figure 45–4 has a vertical head movement of 22" (560mm) and a quill diameter of 6" (152.4mm). The combination of longitudinal, cross slide, and vertical travel of this machine provides it with a work capacity of 51" × 20 5/8" × 26 3/4" (1300mm × 525mm × 680mm).

It is evident that the bed-type vertical milling machine with its #50 NS spindle taper is used on large workpieces where great power and heavy chip removal capacity are required.

NUMERICALLY CONTROLLED VERTICAL MILLING MACHINE

Vertical milling machine control movements may be actuated from information that is stored or punched on tape (numerically controlled, NC) or that is recorded in a computer program (computerized numerically controlled, CNC). (Concepts and programming of numerical control for two- or three-axes processes and four- or five-axes centers are examined in detail in Part Twelve.)

A *numerically controlled (NC) or computerized NC (CNC) vertical milling machine* may be used for point-to-point or contour milling with two- or three-axes continuous path controls.

MAJOR COMPONENTS OF STANDARD VERTICAL MILLING MACHINE

The six major design components of a standard vertical milling machine are as follows:

- Column and base;
- Knee, saddle, and table;

- Turret head and ram;
- Spindle (tool) head;
- Manual, power, and rapid traverse feed drives and controls;
- Cutting fluid and lubricating systems.

The functions of each component and of the control levers for speeds and feeds are similar on standard vertical and horizontal milling machines. The exceptions are the turret head, ram, and spindle (tool) head.

FEATURES OF THE TURRET HEAD AND RAM

The turret swings on a machined base on top of the column. The turret head assembly is graduated through 360°. The spindle (tool) head is secured to the outer end of the ram. The ram is dovetailed to slide in the turret base. The movement of the ram, toward or away from the head, is controlled by a rack-and-pinion device.

FEATURES OF THE SPINDLE HEAD

The spindle head serves as a cutter holding unit, the power source for cutting, and houses the cutter feed control mechanism. To serve these functions, the spindle head consists of the power source, speed and feed controls, a quill, and the spindle. The main design elements of the spindle head are shown in Figure 45–5.

The spindle is usually splined to deliver maximum power without slippage between the cutter or holder shank and the spindle. A spindle lock prevents the spindle from turning when holders or tools are being inserted or removed.

Many standard taper systems and taper sizes are used by different manufacturers. The number of the taper varies with the size of the machine and spindle. The taper systems include Brown & Sharpe, American (Morse), and American Standard (NS) milling machine taper.

In addition, a Bridgeport® quick-change spindle nose is available for production purposes. By simply turning a collar, cutting tools may be changed quickly. Adapters are fitted to some spindle-nose tapers. These adapters accommodate Brown & Sharpe or Morse taper end mills. Similarly, some spindle tapers accommodate collets for holding straight-shank cutting tools.

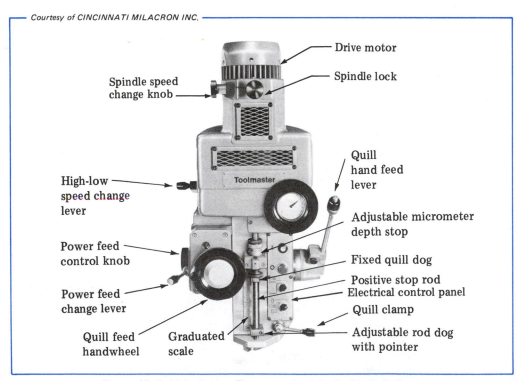

Courtesy of CINCINNATI MILACRON INC.

Figure 45–5 Main Design Elements of a Spindle (Swivel) Head

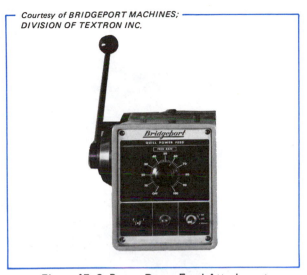

Figure 45–6 Power Down-Feed Attachment

VERTICAL MILLING MACHINE ATTACHMENTS

POWER DOWN-FEED

The *power down-feed attachment* (Figure 45–6) controls the quill feed rate. The up-down feeds are infinitely variable from 0.2 to 2.5 ipm or 5 to 64 mm/min. The feed direction is controlled by the up-off-down switch. A pilot light indicates when the control is in operation.

VERTICAL SHAPER HEAD

Some toolrooms and jobbing shops use a *vertical shaper head* as a milling machine attachment. The vertical tool motion permits shaping processes that normally require special machines or broaches.

The vertical shaper head shown in Figure 45–7 may be used to form shapes in blind holes where broaches cannot be used, cut gear teeth and racks, and produce sharp internal corners. Shaping may be done at a right angle or at any simple or compound angle to the table.

ATTACHMENTS FOR VERTICAL MILLING MACHINE SPINDLE HEADS

HIGH-SPEED ATTACHMENT

The *high-speed attachment* to the standard spindle head provides for maximum rigidity and

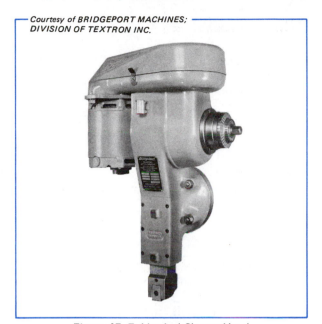

Figure 45–7 Vertical Shaper Head

rotational speeds up to 50% higher than the usual speed range. The Quill Master® high-speed head attachment (Figure 45–8) is designed for operations requiring the efficient use of small end mills and drills. The Quill Master is available with a 1/8″ (3mm) collet, a 3/16″ (5mm) spring collet, and a 3/16″ (5mm) solid end mill holder.

One manufacturer's design of the high-speed attachment is capable of operating at 9,000 RPM.

Figure 45–8 Vertical Miller High-Speed Head Attachment

Courtesy of BRIDGEPORT MACHINES; DIVISION OF TEXTRON INC.

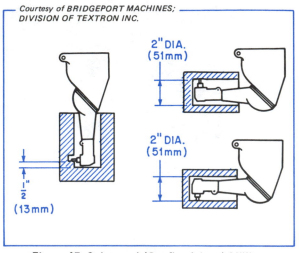

Figure 45-9 Internal (Confined Area) Milling, Using High-Speed and Right-Angle Attachments

Courtesy of DoALL COMPANY

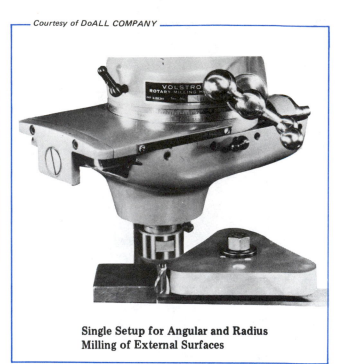

Single Setup for Angular and Radius Milling of External Surfaces

Figure 45-10 Application of Rotary Cross Slide Milling Head

Mounted on the quill, the attachment requires one-sixth of the spindle speed. Operating at this lower spindle speed eliminates vibration. At the same time, the high speed permits small size drills and end mills to be run at proper cutting speeds. Thus, feed rates are increased over the rates used for slower speeds.

RIGHT-ANGLE ATTACHMENT (CONFINED AREAS)

Special tooling and fixtures, particularly for milling out pockets and cavities, may be eliminated by using a *right-angle attachment*. Figure 45-9 shows a vertical and two horizontal setups for end milling inside a workpiece. The right-angle attachment shown in the figure works equally well on inside or outside cuts on regular parts or irregularly shaped castings. The unit has permanently lubricated bearings and a gear housing. The minimum working space is 2″ (51mm) diameter. Milling cuts may be taken to within 1/2″ (13mm) of a walled surface.

ROTARY CROSS SLIDE MILLING HEAD

The *rotary cross slide milling head* mounts on the vertical head quill. The attachment provides mechanical control of the cutting tool through straight, angular, and radial movements. For example, combined operations such as angles and radii, angles tangent to radii,

and one radius blending into another are possible (Figure 45-10).

The rotary cross slide milling head may be rotated in a 360° planetary motion. The degrees are subdivided into five-minute graduations. Operation of the head may be by manual or power feed.

OPTICAL MEASURING SYSTEM

Positive, accurate, and fast determination of the table position is possible by a *direct-reading optical measuring system*. Reading accuracies are within 0.0001″ with the inch-standard attachment. The operator reads a single line on a scale that is calibrated every 0.010″. Settings in 0.001″ are obtained through a drum dial that is calibrated in 0.0001″ increments. The metric-unit scales are calibrated every 0.10mm; the drum dial, in 0.002mm increments. The optical measuring system is used for longitudinal and cross travel measurements.

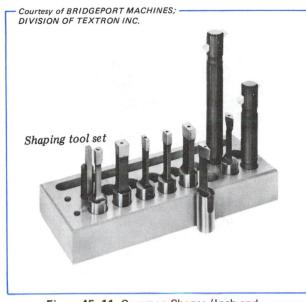

Shaping tool set

Figure 45–11 Common Shapes (Inch and Metric) of Shaping Tools

Figure 45–12 Precision Boring Chuck with Direct 0.0001″ Vernier Reading on Bore Diameter

VERTICAL MILLING MACHINE TOOLING ACCESSORIES

QUICK-CHANGE TOOLING SYSTEM

The *quick-change tooling system* is designed for fast tooling changes during production. The system is available with #30 or #40 NS taper holders. One manufacturer has the following cutter holders or adapters available for quick-change tooling (the collets are available in both inch and metric sizes):

- Quick-change spindle,
- Chucks for drills and end mills,
- Drill extension chucks,
- Non-pullout end mill collets,
- Floating reamer holders,
- Tenthset boring heads and boring bars,
- Morse taper adapters,
- Jacobs taper adapters,
- Shell end mill adapters,
- End mill adapters,
- Tap holders,
- Preset locking fixture for chucks,
- Spade blade holders,
- Spade blades,
- Fly-tool cutter holders.

SHAPING TOOL SET

Vertical shaping processes often require the use of a *shaping tool set*. The seven common shapes of cutting tools in such a set are illustrated in Figure 45–11.

BORING HEAD SETS

Precision boring is done on a vertical milling machine by using a *boring head set*. This set includes a boring head, a number of boring bars, other solid boring tools, and a container.

The shanks of boring heads are ground to #30, #40, or #50 NS tapers or to the straight style (R–8). The boring bit capacity ranges from 1/2″ for the R–8 and #30 NS shank sizes to 5/8″ for #40 NS and 3/4″ for #50 NS.

Boring heads have micrometer dials that permit settings to accuracies of 0.001″ and finer (0.0005″). Metric divisions read to 0.01mm. The tenthset boring head (Figure 45–12) has a direct 0.0001″ (0.002mm) vernier-reading adjustment of the bore diameter.

END MILL HOLDERS, COLLETS, AND ADAPTERS

While there are slight variations in construction details among manufacturers of end mill

Figure 45-13 Collet Design for Vertical Milling Machines

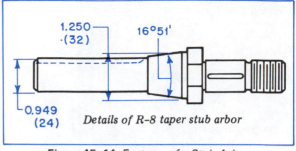

Details of R-8 taper stub arbor

Figure 45-14 Features of a Stub Arbor (R-8 Taper Type)

holders, collets, and adapters, three features are standardized: (1) the shank with its taper, (2) the threaded end to receive the drawbar, and (3) the hole to accommodate either straight or taper-shank end mills.

The shank tapers may be #30, #40, or #50; Brown & Sharpe tapers #7, #9, #10, #11, and #12; or the R-8 combination of a straight body and 16°51' included angle taper nose.

Collets for vertical milling machines are available for straight-shank tools in R-8 and N-2 types, Brown & Sharpe tapers of 0.500" (13mm) taper per foot, and Morse tapers of 0.599" (15.2mm) taper per foot. One type of collet is shown in Figure 45-13.

Adapters are used for taper-shank end mills and drills. These adapters are finished with an R-8 taper. Hole sizes are available for Morse or Brown & Sharpe shanks.

Shell end mill holders extend the tooling capability to the use of shell end mills for face and side milling in one operation. These holders are manufactured for a number of different pilot diameters. The shell end mill is secured on the pilot diameter and shouldered by means of an arbor screw and end mill wrench.

Stub arbors, as the name implies, are short-length spindle arbors. Figure 45-14 shows the features of an R-8 taper stub arbor. The arbor is provided with narrow-width spacers. Details are illustrated by the line drawing.

DIGITAL READOUT

Table movements along the horizontal (X) axis and the cross slide (Y) axis may be viewed rapidly on a *digital readout* accessory. The dual readout panel has a visual readout (separate display) for the X and Y axes (Figure 45-15). The readouts may be in the inch or metric system. Inch/metric conversion is optional.

Digital readout units are accurate within ±0.0005" (0.01mm). Readings are displayed in color for ease of reading and viewing from any working position at the machine. The axis display shows a + or – directional sign, a decimal point, and positions for six-digit readings.

Safe Practices Using Vertical Milling Machines, Attachments, and Accessories

- Check to see that all machine guards are in place before starting the machine.
- Secure assistance for mounting heavy machine attachments and/or difficult-to-handle workpieces. Keep the back straight and vertical. Lift with the force exerted by the legs.
- Exercise care when swiveling the spindle head for angular cuts. Unless the head is counterbalanced or adjusted through gearing, the clamping bolts should be retightened lightly so that the spindle head may be moved when a force is applied to it.

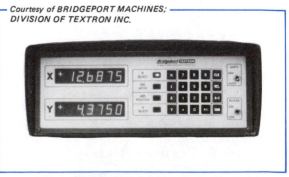

Figure 45-15 Digital Readout Panel with Display for X and Y axes

- Shut down the spindle to check the bore for nicks or burrs. Stone, if necessary.
- Check the mating taper of the spindle and the shank taper of an adapter, collet, or other tool-holding accessory. Remove any burrs.
- Select the cutting speed that is appropriate for the operation, size of cutter, shape of the workpiece, and results in the required dimensional accuracy. Set the spindle RPM to the closest cutting speed.

- Determine and set the power down-feed or up-feed that is correct for the setup and operation. The feed must permit machining to the required dimensional and surface finish accuracies.
- Shut down the spindle before any measurements are taken.
- Select and use comfortably fitting safety goggles and/or a protective face shield.
- Use protective footwear.

UNIT 45 REVIEW AND SELF-TEST

1. List four significant design changes in vertical milling machines that have increased its versatility.

2. Give three functions of the quill on a vertical milling machine.

3. State three functions of the spindle on a vertical milling machine.

4. a. Identify two different types of attachments for vertical milling machine spindle heads.
   b. Describe briefly the function of each attachment.

5. Name six different cutter holders or adapters.

6. Tell what the difference is between an NS shank taper holder and an R-8 shank holder.

7. a. Explain briefly the purpose served by a digital readout system.
   b. Cite two advantages of a digital readout system.

8. State two vertical milling machine safety precautions to observe before turning on the power for the spindle.

SECTION SIX
Cutting Tools and Basic Processes

The first unit in this section deals with the cutters used for basic vertical milling processes. End mills are covered, as well as the step-by-step procedures and safe practices for the milling of flat, angular, beveled, and round surfaces; stepped surfaces; and slots, dovetails, and keyways. The second unit in the section deals with the cutting tools, machines, attachments, and setups for general hole-forming and shaping processes on the vertical milling machine when ordinate, tabular, or baseline dimensioning is used.

UNIT 46

Cutting Tools, Speeds and Feeds, and Basic Processes

The cutters for the basic vertical milling processes are the standard cutters used for horizontal milling and other general machining processes. While high-speed steel cutters are common, cobalt high-speed steel, solid carbide, and carbide-tipped tools are widely used. The carbide tools provide for maximum production, efficient chip removal, and high abrasion resistance.

DESIGN FEATURES OF END MILLS

Like all other cutting tools, the main design features of end mills are designated by technical terms, as shown in Figure 46–1. Here, the tooth face is ground with a *positive radial rake angle*. End mills are also designed with a *negative radial rake angle*. End mills are formed with *straight* or *spiral teeth* and with *right-* or *left-hand helix angles*. The angles may be *standard* for general-purpose milling or *high* for heavy-duty production milling.

CONSIDERATIONS IN SELECTING END MILLS

Single-end end mills are general-purpose mills. *Double-end* end mills are more economical production milling cutters. Fractional sizes of end mills up to 1″ (25mm) are standard stock items. Large sizes are also commercially available, but in limited sizes. Figure 46–2 illustrates *stub*, *regular*, and *long-length* double-end end mills.

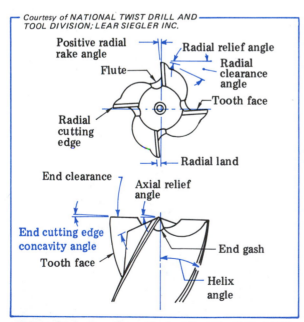

Courtesy of NATIONAL TWIST DRILL AND TOOL DIVISION; LEAR SIEGLER INC.

Positive radial rake angle
Radial relief angle
Flute
Radial clearance angle
Tooth face
Radial cutting edge
Radial land
End clearance
Axial relief angle
End cutting edge concavity angle
Tooth face
End gash
Helix angle

Figure 46–1 Four-Flute (Gashed End Type) End Mill Showing Design Features and End Mill Terminology

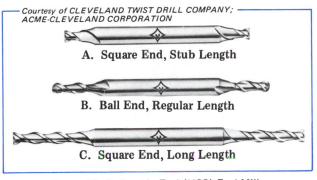

A. Square End, Stub Length

B. Ball End, Regular Length

C. Square End, Long Length

Figure 46-2 Double-End (HSS) End Mills

SUB-LENGTH, SMALL-DIAMETER END MILLS

Stub-length, small-diameter end mills combine positive rake angles on the flute face, primary relief angles, and short-length teeth. These end mills are designed to machine efficiently at the higher speeds that are required for their small diameters.

A feed of approximately 0.0005″ (0.01mm) per tooth is recommended for general-purpose end milling with small-diameter, high-speed steel end mills.

SQUARE- AND BALL-END MULTIPLE-FLUTE END MILLS

Square-end, *four-flute*, single- and double-end end mills are used for general-purpose end milling, slotting, stepping, slabbing, shallow pocketing, tracer milling, and die sinking processes. When ground to a *ball end*, these end mills are adapted to center cutting for processes such as die sinking, fillet milling, tracer milling, and other processes requiring a radius (ball) to be formed.

Long and *extra-long*, square- and ball-end, multiple-flute end mills permit deep cavity milling due to the long flute (cutting tooth) length.

HIGH-HELIX END MILLS

High-helix end mills provide the shear cutting action necessary for producing fine finish surfaces. These general-purpose end mills are used primarily for milling nonferrous and harder alloy workpieces.

When these end mills are used as *slabbing mills* for heavy stock removal, additional support for slabbing cuts is sometimes obtained by using a center in the center hole of the end mill.

A. Ball End, Center Cutting (Short, Multiple Flute)

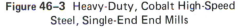

B. Multiple Flute, Roughing (Small-Radius Thread-Form Edges)

Figure 46-3 Heavy-Duty, Cobalt High-Speed Steel, Single-End End Mills

KEYWAY-CUTTING END MILLS

Keyway-cutting end mills have the same construction details as general-purpose end mills. One difference is that the outside diameter is ground to a tolerance of +0.0000″/−0.0015″ (+0.00mm/−0.04mm). The undersize diameter compensates for the tendency of the end mill to cut oversize. The −0.0015″ (−0.04mm) permits the end mill to cut a keyway to a precise, nominal size.

TAPERED END MILLS

Tapered end mills are constructed with cutting edges formed at standard included angles of 10°, 14°, and 15°. The ends are usually ball shaped. The taper end mill is adapted to machining angle surfaces at 5°, 7°, and 7 1/2° and for die sinking processes that require the sides of the relieved surfaces to be cut at an angle.

COBALT HIGH-SPEED STEEL END MILLS

Cobalt high-speed steel end mills are used for heavy-duty center cutting of high-tensile steels and high-temperature alloys. The extra-heavy-duty end mill in Figure 46-3A has a short flute length, which allows for greater rigidity and less deflection during the cutting of difficult-to-machine materials.

One of the newer designs for a multiple-flute *roughing end mill* is illustrated in Figure 46-3B. This cobalt high-speed steel end mill is designed for machining low-alloy, high-strength steels and die steels at accelerated feeds and speeds.

Cobalt high-speed steel roughing end mills have the following advantages:

- High material removal rate,
- Minimum horsepower requirement,
- Ability to take heavier cuts at high speeds and with less chatter and vibration,
- Less deflection for the same stock removal rate than conventional end mills,
- Excellent heat dissipation (because of the tooth-form design) under heavy machining operations,
- Fast chip-breaking characteristics,
- Simplified cutter resharpening without affecting the tooth profile.

TYPES OF END TEETH

Three widely used tooth forms for end mills include the two-flute center cutting, the four-flute gashed end, and the four-flute center cutting.

The center-cutting end mills may be used for plunge cutting as the feeding method in *slabbing*, *pocketing*, and *die sinking* operations. The gashed-end form is primarily used for conventional milling and slotting.

CUTTING FLUIDS FOR END MILLING

Mineral, sulphur-base, and soluble oils, as well as water-base synthetics, are common cutting fluids. Water-base cutting fluids have excellent cooling qualities. Oil-base cutting fluids produce a high-quality surface finish.

Cutting fluids are especially necessary to clear the large volume of chips produced by carbide cutters.

Note: The teeth on a carbide end mill will chip when milling cast iron or steel unless there is a *continuous supply* of cutting fluid *under pressure.*

In the case of cast iron, compressed air is used. Brass and plastics are also machined dry. The air supply disperses the chips from the cutting area and cools the end mill.

Note: Safety guards or protective chip shields must be placed around the table and cutting area to prevent chips and other foreign particles from flying.

INTERPRETING CUTTING SPEED AND FEED DATA FROM HANDBOOK TABLES

The nature of the technical information contained in reference tables of starting cutting speeds, RPM, and feeds for end mills for selected materials is illustrated by Table 46–1.

The feed rate (F) for milling operations is the product of the chip load per tooth (f) multiplied by the number of teeth (n) in the cutter and by the RPM. The feed rate is in inches per minute (ipm) or mm/min.

BASIC VERTICAL MILLING MACHINE PROCESSES

The greatest application of the vertical milling machine is in the machining of flat external and internal surfaces. These surfaces may be milled parallel, perpendicular, or at a single or compound angle to the face of the table. Grooving, slotting, and the cutting of T-slots, dovetails, and keyways are common vertical milling processes. The setups and procedures for each of these basic processes follows.

How to Mill a Flat Surface with an End Mill

End Milling a Horizontal Plane Surface

STEP 1 Select an end mill with a diameter and cutter direction appropriate to the job requirements.

Note: It may be necessary to take more than one cut if the width of the surface is greater than the cutter diameter. Large-width surfaces are usually milled with a shell end mill.

STEP 2 Secure the end mill in an end mill holder.

STEP 3 Set the spindle head at the 0° vertical position.

STEP 4 Set the spindle RPM and the horizontal table feed travel rate.

Table 46–1 Examples of Suggested Starting Speeds and Feeds
for High-Speed Steel End Mill Applications

Diameter of End Mills (in Inches)	Machine Steel, Hard Brass and Bronze, Electrolytic Copper, Mild Steel Forgings (20-30C)		Brass, Bronze, Alloyed Aluminum, Abrasive Plastics	
	High-Speed Steel End Mills, 2 or More Flutes		High-Speed Steel End Mills of High-Helix Type, 1 or 6 Flutes	
	Speed 60–80 sfpm	Feed	Speed 100–200 sfpm	Feed
	RPM	Chip Load per Tooth	RPM	Chip Load per Tooth
	∠1 ∠2	∠2 ∠1	∠1 ∠2	∠2 ∠1
1/16	3667–4888	.0002–.0005	6111–12222	.0002–.0005
3/32	2750–3259	.0002–.0005	4073–8146	.0002–.0005
. . .	. . .	. . .	. . .	. . .
1/2	458–611	.001–.003	764–1528	.0005–.003
9/16	412–543	.001–.004	678–1356	.0005–.004
. . .	. . .	. . .	. . .	. . .
2	115–153	.001–.004	191–382	.0005–.004
2 1/8	108–144	.001–.004	179–358	.0005–.004
. . .	. . .	. . .	. . .	. . .

∠1 Roughing cuts, ∠2 Finishing cuts

STEP 5 Position the workpiece and cutter so that a minimum number of cuts are taken.

STEP 6 Lock the spindle for maximum rigidity. Start the spindle and flow of cutting fluid.

STEP 7 Raise the knee until the cutter just grazes the workpiece. Set the micrometer collar on the knee at zero.

STEP 8 Clear the end mill and workpiece. Raise the table and take a trial cut.

STEP 9 Move the cutter away from the workpiece. Measure the overall height to check the dimension.

STEP 10 Make whatever adjustment is needed to cut to the required depth. Then, engage the power feed and take the first cut.

STEP 11 Take successive cuts until the surface is machined to the required dimension.

How to End Mill an Angular Surface

Positioning the Workpiece at an Angle

When the angle is layed out, the workpiece is positioned in a vise so that the layout line is horizontal with the table. A surface gage is usually used. Sine plates and tables are used for positioning a workpiece at a precise angle.

Setting the Spindle Head at an Angle

STEP 1 Position and lock the spindle head at the required angle (Figure 46–4).

STEP 2 Move the spindle head so that the center line of the cutting edges of the end mill falls near the center line of the angular surface to be milled.

Figure 46-4 Spindle Head Setup for End Milling a Required (45°) Angle

Figure 46-5 Elongated Slot Milled to Size with an End Mill

STEP 3 Turn the quill handwheel to feed the end mill to depth.

STEP 4 Engage the power feed. Take the first cut. Measure the angle. Make angle adjustments when needed.

STEP 5 Take successive cuts until the angular surface is machined to the required dimension.

How to End Mill a Slot or Keyway

STEP 1 Center the keyway end mill (or regular end mill for slotting) with the centerline location on the workpiece.

STEP 2 Feed the end mill to depth.

STEP 3 Take a trial cut. Make adjustments as needed.

STEP 4 Proceed to end mill the keyway or slot to the required dimensions. Figure 46-5 shows an elongated slot milled to size with an end mill.

How to Cut a T-Slot on a Vertical Miller

STEP 1 Select an end mill for first milling a groove to the width of the T-slot opening.

STEP 2 Replace the straight end mill with the T-slot cutter. Center the cutter with the slot. Figure 46-6 shows the cutter and workpiece setup for machining a T-slot on a vertical milling machine.

STEP 3 Move the cutter down to the required dimension for the depth of the T-slot.

STEP 4 Feed the cutter into the workpiece by hand to be sure the cutting action is correct. Engage the power feed.

Safe Practices for Basic Vertical Milling Processes

• Flow the chips rapidly and with a sufficient volume of cutting fluid to move them from the cutting area.

Figure 46-6 Setup for the Vertical Milling of a T-Slot

- Supply a continuous flow of cutting fluid to carbide end mills when cast iron or steel is milled.
- Feed dovetail cutters slowly into a workpiece. Any sharp contact between the fragile ends of the cutter teeth and workpiece may cause the teeth edges to fracture.
- Place safety guards on the table around the workpiece area if air is used as a coolant.
- Use a stiff brush and avoid handling the long, needle-like sharp chips that are produced by an end mill.
- Use safety goggles and/or a transparent shield as protection against flying chips and particles.
- Stop the spindle before cleaning a workpiece. Remove burrs before measuring.

UNIT 46 REVIEW AND SELF-TEST

1. Name six design features of standard end mills.
2. a. Identify three design features of stub-length, small-diameter end mills that permit them to cut efficiently at the required higher speeds.
   b. Give the feed rate for general-purpose end milling with stub-length, small-diameter end mills.
3. Give three advantages of high-helix end mills over standard-helix end mills.
4. Explain why a square-end, multiple-flute end mill is a higher production milling cutter than a two-flute end mill.
5. Tell why a keyway-cutting end mill is preferred over a standard end mill for machining a keyway to a precise, nominal dimension.
6. Give two reasons for using a cobalt high-speed steel end mill instead of a conventional high-speed steel end mill for production milling a high-tensile material.
7. Differentiate between a center-cutting and a gashed-end end mill.
8. Describe briefly how a collet chuck holder operates to hold an end mill solidly in the spindle.
9. Use a table to establish (a) the recommended starting spindle speed range for taking roughing and finish cuts and (b) the corresponding feed range for end milling a step in a brass/bronze block. A 1/4" (6mm) diameter high-speed end mill is used.
10. Give two personal safety precautions to take when performing end milling operations.

Hole Forming, Boring and Shaping

Design features of the vertical milling machine permit positioning a workpiece precisely by the longitudinal and cross slide movements of the table. These movements fall along the X axis and the Y axis. The vertical movement of the knee makes it possible to set and machine a workpiece according to dimensions on the Z axis. The graduated micrometer collars on the cross slide, table, and knee provide a quick, accurate, precision method of locating cutting tools and a workpiece. The micrometer collars permit direct measurements to accuracies within 0.001″ (0.02mm). Vernier, bar, and indicator attachments extend the accuracies to 0.0001″ (0.002mm).

Hole and other surface locations and dimensions for milling processes are often specified by ordinate, tabular, and baseline dimensioning. Essential layout and dimensional information is given in relation to the X, Y, and Z axes (Figure 47–1).

WORK- AND TOOL-HOLDING DEVICES FOR BASIC HOLE-FORMING PROCESSES

In drilling, reaming, counterboring, countersinking, and tapping processes, the same construction and sizes of drills, machine reamers, counterbores, countersinks, and machine taps that are applied to drilling machines, horizontal milling machines, lathes, and screw machines are used on a vertical milling machine. These fixed-diameter, multiple-edge cutting tools may be held in a chuck, adapter, or collet. Taps are turned for threading and removing from a workpiece by a tap adapter or a reversing tapping attachment.

FACTORS AFFECTING HOLE-FORMING PROCESSES

In vertical milling, the material of which a hole-forming cutting tool is constructed, its cutting speed for a particular process and workpiece, feed rates, and cutting action are all influenced by the same factors that apply to drill press, horizontal miller, and lathe work. However, the following additional factors must be considered:

- The vertical head may be positioned at a single or compound angle. Holes may be formed with the spindle axis (and the axis of the cutting tool) set to cut at a required angle.
- The table may be moved longitudinally or transversely to any precise dimension. Hole locations to tolerances of 0.0001″ (0.002mm) are possible by using measuring rods and a tenth dial indicator.

THE OFFSET BORING HEAD AND CUTTING TOOLS

Offset boring heads are available in a number of different models and sizes. Adjustments for machining diameters in increments of 0.001″ (0.02mm) are made rapidly by moving the cutting tool laterally in the holder. The amount of movement is read directly on a micrometer dial on the boring head. Microboring heads permit diametral adjustments within 0.0001″ (0.002mm).

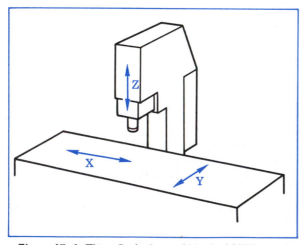

Figure 47–1 Three Basic Axes of Vertical Milling and Numerically Controlled Machines and Machining Centers

FEATURES OF THE OFFSET BORING HEAD

The major construction features of an offset boring head are illustrated in Figure 47–2. The micrometer dial is graduated into 50 divisions that permit adjustments to 0.0005″ (0.01mm). There is a positive lock when the tool slide is set at a required size.

The boring head set includes three *boring bars*. Two of these bars permit the use of standard-size, round boring tools. The third bar holds the different sizes of solid-shank boring bars in a set. The two *reducer bushings* accommodate two different sizes of boring tools. Like the boring bars, the reducer bushings are held securely in a particular cutting-tool location in the tool slide by a setscrew.

SPECIFICATIONS OF SELECTED BORING HEAD FEATURES

The boring head in Figure 47–2 is furnished with shanks that fit R–8 and #30, #40, and #50 NS spindles. The diameters that may be bored range from 1/4″ to 6″ (6mm to 150mm) for the R–8 size. The largest #50 NS shank boring head accommodates hole sizes from 5/16″ to 11 1/2″ (8mm to 292mm). The total travel of the tool slide varies for each head. The N–8 size has a travel of 31/32″ (24mm). The #50 NS shank boring head has a travel of 1 11/16″ (42.8mm).

The R–8 and the #30 NS shank boring heads hold 1/2″ (12.5mm) diameter boring bars. The #40 NS and #50 NS heads accommodate 5/8″ (15.8mm) and 3/4″ (19mm) diameter bars, respectively.

SHAPING ON A VERTICAL MILLING MACHINE

The vertical shaper head attachment is used to cut internal keyways, slots, and splines. The sides of openings and other intricate forms may be shaped on jigs, fixtures, punches, dies, and other workpieces that are to be precisely machined. Gear teeth and racks may be cut with the shaper head. Some heads are adapted to blind-hole machining processes. The shaper head often replaces a second machine—for example, a broaching machine for broaching.

The cutting tools are formed for side and end cutting, with corresponding rake angles and clearance. The tools are mounted in the tool head. A cutting tool is positioned by the table and cross slide handwheels.

Vertical positioning of a workpiece is controlled by elevating or lowering the knee. The length of stroke is adjusted directly on the shaper head. The attachment may be set for shaping at a 0° vertical angle or at any other single or compound angle. The settings are made directly. The angle is read on the graduated shaper head and/or the ram, depending on the job specifications.

Surfaces that require shaping in a circular direction are positioned, mounted, and secured on a rotary table or by using a dividing head. The accessories permit workpieces to be indexed or turned in relation to other vertical milling processes.

The shaper head (in general) has a stroke capacity of 0″ to 4″ (0mm to 102mm). The length of stroke can be dialed in increments of 1/8″ (3mm). The head is available with a slow speed and a standard speed range (strokes per minute). The slow speed range is 35, 50, 75, 100, 145, and 210 strokes per minute. The

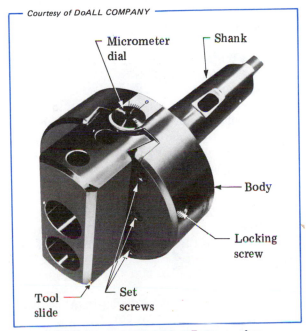

Courtesy of DoALL COMPANY

Micrometer dial — Shank — Body — Locking screw — Tool slide — Set screws

Figure 47–2 Construction Features of an Offset Boring Head

standard speed range is 70, 100, 145, 205, 295, and 420 strokes per minute.

How to Machine Holes by Using a Conventional Vertical Milling Machine Head

STEP 1 Select an appropriate cutter holder, adapter, collet, or tapping unit. Secure the tool-holding device in the vertical spindle.

STEP 2 Mount and secure the cutting tool.

STEP 3 Determine the correct cutting fluid, if needed. Set the nozzle to direct the quantity and flow of cutting fluid.

STEP 4 Position the cutting tool at the center of the hole. Lock the knee, cross slide, and table. Set each of the micrometer collars at zero.

Note: Ordinate, tabular, and baseline dimensioned drawings give each dimension from a zero position on the X, Y, and/or Z axes. The operator usually sets the table, cross slide, and knee micrometer collars at zero. All table movements to position the cutting tool relate to the zero settings.

STEP 5 Proceed to carry on the hole-forming processes by following the steps that are used on a drilling machine, lathe, or screw machine.

How to Bore Holes with An Offset Boring Head

Boring a Predrilled or Formed Hole Vertically (0°)

STEP 1 Align the vertical spindle head at 0°.

STEP 2 Select an offset boring head to accommodate the size and type of tool bit needed for the job. Check for and remove any burrs. Mount and secure the boring head in the spindle.

STEP 3 Mount the boring tool so that the cutting face is aligned with the center line of the head. Tighten the cutter setscrew.

STEP 4 Align the spindle (and boring head) axis with the zero coordinates on the X and Y axes. Set the cross slide (saddle) and table feed micrometer collars at zero.

STEP 5 Move the table horizontally and transversely to the required center line dimension.

STEP 6 Adjust the boring tool for a trial cut. Take the cut for about 1/16″ (1mm to 2mm) deep.

Note: The cut should clean up to more than half the diameter to permit taking an accurate diametral measurement.

STEP 7 Set the micrometer dial on the boring head at zero. Turn the micrometer dial adjusting screw on the boring head to feed the cutter for a roughing cut. Lock the tool slide on the boring head.

STEP 8 Bring the cutter into contact with the workpiece. Continue feeding by engaging the power quill feed.

Note: The proper cutting fluid and volume must be used to flow the chips out and away from the area being bored.

STEP 9 Return the boring tool to the starting position. Stop the machine. Remeasure the bored diameter.

STEP 10 Set the boring tool for subsequent roughing and/or finish cuts.

Note: The feed rate is decreased to obtain a higher-quality surface finish. Some operators use the down-feed and, at the end of the cut, change to an up-feed without changing the finish cut position of the boring tool.

Note: The cutting edge must be moved clear of the workpiece if it is brought up and out of the bored hole on the final cut.

Drilling and Boring a Hole

STEP 1 Select a drill that is smaller than the diameter of the hole to be bored.

STEP 2 Mount the drill in a tool-holding device. Secure it to the spindle.

Figure 47–3 Use of Boring and Facing Head to Machine a Stepped Circular Area

STEP 3   Set the spindle RPM and feed rate for drilling the first hole. Position the drill at the center of the hole. Drill.

STEP 4   Remove the drill and adapter. Replace with an offset boring head and boring tool.

STEP 5   Proceed to bore the hole as for conventional boring.

Boring a Hole at a Simple or Compound Angle

STEP 1   Position and lock the vertical milling machine head at the required angle in the plane that is parallel to the column face. Read the angle setting on the graduated quadrant of the head.

STEP 2   Proceed to bore the hole to size.

Boring a Radius or Stepped Circular Area (Figure 47–3)

STEP 1   Insert a centering rod in a boring head bushing or a chuck. Move the table longitudinally and transversely until the spindle axis is aligned at the layed out center line at the end (Y datum) of the workpiece.

STEP 2   Move the table longitudinally on the center line. The distance moved along the X axis should center the boring head at the center line of the radius to be bored.

STEP 3   Insert a tool bit, with correctly ground rake and clearance angles, in a boring bar. Mount the tool bit so that the cutting edge is aligned with the center axis of the boring head tool slide block.

STEP 4   Move the boring cutter until it grazes the end of the workpiece. Set the micrometer adjusting screw at zero.

STEP 5   Feed the tool outward for a trial cut. Take the cut for approximately 1/16″ (1mm to 2mm) deep.

STEP 6   Stop the spindle. Measure the diameter.

STEP 7   Increase the depth of cut, if possible. Use a coarse feed for a roughing cut.

> **Note:** The depth of cut (amount the boring tool is moved on the tool slide) is established by the micrometer dial reading.

STEP 8   Reset the cutting tool for subsequent roughing and finish cuts.

Boring a Counterbored Area or a Flat Shouldered Surface

STEP 1 Mount a boring and facing head on the quill so that the boring tool is driven by the spindle.

STEP 2 Select a boring tool that is ground to produce the specified square, round, or angular corner for the shoulder.

STEP 3 Take a series of cuts to bore the outside diameter to depth and to remove the excess material within the area to form the flat face.

Note: The cutting tool is fed downward to depth by the quill feed handwheel. The flat shouldered surface is cut by turning the boring and facing head handwheel to feed the boring tool inward.

Safe Practices for Using Boring and Shaper Tools and Attachments

- Start with the recommended starting cutting speed and feed. Adjust the speed and feed rates, as the cutting action may require, after all factors are considered.

- Check the workpiece for effects on surface texture and dimensional accuracy caused by any eccentric forces of the boring head. Reduce the spindle RPM if vibration or chatter occurs.

- Set the depth of cut for a boring tool or shaper tool so that the cutting forces will not deflect the cutter and cause an incorrect work surface. The feed may also need to be reduced.

- Rotate the boring tool and head slowly and carefully when the cutting edge is set to take the first cut in forming a radius.

- Move the cutting tool away from the finish bored surface before it is withdrawn from a workpiece.

- Place a wooden pad over the table surface to protect it from possible damage when removing or replacing a vertical spindle head with a shaper head. Two persons should be involved in removing any cumbersome and heavy attachment, accessory, or workpiece.

- Set the cutting stroke length to permit the shaping tool to clear the workpiece at the end of the cut. Adequate clearance must be provided in the starting position to take measurements safely.

UNIT 47 REVIEW AND SELF-TEST

1. a. Identify three basic hole-forming processes that are common to vertical milling machine work.
   b. Indicate the cutting tool that is used for each of the three processes.
   c. Name the tool-holding device for each cutting tool.

2. Explain how the T-slot grooves on a vertical milling machine table may be used to position a rectangular workpiece for work processes that are parallel to the table axis.

3. State three advantages of using a vertical milling machine instead of using a drilling machine for drilling, reaming, counterboring, countersinking, and tapping.

4. a. Name three design features of an offset boring head.
   b. Explain the function served by each feature.

5. a. Secure the specifications of an offset boring head.
   b. Furnish technical information about the following design features: (1) off-set boring head manufacturer, (2) shank style and size, (3) draw-in bar thread specifications, (4) boring diameter range, (5) boring bar set lengths, (6) diameter of boring bar, and (7) total tool slide travel.

6. State two practices to observe to ensure maximum rigidity in setting up a boring head and boring tool.

7. Give (a) two possible causes of deflection in a boring tool and (b) the corrective steps that the operator may take.

8. State two unique applications of the shaper head that save second machine setups or special tooling.

9. Tell how ordinate dimensioned drawings are used by the vertical milling machine operator to position the table for drilling and reaming.

10. State two tool or machine safety factors to consider to ensure dimensional accuracy when shaping is performed.

SECTION ONE

Band Machine Technology and Basic Setups

Vertical band machines are used for intricate and precise sawing and filing processes and in other processes requiring fast, economical removal of material. This section deals with:

- Principles of band machining and types of machines;
- Functions of major band machine components;
- Factors that affect cutting speeds and feeds;
- Machine and saw band setups for preparing the saw band, setting up the band machine, and regulating the speeds and feeds.

UNIT 48

Band Machine Characteristics, Components, and Preparation

TYPES OF BAND MACHINES AND CUTS

Band machines are used primarily for sawing functions such as cutting off, shaping, and slotting. Figure 48–1 illustrates how, by combining internal and external straight and corner radius cuts in two planes, it is possible to produce a part that otherwise would be complicated to machine. In this illustration, the term *three-dimensional cutting* is used. This term means that sawing cuts are made in two or more planes to alter the shape of a part.

Slicing refers to a cutoff operation that is usually performed with a knife-edge blade. *Slabbing* is another cutoff operation. It refers to cutting a thin section or several thin pieces from heavy workpieces.

Band machine shaping refers to cuts that alter the shape of a part by removing excess portions.

SIZES AND TYPES OF BAND MACHINES

General-purpose band machines (Figure 48–2) usually have nonpower-feed work tables. The table can be tilted for angle cutting. Angle cuts up to 10° may be taken when the table is tilted 10° counterclockwise. On large models, the cutting angle is limited to 5°. The table is designed for settings up to 45° clockwise. The workpiece may be fed by hand or power. A *hydraulic tracing accessory* is also available. The cutting path is guided by a stylus as it follows a pattern or template.

Heavy-duty band machines have power-feed work tables. The table is hydraulically fed. The path of the cut is produced according to a steering mechanism setup. The operator guides and controls the movement by turning a handwheel.

Courtesy of DoALL COMPANY

Figure 48–1 Three-Dimensional Band Sawing to Produce a Complicated Machined Part

Heavy-duty production band machines are developed to accommodate special needs.

High-tool velocity band machines (also called high-speed band machines) are adapted to the cutting and trimming of nonmetallic products such as nonferrous metals, plastic laminates, paper, wood, and other fibrous materials. Band speeds range as high as 15,000 feet per minute (fpm) to cut these comparatively soft materials. High-speed band machines are also adapted to continuous friction sawing applications. These machines are designed for vibration-free operation at high tool speeds.

Manufacturers' Recommendations. Table 48–1 is a partial table of band machine tool and cutting tool manufacturer's recommendations for conventional sawing. Data is provided for selected band machining high-speed steel and water-hardening carbon tool steel parts. The working data relates to information the machine operator must know for each of the five basic processes. This information is provided for parts ranging in thickness from 0″ to 12″ (0mm to 304mm).

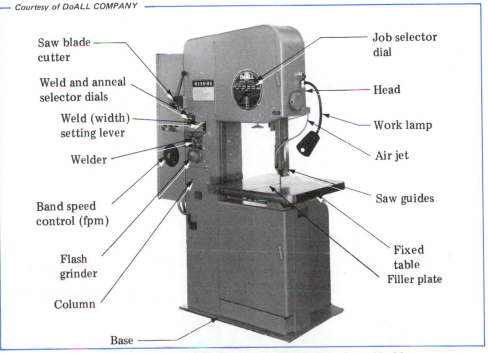

Courtesy of DoALL COMPANY

Saw blade cutter

Weld and anneal selector dials

Weld (width) setting lever

Welder

Band speed control (fpm)

Flash grinder

Column

Job selector dial

Head

Work lamp

Air jet

Saw guides

Fixed table

Filler plate

Base

Figure 48–2 Basic Design Features of the Fixed-Table Band Machine

Table 48-1 Partial Table of Manufacturers' Recommendations for Band Machining
High-Speed Steel and Carbon Tool Steel Parts

| Material | Work Thickness | | Conventional Sawing | | | | | | | | Coolant/ Lubricant# |
| | | | High-Carbon Saw Bands | | | | High-Speed Steel Saw Bands | | | | |
	(in)	(mm)	Tooth Form	Pitch	Band Speed (fpm)	Band Speed (m/min)	Tooth Form	Pitch	Band Speed (fpm)	Band Speed (m/min)	
High-speed steel											
	0-1/4	0-6.4	P	18	140	43					240
	1/4-1/2	6.4-12.7	P	14	110	34	P	10	160	49	240
Tungsten base	1/2-1	12.7-25.4	P	10	90	27	P	8	135	41	240
types T_1, T_2	1-3	25.4-76.2	P	8	70	21	P	6	110	34	240
	3-6	76.2-152.4	C	3	55	17	P	4	85	26	240
	6-12	152.4-304.8	C	3	50	15	C	3	60	18	240
	...				...				...		
Carbon tool steel											
Water	0-1/4	0-6.4	P	18	175	53					240
hardening	1/4-1/2	6.4-12.7	P	14	150	46	P	10	270	82	240 HD-600
W-1 special	1/2-1	12.7-25.4	P	10	125	38	P	8	225	69	240 HD-600
	...				...				...		

Note: mm values are rounded to one decimal place; m/min values, to the nearest whole number
Tooth Form   P = Precision   B = Buttress   C = Claw tooth

FUNCTIONS OF MAJOR MACHINE COMPONENTS

BASIC DRIVE SYSTEMS

Fixed-speed machines have a belted or a direct drive. Intermediate speeds may be set in steps of 1,000 feet per minute (fpm) between low and high range. Variable-speed machines may be set for intermediate band speeds within either a low-speed or a high-speed range.

The controls for setting the band speed are shown in Figure 48-3. The band speed is set by first positioning the *gearshift control lever* to the range of speeds appropriate for the job. The *variable-speed control handwheel* is then turned to increase or decrease the speed within the range. The band fpm (m/min) to which the controls are set is read directly on the *tachometer*.

WORK TABLES

The work table is mounted on and secured to the base by a *trunnion*. A protractor scale and a pointer are attached to the trunnion cradle for angle settings of the table (Figure 48-4).

Power-table band machines are designed with forward and reverse table workstops. The work table rests on hardened and ground steel *slide rods*.

Table feed controls (Figure 48-5) are located in the column. The controls include a *vernier control knob* (feed-pressure) and a *positioning lever*. The control knob regulates feeds from 0 to 12 fpm (0 to 3.6 m/min) on lighter machines and 0 to 8 fpm (0 to 2.4 m/min) on heavy-duty models. Table movement is controlled by turning a directional handle to the power-feed or stop position. The forward position is used for rapid forward traverse; reverse is for rapid reverse traverse away from the work.

CARRIER WHEELS AND BAND ASSEMBLY

The saw band, as the cutting tool, forms an endless band that rides on the *upper* and *lower carrier wheels*. The band is driven by the lower carrier wheel. The upper wheel is free wheeling.

The wheels are flanged and tapered and have a replaceable flanged back. The tapered face permits *tracking*. When correctly adjusted, the

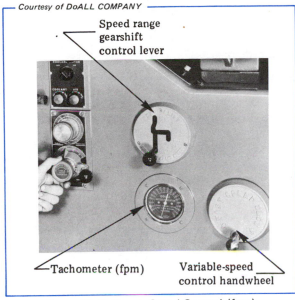

Speed range gearshift control lever

Tachometer (fpm)

Variable-speed control handwheel

Figure 48–3 Range, Speed Control (fpm), and Speed Indicator

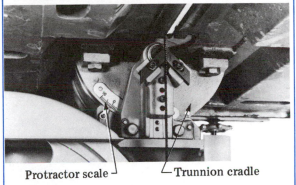

Protractor scale

Trunnion cradle

Figure 48–4 Trunnion Cradle and Protractor Scale Used for Angle Settings of the Table

Figure 48–5 Table Feed Controls

saw band rides up the taper, onto the rim, and against the flange at the back of the wheels.

Filing, polishing, and special bands work only on rubber-tired wheels. These wheels are standard equipment on fixed-table and small sizes of band machines. They are available for power-table machines. The rubber tire has a convex shape and is carried toward the center. As the carrier wheels revolve, the band moves toward and correctly rides on the top of the crown.

Heavy-duty production machines require steel-rimmed, flat-steel, flanged wheels. These wheels permit the saw band to be tightened for more positive drive and nonslip gripping.

Tracking and Saw Band Adjustments. The upper carrier band wheel tilts in or out to position and track a saw band. The tilt angle is adjusted by turning the tilt adjusting screw until the belt tracks. The upper carrier band wheel may also be moved up or down to change the band tension. A removable hand crank is turned clockwise to move the carrier band wheel upward to tighten the tension of the band.

Accurate band machining partly depends on band tension. The amount of and changes in band tension are given on the *band tension*

indicator (Figure 48–6). The indicator face shows three sets of values.

SAW GUIDE POST AND SAW BAND GUIDES

The *saw guide post* serves two prime functions. It supports the upper saw band guide. The post is adjustable up or down to permit machining workpieces of different sizes. Some posts have a calibrated scale to indicate the distance between the table and the section of the saw guide. The post is positioned close to the width of the workpiece in order to provide maximum support for the saw band.

As its name implies, the *saw band guide* functions to provide a bearing for a saw band. During sawing, the saw band is subjected to the force applied by feeding the work into the band. A second

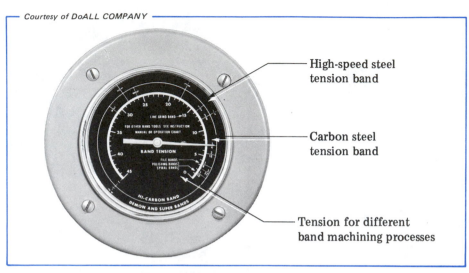

High-speed steel
tension band

Carbon steel
tension band

Tension for different
band machining processes

Figure 48–6 Band Tension Indicator

force is produced by the resistance of the work-piece to the cutting action. This force tends to bend (laterally deflect) the saw band sideways. The saw band guides help to overcome these two forces.

The guide must be able to support the back edge of the saw band without causing damage. Therefore, the guide is designed with a *backup bearing*. This bearing takes up the back feed thrust and permits the band to move at high speed. The lateral deflection is controlled by

Figure 48–7 Adjusting Roller-Type Saw Band Guide to Accommodate Saw Band Thickness

saw band guide inserts. They are brought in close to the saw band, but with enough space left to permit the saw band to travel freely during a cut. The two common types of saw band guides are *roller* and *insert*.

Roller-Type (High-Speed) Saw Band Guides. Roller-type saw band guides (Figure 48–7) are preferred for continuous high-speed sawing. Rollers are available for use with different widths of saw bands from 1/4″ to 1″ (6.35mm to 25.4mm) and for speed ranges from 6,000 to 10,000 fpm (1,820 to 3,048 m/min).

Insert-Type Saw Band Guides. The *light-duty* saw guide insert is a precision sawing guide for bands from 1/16″ to 1/2″ (1.5mm to 12.7mm) wide. It operates at comparatively low band speeds not exceeding 2,000 fpm (610 m/min).

The *heavy-duty* carbide-faced saw guide insert (Figure 48–8) is used with bands from 5/8″ to 1″ (15.8mm to 25.4mm) wide for heavy-duty continuous sawing.

The *high-speed* saw guide insert is used for band sizes from 1/16″ to 1/2″ (1.5mm to 12.7mm) wide and for band speeds in the 2,000 to 6,000 fpm (610 to 1,830 m/min) range. Guides are available for all different band sizes. The guides on the high-speed type are constructed with a large antifriction bearing and

Figure 48-8 Heavy-Duty Model of Insert-Type Saw Band Guides

a hardened, wear-resistant steel, thrust roller cap.

COOLANT SYSTEMS

Heat generated in band machining may be dissipated by either reducing the band speed and cutting feed or using a cutting fluid. In some applications where cutting oils are not required, the chips may be removed by a jet air stream.

Where a mist is preferred, the coolant and air are mixed in the same manifold. The controlled mist is forced by air pressure onto the saw band teeth and workpiece. The nature and amount of coolant flow is regulated by turning the knob on the spray manifold.

SAW BAND WELDING

Four separate components are needed to weld a saw blade. They include a *cutoff shear*, *butt welder*, *annealing unit*, and *grinder*.

CUTOFF SHEARING

The manufacturer's name plate on the machine provides information about the required length of band. This length is layed out and accurately marked. The saw blade is placed (teeth facing outward) in the throat of a cutoff shear. A continuous, firm force is applied to the handle to make a sharp, clean cut.

With the blade cut to length, it is necessary to grind off one or more teeth on each side of the cut. This grinding is required on blades of 4 to 10 pitch because about 1/4″ (6.35mm) of

Table 48-2 Position of Shear Cut and Required Teeth to Be Ground for Proper Spacing after Butt Welding (Examples: 4 and 10 pitch saw blades)

Saw Blade Pitch	Equivalent Teeth to Be Ground to Gullet Depth*	Position of Cut Relative to the Gullet of the Teeth
4	1	1/2 tooth ← → 1/2 tooth Area of blade consumed during butt welding 1 tooth ground off
10	2 1/2	1 1/4 teeth ← → 1 1/4 teeth 2 1/2 teeth ground off

*Tooth form is ground equally 1/8″ (3mm approximately) on both sheared blade ends.

Table 48–3 Control Setting Chart for Jaw Gap, Jaw Pressure, and Annealing Settings

Carbon Alloy Steel Bands

Width (mm)*	(inch)	Gage (mm)*	(inch)	Jaw Pressure	Jaw Gap	Annealing Setting
1.6	1/16	.6	.025	2	1	1
6.4	1/4	.6	.025	3	4	1
9.5	3/8	.6	.025	3	4	1
12.7	1/2	.6	.025	3	4	1
15.9	5/8	.8	.032	3	4	2
19.1	3/4	.8	.032	3	5	2
25.4	1	.9	.035	4	5	3
25.4	1	1.0	.050	5	6	3

High-Speed Steel Saw Bands

Width (mm)*	(inch)	Gage (mm)*	(inch)	Jaw Pressure	Jaw Gap	Annealing Setting
6.4	1/4	.6	.025	2 turns less than maximum	6	1
12.7	1/2	.6	.025	maximum	6	1

Friction Saw Bands

Width (mm)*	(inch)	Gage (mm)*	(inch)	Jaw Pressure	Jaw Gap	Annealing Setting
12.7	1/2	.8	.032	3	5	2
19.1	3/4	.9	.035	4	5	2
25.4	1	.9	.035	4	5	2

Spiral Band/Rod Welding

Diameter (mm)*	(inch)	Jaw Pressure	Jaw Gap	Annealing Setting
1.6	1/16	2	1	1
2.4-3.2	3/32-1/8	3	3	1
4.8	3/16	4	4	2
6.4	1/4	6	6	3
7.9	5/16	6	6	3

*mm values rounded to one decimal place

Figure 48–9 Positioning the Jaw Pressure Selector at a Required Control Chart Number

sitioned to hold the saw blade segments with the teeth facing inward toward the column.

During the butt welding process, particles of excess metal are blown in the area of the welding jaws and inserts. Therefore, the jaws, clamping assembly, and inserts must be cleaned after every weld.

Using a Control Setting Chart. A *control setting chart* (Table 48–3) is provided by band machine manufacturers. The chart gives the jaw gap and jaw pressure for the width, gage, and material on the saw band. The setting for annealing the welded area is also given.

A number on the *jaw pressure selector* corresponds to a number on the chart for the blade to be welded. Figure 48–9 shows the jaw pressure selector being turned to a particular setting.

The control knob on the opposite side of the butt welder is called the *jaw gap control*. The control knob is turned for the proper jaw gap according to the width, gage, and material in the saw blade.

Making the Weld. If an internal cut is to be made, the blade is moved through the workpiece. The blade is aligned so that the ends are clamped as they touch and are centered between the two jaws.

material is consumed in making the butt weld. Table 48–2 shows the position of the cut relative to the gullet and the number of teeth to grind off for blades of 4 and 10 pitch.

BUTT WELDING

The edges of the upper and lower welding jaw inserts are beveled. The narrow beveled edge is used for blades up to 1/2″ (12.7mm) wide and fine-pitch blades 3/4″ (19mm) and wider. The wide, beveled edge accommodates coarse-pitch blades of 3/4″ (19mm) or wider. The upper and lower jaw inserts, with matching bevels, are po-

Caution: Use safety goggles or a shield. Step to one side to avoid the welding flash.

The weld switch is pressed in to its stop. The weld is made automatically. The movable jaw of the butt welder advances the jaw and blade.

At the time welding takes place, there is a consistent flash. It is sharp and bright in color. A dull color or *sputtering* indicates an error in setting the welder controls.

The *reset lever* is raised when the weld is completed. This lever releases the spring tension. The saw band clamp handles are turned down to release the clamping pressure. The saw band is then removed and inspected at the weld.

Inspecting and Treating the Weld. The *flash* is the buildup material that is compressed during heating and flows on both sides of the band at the weld position. The color of the upset material should be blue-gray and of equal intensity across the flash. The spacing of the teeth must be uniform. The weld should be located in the center of the gullet. After grinding, the straightness of the band is checked with a straight edge. If the blade sections are misaligned, the band must be broken at the weld. The edges are ground square and the ends are rewelded. Common saw band welding problems, causes, and corrective action are given in Table 48–4.

ANNEALING THE WELD

During saw band welding, the metal around the joined area hardens and becomes brittle. The band must be annealed in the area. Annealing restores the joint to its original heat-treated hardness.

The butt-welded area is clamped in the butt welder jaws. The teeth are positioned to the rear. The welded section is in the gap midway between the two jaws. The reset lever is moved to the anneal position and then back toward the weld position. This movement positions the movable jaw about 1/16″ (1.5mm) toward the stationary jaw and allows for expansion of the band upon heating.

The *anneal selector switch* is set to the annealing heat number found in the control setting chart. The anneal switch button (Figure 48–10) is jogged intermittently until the saw band reaches a dull cherry red for both carbon and high-speed steel bands.

Note: Overheating, which causes the band to harden and become brittle, must be avoided.

The switch button is then released. As the welded area starts to cool, the switch button is again jogged and released. The cooling process slows down and the annealed area is thus produced.

Table 48–4 Common Saw Band Welding Problems, Probable Causes, and Corrective Action

Problem	Probable Cause	Corrective Action
Overlapped or crooked welds	—Misalignment	—Check and adjust jaw alignment
	—Dirty or worn jaw inserts	—Clean or replace inserts
	—Wrong jaw pressure setting	—Recheck control setting chart
Brittle welds; hard spots in weld; spots of lighter color	—Incorrect jaw gap setting	—Increase jaw gap setting by 1/2 notch
	—Low voltage	—Check incoming voltage
Incomplete or partially joined weld	—Low voltage	—Check incoming voltage
	—Burned or pitted weld switch contacts	—Replace contacts
	—Wrong electrical cutoff timing	—Check timing control
	—Wrong jaw pressure setting	—Reset according to control setting chart

Figure 48–10 Jogging the Anneal Switch Button for Gradual Temperature Buildup and Slow Cooling Rate

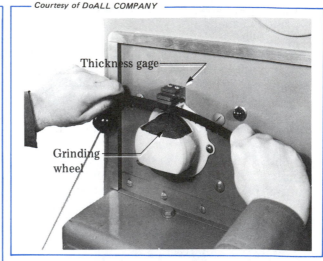

Figure 48–11 Band Grinder Unit

GRINDING THE WELD FLASH

A grinder unit is included on the band machine. The unit has a grinder wheel and a thickness gage (Figure 48–11). After annealing, the excess flash is removed. The welded area is ground to the blade thickness. The band is held with the teeth facing outward. Care must be taken to prevent grinding into the teeth. The band is moved continuously to prevent overheating or burning the ground area. If the saw band is burned, it will need to be rehardened.

Grinding starts at the tooth gullet. The band is inclined at a slight angle with the teeth higher than the back edge. The band is brought into contact with the bottom of the grinding wheel on its front edge. The band is moved forward with a light, uniform force. The positioning and grinding are repeated until the flash is removed.

The band is then flipped over and turned around. The flash is ground off the other side. The blade weld is checked for thickness as the ground area approaches the band thickness. The blade is checked with the thickness gage.

Safe Practices in Setting Up the Vertical Band Machine

- Close the guard on the saw guide post and upper and lower wheel door guards before applying tension or starting the machine band.
- Position the upper and lower saw guides close to the work to provide maximum support against cambering or deflection during machining.
- Adjust the cutting feed to prevent work hardening, vibration, and bending of the saw band.
- Maintain a feed rate that produces a chip with a tight curl.
- Set the band speed within the manufacturer's recommendations for the job at hand.
- Set the forward and reverse table workstops on power-table machines. These workstops prevent table movement beyond fixed limits.
- Point the teeth of a saw band down toward the machine table in the direction for cutting. The saw teeth dull quickly if reversed.
- Track the saw band to ride centrally on center-crowned wheels or against the rim of the upper carrier wheel.

- Keep the band tension constant as recommended. Read the tension indicator during machining. Adjust as needed.
- Check each flash area after welding and grinding. The butt weld must be solid across the width of the blade. The blade color must remain unchanged. A crack or incomplete weld or softening of the blade is dangerous and requires corrective action.
- Use protective gloves for handling saw stock or bands. Safety goggles or eye shields are required during machining.

UNIT 48 REVIEW AND SELF-TEST

1. Inspect a vertical band machine. Identify three safety features that are incorporated in the design of the machine.

2. a. State the purpose served by grinding off teeth at both ends of a blade where the blade is to be butt welded.

   b. Tell how many teeth are to be ground on a (1) 6-pitch blade, (2) 8-pitch blade, and (3) 10-pitch blade in preparation for butt welding.

3. a. State one function of a roller-type and another function of an insert-type saw band guide.

   b. Identify two construction features that distinguish the roller-type from the insert-type saw band guide.

4. Tell what purpose is served by annealing the weld joint area of a saw band.

5. The four parts listed below are to be band sawed and filed. The material and thickness of each part is also listed.

	Material in Part	Thickness of Part
Part A	High-speed steel T-2	20mm
Part B	High-speed steel M-2	1 1/2"
Part C	Carbon tool steel W-1	8mm
Part D	Carbon tool steel W-2	1 1/2"

   a. Prepare a chart with five vertical columns in addition to the three above. Three of the new columns relate to conventional sawing with HSS saw bands. Mark these three columns: Tooth Form, Pitch, and Band Speed. The two remaining new columns are for recording the Band Tool and Band Speed for filing the parts.

   b. Refer to a manufacturer's table of recommendations for band machining high-speed and carbon tool steel parts to complete the chart.

6. Describe briefly how a mist spray is produced and regulated.

7. State four precautions to take for grinding the flash on a butt-welded saw band.

Band Machining: Technology and Processes

This section applies the technology, basic principles, and setup procedures of vertical band machining to straight and contour sawing processes, filing, polishing, and grinding. Spiral band cutting, the use of diamond-edge saw bands, friction sawing, line grinding, and electroband machining are also described. Tables are included on common band machining problems; setups for cutting internal and external sections in one operation; and band velocity, pitch, and cutting rates according to work thickness.

UNIT 49

Band Machine Sawing, Filing, Polishing, and Grinding

EXTERNAL SAWING

Before sawing to a layout line, the machine operator must read the part print to determine the degree of accuracy of the sawed surface. If the specified quality can be met by sawing alone, the cut is taken to the layout line. Allowance is made for the saw kerf from the waste side (Figure 49–1). If the surface is to be further ma-chined, an allowance of about 1/64″ (1.5mm) is made for finish filing; 0.010″ to 0.015″ (0.25mm to 0.4mm), for grinding. Allowance for the width of the saw kerf and an additional width for machine finishing are taken on the waste side.

For cutting corners that have a radius, a hole is drilled to match the radius. If the hole size is large enough, the workpiece may be fed to cut to the first layout line. The workpiece is turned at the corner formed by the radius of the drilled hole.

A square corner may be cut to shape by either drilling a hole or making straight cuts. The excess stock at the corner radius is re-moved by *notching*. The same square corner may also be produced by cutting along one lay-out line almost to the second line. The workpiece is carefully removed and fed so that a cut is taken along the second layout line.

The width of blade to use when a hole is first drilled is determined by the size of the radius to be cut and the thickness of the work-piece. The widest possible blade should be used.

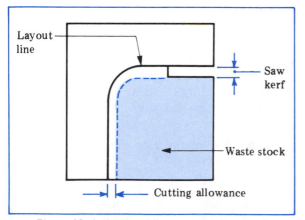

Figure 49–1 Cutting Allowance for Saw Kerf

How to Make External Cuts

STEP 1 Determine the correct type of saw band teeth, pitch, blade material, and saw width.

STEP 2 Mount, track, and apply the necessary tension to the saw band.

STEP 3 Lower the upper saw guide until it clears the workpiece by approximately 1/4" (6mm). Secure the guide at this position.

STEP 4 Check the job selector for the saw band speed and feed. Set the speed at the required fpm (m/min).

STEP 5 Determine the correct cutting fluid, if required. Position the air jet and the coolant nozzle.

Note: The air stream blows the chips away from the layout lines. The cutting fluid keeps the blade and workpiece cool and adds to cutting efficiency.

STEP 6 Move the workpiece up to the moving saw band.

STEP 7 Start the cut on the waste material side.

STEP 8 Feed the workpiece into the saw band.

Caution: Use a pusher block on small parts. Slow down the feed rate toward the end of the cut.

STEP 9 Stop the machine. Use a brush to remove chips. Burr the workpiece.

INTERNAL SAWING

Internal sawing refers to the removal of an internal section of a workpiece. The cut begins and ends on the inside. The steps in contour sawing inside a workpiece are shown in Figure 49–2. One or more starting holes are drilled tangent to the layout lines. The drill diameter must be large enough to permit the correct width blade to be used.

The saw blade is cut and threaded through a starting hole. The blade ends are butt welded and the flash area is ground to saw band thickness. The saw band, which is free to move through the workpiece, is tracked and then tensioned.

The saw cut is started parallel with the layout line. In cases of internal contour sawing (Figure 49–2), the stock is first notched out. Notching permits feeding the workpiece so that the cut follows the curved layout line.

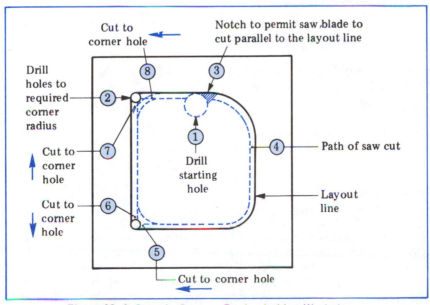

Figure 49–2 Steps in Contour Sawing Inside a Workpiece

Table 49–1 Selected Common Band Machine Sawing Problems, Probable Cause, and Corrective Action

Problem	Probable Cause	Corrective Action
Saw band vibrating in cut	—Velocity of saw band is too slow or too fast	—Adjust fpm (m/min) control according to job requirements
	—Feed rate is too light or too heavy	—Adjust rate of speed
Saw band breaking prematurely	—Velocity of saw band is too slow	—Adjust fpm (m/min) control according to job requirements
	—Feed rate is too heavy	—Adjust rate of speed
	—Saw band tension is too great	—Turn tension adjusting handwheel to correct tension
Teeth dulling prematurely	—Velocity of band is too fast	—Adjust fpm (m/min) control according to job requirements
	—Feed rate is too light	—Adjust rate of speed
	—Saw band pitch is too coarse	—Use finer-pitch saw blade
Saw band teeth chipping or breaking	—Velocity of band saw is too slow	—Adjust fpm (m/min) control according to job requirements
	—Feed rate is too heavy	—Adjust rate of speed
	—Saw band pitch is too coarse	—Use finer-pitch saw blade
Loading of gullets of saw band teeth	—Velocity of saw band is too fast	—Adjust fpm (m/min) control according to job requirements
	—Saw band pitch is too fine	—Use coarser-pitch saw blade
Welding of chips to saw band teeth	—Feed rate is too heavy	—Adjust rate of speed
	—Wrong type of cutting fluid and/or improper flow	—Change cutting fluid to meet job requirements
Wandering of saw band	—Feed rate is too heavy	—Adjust rate of speed
	—Saw band improperly tracking	—Adjust carrier wheel tracking nut

The corners are then notched to remove the excess material. A stock allowance is made for machine finishing the inside surface. When the machine is stopped, the tension is released and the saw band is removed. The saw band is cut at the point of the weld to completely remove the weld.

COMMON BAND MACHINE SAWING PROBLEMS

Factors which may cause premature dulling of the teeth, vibration in the cut, teeth chipping or loading, welding of chips, and other damage to the saw band or the workpiece are given in Table 49–1. Probable causes and corrective action the machine operator may take are included.

SLOTTING

Slotting is an adaptation of the band sawing process. It requires the use of a saw band having a kerf equal to the required width of the slot. Slots may be sawed to a tolerance of +0.002″ (+0.05mm). The advantage in slotting on a band machine over a milling machine is the shorter distance the table travels to complete the cut.

SLITTING

Slitting is another typical band sawing process. It is used widely to separate parts of castings or forgings. The separated parts are then further machined and held together with fasteners. Other parts such as bearings and bushings are first machined as a single piece and then slit. Fixtures

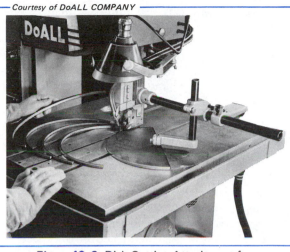

Figure 49-3 Disk-Cutting Attachment for Band Machining Radii

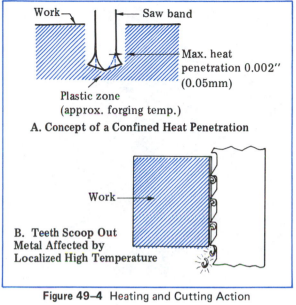

Figure 49-4 Heating and Cutting Action in Friction Sawing

are used in such applications to seat and locate the workpiece and hold it securely.

RADIUS AND OTHER CONTOUR CUTTING

Contours may be sawed by using any one of three common methods: (1) hand-feeding, (2) sawing with a disk-cutting attachment, and (3) cutting with a contour sawing accessory.

The *disk-cutting attachment* (Figure 49-3) simplifies single-radius contour sawing. This attachment has an adjustable center point. The center is positioned at the radius distance from the saw band. The workpiece center provides a bearing surface for the attachment center. The workpiece is fed at the radius distance past the saw band.

The *contour sawing accessory* combines the forward hydraulic table motion and the rotary motion produced by the turning of the control handwheel.

FRICTION SAWING

PRINCIPLES OF FRICTION SAWING

Friction sawing requires the instant generation of heat immediately ahead of the saw teeth. The heat produces a temperature that causes a breakdown of the crystal structure of the metal. The cutting edge of the saw band then removes the material affected by the friction heat. The

saw teeth, traveling at extreme velocities between 6,000 to 15,000 fpm (1,828 to 4,572 m/min), generate the heat. During friction sawing, sparks and extremely hot metal chips are produced.

Figure 49-4 illustrates the heating and cutting action that occur in friction sawing. Figure 49-4A shows the confined heating area. The maximum heat penetration into the sides of the cut is 0.002″ (0.05mm). The heat is further confined to the small area ahead of the kerf. The metal immediately ahead of the saw teeth is referred to as the *plastic zone*. In this zone, the approximate forging temperature of the workpiece is reached. The action of the saw blade in forming the chips is shown in Figure 49-4B.

APPLICATIONS OF FRICTION SAWING

Friction sawing is used to cut hard-to-machine steel alloys without annealing them. Other alloys that work-harden during conventional sawing may be cut efficiently by friction sawing. Friction sawing also replaces slower grinding and chemical machining processes on ferrous metals having a high hardness rating.

Friction sawing often replaces cutting methods that produce distortion in the workpiece. Friction sawing may be used on thin-walled

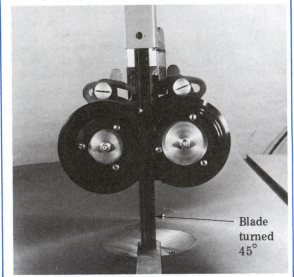

Blade turned 45°

Figure 49–5 High-Speed, Roller-Bearing Angle Saw Band Guides

tubes, thin parts, and thin-sectioned workpieces. This sawing process may be applied to straight, radius, and intricate contour shapes.

LIMITATIONS OF FRICTION SAWING

Friction sawing does produce a 1/32" to 1/16" (0.8mm to 1.6mm) thick, sharp burr on the underside of the work.

Some metals, such as aluminum, copper, and brass, most thermoplastic materials, steels containing tungsten, and most cast irons will not friction saw.

Work thickness is also a limiting factor. Friction sawing is practical for workpieces 1" (25.4mm) and smaller in thickness. Within this range, it is easy to control the high unit forces between the saw band and the work. Larger sizes may be accommodated if the workpiece may be rocked.

FRICTION SAWING BAND MACHINES
DESIGN OF SAFETY FEATURES

The high band speeds required for friction sawing require the saw band to be completely guarded to the point of the cut and at all times. Heavy-constructed telescoping saw band guards are designed and provided to cover the space between the upper carrier wheel and the work table.

Hydraulic brakes are included on both the upper and lower carrier wheels. The saw band guides, which provide adequate backing and side support for the band tool, are mounted on a heavy, precision-ground post. A slide block provides the bearing on which the post is mounted. A set of high-speed, roller-bearing saw guides is shown in Figure 49–5.

FRICTION SAW BAND CHARACTERISTICS
BLADE SELECTION

The features of the friction saw blade (Figure 49–6) enable it to overcome the extreme flexing action, stresses at high speeds, and heavy feed forces.

The standard widths for friction saw blades are 1/2", 3/4", and 1" (12.7mm, 19mm, and

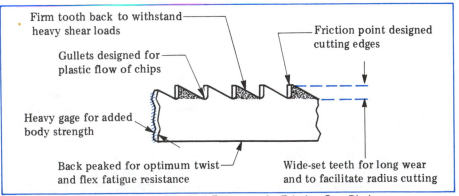

Firm tooth back to withstand heavy shear loads

Gullets designed for plastic flow of chips

Heavy gage for added body strength

Back peaked for optimum twist and flex fatigue resistance

Friction point designed cutting edges

Wide-set teeth for long wear and to facilitate radius cutting

Figure 49–6 Principal Features of a Friction Saw Blade

Table 49-2 Examples of Band Velocity, Pitch, and Cutting Rate for Friction Sawing According to Work Thickness for Selected Metals

Material (AISI–SAE Designation)	Work Thickness in Inches (mm)										
	1/16–1/4 (1.6–6.4)	1/4–1/2 (6.4–12.7)	1/2–1 (12.7–25.4)	1/16–1/4 (1.6–6.4)	1/4–1/2 (6.4–12.7)	1/2–1 (12.7–25.4)	1/16 (1.6)	1/4 (6.4)	1/2 (12.7)	3/4 (19.1)	1 (25.4)
	Saw Band Velocity			Saw Band Pitch			Approximate Linear Cutting Rates in/min (m/min)*				
Carbon steel (plain) 1010–1095	6,000	9,000	12,500	14	10	10	1,400 (35.6)	60 (1.5)	30 (0.8)	8 (0.2)	6 (0.2)
Free-machining steel 1112–1340	6,000	9,000	12,500	14	10	10	1,400 (35.6)	60 (1.5)	25 (0.6)	7 (0.2)	3 (0.1)

*Rounded to nearest tenth of a meter

25.4mm). A standard carbon alloy, precision-type blade is used for applications requiring a narrower saw band width than 1/2″ (12.7mm).

PITCH

Most friction sawing is done with the 10-pitch blade. A 14-pitch blade is used for thin work. The teeth are *raker set*. The pitch is affected by the work thickness and the amount of frictional heat to be generated in relation to the number of teeth that are removing the chips.

Table 49–2 provides examples of the kind of information the operator needs for friction sawing.

FRICTION SAWING SETUP PROCEDURES

Procedures for setting up the friction sawing band machine are similar to the procedures used in conventional sawing. There are two changes, however. First, no coolant is used. Second, roller guides are required unless only an occasional single part is to be sawed.

Extra safety precautions must be observed. The plastic guard must be positioned around the saw band and guides. The machine operator should wear a helmet with transparent visor. A pusher block should be used to feed small pieces. The friction-sawed chip is stubby, wrinkled, and has numerous cracks. The chips differ from the uniformly curled chips produced by conventional sawing.

HIGH-SPEED SAWING

High-Speed sawing refers to the speed of the saw band, usually within the range of 2,000 to 6,000 fpm (610 m/min to 1,830 m/min).

APPLICATIONS OF HIGH-SPEED SAWING

High-speed sawing is practical for such soft materials as plastics, wood, paper, and other fibrous products. It is also widely used on non-ferrous metals, such as aluminum, brass, bronze, and magnesium. Common forms that are cut include bar stock and plate and sheet stock, as well as extruded, cast, and tube forms.

Insert-type guides are used for short runs of high-speed sawing. These guides permit the greatest accuracy. However, production high-speed sawing requires the use of roller-type guides. Since high-speed sawing is used largely in production, special holding, positioning, and feeding fixtures are used.

SPIRAL-EDGE BAND SAWING

Spiral-edge band sawing requires the use of a round, continuous spiral-tooth band of small diameter. The spiral-edge saw band is unique. A continuous tooth is formed by a spiral cutting edge. The saw band is round as opposed to conventional flat-back blades. The spiral tooth design provides 360° of cutting. Intricate-pattern contours may be formed in thin-gage metals and

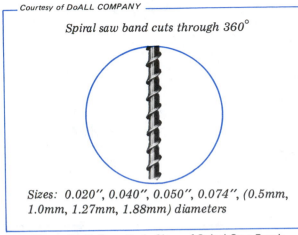

Spiral saw band cuts through 360°

Sizes: 0.020", 0.040", 0.050", 0.074", (0.5mm, 1.0mm, 1.27mm, 1.88mm) diameters

Figure 49–7 Four Basic Sizes of Spiral Saw Bands

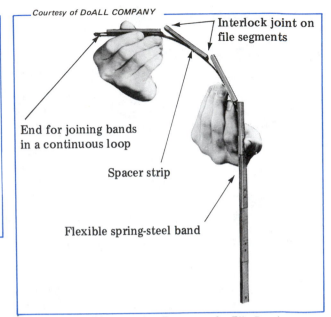

Interlock joint on file segments

End for joining bands in a continuous loop

Spacer strip

Flexible spring-steel band

Figure 49–8 Design Features of a File Band

other soft materials. Spiral-edge saw bands are practical for cutting small-radius contours that cannot be sawed by a conventional saw band because of the saw band width.

SIZES AND DESIGN FEATURES OF SPIRAL SAW BANDS

There are four basic sizes for spiral saw bands. The outside diameters are 0.020", 0.040", 0.050", and 0.074" (0.5mm, 1.0mm, 1.27mm, and 1.88mm). A close-up of the tooth form is illustrated in Figure 49–7. Spring-tempered spiral-edge saw bands are used for plastics, wood, and soft materials. Hard spiral-edge saw bands are adapted to cutting metals. The sharp, spiral cutting edges permit the cutting of true, precise contours to the minimum radius of the spiral band.

Spiral-edge bands are used only with center-crowned, rubber-tired carrier wheels. A special saw guide design is required.

DIAMOND-EDGE BAND SAWING

Diamond-edge saw bands, like other diamond-impregnated cutting tools, are designed to cut superhard materials. Straight, angular, and contour cutting is done on abrasive plastics, ceramic products, hard carbons, quartz, glass, granite, and other difficult-to-cut materials.

PREPARATION FOR DIAMOND-EDGE SAWING

Diamond-edge saw bands operate at maximum velocities between 2,000 and 3,000 fpm

(600 to 900 m/min). The saw bands are available in 1/4", 1/2", 3/4", and 1" (6.4mm, 12.7mm, 19.1mm, and 25.4mm) widths. Tapered, rubber-tired, flanged carrier wheels and roller-type saw band guides are required.

The diamond band is welded in a manner similar to a conventional saw blade.

BAND FILING

The vertical band machine is used extensively for band filing. *Band filing* refers to the filing of internal or external surfaces to a uniformly smooth, accurate surface. Band filing may be done on almost any kind of metal.

The cutting tool consists of a series of file segments. These segments are riveted on one end to a flexible spring-steel band (Figure 49–8). As the flexible band travels around the carrier wheels, the loose ends of the file segments lift away. When the flexible band returns to the vertical plane, the loose end of one file segment fits into the next segment. A special interlocking joint between file segments permits the formation of a continuous, flat filing band.

KINDS OF FILE BANDS

Cut, as related to file bands, refers to the short angle or bastard pattern; grade (such as coarse,

Table 49–3 Sample File Band Segment Cuts and Applications

Courtesy of DoALL COMPANY

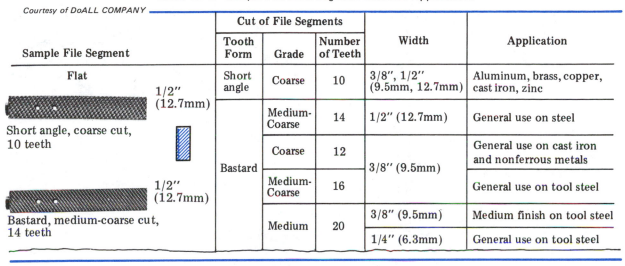

Sample File Segment		Cut of File Segments			Width	Application
		Tooth Form	Grade	Number of Teeth		
Flat Short angle, coarse cut, 10 teeth	1/2″ (12.7mm)	Short angle	Coarse	10	3/8″, 1/2″ (9.5mm, 12.7mm)	Aluminum, brass, copper, cast iron, zinc
		Bastard	Medium-Coarse	14	1/2″ (12.7mm)	General use on steel
Bastard, medium-coarse cut, 14 teeth	1/2″ (12.7mm)		Coarse	12	3/8″ (9.5mm)	General use on cast iron and nonferrous metals
			Medium-Coarse	16		General use on tool steel
			Medium	20	3/8″ (9.5mm)	Medium finish on tool steel
					1/4″ (6.3mm)	General use on tool steel

medium-coarse, or medium); and the number of teeth. The job selector is used to identify the type of file. For example, a short-angle file is used for aluminum, brass, cast iron, copper, and zinc products.

Table 49–3 identifies information which is available for flat file band segments. The two other common segments are the oval and half-round. The widths of these files are 1/4″, 3/8″, and 1/2″ (6.3mm, 9.5mm, and 12.7mm).

The file cuts illustrated in Table 49–3 are for 10, 12, 14, 16, and 20 teeth. The short-angle coarse cuts with 10 teeth are for band filing soft materials; the medium-coarse cuts, for general use on mild steels; and the medium cuts for filing tool steels.

BAND FILING SPEEDS AND FEEDING FORCE

Band speeds between 50 to 100 fpm (15.2 to 30.5 m/min) are used for all-purpose band filing applications. However, recommendations on the job selector and in manufacturers' tables on band filing should be followed.

BAND GRINDING AND POLISHING

Band grinding and/or *polishing* produces an excellent grained or polished surface in contrast to the coarser surface produced by band sawing or filing. Abrasive polishing bands are used. The bands are 1″ (25.4mm) wide.

APPLICATION OF GRIT SIZES

The 50 grit size of aluminum oxide abrasive bands is adapted to heavy stock removal operations. The operating band speed should be checked on the job selector. The 80 grit size is for general surface finishing or coarse polishing operations. Band speeds for these operations run to 1,000 fpm (305 m/min). The 150 grit size is for light stock removal and high polish applications. Band speeds range between 800 and 1,500 fpm (244 and 457 m/min) for these abrasive bands.

ABRASIVE BAND GUIDE COMPONENTS

Abrasive bands require a rigid backup support and an adapter. The abrasive band adapter replaces the grooved face adapter used for the file band support. Figure 49–9 shows how the band polishing guide (which is secured in the upper mounting bracket) rides against the polishing guide support. The guide usually has a graphite-impregnated facing, which serves as a dry lubricant. A graphite powder is usually rubbed in the polishing band fabric to lubricate it and to increase wear life.

SILICON CARBIDE ABRASIVE GRINDING BAND

Silicon carbide abrasive bands are used for grinding tough, hard metals such as carbide-tipped cutting tools. Grit sizes of 50, 80, and 150 are

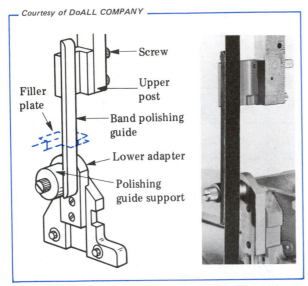

Screw

Upper post

Filler plate

Band polishing guide

Lower adapter

Polishing guide support

Figure 49–9 Guide Components for Band Polishing

common. The band width is 1″ (25.4mm). A coolant is generally used.

LINE GRINDING

Line grinding requires a line-grind band. This band tool has a continuous abrasive cutting edge bonded to a steel band. The abrasive coating may be silicon carbide or aluminum oxide. Carbide-coated bands are recommended for grinding quartz, granite, ceramic, glass, and other hard nonferrous materials. Bands coated with aluminum oxide are best for grinding metals, heat-treated alloy steels, heat-resistant steels, wear-resistant nonferrous alloys, and other difficult-to-machine materials.

QUALITIES OF LINE-GRIND BANDS

Line-grind bands produce a clean, smooth finish surface. The grinding capacity (width) is limited only by the working depth of the machine.

Tapered, rubber-tired carrier wheels and roller-type saw guides are used with line-grind bands. The bands are welded the same as diamond bands. The coated edge is protected by the beveled jaw inserts that face the operator. Before welding, 1/8″ (3.2mm) of abrasive is removed from each end of the band. When this amount of the metal band is consumed in welding, the abrasive edge forms a continuous band.

HEAT GENERATION AND BAND SPEEDS

Considerable heat is generated by the cutting action of the abrasive grains. Therefore, a flood coolant application is needed. Special qualities are required in the coolant. It may be necessary to flush the system to change to the coolant recommended for line grinding a particular metal.

Band speeds for line grinding between 3,000 to 5,000 fpm (900 to 1,500 m/min) are best. Bond abrasives are sharpened by dressing the band with a diamond dressing stick. The stick is moved lightly across the abrasive cutting edge.

ELECTROBAND MACHINING

Electroband machining operates on the scientific principle of disintegrating the material in front of a cutting edge without having a knife-type band ever touch the material. A low-voltage, high-amperage current is fed into a knife-type band. The electrical input causes the knife edge to discharge a sustained arc into the material being cut. The arc disintegrates the material at the same instant a flood of coolant quenches the arc. The coolant prevents burning and damaging the material.

ELECTROBAND TOOL CHARACTERISTICS, OPERATION, AND FEED RATES

The band tool used in electroband machining must have properties to withstand arcing without cracking and to maintain long flex life. Electrobands are welded in the same manner as knife-edge blades.

The electroband machine tables are hydraulically powered. The band speed is fixed at 6,000 fpm (1,828 m/min). Splash guards contain the coolant without obstructing the operation.

Surfaces requiring a fine finish may be electroband machined at a rate ranging from 5 to 50 square inches (32.2 to 322 sq. cm) per minute. This rate can be increased (when the quality of the work edge is not important) to 150 to 200 square inches (968 to 1,290 sq. cm) per minute.

KNIFE-EDGE BANDS AND PROCESSES

Thus far, blades with cutting teeth have been described and applied primarily to metal-working jobs and processes. *Knife-edge blades*, as the name implies, have a knife edge. The edge may be straight (knife), wavy, or scalloped. When formed into a band, the knife-edge blade is especially adapted for cutting soft, fibrous materials. Other types of cutting edges would tear or fray such materials.

The knife-edge saw blade is produced with a single- or double-bevel cutting edge. The wavy-edge and scallop-edge blades have a double-bevel cutting edge.

The three types of saw blades come in different widths and gages. The sizes of the knife-edge blades range from 1/4″ to 1 1/2″ (6.4mm to 38.2mm) wide and from 0.018″ to 0.032″ (0.46mm to 0.8mm) gage thickness. The sizes of the scallop- and wavy-edge saw blades are from 0.015″ to 0.032″ (0.4mm to 0.8mm) thick and from 1/4″ to 2″ (6.4mm to 50.8mm) wide.

Knife-edge bands cut efficiently at relatively high band speeds. The bands must withstand maximum *flexation* (bending) over the carrier wheels. In addition, the blades must be ground to razor sharpness and be able to maintain it over long usage. Most knife-edge cutting is done dry and at high speeds. The knife-edge bands slice through to separate material without producing a kerf. As a result, no chips are produced.

Wavy- and scallop-edge forms are used on materials that offer increased resistance to cutting.

Safe Practices in Setting Up and Performing Basic Band Machining Operations

- Use the machine roller guide shield on high-speed saw cuts. In addition, wear protective safety glasses when working at or near a band machine.
- Replace the table filler plate with the plate designed for the operation (sawing, filing, grinding, and so on) to be performed.
- Wear gloves for handling saw bands during mounting and dismounting. Then, remove the gloves unless the condition of the workpieces or operation justifies their use.
- Check each cutting band before mounting for possible weld defects, the condition of the teeth or cutting edge, and signs of band fatigue.
- Recheck band tracking and the application of the recommended band tension on all band machine cutting tools. Close all doors and secure all guards before starting the machine.
- Adjust the automatic hydraulic carrier wheel brakes for the band tool being used.
- Use only center-crowned, rubber-tired carrier wheels for driving spiral-edge saw bands.

UNIT 49   REVIEW AND SELF-TEST

1. Referring to a job selector on a band machine, list the kind of information an operator needs to know about conventional sawing a 25.4mm thick block of free-machining steel.

2. List three differences between high-speed sawing and friction sawing.

3. a. Identify two distinguishing design features of a spiral-edge saw band.

   b. Cite two types of jobs for which spiral-edge band sawing is uniquely adapted.

4. Compare electroband machining with conventional band sawing in terms of (a) principles of machining and (b) a typical application.

5. a. Study the following working drawing of a form guide. This part is to be band machined.

    b. List the work processes required (1) to saw the square corner, relief, and the contour; (2) to file; and (3) to grind the 0.750 ±0.001 area. Assume the workpiece has the one edge ground to $\frac{32}{}$, the plug end is square, and the part is layed out.

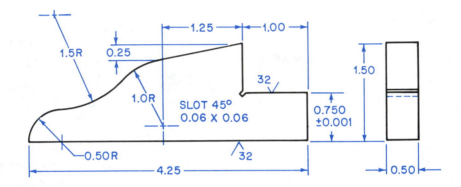

6. a. State one similarity of a line-grind band and an abrasive band.

    b. Give two major differences between the two bands.

7. Give three personal safety precautions to take prior to operating a band machine.

SECTION ONE

Machine Tool Technology and Processes

Shapers and planers are machine tools that are used for machining internal and external plane and contour surfaces. Shapers and planers generate a required linear surface or form by a reciprocating straight-line cutting motion. Material may be removed to produce a horizontal, vertical, or angular plane surface or a varying-contour form. This section provides an overview of shaping and planing machine design features, fundamental tool geometry, cutting speed and feed computations, and basic setups and machining processes.

UNIT 50

Shapers, Slotters, and Planers: Machines and Setups

BASIC TYPES OF SHAPERS

There are two basic types of shapers: *vertical* and *horizontal.* As the names indicate, the movement of the cutting tool is in either a vertical plane or a horizontal plane.

THE VERTICAL SHAPER (SLOTTER)

The vertical shaper is sometimes called a *vertical slotter.* It is designed primarily to shape internal and external flat and contour surfaces, slots, keyways, splines, special gears, and other irregular surfaces. The design features of a vertical shaper are shown in Figure 50–1.

A rotary table is mounted on the carriage. The table is also movable longitudinally. The combination of three directions of movement permits the table and workpiece to be positioned and fed for straight-line and rotary cuts.

Motion for cutting is produced by the reciprocating movement of the *vertical ram.* The ram and mechanism for producing motion are housed in the column. Cutting tools are secured in a toolholder on the forward section of the ram. The ram position is adjustable. The cutting tool is set in relation to the starting and ending of a cut. The distance the ram travels is called the *length of stroke.* The ram on some machines may be adjusted at an angle to take angular shaping cuts on vertical surfaces. Power feed is

THE HORIZONTAL SHAPER

In contrast to the reciprocating cutting movement of the vertical shaper, cutting action takes place on the *horizontal shaper* by feeding the workpiece for the next cut on the noncutting return stroke. The cutting tool removes metal on the forward longitudinal stroke of the *horizontal*

403

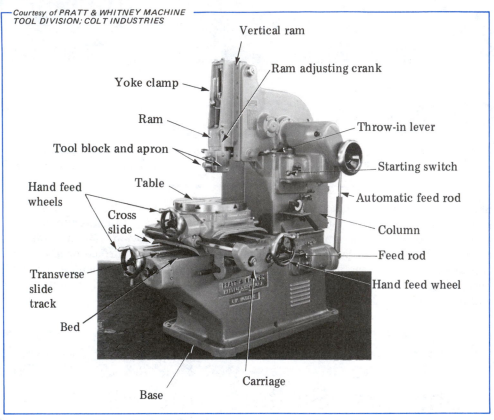

Courtesy of PRATT & WHITNEY MACHINE
TOOL DIVISION; COLT INDUSTRIES

Figure 50–1 Design Features of a Vertical Shaper (Slotter)

ram. A surface is shaped by taking a series of successive cuts. For each cycle of the ram, cutting is done on the forward stroke; feeding, on the return stroke. Shaping of small workpieces is usually done on *bench shapers.*

Positioning, Feeding, and Cutting Movements. The different positioning, feeding, and cutting movements on a universal horizontal shaper are shown in Figure 50–2. Note on this model that the *universal table* may be rotated through a 90° arc.

The *tool head* may be swiveled to feed the cutting tool at a required angle. The angle settings are read directly from the angle graduations on the end of the ram. The clapper box is adjustable to permit the cutting tool to clear the workpiece on the return stroke.

The table is movable vertically and transversely. The cutting tool may be moved up or down by turning the handwheel on the vertical tool-head slide.

BASIC TYPES OF SHAPER CUTS

SHAPING VERTICAL SURFACES

The tool head and clapper box are positioned vertically for machining flat and regular curved surfaces. However, the clapper box and tool head must be swiveled for angular and vertical cuts. Otherwise, the toolholder and/or the cutting tool may not clear the workpiece. Swiveling the head prevents the cutting tool from dragging along and scoring the machined surface during the return stroke.

SHAPING SIMPLE AND COMPOUND ANGLES

Angular surfaces are shaped either by positioning the workpiece or the work-holding device or by setting the tool head at the required angle.

Some parts require the machining of surfaces at a compound angle. The vise is set at the first angle; the standard table, at the second angle; and the universal table, at the third angle. The

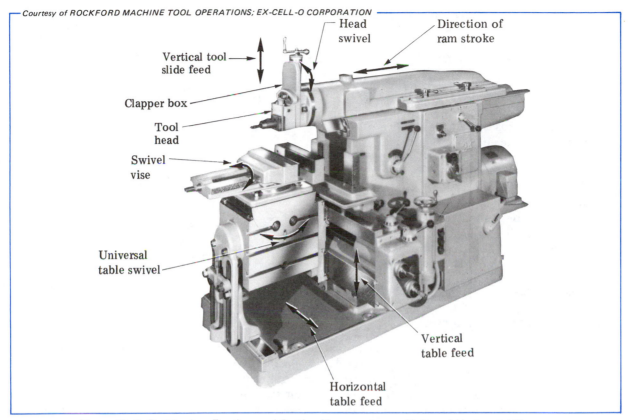

Figure 50-2 Positioning, Feeding, and Cutting Movements on a Universal Horizontal Shaper

compound angle is then produced by horizontal shaping.

In addition to standard straight and offset shaper toolholders, an adjustable-type universal cutting-tool holder is available (Figure 50-3). The design permits setting a tool bit for horizontal, vertical, angular, and contour cuts.

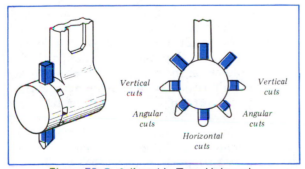

Figure 50-3 Adjustable-Type Universal Cutting-Tool Holder

SHAPING DOVETAILS

When dovetails are cut on shapers, the tool head is swiveled at the angle of the dovetail. The cutting tool is ground at a smaller included angle than the angle of the dovetail. The tool is ground for cutting on the front face and either the left- or right-hand edges, depending on whether a right- or left-hand dovetail is to be cut. The front cutting edge is formed to the required corner size and shape. Usually, the corners of an internal dovetail are relieved.

SHAPING CONTOUR SURFACES

Contour surfaces may be shaped by using a shaper tracing attachment or by combining the vertical feed on the tool slide and the longitudinal feed of the table and workpiece. In general practice, power transverse feed and manual vertical feed are used for roughing cuts. Finish cuts are taken by synchronizing the table and cutter movements. Finish cutting tools and finer feeds are used.

SHAPING INTERNAL FORMS

One common practice in machining a regular form inside a workpiece is to use a broach. The process is fast. The size and shape of the form may be held to close tolerances. When a single part is to be produced, the shaper is a practical machine tool to use for the occasional shaping of regular or other internal forms. One type of general-purpose toolholder for internal shaping cuts includes an extension bar and cutting tool which may be turned for vertical and angular shaping cuts. The bar is adjustable, depending on the length of the cut. A slower ram speed is used than for external shaping. The internal toolholder is not as rigid as a solid tool or toolholder.

CARBIDE CUTTING TOOLS

Carbide cutting tools permit machining at faster cutting speeds and about the same feed rates as the speeds and feeds used for shaping with most high-speed steel tool bits. However, the greater cutting efficiency depends on three conditions:

- The cutting speed of the shaper must exceed 100 fpm (30 m/min);
- A constant speed and feed rate must be maintained;
- The tool head should be fitted with a tool lifter to permit the tool to clear the work on the return stroke.

Cutting speeds of carbide cutting tools are usually two or three times faster than for high-speed steels. In using carbides, it is better practice to take heavy, deep cuts and to use a lighter feed. This practice distributes the chip force over a longer cutting area.

TOOL GEOMETRY FOR CARBIDE TOOLS

The tool geometry relating to the shape and angles of a carbide shaper cutting tool is shown in Figure 50-4. Tool manufacturers furnish tools having specified angles and tool forms. These tools produce efficiently, machine to particular surface finishes, and have maximum tool life.

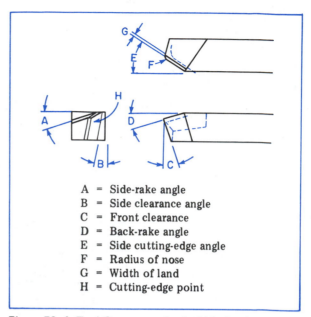

A = Side-rake angle
B = Side clearance angle
C = Front clearance
D = Back-rake angle
E = Side cutting-edge angle
F = Radius of nose
G = Width of land
H = Cutting-edge point

Figure 50-4 Tool Geometry of a Carbide Cutting Tool

Like the high-speed steel tool bits, the front clearance on carbide cutters is 4°. A 0° to 20° negative back-rake angle is used for roughing cuts on steel and cast iron. The back-rake angle permits the cut to start above the cutting edge. This action protects the cutting edge (as the weakest part) from the shock of starting or intermittent cuts. The back-rake angle is increased to a positive angle on finish cutting tools. The positive angle prevents tool chattter when light finish cuts are taken. A side cutting-edge angle of 30° to 40° with a large nose radius should be used on roughing tools.

MODERN PLANER DESIGN

Some planers are designed with their upright columns joined together at the top to increase machine rigidity. Such planers are called *double-housing planers*. Other planers are designed with a single column on the right side. These *open-side planers* permit parts to be machined that are wider than the width of the platen. However, the size is still specified by the width, height (depth), and length measurements on planers with two permanent columns. Some open-side planers are designed with a support column accessory to provide additional machining rigidity. Figure 50-5

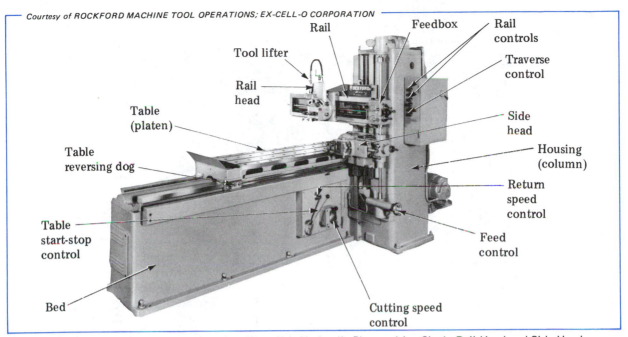

Figure 50–5 Major Components of an Open-Side Hydraulic Planer with a Single Rail Head and Side Head

shows the major components of an open-side hydraulic planer with a single rail head and side head.

One machine tool manufacturer identifies open-side planers that are under 42″ × 42″ (1m × 1m) as *shaper-planers.* They are available with single or double rail heads and with or without a side head mounting. Typical horizontal and vertical roughing cuts may be taken by both rail and side heads simultaneously.

Milling and grinding heads are available as accessories. These heads provide the added capability to perform milling and grinding processes, respectively. Template and contour following, indexing, and footstock accessories are also available. Cutting tools are sometimes set up on multiple heads for gang planing. Some cutters are actuated by a stylus to permit two- and three-axis duplicator processes to be carried on.

One of the greatest advantages of a planer is its versatility and lower costs for certain machining operations. Single-edge cutting tools that are primarily used for planer operations may be easily modified. These tools often replace more highly expensive multiple-point cutting tools when one part or a small number of parts are to be machined.

Universal planers permit cutting during both the forward stroke and the return stroke of the platen. On the forward stroke, one cutting tool edge is firmly held against the tool block during the cut. The second tool edge is then automatically indexed at one end of the stroke. With the second cutting edge in position, a cut is taken during the return stroke. Machining time is reduced as the lost return stroke time is eliminated.

MACHINE AND WORKPIECE SETUPS IN PREPARATION FOR PLANER WORK

PLACING THE WORKPIECE

Single or multiple workpieces are placed on the table so that planer stops may be inserted in holes that are just in front of the workpieces to be machined (Figure 50–6). Poppets and planer pins are placed behind the workpieces. The poppets and pins force the workpieces to the front stops. The parts are then secured by using T-bolts and appropriate shapes and sizes of straps.

ADJUSTING THE RAIL

The rail is set at a height that permits the greatest support for the cutting tool with minimum overhang of the tool head. There must,

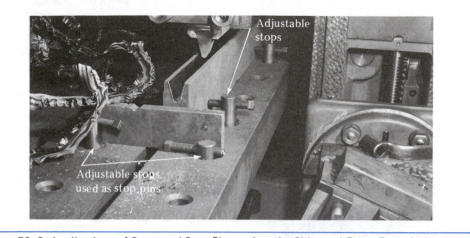

Figure 50–6 Applications of Stops and Stop Pins against the Sides and Front End of a Workpiece

however, be adequate, safe clearance between the workpiece and the cutter. When a rail height needs to be changed, the rail locking clamps and bolts need to be tightened. Open-side shaper-planers have back rail clamps that are loosened for rail adjustment and tightened to rigidly support the cross rail apron to the front of the column. Other planing machine designs require the tightening of bolts to lock the cross slide apron at a fixed height. Consideration is given to taking up backlash on the elevating screw. The rail is lowered to slightly lower than the required height and then raised to position and locked.

ADJUSTING THE TOOL HEAD

The tool head is moved downward almost to the surface to be planed. Once the table stops and cutting speeds are set, the tool slide is moved to locate and to feed the cutting tool to the depth of the first cut. The same procedure is followed to position the cutting tool on a second tool head.

Somewhat similar steps are taken to mount, secure, position, and locate the cutting tool on a side head. The difference is that the cutting action will be vertical and with down-feed. This vertical cut is usually combined with a right (hand) horizontal cut.

SETTING THE CUTTING SPEEDS AND TABLE STOPS

With the table stops positioned with adequate clearance for the starting and ending of each cut, the cutting and return speeds are next set. Some hydraulic planers have adjusting valves on two sides of a hydraulic pump. The valve on one side controls the rate of the cutting stroke. The independent valve on the other side regulates the return stroke. Other hydraulic planers have a single cutting-speed control lever.

ADJUSTING THE FEED RATE

Feed rate on a hydraulic planer is controlled by a valve. The feed control valve on the side of the machine column is adjusted to the required feed rate.

PLANING FLAT SURFACES

Roughing cuts are taken to within 0.005″ (0.12mm) of the finished dimension. As the roughing cuts are completed, the tool slides are raised. The cutting tools are replaced by finish cutting tools that have a broad contact surface.

Finish cuts at a depth of about 0.005″ (0.12mm) are taken. The feed rate is increased to about two-thirds of the cutting face width. Usually, the same table speed is used for roughing and finish cuts. If a cutting oil is to be used, the entire surface is coated before there is any table movement. The combination of a flat tool form, fine depth of cut, and wide feed rate results in a smooth, finely finished plane surface.

SIMULTANEOUS PLANING WITH A SIDE HEAD

When a side head is used, steps must be taken to position the workpiece against side stops. The surface to be machined must be parallel to the edge (side face) of the planer table. The front stops are used to prevent any forward movement under heavy cutting forces. The side stops position the workpiece in relation to the cutting action of the side head cutting tool. Once the workpiece is positioned, the side head cutting tool feeds and speeds are similar to the feeds and speeds used with the rail heads.

PLANER CUTTING TOOLS AND HOLDERS

Tool geometry for planer cutting tools is similar to shaper tools. Clearance angles are reduced to a minimum of 4° to 5°. This angle provides maximum support behind the cutting edge. Carbide tools are ground from 0° to negative top rake angles away from the working area of the cutting tool edge.

Planer toolholders are generally drop forged of nickel chromium steel. The holders are heat treated for maximum toughness. Some holder designs have a separate *taper seat*. The seat provides a reference surface for holding, adjusting, and replacing a tool bit as the cutting tool. The seat is adjusted to pull the tool bit into position until it is securely locked on three sides: back, top, and bottom. The seat is *serrated* with a series of evenly spaced V-grooves.

Straight planer toolholders assure rigidity and permit the heaviest cuts to be taken. *Gooseneck* planer toolholders (Figure 50–7A) provide an "underhung" tool height. This design of toolholder permits the cutting tool to lift under severe cutting forces and to minimize chatter.

Spring-type planer toolholders are used for finishing operations. A chatter-free finish is produced because the holder design absorbs vibration. *Universal clapper box* planer toolholders (Figure 50–7B) are adjustable to any angle.

Other planer toolholders, classified as special-purpose holders, are designed for specific processes. *Adjustable* toolholders are used for planing T-slots. *Multiple* toolholders permit gang planing or simultaneously planing machine bed V-ways. *Double* toolholders provide for cutting at two depths at the same time. *Reversing* toolholders are used for planing on the forward and return strokes on planers designed for such cutting action.

ADJUSTABLE, REPLACEABLE, SERRATED PLANER TOOL BIT GEOMETRY

The tool bits for planer toolholders with serrated seats have a corresponding series of grooves. The grooves permit a tool bit to be adjusted sideways in steps of 0.060″ (1.5mm) and to compensate for wear. Also, the tool bit can be changed without disturbing the original setup.

Tool bits are available in high-speed steel, cobalt high-speed steel, cast alloys, and all grades of carbides. Standard steel serrated tool bits are hardened, tempered, and ground. The end of each tool bit is preformed to a particular shape. Tool bits are supplied in sets in sizes such as #4, #6, #7, #8, and so on. The higher the number, the larger the features are proportionally. The design of each numbered size provides for interchangeability with all styles of holders of the same size. Each tool bit may be inserted and held

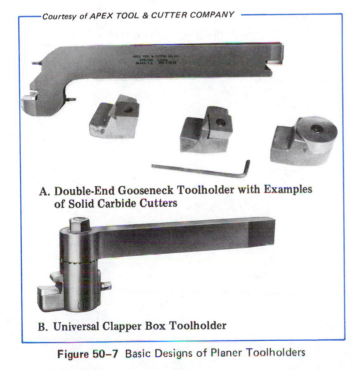

Courtesy of APEX TOOL & CUTTER COMPANY

A. Double-End Gooseneck Toolholder with Examples of Solid Carbide Cutters

B. Universal Clapper Box Toolholder

Figure 50–7 Basic Designs of Planer Toolholders

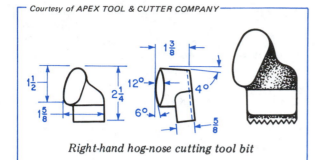

Right-hand hog-nose cutting tool bit

Figure 50–8 Manufacturer's Specifications for a Selected Replaceable, Serrated Planer Tool Bit

in a standard straight, gooseneck, or spring-type holder. The cutting tools in a set include right- and left-hand forms, a straight parting tool; full-width, flat-nose tool; diamond-point tool; and straight round-nose tool.

The manufacturer's specifications for three selected size #8, replaceable, serrated-base planer tool bits are shown in Figure 50–8.

Full-radius and other special-form tool bits are available in high-speed steels or cobalt steels or as carbide-tipped tool bits. The term *plug* is used with carbide inserts (cutting tool bits). The plugs are generally square or round and are designed with either positive or negative clearance.

Safe Practices in Tooling Up and Preparing for Shaping or Planing

- Stop the machine. Under no circumstances are workpiece adjustments or checking to be made while the shaper ram or planer table is moving.
- Stand clear of a heavy or hard-to-handle part that is being moved to or set on the platen. Keep hands away from the underside of the workpiece. Be ready to instantly move safely away from the setup in the event of danger.
- Use stops against the front end of a workpiece to take up the cutting thrust.
- Protect the platen, table, or other work-holding device against scratching, burring, or denting.
- Position the safety dogs on a planer to keep the table from moving past the table length.
- Move the shaper ram or planer table manually (or at an extremely slow speed) through at least one forward and reverse stroke.
- Check for excessive cutting speeds.
- Use a protective screen for shaper and planer operations.
- Wear safety glasses or a protective shield during setup and machining processes.

UNIT 50 REVIEW AND SELF-TEST

1. List three operations that are best performed on each of the following three machine tools: (a) shaper, (b) planer, and (c) slotter.

2. a. Name three major parts of a tool (swivel) head on a shaper.
   b. List the steps to follow in swiveling a tool head to cut a 30° angle surface.

3. Explain why most carbide tools used for roughing shaper and planer cuts have a negative back-rake angle.

4. a. Prepare a work production plan for shaping the cast iron angle form block. A standard shaper is used. The workpiece is to be positioned and held in a swivel vise.
   b. Determine the angle for setting the vise.
   c. Calculate (1) the cutting speed in fpm and (2) the ram strokes per minute.

   The casting size allows for one approximately 3/16″ deep roughing cut and a 0.010″ finishing cut on each surface. The feed for the roughing cut is 0.060″. The feed factor of 1.00 may be used for computing the cutting speed of the final cut. High-speed roughing and finish cut cutting tools are used.

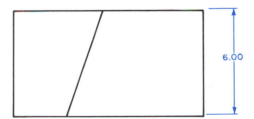

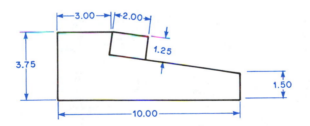

5. a. Name the four principal construction features of a shaper-planer.
   b. Describe the function of each feature.

6. a. Identify two different types of planer toolholders that may be used to prevent chatter.
   b. Provide a brief explanation of how chatter is overcome with selected toolholders.

7. Give two advantages for using replaceable, serrated planer tool bits instead of older, solid-forged cutting tools.

8. Cite one machine or tool safety precaution to take for each of the following conditions: (a) setting the ram or platen stroke, (b) setting a solid-shank tool or toolholder, (c) swiveling a tool slide at a steep angle on a shaper, and (d) machining at an angle on a shaper.

9. State three personal safety precautions the operator follows after completing a shaper or planer job.

SECTION ONE

Grinding Machines, Abrasive Wheels, and Cutting Fluids

This section deals with the types and construction features of grinders used in abrasive machining and precision grinding processes. The characteristics of abrasives and feeding techniques to accommodate a wide variety of work parts are described. Principles of cutting action relating to surface, centerless, cylindrical, form, plunge, and tool and cutter grinding are examined. Microfinishing, honing, lapping, and superfinishing processes are treated.

Also covered is the technology of grinding wheels, ANSI representation and dimensioning, and wheel speeds and feeds. General grinding problems, causes, and corrective action are summarized. Further, a description of cutting fluids and the selection, dressing, truing, and maintenance of grinding wheels is provided.

UNIT 51

Grinding Machines and Grinding Wheels

A. GRINDING MACHINES: FUNCTIONS AND TYPES

CHARACTERISTICS OF ABRASIVES AND GRINDING

Abrasive cutting tools have thousands of cutting edges (grains). Each sharp grain that comes in contact with the workpiece cuts away a chip. Chips magnified 350X are shown in Figure 51–1. The many cutting edges provide continuous cutting points. Abrasive cutting tools are capable of generating a smoother surface than any other machine tool. Also, the surface generated is dimensionally precise; therefore, grinding processes are widely used in finishing workpieces.

Grinding wheels operate at speeds of 6,000 to 18,000 sfpm (1,830 to 5,490 m/min). These speeds are roughly equivalent to 1.1 to 3.4 miles (1.8 to 5.4km) per minute. Usually, the distance between abrasive grains is less than 1/8″ (3mm). On a 1″ (25mm) wide grinding wheel, over 4,500,000 (at 6,000 sfpm) tiny, sharp cutting edges may come in contact with the workpiece each minute.

ABRASIVE MACHINING

Abrasive machining refers to heavy metal-removing operations that previously were performed with conventional cutting tools and

412

machines. Abrasive machining may be done using abrasive grinding wheels or coated abrasive belts.

ADVANTAGES OF ABRASIVE MACHINING

Six advantages are generally cited for abrasive machining:

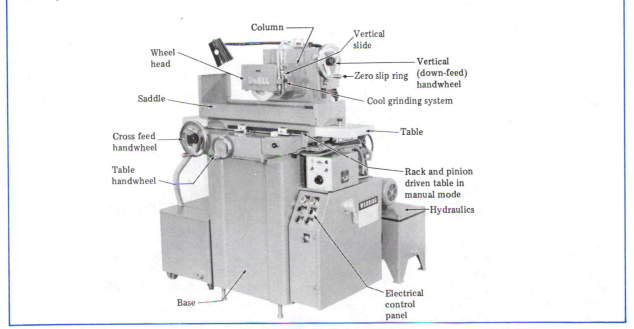

Courtesy of BAY STATE ABRASIVES; DRESSER INDUSTRIES INC.

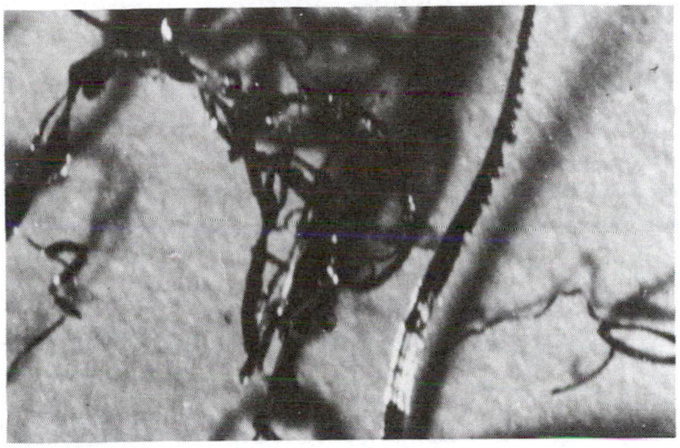

Figure 51-1 Chips Produced during Grinding (Magnified 350X)

- Machine tools are designed with *automatic sizing* of the abrasive wheel to compensate for cutting tool wear.
- Parts handling, loading, and unloading costs are reduced.
- The costs of fixtures may be reduced or eliminated.
- Abrasive wheels cut through the outer scale of castings and forgings, hard spots, and burned and rough edges due to welding processes.
- Thin-walled areas of workpieces may be abrasive machined with less springing or breakage.
- Casting and forged part design may be simplified. Normal allowances of extra stock in order to cut under the scale may be reduced.

BASIC PRECISION GRINDING

Precision grinding processes are generally grouped into six categories: surface, cylindrical (external and internal), form, plunge, centerless, and tool and cutter grinding.

Courtesy of DoALL COMPANY

Column
Vertical slide
Wheel head
Vertical (down-feed) handwheel
Zero slip ring
Cool grinding system
Saddle
Table
Cross feed handwheel
Table handwheel
Rack and pinion driven table in manual mode
Hydraulics
Base
Electrical control panel

Figure 51-2 Basic Construction Features of a Saddle-Type, General Purpose 6" X 12" (Type I) Horizontal Surface Grinder

SURFACE GRINDING MACHINES

Essentially, *surface grinding* deals with the generating of an accurate, finely finished, flat (plane) surface.

The four basic types of *surface grinders* are:

- *Type I* (horizontal spindle, reciprocating table);
- *Type II* (horizontal spindle, rotary table);
- *Type III* (vertical spindle, reciprocating table);
- *Type IV* (vertical spindle rotary table).

TYPE I (HORIZONTAL-SPINDLE, RECIPROCATING-TABLE) SURFACE GRINDER

The type I surface grinder (Figure 51–2) is the most commonly used surface grinding machine. Flat, shoulder, angular, and formed surfaces are ground on the machine. Workpieces are generally held by a magnetic chuck on the table. The table is mounted on a saddle. Cross feed for each stroke is provided by the transverse movement of the table. The workpiece is moved longitudinally, under the abrasive wheel on the reciprocating table. The grinding wheel downfeed is controlled by lowering the wheel head. The amount of feed is read directly on the graduated micrometer collar.

Depending on the material used and nature of the operation, grinding may be done dry or wet. All general-purpose machines are designed either with or for the attachment of a coolant supply system. Dry grinding requires an exhaust attachment to carry away dust, abrasives, and other particles.

Most type I surface grinders (Figure 51–2) are 6″ × 12″ (150mm × 300mm). This designation indicates that workpieces up to 6″ (150mm) wide × 12″ (300mm) long may be surface ground. Standard heavy-duty type I machines are produced to grind surfaces 16′ and longer.

By dressing the wheel, the face and edges may be cut away in reverse form to grind to a required shape. Radius- and other contour-forming devices and diamond-impregnated formed blocks are used. The formed wheel then reproduces the desired contour.

TYPE II (HORIZONTAL-SPINDLE, ROTARY-TABLE) SURFACE GRINDER

The type II surface grinder produces a circular scratch-pattern finish. This finish is desirable for metal-to-metal seals of mating parts. Workpieces are held on the magnetic chuck of a rotary table. The parts are nested so that they rotate in contact with one another. The grinding wheel in-feed is measured by direct reading on the wheel feed handwheel micrometer collar. The traverse movement of the wheel head feeds the grinding wheel across the workpieces. The principle of grinding on a type II surface grinder is shown in Figure 51–3. The table may be tilted to permit grinding a part thinner either in the middle or at the rim.

TYPE III (VERTICAL-SPINDLE, RECIPROCATING-TABLE) SURFACE GRINDER

The type III surface grinder, designed with a *vertical spindle* and a *reciprocating table*, grinds on the face of the wheel. The work is moved back and forth under the wheel. The positioning of workpieces on a magnetic table and the cutting action of a type III surface grinder are pictured in Figure 51–4.

Type III grinders are capable of taking comparatively heavy cuts. The wheel head is designed so that it may be tilted a few degrees from a vertical position. This design permits faster cutting and metal removal as the rim of the grinding wheel contacts the workpiece. In a vertical position, the surface pattern produced is a uniform

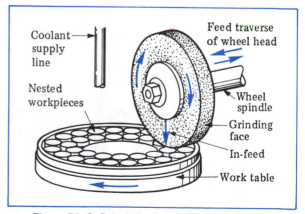

Figure 51–3 Principle of Grinding on a Type II Surface Grinder

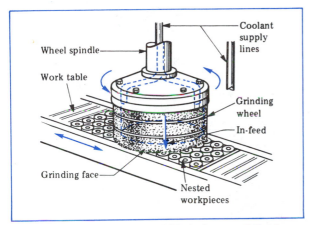

Figure 51–4 Positioning of Workpieces and Cutting Action of a Type III Surface Grinder

series of intersecting arcs. A semicircular pattern is generated when the wheel head is tilted.

ADAPTATIONS OF TYPE III SURFACE GRINDERS

Figure 51–5 shows examples of surface grinding processes using auxiliary spindles positioned vertically, at an angle, or with the wheel dressed. Straight, angular, and dovetailed surfaces may be ground. Auxiliary spindles are used largely with heavy-duty or special multiple-operation surface grinders.

TYPE IV (VERTICAL-SPINDLE, ROTARY-TABLE) SURFACE GRINDER

The type IV surface grinder, designed with a *vertical spindle* and a *rotary table*, also grinds on

the face of the wheel to generate a flat (plane) surface.

GENERAL-PURPOSE CYLINDRICAL GRINDING MACHINES

Cylindrical grinding basically involves the grinding of straight and tapered cylindrical surfaces concentric or eccentric in relation to a work axis. Such grinding is generally performed on either *plain* or *universal cylindrical grinding machines*.

Form grinding refers to the shape. The grinding of fillets, rounds, threads, and other curved shapes are examples. Straight form grinding processes are performed on surface grinders.

Cylindrical grinders are also used for producing straight, tapered, and formed surfaces by moving the grinding wheel into the workpiece without table traverse. The process is referred to as *plunge grinding.*

UNIVERSAL CYLINDRICAL GRINDING MACHINES

The universal cylindrical grinder has the capability to perform all the operations of the plain cylindrical grinder. In addition, both external and internal cylindrical grinding operations, face grinding, and steep taper and other fluted cutting tool grindings may be performed.

The principal parts of a standard universal cylindrical grinder are shown in Figure 51–6. A strong, heavy base adds to the stability of the machine. The table may be swiveled for taper grinding. The headstock and wheel spindle head may also be swiveled for grinding steep angles and

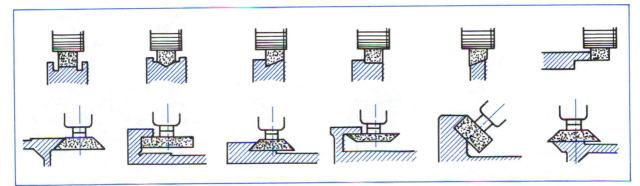

Figure 51–5 Examples of Surface Grinding Processes Using Auxiliary Spindles Positioned Vertically, at an Angle, or with the Wheel Dressed

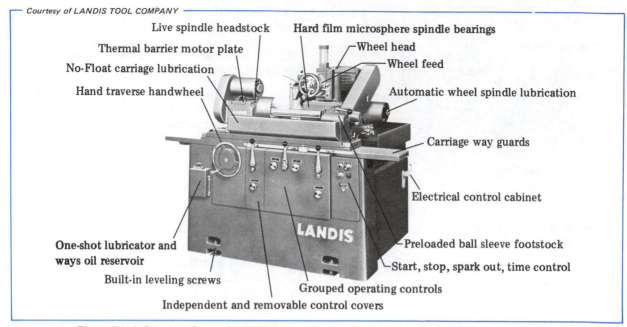

Live spindle headstock
Hard film microsphere spindle bearings
Thermal barrier motor plate
Wheel head
No-Float carriage lubrication
Wheel feed
Hand traverse handwheel
Automatic wheel spindle lubrication
Carriage way guards
Electrical control cabinet
One-shot lubricator and ways oil reservoir
Preloaded ball sleeve footstock
Start, stop, spark out, time control
Built-in leveling screws
Grouped operating controls
Independent and removable control covers

Figure 51-6 Principal Parts of a Universal Hydraulic Traverse and In-Feed Cylindrical Grinder

faces. The footstock is provided with a center for holding work between centers. On some machines the internal grinding unit is mounted directly above the wheel spindle head. The versatility of the machine is extended by a number of work-holding devices and machine accessories.

Plain and universal toolroom and commercial shop cylindrical grinders are capable of grinding diameters to within 25 millionths of an inch (0.0006mm) of roundness and to less than 50 millionths of an inch (0.001mm) accuracy. Surface finishes to 2 microinches and finer are pro-

duced in the cylindrical grinding of regular hardened steel as well as tungsten carbide parts.

INTERNAL CYLINDRICAL GRINDING MACHINE

The setup for and principle of internal cylindrical grinding are shown in Figure 51-7.

Concentric workpieces may be held in collets and chucks. Irregularly shaped parts, such as an automotive connecting rod, may be positioned and held in a special fixture. Internal grinding may be done on small-diameter holes with mounted grinding wheels. The diameter should be at least three-quarters of the finished hole diameter. Frequent dressing is required unless a *cubic boron nitride* abrasive is used. This abrasive is a recently developed product of the Specialty Materials Department, General Electric Company. The abrasive bears the trademark of BOROZON™ CBN. While cubic boron nitride abrasives are almost as hard as diamonds, they are an exceptionally good abrasive on hardened steels. Cubic boron nitride wheels require infrequent dressings.

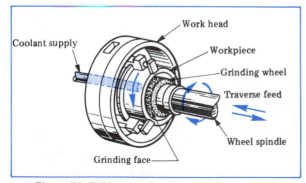

Work head
Coolant supply
Workpiece
Grinding wheel
Traverse feed
Wheel spindle
Grinding face

Figure 51-7 Machine Features and Principle of Internal Cylindrical Grinding

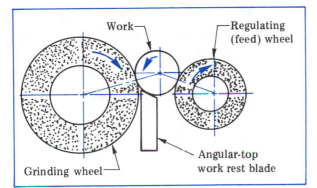

Figure 51–8 Principle of Controlling RPM of Workpiece on a Centerless Grinder (Work Set above Center)

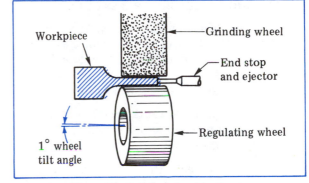
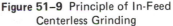

Figure 51–9 Principle of In-Feed Centerless Grinding

CENTERLESS GRINDING METHODS AND MACHINES

Centerless grinding machines are primarily used for production machining of cylindrical, tapered, and multiple-diameter workpieces. The work is supported on a *work rest blade.* The blade is equipped with guides suited to the design of the workpiece.

The principle of centerless grinding requires the grinding wheel to rotate and force the workpiece onto the work rest blade and against the regulating wheel (Figure 51–8). This wheel controls the speed of the workpiece. When the wheel is turned at a slight angle, it also regulates the longitudinal feed movement of the workpiece. The rate of feed is changed by changing the angle of the regulating wheel and its speed.

The center heights of the grinding wheel and the regulating wheel are fixed. The work diameter is, thus, controlled by the distance between the two wheels and the height of the work rest blade. There are three common methods of grinding on centerless grinders:

- In-feed centerless grinding,
- End-feed centerless grinding,
- Through-feed centerless grinding.

IN-FEED CENTERLESS GRINDING

In-feed centerless grinding is used to finish several diameters simultaneously or to grind tapered and other irregular profiles. The principle of in-feed centerless grinding is illustrated in Figure 51–9.

END-FEED CENTERLESS GRINDING

End-feed centerless grinding is used to grind tapered parts. The principle of end-feed centerless grinding is shown in Figure 51–10.

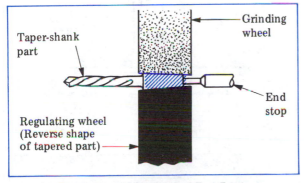

Figure 51–10 Principle of End-Feed Centerless Grinding

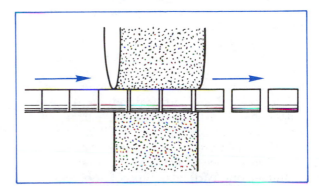

Figure 51–11 Principle of Through-Feed Centerless Grinding on a Cylindrical Grinder

THROUGH-FEED CENTERLESS GRINDING

Through-feed centerless grinding involves feeding the work between the grinding and regulating wheels, as illustrated in Figure 51–11.

ADVANTAGES OF CENTERLESS GRINDING

The four prime advantages of using centerless grinding processes and machines over other methods are as follows:

- Long workpieces that would otherwise be distorted by other grinding methods may be ground accurately and economically by centerless grinding. Axial thrust is eliminated;
- Generally, less material is required for truing up a workpiece. The work *floats* in a centerless grinder, in contrast to the eccentricity of many workpieces that are mounted between centers;
- Greater productivity results as wheel wear and grind time are reduced (with less stock removal);
- Workpieces of unlimited lengths may be ground on a centerless grinder.

TOOL AND CUTTER GRINDERS

Tool and cutter grinders are especially designed for grinding all types of milling cutters, reamers, taps, drills, and other special forms of single- and multiple-point cutting tools.

UNIVERSAL AND TOOL AND CUTTER GRINDER

The *universal and tool and cutter grinder* (or *universal tool and cutter grinder*) is a general-purpose grinding machine.

Many attachments are available for the machine. Magnetic chucks, collet chucks, universal scroll chucks, index centers, and straight and ball-ended mill sharpening attachments are produced for tool and cutter grinding. While many cutter sharpening operations are performed dry, wet grinding attachments are available.

The general features of grinding machines are modified to meet special job requirements. For example, one particular type of tool and cutter grinder is capable of generating helical surfaces and of sharpening ball-ended mills.

GEAR AND DISK GRINDING MACHINES

Gear grinding machines are usually of two types. The first type, a *form grinder*, has the grinding wheel dressed to a slope opposite to the slope of the gear tooth form. The second type of gear grinding machine generates the tooth form. The form is produced by the combination movement of the workpiece and the cutting action of the abrasive wheel. Gear grinding machines are designed for the range of general straight-tooth spur gears to hypoid gears.

OTHER ABRASIVE GRINDING MACHINES

Many surfaces are finished by using abrasive belts. Workpieces, as pictured in Figure 51–12, are moved under the abrasive belts by a conveyor drive unit. The abrasive finishing process is efficient on soft metal, plastic, and other parts requiring a smooth surface.

VIBRATORY FINISHING (DEBURRING) MACHINE

In addition to the bonding of abrasives into sheet, roll, grinding wheel, and other forms, the grains themselves are used in many production finishing and deburring processes. For example, machined parts may be deburred in a *vibratory finishing machine*. The parts and abrasive grains are vibrated. During this process, the tumbling abrasive reaches and deburrs inaccessible areas of workpieces.

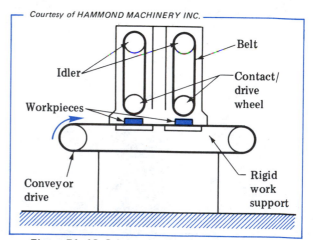

Courtesy of HAMMOND MACHINERY INC.

Figure 51–12 Schematic of Simple, Double-Belt, Coated-Abrasive Finishing Process

FREE-ABRASIVE GRINDING MACHINE

Flat surfaces may be generated by using free abrasive grains in a free-abrasive grinding machine. The grains are made into a slow moving *slurry* and fed onto a hardened steel plate. The plate is water cooled. The grains circulate along the plate. As the grains roll around the workpiece, they cut into the material.

MICROFINISHING PROCESSES AND MACHINES

The term *microfinishing* is used here to identify four selected high-quality surface finishing processes. Each process requires the removal of limited amounts of material. The ultimate finely finished and highly accurate surface is specified in microinches.

Honing, lapping, and superfinishing processes and machines are discussed. Polishing is included although the process is not intended to control the size or shape of a finished part.

HONING PROCESSES AND MACHINES

Honing is a grinding process in which two or three different low-pressure cutting motions of a honing stone produce a microfinished surface. The stock removal rate of honing, up to 0.030" (0.8mm), is the highest of the microfinishing process rates. Honing can correct out-of-roundness and taper and axial distortion. Usually, from 0.004" (0.1mm) to 0.010" (0.25mm) is left for honing. The general range of honing finishes is from 8 to 10 microinches. However, a fine surface finish to 1 microinch is obtainable. Honing may be employed on soft or hardened materials.

HONING STONE HOLDERS AND ABRASIVES

Honing requires the use of bonded abrasive *honing stones*. Some honing stones are held loosely in holders. They may be regularly spaced or interlocked. The interlocked design provides a continuous honing surface. An example of a typical honing tool (holder and stones) is shown in Figure 51-13.

The abrasive stones are expanded against a bored hole. This action is produced mechanically or by hydraulically operated mechanisms. In production, automatic controls to within 0.0001"

(0.0025mm) to 0.0003" (0.0075mm) of a specified size are used with hydraulic types of honing tools.

Aluminum oxide, silicon carbide, and diamond grain abrasives are used. The usual bonds are resinoid, shellac, vitrified, or metallic. The cutting action and wear life of the stones are increased and controlled by using a "fill" of sulfur, resin, or wax.

A coolant is required in honing. The general coolants include kerosene, turpentine, lard oil, and manufacturers' trade brand names. Water-soluble cutting oils are not recommended.

Completely round holes or interrupted holes such as keyways may be honed, provided the stones overlap the cutaway area. Honing may be used on ferrous and nonferrous parts, carbides, bronzes and brass, aluminum, ceramics, and some plastics.

LAPPING PROCESSES AND MACHINES

Lapping is a process of *abrading* (removing) material by using a soft-metal, abrasively charged lap against the surface to be finished. Loose-grain *abrasive flows* are used.

Courtesy of MICROMATIC HONE CORPORATION

Figure 51-13 Typical Honing Tool (Holder and Stones)

CHARACTERISTICS OF LAPS AND LAPPING

Laps are made of a material that is softer than the parts to be lapped. Soft cast iron, brass, copper, lead, and soft steels are commonly used materials. They easily receive and retain the abrasive grains. Soft, close-grained gray cast iron laps are common.

An open-grained lap will cut faster than a close-grained lap. The harder the lap, the slower the cutting action. Also, the harder the lap, the duller the finish and the faster the wear. The faces of laps are serrated (grooved) to permit the working surface to remain clean and free of abrasive and other particles.

Lapping is used as a finishing process, not as a metal-removing process alone. The amount of material removed in rough lapping is only 0.003" (0.08mm). In finish lapping, the material removed can be as little as 0.0001" (0.002mm). The common practice in commercial lapping processes is to grind to within 0.001" (0.02mm) of the basic size.

ABRASIVES AND WORKHOLDERS (RETAINERS OR SPIDERS) FOR LAPPING

The three most commonly used abrasives for lapping include aluminum oxide, silicon carbide, and diamond flours. Emery, rouge, levigated alumina, and chromium oxide are also used. Aluminum oxide is employed for soft metals. Silicon carbide is a practical abrasive for hardened parts. Diamond flours are used for small precision work and extremely hard workpieces.

The abrasives are used with a *vehicle*, or lubricant. The lapping workholder is known as a *spider* or *retainer*. The workholder is usually a flat plate designed with holes to accommodate workpieces of a particular shape. In addition to nesting workpieces, the retainer confines them.

The upper wheel (lap) on the lapping machine and the bottom wheel rotate in opposite directions. The wheel speeds are approximately one-tenth the surface feet per minute of regular grinding wheels. The combination of movements produces an ever-changing cutting pattern and an even wear on the laps. On some machines, the retainers provide a planetary motion to the workpiece.

SUPERFINISHING PROCESSES AND MACHINES

Superfinish™ is the trade name of the process that was invented and developed by the Chrysler Corporation. *Superfinishing* is a refined honing process in which bonded abrasive stones at low cutting speeds and constantly changing speeds produce a scrubbing action. Forces applied on the abrasive are low. From three to ten different motions are involved. As a result, a random, continuously changing surface pattern is produced. Superfinishing to 2 and 3 microinches is common practice.

SUPERFINISHING MACHINES

The three major features of a superfinishing machine include a headstock, tailstock, and a vertical spindle. The workpiece is rotated around a horizontal or vertical axis, as required. The abrasive stone is mounted above the work. It is oscillated and moved by short strokes across the workpiece. The abrasive stone may be shaped with a diamond-type boring bar. The vertical spindle superfinishing machine is usually used for flat surfaces on round parts.

POLISHING AND BUFFING AS SURFACE FINISHING PROCESSES

POLISHING

Polishing wheels are used in production work. These wheels are made of leather, canvas, felt, and wool. The abrasive grains are glued to the wheel face. Workpieces are held against the revolving polishing wheels and rotated until the desired polished surface is obtained.

BUFFING

Buffing produces a high luster and removes a minimum amount of material. Buffing follows polishing. The luster is produced by feeding and moving the surface areas of a workpiece to a buffing wheel revolving at high speed.

Buffing wheels consist of a number of layers of soft fabric. They may be linen, cotton, wool, or felt. The wheels are charged with extremely fine (flour size) grains of rouge or other abrasives. The abrasive and a wax are formed into a stick or cake. In this form, the abrasive is applied by force against the revolving buffing wheel. Buffing is not a widely used jobbing shop process.

B. GRINDING WHEEL CHARACTERISTICS, STANDARDS, AND SELECTION

CONDITIONS AFFECTING A GRINDING WHEEL

Conditions of use effect the nominal grade of a grinding wheel. Conditions of use cause a grinding wheel to act harder or softer than its nominal grade. If the cutting forces on the individual abrasive grains increase, the wheel acts softer, and the grains tear away more rapidly. If the cutting force is decreased, the wheel retains the grains longer and thus acts harder than the nominal grade. The cutting force is partly controlled by the grain depth of cut.

The rate of feed also causes changes in the action of the wheel. As the feed increases, a greater force is applied by the grains as they cut. Again, this force produces a softer-acting wheel.

The diameter of the grinding wheel also affects the cutting action. If constant spindle RPM is maintained, as the wheel wears the sfpm decreases. The cutting force on each grain then increases. The grinding wheel acts softer.

Conditions of use, as they affect cutting forces and cutting action, are illustrated in Figure 51–14. The depth of cut, chips, and feed are greatly enlarged. An abrasive grain, represented at A, moves from A to B during its cutting action. The distance is called the *arc of contact*. The *grain depth of cut* is from C to the arc of contact.

The feed of the workpiece during the arc of contact is from C to B. The shaded area represents the chips taken by the force of the grain against the workpiece. The *wheel depth of cut* is from C to D.

OTHER CONDITIONS AFFECTING THE GRADE

Four other conditions must be considered by the worker as follows:

- Area of contact (area represented by the arc of contact and the width of the cut);
- Kind and application of a coolant;
- Hardness, softness, and composition of the workpiece;
- Required surface finish.

GRADE AND ABRASIVE TYPE

No one factor may be isolated when determining the wheel characteristics for a particular job. Grade must be related to the kind of abrasive. Grain size, bond, and structure must also be considered.

Iron, steel, steel alloys, and other ferrous metals usually require wheels with aluminum oxide. Copper, brass, aluminum, and other soft nonferrous metals, and ceramic parts are ground with silicon carbide wheels. Cemented carbides and nonmetallic materials such as granite, glass, quartz, and ceramics are ground with diamond wheels.

Cubic boron nitride wheels are efficient for grinding hard steel and other high-speed steels. Cubic boron nitride wheels are usually more expensive than aluminum oxide wheels. Therefore, they are used only when there is a demonstrated advantage, as for example, the grinding of very hard steel alloys. Many cubic boron nitride wheels are made of a CBN layer around a core of other material.

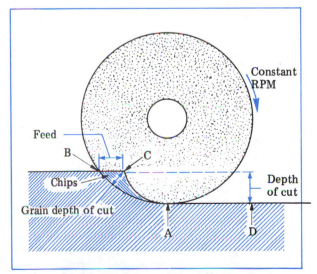

Figure 51–14 Conditions of Use Affecting Cutting Forces and Cutting Action

WHEEL TYPES, USES, AND STANDARDS

HARDNESS OF MATERIAL

Coarse and hard-grade abrasive grains are used on ductile materials. The grain size for general-purpose grinding is usually in the 36 to 60 range. Finer grain sizes in the 80 to 120 range and softer grains are used for hard materials. The hardness range of soft wheels for ordinary grinding is from F to I. The medium-hard wheel range includes J, K, L, and often M and N. Grade P, Q, and R are hard wheels.

Grinding dry or wet influences the grain hardness. Generally, when a coolant is applied, a one-grade-harder grinding wheel may be used. However, if a wheel is too hard, the workpiece may be *burned* due to overheating. Burning produces surface discoloration and may cause internal stresses, distortion, or checking.

REQUIRED SURFACE FINISH

Grinding wheel selection is based on the amount of stock to be removed and the required degree of surface finish. For heavy stock removal, a very coarse, resinoid-bonded wheel is recommended. Where fine production finishes are specified, and a coolant is used, a fine-grain, rubber- or shellac-bonded wheel is suggested. It is practical, with proper dressing, to produce an excellent surface finish by using a coarser grain.

AREA OF WHEEL CONTACT

Grinding wheels are used in both *side grinding* (grinding with the *side* of the wheel) and *peripheral grinding* (grinding with the *face* of the wheel). The area of wheel contact, however, varies. Cup wheels and segmented wheels are side grinding wheels. Coarser and softer grain sizes—for example, 301 and 401—are used.

In peripheral grinding, the face of the wheel makes (compared to side grinding) a small line contact with the workpiece. Cylindrical, tool and cutter, and surface grinders provide examples of peripheral grinding. The area of contact, however, varies in each instance. A small area of contact is made in cylindrical (peripheral external) grinding. The area is generated by the arc of contact on the periphery of the workpiece and

the wheel. Grain sizes from 54 to 80 and hardness grades of K, L, and M are used in general cylindrical grinding processes.

There is a larger area of contact in surface grinding than in external cylindrical grinding. The area is produced by the arc of the wheel in relation to a flat surface. A coarser grain size, such as 46, and a softer grade, such as I or J, may be used.

On internal cylindrical grinding, the area of contact is formed by the inside diameter of the workpiece and the grinding face of the wheel. The area of contact is larger than in external grinding. A softer grinding wheel is usually used.

WHEEL SPEED

In all instances, wheel speed is a factor. Every grinding wheel is marked with the safe limits of sfpm or RPM. The safety range of vitrified wheels is around 6,500 sfpm. Resinoid-, shellac-, and rubber-bonded wheels may be operated at speeds up to 16,000 sfpm.

WHEEL MARKING SYSTEM

The grinding wheel marking system incorporates a series of letters and numbers. The standard marking system for abrasive (wheel) types A, B, and C is summarized in Table 51–1. The kind of information a manufacturer generally gives appears on the blotter of a grinding wheel. This information relates to operator safety checkpoints, the safe operable RPM, wheel composition (according to ANSI codes), wheel dimensions, and a repeated safety precaution.

PERIPHERAL GRINDING WHEELS

Types 1, 2, 5, and 7 are straight wheels used primarily for peripheral grinding. A number of standard straight *wheel faces* are obtainable from grinding wheel manufacturers. Twelve standard shapes of wheel faces are pictured in Figure 51–15. The standard code of letters which is used in relation to dimensional features on all types of grinding wheels appears in Table 51–2.

SURFACE GRINDING WHEELS

The design features of three types of peripheral grinding wheels used for surface grinding are

Table 51–1 Standard ANSI Marking System for Abrasive Types A, B, and C

1	2	3	4	5	6
Abrasive Type	Grain Size	Grade (Hardness Scale)	Grain Structure	Bond Type	Manufacturer's Record
Prefix					
51	**A**	**24**	**M**	**6**	**V**	**32**

Manufacturer's symbol indicating exact kind of abrasive (optional)

Aluminum oxide = A
Silicon carbide = C
Cubic boron nitride = B

Coarse	Medium	Fine	Very Fine
10	30	70	220
12	36	80	240
14	46	90	280
16	54	100	320
20	60	120	400
24		150	500
		180	600

Dense to Open

1	9
2	10
3	11
4	12
5	13
6	14
7	15
8	etc.

(Use optional)

V = Vitrified
R = Rubber
RF = Rubber reinforced
B = Resinoid
BF = Resinoid reinforced
E = Shellac

Manufacturer's symbol identifying bond composition (optional)

Soft — Medium — Hard

A B C D E F G H I J K L M N O P Q R S T U V X Y Z

Table 51–2 Standard Code of Letters for Dimensions on All Types of Grinding Wheels

Letter Code	Dimension
A	Radial width of flat at periphery
B	Depth of threaded bushing in blind hole
D	Outside or overall diameter
E	Thickness of hole
F	Depth of recess on one side
G	Depth of recess on second side
H	Diameter of hole (bore)
J	Diameter of outside flat
K	Diameter of inside flat
N	Depth of relief on one side
O	Depth of relief on second side
P	Diameter of recess
R	Radius
S	Length of cylindrical section (plug type)
T	Overall thickness
U	Width of edge
V	Face angle
W	Wall (rim) thickness of grinding face

shown in Figure 51–16. The wheel types include types 1, 5, and 7.

CYLINDRICAL GRINDING WHEELS

Many grinding wheels have one or two sides *relieved*—that is, the side of the wheel is cut away. The side tapers from the diameter of the inside flat (K) to the radial flat at the periphery (A). The design features of four types of peripheral

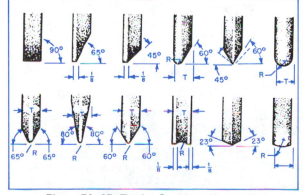

Figure 51–15 Twelve Standard Shapes of Wheel Faces

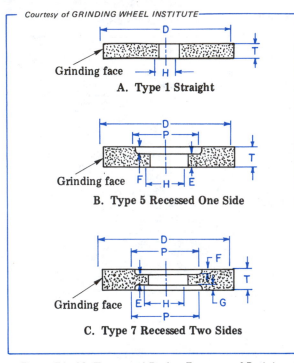

A. Type 1 Straight

B. Type 5 Recessed One Side

C. Type 7 Recessed Two Sides

Figure 51–16 Types and Design Features of Peripheral Grinding Wheels Used for Surface Grinding

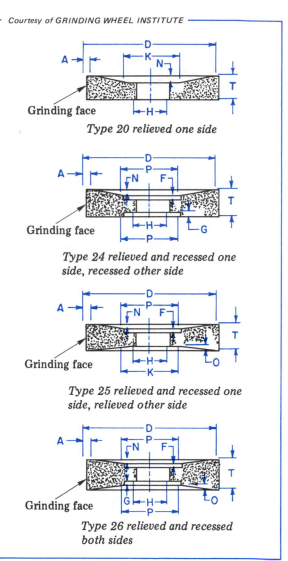

Type 20 relieved one side

Type 24 relieved and recessed one side, recessed other side

Type 25 relieved and recessed one side, relieved other side

Type 26 relieved and recessed both sides

Figure 51–17 Types and Design Features of Selected Peripheral Grinding Wheels Used for Cylindrical Grinding

grinding wheels used for cylindrical grinding are shown in Figure 51–17. The wheel types include types 20 through 26. Note that some wheels are relieved on two sides. Other wheels are relieved and recessed.

SIDE GRINDING WHEELS

Side grinding, to repeat, means grinding with the side of the wheel in contrast to grinding with the face as in peripheral grinding. Three of the most widely used side grinding wheels used primarily on vertical-spindle surface grinders are identified as types 2, 6, and 11. The design features of these types of side grinding wheels are shown in Figure 51–18.

COMBINATION PERIPHERAL AND SIDE GRINDING WHEELS

The type 12 dish wheel, illustrated in Figure 51–19, is designed for straight peripheral (face) grinding in addition to side (wall) grinding. This wheel is similar to a flaring-cup wheel. However, there are two essential differences. The

dish wheel is shallower. The periphery is dressed as a secondary cutting face.

MOUNTED WHEELS (CONES AND PLUGS)

Types 16 through 19 are referred to as *cones* and *plugs*. They are manufactured to be threaded on a bushing or they are made with a standard-diameter shank. The cones and plugs may be formed to any shape. Figure 51–20A shows the

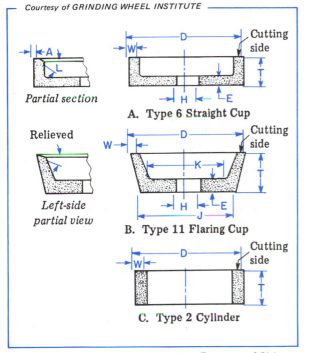

Courtesy of GRINDING WHEEL INSTITUTE

A. Type 6 Straight Cup

B. Type 11 Flaring Cup

C. Type 2 Cylinder

Figure 51–18 Types and Design Features of Side Grinding Wheels (Vertical-Spindle Surface Grinding)

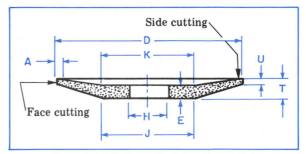

Figure 51–19 Face (Peripheral) and Side Cutting Type 12 Dish Wheel

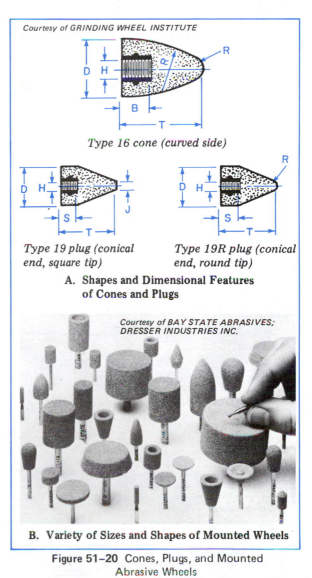

Courtesy of GRINDING WHEEL INSTITUTE

Type 16 cone (curved side)

Type 19 plug (conical end, square tip)

Type 19R plug (conical end, round tip)

A. Shapes and Dimensional Features of Cones and Plugs

Courtesy of BAY STATE ABRASIVES; DRESSER INDUSTRIES INC.

B. Variety of Sizes and Shapes of Mounted Wheels

Figure 51–20 Cones, Plugs, and Mounted Abrasive Wheels

shapes and dimensional features of cone and plug types 16 through 19. An assortment of mounted wheels used for burring and general grinding are illustrated in Figure 51–20B. Cutting is done on any face. Some cones and plugs are used for internal grinding. Most often, they are used in portable grinders for deburring, breaking the edges, or rough grinding.

COMPUTING WORK SPEED AND TABLE TRAVEL

WORK (SURFACE) SPEED

Work (surface) speed relates to the feet (or meters) per minute that a workpiece moves past the area where cutting takes place.

The RPM of the headstock for cylindrical grinding processes is equal to the work speed (fpm) divided by the circumference of the work in feet:

$$RPM = \frac{\text{Work Speed (fpm)}}{\text{Work Circumference (in feet)}}$$

Table 51–3 Grain Sizes for Grinding Selected Metals

Metal	Recommended Coarse-to-Medium Grain Sizes
Soft steel	20 to 46
Hardened steel	16 (roughing cuts)
	36 to 46 (finishing cuts)
High-speed steel	36 to 46
Cast iron	16 to 36
Aluminum	20 to 46
Brass	20 to 24
Copper	36 to 54

If the work speed is given in meters per minute,

$$RPM = \frac{\text{Work Speed (m/min)}}{\text{Work Circumference (in meters)}}$$

RATE OF TABLE TRAVEL

Table travel rates of grinding machines are generally given in inches per minute (ipm). The rate of table travel varies from two-thirds of the width of the grinding wheel down to one-eighth. The smaller rate of travel is used to produce a very smooth surface finish. Longitudinal table travel rate changes are made by adjusting the table speed-selector knob.

Table 51–4 Common Grinding Wheel and Workpiece Problems, Probable Causes, and Corrective Action

Probable Cause and/or Corrective Action	Surface Texture					Dimensional Accuracy			Operating Conditions Wheel		Spindle
	Chatter Marks	Scratches	Spiral Marks	Burning/Checking	Burnishing	Not Ground Flat	Out-of-Parallel or round	Not Sizing Uniformly	Loading	Glazing	Spindle Running Too Hot
Grinding Wheel											
Grain size											
Too fine				X	X				X	X	
Too coarse		X									
Structure											
Too dense				X					X	X	
Too hard	X			X	X	X			X	X	
Grade too soft	X	X									
Dressing too fine				X		X			X	X	
Out-of-balance	X										
Dull, glazed, loaded	X			X							
Not trued properly	X	X									
Speed too high				X							
Workpiece											
Work speed too slow				X							
Out-of-balance	X										
Not adequately supported	X										
Centers worn or require lubrication	X						X	X			
Insufficient number or improper adjustment of back rests	X						X	X			
Work speed or table travel too high	X	X									

Table 51–5 Selection Factors and Wheel Characteristics Affected

Factors Influencing Grinding Wheel Selection	Wheel Characteristics Affected				
	Abrasive Type	Grain Size	Grade	Structure	Bond Type
Type of grinding machine and operation (surface, cylindrical, tool and cutter; form, shoulder, and so on)	X	X	X	X	X
Characteristics of material to be ground (tensile strength and hardness)	X	X	X	X	
Machine features and condition (heavy-duty, light; worn bearings, loose fitting parts)			X		
Operator work practices (tendency toward light or heavy cuts)			X		
Area of contact		X	X	X	
Stock removal (severity of heavy or light cuts)		X	X	X	X
Wheel speed (sfpm)			X	X	X
Work speed (lpm)			X		
Feed rate		X	X	X	
Quality of surface finish required		X	X	X	X
Wet or dry grinding process			X	X	X

The rate of table travel in inches is determined (for cylindrical grinding processes) by multiplying the desired work speed in RPM by the distance (in inches) the workpiece should travel each revolution:

$$\text{table travel} = \frac{\text{work speed (RPM)} \times \text{table travel per revolution}}{}$$

Manufacturer's tables are used for selecting the grain sizes of grinding wheels. Recommended coarse-to-medium grain sizes for commonly machined metals are given in Table 51–3.

RECOGNIZING GRINDING PROBLEMS AND TAKING CORRECTIVE ACTION

Three groups of conditions must be recognized as contributing to grinding problems:

- *Machine systems.* The surface texture quality and dimensional accuracy of a workpiece are partially controlled by the correct functioning and precision of the pneumatic, electrical, and mechanical systems;

- *Machine processes.* A number of problems are grouped around the condition of the grinding wheel, coolant system, removal of chips, and basic processes;
- *Operator responsibilities.*

Many of the common problems encountered in grinding, their probable causes, and corrective action are summarized in Table 51–4.

SUMMARY OF FACTORS INFLUENCING THE SELECTION OF GRINDING WHEELS

Table 51–5 indicates a number of major factors influencing the selection of grinding wheels and the grinding wheel characteristics that are affected.

Safe Practices for Machine Grinding

- Store grinding wheels in a storage rack. Use protective corrugated shims between wheels. Place each wheel in its proper compartment.

- Handle wheels carefully. Avoid hitting a wheel, particularly on its edges.
- Check the outside diameter of the wheel (if worn) and the recommended maximum sfpm on the wheel blotter. Adjust the spindle to keep within the manufacturer's range.
- Use a guard that covers at least half the diameter of the wheel. Adjust the spark guard and work rest on a high-speed grinder.
- Stand to one side of the grinding wheel, particularly during start-up.
- Bring the wheel speed up to operating speed. Run at this speed for about a minute before taking a cut.

- Use only the recommended face(s) of a grinding wheel as designed and specified for each process.
- Shut off the coolant at the end of the operation before stopping the spindle to prevent the coolant from collecting at the bottom of the wheel. Such a collection produces an out-of-balance condition.
- Use safety goggles or a protective shield. Position safety glass shields (when they are provided) on the grinding machine.

UNIT 51 REVIEW AND SELF-TEST

A. GRINDING MACHINES: FUNCTIONS AND TYPES

1. State two differences in the cutting action of abrasive grains as compared to the cutting action of multiple-tooth milling cutters.

2. a. Differentiate between in-feed and end-feed centerless grinding.
   b. Give one application of in-feeding and another of end-feed centerless grinding.

3. Describe briefly what is meant by (a) a soft, abrasively charged lap and (b) the lapping process.

4. Refer to the Appendix Table on the microinch range of surface roughness for selected manufacturing processes.
   a. Indicate the general manufacturing surface finish roughness height for the following processes: (1) finish grinding, (2) honing, (3) lapping, and (4) superfinishing.
   b. Compare the surface textures produced by finish turning and by commercial grinding in terms of (1) measurement and (2) quality of the surface texture.

5. State two machine and/or cutting tool precautions the operator must take when starting up the spindle of a grinding machine.

B. GRINDING WHEEL CHARACTERISTICS, STANDARDS, AND SELECTION

1. State three conditions of use that affect the selection of a grinding wheel. (As a result, the wheel for a particular job may differ from a manufacturer's standard recommendations.)

2. Tell what effect the area of contact has in selecting a grinding wheel for cylindrical grinding as compared to surface grinding.

3. a. List the kind (category) of information that is covered by each of three major symbols in ANSI standards for grinding wheel selection.
   b. Describe the wheel construction features for each symbol.

4. a. Compute the wheel speed (in meters per minute) of a 450mm diameter grinding wheel that is traveling at 1,500 RPM. Use $\pi = 3.14$.
   b. Determine the spindle speed for a 12″ diameter grinding wheel that is traveling at 6,200 fpm. Use $\pi = 3.14$.

5. a. Compute the RPM at which a 6″ diameter workpiece is to be turned on a cylindrical grinder to produce a work speed of 84 fpm. Use $\pi = 3.14$.
   b. Determine the table travel rate for a 1 1/2″ wide wheel that is operating at 400 RPM. The table advances one-sixth of the wheel face width each revolution.

6. State three machine and/or personal safety precautions to follow while operating a grinding machine.

Grinding Wheel Preparation and Grinding Fluids

The grinding wheel must be moutned, trued, dressed, and balanced. This unit examines the principles and practices related to this preparation. Several wet coolant systems and procedures for maintaining the quality of the grinding fluid are described. Many key items in the ANSI "Safety Code for the Use, Care, and Protection of Abrasive Wheels" are summarized.

WHEEL DRESSING TECHNIQUES AND ACCESSORIES

FORM DRESSING

More and more form work is being precision ground to shape and size. Slots, grooves, and contours may be ground by *form dressing* the grinding wheel.

Wheels may be form dressed by a crush roll or with diamond rolls or dressing blocks.

CRUSH FORMING

A *crush roll* is a duplicate form of the required shape to be ground. The crush roll is brought against a slowly revolving grinding wheel (100 to 300 sfpm) with considerable force. The roll *crushes the form* into the grinding face of the wheel. Figure 52–1 illustrates the setup for forming a surface grinder wheel with a crush roll.

DIAMOND ROLL FORM AND PLATED-BLOCK DRESSER

The action of a *diamond roll* form dresser is not as severe as a crush roll. As a result of less force, the abrasive grains are cut instead of crushed. The form generated on the wheel by using a diamond roll form dresser produces a finer-quality surface finish.

A *diamond-plated dressing block* is often used for form dressing a wheel. The dressing block is preformed to the desired shape. It is then positioned on the table of the surface grinder. The revolving grinding wheel is brought down to depth while being traversed forward and backward over the block.

CONCENTRIC AND PARALLEL FACE DRESSING (TRUING)

SINGLE-POINT AND DIAMOND CLUSTER DRESSERS

The *single-point diamond dresser* is widely used to produce a true wheel face. The face is usually concentric and parallel with the spindle centerline. The *single-point diamond dresser* may be mounted in a simple holder (Figure 52–2A). Some surface grinders have the diamond mounted in an *overhead dresser* (Figure 52–2B). The diamond is fed to depth by means of a micrometer dial control. A dresser traverse handle is moved to the right and left to produce a true and parallel face.

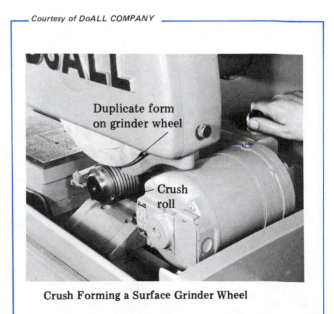

Courtesy of DoALL COMPANY

Duplicate form on grinder wheel

Crush roll

Crush Forming a Surface Grinder Wheel

Figure 52–1 Application of a Crush Roll

Courtesy of DoALL COMPANY

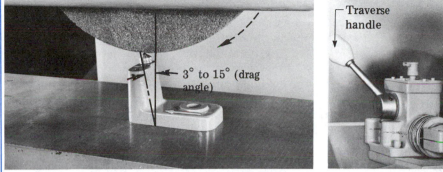

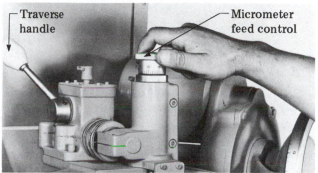

A. Diamond Dresser Mounted in a Simple Holder

B. Overhead Dresser with Micrometer Feed Control and Traverse Handle

Figure 52–2 Applications of a Single-Point Diamond Dresser

GENERAL PROCEDURES FOR USING DIAMOND DRESSERS

How to True and Dress a Surface Grinder Wheel

STEP 1   Select a sharp single-point diamond.

STEP 2   Examine the diamond to see that it is not flat, cracked, or loose in its setting.

STEP 3   Mount the diamond in a block or wheel truing fixture. The diamond mounting should bring the center line of the diamond and wheel at a 10° to 15° intersecting angle.

STEP 4   Mount the grinding wheel (or combination unit) on the spindle.

STEP 5   Position the fixture or diamond holder on a magnetic chuck.

STEP 6   Energize the magnetic chuck. Test to see that the holder is properly seated and secure.

STEP 7   Replace and secure the wheel guard.

> **Caution:** Be sure to use safety goggles or a protective shield.

STEP 8   Start the grinder. Run it free for a minute or so. Position the grinding wheel so that the center line is 1/8″ (3mm) to 1/4″ (6mm) to the right of the diamond.

STEP 9   Bring the grinding wheel down slowly until contact is made at the high point.

> **Note:** In wet grinding, use a coolant. In dry grinding, be sure the diamond is air cooled after each two or three passes.

STEP 10   Feed the diamond across the wheel face by using the cross feed handle.

STEP 11   Continue to feed the diamond 0.0005″ (0.01mm) for each pass across the wheel face. The final feed for precision grinding may be reduced to 0.00015″ (0.004mm).

> **Note:** In the last pass or two across the wheel face, there is no wheel feed—that is, the wheel is allowed to *spark out* as it is traversed over the diamond.

STEP 12   Dress the wheel corners with a hand dresser.

STEP 13   Stop the coolant flow a minute or more before the machine.

How to True and Dress a Cylindrical Grinder Wheel

STEP 1   Secure the diamond holder on the table or footstock.

STEP 2   Set the center of the holder on the horizontal center line of the wheel. The vertical angle of the holder should be between

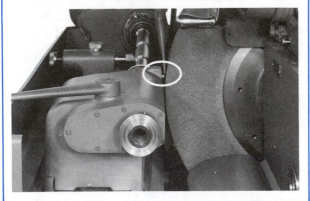

Courtesy of LANDIS TOOL COMPANY

Figure 52–3 Diamond Dresser Setting in Footstock for Dressing a Cylindrical Grinder Wheel

3° to 15°. The position of a diamond holder in a footstock is shown in Figure 52–3.

STEP 3 Replace all guards. Start the machine.

Caution: Use an eye protection device. Stand to one side. Allow the bearings and machine to reach operating temperature.

STEP 4 Adjust the wheel until the highest point is in contact with the diamond.

Note: The coolant supply must be flowing before there is any contact. Otherwise, it is possible to crack and damage the diamond.

STEP 5 Feed the diamond 0.0005″ (0.01mm) each pass.

STEP 6 Continue to feed and traverse the wheel each pass.

EFFECT OF TRAVERSE RATE ON WHEEL CUTTING ACTION

The same wheel may be dressed for rough grinding and finish grinding. The faster the dresser traverses the face of a revolving grinding wheel, the sharper and more open the grains that are produced. Sharp grains are better suited for fast cutting action and heavy cuts. The slower the traverse, the finer the grains. With a slow traverse, the diamond dresser cuts the abrasive and dulls the grains slightly. Such grains cut less and produce a finer surface finish.

In general practice, traverse rates from 12 to 25 ipm produce sharp cutting grains. A medium cutting grain results from using traverse rates from 6 to 12 ipm. A fine cutting grain is produced at a traverse rate of 2 to 6 ipm.

PROPERTIES AND FEATURES OF CBN AND DIAMOND WHEELS

CUBIC BORON NITRIDE (BORAZON CBN) WHEELS

The hardness range of CBN is between silicon carbide and the diamond. CBN has a Knoop hardness reading of 4,700. This reading compares to 7,000 for the diamond, which is the only material harder than CBN. The extremely hard CBN crystals have sharp corners, remain sharper longer, and normally do not load. Therefore, in many machine grinding operations where CBN wheels are used, the following note appears: **USE AS IS. DO NOT DRESS THE WHEEL.** The cooler cutting action of CBN produces less thermal damage to the workpiece.

Like diamond grinding wheels, CBN wheels consist of a core. This core is surrounded by a bond in which abrasive crystals are impregnated. The core is precision machined to fit the spindle within a closer tolerance than is allowed for conventional wheels. The close fit helps the wheel to stay in balance.

DIAMOND GRINDING WHEELS

Diamond grinding wheels are generally produced in resinoid, metal, and vitrified bonds. The core is molded into one of the wheel type forms. The core material may be steel, an aluminum-filled resin composition (aluminoid); or bronze, copper, or plastic. Three common bonds are generally used in the manufacture of diamond grinding wheels:

• *Resinoid bond*, which produces fast and cool cutting action for grinding carbide cutters and for other precision grinding processes,

- *Metal bond*, which is a harder bond that resists the wearing effect produced, for example, by offhand grinding single-point parts and tools;
- *Vitrified bond*, which combines grinding characteristics of both the resinoid bond and the metal bond to produce an intermediate fast and cool cutting action and a high resistance to wear.

DIAMOND WHEEL MARKINGS

The depth of the diamond section varies from 1/16″ (1.6mm) to 1/4″ (6.4mm). The greater depths are used for heavy stock removal, large areas of wheel contact, and heavy production demands. Diamond and cubic boron nitride wheel markings differ from the markings used on conventional wheels. Diamond and CBN grinding wheel markings include a *concentration number* which represents the ratio between the diamond grains and the bonding material. A high concentration (100) is desirable for heavy stock removal. A low concentration (25) is adequate for light cuts and where low heat generation is a requirement.

CONDITIONING DIAMOND AND BORAZON (CBN) GRINDING WHEELS

Great care must be taken in conditioning diamond grinding wheels. If the wheel is not concentric, the edges tend to chip, greater wheel wear occurs, and surface finish and dimensional accuracy are impaired. Therefore, diamond wheels—and borazon(CBN) wheels—must run true. As a general rule, the tolerance for outside diameter wheel runout in cylindrical grinding must be within 0.0005″ (0.01mm). The tolerance for extremely precise work is 0.00025″ (0.006mm). The grinding face of a cup wheel should be held to 0.0001″ (0.002mm) minimum. Testing the trueness requires the use of a *tenths dial indicator*.

Diamond and cubic boron nitride wheels are never trued or dressed with a diamond dresser. A coolant should always be used when truing a diamond wheel. A small amount of water as a spray is applied to CBN. Three common devices for truing diamond wheels are used. A standard dressing stick or brake-controlled dresser (Figure 52–4) is used on surface grinder applications.

Courtesy of DoALL COMPANY

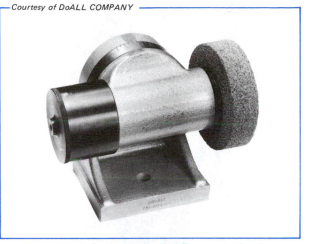

Figure 52–4 Brake-Controlled Dresser for Truing Diamond Wheels

On cylindrical and tool and cutter grinding, a toolpost grinder is occasionally used for truing the wheel.

How to Prepare Borazon (CBN) and Diamond Abrasive Wheels

Using a Brake-Controlled Device to True Diamond and CBN Wheels

STEP 1 Select the appropriate grade and grain size to accommodate the diamond or CBN wheel to be trued and dressed.

> Note: A metal- or vitrified-bonded diamond wheel is trued with a 100 grain, M grade, silicon carbide vitrified wheel. A resinoid-bonded diamond wheel is trued with a 100 grain, M grade, aluminum oxide vitrified wheel. A resinoid-bonded borazon (CBN) wheel is trued with an aluminum oxide or silicon carbide stick or wheel. A fine grit (400 or finer) and G hardness bond are recommended.

STEP 2 Position the brake-type truing device. The holder must have about a 30° angle with the vertical center line of the wheel.

STEP 3 Run the spindle for a minute. Move the diamond or CBN wheel into contact with the brake-controlled dresser wheel.

Courtesy of DoALL COMPANY

Figure 52–5 Parallel Balancing Ways

STEP 4 Retard the dressing wheel speed by applying the spindle wheel brake. This action results in scrubbing and truing of the diamond or CBN wheel face.

Note: This truing is usually followed by light stick dressing. This dressing improves the cutting action of the wheel.

FUNCTIONS AND TECHNIQUES OF WHEEL BALANCING

PARALLEL AND OVERLAPPING DISK BALANCING WAYS

Parallel and overlapping disk balancing ways are two common devices used in balancing grinding wheels. The *parallel balancing ways* (Figure 52–5) consist of two parallel knife-edge surfaces. These ways are part of a solid frame. The device is leveled so that the two parallel ways are on a horizontal plane.

A grinding wheel is balanced by placing the wheel and its adapter on a *balancing arbor*. The assembly is placed carefully on the ways. The wheel is allowed to turn slowly until it comes to rest. The heavy spot is on the underside of the arbor. This spot is marked with chalk. Two other lines are then drawn at a right angle to the heavy spot.

Courtesy of DoALL COMPANY

Wheel mount

Balancing weights

Mount with Two Balancing Weights

Figure 52–6 Positioning of Weights to Balance a Grinding Wheel

Balancing weights in the flanges of the adapter (Figure 52–6) are positioned for a *trial balance*. The wheel is retested. Further adjustments of the balancing weights are made until the wheel remains at rest in any position on the balancing ways. Wheel mounts are available with two, three, or more balancing weights.

The *overlapping disk balancing ways* consist of four perfectly balanced and free-turning overlapping disks. The grinding wheel assembly and balancing arbor are placed across the overlapping edges of the disks. The assembly is allowed to turn until the heavy spot comes to rest at the lowest point. The balancing weights are then adjusted the same as when parallel balancing ways are used.

GRINDING PROBLEMS RELATED TO TRUING, DRESSING, AND BALANCING

Any one or combination of grinding wheel, machine, and operator variables affects grinding. Eight common truing, dressing, and balancing problems with probable causes and corrective actions appear in Table 52–1. If the problem is not corrected, other variables must be checked. These variables deal with the characteristics of the grinding wheel, the workpiece, cutting fluid, machine condition, and work processes.

Table 52–1 Truing, Dressing, and Balancing Problems with Probable Causes and Corrective Action

Problem	Probable Cause	Corrective Action
Chatter marks	—Out-of-balance	—Rebalance after truing operation —Balance carefully on own mounting
	—Out-of-round	—True before and after balancing
	—Acting too hard	—Use faster dressing feed and traverse
Scratching of the work	—Foreign matter in wheel —Coarse grading	—Dress the wheel
Checking or burning of the work	—Wheel too hard	—Use a faster dress rate to open the wheel face
Diamond lines in work	—Dressing too fast	—Slow dress the wheel face
Inaccuracies in work	—Improper dressing	—Check alignment of dressing process
Rate of cut too slow	—Dressing too slow	—Increase the dressing rate to open the wheel face
Wheel acting too soft (not holding size)	—Improper dressing	—Slow down the traverse rate —Use lighter dressing feed
Wheel loading	—Infrequent dressing	—Dress the wheel more often

GRINDING FLUIDS

KINDS OF GRINDING FLUIDS

The terms *grinding fluids*, *grinding coolants*, and *cutting coolants* are used interchangeably in the shop.

Three general kinds of grinding fluids are used. *Water-soluble chemical grinding fluids* are the most common. They are particularly useful in medium to heavy stock removal processes. These fluids are transparent (a valuable aid to the operator) and have excellent cooling properties.

Rust inhibitors and detergents are added to the solutions to provide good adhesion and cleaning qualities. The lubricating properties are improved by adding water-soluble polymers. Fluid life is increased by adding bactericides and disinfectants in the solutions. They control bacteria growth and retard rancid conditions in the fluids.

Water-soluble oil grinding fluids combine the cooling qualities of water and the lubricating advantages of oil. These oil/water fluids form a milky solution. An emulsifying agent is added to the water and oil (either natural or synthetic).

Caution: The additives must be controlled. Too strong a disinfectant solution may produce skin irritation or possible infection.

Water-soluble oil grinding fluids are used for light to moderate stock removal processes. Parts may be held to reasonably accurate dimensional and surface finish tolerances.

Straight oil grinding fluids are the most costly of the three fluids. These fluids provide excellent lubrication. However, they are not as effective as the water-soluble fluids in dissipating heat. Straight oil grinding fluids are used primarily on very hard materials requiring high dimensional accuracy and surface finish and long wheel wear life. Oil grinding fluids are especially practical in thread and other heavy form grinding processes.

COOLANT SYSTEMS

WET GRINDING METHODS

There are three principal methods of supplying grinding fluids at the point of cutting action between the workpiece and the grinding wheel. These methods of wet grinding include the *flood*, *through-the-wheel*, and *mist coolant systems*.

Flood Coolant System. The grinding fluid is forced from a reservoir through a nozzle. The fluid floods the work area and is then recirculated through the system. Both the volume and the force of the coolant are controlled to overcome air currents around the wheel. Otherwise, these currents tend to force the fluid away from

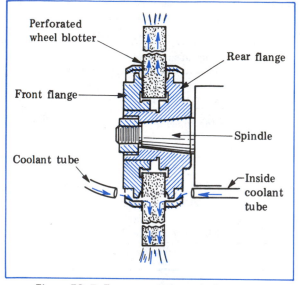

Figure 52–7 Features of Through-the-Wheel Coolant System

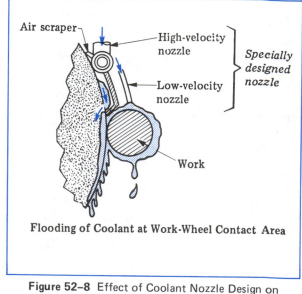

Figure 52–8 Effect of Coolant Nozzle Design on Coolant Application at Point of Contact

the required point of contact. The fluid system provides for straining out foreign particles and producing a recycled, clean coolant.

Through-the-Wheel Coolant System. The features of the through-the-wheel coolant system are illustrated in Figure 52–7. The grinding fluid is pumped through two coolant tubes. These tubes are on both sides of the grinding wheel. The wheels have flanges. As the fluid is discharged into a recess on each flange, the centrifugal force of the revolving wheel carries the fluid through a series of holes in the flanges, washers, and blotters to the center of the wheel. The flow continues with force to the periphery of the wheel. The porous feature of vitrified wheels is applied in flowing the coolant through the wheel.

This system keeps the wheel face flushed free of chips and abrasive particles so that there is a cool cutting action.

Mist Coolant System. This system works on an atomizer principle. Air at high velocity flows through a T-connection. One branch includes the coolant supply. By passing the air through the connection, a small volume of liquid is atomized into a large mass with air. The cool-

ant stream emerges from the nozzle as a mist. Cooling action is produced by the air and the evaporation of vapor. The material and abrasive particles are blown from the grinding area. One advantage of this system is that the operation is unobstructed.

IMPORTANCE OF COOLANT NOZZLE DESIGN

With a standard nozzle the air flow produced by a highly revolving grinding wheel causes an air pocket or bubble to form. This bubble causes the fluid to be blown away from the area where it is needed. With a specially designed nozzle consisting of a high-velocity nozzle, a low-velocity nozzle, and an air scraper (Figure 52–8), flooding of the work-wheel interface and the immediate adjacent area occurs. This condition is needed for effective cooling, lubricating, and chip removal.

EXHAUST SYSTEMS FOR DRY GRINDING

Dry grinding refers to operations that are carried on without a liquid fluid. An exhaust system is needed to remove grinding dust, abrasive material, and chips from the work, grinding wheel, and surrounding area. This system is important for the protection of all machines in the work area. Removal of the dust from the air also constitutes an industrial safety requirement to protect the operator.

The exhaust system usually consists of a nozzle attached to the wheel guard. A flexible tubing leads from another part of the guard to the exhaust attachment. The foreign particles that are removed (drawn) from the grinding area and around the wheel and guard are sucked through a spiral separator.

Safe Practices in Grinding Wheel Preparation

The following safe practices are summarized from the ANSI "Safety Code for the Use, Care, and Protection of Abrasive Wheels."

- Fit the wheel to the adapter mounting flange or spindle. A close sliding fit is needed for close wheel balance. All contacting surfaces of the wheel and mounting must be burr free and perfectly clean.

- Place a compressible washer wheel blotter of proper thickness and diameter between each flange and each side of the wheel.
- Tighten any retaining screws on an adapter so that they hold the wheel securely.
- Secure the wheel guard. It must cover more than half the wheel.
- Ring test each grinding wheel. Check the recommended sfpm. Be sure the spindle RPM does not exceed the speed limitations.
- Stand to one side and out of line with the wheel when starting a grinding wheel.
- Wear only approved safety glasses or other face protective devices.
- Grind only on the face(s) that is specifically designed for a particular grinding operation.
- Stop the coolant flow a short time before the grinding wheel is stopped to prevent a heavy side of the wheel from developing and throwing the wheel out of balance.
- True and dress a wheel with a diamond or other dresser pointed away from the centerline of the wheel.
- Start to true and dress a wheel by carefully bringing the wheel face so that its high point makes contact. Successive cuts are then taken to produce a concentric wheel face.

UNIT 52 REVIEW AND SELF-TEST

1. Explain briefly why cubic boron nitride and diamond wheels require less frequent dressing than aluminum oxide and silicon carbide abrasive wheels.

2. State the difference between crush forming and preformed diamond dressing block contour forming on a grinding wheel face.

3. List the steps for truing and dressing an aluminum oxide abrasive wheel for grinding flat surfaces with a horizontal surface grinder.

4. Describe briefly how the grinding machine operator may change the cutting action of an abrasive wheel during dressing.

5. a. State a probable cause of each of the three following grinding problems:
   (1) scratches in the surface finish, (2) dimensional variations of the finished surface (the dimension is not held uniformly), and (3) chatter marks.
   b. Give the correct action to take for each problem.

6. a. Name three additives to cutting fluids used in grinding.
   b. Make a general statement about the functions performed by the cutting fluid additives.

7. State three checks the surface grinder operator makes to prevent damage to a workpiece due to grinding wheel faults.

Surface Grinders and Accessories: Technology and Processes

This section deals with the major construction and control features of standard and automated precision surface grinders. The technology and shop practices are related to work-holding accessories, machine setups, and flat grinding processes. Accessories and procedures for grinding angular and vertical surfaces and shoulders, truing and dressing devices, and methods of form (provile) grinding and cutting off, are covered in detail.

UNIT 53

Design Features and Setups for Flat Grinding

DESIGN FEATURES OF THE HORIZONTAL-SPINDLE SURFACE GRINDER

COLUMN

The column serves two main purposes. It supports the spindle housing and wheel head. It permits vertical movement for positioning and controlling the depth of cut and performing all grinding operations.

TABLE AND CROSS FEED FEATURES

The main components and operating handwheels on a horizontal-spindle, reciprocating-table surface grinder are identified in Figure 53–1. The newer 612 and 618 models of surface grinders employ both inch and SI metric units of measure.

The usual table travel of a 612 model is 14″ (356mm). Table feeds with hydraulic controls for this size of machine are infinitely variable from 0 to 78 sfpm (0 to 21.3 m/min). The range of automatic cross feed increments is from 1/64″ to 1/4″ (0.4mm to 6mm). Cross feed reference graduations are usually in 0.001″ (0.025mm). However, a fine-feed cross feed attachment is available. The attachment is graduated in 0.0001″ (0.002mm) increments.

WHEEL HEAD

The down-feed handwheel on toolroom grinders is graduated in 0.0001″ or 0.002mm (Figure 53–2). A handwheel may be fitted with a *zeroing slip ring* or mounted pointer, which permits the operator to set the handwheel to a zero reference point. This setting simplifies the dimensional information the operator must remember. The zeroing ring also reduces setting errors.

OPERATING FEATURES OF THE FULLY AUTOMATIC PRECISION SURFACE GRINDER

Plunge and straight surface grinding may be done automatically on production runs. Dimensional accuracies may be held to within ±0.0001″ (±0.002mm) with finishes of 8 to 10 microinch a.a. The five basic features included in an automatic precision surface grinder are as follows:

- A complete digital control system,
- Automatic down-feed (plunge grinding),
- Automatic surface grinding,
- Automatic wheel dressing,
- Automatic digital position readout.

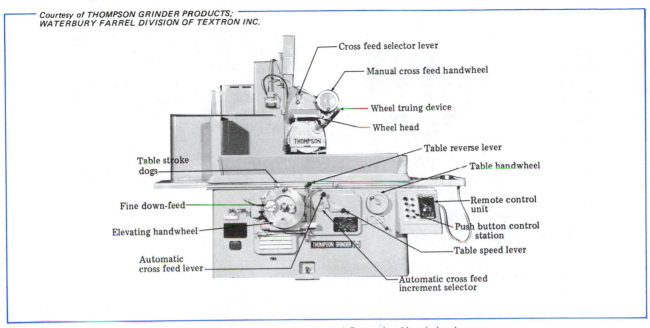

Figure 53–1 Main Components and Operating Handwheels on a
Horizontal-Spindle, Reciprocating-Table Surface Grinder

Figure 53–2 Down-Feed Handwheel and Zeroing
Slip Ring (0.0001″ Graduations)

DIGITAL CONTROL SYSTEM

The digital control system panel on a fully automatic precision surface grinder (Figure 53–3) provides the following controls:

- Vertical feed, a four-position switch for manual, automatic, down, and up movements;
- Single step, a control that provides up or down wheel head movement at 0.000050″ (0.0001mm) increments;
- Full feed down, a control that permits the operator to conserve on setups and on productive machine time;
- Down-feed table, the down-feed selector control for plunge grinding;
- Down limit set, a control used for referencing the down-feed counter to the zero plane;
- Auto dresser, an off-on selector switch;
- Single step offset, a control that moves the wheel head in single 0.000050″ (0.0001mm) increments to correct any accumulated error;
- Reset up, a control that raises the wheel head to the up limit;
- Cycle start and cycle stop controls.

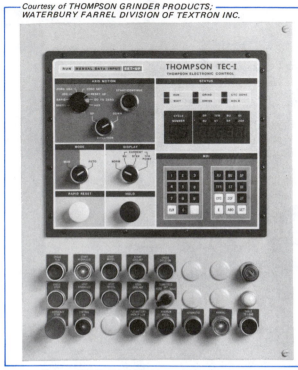

Figure 53–3 Digital Control System Panel on Fully Automatic Precision Surface Grinder

Other switches on the control panel are used with the coolant system, lubricant control, and for an emergency stop.

AUTOMATIC DOWN-FEED

Down-feed may be controlled for each table reversal or every other table pass. The feed increments may range from 0.0001″ (0.025mm) to 0.0099″ (0.25mm).

The grinding wheel is first referenced to the zero plane. The up limit thumb switches on the console panel are set to the desired dimension. The coarse-feed increment (0.0001″ to 0.0099″ or 0.02mm to 0.25mm) is set. Then the fine-feed increment (0.0001″ to 0.0009″ or 0.02mm to 0.025mm) and the point at which the fine feed is to begin are set.

Additional thumb switches are set for the number of spark-out passes, automatic wheel dressing and compensation (if required), and when the down-feed is to occur. The vertical feed selector is set to auto. Finally, the cycle start button is pushed.

AUTOMATIC SURFACE GRINDING

Selector switches are used for machine cycling. The down-feed increments and maximum depth are established as for plunge grinding. Both the cross feed reversal and wheel down-feeds are controlled automatically. They may be set for each or every other reversal.

AUTOMATIC WHEEL DRESSING

The operator selects when the grinding wheel is to be automatically dressed. The digital control is set to the number of feeds to wheel dress. The dressing may take place at any given number of grinding passes from 1 to 999.

The diamond dressing tool in the over-the-wheel dresser is fed automatically down to a preselected depth. Following this dressing pass, the wheel height is set to lower automatically. This lowering is to compensate for the reduction of wheel diameter resulting from dressing. Normally, increments of from 0.0002″ (0.005mm) to 0.001″ (0.02mm) are used.

DIGITAL POSITION READOUT

The position of the freshly dressed wheel above the table top or any other reference surface is shown on the digital position readout. The accuracy of a readout is within 0.0001″ (0.002mm).

WHEEL AND WORK-HOLDING ACCESSORIES

General grinding wheel accessories include wheel adapters, a wheel balancing arbor, and parallel ways or an overlapping disk balancing stand.

Magnetic chucks are the most versatile work-holding devices for surface grinding ferrous metals. Magnetic chucks may be of the *permanent magnet* or *electromagnetic* type. The common chuck shape for horizontal-spindle surface grinders is rectangular. However, a rotary magnetic chuck is adapted to operations that require a circular, scratch pattern. This chuck has an independent power and magnetizing source. Applications of magnetic chucking principles are to be found in magnetic V-blocks, vises, and parallel blocks, and sine chucks.

Magna-vise clamps, the toolmaker's vise, and the three-way vise are commonly used in combination with magnetic accessories. The three-way vise may be set quickly at any compound angle.

ELECTROMAGNETIC CHUCKS

Electromagnetic chucks stand higher on the table than permanent magnet chucks. The top face is laminated and usually has holding power across its entire area. Ferrous metal workpieces may be held directly on the face or by using other accessories such as magnetic V-blocks and parallels and vises.

Electromagnetic forces are controlled by an on-off position switch. Direct current is used to produce the electromagnetic effect. Residual magnetism remaining in the workpiece is removed on some machines by using a *selective chuck control.*

MAGNA-VISE (PERMA-)CLAMPS

Nonmagnetic workpieces that do not have a large bearing surface may be held for grinding with magna-vise clamps. Each clamp is made of a flexible, comb-like, thin, flat bar. The bar is attached to a solid bar by a spring-steel hinge.

The clamps, also called *perma-clamps*, are used in pairs (Figure 53–4). The solid edge of one clamp is positioned against the backing plate of the magnetic chuck. The work is brought against the toothed edge of the clamp. The toothed edge of the second clamp is fitted snugly against the side of the workpiece. As the magnetic chuck is energized, the toothed edges of the perma-clamps pull down. The force exerted by the perma-clamps holds the workpiece against the chuck face.

LAMINATED PLATES, PARALLELS, AND BLOCKS

Magnetic chuck parallels are referred to as *laminated blocks.* Usually, narrow strips of low-carbon steel, alternated with thin separators (spacing strips), are welded into a positive non-shift unit. The chuck parallels are then precisely ground for parallelism.

Round, square, and irregular-shaped workpieces are sometimes held on precision laminated V-blocks, also called *chuck V-blocks.* An adapter (auxiliary top plate) is used to hold small and thin workpieces securely.

Backing plates on the sides and ends of magnetic chucks provide ideal locating surfaces for many grinding setups.

VACUUM CHUCKS

Vacuum chucks hold work against a plane surface by exhausting the air from between the part and the chuck face. This technique of holding applies to magnetic and nonmagnetic materials. The vacuum chuck has been found to be a practical device for holding paper-thin (0.002″ to 0.003″) workpieces.

PROBLEMS ENCOUNTERED IN PRODUCING FLAT SURFACES

DISCONTINUOUS SURFACES

Flat surfaces are *discontinuous* if they include holes, ridges, grooves, or slots. When many work-

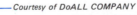

Comb-like clamp bars

Spring-steel hinge — Solid backing bar

Figure 53–4 Magna-Vise (Perma-) Clamps

Figure 53–5 Discontinuous Surface Grinding Setup Requiring Operator Judgment and Control

pieces having a space between each one are ground at the same setting, the flat surface is interrupted. A discontinuous surface grinding setup is shown in Figure 53–5. As the grinding wheel approaches an interrupted section, the reduced area of wheel contact permits the wheel to cut slightly deeper. This dipping of the wheel is due to the fact that a grinding wheel normally exerts force during the cutting action. Dimensional and surface flatness inaccuracy may be produced.

The problem of discontinuous surfaces is overcome by (1) keeping the wheel sharp and cool cutting at all times and (2) taking light finish cuts.

WORK STRESSES AND DISTORTION

Castings; forgings; and heat-treated, cold-formed, extruded, or other manufactured parts have internal stresses and strains produced by chilling or mechanically working the material. Such stresses and strains are relieved, or unlocked, when the top surface or skin is machined away. A relieved surface changes shape. The grinder operator needs to recognize that the workpiece is *distorted* (warped or bent) away from its nominal shape. Compensation must be made to overcome this distortion (warp or bend). This may be done at the machine or by subjecting the part to heat treatment. Severely distorted parts require straightening by force on arbor and other presses.

WARPED WORKPIECES

Thin shims are often used to facilitate holding and grinding a warped piece. The top surface is carefully ground to flatness. The part is then turned over and held without the use of shims. The second surface, when ground, should be parallel to the flat top surface.

One of the easiest ways of checking the flatness of a surface is to place a precision straight edge on the ground surface.

Often, the workpiece is placed on a surface plate. If a feeler gage can be inserted at any point between the workpiece and the surface plate, the ground surface is not perfectly flat.

DOWN-FEEDS AND CROSS FEEDS FOR ROUGH AND FINISH GRINDING CUTS

Coarse, rough cuts are taken at the beginning of the grinding operation. The depth of successive cuts is reduced. Sufficient stock is left for finish grinding. The surface is generally rough ground by using cross feed increments of from 0.030″ to 0.050″ (0.8mm to 1.2mm). The rough-cut down-feed per pass on general jobbing shop work is 0.003″ to 0.008″ (0.1mm to 0.2mm) for average rough grinding. A heavier down-feed may be used on cast iron and soft steel. The down-feed may be increased still further if the work area being ground is small.

Fine down-feeds are used for the finishing passes. An excessive number of fine down-feed passes tends to load the wheel face. The cutting action is then inefficient and overheating may result. It is important to redress the wheel for light finish cuts of 0.0005″ (0.01mm) and finer. Where the down-feed is very light, fast table speeds and large cross feed increments are used. The workpiece must be passed under the grinding wheel fast enough to eliminate the possibility of overheating.

GRINDING EDGES SQUARE AND PARALLEL

In addition to having flat surfaces ground, most rectangular workpieces require that the edges be ground parallel and square. Such workpieces are usually rough machined slightly oversize. Stock is left to correct any inaccuracy in squareness, parallelism, and machining surface irregularity and to permit the sides to be ground square and to the required dimensional accuracy for surface finish and size. The general shop practice is to allow 0.010″ (0.2mm) on each side for rough and finish grinding.

A common method of grinding a workpiece square involves the use of a precision vise. The vise base surface is first checked for parallelism in relation to the spindle. The dimensional accuracy requirements (length, width, and height) for grinding the edges of a workpiece square and parallel are illustrated in Figure 53–6. The workpiece is placed on a parallel on the vise base. The first edge (1) is ground. The adjacent edge (2) is ground next. The squareness of its posi-

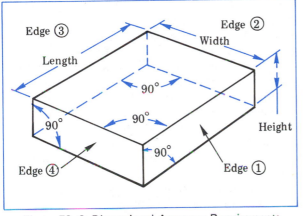

Figure 53–6 Dimensional Accuracy Requirements for Grinding Edges Square and Parallel

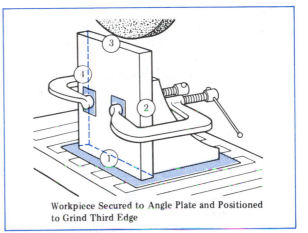

Workpiece Secured to Angle Plate and Positioned to Grind Third Edge

Figure 53–7 Position and Holding of Workpiece to Grind Third and Fourth Edges

tion in the vise is checked vertically (at 90°) with a dial indicator.

After these two edges are ground square, each edge is used as a reference plane. Edge ① is placed on a parallel. The workpiece is secured and edge ③ is ground to the required dimension. Then edge ② is positioned on a parallel and clamped securely. Edge ④ is then ground parallel, square, and to the specified dimension. The squareness of edges ① and ② may be checked with a solid-steel square. A more precise measurement may be taken by using a cylindrical square.

A second method of grinding square edges is to use an angle plate (Figure 53–7). The large, flat surfaces are first ground parallel. The faces then act as plane reference surfaces. The use of a V-block and angle plate for holding a round part is illustrated in Figure 53–8.

HOLDING THIN WORKPIECES

USE OF MAGNA-VISE CLAMPS

Serrated magna-vise clamps provide a convenient work-holding device for small, thin parts (that are slightly thicker than the height of the clamps). A thin part should be shimmed, if necessary. With this setup, the workpiece is mounted parallel to the side faces of the magnetic chuck. Nonferrous metals are also held securely in a magnetic chuck with magna-vise clamps.

USE OF ADAPTER PLATES AND FINE POLE MAGNETIC CHUCKS

An adapter plate is specially designed to hold small, thin workpieces. The adapter plate is a

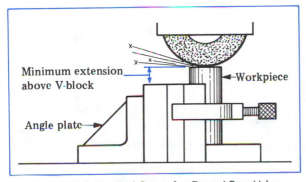

Figure 53–8 Vertical Setup for Round Part Using V-Block and Angle Plate on a Magnetic Chuck

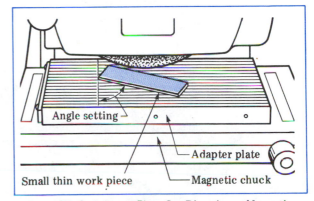

Figure 53–9 Adapter Plate Set Directly on Magnetic Chuck for Holding Small, Thin Wirkpieces

precision ground rectangular plate. It is set directly on a magnetic chuck as illustrated in Figure 53–9. The laminations run lengthwise and are finely spaced to allow more, but weaker, lines of force to act on the workpiece. The adapter plate is energized by turning the magnetic chuck on. Small, thin workpieces are positioned at an angle of 15° to 30° to minimize the heat generated during each pass.

Safe Practices in Grinding Flat Surfaces and Thin Workpieces

- Add a square block or angle plate in front of a V-block to give added support against the cutting forces of a grinding wheel. Nest small workpieces against end and side blocks.

- Check the clearance between the workpiece, mounting device, wheel, and other machine parts before starting the machine.
- Bring the grinding wheel into contact at the high spot of a surface to be ground by slowly feeding the grinding wheel.
- Recheck how securely the diamond holder is held on the magnetic chuck for dressing and truing operations. Maintain a sharp diamond dresser by rotating the point from 10° to 20° after each dressing. Use a coolant during dressing with a diamond.
- Replace the machine guard and wear protective goggles or a shield. Observe general machine safety precautions. Stand out of line of the wheel travel.
- Check the position of splash guards. The coolant must be contained on the machine and flow back to the reservoir. Any spilled grinding fluid or other oils are to be wiped from the machine area.

UNIT 53 REVIEW AND SELF-TEST

1. a. List three design features of a fully automatic surface grinder, exclusive of the digital control system.
    b. Describe briefly the function of each of the design features listed.

2. a. Tell what common practices are followed with table speeds and traverse feeds when fine finish cuts of 0.0005″ (0.01mm) are taken.
    b. Identify two problems encountered by the machine operator (1) when an excessive number of fine cuts (passes) are taken and (2) what corrective steps may be followed.

3. Identify the steps for setting up and grinding the two ends square and to accurate length for a short-length die block. An angle plate and magnetic chuck setup are to be used.

4. a. Tell what effect relieving part of a wide-face abrasive wheel has on the generation of heat during grinding.
    b. Explain how the opposite faces of thin parts that are bent or warped may be ground parallel.

5. State two machining techniques that may be used to prevent the distortion that may be produced when thin workpieces are ground.

6. List six safety precautions the surface grinder operator must check for the setup, prior to taking the first cut.

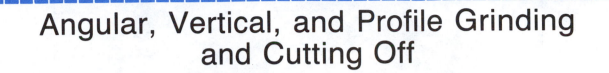

Angular, Vertical, and Profile Grinding and Cutting Off

A. GRINDING ANGULAR AND VERTICAL SURFACES

Common accessories that are used to hold workpieces at an angle and to generate a flat (angular) surface are grouped as follows:

- Vises (plain, adjustable, universal, and swivel);
- Angle plates (adjustable and sine angle plates);
- Magnetic work-holding accessories (V-blocks, toolmaker's knee, swivel chucks, and sine table);
- Grinding fixtures for production work.

The first part of this unit describes these accessories with applications to angular surface grinding. Step-by-step procedures are given for machine setups and the actual grinding of straight, filleted, and angular shoulders and vertical surfaces.

MAGNETIC WORK-HOLDING ACCESSORIES

MAGNETIC TOOLMAKER'S KNEE

The magnetic toolmaker's knee (Figure 54–1) is especially designed to hold odd-shaped parts for grinding. The model illustrated is a permanent magnetic tool. The magnetic surface is 4″ × 6 1/2″ (100mm × 165mm) and is parallel within 0.002″ (0.05mm) overall. The tolerance for squareness is 0.00005″ per inch.

SWIVEL-TYPE MAGNETIC CHUCK

This horizontal magnetic chuck has a swivel feature. This feature makes it possible to position the chuck at an angle from the usual horizontal plane. The top face on which the workpiece is positioned and held by magnetic force is set at the required angle. The accuracy of the angle may be checked with a vernier bevel protractor. Finer angle setups are checked against an angle sine block combination.

MAGNETIC SINE TABLES

Plain and compound electromagnetic sine chucks are shown in Figure 54–2. A plain electromagnetic sine chuck (Figure 54–2A) has two hinged sections. The first section includes a base and angle plate. Angles from 0° to 90° may be set. Gage blocks are used to establish the vertical height for a required angle.

The second section is hinged to the accessory table. Again, gage blocks are used to set the second table (a magnetic table) at a specified angle from 0° to 90°. Workpieces are positioned directly on the magnetic chuck face.

A compound electromagnetic sine chuck (Figure 54–2B) includes a third hinged section. This design feature permits setting the electromagnetic chuck face at a compound angle. Each section is set precisely at one of the required angles by using gage blocks.

Accurate → serrated holding surface

Figure 54–1 Permanent Magnetic Toolmaker's Knee for Inspection, Layout, and Grinding Setups

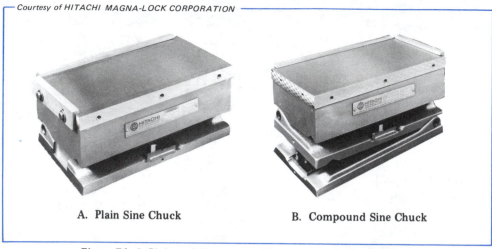

A. Plain Sine Chuck   B. Compound Sine Chuck

Figure 54–2 Plain and Compound Electromagnetic Sine Chucks

How to Grind Angular Surfaces

Using a Right Angle Plate

STEP 1 Select an angle plate of a suitable size. Set the angle plate on the chuck. Align the vertical face.

> **Note:** The surface to be ground should be set parallel with the table travel.

STEP 2 Place the surface of the work against the one face of the right angle. Position the work at the required angle. Use a vernier bevel protractor, gage blocks, or other measuring device. (The angle plate setup for grinding an angle is shown in Figure 54–3.)

STEP 3 Clamp securely. Check to be sure the wheel clears any clamps or other accessory.

STEP 4 Set up the machine. Take roughing and finishing cuts. Check the angular measurements. Grind to size and surface finish.

Using a Magnetic V-Block

STEP 1 Select the magnetic V-block that accommodates the part to be ground. Clean the base and the work contacting surfaces.

STEP 2 Locate the magnetic V-block centrally on the chuck. Position it parallel with the side of the chuck. (The setting and use of a magnetic V-block to position and hold a workpiece are illustrated in Figure 54–4.

STEP 3 Nest the workpiece in the magnetic V-block. Turn the magnetic chuck control lever to the on position.

> **Caution:** The side of the wheel must clear the vertical surface. During the last pass or two, disengage the automatic traverse feed to provide better machine operator control in feeding the wheel as it contacts the side of the workpiece at the shoulder.

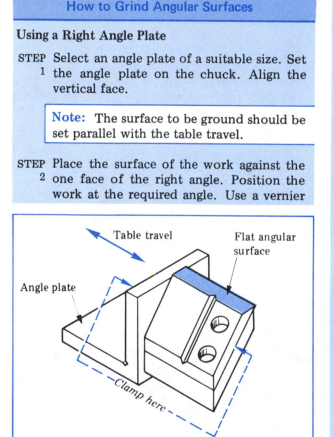

Table travel

Flat angular surface

Angle plate

Clamp here

Figure 54–3 Angle Plate Setup for Grinding an Angle

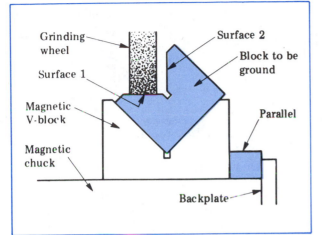

Figure 54–4 Setting and Use of Magnetic
V-Block to Position and Hold Workpiece

How to Grind an Angle With a Right Angle Plate, Sine Bar, and Gage Block Setup

STEP Place the angle plate on a surface plate.
1 Set up the angle sine bar combination for the required angle. (Figure 54–5 shows the sine bar and angle plate setup commonly used for laying out, holding, and grinding a precision angular surface. As shown, the workpiece is set at a 15° angle.)

STEP Mount the workpiece on the angle sine
2 bar. Clamp in position.

UNDERCUTTING FOR SHOULDER GRINDING

Wherever practical, the juncture of the two flat planes of a square shoulder should be undercut by prior machining processes to eliminate a sharp corner (Figure 54–6). Undercutting is important for the following reasons:

- To preserve the shape of the wheel at the edge.
- To permit the wheel to produce a true surface over the entire area without leaving a slight ridge in the corner.
- To help to eliminate wheel stresses at the corner contact area.
- To increase the life of the abrasive wheel and thus decrease the time consumed in truing and dressing operations.

TRUING AND DRESSING DEVICES FOR GROUND SHOULDERS

The straight-face and the dish-type wheels illustrated in Figures 54–7 and 54–8 are commonly used for grinding square and vertical face shoulders. The redressing or relieving of the side of the wheel may be done by using an abrasive stick offhand. A diamond dresser or other truing and

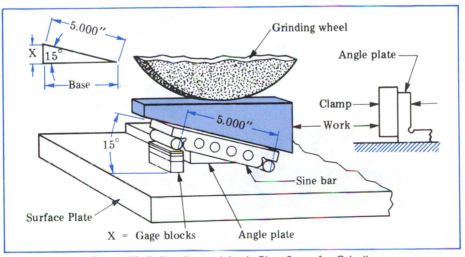

Figure 54–5 Sine Bar and Angle Plate Setup for Grinding
a Precision Angular Surface

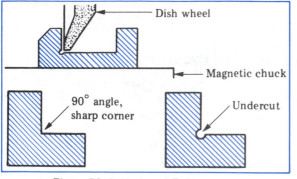

Figure 54-6 Undercut Features on a Square Shoulder

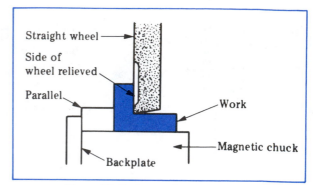

Figure 54-7 Setup of Workpiece for Square Shoulder Grinding

dressing tool, mounted in a suitable holding device, is used for more accurate truing and dressing.

Figure 54-9 illustrates the grinding of square, angular, and filleted shoulders by using straight wheels dressed to shape.

Similarly, one or both sides of a straight-face wheel may be dressed to the required angle for an angular shoulder. The correctness of the angle is usually checked with a bevel protractor or angle gage. When a wide horizontal surface and angle are being ground simultaneously, the face (periphery) is sometimes relieved as shown in Figure 54-9.

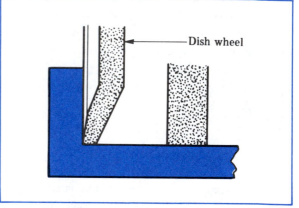

Figure 54-8 Use of Dish Wheel to Grind Vertical Surface

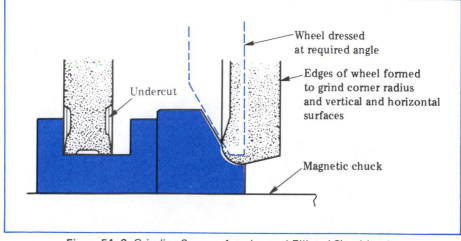

Figure 54-9 Grinding Square, Angular, and Filleted Shoulders by Using Straight Wheels Dressed to Shape

How to Form Dress a Grinding Wheel for Angular Grinding

STEP 1 Select the wheel dressing tool. Mount it in a fixture or other holding device on the magnetic chuck.

> **Note:** The dressing tool or stick is set in a fixture at the designated angle. The angular side of the wheel is formed by feeding the wheel truing diamond or other abrasive cutter across the side face.

STEP 2 Increase the dressing tool contact a few thousandths of an inch at a time. Continue to dress the side of the wheel until the correct angle or radius is formed and the cutting faces of the wheel are trued.

STEP 3 Stop the wheel spindle. Check the radius with a radius gage or template. Check the angular side face with a bevel protractor or other gage.

How to Grind Shoulders

Rough Grinding Shoulders

STEP 1 Start the table reciprocating. The coolant is turned on, if the work is to be wet ground.

STEP 2 Feed the wheel down to start rough grinding the horizontal surface. Take from 0.001″ to 0.002″ (0.02mm to 0.05mm) per table stroke as a cut.

> **Note:** This operation requires the manipulation of both the vertical and crossfeed handwheels to set the depth of cut and to traverse the area to be ground, respectively.

STEP 3 Rough grind the vertical surface. Feed the work face toward the grinding wheel. Continue to rough grind the vertical surface to within 0.001″ to 0.0015″ (0.02mm to 0.04mm) of finish size.

> **Note:** A dish-type wheel is preferred over a straight-face wheel when a square shoul-

der with a narrow adjacent horizontal face is ground.

Finish Grinding Shoulders

STEP 1 Traverse the table to the extreme end. True and dress both the face and side of the wheel to true the wheel, produce the necessary relief, and restore sharp abrasive cutting edges.

STEP 2 Repeat the steps used for rough grinding to position the wheel for finish grinding.

STEP 3 Feed the grinding wheel down about 0.0005″ (0.01mm) after the first sparks and grind the horizontal surface.

STEP 4 Feed the vertical face toward the wheel at 0.0002″ to 0.0003″ (0.005mm to 0.008mm) per stroke.

STEP 5 Continue to machine to size and shape according to the required degree of dimensional and surface finish accuracy.

STEP 6 Allow the wheel to spark out in the final finish grinding step.

B. FORM (PROFILE) GRINDING AND CUTTING-OFF PROCESSES

Form grinding, as described in this part of the unit, deals with dresser accessories and dressing tools, machine setups, and form dressing processes. Grinding wheels may be formed by hand. Simple abrasive dressing sticks of aluminum oxide, silicon carbide, and boron carbide are used for simple hand forming and dressing. These hand-guided dressing tools are practical for roughly shaping a wheel to a limited degree of accuracy. Parts that are to be ground within close dimensional tolerances for shape and size require that the wheel be precision dressed.

GENERAL TYPES OF WHEEL FORMING ACCESSORIES

The three general types of wheel forming accessories are as follows:

- Radius truing and dressing devices which form a concave or convex radius;
- Angle truing and dressing devices which form angles, shoulders, and bevels;

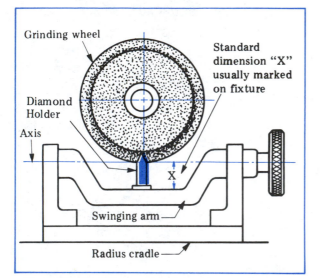

Figure 54–10 Relationship of the Diamond Position and the Axis of a Radius Dresser

- Combination radius and angle truing devices which form both radii and angles as used singly or in combination.

RADIUS DRESSER

While differing in design and construction, all radius truing and dressing devices of the *cradle type* follow the same principles. Each device supports a diamond dressing tool that is used to form a required radius on the grinding wheel. The radius is produced by rotating a curved arm that extends from an upright member on the dresser body. The curved arm may be rotated on its axis to swing the diamond in an arc.

The concave or convex radius produced is determined by the relationship of the diamond nib (point) to the axis of the arm. A concave form is generated in the wheel face when the diamond is set above the axis and is swung in an arc while in this position (Figure 54–10). Conversely, a convex form is produced when the diamond is set below the axis.

The form is produced by feeding the diamond and rotating it as contact is made with the grinding wheel face. A coolant should be used if the form grinding process is to be performed wet.

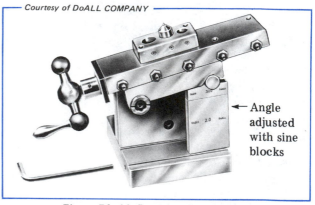

Figure 54–11 Precision Sine Dresser

ANGLE DRESSER

Another method of producing an angular surface is to form the wheel at a required angle with an *angle dresser*. The *precision sine dresser* shown in Figure 54–11 is adjustable to 45° either side of a horizontal (0°) position. This device is set by using gage blocks to accurately obtain the required angle.

COMBINATION RADIUS AND ANGLE DRESSER

A *radius and angle dresser* combines the processes performed by the separate dressers. A radius and angle dresser is designed to form convex and concave outlines and angles on grinding wheels. Moreover, this device makes it possible to generate combinations of radial and angular shapes that are continuous (tangential to each other). On this type of dresser, the diamond dressing tool must be located at the horizontal center line of the wheel. The wheel is formed with the diamond and holder in a right-side position instead of under the wheel.

The principal parts of a precision radius and angle dresser are identified in Figure 54–12. The base carries a sliding platen that is moved horizontally by a micrometer collar. This movement permits the dressing of angles. Concave and convex radii are formed by using a swivel base movement. The stop pins are located to permit dressing each radius and angle according to specified dimensions.

The radius and angle dresser is adaptable to the precision grinding of punches, sectional dies,

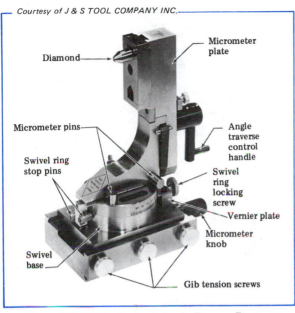

Diamond

Micrometer plate

Micrometer pins

Swivel ring stop pins

Angle traverse control handle

Swivel ring locking screw

Vernier plate

Micrometer knob

Swivel base

Gib tension screws

Figure 54–12 Radius and Angle Dresser Features

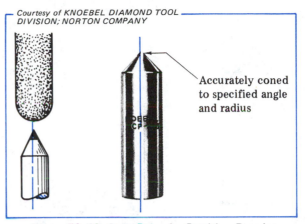

Accurately coned to specified angle and radius

Figure 54–13 Cone Point for Precision Dressing Intricate Radii and Combination Forms

cutting tools, special forming tools, and other work involving hardened steels.

FORMING A WHEEL WITH CRUSH ROLLS

In addition to the single high-speed steel or carbide crush roll described in Unit 52, two crush rolls are preferred for quantity production. One roll serves as a work roll. The second roll, a *reference roll* is used to form dress the work roll when it becomes worn.

CONSIDERING GRINDING WHEEL CHARACTERISTICS

All grinding wheels cannot be crush formed. It is impractical to apply this process to shellac-, rubber-, or resinoid-bonded wheels. The grains tend to break out in clusters. Thus, crush forming is possible only with vitrified-bonded aluminum oxide and silicon carbide grinding wheels. Organic-bonded wheels are too elastic to permit building up enough force between the roll and the wheel to crush form them.

The radius of the form is important in selecting the size of grit. The grit size must be smaller than the smallest radius. The general grit size of grinding wheels falls within the 120 to 400 medium-grade range.

Crush dressing produces a sharp cutting wheel. The grains are sharp pointed for fast removal of material. However, the quality of the surface finish is not as high as the finish produced by diamond dressing a wheel.

DIAMOND DRESSING TOOLS FOR FORMING GRINDING WHEEL SHAPES

The size and shape of both the diamond and holder depend on the contour to be generated in the wheel, the wheel diameter and characteristics, and the nature of the application (light, medium, heavy, or extra-heavy). A *cone-point* diamond cutting tool, as pictured in Figure 54–13, is used for the precision dressing of intricate radii and complex combination forms. Cone-point dressing tools are ordered by the type of cone point, the included angle on the diamond, the radius of the diamond, and the application. This information is required by the manufacturer.

The *full-ball* radius dressing tool (diamond dresser) is designed to plunge dress a full 180° concave radius in a grinding wheel. This dressing tool is available in standard radii between 0.010″ to 0.050″ (0.25mm to 1.25mm). Smaller and larger radii dressers are available to meet special requirements. Frequent turning of the tool is necessary to provide longer tool life. When a diamond becomes worn with use, it may be relapped.

There are three common forms of *radius diamond dressing tools*. The first form is

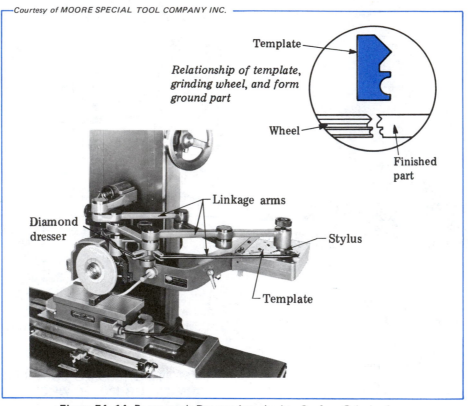

Courtesy of MOORE SPECIAL TOOL COMPANY INC.

Relationship of template, grinding wheel, and form ground part

Template

Wheel

Finished part

Linkage arms

Diamond dresser

Stylus

Template

Figure 54–14 Pantograph Dresser Attached to Surface Grinder for Dressing Intricate Contour on a Grinding Wheel

available in six standard sizes to form concave radii from 0.010″ to 0.250″ (0.25mm to 6.25mm). The second form generates convex radii on small-diameter grinding wheels. There are three general sizes of this dressing tool. The sizes accommodate radii from 0.020″ to 0.500″ (0.5mm to 12.5mm). The third form is applied when dressing half-circle concave radii. Radii of 0.032″ (0.8mm), 0.062″ (0.16mm), and 0.125″ (3mm) are standard sizes. Designs are available for use with radius, angle and combination dressers, and different pantographs.

IRREGULAR FORM DRESSING WITH A PANTOGRAPH DRESSER

The forming of a grinding wheel to produce complex forms or contours in a workpiece requires the use of a *pantograph dresser* (Figure 54–14). This accessory is mounted on a surface grinder. It consists of a tracer that follows a template of the required contour and linkage system that guides the diamond dresser to reproduce the form accurately (resulting from a maximum 10:1 reduction ratio).

FORMING CRUSH ROLLS ON A MICROFORM GRINDER

The *microform grinder* is a precision grinding machine. While it is used extensively to grind complex profiles on crush rolls, other circular and flat forms may be ground to shape. The microform grinder is used to grind form tools, templates, cams, gages, punches, dies, and other intricate forms. Such grinding is done on hardened materials, as well as on carbides. The principal design features of a microform grinder are the viewing screen, scope, pantograph, template, stylus, and drawing table (Figure 54–15A).

The pantograph feature provides for the stylus to follow the outline of a profile drawing.

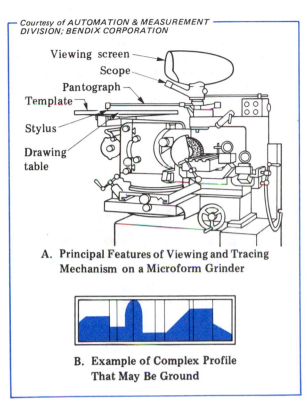

A. **Principal Features of Viewing and Tracing Mechanism on a Microform Grinder**

B. **Example of Complex Profile That May Be Ground**

Figure 54–15 Microform Grinder and Application in Grinding Intricate Forms

The pantograph is mounted on the drawing table. The arms permit the transfer, through a control mechanism, to form dress the crush roll to the required shape. The combination of viewing screen and scope permits the operator to follow the profile lines on the layout drawing and to control the form grinding process. The ratio between the enlarged drawing size and the reproduced profile on the crush roll is 50:1 on the model illustrated. This ratio helps to control the dimensional accuracy and the reproduced profile on a ground crush roll. An example of a complex profile that may be ground is shown in Figure 54–15B.

ABRASIVE CUTTOFF WHEELS AND CUTTING-OFF PROCESSES

Cutting off (while not a widely used surface grinding process) means severing a material at a specific dimension. The cutting is done with a cutoff abrasive grinding wheel or abrasive saw.

The process is particularly suited to cutting off hardened steel, ceramic, stainless steel, new high-temperature metals, and other difficult-to-cut materials. Unusual-shaped parts and extruded forms are held and positioned in fixtures for cutting-off processes.

The process is fast cutting. The temperature may be controlled so that the parts are not work hardened. Cutting-off accuracy is better controlled than heat (flame) or arc cutting processes. The quality of surface finish is also far superior.

Cutoff wheels vary in thickness from 0.006″ to 0.187″ (0.15mm to 4.8mm) for general operations.

NONREINFORCED CUTOFF WHEELS

Cutoff wheels are available with or without reinforcing. The severity of the forces and the conditions under which parts are cut off determine whether a *reinforced* or *nonreinforced* wheel is used. The *nonreinforced cutoff wheel* is designed for machines where the work is clamped securely, the wheel operates in a controlled plane, and the wheel is adequately guarded. The surface grinder meets these requirements.

The nonreinforced wheel does not contain a strengthening fabric or filaments throughout the wheel diameter or reinforcement in special areas of the wheel. General cutoff wheels are made of aluminum oxide or silicon carbide for cutting ferrous metals, except titanium. Silicon carbide cutoff wheels are used for titanium and nonmetallics such as ceramics, plastics, and glass.

SPECIFIC BOND RECOMMENDATIONS

All cutoff wheels are type 1. They are furnished for both wet and dry grinding operations. Resinoid, rubber, and shellac cutoff wheel bonds are available for a wide variety of operations. For example, resinoid wheel bonds permit applications that range from the high production cutting of bar stock to the cool cutting action on nonproduction, heat-sensitive materials. Resinoid wheels are adapted to dry grinding.

Rubber wheel bonds range from free-cutting ferrous and nonferrous metals (either wet or dry) to the soft cutting action required for cutting glass and ceramics. Rubber-bonded wheels produce a good surface finish with a minimum of burrs and discoloration. Rubber wheels produce smooth side (vertical) surfaces and square corners. Thus, they are good to use for grooving and slotting operations.

Shellac wheel bonds are adaptable to highly critical applications where no burrs, checking from heat, or discoloration is permitted. Some shellac-bonded wheels are adaptable to extreme applications requiring fine grit sizes and soft grades. Shellac wheels are generally used in toolroom applications where the quality of cut and versatility are the main requirements.

Diamond cutoff wheels are recommended for cutting ceramic, stone, and cemented carbide materials. A *metal bond* is used because of its strength and ability to hold diamond abrasives securely in the wheel. Cutoff wheels are marked with letters to designate the abrasive: *A* designates aluminum oxide as recommended for metal cutoff operations; *C* represents silicon carbide for nonmetallic and masonry materials; *D* indicates a diamond wheel.

REINFORCED CUTOFF WHEELS

The *reinforced cutoff wheel* is termed *fully reinforced* when internal or external strengthening fabrics are incorporated throughout the entire diameter. When the strengthening material covers just the hole and flange area, the wheel is termed *zone reinforced.*

Reinforced wheels are designed for use where the grinding machine is hand guided or the workpiece is not clamped securely. Reinforced wheels resist breakage that normally results from severe cross bending. Reinforced cutoff wheels are manufactured in aluminum oxide or silicon carbide. These wheels may be used for surface speeds up to 16,000 sfpm for wheel diameters up to 16″ (406mm) and 14,200 sfpm for wheels over 16″ (406mm) diameter.

A. Safe Practices in Grinding Angular and Vertical Surfaces and Shoulders

- Disengage the automatic transverse feed upon approaching a vertical, filleted, or angular shoulder. Disengagement permits hand control in order to take fine, precise cuts in forming a corner and the adjacent faces.
- Judgment is needed to avoid excessive side force on the grinding wheel. Such a force may cause distortion, uneven wheel rotation, and even fracturing.
- Avoid a sudden tap or hitting the side of the grinding wheel.
- Work from a side position and never directly in line with the revolving wheel. When a grinding wheel fractures, the spindle speed and the centrifugal force cause the broken pieces to fly at a terrific speed. These pieces can cause serious injury to the operator and damage to the machine.
- Check each setup to ensure that the work is held securely, the grinding wheel is free to rotate and travel without interference, and the trip dogs are tightly fastened.

B. Safe Practices in Using Form Dressers and in Crush Dressing and Form Grinding

- Remove all dust and abrasive particles during the form grinding of the wheel. The air exhaust system on the wet grinding system on the machine must be adequate to remove the volume of dust and particles produced when form grinding the wheel.
- Crush dress only on a surface grinder that is designed with a spindle that compensates for the severe forces required during the process.
- Select a grinding wheel with appropriate characteristics for crush grinding without damage to the wheel. Also, the wheel must have properties that permit it to reproduce the form accurately with maximum operator safety and wheel wear life.
- Select a cutoff wheel with adequate reinforcement to withstand the cutting and bending forces required for a specific set of conditions.
- Analyze the cutting action, burring, and surface finish produced with a cutoff wheel.
- Extend the workpiece (with a minimum overhang) so that the section to be cut off drops away from the grinding wheel onto a protective tray.

UNIT 54 REVIEW AND SELF-TEST

A. GRINDING ANGULAR AND VERTICAL SURFACES

1. State three different work-positioning requirements for setting a workpiece to grind a compound angle. A toolmaker's universal vise is to be used.

2. Prepare a work production plan for grinding an angular surface on a workpiece. A standard angle plate setup is used. The workpiece is positioned by using a sine bar and gage blocks.

3. State two reasons why the use of a dish wheel is often preferred to the use of a type 1 wide-face abrasive wheel when grinding two surfaces of a shoulder simultaneously.

4. List the steps for form dressing a grinding wheel to a required cutting angle.

5. State three personal safety precautions the surface grinder operator must observe, particularly when grinding vertical shoulders and angular surfaces (with the side face of a grinding wheel).

B. FORM (PROFILE) GRINDING AND CUTTING-OFF PROCESSES

1. List the steps that must be taken to form grind hardened workpieces that have an internal corner radius of 0.75″ (18.8mm). The grinding wheel is to be formed with a cradle-type radius dresser. Stock is left on the part for a series of roughing cuts and a finish cut.

2. a. Cite two advantages to using a sine dresser for dressing a grinding wheel at angles up to 45° from either side of a 0° horizontal plane.
   b. Calculate the gage block height for setting the sine dresser at a 7°15″ angle above the 0° horizontal plane.
   c. Select a set of gage blocks that will produce the required height.

3. a. Explain why it is impractical to crush form rubber- or resinoid-bonded grinding wheels.
   b. Compare (1) the cutting action and (2) the surface finish of a crush formed wheel with a contour formed on an abrasive wheel when a diamond nib and wheel dresser are used.

4. Secure an abrasive wheel manufacturer's technical manual on cutoff wheels. Also, check the specifications of a surface grinder in the shop with respect to spindle features, size, and speed range.
   a. Select the wheel size and determine the specifications for a wheel that is adapted to cutting off 20 pieces 4mm thick from a 25mm square bar of SAE 1020 mild steel.
   b. Give the wheel size and specifications.
   c. Compute the spindle speed (RPM) and indicate the rate of table feed.

5. State two personal safety practices the grinder machine operator must follow (a) before crush dressing a wheel and (b) when mounting a cutoff wheel.

Cylindrical Grinding:
Machines, Accessories, and Processes

This section begins with a general description of basic types of cylindrical grinding machines, design features and their functions, work-holding devices, and accessories. Peripheral speeds, problems and corrective action, and basic cylindrical grinding process data are applied.

Procedures are given for setting up and operating the machine, grinding external straight and taper cylindrical surfaces, angle grinding, face grinding, the internal grinding of straight and slight and steep taper cylindrical surfaces, and using angle and radius truing and dressing devices.

UNIT 55

Machine Components:
External and Internal Grinding

GENERAL TYPES OF CYLINDRICAL GRINDING MACHINES

The *plain and universal cylindrical grinders* continue to be the basic types. The plain cylindrical grinder is primarily used for straight and taper cylindrical grinding. The machine may be adapted to plunge grinding and to the producing of contour-formed cylindrical surfaces.

The term *universal* indicates that multiple processes may be performed in addition to plain cylindrical grinder processes. Slight and steep tapers (angles), faces, and combined internal and external grinding processes are performed on the universal cylindrical grinder.

DESIGN FEATURES OF THE UNIVERSAL CYLINDRICAL GRINDER

The versatility of the universal cylindrical grinder makes this machine ideal for identifying its design features, operating controls, and basic grinding processes. The six principal units of the universal cylindrical grinder (Figure 55–1) are as follows:

- Base, which houses motors for the coolant system, lubrication system, and drive components for the table. The machined ways accommodate the saddle;
- Saddle, swivel table, and drive mechanism for longitudinal travel;
- Headstock;
- Footstock;
- Wheel head, carriage, and cross feed mechanism;
- Operator control panel.

Toolroom universal cylindrical grinders are capable of grinding perfectly round cylinders to less than 25 millionths of an inch (0.0006mm) of roundness. Diametral sizes may be held to less than 50 millionths (0.0013mm). Surface finishes may be ground to less than 2 microinches and metric equivalent.

The universal cylindrical grinder is usually identified by size and whether the table traverse and/or the wheel feed is hand or hydraulically operated. Two popular toolroom sizes are 6″ × 12″ and 10″ × 20″ (152mm × 305mm and

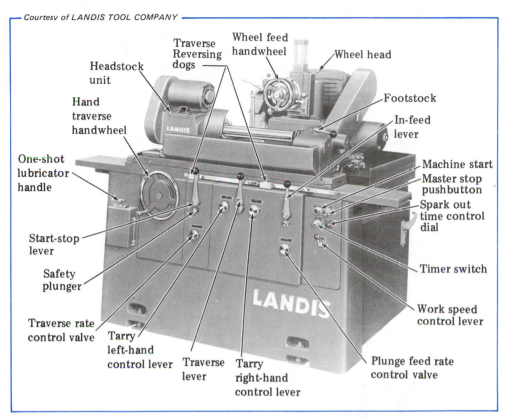

Figure 55–1 Major Components of a Universal Cylindrical Grinder
(Hydraulic Traverse and In-Feed Model)

254mm × 508mm). The grinder may be designed for:

- Hand traverse and hand wheel feed,
- Hydraulic traverse and hand wheel feed,
- Hand traverse and hydraulic in-feed,
- Hydraulic traverse and hydraulic in-feed.

SWIVEL TABLE

The table may be swiveled to position the work axis at a short angle to the grinding wheel face. The swivel table is adjustable from 0° to 20° included angle and up to 4″ per foot or metric equivalent.

The table may be moved by hand for positioning the workpiece in relation to the grinding wheel or for traversing the workpiece past the wheel face. Trip dogs are provided to control the length of table movement and to actuate the table reversing lever. Angle settings are made by turning an adjusting screw to position the table

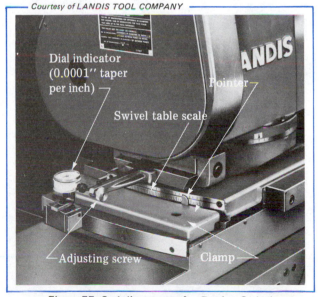

Figure 55–2 Adjustments for Setting Swivel
Table for Grinding a Taper

at the required angle indicated on the swivel table scale and tightening the table clamping screws (Figure 55–2). It should be noted that the taper angle and the table setting depend on the location of the taper on the workpiece, the work diameter, the length to be ground, and the grinding wheel diameter.

LONGITUDINAL TABLE TRAVEL

Table travel is controlled manually or automatically. The table travel rate is selected with a table speed selector knob. On some 10″ × 20″ (254mm × 508mm) models, the traverse speed is from 2″ to 240″ (50mm to 6,096mm) per minute. Models are designed so that the table may *dwell* from 0 to 2 1/2 seconds at the end of one or both table reversals.

HEADSTOCK

Figure 55–3 illustrates the general design features of a headstock. On the model illustrated, there are six work speeds of 60, 90, 150, 225, 400, and 600 RPM. The work speeds are controlled by the headstock speed control knob.

The headstock is powered by its own motor. The headstock also includes a work-driving plate

and arm that are mounted on the spindle. The base is graduated so that the headstock may be swiveled through a 180° arc. The swivel feature permits steep angle grinding and face grinding on workpieces that are held in a mechanical or magnetic chuck or collet or that are mounted in a fixture on the headstock spindle.

The rotation of the headstock spindle (and workpiece) is controlled by a lever or a handwheel control. The start, stop, jog, and brake of the headstock spindle (workpiece) is controlled by a spindle control lever. The headstock spindle for the model illustrated has a 1″ (25.4mm) diameter hole through the spindle. The headstock work center has a #10 Jarno taper; the footstock, a #6 Jarno taper.

WHEEL HEAD AND CROSS FEED MECHANISM

The wheel head is a self-contained unit. It includes the spindle and a swivel base on which are mounted the drive motor, spindle, wheel feed, lubricating pump, oil reservoir, and pressure switch.

The handwheel is graduated for a work diameter reduction of 0.100″ (2.5mm) per revolution. The work diameter reduction per graduation of the regular feed knob is 0.001″ (0.025mm). The fine wheel feed graduations are in increments of 50 millionths (0.000050″) of an inch (0.00125mm) on the work diameter.

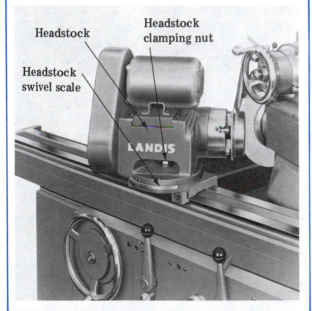

Figure 55–3 General Design Features of a Headstock

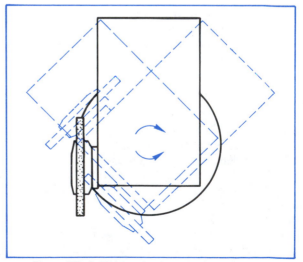

Figure 55–4 Universal Wheel Head Maximum Angular Settings at 45° Each Side of Center

Graduated Base. The wheel feed unit on the universal cylindrical grinder may be swiveled when the wheel head is swiveled. The wheel head may be swiveled 90° to position it at a required angle either side of center. Figure 55–4 shows universal wheel head settings at 45° each side of center. The large, circular graduated swivel base is graduated in degrees. Graduated machine scales are used only for general reference. Other precision instruments and gages are employed to measure within highly precise limits of accuracy.

CROSS FEED MECHANISM

The cross feed mechanism provides for feeding the grinding wheel into the workpiece and automatically stopping the in-feed movement when the required diameter is produced. In-feeding is controlled manually or by power feed.

The cross feed mechanism is started by the cross feed control lever. Positive stop mechanisms differ among machine-tool builders. However, the function in each case is to stop cross feeding when the workpiece is ground to a required diameter. The positive stop is set for the desired diameter by in-feeding the grinding wheel until it grazes the work surface. The handwheel disengaging knob is pulled out to free the handwheel. The handwheel is then turned until the fixed stop pin comes in contact with the right side of the switch operating slide. The cross feed is engaged so that the work is ground to the size to which the stop is set.

When the grinding has sparked out at the footstock end, the cross feed is disengaged. The work diameter is then measured. The index dial is set for any material that still is to be removed. The cross feed is reengaged to take successive cuts until the cross feed is stopped by the positive stop. The cylindrical grinder operator usually sets the stop to grind from 0.001″ to 0.002″ (0.025mm to 0.05mm) oversize. Further adjustments are made in a second setting to bring the workpiece to the required size and within surface texture limits.

BASIC CYLINDRICAL GRINDER ACCESSORIES

Standard jaw and magnetic chucks and collets are accessories that are mounted on and secured to the headstock spindle. Although chucks are primarily used to hold parts for internal grinding, such chucking permits external grinding. Face grinding operations require holding the workpiece in a chuck or on a magnetic faceplate or fixture.

Other accessories include work-supporting *universal spring back rests* (Figure 55–5) and a series of wheel truing, dressing, and forming devices with functions similar to surface grinding devices. Main differences relate to design modifications of the device for mounting on the table and wheel forming from a different position.

OPERATIONAL DATA FOR CYLINDRICAL GRINDING

The following *operational data* is commonly found in handbook tables and other technical manuals. For cylindrical grinding purposes, basic process data must be known for:

- The work surface speed in feet or meters per minute,
- The in-feed in fractional parts of an inch or millimeter per pass for roughing and finish grinding,
- The fractional part of the wheel width that traverses the workpiece for each revolution.

Table 55–1 provides operational data for traverse and plunge grinding of common workpiece materials on a cylindrical grinder. Using the same work surface speed for a specific material, the information needed for plunge grinding deals with the amount of in-feed.

Figure 55–5 Universal Spring Back Rest for Work Support

Table 55–1 Operational Data for Traverse and Plunge Grinding of Common Workpiece Materials

Traverse Grinding

| Workpiece Material | | Surface Speed of Work | | In-Feed per Pass | | | | Fractional Wheel Width Traverse per Revolution of Workpiece | |
| Kind | Machinability (Hardness) Characteristics | | | Roughing | | Finishing | | | |
		(fpm)	(m/min)*	(0.001")	(0.01mm)	(0.0001")	(0.001mm)	Roughing	Finishing
Carbon steel (plain)	Annealed	100	30	0.002	0.05	0.0005	0.013	1/2	1/6
	Hardened	70	21	0.002	0.05	0.0003 to 0.0005	0.008 to 0.013	1/4	1/8
Tool steel	Annealed	60	18	0.002	0.05	0.0005 max.	0.013 max.	1/2	1/6
	Hardened	50	15	0.002	0.05	0.0001 to 0.0005	0.003 to 0.013	1/4	1/8

Plunge Grinding

| Workpiece Material | | In-Feed of Grinding Wheel per Revolution of Workpiece | | | |
| Kind | Machinability (Hardness) Characteristics | Roughing Cuts | | Finishing Cuts | |
		(0.0001")	(0.001mm)**	(0.0001")	(0.001mm)
Steel	Soft	0.0005	0.013	0.0002	0.005
Alloy and tool steel	Hardened	0.0001	0.003	0.000025	0.0006

*Values rounded to closest meter.
**Values rounded to closest 0.001mm.

AUTOMATED CYLINDRICAL GRINDING MACHINE FUNCTIONS

Automation is possible in at least seven different categories, briefly described as follows:

- *Work loading and positioning*, which involves work-feeding devices, loading and unloading, and work positioning;
- *In-feed*, which involves automatic presetting of advance feed rates, cutoff points, and duration of time-process functions. In-feed functions relate to rapid approach rates, feeds for roughing and finishing cuts, and the sparkout period that follows;
- *Table traverse tarry (dwells)*, which involves continuation of wheel and workpiece contact (exposure) at the ends of a cut for a preset time to ensure uniform cutting conditions;
- *Size control*, which provides for automatically adjusting the cutoff points of the in-feed to control the advance movement of the grinding wheel. Provision is also made to automatically compensate for wheel size variations through signals originating with a master gage;
- *Automatic cycling*, which relates to the actuating and shutting down of all mechanisms dealing with work rotation, in-feed, table traverse, and coolant supply as required through a complete machining cycle. At the completion of the operation, cycling refers to the retracting of the wheel slide and grinding wheel and the stopping of the table movement, work rotation, and coolant supply;
- *Wheel dressing*, which specifies wheel dressing at a preset frequency. Compensation is

Table 55–2 Selected Common Cylindrical Grinding Problems, Probable Causes, and Corrective Action

Cylindrical Grinding Problem	Probable Cause	Corrective Action
Chatter Wheel	—Out-of-balance	—Run the wheel without coolant to remove the excess or unevenly distributed fluid in the wheel
		—Rebalance the wheel before and after truing and dressing
	—Out-of-round	—True the wheel face and sides before and after balancing
	—Too hard	—Change to coarser grit, finer grade, and a more open bond
	—Improperly dressed	—Check the mounting position and rigidity of the diamond dressing tool and its sharpness
. . .	. . .	. . .
Work Defects Check marks	—Improper cutting action	—Dress the grinding wheel to act softer
		—Increase the coolant flow and position the nozzle for maximum cooling and cutting efficiency
	—Incorrect wheel	—Use a softer-grade wheel and check the correctness of grain size, abrasive, and bond
. . .	. . .	. . .
Improperly Operating Machine Dimensional inaccuracies	—Out-of-roundness	—Relap the work centers
		—Check the condition of the machine centers, accuracy of mounting, and the footstock force against the workpiece
. . .	. . .	. . .
Grinding Wheel Conditions Wheel defects	—Faulty wheel dressing	—Incline the dressing tool at least three degrees from the horizontal plane
		—Reduce the depth of the dressing cuts
		—Round off the wheel edges
	—Wheel acting too hard (problems of loading, glazing, chatter, burning, and so on)	—Increase the rate of in-feed, the RPM of the work, and the traverse rate
		—Decrease the wheel speed or width
		—Select a softer wheel grade and coarser grain size or dress the wheel to cut sharper

made for wheel size reduction through automatic in-feed movement;

• *Computer numerical control operations*, which involve advanced levels of automating cylindrical processes and are designed to improve the speed, accuracy, and efficiency of the setup and to check out the program and troubleshoot problems.

CYLINDRICAL GRINDING PROBLEMS AND CORRECTIVE ACTION

The four fundamental cylindrical grinding problems that confront the machine operator are as follows:

• Chatter as related to the wheel, workpiece, and machine units;

Table 55–3 Wheel Recommendations for Cylindrical Grinding of Materials Generally Used in the Custom Shop

Material	Wheel Marking*
Aluminum parts	SFA 46–18 V
Brass	C 36–K V
Bronze	
Soft	C 36–K V
Hard	A 46–M 5 V
Cast iron (general, bushings, pistons)	C 36–J V
Cast alloy (cam lobes)	
Soft	
Roughing	BFA 54–N 5 V
Finishing	A 70–P 6 B
Steel	
Forgings	A 46–M 5 V
Soft	
1″ (25.4mm) diameter and under	SFA 60–M 5 V
Over 1″ (25.4mm) diameter	SFA 46–L 5 V
Hardened parts	
1″ (25.4mm) diameter and under	SFA 80–L 8 V
Over 1″ (25.4mm) diameter	SFA 60–K 5 V
Stainless (300 series)	SFA 46–K 8 V
General-purpose grinding	SFA 54–L 5 V

The prefix **BF** with the aluminum oxide designation (A) indicates *blended friable* (a blend of regular and friable). SFA indicates *semifriable*.

*The grain size, grade, and structure are the recommendation of one leading manufacturer.

- Work defects;
- Improperly operating machine;
- Grinding wheel conditions.

Examples of the probable causes of these major problems and recommended corrective action are provided in Table 55–2.

WHEEL RECOMMENDATIONS FOR CYLINDRICAL GRINDING (COMMON MATERIALS)

Table 55–3 provides an example of wheel recommendations of one manufacturer for the cylindrical grinding of common metals. In some instances, grinding processes are identified.

BASIC EXTERNAL AND INTERNAL CYLINDRICAL GRINDING PROCESSES

Once the cylindrical grinder is set up, five basic processes are performed in toolroom work or for small production runs. These processes include:

- Grinding parallel surfaces,
- Grinding tapered surfaces,
- Grinding contour-shaped (curvilinear) surfaces,
- Grinding at a steep angle to form beveled surfaces and shoulders up to 90° to the work axis.
- Face grinding.

While most surface forming on a cylindrical grinder is on external surfaces, the same applications are made to internal grinding.

How to Grind Straight Cylindrical Work

STEP 1 Select the appropriate cylindrical grinding wheel. Check it for balance. Mount the wheel on the spindle and true and dress it.

STEP 2 Set the wheel head at zero.

STEP 3 Attach a driving dog on the workpiece so that there is clearance for grinding.

STEP 4 Check the center holes (on center work) for quality of surface and accuracy of the included angle.

> **Note:** Some workpieces may be held in a chuck when there is limited overhang. Longer chucked pieces may be supported by the footstock center. Collets and fixtures may also be used in the headstock.

STEP 5 Position the grinding wheel. Set the table trip dogs for the length to be ground. Set the dwell control to spark out each cut. Adjust the traverse rate. Set the spindle speed at the RPM required for rotation of the workpiece.

STEP 6 Feed the grinding wheel into the revolving workpiece until it just grazes the surface. Set the cross feed handwheel at zero.

STEP 7 Set the cross feed in-feed trip to permit grinding to the required finished dimension.

STEP 8 Take successive roughing and finishing cuts.

SUPPORTING LONG WORKPIECES

Long, slender workpieces need to be supported by *spring back rests* or *steady rests.* The number and placement of the back rests depend on the work diameter and the length and nature of the grinding process. Universal spring back rests (Figure 55–5) are designed to be positioned along and secured to the table. The *shoes* (bearing surfaces) are usually made of bronze to protect the work surfaces. Shoes are available in increments of 1/16″ (1.5mm) for work sizes beginning at 1/8″ (3mm) through 1″ (25.4mm). Each shoe remains in contact with the workpiece automatically, giving constant support.

In general practice, long workpieces that are 1/2″ (12.7mm) in diameter are supported at 4″ to 5″ intervals along the length. Fewer spring back rests are needed to support stronger, larger-diameter workpieces. Steady rests are used primarily as an intermediate work support.

How to Grind Small-Angle and Steep-Angle Tapers

Grinding Small-Angle Tapers up to 8°

STEP 1 Release the swivel table clamping bolts.

STEP 2 Swivel the table to the required angle or taper per foot (or metric unit). Read the setting on the graduated scale on the end of the table.

STEP 3 Tighten the table clamping bolts to secure the table at the angle setting.

STEP 4 Set the headstock spindle RPM, the rate of table traverse, in-feed rate, and dwell time. Adjust the flow rate of the cutting fluid and the position of the nozzle. Secure the wheel and splash guards.

> **Caution** Use safety goggles or a protective shield.

STEP 5 Bring the grinding wheel into position to take a first cut. Set the wheel head positive-stop mechanism to automatically trip the in-feed at the required position.

STEP 6 Take several cuts until the ground taper length is sufficient to be measured.

STEP 7 Adjust the table setting to compensate for any variation between the taper being ground and the required one.

STEP 8 Take successive roughing cuts by using power table traverse. Leave the workpiece from 0.001″ to 0.002″ (0.025mm to 0.050mm) oversize.

STEP 9 Set the feed rate on the fine-feed dial to remove the remaining material. Take several light finish cuts.

STEP 10 Spark out at the footstock end so that the grinding wheel clears the workpiece. Then, stop the table travel and rotation of the part.

Grinding Steep-Angle Tapers

Steep tapers may be ground by three fundamental methods:

- Swiveling the universal wheel head on its swivel (angle-graduated) base to the required angle;
- Swiveling the headstock for workpieces that are held directly in a chuck, fixture, collet,

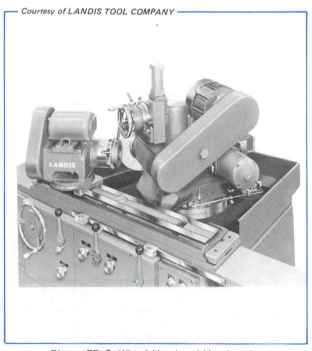

Courtesy of LANDIS TOOL COMPANY

Figure 55–6 Wheel Head and Headstock Swiveled to Grind a Steep Angle

or the tapered spindle. The angle setting of the headstock is read directly from graduations on the swivel base;

- Swiveling the universal wheel head and headstock spindle axes (as shown in Figure 55–6).

FACE GRINDING ON A UNIVERSAL CYLINDRICAL GRINDER

Face grinding refers to the grinding of a plane surface at a right angle to the axis of the part. Face grinding on a universal cylindrical grinder requires the workpiece to be set in a plane that is parallel to the axis of the grinding wheel spindle. The headstock is swiveled 90° for face grinding. The grinding wheel is hand fed to the depth of cut against the revolving part. The part is fed across the face of the grinding wheel either by hand or automatically by using the table traverse feed. The wheel is fed into the part for successive cuts.

GRINDING INTERNAL SURFACES ON A UNIVERSAL CYLINDRICAL GRINDER

STRAIGHT (PARALLEL) GRINDING

An internal grinding attachment permits grinding outside diameters and inside diameters without removing the part, grinding wheel, or spindle. The attachment design permits it to be mounted on the wheel head. It may be swung to a down position for internal grinding (Figure 55–7).

The spindle setup in Figure 55–7 permits grinding a hole to 4 3/8″ (110mm) diameter × 2 1/2″ (63mm) deep without removing a standard 12″ (304mm) diameter external grinding wheel. Spindle quills extend the depth to which an internal surface may be ground.

On the model illustrated, the belted motor drive produces spindle speeds up to 16,500 RPM. Smaller machines are designed to deliver infinitely variable spindle speeds to 45,000 RPM through a dial control on an electrical panel.

INTERNAL GRINDING PROBLEMS

Internal grinding spindles, by the nature of their size and other design limitations, are not as strong or rigid as spindle assemblies for external grinding. As a result, three common grinding problems may occur. Unless the wheel is dressed correctly and run at the required cutting speed with proper in-feed and traverse feed rate, the internally ground hole may have one or more of the following dimensional defects:

- The hole may be bell-mouthed. This defect is caused by the wheel overlapping the hole too much and may be corrected by reducing the overlap at each end of the hole. The bore length may also be too long and outside the capacity of the spindle to accurately machine within the required tolerance limits.
- The hole may be out-of-round. This defect may be caused by overheating during the grinding process or through distortion in chucking or holding the workpiece.
- The hole may be tapered. This defect may be caused by using a wheel that is too soft to hold its size, too fast a feed, or the setting of the workpiece axis at a slight angle.

Other general grinding problems, probable causes, and recommended corrective action were summarized earlier in Table 55–2.

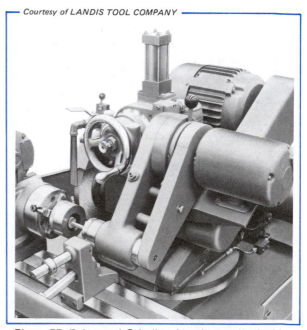

Courtesy of LANDIS TOOL COMPANY

Figure 55–7 Internal Grinding Attachment Locked in Operable Position (Setup for Internal Grinding)

How to Set Up for and Cylindrically Grind Internal Angular Surfaces

Grinding Small-Angle Internal Tapers

STEP 1 Check the wheel head to be sure it is set at 0°.

STEP 2 Check the 0° setting of the headstock.

STEP 3 Set the swivel table at the required taper angle, taper per foot, or metric equivalent.

STEP 4 Proceed to grind internally following the same steps as for grinding a straight internal cylinder.

> **Note:** The correctness of taper is checked with a taper plug gage or the mating part.

Grinding Steep-Angle Internal Tapers

STEP 1 Check the wheel head to be sure it is set at 0°.

STEP 2 Check that the swivel table is set at the zero position.

STEP 3 Loosen the clamping bolts on the headstock base.

STEP 4 Swivel the headstock to the required angle. Secure the headstock in position.

STEP 5 Proceed to grind the steep taper. Follow the same steps as were used for small-angle internal taper grinding and measuring.

CYLINDRICAL GRINDER WHEEL TRUING AND DRESSING DEVICES

There are three basic types of wheel truing and dressing devices for cylindrical grinders. The most common are the wheel truing and dressing fixtures for generating a flat wheel face. These fixtures are designed for mounting either on the footstock or on the table.

Grinding wheels may be dressed at an angle by using a table-mounted *angle truing attachment*. Concave and convex radii are formed by using a *radius wheel truing attachment* (Figure 55–8). The *combination angle and radius wheel truing attachment* is used to accurately form wheels to different radius and angle combinations.

Courtesy of LANDIS TOOL COMPANY

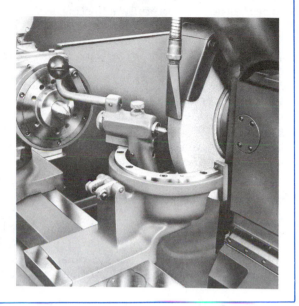

Figure 55–8 Radius Wheel Truing Attachment

The principles for setting these attachments are the same as described for similar surface grinding attachments.

Safe Practices in Setting Up and Operating Cylindrical Grinders

- Observe the ANSI "Safety Code for the Use, Storage, and Inspection of Abrasive Wheels." Visual inspection is accompanied by ring testing for possible cracks.

- Examine the construction and clearance between the guard and the grinding wheel. The clearance should be adequate to avoid interference with the grinding operation.

- Check the balance of the grinding wheel.

- Establish from the manufacturer's specifications what the safe operating (fpm or m/min) speed is for the grinding wheel. Convert the surface speed into the RPM at which the grinding wheel is to revolve.

- Position splash guards to contain the cutting fluid. Immediately wipe up spilled cutting fluid and use a floor surface compound to keep the machine area clean.

- Control the in-feed of the grinding wheel (wheel head) by carefully adjusting with the in-feed handwheel. Power in-feed is not recommended until the first cut is set by hand, clearances are checked, and everything is operating correctly.
- Recheck the settings of the trip dogs by moving the table longitudinally by hand. Finer adjustments are made before the power feed is engaged.
- Lock the internal grinding attachment when it is swung out of operating position.
- Set the wheel head travel movement so that it is automatically stopped when the required ground diameter is reached.
- Review the table of common cylindrical grinding problems, probable causes, and corrective action (Table 55–2). Study the potential machining problems and surface and dimensional defects that may result from unsafe practices.
- Use eye protection devices against flying chips and abrasive particles.
- Leave the machine tool in a safe, operating condition.

UNIT 55 REVIEW AND SELF-TEST

1. a. State when (1) regular feed knob graduations and (2) fine wheel feed graduations are used in cylindrical grinding.
   b. Indicate the customary inch and metric graduation increments for reducing the work diameter on (1) a regular feed knob and (2) the fine-feed graduated handwheel.

2. Identify the function of each of the following: (a) sparkout timer dials, (b) positive-stop cross feed mechanism, and (c) universal spring back rests.

3. Cast steel parts are to be rough and finish ground using the table traverse method.
   a. Refer to a table of operational data for traverse and plunge grinding.
   b. Establish (1) work surface speed in m/min, (2) roughing in-feed per pass in mm, (3) finishing cut in-feed per pass, and (4) fractional wheel width traverse per revolution of the work for roughing cuts, and (5) fractional wheel width traverse for finishing cuts.

4. Calculate the headstock spindle RPM at which a 3.625″ diameter hardened tool steel punch may be ground. Round off the work speed at which the headstock speed control is set.

5. List the steps or set up a cylindrical grinder and grind the mandrel to the required taper and surface finish.

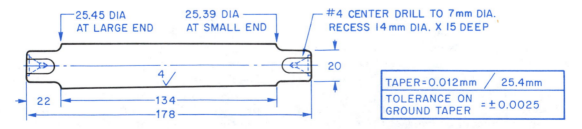

6. List the steps or dress a type 1 wheel for parallel (straight) grinding on a cylindrical grinder.

7. State three personal safety precautions the cylindrical grinder operator must observe with respect to the grinding wheel.

SECTION FOUR

Cutter and Tool Grinding:
Machines, Accessories, and Processes

This section covers the functions of the major components of the cutter and tool grinder, attachments and accessories, and basic machine setups in preparation for grinding straight and helical teeth on a plain milling cutter; slitting saws; staggered-tooth side milling cutters; shell and small end mills; inserted-tooth face milling cutters; angle cutters; and form relieved cutters. Calculations are required to establish the offset for grinding relief, clearance, and rake angles. Considerations for grinding solid and adjustable machine reamers and starting teeth on taps are also included.

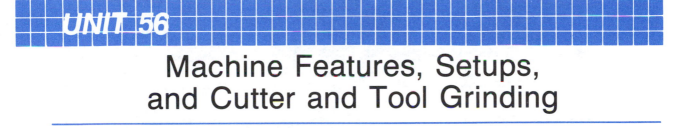

UNIT 56

Machine Features, Setups,
and Cutter and Tool Grinding

A. CUTTER AND TOOL GRINDER FEATURES AND SETUPS

MACHINE DESIGN FEATURES OF THE UNIVERSAL CUTTER AND TOOL GRINDER

One popular size of toolroom universal cutter and tool grinder has a 10″ (254mm) swing capacity over the table and a 16″ (406mm) longitudinal table movement. The major components and controls of a cutter and tool grinder are illustrated in Figure 56–1.

TABLE

The cross (traverse) movement of the table is 10″ (254mm). The table swivels 180°. A taper setting device permits taper settings toward or away from the wheel head up to 5″ (127mm) per foot on the work diameter. The swivel table is graduated in degrees from either side of center.

The table is fitted with spring-cushioned table dogs (Figure 56–2). The dogs govern the length of table traverse, absorb the shock of table reversal, and provide positive table stop, where required.

"Tange Bar" Taper Setting Device. The "tange bar" taper setting device (Figure 56–3) uses a standard setting block and a gage block. The center distance from the pivot point of the table to the center of the block setting combination is fixed at 12″ (304.8mm). The gage block size is determined by the height measurement of 1/2 the included angle (Figure 56–3A). This method is used to set the swivel table to the required angle for precision taper grinding (Figure 56–3B).

WORK HEAD

The standard universal work head (Figure 56–4) is designed with two tapered holes running through the spindle. One end accommodates a Morse or Brown & Sharpe taper. The other end is designed for a standard—for example, a #50—National Series taper.

WHEEL HEAD

Wheel heads are designed with a graduated base that permits settings through 360°. Wheel

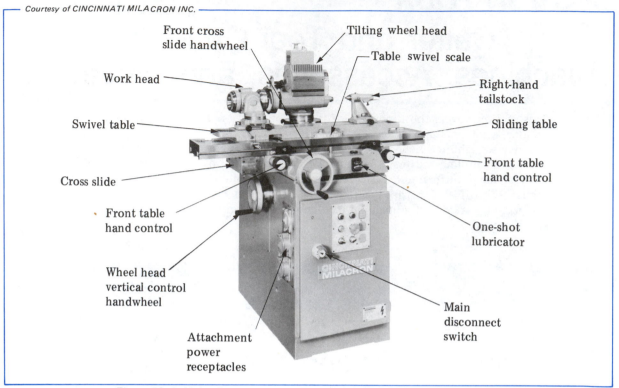

Figure 56-1 Major Components and Controls of a Cutter and Tool Grinder

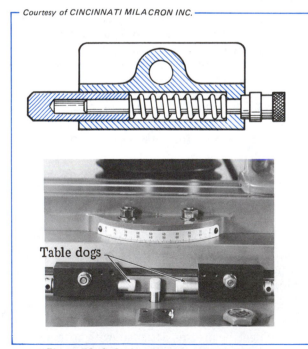

Table dogs

Figure 56-2 Spring-Cushioned Table Dogs

heads may also be swiveled to an angle in the vertical plane. An added feature on some machines is called an *eccentric column*. The eccentric column is also graduated through 360° of arc.

Standard toolroom model wheel heads are operable at spindle speeds of 3,834 RPM for 6″ (152mm) diameter wheels having a maximum wheel surface speed of 6,022 fpm (1,835 m/min). The spindle speed of 6,425 RPM is used for 3 1/2″ (89mm) diameter grinding wheels at a maximum surface speed of 5,887 fpm (1,794 m/min).

CUTTER AND TOOL GRINDER ATTACHMENTS

SURFACE GRINDING ATTACHMENT

The *surface grinding attachment* is applied to the grinding of flat forming tools for turning and other machine tool processes. This attachment consists of a swivel base, a vise, and an intermediate support between these two units. The base and intermediate support are graduated

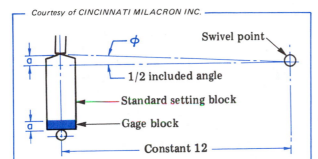

A. Determination of Gage Block Height

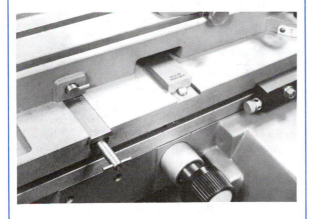

B. Swivel Table Set to Required Angle

Figure 56-3 "Tange Bar" Taper Setting Device

and are designed to be swiveled 360° in both the horizontal and vertical planes.

CYLINDRICAL GRINDING ATTACHMENT

The *cylindrical grinding attachment* provides for the grinding of outside diameters on work that is held in a chuck or fixture or between live or dead centers. The attachment is adapted to straight and taper cylindrical grinding and for facing operations.

INTERNAL CYLINDRICAL GRINDING ATTACHMENT

Internal diameters may be ground by the addition of an *internal grinding spindle*. The spindle is used with the work head drive unit of the cylindrical grinding attachment (Figure 56–5). Spindle RPM up to 23,000 is within the range of this attachment. This speed provides for the use of small-diameter grinding wheels to grind smaller-diameter holes than are normally produced.

GEAR CUTTER SHARPENING ATTACHMENT

The *gear cutter sharpening attachment* (Figure 56–6) is designed for sharpening form relieved cutters by grinding the face of the teeth. The cutter is supported on a bracket. The bracket

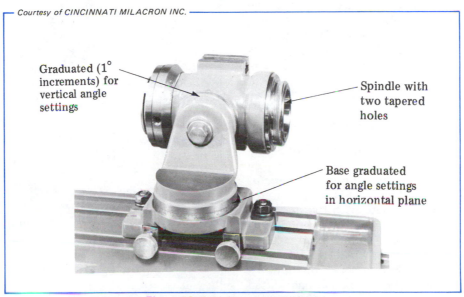

Graduated (1° increments) for vertical angle settings

Spindle with two tapered holes

Base graduated for angle settings in horizontal plane

Figure 56–4 Universal Work Head

Figure 56-5 Application of Work Head Drive Spindle and Internal Grinding Spindle to Internal Cylindrical Grinding

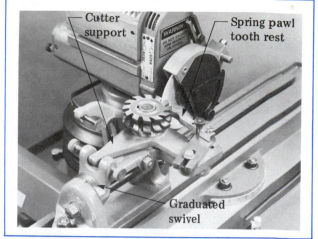

Figure 56-6 Application of Gear Cutter Sharpening Attachment for Grinding the Tooth Face on Form Relieved Cutters

may be swiveled to accommodate the angle of the tooth face. The spring pawl tooth rest lies in a horizontal position so that the edge of the spring locates against the back of the tooth.

SMALL END MILL GRINDING ATTACHMENT

For cylindrical grinding purposes, small end mills are held in collets. The *small end mill grinding attachment* (Figure 56-7) includes an intermediate support, a 24-division master index plate, and a plunger-type indexing mechanism. The spindle is designed to take straight cylindrical and taper collets.

RADIUS GRINDING ATTACHMENT

Ball-shaped end mills with straight or helical flutes and cutters that are to be ground to an accurate 90° radius require the use of a *radius grinding attachment* (Figure 56-8). The design features include a base plate with two mounted table slides and a swivel plate. Movement of each slide is controlled by a micrometer adjustment knob (or graduated collar). A 24-notch index plate may be mounted at the back of the work head to permit direct indexing of straight-fluted cutters.

The starting position is established by a micrometer gage. Radius grinding attachments are available with capacity to position cutters for grinding radii from 0″ to 2″ (0mm to 50mm), cutter diameters up to 12″ (305mm), and to a maximum width of cutter face of 3″ (76mm).

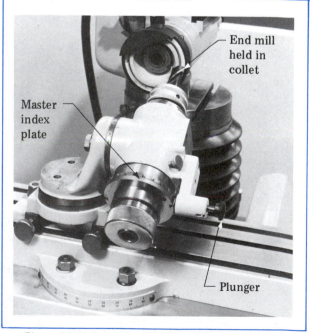

Figure 56-7 Small End Mill Grinding Attachment

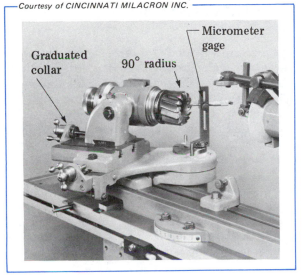

Graduated collar — 90° radius — Micrometer gage

Figure 56—8 Radius Grinding Attachment for a Universal Cutter and Tool Grinder

ADDITIONAL ATTACHMENTS

A number of other attachments are optional, depending on the variety of toolroom or small manufacturing cutter and tool grinding operations. For example, the standard work head may be equipped with an *indexing attachment*. This attachment eliminates the tooth rest as the teeth may be indexed according to the notches on an index plate.

A *micrometer table positioning attachment* adds the feature of a precision lead screw for the machine table. A *sine bar attachment* permits the work head to revolve at a predetermined lead without a tooth rest. A *heavy-duty tailstock* increases the swing capacity over the table. A special *extended grinding wheel spindle* is interchangeable with the conventional spindle. The extended spindle provides added range. A *draw-in collet attachment*, having similar design features to draw-in attachments for turning machines, may be used with the work head. Collets are available in sizes from 0.125″ to 1.125″ in increments of 1/64″. Metric sizes range from 3mm to 28mm in increments of 1mm. Small, straight-shank cutters are conveniently held in draw-in collets.

CENTERING GAGE, MANDREL, AND ARBOR ACCESSORIES

CENTERING GAGE

The *centering gage* consists of a base, arm, and center. As the name indicates, it is used to position the wheel head (and spindle axis) at the center height of the work head and tailstock centers. It is also used to locate the cutter teeth to coincide with the center axis (when this setup is required).

GRINDING MANDREL

The *grinding mandrel* serves to hold a cutter so that it may be accurately mounted between centers. The cutter grinding mandrel has a slight taper for about one-third of its length. When pressed against this slight taper, the cutter is held securely for grinding.

CUTTER GRINDING ARBOR

A *cutter grinding arbor* is used when there is considerable grinding to be done. The general design of the arbor includes a centered, ground shaft to accommodate the bore diameter of the cutter and a series of spacing collars, washer, and nut.

FUNCTIONS AND TYPES OF TOOTH RESTS AND BLADES

FUNCTIONS OF TOOTH RESTS

A tooth rest serves three functions:

- To position the tooth or surface to be ground in a fixed relationship to the grinding wheel,
- To provide a support (the top face of the tooth rest blade) for the tooth during the grinding process,
- To permit the tool or cutter to be indexed to the next tooth. The tooth rest springs back into working position after each index.

TYPES OF TOOTH RESTS

Tooth rests are of two general types: (1) stationary and (2) adjustable. The adjustable type is designed with a micrometer adjustment. The blade (and tooth rest) may be adjusted to more

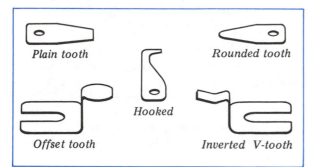

Figure 56-9 Five Common Shapes of Tooth Rest Blades

accurately position the cutter for grinding the relief angle. The two types of cutter tooth rests may be fastened to the wheel head or the table.

FORMS OF TOOTH REST BLADES

Different forms of tooth rest blades are available. The blade form depends on the shape of the cutter teeth that rest against the blade. Five common shapes of tooth rest blades are shown in Figure 56-9.

Straight-tooth milling cutters are supported by a *plain tooth rest blade*. Shell end mills, small end mills, taps, and reamers are ground with the support of a *rounded tooth rest blade*. An *offset tooth rest blade* is applied in the grinding of coarse-pitch helical milling cutters and large face mills with inserted cutter blades. A *hooked blade* is used with straight-tooth plain milling cutters having closely spaced teeth, end mills, and slitting saws. An *inverted V-tooth blade* is used for grinding staggered-tooth cutters.

TOOTH REST APPLICATIONS AND SETUPS

Cutters are divided for sharpening purposes into two general groups. The first group includes cutters that are sharpened by grinding primary relief and/or secondary clearance angles behind the cutting edge of each tooth. Examples of cutters that are ground on the periphery are plain and helical milling cutters, cutoff saws, and reamers. Cutters that are ground on the sides or ends are also included.

The second group of form-relieved cutters are sharpened by grinding the cutting faces of the teeth. Such cutters have a definite profile for form machining. Gear tooth cutters, combination radius and angle form cutters, and taps are examples.

CUTTER GRINDING WITH TOOTH REST MOUNTED ON TABLE

The height of the tooth rest depends on the type of grinding wheel and its direction of rotation in relation to the cutting edge of the cutter and the location of the cutter and machine centerline. One typical setup of a tooth rest for hollow grinding a plain milling cutter with a straight-type wheel is illustrated in Figure 56-10. The direction of the grinding force is clockwise. The grinding process produces a counterclockwise force on the cutter to hold it against the tooth rest.

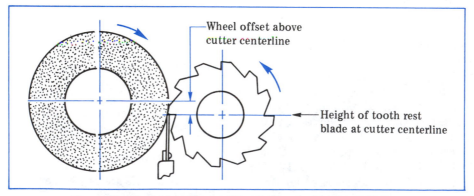

Figure 56-10 Typical Setup of Tooth Rest for Hollow Grinding Plain Milling Cutter with Straight Wheel

CUTTER GRINDING WITH TOOTH REST MOUNTED ON WHEEL HEAD

When the tooth rest blade is positioned above the cutter, the centers of the wheel and cutter are offset to produce the required clearance angle. Here, the wheel is offset below the center line of the cutter. A caution is in order when using this setup. The cutter must be held securely against the tooth rest. The cutting action tends to move the cutter away from the tooth rest. Thus, there is the possibility of personal, wheel, and/or cutter damage.

With the wheel and cutter properly set up, there are two advantages of using this method. First, burr-free cutting edges are produced. Second, the possibility of overheating the cutting edges is reduced.

POSITIONING OF FLARING-CUP WHEELS

Flaring-cup wheels are widely used for cutter and tool grinding. The two general methods of positioning the tooth rest are similar to the setups used with standard straight grinding wheels. The main difference is that the axes of the cutter and wheel fall on the same center line. The clearance angle for grinding the cutter teeth is set by adjusting the tooth rest.

The setups for grinding the primary relief and/or secondary clearance angles with a flaring-cup wheel are shown in Figure 56–11. The positioning of the tooth rest below or above center at the required angle depends on whether the tooth rest is mounted on the table or the wheel head.

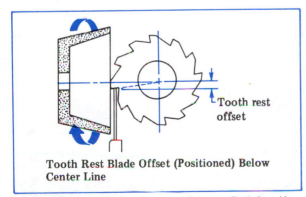

Tooth Rest Blade Offset (Positioned) Below Center Line

Figure 56–11 Setup for Grinding Primary Relief and/or Secondary Clearance Angles with Flaring-Cup Wheel

There usually is a greater area of contact when grinding with cup wheels. Therefore, the cuts are lighter than cuts taken with a straight type 1 wheel.

GRINDING CONSIDERATIONS AND CLEARANCE ANGLE TABLES

The clearance angle produced by a straight grinding wheel depends on the diameter of the *wheel*. When a cup wheel is used, the diameter of the *cutter* determines the cutting angle to be ground. The distance the tooth rest is set above or below center determines the clearance angle.

PRIMARY RELIEF AND SECONDARY CLEARANCE ANGLE CONSIDERATIONS

Summary tables provide a guide for grinding clearance angles. Table 56–1 gives recommended clearance angles for high-speed steel and cemented carbide cutters. Note that different cutting angles depend on the type of cutter and whether the cutting edge is on the periphery, corner, face, or a combination.

ESTABLISHING THE OFFSET

Table 56–2 lists the offset for sharpening milling cutters with straight (type 1) wheels. The amount of offset shown is for standard clearance angles from 3° to 7° for wheel sizes ranging from 3″ (76.2mm) to 7″ (177.8mm).

Table 56–3 gives the offset for sharpening milling cutters with flaring-cup wheels. Here, the cutter diameter influences the offset. While the table covers cutter diameters to 3 1/2″ (88.9mm), the values given in Table 56–2 for larger straight-wheel diameters may be substituted.

AMOUNT TO RAISE OR LOWER THE WHEEL HEAD

When the wheel head is raised or lowered (depending on the grinding method) to produce the required primary relief angle, the amount may be calculated by formula. The cutter tooth level must be on center. The wheel head offset measurements originate from the same spindle and work center axes. The formula is as follows:

Table 56–1 Recommended Clearance Angles for High-Speed Steel and Carbide Milling Cutters

Recommended Relief and Clearance Angles for High-Speed Steel Milling Cutters		
Material to Be Machined	Primary Relief Angle	Secondary Clearance Angle
Gray cast iron Malleable cast iron	4° to 7°	9° to 12°
Plain carbon steel Cast steel Tool steel High-speed steel Alloy steel	3° to 5°	8° to 10°

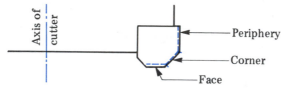

Type of Milling Cutter	Periphery Angle			Corner Angle			Face Angle		
	Steel	Cast Iron	Aluminum	Steel	Cast Iron	Aluminum	Steel	Cast Iron	Aluminum
				Primary Clearance Angle					
Face or side	4° to 5°	7°	10°	4° to 5°	7°	10°	3° to 4°	5°	10°
Saw and slotting	5° to 6°	7°	10°	5° to 6°	7°	10°	3°	5°	10°

Table 56–2 Examples of Offsets for Sharpening Milling Cutters with Straight (Type 1) Wheels

Grinding Wheel Diameter		Cutter Clearance Angle								
		3°		4°		5°		6°		7°
		Offset Required to Grind Cutter Teeth to Clearance Angle								
(Inches)	(mm)	(0.001″)	(mm)	(0.001″)	(mm)	(0.001″)	(mm)	(0.001″)	(mm)	(0.001″)	(mm)
3	76.2	0.079	2.01	0.105	2.67	0.131	3.33	0.158	4.01	0.184	4.67
4	101.6	0.105	2.67	0.140	3.56	0.175	4.45	0.210	5.33	0.245	6.22
6	152.4	0.157	3.99	0.210	5.33	0.262	6.65	0.315	8.00	0.368	9.35

Table 56–3 Offset for Sharpening Milling Cutters with Flaring-Cup Wheels

Cutter Diameter		Cutter Clearance Angle								
		3°		4°		5°		6°		7°
		Offset Required to Grind Cutter Teeth to Clearance Angle								
(Inches)	(mm)	(0.001″)	(mm)	(0.001″)	(mm)	(0.001″)	(mm)	(0.001″)	(mm)	(0.001″) (mm)
1/2	12.7	0.013	0.33	0.017	0.43	0.022	0.56	0.026	0.66	0.031 0.79
3/4	19.1	0.019	0.48	0.026	0.66	0.033	0.84	0.040	1.02	0.046 1.17
1	25.4	0.026	0.66	0.035	0.89	0.044	1.12	0.053	1.35	0.061 1.55
1 1/2	38.1	0.039	0.99	0.053	1.35	0.066	1.68	0.079	2.01	0.092 2.34
2	50.8	0.052	1.32	0.070	1.78	0.087	2.21	0.105	2.67	0.123 3.12
2 1/2	63.5	0.065	1.65	0.087	2.21	0.109	2.77	0.134	3.40	0.153 3.89
3	76.2	0.079	2.01	0.105	2.67	0.131	3.33	0.158	4.01	0.184 4.67
3 1/2	88.9	0.092	2.34	0.122	3.10	0.153	3.89	0.184	4.67	0.215 5.46

$$\text{Wheel Head Offset} = \text{Sine of Required Primary Relief Angle} \times \text{Grinding Wheel Radius}$$

AMOUNT TO RAISE OR LOWER THE TOOTH REST

The distance the tooth rest is lowered or raised to position the edge of the cutter in relation to the center line of a flaring-cup wheel used to grind the required primary relief angle may be calculated by formula:

$$\text{Tooth Rest Offset} = \text{Sine of Required Primary Relief Angle} \times \text{Cutter Radius}$$

METHODS OF CHECKING ACCURACY OF CUTTER CLEARANCE ANGLES

DIAL INDICATOR "DROP" METHOD

Cutter clearance angles may be checked by using the *dial indicator "drop" method.* The term *drop* is derived from the fact that when a dial indicator is used to measure the ground clearance angle, the pointer end "drops" (moves downward) as the cutter is revolved. The pointer measures the movement from the front to the back of the land. Tables are available that give the dial indicator movement for different cutter diameters, land widths, and relief angles.

Figure 56–12 illustrates how the dial indicator "drop" method is used to check a primary relief angle. As a general rule of thumb, for each degree of relief on a 1/16″ (1.5mm) land, there is a 0.001″ (0.025mm) movement (drop) on a dial indicator.

CUTTER CLEARANCE GAGE METHOD

Cutter clearance angles may also be checked by using the *cutter clearance gage method.* There are two common designs of cutter clearance angle gages. One design consists of two hardened steel arms that are at right angles to each other. The arms are placed on top of two teeth on the cutter. A hardened sliding center blade, ground on the end with the required angle, is brought into contact with the face of a ground tooth. The tooth is ground to the required angle when the ground angle and the blade angle coincide.

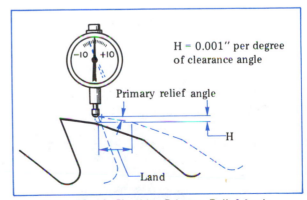

Figure 56–12 Checking Primary Relief Angle by Dial Indicator "Drop" Method

The second design of cutter clearance angle gage consists of a graduated frame with a fixed foot and beam. The gage is set by positioning the feet (fixed and adjustable feet) on two alternate teeth of the cutter and at a right angle to the tooth face. The blade is set to the clearance angle of the cutter. The reading is taken directly from the graduated scale on the frame. The gage may be used to read angles from 0° to 30° on straight, helical, side, or inserted-tooth milling cutters; straight or helical end mills; saws and slitting cutters; and T-slot cutters. The gage is practical for measuring or checking cutter diameters from 2″ (50mm) to 30″ (762mm). Angles may be measured on end mills (if the teeth are evenly spaced) in diameters that range from 1/2″ (12.7mm) to 2″ (50mm).

B. GRINDING MILLING CUTTERS AND OTHER CUTTING TOOLS

SLITTING SAWS

How to Sharpen Peripheral Teeth on a Slitting Saw

STEP 1 Set the edge of the tooth rest blade and the wheel head at center height with the centering gage.

STEP 2 Position the micrometer collar on the wheel head at zero.

STEP 3 Calculate the wheel head offset required to produce the primary relief angle.

> **Note:** The following formula is used:
>
> $$\begin{array}{l}\text{wheel} \\ \text{head} \\ \text{offset}\end{array} = \begin{array}{l}\text{sine of required} \\ \text{relief or clearance} \\ \text{angle}\end{array} \times \begin{array}{l}\text{grinding} \\ \text{wheel} \\ \text{radius}\end{array}$$

STEP 4 Raise the wheel head the amount of offset.

STEP 5 Mount the arbor and cutter between centers.

STEP 6 Position the tooth rest on the table so that the blade supports the face of the tooth to be ground. Secure the tooth rest to the table.

> **Note:** Hook or L-type rest blades are adapted to the grinding of slitting saws. A flicker-type tooth rest support permits each tooth to be ratcheted into position for grinding.

STEP 7 Start the spindle and run it at operating speed for a minute.

> **Caution:** Stand clear and work from the side of the revolving grinding wheel.

STEP 8 True and dress the wheel face.

STEP 9 Bring the grinding wheel into position for the first cut.

> **Caution:** The cutter must be held securely against the tooth rest blade throughout the grinding of each tooth.

STEP 10 Grind each tooth by traversing the table and cutter past the grinding wheel. Use the back machine controls for table traverse.

Grinding the Secondary Clearance Angle

STEP 1 Calculate the height needed to produce the secondary clearance angle.

STEP 2 Raise the wheel head the additional amount.

> **Note:** A smaller-diameter wheel may be needed to grind the secondary clearance angle clear of the primary relief angle.

> **Note:** Many times, just the primary relief angle is ground.

STAGGERED-TOOTH MILLING CUTTERS

The functions of the alternate right- and left-hand helical teeth on staggered-tooth milling cutters and on staggered-tooth slitting saws are similar.

An inverted V-tooth rest blade, with a slightly smaller included angle, is used to support the cutter teeth. The blade requires accurate positioning at the center of the narrow, flat area on the grinding wheel. Unless the V is centered, every other tooth will be ground slightly higher or lower than the adjacent tooth. The tooth rest support is mounted on the wheel head assembly. The relief angle and the secondary clearance angle are produced by lowering the wheel head. The wheel head offset equals the sine of the primary relief angle multiplied by the cutter radius.

CHECKING THE CUTTER TEETH FOR CONCENTRICITY

The teeth must be checked for concentricity on both helixes of the cutter. The position of the apex of the V-shaped tooth rest blade determines concentricity. Concentricity is checked with a tenth dial indicator (0.0001″ or 0.0025mm). A tolerance of 0.0003″ (0.0075mm) is accepted for general milling purposes. When the difference between two teeth exceeds this amount, it is necessary to adjust the location of the blade apex. The apex (blade) is moved in the direction of the helix with the higher tooth. The blade adjustment moves the tooth slightly higher. The result is that a slight amount of additional material is ground away.

GRINDING SECONDARY CLEARANCE

The secondary clearance may be ground by following steps similar to the steps used for grinding the primary relief angle. The wheel head is lowered to permit grinding at the required secondary clearance angle.

Another method is to use a tooth rest with a micrometer adjustment. The blade is lowered for the required offset. The table is swiveled for grinding the peripheral teeth at the right- and then the left-hand helix. Grinding the secondary clearance at the helix angle also produces a uniform land width for the primary relief angle.

CONSIDERING SIDE TEETH AND LAND WIDTH

Generally, the smaller the side tooth land width, the less heat is generated through contact. A slight (about $1/2°$) back clearance further reduces the contact area of the land. The design of land width, small back clearance angle, and minimal relief angle produces a high-quality surface finish on the sides of deeply milled slots. A flaring-cup wheel is used for grinding the side teeth. The wheel face is narrowed to a small, flat width.

CONSIDERATIONS FOR GRINDING CARBIDE-TOOTH SHELL MILLS

Similar setups and procedures are used for grinding primary relief and clearance angles on the side, bevel, and face on a carbide-tipped (inserts) shell mill. A diamond-grit cup wheel is generally used for sharpening purposes. Carbide cutters may be ground to positive or negative radial rake. A coolant is used to prevent overheating and checking at the cutting edge. The depth of each cut is usually restricted to 0.001″ (0.025mm) or less.

END MILLS

RADIAL RELIEF ANGLES

Table 56–4 provides recommended radial relief angles for high-speed steel end mills (for machining mild carbon steels, tool steels, and nonferrous metals). The angles are for the

Table 56–4 Recommended Radial Relief Angles for High-Speed Steel End Mills

Material to Be End Milled	End Mill Diameter							
	1/8″ (3.2mm)	1/4″ (6.4mm)	3/8″ (9.5mm)	1/2″ (12.7mm)	3/4″ (19.0mm)	1″ (25.4mm)	1-1/2″ (38.1mm)	2″ (50.8mm)
	Radial Relief Angle* (in Degrees)							
Carbon steel	16	12	11	10	9	8	7	6
Tool steels	13	10	9	8	7	6	6	5
Nonferrous metals	19	15	13	13	12	10	8	7

*For secondary clearance angles, multiply the radial relief angle by 1.33.

conventional grinding of radial relief and accompanying secondary clearance angles.

How to Sharpen an End Mill

Grinding the Side Cutting Edges (Conventional Method)

STEP 1 Select, mount, and true a straight grinding wheel. Dress the wheel and narrow the face to a width of about 1/16'' (1.5mm).

STEP 2 Mount the small end mill grinding attachment.

STEP 3 Select a narrow-width blade to match the helix of the end mill. Install the tooth rest and blade assembly on the wheel head.

STEP 4 Secure the end mill in a straight or taper collet, depending on the body shape.

STEP 5 Adjust the wheel head, tooth rest blade, and mounted end mill to the same center height.

> **Note:** Each end tooth must be in a horizontal plane.

STEP 6 Set the micrometer collar on the wheel head column at zero.

STEP 7 Calculate the required offset for lowering the wheel head:

$$\text{wheel head drop} = \text{sine of primary relief angle} \times \text{cutter radius}$$

STEP 8 Lower the wheel head and tooth rest to grind the required primary relief angle.

STEP 9 Move the fixture away from the grinding wheel to clear the end mill.

STEP 10 Advance the cutter forward along the tooth rest blade until the cutting edge at the shank end is in position for the start of the cut.

> **Note:** While an experienced operator may start the cut from the end of the shank area, it is safer (in terms of possible damage to the face end of the end mill if there is accidental contact) to start at the shank end.

STEP 11 Take a light first cut on each flute. Increase the feed increments until the primary relief

area of the land is ground to correct any cutter wear or damage.

STEP 12 Spark out the final cut.

STEP 13 Determine the amount of wheel head offset required to grind the secondary clearance angle.

STEP 14 Lower the wheel head the additional amount.

STEP 15 Grind the secondary clearance until the land width is ground to the required size.

Grinding the End Teeth (Universal Work Head Method)

STEP 1 Select a small-diameter flaring-cup wheel. True the wheel and dress it to produce a narrow 1/8'' to 3/16'' (3mm to 5mm) grinding area. Stop the spindle.

STEP 2 Swivel the workhead counterclockwise past the 90° graduation on the base to 88°.

> **Note:** An additional 2° to 3° is recommended so that the teeth are ground slightly lower at the center than at the outside edge.

STEP 3 Tilt the work head spindle to the number of degrees specified for the relief angle of the axial end teeth.

STEP 4 Attach a flicker-type tooth rest support with micrometer adjustment to the work head to provide ratchet indexing.

STEP 5 Insert the end mill in a collet and level the tooth. Adjust the tooth rest support in relation to the master index plate.

STEP 6 Traverse the first end tooth so that the grinding wheel completes the cut at the center of the end mill.

> **Note:** The table stop is set to prevent feeding with possible damage to the opposite tooth face.

STEP 7 Return the grinding wheel to clear the cutter. In-feed about 0.003'' (0.08mm). Traverse the cutter across the wheel face.

STEP 8 Index (ratchet) the spindle to position the next tooth. Continue the grinding and indexing until all teeth have been ground to correct cutter wear or damage.

STEP 9   Spark out the final cut without additional in-feed.

STEP 10   Calculate the additional offset to which the work head is to be set to grind the secondary clearance angle.

STEP 11   Adjust the work head angle and wheel to permit grinding the secondary clearance.

STEP 12   Take successive cuts until the required land width is reached for the primary relief angle.

DESIGN FEATURES AND GRINDING OF FORM RELIEVED CUTTERS

Form cutters are usually marked with the radial rake. Some of the most frequently used form cutters that require grinding on the tooth face include gear tooth milling cutters, convex and concave cutters, and multiple-form gang milling cutters. Before the faces of the cutter teeth are ground, it is necessary to check the uniformity of each tooth from the cutting face to the

Courtesy of AMERICAN TOOL & GRINDING COMPANY INC.

Roughing depth

Figure 56–13 Wheel Head (and Wheel) Set for Depth of Down-Feed Roughing Cut

back of the tooth form. If there is a variation in micrometer measurement, it is necessary to accurately grind a reference area on the back of each tooth for indexing. Dish (type 12) wheels (Figure 56–13) are used to grind the radial face of a cutter tooth down to and including part of the tooth gullet.

The two general methods of grinding the face of the teeth include (1) mounting the cutter between centers and (2) positioning the cutter horizontally in a gear cutter attachment or form relieved cutter holder.

GRINDING SINGLE- AND DOUBLE-ANGLE MILLING CUTTERS

Angle cutters are ground with a flaring-cup wheel. The teeth are set parallel to the table. Usually, the cutter is mounted on an arbor and held in the work head. The wheel head and grinding wheel spindle are set at the back clearance angle for grinding the vertical face of a single-angle milling cutter. The work head is tilted to the primary relief or secondary clearance angle for the vertical face teeth.

The teeth on the angular surfaces of single- and double-angle cutters are ground by setting the work head at 0° and swiveling the table to each required angle setting. The blade of the tooth rest is aligned at center with the angular cutter. The work head is tilted to the required primary relief or secondary clearance angle when the cutter is mounted on the work head. On cutters that are mounted between centers, either the spindle is tilted or the cutter is offset to produce the relief and clearance angles.

DESIGN CONSIDERATIONS FOR GRINDING MACHINE REAMERS

When the lands on solid reamers become worn, they are cylindrically ground to a smaller size. The lands are ground to a slight taper of 0.0002″ (0.005mm) per inch to provide clearance (longitudinal relief), especially on deep-hole reaming.

One caution must be observed in cylindrically grinding of the lands. The direction the reamer turns must be opposite to the direction of normal rotation. The back of the land margin contacts the grinding wheel first. A second caution requires that the design of the cutter grinder must permit the direction of spindle rotation to be reversed and operated safely without loosening the wheel setup.

Secondary clearance is primarily provided to reduce the margin area along the flutes of a reamer. The procedure is basically the same as for grinding secondary clearance on a plain milling cutter or end mill. Either a straight wheel or a flaring-cup wheel may be used. The amount of offset (using the type 1 straight wheel method) is equal to the sine of the secondary clearance angle multiplied by the radius of the grinding wheel. In the case of solid-type machine reamers, there is no primary clearance angle to the ground narrow band margin.

To grind an adjustable machine reamer to ream accurately, the reamer blades are adjusted oversize. Enough stock is allowed to grind the blades to the correct diameter, starting taper, and correct margin (width) of the land.

DESIGN FEATURES OF ADJUSTABLE HAND REAMERS

Table 56–5 gives the primary relief and secondary clearance angles for reamer diameters ranging from 1/2″ (12.7mm) to 2″ (50.8mm). It shows a range of margins from 0.005″ (0.12mm) for reaming steel to a maximum width of 0.030″ (0.75mm) for bronze and cast iron. It is important to grind the land width to the recommended margin. Otherwise, the primary relief may be insufficient to prevent land interference when reaming.

Table 56–5 Recommended Primary Relief and Secondary Clearance Angles for Hand Reamers

Diameter of Reamer		Material to Be Reamed			
		Steel		Cast Iron and Bronze	
		Width of Margin			
		0.005″ to 0.007″ 0.12mm to 0.17mm		0.025″ to 0.030″ 0.61mm to 0.75mm	
		Grinding Angle			
(in.)	(mm)	Primary Relief	Secondary Clearance	Primary Relief	Secondary Clearance
1/2	12.7	3°	12°	7°30′	17°
1	25.4	1°30′	10°30′	4°30′	14°
1 1/2	38.1	1°	10°	3°30′	13°
2	50.8	45′	10°	3°15′	12°30′

GRINDING CUTTING, CLEARANCE, AND RAKE ANGLES ON FLAT CUTTING TOOLS

Flat cutting tools such as thread chasers often require the grinding of a chamfer or face, top rake, lip rake, and clearance angles. While these processes are often performed on surface grinders, the use of a surface grinding accessory extends the versatility of the cutter and tool grinder to cover the grinding of flat cutting tool forms.

CONSIDERATIONS FOR SHARPENING TAPS

Where precision tapping is required, using either hand or machine taps, the taper cutting portion is sharpened by machine grinding. Usually a grinding fixture is used. The tap is mounted between centers. The finger rest is set behind each flute. Small taps are held in a chuck and are positioned with a ratchet stop. The taper at the front end of the tap and the clearance are produced by adjusting the attachment and turning the tap by a handle on the tap grinding attachment.

A type 1 grinding wheel is used. The wheel head is set at 1° more than the angle of the attachment setting to reduce the width of the grinding face. The taper section of each flute is indexed and secured in position by the finger rest.

A. Safe Practices for Setting Up and Operating a Cutter and Tool Grinder

- Make standard wheel checks for soundness, balance, truing, and dressing.
- Test the wheel fit on the spindle, correct mounting, and condition of the wheel for grinding.
- Check the grinder stops and tripping devices.
- See that the blade and tooth rests are securely attached to the table or wheel head column.
- Examine the work-holding device and the setup. Adequate, safe working space must be provided. The operator must be sure the cutter is held securely against the work rest during grinding.
- Test the condition of the tailstock center and the tension (force) of the adjustable center against the arbor. The force exerted must be sufficient to hold the work securely.

- Check the diameter of the grinding wheel and the spindle speed before starting the spindle. The fpm or m/min must be within the maximum recommended wheel speed. In the case of small-diameter wheels, it is equally important to bring the fpm up to maximum wheel speed in order to grind efficiently.
- Regulate the amount of in-feed per cut. Avoid undue force on the grinding wheel, particularly when flaring-cup, saucer, and relieved types of wheels are used.
- Feed the cutter carefully so that the wheel is not brought into sharp contact with the workpiece.
- Protect machine surfaces when dressing the wheel.

B. Safe Practices for Machine Setups and Cutter Grinding Processes

- Apply a continuous force against the cutter during grinding to hold it securely against the tooth rest blade.

- Allow the wheel spindle to reach its operating speed when starting up for a grinding operation. Stand clear of the revolving grinding wheel at all times.
- Make a test cut on wide cutters to correct for unusual wear. It may be necessary to reduce the depth of the first cut.
- Stone and hone the razor-sharp grinding burrs produced by grinding.
- Use a cloth as protection against burrs when removing a cutter from the setup or during burring or other handling.
- Grind toward the cutting edge of carbide-tipped cutting tools. Use a coolant to avoid overheating and checking.
- Check the safety features of the spindle head when it is necessary to change the direction of wheel rotation. The direction of rotation of a reamer (when grinding the margin) must be opposite to its normal rotation.
- Replace guards after a wheel is mounted and before the spindle is started.
- Use safety goggles and/or a protective shield at all times.

UNIT 56 REVIEW AND SELF-TEST

A. CUTTER AND TOOL GRINDER FEATURES AND SETUPS

1. State three functions of tooth rests.

2. Make a simple freehand sketch of a setup to flat grind a primary relief angle on a cutter. A flaring-cup wheel is to be used. The cutting action is from the face of the cutting edge back across the land.

3. Calculate the tooth rest offset required to grind (a) a 6° primary relief angle and (b) an 11° secondary clearance angle on a 100mm diameter helical-tooth milling cutter.

4. List the steps for measuring a ground secondary clearance angle on a helical-tooth milling cutter by using a cutter clearance angle gage.

5. a. Select a straight face milling cutter from the toolroom.
   b. Use a cutter clearance angle table. Establish the smallest primary relief angle required for the machining of annealed gray cast iron workpieces.
   c. Calculate the offset between the centers of the cutter and a straight (type 1) grinding wheel.
   d. Set up the cutter grinder to use a type 1 wheel to hollow grind the primary angle. The grinding wheel is to rotate in a direction toward the cutting edge of the cutter teeth.

6. Give two reasons why it is important to maintain a constant force to hold the cutter securely against a tooth rest blade when grinding.

B. GRINDING MILLING CUTTERS AND OTHER CUTTING TOOLS

1. Give the formulas for calculating the offset of the wheel head to grind the primary relief (or secondary clearance angle) on the peripheral teeth of (a) slottings saws and (b) staggered-tooth milling cutters.

2. a. Describe briefly what effect the positioning of the tooth rest blade apex has on the concentric grinding of the peripheral cutting teeth edges on a staggered-tooth milling cutter.
   b. Tell how nonconcentric grinding of the cutter teeth may be corrected.

3. Refer to a manufacturer's table of wheel recommendations for grinding selected materials.
   a. Look up and record the recommendations for sharpening high-speed steel milling cutters with the following wheels: (1) type 1 straight wheel (dry), (2) type 11 flaring-cup wheel (wet), and (3) type 12 dish wheel (dry).
   b. Give the recommendations for sharpening cemented carbide cutting tools with (1) a type 11 cup wheel for offhand grinding (roughing cuts) single-point cutting tools and (2) a cup wheel for backing-off (finishing cuts) cutters.

4. a. Select a staggered-tooth side milling cutter that requires sharpening of the peripheral and side teeth.
   b. Determine the most suitable types, sizes, and specifications of the required grinding wheels.
   c. Establish the primary relief and secondary clearance angles for the cutter and the amount of offset required.
   d. Set up the workpiece, machine, and attachment to sharpen the peripheral and side teeth. Grind the lands to the specified size.
   e. Sharpen the cutter. Test for concentricity. Stone and hone the cutting edges. Check the angles.

5. a. Select a dull adjustable hand reamer or one with unevenly worn or out-of-parallel blades.
   b. Grind the outside diameter.
   c. Regrind the primary relief and secondary clearance on each blade. The reamer is to be used for reaming holes in cast iron plates.
   d. Regrind the starting taper.
   e. Check the concentricity and parallelism of the blades, margin width, and starting taper.

6. Identify two personal safety factors the cutter grinder operator observes in relation to the use of tooth rests and blades.

SECTION ONE

Precision Machine Tools for Production Tooling

Three of the most widely used machine tools for the laying out, machining, and measuring of production tooling include jig boring, jig grinding, and universal measuring machines. Older models of these high-precision machines are being retrofitted for NC and CNC modes of control, operation, and readout. Tool-making techniques are examined in this section for the building and measuring of tooling commonly used on conventional and NC machine tools.

UNIT 57

Jig Boring, Jig Grinding, and Universal Measuring Machines and Processes

PRECISION JIG BORING MACHINE TOOLS, PROCESSES, AND ACCESSORIES

The jig borer is especially adapted to machine tool operations requiring extreme dimensional measurement accuracies. Linear and machining accuracies are expressed in terms of millionths of an inch (0.025µm). Repeatability in maintaining dimensional accuracy and complete positioning and machining controls make the jig borer an ideal machine tool for precision machining.

Jig borers are widely used in toolrooms for the construction of jigs and fixtures, punches and dies, special cutting tools, gages and other fine machine work. The basic processes include centering, drilling, reaming, through and step boring, counterboring, and contouring. Holes as small as 0.013″ (0.33mm) may be drilled without spot drilling. The spindles are rigidly con-

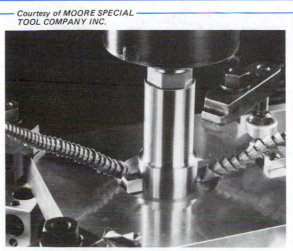

Courtesy of MOORE SPECIAL TOOL COMPANY INC.

Figure 57–1 Hole-Hog Tool Taking a 1/2″ (12.7mm) Cut on a Jig Borer

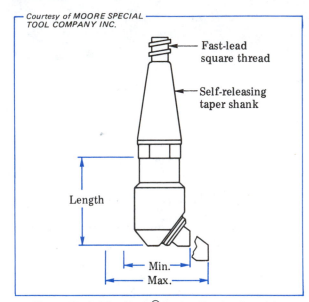

Fast-lead
square thread

Self-releasing
taper shank

Length

Min.

Max.

Figure 57–2 Microbore® Boring Bar for Roughing
and Precision Finishing Operations on a Jig Borer

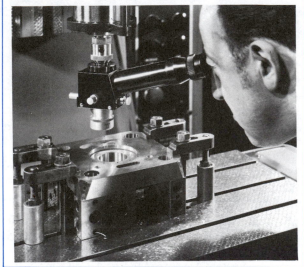

Figure 57–3 Application of a Locating
Microscope on a Jig Borer

structed for power and strength in taking heavy
hogging cuts. For example, the power, strength,
and rigidity of a hole-hog tool in taking a 1/2″
(12.7mm) cut on a jig borer is shown in Figure
57–1.

Since a great variety and number of toolroom
applications require constant tool changing, a
special shank design permits easy removal and
replacement of cutting tools and noncutting,
locating, and positioning devices. The Moore
taper-shank, fast-lead square thread tool illustrated in Figure 57–2 ensures accurate, quick-
change tooling and permits the removal and
replacement of a tool to repeat the hole size on
any diameter. Most jig borers have an infinitely
variable spindle speed. Speed is read directly on
a tachometer.

JIG BORING TOOLS AND ACCESSORIES

Precision Boring Chuck.   The most common
accessory on the jig borer is the precision boring
chuck. Hole diameters may be increased 0.001″
(0.02mm) per graduation. Vernier settings permit
hole diameter increments of 0.0001″ (0.002mm)
per graduation.

Microbore® Boring Bars.   A Microbore®
boring bar series is available for jig boring work.

The Microbore® boring bar, illustrated in Figure
57–2, is used for roughing and finishing operations on a jig borer where close dimensional
limits and accuracy are to be maintained. The
micrometer dial permits size changes to be read
easily. These boring bars are available in overlapping sizes from 3/4″ to 4 5/16″ (19mm to
109.5mm) diameter. Complete bars are furnished in high-speed steel and carbide.

Chucks and Collets.   Key and keyless chucks
are used. Chuck shanks may be straight fitted
for insertion in collets or tapered for easy mounting in the quill. Collets are designed for use with
spotting drills, conventional drills, reamer drills,
and end reamers. The hole tolerance on standard-
quality collets is +0.0005″ (0.01mm) and
–0.0000″. The hole concentricity of the shank
to the collet hole is 0.001″ (0.02mm). The precision quality is finer. The hole tolerance is
+0.00005″ (0.001mm); the hole concentricity,
0.0002″ (0.005mm). Collet hole sizes range
from 1/8″ to 49/64″ diameter. Collet sets are
also available in metric sizes in standard and precision quality.

Tapping Heads.   One widely used model of
tapping head permits the tapping of holes up to
5/16″ (6.4mm) in mild steel and 1/4″ (6.4mm)

in tool steel. The tapping head permits tooling without changing the position of the workpiece. This instant-reversing, speed-reducing tapping attachment is held in a standard collet or drill chuck.

Locating Microscopes. Jig borer setups and other workpiece setting needs require the use of a locating microscope. The microscope is readily adapted to locating edges, irregular shapes, and holes that are too small to be located with an indicator.

The microscope (Figure 57–3), with suitable attachments, is interchangeable for application on the jig borer, jig grinder, and universal measuring machine.

Digital Readout and Printer. The digital readout and printer accessory verifies measurement accuracy for the X and Y axes to assure precise hole location. Readout modes are available for inch-standard screw, metric-standard screw, or switchable inch/metric screws. Readout accuracies to 0.00005″ (0.0012mm) are provided. The X and Y channel readouts are ± through seven digits.

The digital readout accessory includes positive backlash take-up, manual zero reset, and entering of slide travel distance for reference to readout. Visual signals indicate problems of digital readout overflow and loss of power, when all numbers return to zero and there is no counting until the unit is reset.

TABLE MOUNTING ACCESSORIES

Rotary Tables. Standard, precision, and ultraprecise rotary tables permit indexing within tolerances of ±12 seconds, ±6 seconds, and ±2 seconds of arc, respectively. Rotary tables are available for hand-operated mechanical indexing. Rotary tables for NC and CNC machine tools are driven by stepping motors that eliminate mechanical indexing.

A *tilting rotary table* (Figure 57–4) is available with a tilt axis dial and rotary axis dial capable of direct readings to 1 second of arc. A disengageable worm permits quick alignment during setup and engagement for precise angle positioning.

Courtesy of MOORE SPECIAL TOOL COMPANY INC.

Figure 57–4 Tilting Rotary Table for Jig Boring Work

Center and Micro-Sine Table. A tailstock and adjustable center permit work on vertically mounted rotary tables to be held between centers. Boring, grinding, inspection, and other operations may be performed. The tailstock is adjustable laterally and vertically.

A *micro-sine table* is a jig borer accessory used to position and hold workpieces for accurately machining or measuring angles. A set of adjusting rods permits angular adjustments for fine settings from 0° through 95°.

Precision Index Center. This table mounting accessory facilitates jig boring, milling, grinding, and other work processes requiring the workpiece to be rotated accurately on centers.

Measuring and Locating Accessories. The standard indicator set consists of an indicator, indicator holder, edge finder, and line finder. Indicators are of the dial indicator type with readings to within 50 millionths of an inch (0.00005″). Metric dial indicators are also available.

DESIGN FEATURES OF A LOCATING MICROSCOPE

A locating microscope is used when small or partial holes, slots, or irregular contours serve as reference points. A locating microscope with magnification power of 20X or 50X locates edges, contours, and holes that are too small to be indicated with a dial indicator.

The reticle on the microscope has a number of concentric circles and two pairs of cross lines. These vary between 0.0025″ on the 20X magnification to 0.001″ on the 50X magnification. On the 20X microscope, six concentric circles range in diameter from 0.005″ to 0.030″ in increments of 0.005″. Another seventeen circles continue from 0.030″ to 0.200″ in increments of 0.01″. Finally, eight additional concentric circles are spaced 0.020″ apart from 0.200″ to 0.360″.

How to Set Up and Use a Jig Borer

Positioning the Workpiece and Drilling

STEP 1 Place the work on parallels. Align the work with respect to X and Y coordinates.

STEP 2 Align the axis of the machine spindle with the starting reference point.

STEP 3 Set the longitudinal and cross feed micrometer collars at zero.

STEP 4 Determine the machining sequence steps and tooling changes in advance.

STEP 5 Position the spindle at the X and Y axes for the first hole. Center drill or use a spotting drill.

STEP 6 Position the spindle and center drill at the location of each hole. Center or spot drill each hole location.

STEP 7 Return the spindle to the starting point. Make a tool change.

STEP 8 Position and drill all holes of a particular size. Change drill sizes as may be required.

STEP 9 Check the work alignment and the setup to be sure the workpiece does not shift if any heavy rough machine drilling takes place.

Boring Concentric, Parallel Holes

STEP 1 Select a Microbore® boring head or a boring chuck and a straight or offset boring tool. Replace the drill chuck with the boring chuck and cutting tool or boring bar setup.

STEP 2 Set the cutting speed (spindle RPM) according to the job requirements for rough boring.

STEP 3 Take a series of roughing cuts. Rough bore all holes to within 0.003″ to 0.005″ (0.08mm to 0.1mm) of finish size.

STEP 4 Replace the rough boring tool with a finish boring tool for the final cut. Increase the spindle speed. Decrease the tool feed.

STEP 5 Use a solid plug, leaf taper gage, or other measuring instrument to check the bore size. The surface finish must meet the specified requirements.

Note: Measurements must be taken at room (a constant 68°F, 29°C) temperature, particularly when the workpiece must be held to extremely precise dimensional tolerances.

Note: Leaf taper gages may be used for measuring trial cuts when boring or grinding holes to size. The leaves are short enough to permit inserting into the hole when the cutter is withdrawn, without changing the work position. Leaf taper gages may also be used to check a through hole for bell-mouth, taper, and out-of-roundness.

PRECISION JIG GRINDING MACHINE TOOLS AND PROCESSES

The jig grinder is used to grind straight and tapered holes and to grind contour forms. Such forms combine straight, angle, round, radii, and tangent surfaces. Since a jig grinder is an abrasive machining machine tool, the process may be used on soft or hardened metals.

Jig grinders are widely used in toolrooms for finishing punches, dies, jigs, fixtures, special cutting tools, and gages to exact size and high-quality surface finish. Jig grinding eliminates many earlier hand fitting processes.

GRINDING IN A HORIZONTAL PLANE

Jig grinders provide versatility in planetary jig grinding in a horizontal plane. They are used for out-feed, wipe, chop, plunge, taper, and shoulder grinding.

In *out-feed grinding* (Figure 57–5), the grinding wheel spindle moves in a planetary path at a slow rate of rotation. The axis of the wheel

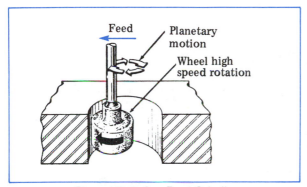

Figure 57–5 Out-Feed Grinding

Courtesy of MOORE SPECIAL TOOL COMPANY INC.

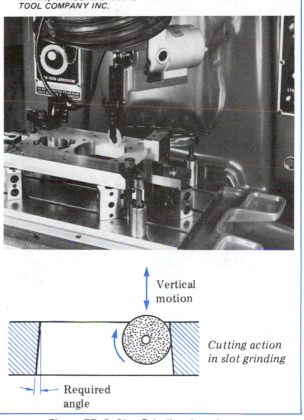

Cutting action in slot grinding

Figure 57–6 Slot Grinding Attachment Positioned for Slot Grinding Operation

spindle moves at a high grinding wheel speed within the planetary path. The diameter of a hole is enlarged by continually *out-feeding* the wheel while grinding.

In *wipe grinding*, the workpiece is fed past the wheel without any oscillating motion. By contrast, in *chop grinding* the grinding wheel has an oscillating movement. Metal is ground as the fast-revolving wheel oscillates and the work is fed past it. Chop grinding is used in contour grinding for fast stock removal.

Plunge grinding is done with the bottom edge of the grinding wheel and is used for rapid stock removal. The wheel spindle travels in a planetary path with the wheel traveling at a high rate of speed. The wheel is set radially at the required diameter. Feeding is axially into the workpiece.

Taper grinding requires the wheel axis to be inclined at the required taper angle. Straight-sided wheels are usually used in jig grinding tapers.

Shoulder grinding requires grinding with a concaved-end grinding wheel. Increments of down-feed are controlled by a positive stop or precision depth stop on the machine.

SLOT GRINDING IN A VERTICAL PLANE

Many tooling requirements call for the grinding of angles, slots, and corners that cannot be generated by vertical-spindle grinding wheel action. A *slot grinding attachment* (Figure 57–6) is used in a slot grinding operation to produce corners, radii, slots, and concave sections.

The wheel motion is vertical as contrasted with planetary and horizontal motion used to grind straight, cylindrical, and chordal sections of regular shape.

The grinding wheel face is dressed at a double angle. The wheel and attachment are set at the proper angle. A taper-setting feature permits grinding draft on both flanks of the angle at the same time. The inside tapered contour surfaces are ground by vertical oscillating motion of the high-speed revolving grinding wheel.

DESIGN FEATURES OF JIG GRINDERS

The major features on jig grinders, such as the 11″ × 18″ (280mm × 450mm) precision jig grinder are illustrated in Figure 57–7.

JIG GRINDING HEADS

Grinding heads are available for speeds ranging from 9,000 RPM for grinding operations re-

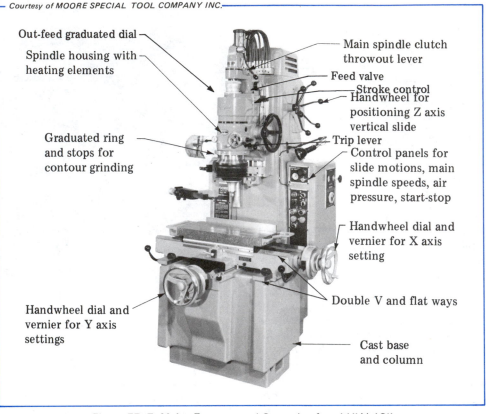

Courtesy of MOORE SPECIAL TOOL COMPANY INC.

Out-feed graduated dial

Spindle housing with heating elements

Graduated ring and stops for contour grinding

Handwheel dial and vernier for Y axis settings

Main spindle clutch throwout lever

Feed valve

Stroke control

Handwheel for positioning Z axis vertical slide

Trip lever

Control panels for slide motions, main spindle speeds, air pressure, start-stop

Handwheel dial and vernier for X axis setting

Double V and flat ways

Cast base and column

Figure 57–7 Major Features and Controls of an 11″ X 18″ (280mm X 450mm) Precision Jig Grinder

quiring power and rigidity for large, deep holes. The 40,000 RPM heads are designed for high thrust capacity for bottom, face, and shoulder grinding. The 175,000 RPM air-driven heads are applied to grinding ultrasmall holes. These heads are self-cooling. The exhaust air tends to keep the work cool. Dimensional errors caused by thermal expansion are thus minimized.

DRESSING ATTACHMENTS FOR JIG GRINDING WHEELS

The radius angle, cross slide, spherical socket, angle, and pantograph dressing attachments are each designed for dressing jig grinding wheels to particular shapes. Each dressing attachment, except for the angle-type dresser, is designed for use with a basic universal wheel dresser unit. Precise micrometer adjustments of 0.001″ to 0.0002″ (0.025mm to 0.005mm) and finer

vernier adjustments of 0.000020″ (0.5μm) are made. The repeatability of positioning is within 0.0002″ (0.005mm).

WHEEL AND DIAMOND-CHARGED MANDREL SPEEDS

As a general rule, grinding wheels operate efficiently at speeds of 6,000 sfm (1,828 m/min). The spindle speed RPM is found by dividing the sfm by the circumference of the wheel diameter. Therefore, the spindle speed range of jig grinders is from 9,000 to 175,000 RPM. Diamond-charged mandrels, which are used for small-hole grinding, are generally operated at 1,500 sfm (457 m/min).

MATERIAL ALLOWANCE FOR GRINDING

General grinding considerations apply to jig grinding. Also, since hardened materials and complex shapes are ground on the jig grinder, adequate

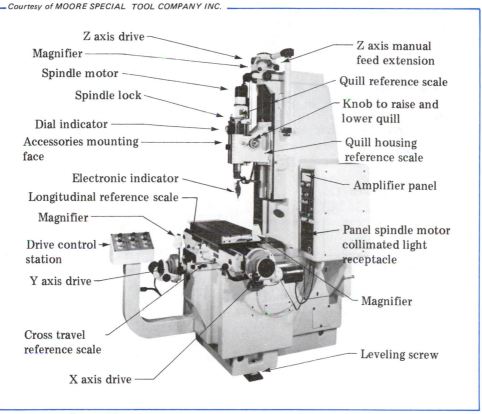

Z axis drive
Magnifier
Spindle motor
Spindle lock
Dial indicator
Accessories mounting face
Electronic indicator
Longitudinal reference scale
Magnifier
Drive control station
Y axis drive
Cross travel reference scale
X axis drive

Z axis manual feed extension
Quill reference scale
Knob to raise and lower quill
Quill housing reference scale
Amplifier panel
Panel spindle motor collimated light receptacle
Magnifier
Leveling screw

Figure 57–8 Major Components of a Universal Measuring Machine with Motorized Lead Screw Drive

material allowance must be made to compensate for distortion resulting from heat-treating processes. The stock allowance for hole sizes smaller than 1/2" (12.7mm) is from 0.004" to 0.008" (0.1mm to 0.2mm). Hole sizes larger than 1/2" are left from 0.008" to 0.015" (0.2mm to 0.4mm) undersize.

UNIVERSAL MEASURING MACHINES AND ACCESSORIES

The fine tolerances between punches and dies; the precise fitting of components in different mechanisms; the precision requirements of fixtures, cutting tools, gages, and gaging systems; and ever higher precision requirements between mating parts of mechanical, pneumatic, and other movements mean that finer measurements are required. Such measurements are in millionths of an inch (0.025μm) and angular dimensioning, within seconds of arc.

The universal measuring machine is a widely used measuring device. The universal measuring machine illustrated in Figure 57–8 is equipped with a motorized lead screw drive. This universal measuring machine may also be equipped for numerical control. Additional accessories, similar to jig borer and jig grinder accessories, are available for preselect positioning and readout and printer information. The standard model is equipped with manually operated lead screws for table positioning.

The extreme accuracies to which these machines measure demand that all components of a machine are held to equally precise tolerances at $68°$F ($20°$C). The total accumulative positioning accuracy of the X and Y axes is 35 millionths of an inch (0.9μm). The greatest amount of positioning error in any 1" (30mm) is 0.000015"

(0.4μm). The straightness of travel longitudinally is 25 millionths of an inch (0.6μm). The cross travel accuracy is 15 millionths of an inch. The spindle accuracy in terms of trueness of rotation is 0.000005″ (0.15μm).

Part drawings, using coordinate systems of measurement, are commonly used in jig boring, jig grinding, and machine measurement. NC and CNC universal measuring machines are programmed for two- or three-axis measurements.

BIDIRECTIONAL GAGING SYSTEM

The bidirectional type of gaging system permits dimensions to be measured without having to rotate the gaging probe. Also, the probe diameter is not subtracted from an observed measurement. The bidirectional gaging system consists of a series of gage head inputs with probe tips, centering controls, and measuring scale. The gaging system is switchable between inch and metric measurements.

TV MICROSCOPES

The *TV microscope* is one of the newer advances in micromeasurement technology. The TV microscope employs closed-circuit television for microscopic observation and measurement. The system provides images that have exceptional fidelity and accuracy. Details may be magnified up to 2500X. The TV microscope is particularly adapted for inspection and measurement of actual components of microminiature parts and circuits.

ANGULAR MEASUREMENT INSTRUMENTS

While conventional angle dividing heads and precision rotary tables are used on universal measuring machines, a *small angle divider* permits greater versatility. For measurement and layout purposes, the small angle divider has the capacity to quickly and accurately divide a circle into 12,960,000 parts. Measurements of ±10 seconds of arc are read directly on a vernier dial.

Other precision angular measurement instruments include the index center, rotary table and tailstock, and the micro-sine table. The precision *spin table* is power operated. The table turns at infinitely variable speeds from 5 to 100 RPM. The spin table is suited for inspection of roundness to tolerances within 0.000005″ (0.15μm).

Safe Practices for Operating Jig Borers, Jig Grinders, and Measuring Machines

- Position the leg of a dial indicator carefully against the edge of a workpiece on a jig borer or a jig grinder. Use care in moving the table to obtain an indicator reading.
- Place parallels and setup blocks so that through holes may be drilled, reamed, bored, or tapped without cutting into these positioning tools.
- Limit the amount of force required to strap workpieces securely on parallels on a jig borer, jig grinder, or accessory table. Excessive or uneven force may cause distortion or cracking if the part is hardened.
- Use the shortest possible shank or diamond mandrel length to reduce wheel overhang.
- Dress the side and bottom wheel faces to a small corner radius. This technique provides a stronger grinding wheel edge and tends to prevent the scoring of ground shoulder surfaces.
- Cycle the NC positioning movements and machining processes for programming accuracy before actual part production.
- Position the jig grinder safety shield for visibility of the grinding process and for eye protection.
- Maintain a constant room and workpiece temperature of 68°F (20°C) for ultraprecise measurements.
- Use machine bibs and aprons to protect housings, tables, slides, graduated collars, and other machine elements from abrasive particles and dust.

UNIT 57 REVIEW AND SELF-TEST

1. List four classifications of toolroom work that are performed on a jig borer.

2. Identify two important design features of drilling tools that are required for jig boring work.

3. a. Identify the function served by a micro-sine table.
   b. Describe briefly how a micro-sine table is set.

4. a. List the steps for setting up a locating microscope on a jig borer to center hole A in the figure shown below. The hole has been drilled and reamed to size.
   b. Tell how hole B is manually positioned for centering, drilling, and boring.

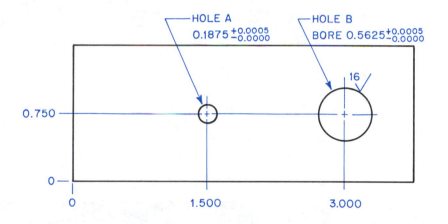

5. State two conditions or factors that differentiate slot grinding from plunge grinding.

6. Identify two methods of manually controlling spindle depth movement on a jig borer or jig grinder.

7. List the general sequence of grinding steps that are followed on a jig grinder.

8. Describe two applications of fixture, punch and die, or gage parts that may be measured on a universal measuring machine.

9. Tell what function a spin table serves when used on a universal measuring machine.

10. State two personal safety precautions to observe when using either a jig borer or a jig grinder.

SECTION ONE

Numerical Control Machine Tools

This section begins with a brief overview of advantages and disadvantages of numerical control (NC) machine tools. Major components of basic numerical control systems and applications on fundamental machine tools are described. The kind of information the machine or tool designer provides for the manufacturer of a part or mechanism is covered. Principles, codes, systems signals, control word language, and the producing of other feed-in information for machining and/or inspecting a part are examined. Three basic format NC programs are considered: tab sequential, work address, and fixed block.

UNIT 58

NC Systems, Principles, and Programming

Numerical control is a complete system of taped or computerized instructions. The basic functions of the system include the following:

- Controlling movements to position cutting and forming tools in relation to a fixed reference point,
- Controlling movements of cutting tools for setups and machining,
- Establishing sequences of operations and time intervals,
- Setting feeds and speeds,
- Monitoring accuracy and cutting tool performance or other machining functions,
- Providing readouts of machining accuracy,
- Changing the nature and sequencing of processes,
- Actuating shutdown or recycling.

ADVANTAGES OF NUMERICAL CONTROL MACHINES

Control functions formerly performed on machine tools by the machine operator are now translated into functions of the numerical control system. The longitudinal distance and direction movement of a machine table (X axis), the traverse cross feed movement (Y axis), and the vertical or angular movement of a spindle (Z axis) may be numerically controlled.

Courtesy of SHELDON MACHINE COMPANY INC.

X axis

Z axis

Figure 58–1 Two Basic Axes of an NC Lathe Equipped with a CNC System

The two basic axes (X, cross slide; Z, longitudinal) of a NC lathe are shown in Figure 58–1. Figure 58–2 illustrates the three basic axes (X, table; Y, saddle; Z, spindle) of a computerized numerical control (CNC) vertical milling machine.

Speed and feed rates for particular cutting tools, cycling for each process, and controls of cutting fluids may also be numerically controlled. In each of these examples, the machine control functions are performed by synchronized motors that respond to pulse commands.

Numerical control machines have many advantages. Five significant advantages include (1) productivity, (2) repeatability, (3) flexibility, (4) reduced tooling, and (5) increased machining capability.

Once programmed, point-to-point positioning accuracies are obtainable for each linear axis to within ±0.0005″ (±0.013mm). The repeatability accuracy is within ±0.0003″ (±0.008mm). Moreover, point-to-point errors are not cumulative.

With tape or computerized programming, it is possible to interrupt a production run, set up to produce an altogether different workpiece, and then return to the machining of the original part. The changeover is accomplished in a minimum of setup time. In many cases, jigs and fixtures are eliminated.

NC machines have a built-in check system. At the end of a block of commands, the control returns the spindle (cutting tool) to a zero starting reference point. On most systems, the oper-

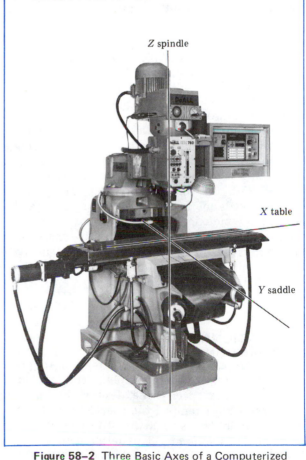

Courtesy of DoALL COMPANY

Z spindle

X table

Y saddle

Figure 58–2 Three Basic Axes of a Computerized Numerically Controlled (CNC) Vertical Milling Machine

ator is alerted by a warning system if the control fails to return to zero.

INCREASED MACHINING CAPABILITY

Tool changers are widely used on NC machine tools. Tool changers serve the function of holding the tooling required for machining a particular part. A *tool storage drum* may be turned until the required tool is in position so that it may be pivoted 90° downward. A *tool changing arm* grasps the pivoted tool and places it in the spindle. Simultaneously, the tool in the spindle is released to the tool changing arm, is returned to the pivot arm, and replaced in the drum.

The tool drum may be located vertically or it may be top mounted. Figure 58–3 shows a tool changing application with a top-mounted tool drum and horizontal spindle.

APPLICATION OF NUMERICAL CONTROL TO STANDARD MACHINE TOOLS

In order to combine a wide variety of machining processes that otherwise would require movement of workpieces among several machines, the whole movement toward numerical control has brought on the concept of the *machining center*. Total machining is performed on a single composite machine tool—that is, a machining center. With the machining center, the several spindle speeds and feed rates; turret indexing and workpiece indexing; positioning of machine components along three, four, and five axes; and continuous-path machine cuts for profile cutting in three dimensions are functions that may be controlled from tape instructions.

FOUNDATIONAL PROGRAMMING INFORMATION

The NC program requires information to position a spindle and work table. The tape instructions then specify the desired machining processes. For example, the specific tape letter code used in one system identifies *machining tasks* as *M functions*. This letter code is followed by a numeric code such as 01, 02, or 06. The following are a few common M functions and their meaning:

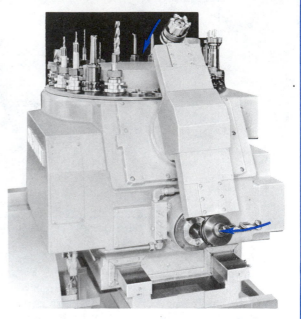

Figure 58–3 Tool Changing with Top-Mounted Tool Drum and Horizontal Spindle

M01	index
M02	end of program (calls for tape rewind)
M06	tool change
M07	coolant #2 on
M08	coolant #1 on
M09	coolant off
M10	clamp
M11	unclamp
M50	spindle up
M51	spindle down
M57	override milling rate

TYPES OF AUTOMATED PROCESSES

All machining requires either *constant* or *intermittent movement (travel)*. Intermittent means the workpiece travels at different lengths of time for a number of processes. When the entire processing is operated by mechanisms, the machine is automated and the workpiece is automatically processed.

CONSTANT CYCLING

Grinding, milling, and turning are operations that are adapted to constant cycling travel. These operations may require straight-line or circular movement. Normally, the workpiece is held in a fixture while one or more operations are performed. In constant cycling processing, the workpieces are removed from moving fixtures. The workpieces move in a designated sequence for machining.

INTERMITTENT CYCLING (TRAVEL AND WORK STATION INDEXING)

Drilling, boring, reaming, and counterboring operations in a single workpiece provide an example of intermittent cycling. Uneven time periods are required for each operation, and, within an operation, holes of different sizes may be machined.

NUMERICAL CONTROL SYSTEMS

CLOSED LOOP NC SYSTEM

In general, there are two numerical control systems: *closed loop* and *open loop*. A signal from the numerical control unit is fed through the machine control unit (MCU) to provide a specific instruction (motion command) to the servo drive unit. The lead screw is actuated by a servomotor. If the signal to the MCU directs the servo drive to feed a machine table 6″, the table is moved this distance. A *sensor* on the servomotor (drive motor) feeds back a signal (through the encoder) to the MCU to indicate the table has moved the instructed distance of 6″. In the closed loop system, the machine control unit is provided with a check on the accuracy of the machine movement.

OPEN LOOP NC SYSTEM

The open loop NC system is used on many installations of numerical control *(retrofitting)* to existing machine tools. In the open loop system, *stepping motors* are used to control the movements of the machine components. There is no feedback system.

The MCU supplies the electric current impulse to the stepping motor. The number of pulses of the MCU is determined by the number of fractions of a revolution required to turn a coupled lead or feed screw to advance or return a table a specific distance within a time limit. Each current pulse causes the motor rotor to turn a fraction of a revolution. Some stepping motors advance a machine table 0.001″ (0.025mm) each pulse. For example, if the table is to advance 1.000″ (25.4mm), the MCU directs 1,000 pulses to the stepping motor.

ADDITIONAL CIRCUITS AND LOOPS

The simplest NC system includes a *control unit*, an *information feeder* (tape or computer), an *actuating/control unit* to supply power to each movement, and a *feedback device (transducer)* that tells how much movement is taking place.

NC RECTANGULAR COORDINATE SYSTEM

Numerical control part drawings are based on what is called the *rectangular (cartesian) coordinate system*. Coordinate dimensions are used on drawings to represent the part.

X, Y, AND Z AXES AND ZERO REFERENCE POINT

Within the rectangular coordinate system, there are two basic axes: X and Y. These axes are perpendicular, lie in the same plane, and are known as *coordinate axes*. A third *spindle axis*, Z, is perpendicular to the X-Y plane. A three-dimensional part (mass) can be described accurately according to relationships with the X, Y, and Z axes. The three axes intersect at a point. The point is identified as the *origin* or *reference point*. The numerical value assigned to this origin point is *zero*. Figure 58–4 illustrates the basic X and Y axes, spindle axis Z, and zero reference point.

ROTATIONAL AXES

Numerically controlled motion around the basic X, Y, and Z axes may be related to *rotational axes*. A rotary table or indexing mechanism may be operated from tape or computer instructions around a basic axis. Information must be provided for (1) direction of rotation

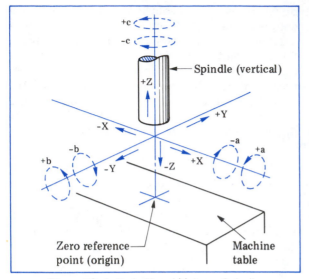

Figure 58–4 Basic X and Y Axes, Spindle Axis Z, Zero Reference Point, and Rotational Axes a, b, and c

and (2) the basic axis around which rotation takes place. Lowercase letters a, b, and c identify the rotational axes (Figure 58–4).

QUADRANTS AND NC SYSTEM POINT VALUES

The X and Y coordinate axes, which are at 90° to each other and are in the same plane, form four quadrants. The quadrants are numbered QI, QII, QIII, and QIV. The origin point is zero. Any point (position or dimension) within a quadrant has a plus or minus value (Figure 58–4) depending on the direction a measurement or distance is taken from the point of origin.

Quadrant I	+X and +Y
Quadrant II	–X and +Y
Quadrant III	–X and –Y
Quadrant IV	+X and –Y

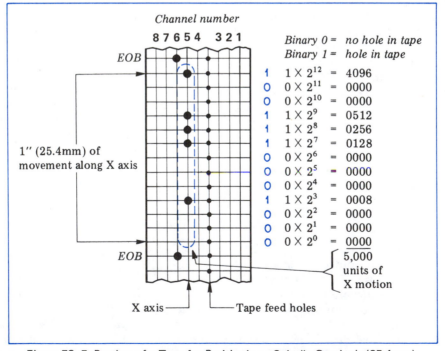

Figure 58–5 Portion of a Tape for Positioning a Spindle One Inch (25.4mm) along X Axis by Using a Straight Binary Number

BINARY SYSTEM INPUT TO NUMERICAL CONTROL

Numerical control is a precise electronic control system. It depends on a two-word numerical vocabulary. The numerals communicate information to "stop" and "go" control pulses. Numerical data are entered on a tape by *binary numbers* and *binary notation*.

BINARY NOTATION

A portion of a tape for positioning a spindle by using a straight binary number is illustrated in Figure 58–5. In this figure, channel 5 is used to program the input information needed to advance a workpiece along the X table axis for 1″ (25.4mm). In this case, each unit of X motion is 0.0002″. Thus, 5,000 units of X motion are required to produce the 1″ of slide motion. A hole in the tape in channel 5 equals a binary 1. The lack of a hole in the channel equals a binary 0. The binary notation for the 5,000 units of movement is expressed as:

$$1\ 0\ 0\ 1\ 1\ 1\ 0\ 0\ 0\ 1\ 0\ 0\ 0$$

Thirteen lines are used in channel 5 on the tape. Binary numerals are identified by a punched hole or no hole. The binary notation for the 5,000 units of motion is punched between the beginning and end-of-block holes in channel 6 of the tape.

BINARY CODE DECIMAL (BCD) SYSTEM

Tape formats for numerical control provide standardized information. The EIA (Electronics Industrial Association) and the ASCII (American Society for Computer Information Interchange) tape formats use the *binary code decimal (BCD) system* of digit coding. Each channel on the tape is assigned a value, as follows:

Channel	Assigned Numerical Value
1	1
2	2
3	4
4	8
6	0

These five channels may be used to designate any number between 0 and 9.

Numerical quantities are expressed in binary notation running along the length of the tape. Each number is expressed as a number of digits, usually six. While a decimal point is not shown, it is understood to be between the second and third digit.

EIA tapes are encoded with numbers (0 through 9), letters (A through Z), signs, and other symbols. The codes are arranged in horizontal rows. Two examples of EIA tape formats are illustrated in Figure 58–6.

NUMERICAL CONTROL WORD LANGUAGE

A *NC word* is a set of letter and numeric characters arranged in a prescribed way. The two basic types of words in a NC word language are (1) *dimension* and (2) *nondimension*.

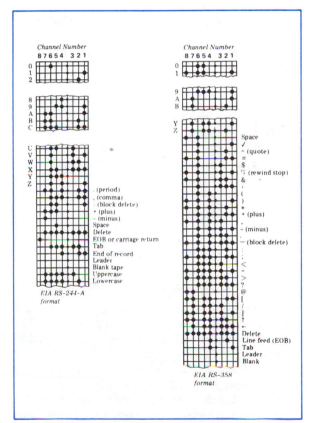

Figure 58–6 Sample EIA Tape Formats (Reduced)

DIMENSION WORDS

Each dimension word begins with the axis address code (letter). Each dimension word has three parts: the axis address code (letter), the appropriate (+ or –) directional sign, and the distance of movement. If there is no + directional sign in the word, the control interprets the word as positive.

NONDIMENSION WORDS AND FUNCTIONS

Nondimension (direction) words fall into categories such as: (1) sequence, (2) preparatory function, (3) feed rate, (4) spindle speed, (5) tool (turret) selection, and (6) miscellaneous function. In each case, an *address character* (designated letter) is followed by a given number of digits.

The *sequence number* is designated by the address character N and three numeric digits. The word indicates the start of a specific sequence of operations. It is the first word for the programming sequence within the block.

The *preparatory function* is designated by the character G and two numeric digits. This word immediately follows the sequence number word. The G word prepares the numerical control unit for a specific mode of operation. Examples of G words include:

G00	point-to-point positioning
G01	linear interpolation
G02	circular interpolation
G03	circular interpolation, Arc CW (arc clockwise)
G03	circular interpolation, Arc CCW (arc counterclockwise)
G04	dwell
G13–G16	axis selection
G33	thread cutting, constant lead
G40	cutter compensation cancel
G80–G89	fixed cycles 1 through 9

Dimension words follow the preparatory function word. For multiple-axis systems, the dimension words follow in order: X, Y, Z; U, V, W; P, Q, R; I, J, K; A, B, C, D, E.

The *feed function* is designated by the letter F and a maximum of eight numeric digits. This word follows the last dimension word. The F is programmed on tape as a coded feed rate number. The EIA feed rate system consists of a series of two-digit code numbers, each representing the linear motion feed rate in inches per minute (ipm).

The programmer follows the same guidelines for establishing feed rates as are applied to conventional machining. Material to be cut, the machining process, and characteristics of the machine tool are considerations. The control units of NC machines are designed with automatic, smooth acceleration or deceleration to a new higher or lower programmed feed rate.

The *spindle speed* is designated by the letter S and three numeric digits. This word follows the last dimension word or feed rate word. The spindle speed is expressed in a coded three-digit number.

The *tool (turret) function* is designated by the letter T and a maximum of five numeric digits. This word immediately follows the spindle speed word. The digits selected must be compatible with the particular numerical control system being used.

The *miscellaneous function* is designated by the letter M and two numeric digits. This word follows the tool function word and immediately precedes the end-of-block (EOB) character.

OTHER SELECTED NC FUNCTIONS

Arc clockwise (Arc CW) specifies the path of curvature generated by coordinating the movement of a cutting tool in a clockwise direction along two axes.

Arc counterclockwise (Arc CCW) specifies the path of curvature generated by a cutting tool whose movements along two axes are coordinated in a counterclockwise direction in the plane of motion.

Automatic acceleration (G08) means accelerating the feed rate from the starting feed rate within a block. A starting feed rate, in any block in which the G08 code is used, of 10% of the programmed feed rate may be accelerated, according to a time constant, to 100%.

Plane selection (G17, G18, or G19) is generally used for X–Y, X–Z, or Y–Z plane selection for circular interpolation and cutter compensation functions.

Program stop is an M00 word that stops the spindle, coolant flow, and feed after completion of all commands in the block. The remainder of the program may continue after the machine operator pushes a button.

Spindle clockwise is an M03 command that starts the machine spindle rotation to advance a right-hand screw into the workpiece. An M04 command starts the spindle to retract a right-hand screw from the workpiece.

Spindle off is an M05 command that stops the spindle as efficiently as possible and turns off the coolant.

FACTORS TO CONSIDER IN NC PROGRAMMING

LINEAR INTERPOLATION

Linear interpolation relates to the control of a travel rate in two directions. The travel rate is proportional to the distance traveled. The axis drive motors must be capable of operating at different rates of speed. Thus, in linear interpolation, the stepping motors on the axes drives permit a cutting tool to move along an angular, circular, or arc path.

CIRCULAR INTERPOLATION

Circular interpolation is defined as the ability of a control unit to generate a circular arc of maximum 99.99° span in one block. A circle or an arc is generated as a continuous curve rather than as a series of straight lines. In circular interpolation, the start and the end of an arc are programmed in only one block of tape. One type of circular interpolation is the EIA standardized method. Some of the newer CNC machines are designed to generate a 360° arc in one block.

DIMENSION COMMAND PULSE WEIGHT

All dimension words must be divisible by the command pulse weight of the numerical control system. The *pulse weight of the system* is the smallest increment of a machine slide movement caused by one single command pulse. For example, if the control unit has a command pulse weight of 0.0002″ (0.005mm), each electronically generated command pulse causes a movement of a machine slide of the same magnitude.

ACCELERATION AND DECELERATION FOR CUTS

The NC programmer must recognize that upon approaching the end of a cut, such as an inside corner, the cutting tool may have to be slowed down (decelerated) to prevent *overshoot*. Overshoot causes the cutter to cut deeper than required and to leave an undesirable indentation in the workpiece. The deceleration is block programmed with a reduced feed in order to machine precision inside corners.

POSITIONING THE SPINDLE

INCREMENTAL MEASUREMENT

Incremental measurement means that the spindle measures the distance to its next location from its last position. Incremental measurement (positioning) utilizes positive and negative directions.

ABSOLUTE OR COORDINATE MEASUREMENT

The spindle of a NC machine tool may also be positioned by *absolute or coordinate measurement*. One system is identified in relation to a fixed zero—that is, all measurements are taken from the same reference point. The advantage of using a fixed zero is that the spindle operates only in quadrant I. All movements for positioning locations have positive values. All coordinate location points are specified in relation to distance from the coordinate axes.

FLOATING ZERO POINT

NC machine tools may also be programmed to permit a *floating zero* to be used as absolute zero. The floating zero may be established as any point that will make programming easier. Spindle positioning using a floating zero point is illustrated. Figure 58–7A is an ordinate drawing of a workpiece that requires the drilling of four holes symmetrically located. Figure 58–7B shows that the absolute zero may be floated to the intersection of X and Y axis center lines. The four holes are positioned by coordinate locations in each of the four quadrants. Thus, the X and Y values change.

Absolute positioning has the same advantages over incremental positioning with respect

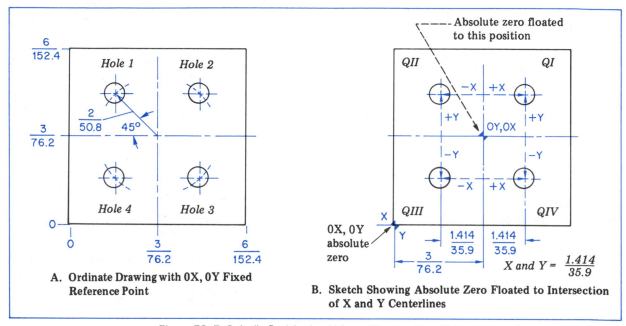

Figure 58–7 Spindle Positioning Using a Floating Zero Point

to machining accuracy. Errors that accumulate from incremental positioning are not a problem in absolute positioning.

NUMERICAL CONTROL TAPE PROGRAMMING, PREPARATION, AND CORRECTION

MANUAL PROGRAMMING

Manual programming requires that all cutter, machine function, and numerical coordinate data be given. The cutter positions must be calculated and specified on the manuscript.

The manuscript is then processed by a *tape preparation unit*. The program for the part is encoded in English letters and Arabic numerals on a tape. The preparation unit also prints a copy or *printout*, of the program. The printout is corrected or changes are made in the tape. Once the tape is accurate, it is fed to the machine control unit (MCU). The MCU feeds control information to the actuating and movement systems, devices, and mechanisms incorporated in the machine tool.

TAPE MATERIAL AND FEATURES

A NC tape is a common method of giving control instructions to a NC machine. NC tape materials include durable paper, paper-plastic, aluminum, and plastic laminates. The nonpaper tapes are able to withstand greater usage with less wear and are not subject to soil from materials used in the shop. Regardless of material, tape sizes are standardized and are manufactured to close tolerances.

TAPE BLOCKS AND REWIND STOP CODE

The information coded on the tape provides input data to the MCU. The MCU directs the machine tool through its various functions. The input coded information on the tape is sectioned in units referred to as *blocks*. Each block represents a complete entity: a machining operation, machine function, or a combination. Each block is separated from a succeeding block by an *end-of-block (EOB) code*. The EOB code is punched on channel 8.

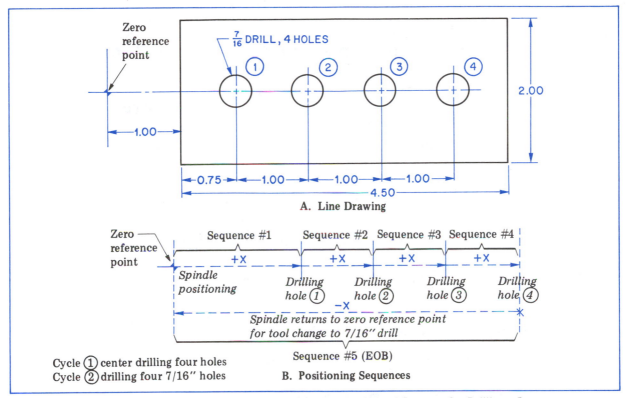

Figure 58-8 Example 1: Tab Sequential Numerical Control Program for Drilling a Part

TAB SEQUENTIAL BASIC FORMAT PROGRAMS

Three basic formats for programming follow. These formats include: tab sequential, word address, and fixed block programs.

Tab sequential is a basic NC format used for point-to-point NC applications. It may also be adapted to continuous-path contour programming. The code produced on the tape is *tabbed* by the tape punch typewriter to give separate information about axis positioning, machining, and other functions. The MCU is able to differentiate in the electrical sections between X, Y, and Z axis positioning, tooling functions, and machining processes.

In tab sequential, the information follows a particular sequence. The first information relates to the positioning or operation step. A tab code follows to separate this information from X axis information.

The next tab code separates X positioning from Y positioning. The third tab separates Y

positioning information from machining M functions. Additional tab codes are used to separate Z positioning and/or rotational axis positioning (if required).

Tab sequential programming is used in the following four examples. These examples apply to drilling and milling.

SINGLE-AXIS, SINGLE MACHINING PROCESS

Example 1: Point-to-Point Program. Drilling processes are a good example of point-to-point machining. The zero reference point for numerical control machining is usually planned to be off the workpiece. The spindle is usually positioned by manual control over the zero point.

An example of a single-axis, single machining process for drilling a part is shown in Figure 58-8. The part, represented by the line drawing, requires the drilling of four holes along the X axis.

Company: C. G. Whitehurst Machine Tool Works

| Dept. 169 | Part DRILLED PLATE | Part # 16 | Oper. #1-A |

Prepared by
Date CTO-86

Ck'd by
Date TPO-86

Tape # 12206

Sheet 1 of 1

Remarks
TOOL MODE SWITCH- AUTO
FEED RATE- HI
BACKLASH- TAKEUP #2

Tools
CENTER DRILL
7/16 DIAMETER DRILL

Seq. #	Tab or EOB	+ or –	(X) Increment	Tab or EOB	+ or –	(Y) Increment	Tab or EOB	(M) Function	EOB	Instructions
									EOB	
0	RWS								EOB	CHANGE TOOL; LOAD; START
1	TAB		1750						EOB	
2	TAB		1000						EOB	
3	TAB		1000						EOB	
4	TAB		1000						EOB	
5	TAB	–	4750	TAB			TAB	02	EOB	

Figure 58–9 Example 1: Tab Sequential Numerical Control Tape Program for Drilling Process

How to Program a Single-Axis, Single-Machining Process

Console Presets

1. The tool mode switch is set to the automatic position. The cutting tool (in this example, the center drill and the 7/16″ diameter drill in turn) is to machine and retract automatically at each position.
2. The tool positioning rate (feed rate) is set at the high (Hi) rate of speed.
3. Lead screw backlash compensation is provided. (In this example, a #2 compensation provides the degree of precise positioning required.)

Beginning of Program, RWS Code, and Sequencing Statements

The NC program (Figure 58–9) begins with the end-of-block (EOB) code. This code is followed by the RWS code. Before there are any program statements, the sequence statement #0 identifies the rewind stop (RWS) code instruction to the tape reader. The tape is stopped at this point after rewinding and the program is recycled.

1. Sequence statement #1 gives the spindle (table) movement along the X axis from the reference point to the center of the first hole to be center drilled. The 1.75″ movement is indicated as 1750. The first hole is center drilled automatically.
2. Sequence #2 gives the spindle movement (1000) along the X axis from the first hole to the second hole. The second hole is center drilled automatically.
3. Sequence #3 provides positioning information or the spindle movement (1000) from the second hole to the third hole. The third hole is center drilled automatically.
4. Sequence #4 prescribes positioning information (1000) to move the spindle from the center of the third center-drilled hole to the fourth hole. The fourth hole is center drilled automatically.
5. Sequence #5 gives the –X increment (–4750). This positioning information

moves the spindle back along the X axis from the fourth center-drilled hole to the reference (origin) point.

Further, the M02 function instructs the tape reader to rewind the tape. The tape does not call for any drilling at the point of origin.

6. The center drill is replaced with the 7/16″ (10.9mm) drill.
7. Sequences #1 through #5 are repeated. The four center-drilled holes are drilled to the required 7/16″ (10.9mm) diameter. At the end of this cycle, the 7/16″ (10.9mm) drill is replaced with the center drill.
8. The M02 function signifies an end-of-program instruction.

X AND Y AXES AND TOOL CHANGES

Example 2: Point-to-Point Program. An example of point-to-point positioning along the X and Y axes, with tool changes for drilling four holes of two different sizes, is illustrated in Figure 58–10. The part drawing shows positioning sequences for a tab sequential program.

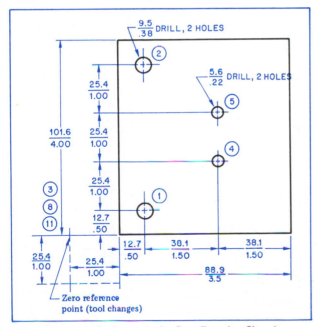

Figure 58–10 Example 2: Part Drawing Showing Positioning Sequences for Point-to-Point Positioning along X and Y Axes, with Drilling Tool Changes

How to Program for Two Axes and Tool Changes

Beginning of Program, RWS Code, and Sequencing Statements

1. After the console presets are made, the instructions start with the loading of the part. Sequence position 0 on the NC tape program RWS for recycling the program.
2. Sequence numbers 1 and 2 give the X axis and/or Y axis increments for positioning and drilling the 9.5mm (0.38″) diameter holes.
3. Sequence number 3 of the program provides information about positioning the spindle at the original zero point. The M06 code M function stops the control and lights the tool change lamp. The instructions column on the tape program indicates that the machine tool operator makes a tool change to the 5.6mm (0.22″) stub drill. –X and –Y increments return the spindle to the starting position.
4. Sequence numbers 4 and 5 position the spindle to drill the two 5.6mm (0.22″) diameter holes.

THIRD AXIS FUNCTION

Example 3: Tab Sequential Program. The third axis function normally controls a spindle component, a Z axis movement of a quill or knee, or an accessory such as a rotary table. Figures 58–11A and B provide an example of a drilling application using a rotary table for the third axis. An angle positioning plate is to be programmed for drilling (Figure 58–11A). Seven sequences are involved in drilling the four holes (Figure 58–11B).

MILLING PROCESSES INVOLVING SPEED/FEED CHANGES

Example 4: Tab Sequential Program. An example of using tab sequential programming in a milling process is provided in Figures 58–12A and B. The part drawing (Figure 58–12A) shows the position of the zero reference point. Three sequences (Figure 58–12B) are involved in the

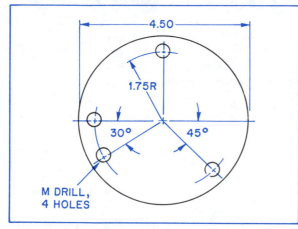

Figure 58–11A Example 3: Part to be Programmed for Drilling Application Using Rotary Table

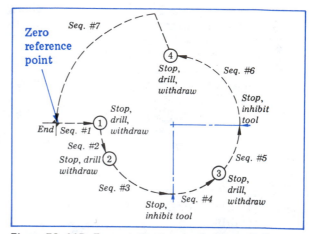

Figure 58–11B Example 3: Schematic Sequence Statements for Drilling Applications Using Rotary Table

procedure for milling the elongated slot. Three new features—M55 (override milling feed rate), M52 (quill down), and M53 (quill up)—are included in the tab sequential program.

NC machine tools are designed to permit the change from rapid traverse in a slow quill (and cutting tool) down-feed and finally to a still different table feed rate.

The milling is completed in sequence #2 and the end mill clears the workpiece. Sequence #3 returns the spindle (end mill) to the starting reference point by using rapid traverse.

How to Program for Milling Processes Involving Speed/Feed Changes

1. Sequence #1 calls for positioning at high feed rate along the X and Y axes to the point where the milling operation begins.
2. Sequence #2 actuates the spindle so that the end mill feeds through the workpiece. The Y axis movement to mill the elongated slot is also identified.

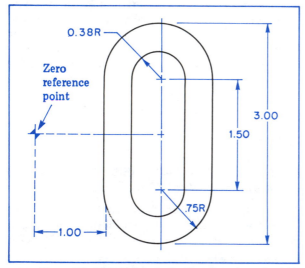

Figure 58–12A Example 4: Part Drawing of Tab Sequential Program for Milling Process

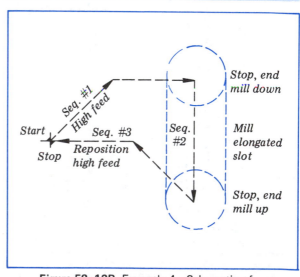

Figure 58–12B Example 4: Schematic of Tab Sequence Statements for Milling Process

3. Sequence #3 provides information for removing the end mill (upward movement of the quill) from the workpiece. Further directions are given for the rapid traverse (or high speed return) table movement to reposition the spindle at the reference starting point. Instructions are indicated to rewind the tape.

WORD ADDRESS FORMAT PROGRAMS

Word address is another NC format. The word address format requires a single letter code (A to Z) to deliver an address. Thus, control information is differentiated by the single letter address. By contrast, tab sequential format provides control information by a specific sequence that is separated by tab codes.

SAMPLE CODED FUNCTIONS

In word address format, specific instructions are given by adding a numeric code to the letter address word. Both the letter and the numeric code are used in word address. For example, a feed rate of 8 ipm is identified in a word address program as F8.

Word address format programs use the letter G for preparatory functions; M, for miscellaneous functions; N, for program sequence numbers; and F, for feed rates. Common letter words of X, Y, and Z for axis distances and + or – directions are included. Examples of coded functions in word address format are as follows:

- A G81 preparatory function calls for the cycling of a milling machine spindle (or the quill of a drilling machine) to perform a milling operation;
- A rapid-traverse feed rate may be programmed into the F address by adding the coded feed rate;
- At the end of a program, the M02 instruction cancels the first sequence so that the tool is not actuated;
- In using the third axis, information is programmed into the Z address;
- Information to indicate circular interpolation for a contouring program is provided through I and J data.

APPLICATION TO CONTINUOUS-PATH MACHINING

Word address may also be used for continuous-path machining. The G01 preparatory function information may be used to call up an interpolation cycle. Radii and angles may be approximated by the cuts in incremental steps.

FIXED BLOCK FORMAT PROGRAMS

Fixed block is a third common type of NC format. The features of a 20-digit fixed block program tape are illustrated in Figure 58–13.

MIRROR IMAGE PROGRAMMING FEATURE

Symmetrical and right-hand and left-hand parts may be efficiently programmed from a single tape by using a *mirror image (axis inversion) feature.*

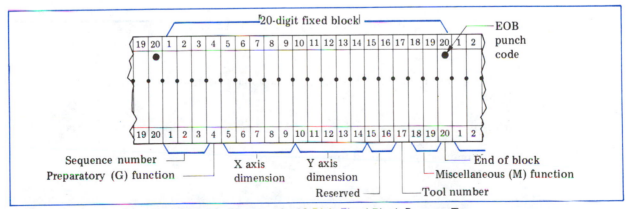

Figure 58–13 Features of a 20-Digit Fixed Block Program Tape

A programmed mirror image produces a reverse duplicate of a part, as illustrated in Figure 58–14. The machine control unit (MCU) is programmed to direct the X and/or Y axis mirror image to electrically reverse the direction of travel. All positive direction (+) movements around a program zero become negative (–). All negative direction movements become positive.

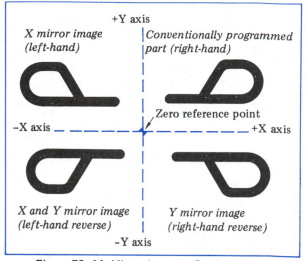

Figure 58–14 Mirror Image to Produce Reverse Duplicate of Part

Safe Practices Relating to Numerical Control Machine Tools

- Reduce the overhang of each cutting tool or the extension of the spindle. A shortened position provides greater rigidity for machining.
- Study tool position heights, especially when multilevel surfaces are to be machined. There must be adequate clearance between cutter, workpiece, and work-holding device to permit safe slide movement between all processes, including loading and unloading.
- Set cutting speeds and feed rates for each tool process within limits recommended for each process. Although processes are tape programmed, the operator still needs to observe the cutting conditions.
- Read the tape printout. Check for axis positioning accuracy (spindle location) and correctness of each machining process.

- Make a dry run for safety checks, spindle positioning, and each machining process.
- Check in advance on how the MCU and machine tool may be stopped immediately in an emergency.
- Use safety goggles or a protective eye shield and follow personal safety precautions.
- Remove burrs carefully. Use standard safety practices for cutters, work-holding devices, and machine tool processes.

UNIT 58 REVIEW AND SELF-TEST

1. Cite three tooling-up and machining functions that may be performed on NC machine tools.

2. Differentiate between a fixed zero reference point and a floating zero point.

3. List the main steps required to manually program a NC machine tool to produce a specific part.

4. a. Prepare a simple sketch to show the programming sequence for drilling the three holes in the drill plate shown in the part drawing below. Locate the zero reference point along the center line, 1.00″ from the left end of the workpiece.
   b. List the console presets.
   c. Prepare the statements for producing the tape according to the tab sequential format.

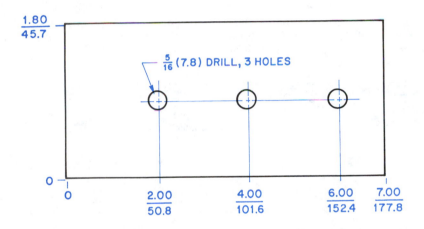

5. a. State the functions performed by the letters G, F, and M in word address format programs.
   b. List examples of code numbers used with each letter.
   c. Give an application of each word format.

6. Describe mirror image.

7. List two personal safety features that apply particularly to the operation of NC machine tools.

NC Applications and Computer-Assisted Programming (CAP) for CNC Machine Tools

This section extends the basic principles of numerical control to on-line machine tool applications, tool gaging and management control, and tool-holding devices. Then, basic descriptions are given of computer-assisted programming (CAP) and computer numerical control (CNC). Step-by-step procedures are included for planning and preparing a computer-assisted program manuscript. The section concludes with descriptions of major components in a sophisticated CNC machining center. In the future, these centers may be adapted to complete computer-aided manufacturing (CAM) systems.

UNIT 59

Basic NC and CNC Tools and Machining Processes

NC MACHINES AND CONTROLS

COMMON CONTROL FEATURES

Regardless of the type of system, there are certain basic control features, as follows:

- Electronic components generate heat and therefore require temperature controls from $50°F$ to $120°F$ ($10°C$ to $48°C$), humidity control up to 95%, and the use of fans and air conditioning;
- The control panel area must include additional space for the machine-tool builder to mount additional or parallel machine function control switches and other accessories;
- The control system incorporates safeguards to ensure an accurate read-in of the tape;
- The NC system is able to control linear and rotary movements in any combination and simultaneously for all axes of machine motion (the smallest increment of motion, de-

pending on the machine tool and processes, ranges from 0.000020″ or 0.0005mm to 0.0001″ or 0.0025mm, and the smallest increment of rotary motion ranges from 1.396 seconds of arc, or 0.000001 of a revolution, to 3.6 seconds of arc);

- The overall accuracy of the control system with an increment of motion of 0.0001″, operating at a constant temperature, should be within ±0.0002″ (0.005mm). Finer accuracies are required for machine tools with smaller increments of motion;
- The feedback system permits the control system to compare the tape command with the actual NC machine positioning or other functions;
- Backlash compensation is provided regardless of direction of motion from the previous position;
- Complete documentation is provided for programming, operating, and maintaining the NC system.

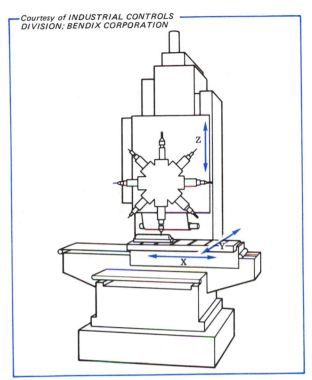

Figure 59–1 Tape-Controlled Drilling Machine
with Turret (Three Axes)

NC DRILLING MACHINES

The NC system for a general drilling machine may include continuous-path NC units for linear or circular interpolation. Arcs up to 90° may be generated within a single quadrant from tape commands in a single block. The following control features may also be included:

- Spindle speed selection, which stores the coded set of digits and then relays the information to the speed changing mechanism on call;
- Tool select feature, which provides for the automatic selection, changing, and storing of tools for multiple-sequence operations;
- Mirror image, which, through axis symmetry switches, permits the sign reversal of dimensional tape information and thus permits a single tape to be used to produce mirror image parts;
- Auxiliary function display, which, through a function selector switch, permits the operator to select an auxiliary function such

as the feed number, preparatory function number, feed rate, spindle speed code, and tool number for display.

Complex NC Drilling Machine. A complex drilling machine includes the control of depth axis (Z) and, if required, a rotary fourth axis. The complex class of NC drilling machines uses combination positioning and contour control systems. A drilling machine with turret and tape control for three axes is illustrated in Figure 59–1.

The third axis is tape controlled so that the spindle is programmed for final depth. In addition, there is a *feed engage point* at which a rapid-traverse feed is changed to a slower programmed feed.

The control system automatically programs preparatory G functions. Thus, the Z axis may be programmed for drilling, boring, reaming, tapping, and other processes.

Since most cutting tools vary in length and may be of a different length than programmed, *tool length compensation* for general machining on some NC machines requires manual setting. The difference in dimensions is set up on switches for each specific tool number. A compensation dial is automatically activated when the tool is selected.

NC TURNING MACHINES
(POINT-TO-POINT POSITIONING)

The engine lathe and the turret lathe (bar and chucking machine) are two basic types of numerical control turning machines. NC principles and machine control units are adaptable to vertical turret and other types of lathes.

NC Engine Lathe (Two Axes). A two-axis NC positioning system may be used for general turning processes that do not require taper turning, thread cutting, or contour forming. Where contours are required, the NC system must have continuous-path capability.

Continuous-Path System NC Engine Lathe. The continuous-path system permits the machining of tapers, threads, radii, and other contour cutting processes. Contour cutting requires the feed rates to be tape controlled. The control unit reads the specified rate from the coded number on the tape and processes it. Other features that are included with a continuous-path system are the following:

- *Linear or circular interpolation*, which provides for straight-line or smooth-curve movement of a cutting tool in machining a contour
- *Internal and external thread cutting*, which is available on numerical contouring control to eliminate manual processes and cumulative errors from thread to thread;
- *Tape-controlled spindle speed selector and auxiliary function display*, a combination of features serving similar functions as described for applications on drilling machines and for drilling processes.

NC Bar and Chucking Machine and Machining Centers. Production turning machines may be tape controlled for two, three, and four axes, depending on the complexity of the machining processes. Some of the common combinations include the following tape controls:

- Simultaneous control of carriage and cross slide along two axes (the turret position selection may also be tape controlled);
- Control of carriage, cross slide, and turret ram (three axes), with turret position selection;
- Simultaneous control of any two of the three axes, with indexing of stations on the turret ram or square tool turret;
- Simultaneous tape control of four axes.

Some typical NC features include a sequence number display, dimension display, manual data input, tool select, spindle speed control, and auxiliary function display. In addition, a *tool offset* feature permits the cutting tool to be moved a preset distance in the automatic tape mode. Tool offset compensates for changes in tool length due to resharpening a cutter or to a position change of the tool in a different holder. The NC machine operator dials in required offsets so that any one tool may be programmed to pick up any given offset.

NC MILLING MACHINES (TWO, THREE, AND FOUR AXES)

When classified according to position of the cutter driving spindle, milling machines are of two types: horizontal and vertical. These types may be programmed for tape control of at least two axes of simultaneous motion. Control of three axes and four axes is general.

A NC system with continuous-path capabilities is practical for angle and other contour milling jobs that are common in jobbing shops. A continuous-path NC system with three axes of control has controls that relate to M functions of table movement, continuous-path controls, and linear and circular interpolation for generating arcs. Additional features, such as sequence number display, dimension display, manual data input, tool selection, and spindle speed control, are available to improve control capability. An auxiliary function display may be included in conjunction with a function selector switch. The operator is able to select any auxiliary function and display it. Auxiliary functions relate to sequence number, preparatory function number, feed rate or spindle speed code, and tool number.

If a considerable amount of drilling, boring, reaming, and tapping is to be done, as in the case of universal vertical turret milling machines, a combination of point-to-point and continuous-path systems is desirable. Multiple-operation milling machines are designed with a combination positioning and continuous-path system.

NC BORING MACHINES

There are two general classifications of boring machines. The first class includes vertical turret lathes and boring mills. The cutting tool, when once positioned for a cut, remains stationary. The workpiece revolves while the cutting tool feeds in to take a cut. In the second classification, the action is reversed. The workpiece is moved into position while the cutting tool revolves and is fed in the direction into the workpiece. Horizontal boring machines and the more widely used jig borer are included in the second class.

NC Jig Borer. The jig borer is regarded by industry as an extremely precise machine tool.

Courtesy of MOORE SPECIAL TOOL COMPANY INC.

Figure 59–2 NC Jig Borer (Equipped with a Rotary Table) and Machine Control Unit

- Complete solid-state integrated circuit system;
- Photoelectric tape reader with capacity of 125 characters per second;
- Temperature-controlled operating range from 50°F to 120°F (10°C to 48°C);
- Auto, single, and manual modes;
- Data input of 1" (25.4mm), 8-track perforated tape (EIA standards RS–227);
- Word address, variable block format (EIA RS–244 and RS–274B);
- Automatic backlash compensation;
- Miscellaneous functions M00 through M99;
- Reference, set, and grid zeroing;
- Absolute programming;
- Test mode and test circuits;
- Incremental feed and programmed feed rates;
- 100% manual feed rate override in 10% steps;
- Milling control of continuous, one-axis-at-a-time capability in a straight line parallel to the X and Y axis;
- Spindle cycle control;
- Multiple depth selection.

The jig borer has program data resolution accuracy of 0.00001" (0.001mm).

NC MACHINING CENTERS

NC machining centers have tape control for three, four, and five axes. It is possible to move the column, spindle head, table traverse, and table rotation and tilt simultaneously under tape control. The NC system may be point-to-point, continuous-path, or a combination system.

The basic control features include simultaneous movements along all axes, programmable feed rates, spindle speeds, and fixed cycle. Typical additional features that increase the control capability or machine operation include: sequence number display, auxiliary function display, manual data input, tool gaging and selection, and tool length compensation. NC machining centers with circular interpolation are capable of generating arcs up to 90° in one quadrant from single-block tape commands. An addition is available on some circular interpolation features for generating 360° of arc from single-block tape

The accuracy of positioning for longitudinal and cross travel is within 0.000030" in any inch or 0.0008mm (0.8μm) in any 30mm. The compound slide is square (along its full travel) to within 0.000060" or 0.0015mm (1.5μm). The alignment of the spindle travel in 5" (127mm) is 0.000090" or 0.0023mm (2.3μm).

The NC jig borer shown in Figure 59–2 has an 11" × 18" (280mm × 450mm) table travel. The jig borer is equipped with point-to-point positioning on X and Y axes. The rotary table is controlled by a preset indexer. The table is actuated by the NC system's auxiliary functions. The rotary table may serve as a fully numerically controlled axis. Variable point-to-point, varying angle positioning, and controlled feed rates are provided. Contour positioning is an optional feature. The manufacturer's specifications identify the following features as common to this particular NC jig borer:

commands. Like the controls on milling and other NC machine tools, the spindle depth axis and the feed engage point are programmed. Specific cyclic preparatory functions are programmed to control the axis for particular operations and to call out the action as required.

NC DATA PROCESSING

PROCESSOR

A *processor* is an NC computer program that performs computations that are *workpiece oriented* or tool offsets. The processor program, as part of the software of numerical control, places the cutting tools on the workpiece. The processor program ignores all control unit information and items that are machine tool oriented.

Additional information required to manufacture a part relates to controls such as spindle speeds, feed rates, spindle direction, coolant requirements, tool selection, and other items that are *machine tool oriented.* The processor assumes that further processing of additional information from another computer program will need to be interlocked to produce the appropriate control tape for the ultimate manufacture of the part.

POSTPROCESSOR

A *postprocessor* is a computer program that accepts a processor program of information about the tool located on a part, machining setups, and machine processes. A postprocessor program makes additional computations to ensure compatibility of information among the MCU, NC machine tool, and the part specifications, including tolerances and machine limitations.

COMPUTER-ASSISTED NUMERICAL CONTROL PROGRAMMING

Thus far, numerical control has been considered in terms of manual programming. All cutter movements and machine setups and machine functions were listed in sequence in the manuscript. All cutter positions were calculated by the programmer and specified on the manuscript. A tape was prepared, a printout was edited, and a control tape was prepared. All of this information was translated in the MCU to command pulses to the NC machine tool. At the end of the program, the machine was reloaded. The processing cycle between the MCU and the NC machine tool was repeated.

Numerical control may also be considered in terms of *computer-assisted programming* (CAP). This method of writing a program is used for continuous-path and other operations requiring complex and time-consuming calculations. CAP eliminates the need to make numerical coordinate data calculations. CAP calculations are made rapidly and are error-free. A number of different processor languages are used for CAP, depending on the complexity of the computations and the combinations of machining processes.

PROGRAM EDITING

A *parity check* permits automatically checking a tape for malfunction errors of the tape punch. Depending on the tape format used, each horizontal row on the tape must have either an odd or an even number of holes. The parity check on the EIA RS–244–A tape format has an odd number of holes; the EIA RS–358 format, an even number of holes. If a row does not have the required odd or even holes, the control system stops to indicate an error.

FUNCTIONS OF THE MACHINE CONTROL UNIT (MCU)

The instructions within the part program are converted in the machine control unit (MCU) into a form of energy that controls the machine tool. Two basic types of machine control units are available: (1) *hard-wired* integrated circuits and (2) *soft-wired* computer numerical control (CNC) units.

MCU DESIGN FEATURES AND FUNCTIONS

An edited punched tape is run through a tape reader in order to enter the program into the computer and to store it in computer memory. One or more part programs may be stored within the computer storage capacity. One feature of a computer is *random access memory* (RAM). This feature permits any stored program to be called up when needed.

When a CNC unit has a *cathode ray tube* (CRT), the part programmer or machine tool operator may visually view positioning and operational information. The CRT is capable of displaying axes positions for all machine slide movements and other machine functions. Messages to the operator may be programmed for display on the CRT screen at a specified time. Other functions displayed on a CRT screen include:

- Part numbers of all programs stored in computer memory;
- Used and available capacity for further storage;
- Compensations for cutter radius, tool lengths, tool offsets, and tool fixture offsets;
- Diagnostic information when the CNC units are able to isolate and identify malfunctions in the numerical control system;
- Part program information when editing.

FUNCTIONS OF A COMPUTER (CPU) IN COMPUTER-ASSISTED PROGRAMMING

The computer used for computer-assisted programming has the computational ability to generate numerical part program data in a useful form by the MCU of a machine. Some computers are located a distance from the MCU. Other smaller computers (often called *minicomputers*) are an integral unit of the MCU.

CNC computers store part programs and process them to generate output signals for the control of a machine tool. Computers use a two-digit binary notation. The two binary digits correspond to two conditions of operation of the electronic components: on or off, charged or discharged, positive or negative charge, and conducting or nonconducting.

The *central processing unit* (CPU) includes all the circuitry controlling the processing and execution of instructions entered into the computer. The circuits relate to basic memory for storage and retrieval and *logic*. The four basic arithmetical processes of addition, subtraction, multiplication, and division are accomplished by adding positive and/or negative numbers. Simple computations are made in billionths of a second.

Input information may be entered into a computer in several forms: punched tape, mag-

netic tape, diskette, or signals from other computers. Similarly, *output information* may be received in these same forms or as printout sheets or electrical signals that control an NC machine tool operation. Buffer storage is provided on some machines so that the stored advance information is available at the point where it is needed.

ROLE OF THE PROGRAMMER IN COMPUTER-ASSISTED PART PROGRAMMING

The Automatically Programmed Tools (APT) system was one of the first computer processor languages. APT, Compact II®, and many other different processor languages are used depending on the machine tool or the complexity of producing parts requiring several machines to be incorporated in a complex machining center.

PROGRAM PLANNING

The part programmer first views the part to be machined in terms of axes, type of cutting tool, the NC machine tool, and the path of the cutting tool. For purposes of providing an example of computer-assisted programming, the part drawing in Figure 59–3 illustrates a workpiece to be turned on a NC lathe. In this instance, the spindle axis (longitudinal feed) is Z and the transverse (in and out feed) axis is X. Plus and minus coordinates are used. The part drawing uses incremental coordinates for dimensioning.

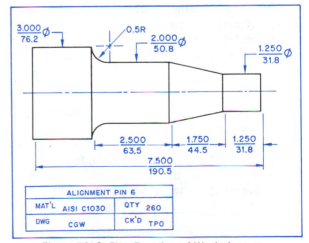

Figure 59–3 Part Drawing of Workpieces to Be Turned on a NC Lathe

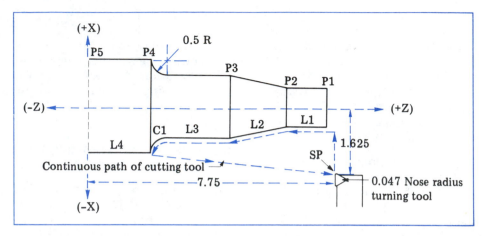

Figure 59–4 Part Programming Layout with Turned Surfaces, End Points, and Starting Points

After viewing the part to be machined, the part programmer prepares a programming layout, as shown in Figure 56–4. The cutting tool has a nose radius of 0.047″ (1.2mm). This layout identifies each process of straight turning (L1), angular turning (L2), straight turning (L3), and radius turning (C1), as well as the reference surface (L4). The starting and ending points (P1, P2, P3, and P4) for each cut are indicated. The reference point (P5) and the starting (SP) point of the cutting tool are given. This reference point is 0.250″ from the end of the workpiece and 1.625″ from the center line.

MANUSCRIPT PREPARATION

After completing a sketch of the programming layout, the part programmer writes the manuscript by preparing a series of statements. An APT program uses four types of statements:

- *Motion* statements to describe cutting tool position,
- *Geometry* statements to describe the features of the part,
- *Postprocessor* statements to identify machine tool and control system data,
- *Auxiliary* statements to provide additional information that is not given in any other statement.

The statements must follow grammatical rules for construction of statements and words, punctuation, and spelling. When the APT system is used, the spelling of an APT word must be exactly as it is spelled in the APT *system dictionary*. This dictionary specifies the only form the computer understands. The computer will indicate an inaccurately spelled word by an undefined symbol.

The following allowable words to be used for programming the turned part shown in Figure 59–3 are selected from and are written according to an APT dictionary. The spelling of each word to be used in programming the part is accompanied with a brief identification of the process, function, or location. An application is then given to show the usage of the word. The program planner first lists postprocessor and auxiliary statements in the manuscript:

PARTNØ	*Part number.* For information. The statement must be the first statement or appear immediately after END.
MACHIN	*Machine identification.* Example: MACHIN/HARDINGE 6. The computer is to make a control tape for a Hardinge turning machine number 6, equipped with a compatible input system.
INTØL	*Inside tolerance.* Example: INTØL/ .001. The computer is to stay within 0.001″ on the inside of curves in making straight-line approximations of curves.

ØUTØL	*Outside tolerance*. **Example:** ØUTØL /.001. The computer is to stay within 0.001″ on the outside of curves in making straight-line approximations of curves.
CUTTER	*Cutter*. **Example:** CUTTER/.94. The nose radius of the cutting tool is designated by the diameter of a theoretical circle. The 0.047″ (3/64″) radius of the cutting tool as shown on the drawing in Figure 59–4 is designated as CUTTER/.094.
CØØLNT	*Coolant*. **Example:** CØØLNT/ØN. Turn the coolant on. It will remain on until CØØLNT/ØFF or STOP command.
CLPRNT	*Print out*. Coordinate dimensions of all end points and straight-line moves are called to be printed out.
FEDRAT	*Feed rate*. **Example:** FEDRAT/4, IPM. Feed rate in all directions, including Z, will be 4″ per minute.
SPINDL	*Spindle*. **Example:** SPINDL/400, RPM. Start spindle at 400 RPM. Spindle stays on until SPINDL/ØFF, STØP, or END.

Postprocessor and auxiliary statements are followed by geometry and then motion statements:

PØINT	*Point*. **Example:** P2 = PØINT/6.25, 2.20, 1.50. **P2 is the point with coordinates X = 6.25, Y = 2.20, and Z = 1.50.**
LINE	*Line*. **Example:** L1 = LINE/P1, P2. Line 1 is the line through points **P1 and P2. Example:** L3 = LINE/P3, RIGHT, TANTØ, C1. Line 3 is a line, tangent to circle 1 (C1), right of P3, through points P3 and P4.

The program and machining cycle are completed by using closing statements:

CØØLNT	*Coolant*. **Example:** CØØLNT/ØFF. Turn the coolant off.
FINI	*Part program is completed*. FINI must be the only word in the statement.

How to Prepare an APT Program Manuscript

Planning the Program

> **Note:** The alignment pin shown in Figure 59–3 is used to illustrate the preliminary steps and the kind of input required to prepare an APT program manuscript. The part is to be turned on a NC lathe.

STEP 1 Study the part drawing in terms of design features, machining processes, tooling, and machine characteristics.

STEP 2 Prepare a sketch to identify the path of the cutting tool in machining the workpiece.

STEP 3 Add the programming layout to the sketch (as shown in Figure 59–4).

STEP 4 Determine the sequence of each APT word to be used in the program.

STEP 5 Check an APT language dictionary for the correct spelling of each word.

Prepare the Program Manuscript

STEP 1 Prepare the computer-assisted program in APT processor language. Start with postprocessor and auxiliary statements—for example,

```
PARTNØ    ALIGNMENT PIN NØ6
MACHIN/HARDINGE 6
INTØL/.001
ØUTØL/.001
CUTTER/.094
CØØLNT/ØN
CLPRNT
FEDRAT/4, IPM
SPINDL/400, RPM
```

STEP 2 Add the geometry statements in APT language—for example,

```
SP  =  PØINT/7.75, –1.625
P1  =  PØINT/7.5, –.613
P2  =  PØINT/6.25, –.613
P3  =  PØINT/4.5, –1.0
P4  =  PØINT/2.0, –1.5
P5  =  PØINT/0, –1.5
```

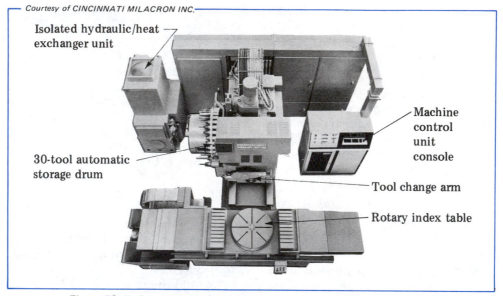

Courtesy of CINCINNATI MILACRON INC.

Isolated hydraulic/heat exchanger unit

Machine control unit console

30-tool automatic storage drum

Tool change arm

Rotary index table

Figure 59-5 Complete CNC Machining Center with Rotary Index Table, 30-Tool Automatic Storage, and Tool Change Arm

```
        L1  =  LINE/P1, P2
        L2  =  LINE/P2, P3
        C1  =  CIRCLE/2.5, –1.5, .5
        L3  =  LINE/P3, RIGHT, TANTØ, C1
        L4  =  LINE/P4, P5
```

STEP 3 Follow the geometry statements with the motion statements—for example,

```
        FRØM/SP
        GØ/TØ, L1
        GØLFT/L1, TØ, L2
        GØLFT/L2, PAST, L3
        GØRGT/L3, TANTØ, C1
        GØFWD/C1, PAST, L4
        GØTØ/SP
```

STEP 4 Conclude the program with two closing statements—for example,

```
        CØØLNT/ØFF
        FINI
```

BASIC COMPONENTS OF A CNC SYSTEM

The maximum productivity of a machining center depends on the versatility of the computer numerical control system. For example, the Acramatic® CNC system is designed with solid-state circuitry and a minicomputer. This combination of hardware and software systems provides flexibility of application to a complete CNC machining center (Figure 59-5). The CNC unit may be adapted at a later date to a complete computer-aided manufacturing center (CAM). The unit is designed to allow for the addition of advanced features, data input/output devices, monitors, and sensors; tool path modification calculations; and tool management information. The CNC system generally has four basic components: (1) tape reader, (2) cathode ray tube (CRT) display, (3) keyboard, and (4) manual and NC panel.

DESIGN FEATURES OF A CNC SYSTEM

TAPE-CONTROLLED PROGRAMMING

Tape-controlled programming features include:

- Tape-controlled feed rates, programmed in ipm or mm/min;
- Tape-controlled spindle speeds, programmed directly in RPM;
- Rotary table index positions, programmed directly in degrees (noninterpolated rotary axis movement is simultaneous with any X-Y-Z plane movement);
- Preselected program pockets or tool numbers, tape controlled for random tool selection;

- Machine slides, programmed for axis dwell that may be varied from 0.01 to 99.99 seconds in 0.01 second increments (the X, Y, Z slides may be kept motionless in a cut to permit a higher-quality surface finish and greater accuracy for certain machining operations);
- Program interruption, permitting a manual tool change.

OTHER STANDARD CNC FEATURES

A few other CNC system features include:

- Slash (/) code, which is an operator selection feature;
- Block delete, which determines whether or not blocks of programmed data, prefixed by the code, will be ignored;
- Two levels of manual control;
- Spindle keyboard override which permits on-site adjustments for material hardness or cutting action through keyboard modification of the programmed spindle speed;
- Gage height feature, which establishes the limit, may be programmed as a six-digit R word;
- Data search, which permits running the tape in forward or reverse direction;
- Manual power axis feed, which permits independent control of each linear axis. Three feed ranges may be manually selected from 0.1 ipm to 1.0 ipm, 1.0 to 10.0 ipm, and 10.0 to 100.0 ipm;
- Incremental jog, which permits independent control of each linear axis;
- Target point align, which provides automatic positioning of X, Y, and Z axes;
- Position set, which controls floating zero capacity;
- Inhibit feature, which permits the operator to interrupt machine activities;
- Push button emergency stop, which initiates stoppage of power to servodrives and feed commands;
- Dry run feature, which allows the program to be cycled through a noncutting tryout at higher than programmed feed rates in order to cycle the numerical control machine tool through all movements and machining processes and check the program before a part is produced.

EXPANDING PRODUCTIVITY AND FLEXIBILITY OF A CNC SYSTEM

A cutter diameter compensation (CDC) feature permits the program to be altered mechanically for variations in cutter diameter. Compensation may be made for undersized or oversized cutter diameters. The keyboard permits compensation in increments of 0.0001" up to a maximum of ±1.0000". The CDC feature is operable in linear and circular interpolation modes, but only in the X–Y plane.

The tool length storage/compensation feature permits accurate tool depth adjustment. This feature eliminates the preset tooling stage. The feature also permits storing a six-digit length of tool dimension in core memory for each available tool pocket. Tool length compensation may be made for variables in increments of 0.0001" up to a maximum of 99.9999".

Variations in setup or irregularities of workpieces may be compensated for by an assignable tool length trim feature. This feature is available for groups of 16 program selectable tool trims. The trim range is ±0.0001" to ±1.0000".

Courtesy of CINCINNATI MILACRON INC.

Figure 59–6 Electronic Tool Gage for Length and Diameter with Digital Readout

PRESETTING NUMERICAL CONTROL CUTTING TOOLS

An *electronic tool gage with digital readout* provides highly discriminating tool length and diameter measurements and adjustments (Figure 59–6). A tool is gaged for length by inserting it into the gage spindle socket. The feeler tip of the tool gage is then brought into contact with the tool point or edge. The enter button is pressed for storage in memory of the CNC system.

Safe Practices for Computer Numerical Control Machine Tools

• Establish whether the MCU console has an emergency work table and/or spindle-positioning stop control. Use the stop control as needed or whenever there is possibility of a personal accident or damage to the cutting tools, machine, table, or other accessories.

• Check the clearance of the work table, vises, clamps, or spindle when manually positioning the work table or spindle. This clearance check is required before positioning data is entered in the MCU console.

• Check the cutter and workpiece clearance before lowering the quill (and cutting tool) on a drill press or vertical mill.

• Reduce the overhang of a quill to ensure maximum rigidity.

• Check the quill movement to be sure a cutter may be raised to clear the workpiece when the quill-up control is actuated by the MCU.

• Verify the tape input by a dry run on the machine tool.

• Observe each operation as it is performed on the NC machine tool, including checks for accuracy.

• Maintain a duplicate, permanent tape record in a separate storage location.

UNIT 59 REVIEW AND SELF-TEST

1. Identify three basic control features that apply to all NC machine tools.

2. Indicate three functions that may be displayed on a control panel.

3. Cite one advantage of using a circular interpolation feature to generate curved surfaces in comparison to a linear interpolation feature.

4. State the reason for parity check.

5. Identify two major features for computers used in computer-assisted programming.

6. The contour on follower cam No. 12 as shown on the part drawing below is to be end milled on a NC vertical milling machine. A 1″ diameter end mill is to be used.
   a. Make a program planning sketch. Indicate the sequence of events.
   b. Write the program in the APT system. Group the statements according to functions.

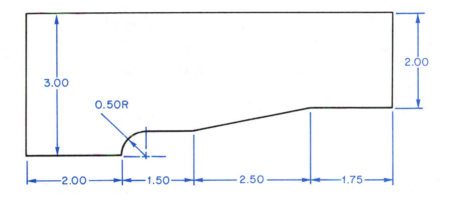

7. State three CNC machine functions that may be tape controlled.

8. Explain the function of a cutter diameter compensation (CDC) feature on a CNC machine tool.

9. State two safety precautions that should be taken with tape information.

PART 13 Production of Industrial and Cutting Tool Materials

Structure and Classification of Industrial Materials

The reduction of iron ores to produce pig iron and processes for producing cast irons and cast steels are described first in this section. Consideration is then given to the characteristics, properties, applications, and manufacture of common ferrous and nonferrous metals and alloys, cemented carbides, and ceramic and diamond cutting tools. Classification systems for iron, steel, aluminum, and alloys are also covered.

UNIT 60

Manufacture, Properties, and Classification of Metals and Alloys

BASIC STEEL MANUFACTURING PROCESSES

Steel-making processes require the impurities within pig iron to be *burned out*. The removal of impurities is done in one of four types of furnaces:

- Open hearth furnace,
- Bessemer converter,
- Electric furnace,
- Oxygen process furnace.

OPEN HEARTH PROCESS

As much as 80% of all steel is produced by the open hearth process, which was introduced in 1908. The dish-shaped *hearth* is charged with limestone and scrap steel. When the scrap is melted, molten pig iron is poured in. The impurities are burned out by sweeping fuel and hot gas over the molten metal. The burned gases are drawn off. The direction of the flames in the furnace is changed at regular intervals. The limestone in the furnace unites with the impurities and rises to the surface of the molten metal.

The process continues for a number of hours. The molten steel is drawn off into a ladle that has the capacity to take the *heat* (the batch of steel to be poured). The slag floating on the metal is drawn off through a spout in the ladle.

Alloying is done by adding fixed amounts of the required alloying metals—for example, tungsten, molybdenum, and manganese—to the molten steel. After mixing, the alloyed steel is poured into *ingot molds.* When the steel within a mold solidifies, the mold is stripped from the solid steel mass, which is called an *ingot.* The ingot is moved into a *soaking pit* so that the complete ingot is slowly brought to a uniform temperature. Next, the heated ingot is rolled into specific sizes and shapes.

BESSEMER CONVERTER PROCESS

The Bessemer converter today produces a very small percent of the steel manufactured as compared to the earlier production of over 90% of all steel manufactured. The converter consists of a lined circular shell that is swiveled to receive a charge of molten iron. The trunnion mountings are designed to provide a blast of air at high velocity. The blast flows through holes in the bottom when the furnace is moved to vertical position. The oxygen in the air burns out the impurities.

After being subjected to the air blast for 10 to 15 minutes, the converter is tilted to pour the metal into a mixing ladle in which alloying elements are added. The alloyed steel is then poured into ingots, soaked, and rolled in much the same manner as the open hearth steel.

ELECTRIC FURNACE PROCESS

The electric furnace permits precise control over the steel-making process. The furnace operates by lowering three carbon electrodes to strike an arc with the scrap metal loaded in the furnace. The quality of scrap steel, the alloying elements, the amount of oxygen used, and the heat are carefully regulated. The heat generated by the electric arc produces the molten steel.

Chromium, tungsten, nickel, or other alloying materials are added to produce the required alloy steel. After the heat (molten steel) is complete, the furnace is tilted in a pouring position. The molten metal is *teemed* (poured from the ladle) into ingots for subsequent soaking and rolling operations.

BASIC OXYGEN PROCESS

The oxygen process is one of the newer steel-making processes. A high-pressure stream of pure oxygen is directed over the top of the molten metal. The furnace is charged with about 30% scrap metal. Then, molten pig iron is poured into the furnace. Finally, the required fluxes are added.

The making of steel in a basic oxygen process furnace is illustrated in Figure 60–1. The furnace is covered with a water-cooled hood. The oxygen lance is lowered to within 5 to 8 feet above the surface of the molten metal. The force of the oxygen at a pressure of 140 to 160 pounds per square inch at a flow rate of 5,000 to 6,000 cubic feet per minute creates a churning high-temperature movement. This action continues until the impurities are burned out. Then, the

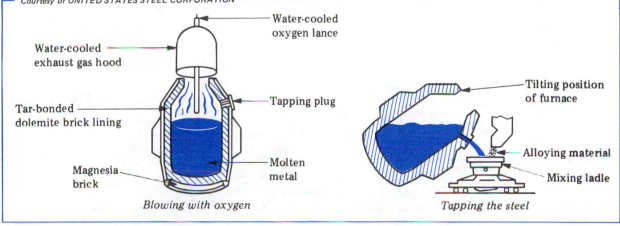

Courtesy of UNITED STATES STEEL CORPORATION

Water-cooled oxygen lance

Water-cooled exhaust gas hood

Tar-bonded dolemite brick lining

Tapping plug

Magnesia brick

Molten metal

Tilting position of furnace

Alloying material

Mixing ladle

Blowing with oxygen

Tapping the steel

Figure 60–1 Making Steel in a Basic Oxygen Process Furnace

flame drops, the oxygen is shut off, and the lance is removed. The entire process takes almost 50 minutes at a steel production rate of about 300 tons per hour.

The furnace is tapped by tilting it. After alloying materials are added to the mixing ladle, the molten metal is poured into ingots. The same processing is followed to form metal stock.

CHEMICAL COMPOSITION OF METALS

Metals are identified according to a natural element. Copper, silver, aluminum, nickel, and chromium are a few examples. Pure metals are alloyed with one or more other elements to provide qualities in a resulting product to meet engineering requirements.

In metallurgical terms, an *alloy* is composed of two or more elements and has metallic properties. One of the elements must be metallic; the other, either metallic or nonmetallic. Commercially, the term *alloy* is used to denote a metallic substance of two of more metallic elements. One of these elements must be intentionally added.

Ferrous alloys contain iron as the base metal and one or more metallic elements. Ferrous alloys such as molybdenum steel, vanadium steel, and nickel steel are all steels that have metallic elements added to change their properties.

Nonferrous alloys such as brass, bronze, and monel metal do not contain iron (except as an impurity). Brass and bronzes are copper-base alloys. Brass is an alloy of copper and zinc; bronze, of copper and tin.

COMPOSITION OF PLAIN CARBON STEEL

The principal elements in plain carbon steel include: *carbon, manganese, phosphorous, silicon,* and *sulphur.* Each element has a decided effect on the properties of steel.

Carbon is the hardening element and exerts the greatest influence on the physical properties. The effect of carbon on certain properties of steels is shown graphically in Figure 60–2. As the percent of carbon increases within a specific range, there is an increase in hardness, hardenability, wear resistance, and tensile strength. The higher the carbon content, the lower the melting point. Any additional carbon above the 0.85% point has limited effect on hardness, although wear resistance increases. The graph shows four general ranges of steel with carbon content within the 0 to 1.3% range. Low-carbon steels contain from 0.02% to 0.30% carbon by weight. Medium-carbon steels contain from 0.30% to approximately 0.60% carbon. High-carbon steels fall within the 0.60% to 0.87% carbon content range. The tool steels begin at the 0.87% carbon content point.

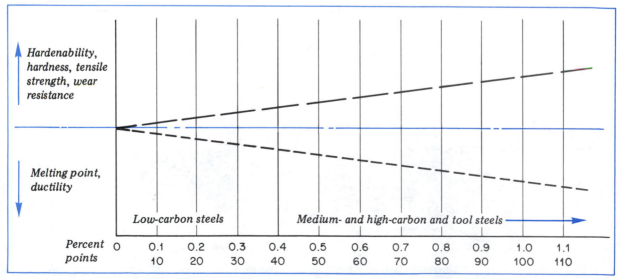

Figure 60–2 Effect of Carbon on Certain Properties of Steels

A small quantity (0.30% to 0.60%) of manganese is usually added during manufacture to serve as a purifier in removing oxygen from the steel. Oxygen in steel makes it weak and brittle. Manganese also mixes with sulphur, which is an impurity. Manganese is added to steel to increase the tensile strength, toughness, hardenability, and resistance to shock. Manganese is added in amounts from 0.06% to over 15% to produce certain physical properties. A 1.5% to 2.0% addition to high-carbon steel produces a deep-hardening steel. Such a steel must be quenched in oil. The addition of 15% manganese produces hard wear-resistant steels.

The addition of small amounts of phosphorous and sulphur to low-carbon steels increases machinability and decreases shrinkage. Amounts above 0.60% are considered undesirable because the phosphorous causes failures due to shock and vibration.

The presence of from 0.10% to 0.30% silicon makes steel sound as it is cast or hot rolled. Between the 0.60% to 2.0% range, silicon becomes an alloying element. Silicon is usually used in combination with manganese, molybdenum, or chromium. As an alloying element, silicon increases tensile strength, toughness, and hardness penetration.

The addition of from 0.08% to 0.30% sulphur to low-carbon steel increases its machinability. Screw stock for automatic screw machine work is a sulphurized free-cutting steel. Otherwise, sulphur is considered an impurity in steel. As an impurity, sulphur causes cracking during rolling or forging at high temperatures.

MECHANICAL PROPERTIES OF METALS

Mechanical properties are associated with the behavior of a metal as it is acted upon by an external force. The properties of a metal determine the extent, if any, to which it can be hardened, tempered, formed, pulled apart, fractured, or machined.

Hardness is defined as the property of a metal to resist penetration. This property is controlled by heat treating in which a machined or forged tool may be shaped while soft and then hardened.

Hardenability represents the degree to which a metal hardens through completely to its center when heat treated. A low hardenability means the surface layer hardens but the metal is softer toward the center. Brittleness is related to hardness. Brittleness refers to the degree to which a metal part breaks or cracks without deformation. Deformation is the ability of a material to flex and bend without cracking or breaking.

Ductility refers to the ability of a metal to be bent, twisted, drawn out, or changed in shape without breaking. For example, a deep draw shell requires sheet metal with high ductility to permit drawing without fracturing the metal. The ductility of a metal is usually expressed as a percentage of reduction in area or elongation.

Malleability relates to the ability of a metal to be permanently deformed by rolling, pressing, or hammering. Toughness of a metal refers to the property that enables it to withstand heavy impact forces or sudden shock without fracturing.

Fusibility is the ease with which two metals may be joined together when in a liquid state. A high-fusibility metal can be easily welded. In such instances, the term weldability is used.

Machinability relates to the ease or difficulty with which a given material may be worked with a cutting tool. After extensive testing of materials under control conditions, machinability ratings have been established. These ratings are published in handbook tables as percentages. AISI 1112 cold-drawn steel is assigned a rating of 100%. Other metals are rated in percentages in comparison to the machinability of AISI 1112 steel. Difficult-to-machine metals are rated below 100%. Easy-to-machine metals are rated above 100%.

Machinability properties need to be considered in terms of the following factors:

- Physical properties of the material (tensile strength, rigidity, hardness, grain structure) and chemical properties;
- Type of chip formation produced during the machining process;
- Cutting characteristics that are influenced by the use or absence of a particular type of cutting fluid;
- Power required in taking a cut.

CLASSIFICATION OF STEELS

There are two general categories of steel: plain carbon steels and alloy steels. The plain carbon

Table 60–1 Effects of Selected Alloying Elements on Certain Properties of Steel

(Examples:) Effect on Properties of Steel	Carbon (C)	Chromium (Cr)	Manganese (Mn)	Molybdenum (Mo)	Nickel (Ni)	Tungsten (T)	Vanadium (V)
Selected Alloying Metallic Element							
Increases							
tensile strength	X	X	X	X	X		
hardness	X	X					
wear resistance	X	X	X		X	X	
hardenability	X	X	X	X	X		X
ductility		X					
elastic limit		X		X			
rust resistance		X			X		
abrasion resistance		X	X				
toughness		X		X	X		X
shock resistance		X			X		X
fatigue resistance							X

steels are grouped in three ranges of carbon content as low-carbon steel, medium-carbon steel, and high-carbon steel.

PLAIN CARBON STEELS

Low-Carbon Steel. The range of carbon by weight is from 0.02% to 0.30%. Low-carbon steels may not be hardened except by adding carbon to permit hardening the outer case while the area below remains soft. The range of carbon is from 0.08% to 0.30% in low-carbon steels that are commonly used for manufacturing parts that do not require hardening. Low-carbon steels such as machine steel and cold-rolled steels are used in manufacturing bolts and nuts, sheet steel products, bars, and rods.

Medium-Carbon Steel. The carbon content is from 0.30% to 0.60%. With this amount of carbon, it is possible to harden tools, such as hammers, screw drivers, and wrenches, that may be drop forged or machined.

High-Carbon Steel. The carbon content range for high-carbon and tool steels is from 0.60% to 1.50%. These steels are adapted for edge cutting tools such as punches, dies, taps, and reamers.

ALLOY STEELS

The addition of alloying elements to steel increases the tensile strength, hardenability, toughness, wear abrasion, red hardness, and corrosion resistance. The most important effects of alloying selected elements on certain properties of steel are indicated in Table 60–1.

SAE AND AISI SYSTEMS OF CLASSIFYING STEELS

Two main systems of classifying steels have been devised. These systems provide for standardization in producing each alloy and in designating each alloy. The systems were developed by the Society of Automotive Engineers (SAE) and the American Iron and Steel Institute (AISI).

FOUR- AND FIVE-DIGIT DESIGNATIONS

A four-digit series of code numbers is used in both the SAE and the AISI systems. Certain alloys are identified by a five-digit series of code numbers. One of two letters is used in cases where an element has been varied from the normal content of the steel. The prefix X denotes a variation of manganese or sulphur. The prefix T denotes a variation of manganese in the 1300-range steels.

The first digit of the SAE and AISI classification systems indicates the *basic type of steel*, as follows:

1 carbon
2 nickel
3 nickel-chrome
4 molybdenum
5 chromium
6 chromium-vanadium
7 tungsten
8 nickel-chromium-molybdenum
9 silicon-manganese

The second digit classifies the steel or alloy within a particular *series*. Table 60–2 shows

Table 60–2 Examples of SAE/AISI Classifications of Standard Carbon and Alloy Steels by Series

Series	Type	Composition or Special Treatment*
Classification of Carbon Steels		
10XX	Plain carbon	Nonsulphurized
11XX	Free-machining	Resulphurized
12XX		Rephosphorized and resulphurized
Classification of Alloy Steels		
13XX	Free-machining, manganese	Mn 1.75%
23XX	Nickel	Ni 3.50%
25XX		Ni 5.00%
31XX	Nickel-chromium	Ni 1.25% Cr 0.65%
30XXX		Corrosion- and heat-resisting
40XX	Molybdenum	Mo 0.20% or 0.25%
. . .	. . .	. . .
93XX	Triple-alloy	Ni 3.25% Cr 1.20% Mo 0.12%
94XX		Ni 0.45% Cr 0.40% Mo 0.12%
98XX		Ni 1.00% Cr 0.80% Mo 0.25%

*Percents represent the average of the range of a particular metal element. Only the predominant alloying elements (excluding carbon) are identified.

the SAE/AISI classifications of standard carbon and alloy steels by series. A steel designated with a 10XX number is in the nonsulphurized plain carbon steel series. An alloy steel numbered 40XX is in the molybdenum steel series that has 0.20% or 0.25% molybdenum content. A zero in the second digit indicates there is no major alloying element.

The third and fourth digits indicate the middle of the carbon content range—that is, the *average percent of carbon*—in the steel. The percent of carbon is indicated by the third and fourth digits as *points*. A point is the same as 0.01% carbon. Thus, a 0.20% carbon content is expressed by the third and fourth digits as 20 points. Figure 60–3 provides two examples of the use of SAE and AISI code numbers.

In a few instances, a five-digit numerical code is used. For example, in the SAE system, a designation of 71250 means the steel is a tungsten steel (7 as the first digit) with 12% tungsten (12 as the second and third digits) and 0.50% carbon (50 points as the fourth and fifth digits).

The letter H after the SAE or AISI code number specifies that the alloy steel meets specific hardenability standards. Steel manufacturers and handbook tables provide craftspersons with additional information about H designations, composition, properties, and applications of steels designated by SAE and AISI codes.

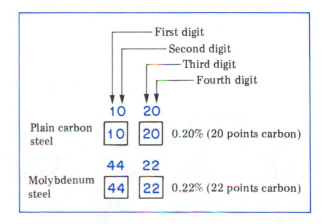

Figure 60–3 Examples of SAE and AISI Code Numbers

Table 60-3 Identification of Major Groups of Tool and Die Steels

Letter Code	Type of Tool and Die Steels		
W	Water-hardening tool steel		
S	Shock-resisting tool steel		
O	Cold-worked tool steel		Oil-hardening
A			Medium alloy, air-hardening
D			High-carbon, high-chromium
H	1-19	Hot-worked tool steel	Chromium types
	20-39		Tungsten types
	40-59		Molybdenum types
M	High-speed tool steel		Molybdenum types
T			Tungsten types
L	Special-purpose tool steel		Low-alloy types
F			Carbon-tungsten types
P	1-19	Mold Steel	Low-carbon types
	20-39		Other types

CODING OF TOOL AND DIE STEELS

There are eleven major groups of tool and die steels. These groups are identified by a letter. In some cases, subgroups are designated by using a numeral following a letter. Table 60-3 provides the codes for identification of the tool and die steel groups.

ADDITIONAL FEATURES OF AISI SYSTEM

A prefix letter is sometimes used with the code number to identify the method used to produce the steel. The four basic prefix letters are: B, for Bessemer carbon steel; C, for general open hearth carbon steel; D, for acid open hearth steel; and E, for electric furnace alloy steel. Where there is no letter prefix to the AISI number given in a table, the steel is predominantly open hearth.

UNIFIED NUMBERING SYSTEM (UNS)

While only the SAE and the AISI systems have been discussed here, various trade associations, professional societies, standards organizations, and private industries have developed different numbering and coding systems. The Unified Numbering System (UNS) represents a joint effort by the American Society for Testing and Materials (ASTM), the SAE, and others to correlate the different numbering systems being used for commercial metals and alloys. The important point to remember about a UNS number is that it provides an *identification, not a specification.* The specifications are to be found in manufacturers' literature, trade journals, and handbooks.

Sixteen code letters are used in the Unified Numbering System. Each letter is followed by five digits. The digits run from 00001 through 99999. Each of the sixteen code letters identifies certain metals and alloys. The letters D, F, G, H, J, K, S, and T designate different cast irons, cast steels, and other ferrous metals and alloys. The letters A, C, E, L, M, P, R, and Z are used for nonferrous metals and alloys.

The numbers in the UNS groups conform, wherever possible, to the code numbers of the other systems. For example, an AISI or SAE 1020 plain carbon steel has a corresponding UNS number of G10200. An SAE or AISI 4130 alloy steel has a UNS identification of G41300. An M33 alloy has a UNS code of T11330. Handbook tables provide the full range of UNS numbers for plain carbon, alloy, and tool steels in the SAE and AISI systems.

ALUMINUM ASSOCIATION (AA) DESIGNATION SYSTEM

A system similar to the SAE, AISI, and UNS classifications for steels and alloys has been developed by the Aluminum Association (AA) for wrought aluminum and aluminum alloys. A four-digit numerical code is used. The first digit identifies the alloy type. For example, a code of 1 indicates an aluminum of 99.00% or greater purity. The numerical code 2 indicates a copper-type alloy; 3, manganese; 4, silicon; 5, magnesium; 6, magnesium and silicon; 7, zinc; and 8, an element other than identified by codes 1 through 7. The number 9 is unassigned at the present time.

The second digit shows the control over one or more impurities. A second digit of zero indicates there is no special impurities control. The third and fourth digits in the 1 series give the amount of aluminum above 99.00%, to the nearest hundredth of a percent. The same digits in the 2 through 8 series are used to identify different alloys in the group. The prefix X is used for experimental alloys. When standardized, the prefix is dropped.

The letters F, O, H, W, and T are used as *temper designations*. These letters follow the four digits (separated by a dash) to designate the temper or degree of hardness, as follows:

F	hardness as fabricated
O	annealed (for wrought alloys only)
H	strain-hardened (for wrought alloys only)
W	solution heat-treated
T	thermally treated

A temper designation may include a numerical code. For example, H designations with one or more digits indicate strain-hardened only (H1), strain-hardened and then annealed (H2), and strain-hardened and then stabilized (H3). A second digit is used to indicate the final degree of strain hardening from 0 to 8—for example, H16. A third digit, when used, gives the variation of a two-digit H temper—for example, H254.

Numerals 2 through 10 have been assigned in the AA system to indicate specific sequences of annealing, heat treating, cold working, or aging, as follows:

T2	annealed (cast products only)
T3	solution heat-treated and then cold-worked
T4	solution heat-treated and naturally aged to a stable condition
T5	artificially aged only
T6	product heat-treated and then artificially aged
T7	product heat-treated and then stabilized
T8	product heat-treated, cold-worked, and then artificially aged
T9	product heat-treated, artificially aged, and then cold-worked
T10	product artificially aged and then cold-worked

VISIBLE IDENTIFICATION OF STEELS

There are two common methods of identifying steels. One method involves the use of the manufacturer's *color code markings*. The other method requires *spark testing*.

COLOR CODE MARKINGS

Some steel producers paint at least one end of steel bars. A color code is supplied by the steel manufacturer. Stock is cut from the unpainted end in order to preserve the painted identification color code.

SPARK TEST

A simple identification test is to observe the color, spacing, and quantity of sparks produced by grinding. The visible characteristics of steel, cast iron, and alloy sparks depend largely on carbon content. A high-carbon tool steel, when spark tested, produces a large quantity of fine, repeating spurts of sparks. These sparks are white in color at the beginning and ending of the stream. By contrast, a spark test of a piece of wrought iron stock produces a few spurts of forked sparks. The color varies from straw color close to the wheel to white near the end of the stream. The spark patterns provide general information about the type of steel, cast iron, or alloy steel.

CHARACTERISTICS OF IRON CASTINGS

Cast iron is an alloy of iron and carbon. The carbon content varies from 1.7% up to 4.5%, with varying amounts of silicon, manganese, and sulphur. The physical properties of cast iron depend on the amount of one of two forms of carbon that is present. *Graphite or free carbon* is one form. *Combined carbon or cementite* (iron carbide) is the second form. The five broad classes of cast iron in general use are as follows:

- Gray cast iron,
- White cast iron,
- Chilled cast iron,
- Alloy cast iron,
- Malleable iron castings.

Ductile iron (nodular cast iron) is a newer type of cast iron used in automotive and industrial applications where factors of high tensile strength and ductility are important.

GRAY CAST IRON

Gray iron castings are widely used in machine tool, farm implement, automotive, and other industries. The American National Standard (ANS) specifications G25.1 and the American Society for Testing and Materials (ASTM) standards A48–64 consider cast iron castings in two groups. The first group of gray iron castings (classes 20, 25, 30, and 35 A, B, and C) have excellent machinability, are comparatively easy to manufacture, and have a low elasticity and a high damping capacity. The second group of gray iron castings (classes 40, 45, 50, and 60 B and C) are more difficult to machine and manufacture, have a lower damping capacity than the first group and have a higher elasticity.

High-strength cast iron castings are produced by the *Meehanite-controlled process*. Some of the more important properties of Meehanite castings were cited earlier in connection with wear-resisting properties on applications such as the cast bed, frame, and column of jig borers, jig grinders, and other machine tools. Meehanite castings may also be produced with heat-resisting, corrosion-resisting, and other combinations of physical properties.

WHITE CAST IRON

White cast iron has a silvery-white fracture. Castings of white cast iron are very brittle, with almost a zero ductility. White cast iron castings have less resistance to impact loading and a comparatively higher compressive strength per square inch than gray cast iron. Nearly all of the carbon in the casting is in the chemically combined carbon (cementite) form. White cast iron is used principally for the production of malleable iron castings and for applications that require a metal with high wear- and abrasive-resistance properties.

A 500X magnification photomicrograph of white cast iron is shown in Figure 60–4 and displays cementite (iron carbide) in the light areas and fine pearlite in the dark areas. Pearlite is comprised of alternating layers of pure iron (ferrite) and cementite.

CHILLED CAST IRON

Chilled cast iron is used for products that require wear-resisting surfaces. The surfaces are designated as chilled cast iron. The hard surfaces are produced in molds that have metal chills for the rapid cooling of the outer surface. Rapid cooling causes the formation of cementite and white cast iron.

ALLOY CAST IRON

Alloy cast iron results when sufficient amounts of chromium, nickel, molybdenum, manganese, and copper are added to cast iron castings to change the physical properties. The alloying elements are added to increase strength; to produce higher wear resistance, corrosion resistance, or heat resistance; or to change other physical properties.

Extensive use is made of alloy cast irons for automotive engine, brake, and other systems; for machine tool castings; and for additional applications where high tensile strength and resistance to scaling at high temperatures are required.

MALLEABLE IRON CASTINGS

Grades and properties of malleable iron castings are specified according to ANS G48.1- and ASTM A47- specifications. These specifications relate to tensile strength, yield strength, and elongation.

Malleable iron is produced by heat treating. In the process, the cast iron is annealed or *graphitized*. Graphitization produces graphite (temper

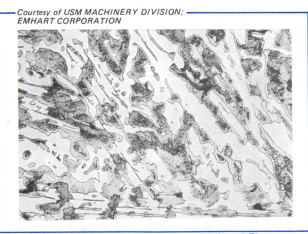

Courtesy of USM MACHINERY DIVISION; EMHART CORPORATION

Figure 60–4 Cementite (Light Areas) and Fine Pearlite (Dark Areas) in White Cast Iron (500X Magnification Photomicrograph)

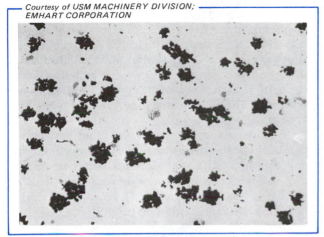

Figure 60–5 Temper Carbon (Graphite) Aggregates in Malleable Cast Iron (100X Magnification)

carbon aggregates). A 100X magnification of temper carbon aggregates in malleable cast iron is shown in Figure 60–5.

Hard, brittle, white cast iron castings are first produced from pig iron and scrap. These castings are then given an annealing heat treatment during which time the temperature is slowly increased over as much as a two-day period up to 1650°F (898.9°C). Then, the temperature is dropped slowly over an equal cooling time.

Malleable iron castings are widely used in industrial applications that require a highly machinable metal, great strength (as compared with other cast irons and nonferrous metals) and ductility, and good resistance to shock.

Cupola Malleable Iron. Malleable iron is also produced by the cupola method. The metal has good fluidity, produces sound castings, and possesses the property of being suited to galvanizing. Therefore, cupola malleable iron is used for making valves, pipe fittings, and similar parts.

Pearlitic Malleable Iron. Pearlitic malleable irons are used where a greater strength or wear resistance is needed than is provided by general malleable iron or in place of steel castings or forgings. There are a number of forms to pearlitic malleable iron. Some forms are engineered to resist deformation. Other forms permit deformation before breaking and are used for crankshafts, camshafts, and differential housings; ordinance equipment; and for other machine parts.

DUCTILE IRON (NODULAR CAST IRON)

Ductile iron is a relatively new type of cast iron. It is also known as *spheroidal graphite* because the graphite is present in ball-like form in contrast to the flake-like form in regular gray cast iron. The spheroidal graphite structure produces a casting of high tensile strength and ductility. The spheroidal structure results from the addition of small amounts of magnesium or cerium-bearing alloys and special processing. A few other advantages of nodular cast iron are as follows:

- Toughness is between the toughness of cast iron and steel;
- Shock resistance is comparable to ordinary grades of carbon steel;
- Melting point and fluidity are similar to the properties of high-carbon cast irons;
- Pressure tightness under high stress is excellent;
- Castings can be softened by annealing;
- Castings can be hardened by normalizing and air cooling or by oil quenching and drawing;
- Wear-resisting surfaces can be cast in molds containing metal chills;
- Surface hardening by flame or induction methods is feasible;
- Parts may be machined with the same ease as gray iron castings.

Nodular cast iron is used in the automotive industry for cylinder heads, crankshafts, and pistons; in the heavy machinery field; and in the machine tool industry for chuck bodies, forming dies, and other industrial applications.

CHARACTERISTICS OF STEEL CASTINGS

Steel castings are used for parts that must withstand shock and heavy loads and must be tougher and stronger than cast iron, malleable iron, or wrought iron.

There are two general classes of steel castings: (1) carbon steel and (2) alloy steel. Carbon steel castings are broken up into three groups: (1) low carbon, (2) medium carbon, and (3) high carbon.

CARBON STEEL CASTINGS

Low-carbon steel castings have a carbon content below 0.20% with most of the castings produced in the 0.16% to 0.19% range. Medium-carbon steel castings have between 0.20% and 0.50% carbon. High-carbon steel castings have a carbon content above 0.50%.

ALLOY STEEL CASTINGS

Alloy steel castings are distinguished from carbon steel castings by the addition of sufficient amounts of special alloying elements to obtain or increase specific desirable qualities. There are two groups of alloy steels: (1) low-alloy steels in which the alloy content is less than 8% and (2) high-alloy steels with more than 8% alloying elements. Alloying elements such as manganese, molybdenum, vanadium, chromium, and nickel are added to steel to affect one or more of the following physical properties:

- Hardenability,
- Distortion reduction and prevention of cracks by decreasing the rate of cooling during the hardening operation,
- Resistance to hardness reduction when being tempered,
- Material strength that may be increased through heat treatment,
- Abrasion resistance at regular and increased temperatures,
- Corrosion resistance at regular and high temperatures,
- Machinability,
- Toughness.

Alloy steels require heat treatment in order to utilize the potential properties of the alloy. There are three categories of alloy steels: (1) construction, (2) special, and (3) tool.

GROUPS OF NONFERROUS ALLOYS

The base metal is used to identify a group of nonferrous metals. A few important groups of nonferrous metals that are commonly used in jobbing shops are described next. The five groups include copper-, aluminum-, zinc-, magnesium-, and nickel-base alloys.

COPPER-BASE ALLOYS

The brasses and bronzes widely used in industry are copper-base alloys. Brasses are pre-dominantly made of copper and zinc. Bronzes are a mixture of copper and tin. Both alloys may be cast, rolled, wrought, or forged. Other alloying elements such as aluminum, manganese, silver, nickel, and antimony are added. High-strength, high-hardness, and age-hardened bronzes are more difficult to machine than free-cutting brasses.

Silicon, manganese, and beryllium are often combined with copper-base alloys. Beryllium in combination with copper forms a hard compound that increases hardness and wear resistance. However, tool wear life is noticeably decreased. Beryllium possesses age-hardening properties.

ALUMINUM-BASE ALLOYS

Alloying elements and heat-treating processes are added to aluminum to increase its tensile strength. Other properties such as machinability, weldability, and ductility are also developed by adding specific alloying elements. Copper, manganese, chromium, iron, nickel, zinc, and titanium are some of the common alloying elements.

NICKEL-BASE ALLOYS

Nickel is noted for its resistance to corrosion from natural elements and many acids. Nickel-base alloys have from 60% to 99% nickel. Nickel is used for parts that are subjected to and must resist corrosive action. In its pure state, nickel is used for plating other metal parts.

One of the common nickel alloys is known as *Monel metal*. This nickel alloy is composed of approximately 65% nickel, 30% copper, and 5% of a few other elements. High-temperature-resistant space age alloys such as Inconel and Waspaloy, depend on nickel as the base metal. Copper, aluminum, iron, manganese, chromium, tungsten, silicon, and titanium are different elements that are alloyed with nickel-base alloys.

ZINC-BASE ALLOYS

Die castings are usually made of zinc-base alloys. The low melting temperature of zinc-base die cast metal alloys of from 725°F to 788°F (385°C to 420°C), the ability to flow easily under pressure into die casting dies, and the quality to which details may be produced make zinc-base alloys exceptionally practical for die casting. Zinc alloys are also rolled into sheets for manufacturing sheet metal products.

Zinc is alloyed with aluminum or small amounts of copper, iron, magnesium, tin, lead, and cadmium to produce the sixteen common zinc-base alloys. Die cast metal alloys fall into six groups: zinc-base; tin-base; lead-base; aluminum-base; magnesium-base; and copper, bronze, or brass alloys. The physical properties of each alloy is affected by the combination and quantity of each element.

APPLICATIONS OF INDUSTRIAL MATERIALS

CAST ALLOY CUTTING TOOLS

Cast alloys are considered as one of the special alloy classifications. The other classification includes *wrought alloys*, which are generally not as hard as the cast alloys and contain up to almost 35% iron. Cast alloys require less than 1% iron.

Because of the extreme hardness of cast alloys, they are used as cutting materials. Cast alloys are known by trade names such as *Stellite*, *Tantung*, and *Rexalloy*. Cast alloys have a cobalt base and are nonferrous (no iron is present, except as an impurity in the raw materials).

One of the wide fields of application of cast alloys is in cutting tools for machining metals. Cast alloys are produced as cutting tool tips to be brazed on tool shanks, as removable tool bits, and as disposable cutting tool inserts. The extreme hardness of cast alloys requires that they be ground to size with aluminum oxide abrasive wheels. The principal elements in cast alloy cutting tools are cobalt (35% to 55%), chromium (25% to 35%), tungsten (10% to 25%), carbon (1.5% to 3%), and nickel (from 0 to 5%).

One big advantage of cast alloy cutting tools is their ability to cut at high cutting speeds and high cutting temperatures. Due to the high red-hardness property, cast alloy cutting tools perform better at temperatures above the temperatures for high-speed cutters. The temperature range is from 1100°F to 1500°F (593°C to 815°C). In terms of cutting speeds, cast alloy cutting tools are capable of machining at speeds above the highest speeds used with high-speed cutters and the lowest practical speeds for carbide cutting tools.

Tool shank support is important. Most cast alloys are brittle and are not capable of withstanding heavy impact forces as are carbon steel and high-speed cutters. There are, however, special grades of cast alloys that have impact-rupture strength comparable to high-speed steel cutters. Cast alloys are also used in noncutting applications as turbine blades, wear surfaces on machines, and for conveyors in heat-treating furnaces.

CEMENTED CARBIDE CUTTING TOOLS

Cemented carbide cutting tools are made of two main materials: tungsten carbide and cobalt. Special properties are obtained by the addition of titanium and titanium carbides. The term *cemented* means that during production cobalt is used to cement the carbide grains together.

Cemented carbides are cast to required shapes. Cemented carbides do not require further heat treatment. The cast form is very hard. The cast shapes are used as disposable inserts, chip breakers, cutting tool tips that are brazed on shanks, and solid smaller cutting tools such as end mills, drills, and reamers.

The properties of cemented carbides are affected by cobalt as a principal ingredient. Cobalt affects hardness, wear resistance, and tool life. Increasing the amount of cobalt increases the hardness of the cutting tool. The brittleness is also increased, producing a decreased ability to resist shock and a lower impact toughness. The alloying elements also affect grain structure. A finer grain structure increases the hardness; a coarser structure decreases hardness.

Carbide manufacturers have established a classifying system for cemented carbide machining applications. Of eight general designations, three are used for machining conditions for cast iron and nonferrous metals and five classifications are used for steel and steel alloys. In addition, the carbide industry has six classifications related to wear and impact applications (Table 60–4).

CERAMIC CUTTING TOOLS

Ceramic cutting tools are produced from metal oxide powders. These powders are formed into shape by cold pressing and sintering or by hot pressing. Regardless of the method of forming, the pressed blanks are not as strong as the carbide-shaped cutting tools. Ceramic cutting

Table 60–4 Carbide Industry Classification System

Material to Be Machined	Cemented Carbide Classification	Application
Cast iron and nonferrous metals	C-1	Medium roughing to finishing cuts
	C-2	Roughing cuts
	C-3	High-impact dies
Steels and steel alloys	C-4	Light finishing cuts at high speeds
	C-5	Medium cuts at medium speeds
	C-6	Roughing cuts
	C-7	Light finish cuts
	C-8	Heavy roughing cuts and general-purpose machining cuts

tools are made of aluminum oxide with a metallic binder to improve impact strength. Silicon oxide and magnesium oxide are added.

Ceramic cutting tools are brittle, have low impact resistance, and shatter easily. The hardness of ceramic cutting tools lies below the hardness of a diamond and above the hardness of a sapphire.

Two great advantages of ceramic cutting tools are: (1) the cutter is not affected by the temperature at the cutting edge and (2) hot metal chips do not tend to fuse to the cutter. Thus, cutting fluids are not required for cutting purposes. However, cutting fluids may be needed to prevent distortion of the workpiece. When cutting fluids are used, a liberal and continuous flow must be provided. Otherwise, any intermittent cooling may cause the cutter to fracture or shatter.

The hardness and wear life properties of ceramic cutting tools make them ideal for machining hard and hardened steels. Cutting speeds up to from two to four times faster than the speeds used for cemented carbides may be used for ceramic cutting tools. The surface finish that may be produced by taking light finishing cuts at high speeds is of a quality that eliminates the need for grinding.

Aluminum oxide/titanium carbide ceramic cutting tools use a newer cutting tool material that combines many features of regular oxide cutting tools with its ability to resist thermal and impact shock. This combination of properties permits this oxide family of cutting tools to be used for milling, turning, and other machine

tool operations that are not possible with other cutting tools.

DIAMONE (POLYCRYSTALLINE) CUTTING TOOLS

Industrial-quality natural diamonds and manufactured diamonds provide the hardest known cutting tool material. Hardness and extreme resistance to heat make the diamond ideal for machining soft, low-strength, and highly abrasive materials. The diamond has low strength and shock resistance. Another limitation is to be found in the geometry of the diamond as a single crystal or as a cluster of single-crystal cutting edges. Since the strength, hardness, and wear resistance is related to the orientation of each diamond crystal, the performance may not be as accurately controlled as desired.

This problem of orientation of the crystal is eliminated by bonding fine diamond crystals to form solid tool shapes. The cutting tools in this diamond family are known as *sintered polycrystalline tools.* They are available in forms for brazing on solid shanks, as tips for bonding to carbide tool inserts, and as solid inserts.

Polycrystalline cutting tool inserts have the advantage over the single-crystal cutting edge in that the diamond crystals are randomly oriented. This feature provides for uniform cutting. Interestingly, the wear-resistance property of diamond cutting tools provides a wear life, based on wear resistance alone, in a ratio of from 10:1 to over 400:1 over carbide cutting tools.

Safe Practices in the Handling and Use of Industrial Materials and Cutting Tools

- Grind sharp fins and edges on castings and tumble if the molding sand is not removed from outside and inside surfaces.
- Check the machining impact conditions against the cutting tool manufacturer's recommendations, especially when carbide, ceramic, cast alloy, and other cutting tool materials that may fracture under severe impact are used.
- Avoid quenching a cast alloy cutting tool after dry grinding. Quenching shock may cause the cutter to fracture or shatter.
- Use a liberal, constant flow of coolant when machining with ceramic cutting tools. Intermittent cooling may cause the cutter to fracture or shatter.
- Select a rigid tool shank on which carbide, ceramic, and cast alloy inserts may be correctly seated and securely held.
- Cut under the hard outer scale of cast iron castings on the first cut.
- Have a chloride-base fire extinguisher available for any possible fire when machining magnesium and zirconium parts that may ignite.
- Take machining cuts at the fpm, depths, and feeds recommended by the manufacturers of carbide, ceramic, cast alloy, diamond, and other cutting tool materials.
- Make provisions in advance—that is, before start-up—for the desired chip formation (continuous or discontinuous), chip disposal (particularly when operating at high speeds), and positioning of guards and protective devices.

UNIT 60 REVIEW AND SELF-TEST

1. Describe briefly how pig iron is produced.

2. State three factors, exclusive of the composition and methods of production, that affect the machinability of metals.

3. Refer to a steel manufacturer's catalog or handbook tables of standard carbon steels.
   a. Give the percent range of the elements for carbon steels (1) 1020 (C1020), (2) 1040 (C1040), and (3) 1095 (C1095).
   b. Rate each of the three carbon steels according to the factors of (1) hardness, (2) hardenability, (3) ductility, and (4) wear resistance. Use A for lowest, B for medium, and C for highest.

4. Identify three different properties of steel that are affected by the addition of (1) chromium, (2) manganese, and (3) tungsten.

5. a. List three different types of tool and die steels.
   b. Indicate the letter (and where applicable the number range) of the selected steels.
   c. Give the (1) hardening temperature range and (2) tempering range for a T4 tungsten high-speed tool steel. Refer to a steel manufacturer's data sheets or a handbook table.

6. a. State two differences between cupola and pearlitic malleable iron.
   b. Give an application of each of the two different malleable irons.
   c. List three physical properties of cast iron castings that are changed by alloys to produce alloy cast irons.

7. State two advantages of using alloy tool steels over plain carbon tool steels.

8. State three safety precautions to take when machining with carbide, ceramic, and cast alloy cutting tools.

PART 14 Basic Metallurgy, Heat Treatment, and Hardness Testing

SECTION ONE

Metallurgy and Heat Treating: Technology and Processes

This section deals with concepts of basic metallurgy and heat treatment of metals. The structure of steel, its behavior, and the transformation products that result from heating and cooling are described. Phase diagrams and transformation curves are used to explain changes in the microstructure and properties of steels and alloys.

Technical material is provided in relation to the critical points and temperature changes in the transformation of grain structures, the relationship of carbon content to hardenability and depth of hardness, the severity and rate of cooling and time on hardness, and the use of quenching media in interrupted quenching processes. Heat treatment of metals is then covered in terms of the following technology and processes:

- Definitions and applications of heat-treating terms;
- Hardening and tempering of carbon tool steels and high-speed steels;
- Annealing, spheroidizing, and normalizing metals;
- Subzero treatments to stabilize precision tooling and instruments;
- Surface hardening through case-, flame-, and induction controlled depth hardening;
- Problems in heat treating;
- Hardness testing equipment and measurement systems;
- A sampling of heat-treating processes for selected nonferrous metals.

UNIT 61

Heat Treating and Metals Technology

Metallurgy is concerned with the technology of producing metals or alloys from raw materials and alloying elements and preparing metals for use. *Heat treatment* is the combination of heating and cooling operations applied to a metal or alloy in the solid state. The purpose of heat treatment is to produce desired conditions or physical properties in the metal or alloy.

HEATING EFFECTS ON CARBON STEELS

Fully annealed carbon steel consists of the element iron and the chemical compound iron carbide. In metallurgical terms, the element iron is known as *ferrite;* the chemical compound *iron carbide* is *cementite.* Cementite is made up of 6.67% carbon and 93.33% iron. There are also traces in carbon steel of impurities such as phosphorous and sulphur.

Depending on the amount of carbon in the steel, the ferrite and cementite are present in certain proportions as a *mechanical mixture.* The mechanical mixture consists of alternate layers or bands of ferrite and cementite. When viewed under a microscope, the mechanical mixture looks like mother-of-pearl. The mixture is known as *pearlite.* With only traces of impurities, pearlite contains about 0.85% carbon and 99.15% iron. Thus, a fully annealed 0.85% carbon steel consists entirely of pearlite. This steel is known as a *eutectoid steel.* A steel having less than 0.85% carbon is a *hypo-eutectoid steel.* This steel has an excess of cementite over the amount that is required to mix with the ferrite to form pearlite. A steel with 0.85% or more carbon content is known as a *hyper-eutectoid steel.* In the fully annealed state, a hyper-eutectoid steel has an excess of cementite so that both cementite and pearlite are present.

CRITICAL POINTS AND HEATING/COOLING RATES

Variations in hardness and other mechanical properties occur when the atomic cell structure of steel is changed through heating and cooling within certain *critical points and rates.* There are two types of atomic structures in steel: *alpha* and *gamma.* Steels that are at a temperature below the critical point have an atomic cell structure of the alpha type. Steels heated above the critical point have a gamma atom structure. Alpha iron *mixes* with carbide. Gamma iron *absorbs* the carbon-iron compound into solid solution. The critical points at which changes take place vary according to the composition of the steel. Hardness and strength variations are produced by changing the atoms and cell structure of the iron atoms from alpha to gamma and then back again to alpha.

The lower critical point for steels is considered to be 1333°F (722.8°C). The upper critical point increases to 1670°F (910°C) for the very low-carbon steels. The lower critical point is used for steels with carbon contents within the 0.8% to 1.7% range. Heat treating takes place when the temperature of a steel is raised above the critical point and the temperature is then reversed and controlled for the required degree rate of cooling.

EFFECTS OF TEMPERATURE AND TIME

Stresses occur in metals at the critical points when a part, or section of a part that is not of equal mass or shape, is heated unevenly or too fast. Usually a part is *preheated* to around 800°F (427°C). Complicated and intricately shaped parts are preheated again to 1250°F (677°C). Preheating permits a better transition from alpha to gamma iron at the critical point.

Steels are heated from 50°F to 200°F (10°C to 93°C) above the critical point to cause all the constituents to go into solid solution. The added temperature also allows the solution to remain in this state a few additional seconds during the cooling cycle between the time the workpiece is removed from the furnace and quenched. This added temperature and short interval of time permit a complete phase change in the metal.

THE IRON-CARBIDE PHASE DIAGRAM

An *iron-carbon phase diagram* contains general information about the full range of hardening temperatures for carbon steels and is used to show the changes in the microstructure of steel that occur during heat treating. Figure 61–1 is an iron-carbon phase diagram with critical temperatures, heat-treating processes, and grain structures. The diagram shows that when carbon steel in a fully annealed state is heated above the lower critical point—between 1335°F to 1355°F (724°C to 735°C)—the alternate bands of ferrite and pearlite that make up the pearlite begin to merge into one earlier.

The process continues above the critical point until the pearlite completely dissolves to form an iron in an *austenite state.* If an excess of ferrite or cementite is present in the steel, they begin to dissolve into the austenite if the temperature of the steel continues to rise. Ultimately, only austenite is present. The diagram shows that above certain temperatures the excess ferrite or cementite is completely dissolved in the austenite. The temperature at which this transformation occurs is known as the *upper critical point*. This temperature is affected by the carbon content of the steel.

EFFECTS OF SLOW COOLING ON CARBON STEELS

The transformation from alpha to gamma iron to the point where the iron is completely austen-itic is reversed upon cooling. However, the upper and lower critical points occur at slightly lower temperatures to change from gamma to alpha iron. At room temperature, the structure of the original fully annealed carbon steel part will return to the same proportions of ferrite or cementite and pearlite that were present before heating and cooling. No austenite will be present.

EFFECTS OF RAPID COOLING ON CARBON STEELS

As the rate of cooling from an austenitic state is increased, the temperature at which the austenite changes to pearlite decreases below

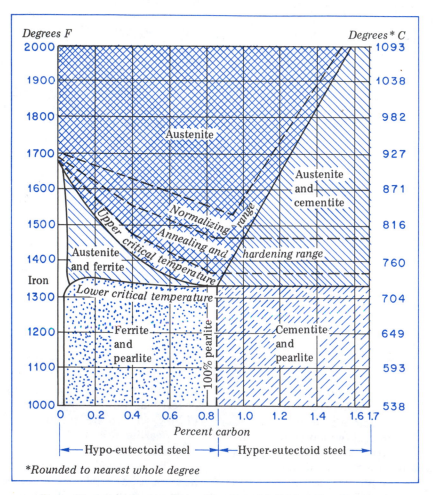

Figure 61–1 Iron-Carbon Phase Diagram with Critical Temperatures, Heat-Treating Processes, and Grain Structures

the slow transformation temperature of 1300°F (704°C). As the cooling rate is increased, the laminations of the pearlite (formed by the austenite as it is being transformed), become finer.

When a carbon steel is quenched at the *critical cooling rate* or faster, a new *martensite structure* is formed. The critical cooling rate is the cooling rate at which there is a sudden drop in the transformation temperature. The martensite structure formed has angular, needle-like crystals and is very hard.

APPLICATION OF IRON-CARBON PHASE DIAGRAM INFORMATION

CHARACTERISTICS OF MARTENSITE

Maximum hardness is produced in steel when the grain structure after quenching is transformed from austenite to martensite in a matter of seconds. It is important to quench carbon steels from hardening temperature to below the 1000°F (538°C) temperature in one second or less to prevent austenite from beginning to transform to pearlite. The hardening process avoids the formation of pearlite grains. If the cooling is slower, any pearlite that is formed mixes with the martensite and reduces hardenability throughout the steel. Martensite is the hardest and most brittle form of steel. Figure 61-2 shows the grain structure of martensite (820X magnification). Martensite in pure form is a supersaturated solid solution of carbon in iron and contains no cementite.

DEPTH OF HARDNESS AND HEAT-TREATING PROBLEMS

Hardenability has been considered in relation to the uniformity of hardness from the outer surface to the center of a part. Hardenability may be improved by adding alloying elements.

When the rate of cooling during quenching is too fast, a number of problems emerge, such as warpage, internal stresses, and different types of fractures. Common heat-treating problems and their probable causes are given in Table 61-1. Usually no problems are encountered with the hardening of workpieces of uniform cross section up to 1/2" (12.7mm) thickness. These workpieces

may be hardened throughout with complete transformation from austenite to martensite. As the thickness increases, additional attention must be paid to the selection of the steel, quenching medium, and cooling time to ensure that there will be uniform hardness or stabilizing of the grain transformation to martensite.

WORKING TEMPERATURES FOR STEEL HEAT-TREATING PROCESSES

Information normally contained on an iron-carbon phase diagram is interpreted into a practical guide in Figure 61-3, which shows the working temperatures for carbon steel heat-treating processes.

Note that the black heat range is from 0°F to 1000°F (538°C); red heat, from 1000°F to 2050°F (1121°C); and white heat, from 2050°F to 2900°F (1593°C). Heat-treating processes for carbon steels are carried on from the subzero range for stabilizing grain structure through normalizing processes starting around 1700°F (927°C) for steels with a minimum of 0.02% carbon. The higher carburizing temperatures are shown graphically above the normalizing range.

AGING AND GROWTH OF STEEL

A small amount of the austenite is sometimes retained when carbon steels are subjected to severe, extremely rapid cooling. All austenite is not transformed into martensite through quench-

Figure 61-2 Grain Structure of Martensite (820X Magnification)

Table 61–1 Common Heat-Treating Problems and Probable Causes

Heat-Treating Problem	Probable Cause
Circular cracks	—Uneven heating in hardening
Vertical cracks; dark-colored fissures	—Steel burned beyond use
Hard and soft spots	—Uneven or prolonged heating; uneven cooling
Hard and soft spots with tendency to crack	—Tool not moved about continuously in quenching fluid
Soft places	—Tool dropped to bottom of fluid tank
Surface scale; rough decarburized surface	—Heated steel surfaces exposed to effects of an oxidizing agent
Excessive strains (workpiece with cavities and holes)	—Formation of steam or gas bubbles as an insulating film in cavities and pockets
Coarse grain; quench cracking; tool fracture in use	—Overheating of tool
Area around tongs softer than remainder of part	—Tongs not preheated
Quench cracks (straight-line fractures from surface to center)	—Overheating during austenite stage —Incorrect quenching medium and/or nonuniform cooling —Incorrect selection of steel —Time delays in a high-stress state between hardening and tempering
Surface cracks formed by high internal stresses	—Unrelieved stresses produced by high surface temperature generated by prior machining operations such as grinding
Stress cracking	—Failure to temper a part before grinding —Reducing surface hardness through subsequent machining —Forming a hardened crust by machining at a high temperature and immediately quenching area with cutting fluid (coolant)
Undersized machining to clean up decarburized surfaces	—Failure to allow sufficient stock to permit grinding all required surfaces to clean up

ing. The tendency over a period of time is for the austenite to be transformed into martensite without further heating or cooling. This process, called *aging*, results in the *growth* of the steel with the increase in volume. Additional stresses are also introduced. Therefore, control rates are established for preheating, heating to the lower and upper critical points, cooling, and quenching to provide optimum hardening conditions.

CRITICAL POINTS IN HARDENING

The two important stages in hardening steel are heating and quenching. The steel is heated above its transformation point to produce an entirely austenitic structure. The steel is then quenched at a rate faster than the critical rate required to produce a martensitic structure. It should be noted that the critical rate depends on the carbon content, the amount of other alloying elements, and the grain size of the austenite. The hardness of steel depends on the carbon content of the martensitic steel.

Pearlite is transformed into austenite as steel is being heated at a transformation point called the *decalescence point*. At this point, while the steel continues to be heated and the surrounding temperature becomes hotter, the steel continues

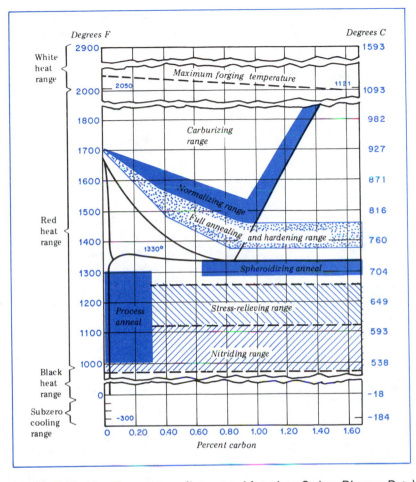

Figure 61-3 Working Temperatures (Interpreted from Iron-Carbon Diagram Data) for Carbon Steel Heat-Treating Processes

to absorb heat without any significant change in temperature.

The process is then reversed in cooling when the austenite is transformed into pearlite at the *recalescence point*. The steel gives out heat at this point so that the temperature rises momentarily instead of continuing to drop.

The control points of decalescence and recalescence have a direct relationship to the hardening of steel. These critical points vary for different kinds of steel, requiring different hardening temperatures. For hardening, the temperature must be sufficient to reach the decalescence point to change the pearlite to austenite. Also, the steel must be cooled rapidly enough before it reaches the recalescence point to prevent the transformation from austenite to pearlite.

HARDENING TEMPERATURES FOR CARBON TOOL STEELS

Tool steel manufacturers' charts contain hardening temperature specifications. As a rule-of-thumb, based on the carbon content in the steel, the following general guidelines are provided:

Carbon Content	Hardening Temperature Range
0.65 to 0.80%	1450 to 1550°F (788 to 843°C)
0.80 to 0.95%	1410 to 1460°F (766 to 793°C)
0.95 to 1.10%	1390 to 1430°F (754 to 777°C)
over 1.10%	1380 to 1420°F (749 to 771°C)

In general, the highest temperature in the range produces deeper hardness penetration and increases strength. Conversely, the lowest temperature in the range decreases the hardness depth but increases the ability of the steel to resist splitting forces.

LIQUID BATHS FOR HEATING

Heating baths are generally used for the following purposes:

- To control the temperature to which a workpiece may be heated,
- To provide uniform heating throughout all sections of a workpiece,
- To protect the finished surfaces against oxidation and scale.

The molten liquid baths that are widely used for steel hardening and tempering operations include sodium chloride, barium chloride, and other metallic salt baths, as well as lead baths. Lead baths are extensively used for quantity heat treating of small tools and for heating below $1500°F$ ($815°C$).

QUENCHING BATHS FOR COOLING

Quenching baths serve to remove heat from the steel part being hardened. The rate of cooling must be faster than the critical cooling rate. Hardness partly depends on the rate at which the heat is extracted. The composition of the quenching bath determines the cooling rate. Therefore, different kinds of baths are used, depending on the steel and the required heat treatment.

The two most common quenching baths are fresh, soft water baths; oils of different classes, or oil-water solutions. Brine and caustic soda solutions are also used. High-speed steels are generally cooled in a lead or salt bath. Air cooling serves as a quenching medium for high-speed steel tools that require a slow rate of cooling. Oil quenching is usually used for applications requiring rapid cooling at the highest temperature and slower cooling at temperatures below $750°F$ ($399°C$).

OIL QUENCHING BATHS

Oil quenching permits hardening to depth while minimizing distortion and the possibility of cracking of standard steels. Alloy steels are normally oil quenched. Prepared mineral oils

have excellent quenching qualities and are chemically stable and cost effective. Vegetable, animal, and fish oils are also used alone or in combination.

One advantage of quenching oils is that they provide fast cooling in the initial stages, followed by slower cooling during the final stages and lower temperatures. This action prevents the steel from cracking. Quenching oils are maintained within a given temperature range of from $90°F$ to $140°F$ ($32°C$ to $60°C$).

WATER QUENCHING BATHS

Carbon steels are hardened by quenching in a bath of fresh, soft water. The bath temperature must be maintained within a $70°F$ to $100°F$ ($21°C$ to $38°C$) range so that subsequent workpieces are cooled at the same rate. The temperature of the water must be kept constant within the range because the water temperature seriously affects the cooling rate and hardness penetration.

Design features in workpieces, such as holes, cavities, pockets, and other internal corners, result in uneven cooling when quenched in fresh water. Gas bubbles and an insulating vapor film in cavities produce uneven cooling, excessive internal stresses, and greater danger of cracking.

The addition of rock salt (8% to 9%) or caustic soda (3% to 5%) to a fresh water quenching bath prevents gas pockets and vapor films from forming. Care must be taken to use a clean, uncontaminated, soft water bath. The quantity must be sufficient to dissipate the heat rapidly and permit the workpiece to be agitated (moved) within the bath. As a general rule, thicker sections of a heated part are immersed first.

MOLTEN SALT QUENCHING BATHS

High-speed steels are generally quenched in a molten salt bath in preference to oil quenching. A molten salt bath produces maximum hardness and minimum cooling stresses. Such stresses result in distortion and possible cracking.

Molten salt quenching baths for high-speed steel are maintained at temperatures of $1100°F$ to $1200°F$ ($593°C$ to $649°C$). After quenching, the hardened part is tempered or drawn in another molten salt bath within a temperature range of 950°F to 1100°F (510°C to 593°C). A general-purpose tempering temperature for high-speed cutting tools is 1050°F (566°C).

QUENCHING BATH CONDITIONS AFFECTING HARDENING

Mention has been made several times about the need to maintain the quenching bath within a specific temperature range. Another condition to consider is the need for movement between the bath and the workpiece so that the temperature throughout the bath remains constant. Under these conditions, cooling proceeds uniformly on all exposed surfaces and completely through the part.

A more desirable practice than agitating the workpiece in the quenching medium is to reverse the process. If the tank permits, the bath is thoroughly agitated while the workpiece is held still. This kind of fluid motion lessens the danger of warping the workpiece during heat treating.

INTERRUPTED QUENCHING PROCESSES

Interrupted quenching processes are used to obtain greater toughness and ductility for a given hardness and to overcome internal stresses that result in quench cracks and distortion. These problems are encountered in general hardening practices.

Austempering, martempering, and *isothermal quenching* are three interrupted methods. The quenching begins at a temperature above the transformation point and proceeds at a rate that is faster than the critical rate. The important fact is that the cooling is interrupted at a temperature above the one at which martensite starts to form.

The steel is maintained at a constant temperature for a fixed time to permit all sections within a part and the external surfaces to reach the same temperature. Transformation of the structure of the steel takes place uniformly for temperature and time throughout the workpiece. Interrupted quenching requires a larger quantity of heat to be absorbed and dissipated without increasing the temperature of the bath.

AUSTEMPERING

Austempering is a patented heat-treating process in which steels (chiefly with 0.60% or higher carbon content) are quenched in a bath at a constant temperature between 350°F to 800°F (176°C to 427°C) at a higher rate than the critical quenching rate. The quenching action is interrupted when the steel temperature reaches the bath temperature.

The steel is held for a specified time at this temperature. The austenitic structure changes to a *bainite structure*, which resembles a tempered martensite structure normally produced by quenching a steel and drawing its temper at 400°F (204°C) or more. The austempered part will have much greater ductility and toughness. In austempering, the steel is quenched rapidly so that there is no formation of pearlite. The steel is also held at the same transformation temperature to ensure that all austenite is transformed to bainite.

MARTEMPERING

Martempering is a heat-treating process that is especially adapted to higher alloyed steels. Martempering produces a high-hardness structure without the problems of internal stresses, which are accompanied in some cases by quench cracks and dimensional changes in the part.

The first rapid quench takes place in martempering at a specific temperature above the transformation point to a temperature above the one where martensite forms. The temperature is held at this point in order to equalize it throughout the part. The workpiece is then removed from the bath and is cooled in air. Martensite begins to form at a uniform rate in a matrix of austenite. The soft austenite tends to absorb some of the internal stresses produced by the formation of martensite.

ISOTHERMAL QUENCHING

Like austempering, isothermal quenching is a process in which steel is rapidly quenched from above the transformation point down to a temperature that is above the one at which martensite forms. The temperature is held constant, usually at 450°F (232°C) or above, until all the austenite is transformed to bainite.

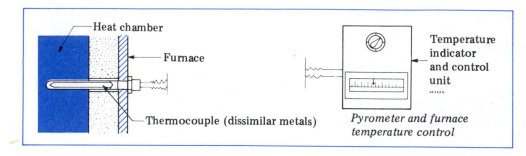

Figure 61-4 Thermocouple and Pyrometer for Measuring, Indicating, and Controlling Temperatures

At this stage, the steel is immersed in another bath and the temperature is raised to a higher specified temperature. After being held at this higher temperature for a definite period, the workpiece is cooled in air. In isothermal quenching, tempering takes place immediately after the steel structure is changed to bainite and before the workpiece is air cooled to normal temperature.

HEAT-TREATING EQUIPMENT

Heat-treating processes require gas, oil-fired, or electrical furnaces that are especially regulated for temperature and, in certain cases, atmosphere. Safety devices are built into the equipment. Exhaust ducts, hoods, close-down valves, and other protective units are requirements of each installation.

Caution: Safety precautions are to be observed for furnace lighting, safety shut off valves and switches, exhausting fumes, performance of all heat treatment processes, and the use of personal protective devices.

HEAT-TREATING FURNACE CONTROLS

Furnace temperatures are usually controlled by a *thermocouple* and a *pyrometer*. A simple installation of an activated thermocouple and an indicator dial pyrometer for measuring, indicating, and controlling temperatures is illustrated in Figure 61-4. The effect of heat in the furnace on the dissimilar metal parts in the thermocouple is to produce small amounts of electrical energy. This varying amount of energy, depending on changes in temperature, is translated on a calibrated temperature scale of the pyrometer.

The pyrometer controls the furnace temperature by setting the instrument at a required temperature.

Pyrometers may have direct temperature sensing and reading controls or a combination of sensing and reading controls and a recording instrument. Control and temperature measurement information provides an important record of heating and cooling processes, particularly when continuous and prolonged temperature controls are required.

BASIC HEAT-TREATING FURNACES

General toolroom hardening, tempering, other heat treatment, forging, and heating processes are performed with small gas-fired or electric heat treatment furnaces. Figure 61-5 shows these two basic types of furnaces with temperature-regulating/indicating controls. The *gas-fired furnace* (Figure 61-5A) is available either for manual regulation or it may be secured with temperature-indicating controls. The *electric furnace* (Figure 61-5B) has its temperatures regulated by electronic controls and readouts on the digital unit. A general temperature control range for these furnaces is from 300°F to 2300°F (148°C to 1260°C). In some furnaces, temperatures are established by using marking crayons, pellets, and liquid coatings. These materials melt at known temperatures and are used to identify the furnace and/or workpiece temperature.

LIQUID HARDENING FURNACES

The *pot-type furnace* is used to heat salt, lead, cyanide, and other baths to a molten state.

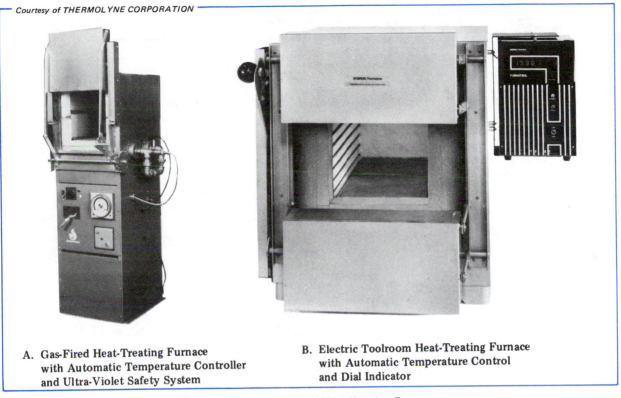

**A. Gas-Fired Heat-Treating Furnace
with Automatic Temperature Controller
and Ultra-Violet Safety System**

**B. Electric Toolroom Heat-Treating Furnace
with Automatic Temperature Control
and Dial Indicator**

Figure 61–5 Basic Types of Heat-Treating Furnaces

This furnace is used for heat-treating processes that require a part to be immersed and heated to a specified temperature. In the case of surface hardening, the outer core that is to be carburized is held in the molten bath for a specified period of time. The time is established by the required depth of the casehardening.

PRODUCTION HEAT-TREATING FURNACES

Two broad groupings of production heat-treating furnaces provide for batch and continuous processing. *Batch furnaces* provide for the heat treatment of quantity parts and process them as a batch. Batch processing is programmed for mass production.

Conveyor furnaces are used for continuous cycling operations. There are different zones of temperature in the furnace to regulate the rate for both heating and cooling. Production furnaces are usually atmospherically controlled—that is, the furnaces are flooded with varying

combinations of carbon dioxide, carbon monoxide, nitrogen, and hydrogen gases. The gases prevent decarburization, scale, and surface rust.

HARDENING AND TEMPERING CARBON TOOL STEELS

HARDENING TEMPERATURE RANGE

The hardening temperature is the maximum temperature to which a steel is heated before being quenched for purposes of hardening. The hardening temperature is above the lower critical point of a given steel.

TEMPERING TEMPERATURES FOR CARBON AND HIGH-SPEED STEELS

Tempering, or *drawing*, reduces the brittleness in hardened steel and removes internal strains produced by the sudden cooling in a quenching bath. A cutting tool or part is tempered by reheating it when the steel is in a fully hardened

Table 61–2 Selected Tempering Temperatures, Colors, and Typical Applications of Carbon Steels

Tempering Degrees °F	°C	Temper Color	Typical Tool Applications
380	193	Very light yellow	Single-point cutting tools and machine centers requiring maximum hardness
430	221	Light straw	Multiple-point milling cutters, drills, reamers, hollow mills; forming tools
450	232	Pale straw-yellow	Twist drills and screw machine centering tools
470	243	Dark straw	Thread rolling dies, punches, stamping and forming dies, hacksaw blades
490	254	Yellow-brown	Taps over 1/2" (12.7mm); threading dies for tool steels; shearing blades
510	266	Spotted red-brown	Machine taps under 1/4" (6.4mm); general threading dies
530	277	Light purple	Hand and pneumatic punches; scribing tools
550	288	Dark purple	Cold chisels and blunt wedge-shape cutting tools
570	299	Dark blue	Cutting/forming tools requiring minimum hardness
590	310	Pale blue	Torque tools: wrenches, screwdrivers, hammer faces
640	338	Light blue	Generally, noncutting tools requiring a minimum hardness

condition and then cooling. Reheating changes the grain structure to one with a reduced hardness.

The general range of temperatures for tempering carbon tool steel is from 300°F to 1050°F (149°C to 565°C). Tempering tables are available for recommended hardness requirements for steel parts, hardening temperatures and time, quenching baths, and tempering temperatures. Table 61–2 gives the tempering temperatures for carbon steels that have been hardened at temperatures from 1350°F to 1550°F (732°C to 843°C).

Alloy tool steels are hardened by heating to temperatures between 1500°F to 1900°F (815°C to 1065°C). High-speed tool hardening temperatures are between 2150°F to 2450°F (1193°C to 1343°C). High-speed parts and tools are tempered at 1000°F to 1100°F (538°C to 593°C).

TEMPERING COLORS

A film of oxide forms on the surface of steel that is heated in an oxydizing atmosphere. As the temperature increases above about 400°F (204°C), the surface of the steel starts to change to a light straw color at 425°F (218°C). As different temperatures are reached, the temper colors change, as listed in Table 61–2. At 640°F (338°C), the temper color of steel is a light blue. Above the tempering color range, the heat colors of steel change from a black red to a white heat around 2400°F (1315°C).

TEMPERING BATHS

Oil Baths. Many tools are tempered in an oil bath that is heated uniformly to a required temperature. Heavy tempering oils may be heated to temperatures between 650°F to 700°F (343°C to 371°C). A partially heated steel part is immersed in the oil. Then, both the steel part and oil are heated to the tempering temperature. The part is next dipped in a tank of caustic soda, followed by quenching in a hot water bath. Oil bath tempering is generally limited to temperatures of 500°F to 600°F (260°C to 316°C).

Salt Baths. Salt baths are recommended for high-speed steel tempering that occurs between 1000°F to 1050°F (538°C to 593°C). Specifications often recommend the use of salt baths above 350°F (173°C) for efficiency and economy. The furnace temperature and part are increased gradually to the tempering range.

Lead Baths. Parts may be tempered by using a lead or lead alloy bath to heat the steel to the required tempering temperature. The workpiece is preheated and immersed in the bath, which has already been brought to the tempering temperature. The part is left in the bath until it reaches the tempering temperature. It is then removed and cooled.

DOUBLE TEMPERING

The tempering operation is often repeated on high-speed steel tools. *Double tempering* ensures that an untempered martensite, which remains in the steel after the first tempering operation, becomes tempered after double tempering. The martensite structure is also relieved of internal strains.

In double tempering, the high-speed steel part is brought to its tempering temperature and held at this temperature for a period of time. The part is then cooled to room temperature. Next, the part is reheated to and held at the original tempering temperature for another period and again cooled to room temperature.

HEAT TREATMENT OF TUNGSTEN HIGH-SPEED STEELS

The hardening temperature for tungsten high-speed steels is from 2200°F to 2500°F (1204°C to 1371°C). Hardening at such a high temperature usually requires one preheating stage. Preheating is particularly important for cutters with thick bodies as compared to the amount of material around cutting teeth. Preheating helps to avoid internal strains. Preheating is done at temperatures below the critical point of the steel, within the 1500°F to 1600°F (816°C to 871°C) range.

At such a temperature, the tool may safely be left in the furnace to heat through uniformly and to bring it up to hardening temperature.

Cutters and other cutting tools that are thicker than 1″ (25.4mm) often require a second preheating.

QUENCHING TUNGSTEN HIGH-SPEED STEELS

High-speed steel tools are generally quenched in oil. A tool is moved in the bath to prevent a poor, heat-conducting gas film from forming on the tool. An oil quench permits uniform cooling at the required rate.

Salt baths are also used for quenching. Salt baths are particularly adapted for tools and parts that have complex sections or areas where hardening cracks may develop.

High-speed tools are sometimes quenched in a lead bath or by air cooling. Small sections may be cooled in still air; large, heavier sections require a stream of dry compressed air.

TEMPERING HIGH-SPEED STEEL TOOLS

The drawing temperature for high-speed tools is from 900°F to 1200°F (482°C to 649°C). The temperature is higher for single-point cutting tools used on lathe work than for multiple-tooth milling cutters and form tools. Cobalt high-speed steel tools are tempered between 1200°F and 1300°F (649°C to 704°C).

Once the tool or part is heated to the required drawing temperature, it is held at this temperature until heated uniformly throughout its mass. The tool is allowed to cool in dry, still air away from drafts. The tool is not quenched for tempering. Quenching produces internal strains that may later cause the part to fracture.

ANNEALING, NORMALIZING, AND SPHEROIDIZING METALS

ANNEALING

Annealing is a heat-treating process that requires heating and cooling to:

- Induce softening,
- Remove internal strains and gases,
- Reduce hardening resulting from cold working,
- Produce changes in mechanical properties such as ductility and toughness and magnetic characteristics,
- Form definite grain structures.

The process is also applied to softer metals to permit additional cold working.

The steel is heated to a temperature near the critical range. It is held at this elevated temperature for a period of time and then cooled at a slow rate. Carbon steels are fully annealed by heating *above the upper critical point* for steels with less than 0.85% carbon content (hypo-eutectoid steels). Steels with more than 0.85% carbon content (hyper-eutectoid steels) are heated slightly *above the lower critical point.*

The heated steel is held at this temperature until the part is uniformly heated throughout. It is then slowly cooled to 1000°F (538°C) or below. The result is the formation of pearlite and a layer-like grain structure.

NORMALIZING

The purpose of *normalizing* is to put the grain structure of a steel part into a uniform, unstressed condition of proper grain size and refinement to be able to receive further heat treatment. The iron-base alloy is heated above the transformation range and then cooled in still air.

Depending on the composition of the steel, normalizing may or may not produce a soft, machinable part. By normalizing low-carbon steel parts, the steel is usually placed in best condition for machining. Also, distortion due to carburizing or hardening is decreased.

SPHEROIDIZING

The *spheroidizing of steels* is defined as a heating and cooling process that produces a rounded or globular form of carbide. Steels are spheroidized to increase their resistance to abrasion and to improve machinability, particulary of high-carbon steels that require continuous cutting operations. Low-carbon steels are spheroidized to increase certain properties—for example, strength—before other heat treatment.

Spheroidizing requires heating steel *below the lower critical point.* The part is held at this temperature for a time and then cooled slowly to around 1000°F (538°C) or below. High-carbon steels are spheroidized by heating to a temperature that alternately rises between a temperature inside the critical range and one that is outside the critical range. Tool steels are spheroidized

by heating the part slightly above the critical range. The part is held at this temperature for a period of time and then cooled in the furnace.

CASEHARDENING PROCESSES

Casehardening relates to the process of increasing the carbon content to produce a thin *outer case* that can be heat treated to harden. Impregnating the outer surface with sufficient amounts of carbon to permit hardening is called *carburizing.* When the carburized part is heat treated, the outer case is hardened, leaving a tough, soft inner core. Casehardening refers to the carburizing and hardening processes. The three general groups of casehardening processes are (1) carburizing, (2) carbonitriding, and (3) nitriding.

CARBURIZATION

Iron or steel is carburized when heated to a temperature below the melting point in the presence of solid, liquid, or gaseous carbonaceous materials. These materials liberate carbon when heated. The steel gradually takes on the carbon by penetration, diffusion, or absorption around the outer surface. This action produces a *case or zone* that has a higher carbon content at the outer surface. When a carburized part is heated to hardening temperature and quenched, the outer core acts like a high-carbon steel and becomes hard and tough.

Casehardening produces a steel having surface properties of a hardened high-carbon steel while the steel below the case has the properties of a low-carbon steel. Thus, there are two heat treating processes. One is suitable for the case; the other, for the core. Following an initial heating and slow cooling, a casehardened part is reheated to 1400°F to 1500°F (760°C to 816°C). It is then quenched in oil or water and given a final tempering.

PACK HARDENING (CARBURIZING)

The purpose of *pack hardening* is to protect the delicate edges or finished surfaces of workpieces and to encourage uniform heating and contact with a carbonaceous material. Pack hardening also prevents scale formation and minimizes the danger from warping or cracking.

Pack carburizing requires the steel part to be enclosed in a box. Carbonaceous material such as carbonates, coke, and hardwood charcoal and oil, tar, and other binders are packed around the part in the box. Since the carburizing materials are inflammable, the box is sealed with a refractory cement to permit the gases generated within the box to escape while air is prevented from entering.

Penetration Time. The box is heated to carburizing temperature between 1500°F to 1800°F (816°C to 962°C), usually within the average temperature range of 1650°F to 1700°F (899°C to 927°C). Naturally, the rate of carbon penetration increases at the higher temperature in the range. The approximate *penetration time* for depths of 0.030" to 0.045" (0.75mm to 1.14mm) is four hours. It generally takes eight hours to penetrate to 1/16" (1.6mm) and 24 hours to penetrate 1/8" (3.2mm).

When the parts are taken out of the box, they are cleaned by wire brush, tumbled, or sand blasted.

Heat Treating Carburized Parts. Once cooled from the carburizing temperature, the parts are reheated to the hardening temperature of the outer case (approximately 1430°F or 777°C) and quenched. This treatment produces a steel with a hardened case and a tough, soft low-carbon steel core.

A finer surface grain structure is produced by *double quenching*, which requires heating to the hardening temperature of the low-carbon steel core (1650°F or 899°C) and quenching. The coarse grain structure produced is then refined by reheating. The part is brought up to the hardening temperature of the case (about 1430°F or 777°C) and quenched. Double quenching combines the good wearing qualities of a hard case and toughness.

LIQUID CARBURIZING

Liquid carburizing has the advantage of not requiring carbonaceous materials to be packed around the workpiece. Liquid baths have a faster and more uniform penetration with minimum distortion. Where sections of a part are to be selectively carburized, the remaining portions may be copper plated. When the whole piece is immersed in the liquid bath, the copper plate inhibits carburization.

Salt bath furnaces are usually designed to be heated by electrodes that are immersed in the bath. The bath is stirred to ensure uniform temperature. The liquid carburizing baths are molten mixtures of cyanides, chlorides, and carbonates. The composition of the mixture depends on the quantity of carbon and/or nitrogen to be absorbed by the steel.

Liquid carburizing temperatures range from 1550°F to 1700°F (843°C to 927°C). In general, it takes about two hours at an average temperature of 1650°F (899°C) to penetrate to a depth of 0.020" (0.5mm). Deeper case depths require proportionally greater periods of time to penetrate.

After carburizing in the molten salt bath, parts may be quenched directly in water, brine, or oil, depending on the job requirements. The parts are then tempered as required.

Caution: A number of safety precautions must be strictly observed. Most liquid carburizing, cyaniding, and nitriding salts are hazardous. Cyanides are extremely poisonous internally and in open wounds and scratches. When heated, cyanide fumes are toxic. Salt baths require careful, direct venting to outdoors.

GAS CARBURIZING

Methane (natural gas), propane, and butane are three gaseous hydrocarbons (carbon-bearing gases) used for *gas carburizing*. In continuous carburizing furnaces, the parts are heated to carburizing temperature in a horizontal rotary type or vertical pit type of gas carburizer. The carbon-bearing gases are mixed with air and other specially prepared dilutent gases. The carburizing gases are fed continuously to the carburizing retort of the furnace. The spent gases are also exhausted.

The parts are soaked in the carburizing chamber. The soaking temperature and time depend on the required depth of case. Gas carburizing temperatures of around 1700°F (927°C) produce an absorption rate for the first 0.020" to 0.030"

(0.5mm to 0.75mm) depth of case during a four-hour period.

CYANIDING (LIQUID CARBONITRIDING)

Cyaniding is a casehardening process that is used generally for limited depth case hardening to about 0.020″ (0.5mm). The amount of carbon required in the surface case establishes the properties of cyanide salts in the salt bath (cyanide, chlorate, and chloride salts).

Cyaniding requires temperatures above the critical range of 1400°F to 1600°F (760°C to 871°C) for steels. Lower temperatures in the range are used for a limited case depth. As a general rule, case depths of from 0.003″ to 0.005″ (0.08mm to 0.13mm) require a half-hour soaking. Depths of 0.005″ to 0.010″ (0.13mm to 0.25mm) take from 60 to 70 minutes. The soaking time at cyaniding temperatures for depths of 0.015″ (0.38mm) is two hours.

Parts are quenched in an appropriate water, brine, or oil bath. Tempering where required, is done at temperatures between 250°F to 300°F (120°C to 150°).

CARBONITRIDING (GAS PROCESS)

Carbon and nitrogen are introduced into the surface of steel by a dry (gas) cyaniding process known as *carbonitriding*. Carbonitrided surfaces possess greater hardenability and are harder and more wear resistant than carburized surfaces.

Parts are soaked at between 1350°F to 1650°F (732°C to 899°C) in a gaseous atmosphere composed of carburizing gas and ammonia. The ammonia produces the nitrogen. Carbon and nitrogen are introduced into the heated parts. Case depths of 0.030″ (0.74mm) require soaking at temperatures of 1600°F (871°C) for four to five hours. A lower temperature is used when a higher proportion of nitrogen is required or for a shallower case. In such instances, a 0.005″ to 0.010″ (0.13mm to 0.25mm) case depth may be produced at a temperature around 1450°F (788°C) within 90 minutes. Generally, carbonitrided parts are quenched in oil to prevent distortion and to gain maximum hardness.

NITRIDING

Nitriding is a casehardening process of producing surface hardening by absorption of nitrogen, without quenching. Special alloy steels are heated in an atmosphere of ammonia or in contact with a nitrogenous material. An exceptionally hard surface is produced on machined and heat-treated parts. Nitriding permits hardening carbon alloy steels beyond conventional hardness readings.

The alloy steels contain chromium, vanadium, molybdenum, and aluminum as nitride-forming elements. Nitriding requires heating the part below the lower critical temperature in a protected nitrogenous atmosphere or a salt bath.

Gas Nitriding. Ammonia gas is circulated through a gas furnace chamber. The parts are heated in an air-tight drum to a temperature between 900°F to 1150°F (482°C to 621°C). The nitrogen in the ammonia gas, which decomposes at this temperature, combines with the alloying elements in the steel. The hard nitrides that form produce a harder surface than may be obtained from other heat treatment processes.

Gas nitriding is particularly adapted to increase the hardness of parts that are hardened and ground. Many high-speed steel cutting tools are nitrided with shallow case depths of 0.001″ to 0.003″ (0.025mm to 0.075mm). The core of these tools is not affected by nitriding.

There is limited, if any, distortion due to the fact that gas-nitrided parts require no quenching. The process is slower than other casehardening processes. Depths of 0.020″ (0.5mm) require two to three days.

Salt Bath Nitriding. Cutting tools such as taps, drills, reamers, milling cutters, and dies are often nitrided. Nitriding increases the surface hardness, fatigue and wear resistance, tool wear life, durability, and corrosive resistance (except on stainless steels).

The liquid salt bath is brought to temperature between 900°F to 1100°F (482°C to 593°C), depending on the tool requirements. The tool (or part) is suspended in the molten nitriding salt for the required period of soaking time.

SPECIAL SURFACE HARDENING PROCESSES

Flame hardening and induction hardening are considered as two special surface hardening processes. These processes require heating without using a furnace. The purpose in each instance is to harden particular surfaces from a skin surface to depths up to 1/4″ (6.4mm).

FLAME HARDENING

Flame hardening requires heating the surface layer of an iron-base alloy above the transformation temperature range.

The high temperature flame is concentrated in the specific area to be hardened. The surface is immediately quenched by quenching jets in back of the torch or burner. A water or compressed-air quenching medium is used, depending on the type of steel. Tempering is recommended close to the hardening process, where practical.

Similar methods of flame heating are used for tempering. A special low-temperature flame head follows immediately behind the quench to localize the heating area, which is brought to the required tempering temperature.

INDUCTION HARDENING

Induction hardening is another localized heat-treating process. Electrical heating is required to bring a piece of steel to the required hardening temperature and subsequently quenching it either in a liquid or air bath. Induction heating is especially adapted to parts that require localized and controlled depth of hardening and/or have an irregularly contoured surface. The major advantages to using induction hardening are as follows:

- A short heating cycle produces the same temperature in seconds that normally requires from 30 to 80 times longer to heat in conventional furnaces;
- There is no tendency to produce oxidation or scaling or decarburization;
- The depth and localized zones of hardening may be exactly controlled;
- The automatic heating and quenching cycles permit close controls of the degree of hardness;
- Warpage or distortion are reduced to a minimum;
- Carbon steels may be substituted for higher-cost alloy steels;

- Stress control is possible by localized heating to relieve internal stresses;
- Welded and brazed design features may be added prior to heat treating particular sections of a workpiece;
- Long parts may be heat treated more efficiently than by conventional furnace methods;
- In gear applications of induction hardening, the teeth may be machined and shaved in a soft-annealed or normalized condition (bushings and inserts can be assembled before hardening the gear teeth).

Principle of Induction Heating and Hardening. The process begins when a metal part is placed inside and close to an applicator coil. The part may be held in a fixed position, turned, or fed through the coil. As a high-frequency electrical current passes through the coil, the surface of the steel is raised to a temperature above the critical temperature in a matter of seconds.

The heated part is then quenched in oil, water, or air. The hardness is localized to the surface. The hardness depth is controlled by the length and intensity of the heating cycle. High-frequency currents are used for localized and surface hardening. Low frequencies are employed for through heating, deep hardening, and large workpieces.

One controlling factor on standard types of steels that may be induction hardened is that the carbon content must permit hardening to the required degree by heating and quenching. Low-carbon steels with a carburized case and plain medium-carbon and high-carbon steels may be induction hardened. Cast irons with a percent of carbon in combined form may also be induction hardened. Induction heat treating of alloy steels is generally limited to shallow hardening types that are not affected by high-stressing and possible cracking induced by the required severe quench.

Quenching After Induction Heating. Heated parts may be quenched by immersing in a liquid bath, by liquid spraying, or by self-quenching. *Self-quenching* is associated with rapid absorption of heat by a large mass of surrounding metal, instead of by using a quenching medium. Self-quenching is generally confined to small, simply designed parts. The exact degree of hardness may be automatically timed for heating and quenching. Induction coils or standard furnaces may also be used to temper parts as required.

SUBZERO TREATMENT OF STEEL

Steels are subjected to a *subzero treatment* in order to stabilize a part and to prevent changes in form or size over a period of time.

How to Stabilize Gages and Other Precision Parts by Subzero Cooling

STEP 1 Cool the hardened and ground gage, sine bar, gage blocks, or other precision part uniformly to −120°F (−84°C).

STEP 2 Place the subzero-cooled part in boiling water, oil, or a salt bath.

Note: Precision tools such as thread and plug gages, gage blocks, and other highly accurate parts are immersed for about two hours.

STEP 3 Repeat the cooling and quenching cycle to eventually transform practically all of the austenite to martensite.

STEP 4 Temper according to standard tempering practices for the specific metal in the part.

STEP 5 Proceed to finish grind or lap the subzero-treated part to the required dimensional size and exact form.

Note: Subzero treatment produces a slight increase in size.

Safe Practices in Basic Metallurgy and Heat Treating

- Avoid bringing the part to a higher temperature than the critical point or quenching at a more severe rate than is recommended by the steel manufacturer. Overheating may cause brittleness, fracturing, or shattering a hardened part.

- Use quenching oils with a flash point sufficiently high to permit safe cooling without the danger of fire.

- Agitate the workpiece in the quenching medium to prevent steam or an insulating vapor film from forming.

- Use safety gloves and face and body protection when working around a furnace with hot metals and when quenching. Stand to one side when plunging and agitating a workpiece in the quenching bath, particularly when using an oil.

- Check the furnace system to see that if the exhaust system fails, there is a shutdown of the main gas/oil supply valve.

- Close the loading door on the jacket enclosure on molten bath furnaces to control and direct the fumes and for protection against spattering.

- Make sure that the workpiece and/or tongs that are to come into contact with molten cyanide are clean and dry. Cyanide, mixtures of nitrate, nitrite, and some other salts spatter and explode violently if any one of them in a molten state comes in contact with water.

- Heat a frozen cyanide bath by electrodes from the top down to ensure that any fumes escape as the bath melts from the top on down through the mass. If electrodes are not used, it is important that a removable wedge be inserted in the bath before it is frozen. The *wide side* is on the bottom. The narrow, tapered end extends beyond the top surface.

UNIT 61 REVIEW AND SELF-TEST

1. Tell how a study of the iron-carbon diagram is related to everyday applications in the heat treating of metals.

2. a. Give two probable causes of quench cracks in hardened metals.
   b. Identify two causes of stress cracks (surface failure) produced by machining a hardened workpiece.

3. State three advantages of using molten baths for heat treatment processes.

4. a. Describe martempering.
   b. Cite two desirable features associated with this process.

5. a. State the function of a thermocouple and pyrometer unit.
   b. List three different types of toolroom furnaces that are used for hardening and tempering processes.

6. a. Name two types of tempering baths and give the heat-treating temperature range of each bath.
   b. Identify the bath materials that are used in each case.

7. a. State the purpose of nitriding.
   b. List three properties of cutting tools that are improved by nitriding.

8. Give three advantages of flame hardening over conventional furnace hardening.

9. a. State the function of the subzero temperature treatment of steel.
   b. Give two applications of this process.

Hardness Testing

A hardened steel is tested to define its capacity to resist wear and deformation. The hardness of steel, alloys, and other materials may also be used to establish other properties and performance.

Hardness testing is performed using measuring instruments of two basic types. The first set of instruments measures the depth of a penetrator for a given load. A few of the instruments that operate under this principle include: *Rockwell, Brinnel, Vickers, Knoop,* and *microhardness testers.* The second set of instruments measures the height of rebound of a given weight hammer of a special shape when dropped from a fixed height. *Scleroscope testers* represent this type of instrument. Each of the two types of instruments is covered in this unit.

Most instrument manufacturers indicate hardness by number scales. The number is related to the size and shape of the penetrator, load, height, and indentation or rebound. Reference tables provide numbers within each major system. These numbers permit conversion of hardness values from one system to another. For example, a Rockwell C (R_C) hardness of 50 is equivalent to 513 on the Vickers scale, 67 on a scleroscope, and BHN 481 on a Brinell scale when a 10mm (0.400″) diameter carbide ball and a 3,000 kg (6,600 lb) load are used. A 75.9

reading is recorded on the Rockwell A (R_A) scale using a diamond penetrator 50 kg (132 lb) load.

Tables for Rockwell C (R_C) hardness numbers 68 through 20 are based on extensive tests with carbon and alloy steels primarily in a heat-treated condition. The hardness numbers may be reliably applied to tool steels and alloy steels in annealed, normalized, and tempered condition. Additional Rockwell B and other hardness scale numbers are provided in handbooks for unhardened or soft temper steels, gray and malleable cast irons, and nonferrous alloys.

ROCKWELL HARDNESS TESTING

Tests are made with the Rockwell hardness tester by applying two loads on the part to be tested. The first load is the *minor load;* the second, the *major load.* The Rockwell tester measures the linear depth of penetration, as shown in Figure

Figure 62-2 Motorized Rockwell Hardness Tester for Applications with Standard, Special, and Superficial Hardness Scales

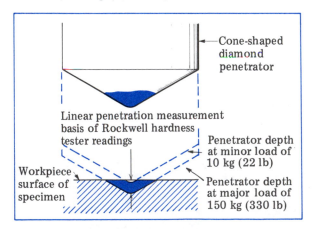

Cone-shaped diamond penetrator

Linear penetration measurement basis of Rockwell hardness tester readings

Penetrator depth at minor load of 10 kg (22 lb)

Workpiece surface of specimen

Penetrator depth at major load of 150 kg (330 lb)

Figure 62-1 Linear Penetration Measured by a Rockwell Hardness Tester

62–1. The difference in depth of penetration produced by applying a minor load and then a major load is translated into a hardness number. This number may be read directly on the indicator dial of a portable tester, a manual or motorized bench model, or on a digital display unit (Figure 62–2).

A shallow penetration indicates a high degree of hardness and a high hardness number. A deep penetration indicates a softer degree of hardness and a low hardness number. In general, the harder the material, the greater its ability to resist deformation.

SIZES AND TYPES OF PENETRATORS

Different penetrator points are used on Rockwell testers, depending on the hardness of the material. Diamond- and ball-point penetrators are used on Rockwell hardness testers. Hard materials such as hardened steels; white, hard cast irons; and nitrided steels require a cone-shaped diamond. The trademark of this form of penetrator is registered as Brale®. A hardened steel ball-shaped point is used for unhardened steels, cast irons, and nonferrous metals.

C and A 120° Diamond Penetrator Points. The C Brale penetrator is used with a major load of 150 kilograms of force (kgf) or 330 pounds of force (lb/f) for hard materials. The *Rockwell C scale* is used with the C Brale penetrator.

Cemented carbides, shallow casehardened steels, and thin steels require the use of the A Brale penetrator and a 60 kgf (132 lb/f). The hardness is read on the *Rockwell A scale*.

Ball Penetrator Points. The second basic type of penetrator has a ball-shaped point and is made of hardened steel. The two diameters are 1/16″ and 1/8″ (1.5mm and 3.0mm). A *Rockwell B scale* is used for hardness readings of unhardened steels, cast irons, and nonferrous metals. The major load with the 1/16″ (1.5mm) diameter ball is 100 kg (220 lb). The minor load is 10 kg (22 lb). The red dial hardness numbers are used for the R_B scale.

HARDNESS SCALES

Table 62–1 furnishes information about standard, special, and superficial Rockwell hardness scales, hardened steel ball-point and conical-shaped diamond-point penetrators, applications, major kg (lb/f) loads, scale symbols, and dial color. The superficial scales are used on parts where only a light, shallow penetration is permitted.

Casehardened parts, nitrided steels, thin metal sheets, and highly finished surfaces are measured for hardness by using a lighter load. Rockwell superficial hardness scales are used to measure such loads. The *Rockwell N scale* requires the use of a special N Brale penetrator; the *Rockwell T scale*, a 1/16″ (1.5mm) diameter hardened steel ball penetrator. The 15, 30, and 45 prefixes indicate the major kilogram load in each case. The minor load is 3 kg (6.6 lb).

FEATURES OF ROCKWELL HARDNESS TESTER

The main features of a Rockwell hardness tester with standard anvil forms are identified in Figure 62–3. A small-diameter plane anvil is shown on the instrument. "V," small- and large-diameter, and roller designs are furnished with this instrument. Cylindrical parts are supported on the V-type centering anvil. Tubing is usually mounted on a mandrel to prevent damage to the walls. Long, overhanging parts are supported by an adjustable jack rest so that the surface of the workpiece is in a horizontal plane with the anvil.

The practice under actual testing conditions is to take readings at three different places along the workpiece. The hardness number is the average of the three readings.

How to Test for Metal Hardness by Using a Rockwell Hardness Tester

STEP 1 Select and mount the appropriate penetrator and anvil in the hardness tester.

STEP 2 Turn the dial on the indicator to position the indicator hand for starting a test.

STEP 3 Place the workpiece or specimen on the anvil. Use a jack rest to support overhanging and long parts.

STEP 4 Raise the anvil or lower the penetrator until it just touches the workpiece.

STEP 5 Apply a 10 kg (22 lb) minor load. This load is shown on the indicator dial. Set the hardness reading (indicator) dial at zero.

Table 62-1 Rockwell Hardness Scales, Penetrators, Major Loads, and Reading Dials for Typical Hardness Test Applications

Scale	Type of Penetrator	Major Load (kgf)	Dial Color (Numbers)	Typical Hardness Test Applications
				Standard Scales
B	1/16″ (1.5mm) ball	100	Red	Soft steels, malleable iron, copper and aluminum alloys
C	Brale (diamond)	150	Black	Steel, deep casehardened steel, hard cast iron, pearlitic malleable iron, other materials harder than B-100
				Special Scales
A	Brale (diamond)	60	Black	Thin steel, shallow casehardened steel, cemented carbides
D	Brale (diamond)	100	Black	Thin steel, medium casehardened steel, pearlitic malleable iron
E	1/8″ (3.0mm) ball	100	Red	Cast iron, aluminum and magnesium alloys, bearing metals
F	1/16″ (1.5mm) ball	60	Red	Thin soft sheet metals, annealed copper alloys
G	1/16″ (1.5mm) ball	150	Red	Malleable iron, phosophor bronze, beryllium copper, other metals within an upper hardness limit of G-92
H	1/8″ (3.0mm) ball	60	Red	Aluminum, lead, zinc
K	1/8″ (3.0mm) ball	150	Red	Harder grades of aluminum, lead, and zinc

L, M, P, R, S and **V** scales with 1/4″ (6.35mm) or 1/2″ (12.70mm) diameter balls and major loads of 60, 100, or 150 kgf are used on very soft and thin materials.

				Superficial Hardness Scales
15 N	Brale (diamond)	15	(N) Green	Casehardened and nitrided parts, thin metal sheets, highly finished surfaces, metal parts requiring lighter loads of 15, 30, or 45 kgf
30 N		30		
45 N		45		
15 T	1/16″ (1.5mm) ball	15	(T) Green	Soft steels, cast iron, nonferrous metals capable of withstanding maximum loads of 15, 30, or 45 kgf
30 T		30		
45 T		45		

STEP 6 Apply the appropriate major load.

STEP 7 Reduce the kgf of the major load to the setting of the minor load.

STEP 8 Read the hardness on the appropriate color scale.

STEP 9 Release the minor load. Move the workpiece or specimen to a second then a third location. Repeat steps 2 through 8 to obtain the second and third hardness readings

STEP 10 Average the three hardness readings.

BRINELL HARDNESS TESTING

The Brinell hardness tester produces an impression by pressing a hardened steel ball under a known applied force into the part to be tested. A microscope is then used to establish the diameter of the impression. The ball is 0.394″

Small pointer

Large pointer

Lever for setting
bezel

Penetrator

Anvil

Dial with Rockwell
scale

Zero adjuster ring

Weights

Capstan handwheel

Zero adjuster ring

Depressor bar

Figure 62–3 Main Features of a Bench Model
Rockwell Hardness Tester

(10mm) diameter. It may be made of hardened steel or carbide, or it may be identified as a Hultgren ball. The main features of a Brinell hardness tester with a measuring microscope are shown in Figure 62–4.

A Brinell hardness number (Bhn) is identified on a Brinell table according to the diameter of the impression and the specified applied load. The Brinell hardness value equals the applied load in kilograms divided by the square millimeter area (impression). The load for hardened steel parts is generally 3,000 kg (6,600 lb). A standard load of 500 kg (1,100 lb) is used for hardness testing nonferrous metals. Lower numbers on the scale are assigned for softer metals and deeper impressions.

RECOMMENDED APPLICATIONS OF BRINELL HARDNESS TESTING

The range of Brinell hardness testing is between Bhn 150 for low-carbon annealed steels to Bhn 740 for hardened high-carbon steels. A carbide ball is required for upper ranges of hardness to Bhn 630. A Hultgren ball accommodates hardness testing up to Bhn 500. The hardened steel ball-point penetrator is adapted to

Load gage

Load and
unload
plunger

Ram

Brinell ball

Handwheel

Dial showing
load in
kilograms

Adjustable air
regulator valve

Air line

Anvil

Screw

*Air-operated metal hardness
tester*

*Microscope for measuring
impression diameter to
establish BHN*

Figure 62–4 Main Features of a Brinell Hardness Tester
with a Measuring Microscope

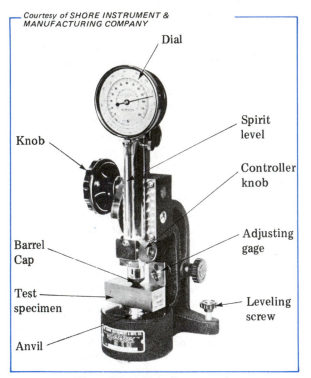

Dial

Knob

Spirit
level

Controller
knob

Barrel
Cap

Adjusting
gage

Test
specimen

Leveling
screw

Anvil

Figure 62–5 Main Features of a
Direct-Reading Scleroscope

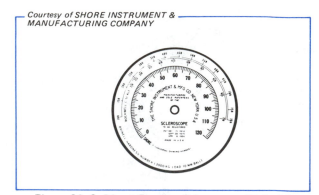

Figure 62–6 Direct-Reading Scleroscope Dial with
Equivalent Rockwell C and Brinell Hardness Numbers

hardnesses to Bhn 450. The Brinell hardness tester works best on nonferrous metals, soft steels, and hardened steels through the medium-hard tool steel range.

VICKERS HARDNESS TESTING

The Vickers hardness test requires a square-based diamond pyramid whose sides are at an angle of 136°. A load of from 5 kg to 120 kg (11 lb to 264 lb) is applied generally for 30 seconds. The diagonal length of the square impression is measured.

The Vickers hardness number equals the applied load divided by the area of the pyramid-shaped impression. The Vickers test is very accurate. It is adapted to large sections as well as thin sheets.

SCLEROSCOPE HARDNESS TESTING

The scleroscope instrument measures hardness in terms of the elasticity of the workpiece. There are two types of scleroscopes. One type has a

direct-reading *scale*. The other type has a direct *dial* recording. Figure 62–5 shows the main features of a direct-reading scleroscope. Scleroscopes are used for hardness testing of ferrous and nonferrous metals. Unlike the Brinell and Rockwell hardness testers, no crater is produced.

A diamond hammer, dropped from a fixed height, rebounds. The rebound distance varies in proportion to the hardness of the metal being tested. The harder the metal, the higher the rebound distance. This movement is either read on a scale or a dial.

The scale is calibrated from the average rebound height of a tool steel of maximum hardness that is divided into 100 parts. The range of rebounds is from 95 to 105. The dial of a direct-reading scleroscope is shown in Figure 62–6. Readings are carried beyond 100 to 120. This range covers super-hard metals. Note that the conversion Rockwell C and Brinell hardness values are given on the dial face; a 3,000 kg (6,600 lb) load and a 10mm diameter ball are used.

MICROHARDNESS TESTING

The Microhardness tester permits hardness testing of minerals, abrasives, extremely hard metals, very small or thin precision parts, and thinly hardened surfaces. Some of these materials and parts cannot be tested by other methods. Microhardness testers are also used to determine the hardness of grains in the microstructure of the material.

Hardness tests are made with a diamond penetrator. The penetrator is pressed into the specimen with extremely light loads of from 25 to 3,600 grams (0.7 oz. to 7.9 lb). Minute impressions are produced under appropriate low loads.

The microscope feature of the tester permits measuring the fine impression. The depth of penetration is determined by the applied load and the hardness of the material. A number value is assigned to each degree of hardness.

KNOOP HARDNESS TESTING

The Knoop hardness test is adapted to the same hard materials and thin parts as the Microhardness tester. The load range in grams is also from 25 to 3,600 grams (0.7 oz. to 7.9 lb). The plane surfaces must be free from scratches.

A *Knoop indentor* is used. The indentor is a diamond in an elongated pyramid form. The indentor acts under a fixed load that is applied for a definite period of time to produce an indentation that has long and short diagonals. The Knoop indentor is used in a fully automatic, electronically controlled Tukon tester.

Knoop hardness numbers are used. The numbers are equal to the load in kilograms divided by the area of the indentation in square millimeters. Tables are available that give the indentation number corresponding to the long diagonal and for a given load.

Safe Practices for Hardness Testing

- Select the size and style of the penetrator and load which are appropriate to the material and condition of the part to be tested.
- Examine the part for any possible fractures caused by heat treatment.
- Check for hardness values at at least three different places on the workpiece. The *average* represents the hardness number.
- Use an adjustable jack for overhanging parts and a tube support for long, thin, pieces. The surface area must rest solidly on the hardness tester anvil to obtain an accurate reading and to prevent damage to the penetrator.

UNIT 62   REVIEW AND SELF-TEST

1. a. Harden and temper the SAE 1041 Aligning Block shown in the illustration. The part is to be drawn to a Bhn 243 hardness. Consult handbook tables to establish the (1) quenching medium, (2) hardening and tempering temperatures, and (3) soaking time.
   b. Test the block after the surfaces are finish ground to check the Bhn 243 hardness. The hardness may be tested on any hardness tester available. Values other than Bhn are to be checked against the required Bhn 243 hardness.
   c. List three personal safety and/or precaution measures to take in carrying out the heat-treating operations.

   Note: Another part may be used or the processes simulated by preparing a step-by-step production plan for the aligning block or the new part.

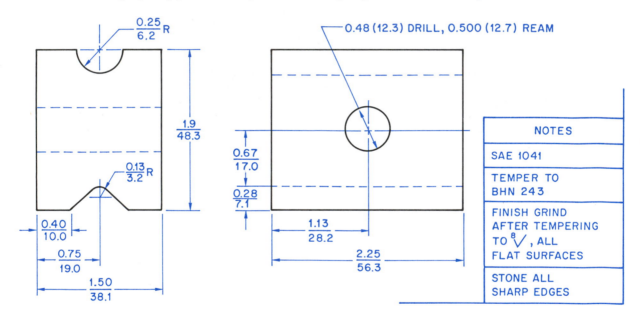

2. a. Make a second hardness test of the Aligning Block shown in the illustration. (A Rockwell hardness test may be substituted, if this process was not used previously.)
   b. Compare the Bhn scale value in test item 1 with the R_C scale reading.
   c. Check the accuracy of the hardness readings against a handbook table.
   d. Establish the comparable scleroscope and Vickers hardness numbers from the table.

3. List three precautions to take to avoid common problems such as soft spots, uneven hardening, and fracturing.

PART 15 New Manufacturing Processes and Machine Tools

SECTION ONE

Nontraditional Tooling and Machines: Technology and Processes

UNIT 63

Nontraditional Machine Tools and Machining Processes

CLASSIFICATION OF NONTRADITIONAL MACHINING PROCESSES AND MACHINES

Nontraditional machining and machine tools, which are currently being used in production, are generally classified within four energy processes groups. The commercially economical nontraditional machining processes, with identifying code letters (acronyms in common industrial use), appear in Table 63–1. The processes are grouped under (1) mechanical energy, (2) electrical energy, (3) thermal energy, and (4) chemical energy machining.

Table 63–1 Nontraditional Machine Tools and (Energy) Machining Processes (with Acronyms)

Mechanical Energy Machining		Electrical Energy Machining		Thermal Energy Machining		Chemical Energy Machining	
AJM*	Abrasive Jet	ECD*	Electrochemical Deburring	EBM*	Electron Beam	CHM*	Chemical
USM*	Ultrasonic	ECG*	Electrochemical Grinding	EDM*	Electrical Discharge Machining	TCM	Thermochemical
AFM	Abrasive Flow	ECM*	Electrochemical Machining	LBM*	Laser Beam	PCM	Photochemical
HDM	Hydrodynamic	ECDG	Electrochemical Discharge Grinding	PBM*	Plasma Beam	ELP	Electropolishing
LSG	Low Stress Grinding	ECH	Electrochemical Honing	EDG	Electrical Discharge Grinding		
		ECP	Electrochemical Polishing	EDS	Electrical Discharge Sawing		
		ECT	Electrochemical Turning	EDWC	Electrical Discharge Wire Cutting		
		ES	Electro-stream	LBT	Laser Beam Torch		
		STEM	Shaped Tube Electrolytic				

*Identifies machines and processes described in this unit

559

Table 63-2 Surface Finish Ranges for Nontraditional
Material Removal Processes

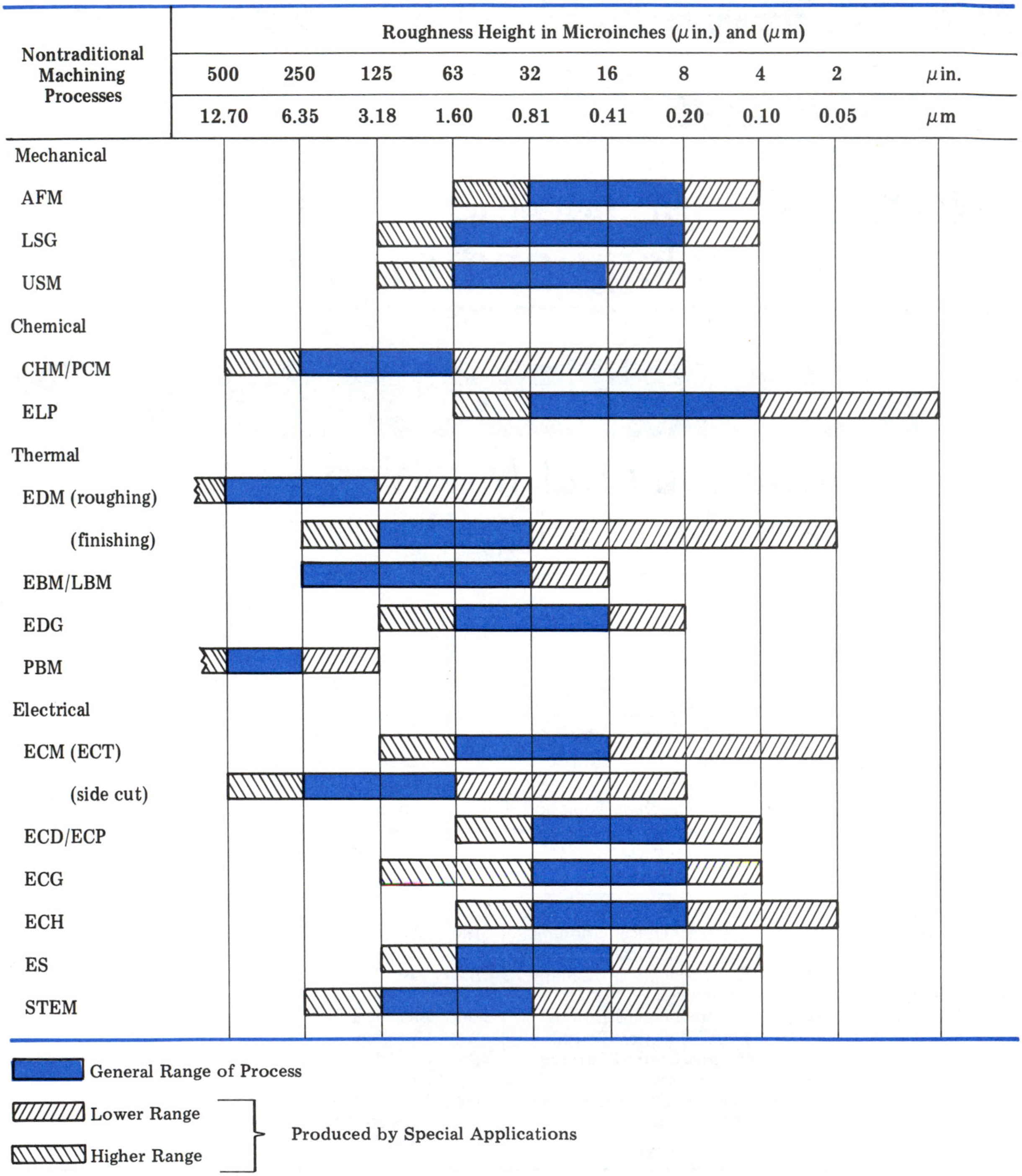

Nontraditional Machining Processes	Roughness Height in Microinches (µ in.) and (µm)									
	500	250	125	63	32	16	8	4	2	µ in.
	12.70	6.35	3.18	1.60	0.81	0.41	0.20	0.10	0.05	µm

Mechanical — AFM, LSG, USM; Chemical — CHM/PCM, ELP; Thermal — EDM (roughing), (finishing), EBM/LBM, EDG, PBM; Electrical — ECM (ECT), (side cut), ECD/ECP, ECG, ECH, ES, STEM

Legend:
■ General Range of Process
▨ Lower Range ⎫
▧ Higher Range ⎬ Produced by Special Applications

Table 63-3 Examples of Metal Removal Rates for Conventional and Nontraditional Machining Processes

Machining Processes	Maximum Metal Removal Rate (in./min.)	Cutting Speed (ft/min, fpm)	Penetration Rate (in/min, ipm)	Machining Accuracy		Typical Horsepower (machine input)
				Finest Accuracy Attainable	At Maximum Metal Removal Rate	
				(inch values)		
Conventional						
Turning	200	250	. . .	0.0001	0.005	30
Grinding	50	10	. . .	0.0001	0.003	25
Nontraditional						
EDM	0.3	. . .	0.5	0.0005	0.005	15
ECM	1.0	. . .	0.5	0.0005	0.005	200
LBM	0.0003	. . .	4.0	0.0005	0.005	20
CHM	30	. . .	0.001	0.0005	0.002	. . .
USM	0.05	. . .	0.02	0.0002	0.001	15
EBM	0.0005	200	6	0.0002	0.001	10
PBM	10	50	10	0.01	0.1	200

Principles, processes, and applications of representative machines are provided in this unit. The machines and processes are indicated by an asterisk (*) in Table 63-1. Many other nontraditional processes have been invented. However, these new processes have yet to emerge as acceptable production processes. Examples of the emerging processes include: Water Jet Machining (WJM, mechanical), Electrovapor Machining (EVM, electrical), High Energy Rate Additive (HERA, thermal), and Electro Gel Machining (EGM, chemical).

SURFACE INTEGRITY: NONTRADITIONAL PROCESSES

The term *surface integrity* relates to the description, control, and alterations in the product or process. Surface integrity in nontraditional machining depends largely upon the *intensity level* during processing. Surface integrity relates, also, to distortion or change in the surface properties of the material being machined. Tables of surface integrity for the four groups of nontraditional machining processes provide recorded data for different types of effects. The effects relate to *surface roughness, mechanical alterations* (such as hardness alteration, cracks, and residual stress), *metallurgical alterations* (such as recrystallization and intergranular attack); and *high cycle fatigue.*

TYPICAL SURFACE ROUGHNESS RANGES

Typical microinch (μin.) and micrometer (μm) ranges of surface roughness for some of the nontraditional machining processes in each of the four process groups are presented in Table 63-2. These may be compared with surface finish ranges produced by conventional machining as given in Table A-23 in the Appendix.

METAL REMOVAL RATES

Table 63-3 provides technical data for turning and grinding by conventional and for other nontraditional machining processes. Note the tremendous variation from accuracies of 0.01" (0.25mm) for plastic beam machining (PBM) as contrasted with accuracies within the 0.0002" to 0.0005" (0.05mm to 0.0125mm) range for other manufacturing processes.

NONTRADITIONAL MACHINING PROCESSES AND MACHINES

ELECTRICAL DISCHARGE MACHINING (EDM)

The *electrical discharge machine* and process were industrially accepted and marketed beginning in 1946. With an electrical discharge machine, holes and other straight, contour, and embossed surfaces are produced without using a movable

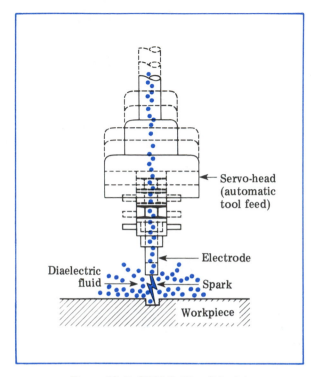

Figure 63–1 EDM Drilling Principle

tool and without tool contact with the workpiece. *Electrical discharge machining* (EDM) uses electrical energy to remove metal of any hardness. Holes and surfaces that have an irregular and intricate contour may be produced. There is no friction-produced heat, so there is no warpage. Tolerances of ±0.002″ (±0.05mm) are practical.

The cutting tool, called an *electrode*, is made of either brass, copper, graphite, or alloys of copper or silver. The tool is hollow to permit pumping a *dielectric fluid* through it. The dielectric fluid covers the workpiece during machining. It helps flush the discharge (chips) and acts as a coolant and as the electrical conductor. The principle of EDM drilling is illustrated in Figure 63–1.

In EDM a good quality surface finish is produced. EDM is especially valuable in producing holes and profiled surfaces where other processes may distort the material and part. EDM is used in machining a cluster of holes and sections having a thin wall thickness, like 0.005″, between the holes. The holes may be through holes or blind holes. Materials such as carbides

and stainless steels, as well as softer and non-ferrous metals, may also be worked by EDM.

Electrical energy is required in EDM. The tool (electrode) is positioned so that there is a gap between it and the work surface. The gap is filled with dielectric fluid. A high-frequency, pulsating electric current creates sparks that jump the gap. The bombardment vaporizes the material under the tool and produces the desired hole or formed surface. The tool is partly consumed during the process.

There are variations of EDM in which there is a constant reversal of polarity that reduces wear on the tool and produces a higher removal rate. However, because the surface finish produced is rougher, these variations are used for roughing operations.

METAL REMOVING (CUTTING) RATES AND SURFACE FINISH

Since EDM metal removing rates are less than conventional machining processes, as much material as possible is removed by conventional tooling. Three conditions which affect EDM cutting rates and surface finish are: *arcing frequency*, the *amperage setting*, and the properties and condition of the *dielectric coolant*. The lower the amperage and the lower the arcing frequency, the slower the cutting rate.

As the amperage (machining current) is increased for a specific discharge (arcing) frequency, there is an increase in the metal removal rate and in the quality of the surface finish. Metal removal rates are expressed in terms of cubic inches (or cubic centimeters) per hour. For example, using a graphite electrode on steel, under ideal conditions 0.05 in³ (0.32 cm³) is removed per hour per ampere. A power supply of 50 amperes (T–50) produces a metal removal rate of 0.05 × 50 or 2.5 in³ (16 cm³) per hour.

Cutting rate, surface finish, and electrode wear life are also affected by the electrode material, the dielectric fluid, the fast removal of metal chips from the working gap, the filtration of the contaminates, and the continuous removal of heat from the dielectric fluid. Common dielectric mediums include silicone and other low-viscosity oils, deionized water, and certain gases.

A. Standard Electrode Holder

B. Square Holder for Electrodes

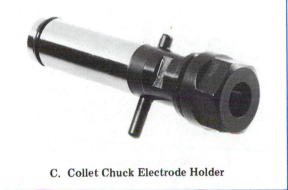

C. Collet Chuck Electrode Holder

Figure 63–2 Three General Types of Electrode Holders

COMMON EDM ACCESSORIES

EDM digital readouts permit inch and metric conversion measurements. These allow for the positioning of X–Y axes of the machine table to within 0.0005″ (0.01mm).

The *electrode holder* is another basic accessory. Three general types are illustrated in Figure 63–2. The mounting end of electrode holder (A) is hardened. The electrode blank is pressed, screwed, soldered, or glued to the unhardened end.

The square holder (B) provides reference surfaces and movable clamping studs to mount electrodes of varying sizes. The collet chuck at (C) is used for holding round electrodes. Both the square holder and the collet chuck may be used with an hydraulic quick-change holder. Collet sets are available to cover a range of electrode sizes: for example, 0.039″ to 0.511″ (1mm to 13mm), or 0.019″ to 0.393″ (0.5mm to 10mm).

There is a through hole in each type of holder, a set of seals, and a clamping key. The hole provides passage for the liquid flushing of the electrode.

TRAVELING WIRE (WIRE-CUT) EDM SYSTEM

A second type of EDM general-purpose machine uses a *traveling wire* as an electrode. Extremely accurate internal or external profiles may be produced vertically or at an angle in metals or other materials of any degree of hardness.

Wire EDM machines may be numerically controlled by manual programming or from computer assisted programming (CAP). CAP is used for NC processes involving long or complicated mathematical computations.

CRT, MDI, AND CNC SYSTEM

A number of machine tool and control panel features are illustrated in Figure 63–3. This model has a *CRT display*, *Manual Display Input (MDI) panel*, and *CNC system*.

Part programs, wire position, power settings, simultaneous four-axis positions, and other operator-needed information may be displayed on the CRT. The MDI panel makes it possible

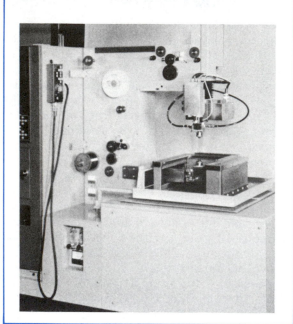

Figure 63–3 High-Speed Wire-Cut Electrical Discharge Machine (EDM) with 4 Simultaneous Controlled Axes and Taper Cutting System

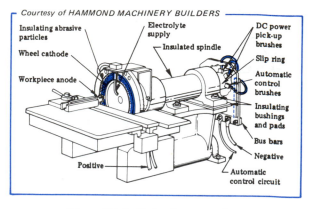

Figure 63–4 Major Features of an ECG Machine

to input additional program information during operation or to edit data. The CNC system accepts input from NC tape, memory cassette, or directly from a programming computer.

Setup time is reduced by high-speed dry run capacity. The programmed path may be checked against machining errors at a rapid advance rate up to 14″ per minute. CNC control permits three-axis contouring, linear and circular interpolation, absolute and incremental programming, Z axis zero return, edge finding, and other often used functions.

ELECTROCHEMICAL MACHINING (ECM)

There are many similarities between *electrochemical machining* (ECM) and EDM. The workpiece may be of any metal capable of conducting electricity. A tool made of material similar to the material of a tool used for EDM is required. The tool acts as a *cathode* (negative terminal). The workpiece serves as an *anode* (positive terminal). An electrolyte is passed through the gap (from 0.001″ to 0.030″ or

0.025mm to 0.76mm) between the tool and the work at a high rate of flow (velocity). The metal is dissolved and removed by using direct current (DC). A low voltage (30 volts) and high current (2,000 amperes per square inch) are required.

ECM, however, is a faster metal-removing process than EDM. Additional operations, such as etching, deburring, and face milling, may also be performed by ECM.

In electrochemical machining (ECM) metal particles from the workpiece are dissolved into the electrolyte by chemical reaction. The particles are flowed away and are filtered out. Electrolyte temperature affects the electrical conductivity, machining rate, and machining accuracy. The temperature is controlled as it passes through a heat exchanger.

Electrodes (tools) are machined undersize to provide a gap or clearance between a final workpiece dimension and the electrode size which is affected by the *particle size*. The gap or clearance is also known as *overcut*.

Electrode Materials and Cutting Rates. Brass, copper, stainless steel (type 316), titanium, and copper alloys are generally used for electrodes. The most commonly used electrolytes consist of water solutions of sodium chloride, potassium chloride, sodium nitrate, and sodium hydroxide. ECM cutting rates range from 0.10 in^3 to 0.27 in^3 (0.65 cm^3 to 1.74 cm^3) per minute per 1,000 amperes. Importantly, no burrs are produced. The range of surface finishes is from 4 μ in. to 30 μ in. (0.1 μ m to 0.75 μ m).

ELECTROCHEMICAL DEBURRING (ECD)

Burrs and sharp edges produced by other machining processes may be removed by electrochemical deburring. The ECD machine is an adaptation of ECM. The workpiece is the anode. The tooling (within an insulated coating) forms the cathode. The electrolyte flows through the gap between the tool and the workpiece. The burrs are removed by electrochemical reaction.

ELECTROCHEMICAL GRINDING (ECG)

Electrochemical grinding (ECG) requires an ECM machine, a grinding head, and conventional grinding wheels. These and other major features of an ECG machine are shown in the line drawing (Figure 63–4).

ECG grinding wheels are made with an electrically-conductive bond. The wheel is the cathode; the work is the anode. Electrical current passes between the work and the revolving grinding wheel. An electrochemical reaction is produced for removing metal. The stock removal rate on a standard ECM tool grinder ranges from 0.030″ to 0.060″ per minute (0.75mm to 1.50mm/min).

The following are major advantages of ECG over conventional grinding.

- Over 75% more metal may be removed in the same period of time.
- Wheel dressing and wheel wear are drastically reduced.
- There is limited loss in dimensional accuracy or the quality of surface finish.
- Cutting edges are smooth and burr free.
- Carbide and other extremely hard alloy cutting tools are ground with greater efficiency and at lower cost, resulting from single-pass grinding.
- Flat surfaces of alloy steel parts may be ground at exceedingly high production rates.

CHEMICAL MACHINING (CHM)

Parts may be *blanked* from thin sheet metals and heavier materials may be *milled* to shape by chemically dissolving unwanted metal. The solutions used for dissolving are generally strongly acid or alkaline

CHEMICAL BLANKING

Chemical blanking is widely used for producing intricately formed, thin sheet metal parts and for the manufacture of printed circuits. Enlarged drawings are made of parts that are to be chemically blanked. The drawing is then reduced to size photographically. The negative produced is used to make a contact printing on the surface of the workpiece which has been given an acid-resisting, photo-sensitive coating.

The light-exposed unwanted portions are dissolved away when the workpiece is placed in an etching solution. The remaining portions form the required blanked workpiece. Less accurate parts are *masked* by using either a silk screen or offset printing process to apply the acid-resistant coatings.

CHEMICAL MILLING

This process is very efficient on very large workpieces which require the removal of considerable amounts of metal, particularly from intricate or complex surfaces. A number of parts may be machined at one time. Delicate parts may be machined without distortion or loss of strength since no cutting forces are involved.

Chemical milling begins when a metal part is thoroughly cleaned in preparation for spraying or flowing on a masking of vinyl plastic or neoprene rubber. The coating is baked to cure it.

Templates are then used to scribe areas from which the masking is to be removed. The remaining coated metal part is then submerged in a chemical solution to dissolve unwanted metal. At that point, the part is removed, rinsed in water, and *stripped (demasked)* to remove the coating.

ULTRASONIC MACHINING (USM)

Ultrasonic machining (USM) is an *impact grinding* process which requires a high-frequency, low-amplitude vibrating tool head; a ductile, tough material formed tool secured to a tool holder; and an *abrasive slurry*. USM is adaptable to machining tough or hard metal, carbide, or nonconducting ceramic, glass, diamond, and less hard gem stone parts. USM is also used for machining plastics and other soft materials.

The USM process is adapted to blanking uniform or irregular sections in thin, hard conducting and nonconducting metals. The process is also used for forming shallow cavities of simple or intricate shapes.

As an impact grinding process, aluminum oxide, silicon carbide, and boron carbide abrasive grains are suspended in a liquid (usually water) slurry. The slurry is pumped into the gap between the formed tool and the workpiece. The tool vibrates a few thousandths of an inch within a 19,000 to 25,000 Hz range of ultrasonic frequencies. Grinding takes place by the action of the abrasive grains impacting at high velocity on the workpiece as the area immediately under the tool is ground to the shape of the tool.

Holes up to 2″ diameter and similar cross-sectional area may be ultrasonically machined, burr free. Machining tolerances may be held within a ±0.001″ to ±0.00025″ (0.02mm to 0.006mm) range. Finer abrasive grit sizes are used to obtain the closer tolerances.

Similarly, the use of finer abrasive grains produces a finer surface finish. For example, an 800 flour size is used to produce a 7 μin. (0.18 μm) surface finish, while a 100 abrasive grit size produces a coarser 30 μin. (0.75 μm) finish.

PLASMA BEAM (ARC) MACHINING (PBM)

Plasma beam (arc) machining (PBM) is primarily a metal cutting process. A high-velocity jet of a super-heated stream of electronically ionized gas is used to form holes and to make other profile cuts in metals. The PBM process is especially adapted to cutting operations on metals which may not be performed efficiently with oxyacetylene equipment. Stainless steels, aluminum alloys, magnesium products, copper-nickel alloys, carbon steels, and other ferrous metals are economically cut by PBM. Plasma beam cutting equipment is displayed in Figure 63–5. Metal thicknesses up to 5″ (127mm) in stainless steel, 6″ (152mm) in aluminum, and 8″ in mild steel, are easily cut by PBM.

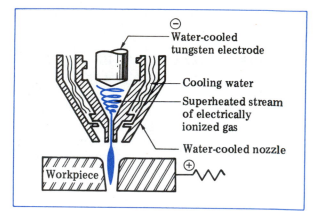

Figure 63–5 Features of Plasma Beam Machining Equipment

The term *plasma* is derived from the extremely hot jet of ionized gas which is produced by a water-cooled plasma arc cutting torch. Cutting temperatures are produced by heating a gas with an electric arc inside a plasma arc torch.

ABRASIVE JET MACHINING (AJM)

Abrasive jet machining (AJM) depends on the cutting force of fine aluminum oxide or silicon carbide abrasive particles. These particles are propelled in a high velocity stream of air to serve as a cutting tool. Accurate AJM cuts as narrow as 0.005″ wide may be made.

AJM is an effective cutting method for internal and external deburring and for cutting ceramic, glass, and other hard semiconductor materials.

AJM and PBM machines and processes make it possible to:

- Produce intricate shapes that cannot be economically machined on other regular or nontraditional machines,
- Penetrate hard materials that are difficult to machine,
- Machine holes and other surfaces to high degrees of dimensional accuracy and quality of surface finish.

ELECTRON BEAM MACHINING (EBM)

Electron beam machining (EBM) is used to cut diamonds, carbides, ceramic oxide, and other

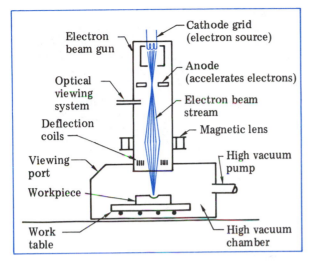

Figure 63-6 Schematic of Major Components in Electron Beam Machining (EBM)

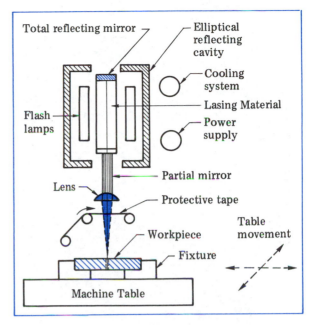

Figure 63-7 Pulsed (Intermittent Bursts) Operation of Solid State Lasers

extremely hard materials. The principles of operation and the major components are pictured in Figure 63-6.

Electron rays from a *cathode (electron) source* are directed by an *electron beam gun*. The electrons are accelerated through an anode to form an electron beam. As the high-speed electron beam passes through a *focusing coil*, electron energy is transformed into an intense heat energy. This energy (focused on the workpiece) is capable of vaporizing and removing the material before it.

Cutting energy is increased by enclosing the workpiece in a high vacuum chamber. The chamber makes it possible to direct full-energy power electron beams to precise locations. *Magnetic deflection coils* provide for movement of the electron beam (heat energy beam) to generate different cutting patterns. Thus, straight line ╱, circle ◯, square ◈, circle-on-circle ◉, and hole offset ⁝⁝ wave forms may be generated by moving the electron beam off-center. Holes and slots that are larger than a few thousandths of an inch in diameter or width; multiple, closely-spaced holes; and various width slots may be cut from one position. Other longer cuts may be taken by rotating the workpiece or moving the X–Y axes positions of the table.

The disadvantages of the system include: high cost of equipment, slow cutting and machining rates, limitations in terms of workpiece capacity, and the need for extensive machine guarding against harmful x-rays. The same principles and type of equipment (except for increased power requirements) are employed in *electron beam welding (EBW)*.

LASER BEAM MACHINING (LBM)

The initial source of energy for *laser beam machining* is a low-intensity light beam. The power density of the light beam is amplified by an *optical pump* to become a *high-intensity laser beam*. This beam is focused through a lens system to an exceedingly small area to produce a high level of heat energy capable of vaporizing any material.

Solid Lasers. Solid lasers produce intermittent bursts of high-intensity power (Figure 63-7). The general-purpose system is used in production work to produce small holes, cut out sections of thin materials, and for spot welding and heat treating processes. Tolerances for hole sizes within ±0.0001″ (0.0025mm) are possible.

Figure 63–8 Major Components of a Multi-Purpose Machining, Cutting, Welding, and Heat-Treating CO_2 Laser System

LBM with solid lasers require the use of a *doped* (neodymium) *glass rod* as the laser medium. The ends of the glass laser rod are formed as optical surfaces having a *reflective coating*. On one end there is a partial reflective coating. This permits a laser beam to escape when the required high intensity is reached. Laser beams may be directed for straight or angle hole drilling up to 0.050″ (1.25mm) diameter on flat, round, or irregular surfaces, and difficult-to-reach internal areas.

Gas Lasers. One advantage of the gas laser system is its capability to produce a continuous laser beam. Gas lasers require a gas such as carbon dioxide (CO_2) to convert electrical energy into laser energy and to maintain the continuous beam. The energy level of gas lasers is further increased by using oxygen to boost the system.

Figure 63–8 shows the major components of a CO_2 laser focusing optic unit with safety enclosure, a CNC panel, and a CAD/CAM interface system. Table movements are along X, Y, and Z axes. A rotary table accessory permits taper and circular machining.

Mild steel plates 0.400″ (10mm) thick are cut at the rate of 30 ipm (inches per minute) or 12 cm/min. Inconel parts 0.060″ (1.5mm) thick may be butt welded with full penetration at 300 ipm (7.6 meters per minute).

Gas lasers are used for marking, embossing, welding, cutting, and heat treating processes. These are in addition to hole forming and other cutting operations.

Safe Practices for Nontraditional Machines and Processes

- Check to see that an operable class B and C fire extinguisher is readily available near EDM equipment in the unlikely event that a gas flame is ignited at the arc gap.
- Shut down the EDM power supply and wait until the voltmeter reads zero before handling the workpiece, tool holder, or electrode. Electrical shock hazards are associated with all EDM.
- Place the plexiglass guard in position to cover the entire EDM work area.
- Wear a face shield or mask to prevent inhaling dusts, mists, and vapors of chemical compounds which may be released by chemical reaction in ECM.
- Make sure the work enclosure is vented away from the work area to remove any gas produced in ECM.
- Use protective clothing when handling any combustible material which may be injurious to body tissues.
- Wear a full-face shield, acid-resisting apron and clothing, and rubber gloves and boots as protection against personal injury when working with acids and other CHM etching solutions.
- Check for the removal of toxic fumes in work areas around CHM equipment and the availability of an emergency shower.

- Check to see that all x-ray shielding panels are securely in place before using EBM equipment.
- Observe all personal, machine, and area safe operating directions for nontraditional equipment which may generate harmful rays and noise levels.

- Wear laser-proof protective shield or safety glasses for eye protection around LBM systems.
- Check that all laser beam paths are enclosed with safety guards which are secured before the system is turned on.

UNIT 63 REVIEW AND SELF-TEST

1. State two design features or properties of regular cutting tools which are not present in nontraditional machining processes.

2. List three materials which may be more readily machined by nontraditional systems than by conventional practices.

3. Identify two functions of digital readouts on an electrical discharge machine (EDM).

4. State three conditions which govern the cutting rate and surface finish in EDM.

5. List three advantages of electrochemical grinding (ECG) of cutting tools in comparison with conventional tool grinding.

6. a. Give a brief explanation of chemical machining (CHM).
   b. Differentiate between the blanking and the milling CHM processes.

7. a. Describe ultrasonic machining (USM).
   b. Determine the effect the use of coarser abrasive grains has on cutting rate and surface quality.

8. List four metals that may be more efficiently cut by plasma beam machining (PBM) than with other gas-cutting equipment.

9. a. Explain briefly the process of abrasive jet machining (AJM).
   b. Give two examples of AJM.

10. Describe how an electron beam in EBM is moved to (a) generate different cutting patterns and (b) how longer cuts may be taken.

11. a. Name two basic systems of lasers used in laser beam machining (LBM).
    b. Identify general applications of each laser system.
    c. State the principles of operation of each system.

12. List one safety precaution which must be followed in operating nontraditional equipment which require: (a) the direction and control of high energy beams (like EBM and LBM); (b) the application of electrical and chemical energy (as for ECM and CHM); and (c), the use of abrasive particles (as for AJM and USM).

PART 16

Flexible Manufacturing Systems (FMS) and Robotics

Automated Flexible Multimachine Manufacturing Systems

A computer-controlled, random-order, flexible manufacturing system (FMS) provides the capability to produce a specified range of parts or components. FMS also represents an advanced stage in the development of a completely automated factory. FMS requires a computer-integrated group of multiple machine tools and/or machining centers, work and storage stations, and an automated handling system. FMS involves subsystems of computer-aided design and drafting (CADD), computer-integrated manufacturing (CIM), variable-computer integrated manufacturing control (CIM/GEN), and others.

Subsystems in FMS are further interlocked with at least one robotic system. Robot functions relate to work handling and tool transfer, sensing, machining and tooling productivity, inspection, gaging, quality control, and others.

This concluding section provides the capstone to all preceding sections and units. Part A of this unit covers functions of automated manufacturing subsystems; Part B, robotics.

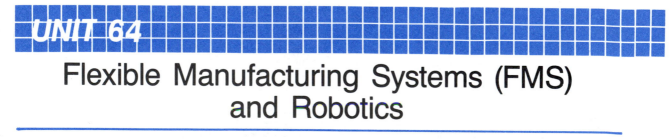

UNIT 64

Flexible Manufacturing Systems (FMS) and Robotics

Computerized subsystems and robots, when fully integrated into conventional and/or traditional CNC and DNC machine tools and machining centers, represent an accelerating movement toward totally automated flexible manufacturing systems.

CAD/CADD, CIM/CAM, AND CIM/GEN COMPONENTS

Descriptions of a few additional components commonly associated with computer-integrated manufacturing follow.

CAD/CADD. CAD refers to automated drafting. Drawings that normally would be prepared by a draftsperson manually are automatically programmed and are graphically displayed. When the CAD system has the added capacity to analyze and to design parts and components

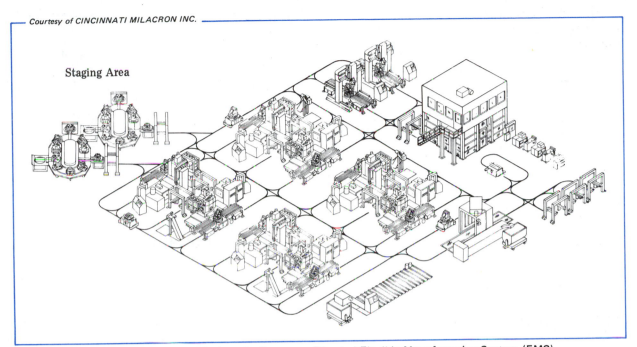

Staging Area

Figure 64-1 Graphic Display of a Computer-Directed Flexible Manufacturing System (FMS)

(using computer graphics to perform such engineering functions as calculations and design analyses), the system is referred to as computer-aided design and drafting (CADD). In CADD, coordinates are calculated and instructions are generated to direct NC operations.

CIM/CAM. This computer-graphics system creates the geometrics of a part, machining data, and other manufacturing input for multiple-axis machining on CAM controlled machine tools (or DNC to remote machine tools). Representative manufacturing processes performed on machine tool units which operate interactively include: milling, turning, punching, nontraditional machining, form cutting, and others.

CIM/GEN. These code letters relate to computer-integrated manufacturing (CIM) with a variable computer-integrated manufacturing control system (GEN). GEN operates on the selection of software routines to build and to control the system. Some modules include manufacturing specifications, inspection and quality control requirements, and inventory control. Other system modules relate to tooling and machine setups, planning material requirements, measurement standards setting, NC programming, production control, and shop floor control.

A. SUBSYSTEMS OF HIGHLY-AUTOMATED MANUFACTURING SYSTEMS

GRAPHIC DISPLAY OF A FLEXIBLE MANUFACTURING SYSTEM

Figure 64-1 visually displays a computer-directed flexible manufacturing system (FMS). This particular system consists of eight machining centers. Each machining center is equipped with a 90 tool-storage matrix to accommodate the range of parts which are to be machined and to automatically replace worn or damaged tools. Also, there are two coordinate measuring machines, one parts cleaning machine, and an automatic chip-removal system.

Workpieces and tools, mounted on pallet carrousels, are transported throughout the system under computer control. Additional carrousels are shown being loaded in the *staging area*. These carrousels are used for continuous parts processing, particularly during unmanned machining periods.

FUNCTIONS OF AUTOMATED SUBSYSTEMS IN FLEXIBLE MANUFACTURING SYSTEMS (FMS)

FMS requires a series of automated subsystems which serve the following functions.

- Complete control of the manufacturing system by the host computer.
- Handling and transporting materials.
- Positioning, adjusting, and changing workpieces.
- Selecting, changing, and handling tooling.
- Continuous cycling of each individual machine.
- In-process and post-process gaging.
- Controlling speeds and feeds.
- Precision multidirectional surface sensing.
- Washing and cleaning workpieces.
- Disposing of chips automatically.

AUTOMATED MATERIALS-HANDLING SUBSYSTEM

This subsystem deals with the storing, control, and retrieval of materials and work that (1) is not actually being processed, (2) finished parts, and (3) work in process. Materials handling is integrated into FMS. The design and development of the subsystem requires a consideration of the following characteristics of materials handling.

- The controlled path or travel of the workpiece.
- The transfer of workpieces from a conveyor line into a precisely piloted position for machining.
- The level(s) for placing machine tools in relation to materials conveyor shuttle levels.
- The need for fixtures and other nesting/holding devices for processing multiple workpieces.
- The advantages and/or problems associated with transporting parts by free or power-driven conveyors, shuttle cars, or other automatically-guided devices.
- Transporter traffic control to establish the number of transporters needed in order to minimize or eliminate machine time lost in waiting for tools or workpieces.

AUTOMATED TOOL-HANDLING SYSTEMS

Manual and computer-assisted loading and unloading of cutting tools and assemblies were described in earlier units on NC and CNC machine tools. In FMS applications, there is the added provision for the withdrawal and replacement of quick-change, cartridge-type, tool drums by a robotized subsystem.

AUTOMATED IN-PROCESS AND POST-PROCESS GAGING

In-process gaging refers to the taking of dimensional and form measurements and parts inspections for a limited number of part features *prior to the removal* of the workpiece. In-process gaging is important in correcting surface problems, maintaining form, and controlling dimensions within allowable tolerances.

New tools are brought up by command of the machine control unit (MCU) to remachine oversize surfaces. Undersize workpieces are rejected and the MCU signals a malfunction.

In-process gaging is carried on by linking a computer-controlled coordinate measuring machine (CMM) program into a DNC system (which remotely controls a group of NC machines). The in-process gaging of intricately detailed parts is exceedingly difficult and complex to achieve.

Post-process gaging, by contrast, does permit checking *all part details*. However, there are two major disadvantages. First, compensation for out-of-tolerance machining takes place *after* the part has been machined. Secondly, the part cannot be returned to the machine in FMS to be reworked.

AUTOMATED PRECISION SURFACE SENSING PROBES

A *sensing probe* is a multidirectional electronic switching device. When applied to a machining center, the sensing probe has the following components.

- A *body* fitted with a standard tapered shank to fit the machine spindle nose and a tool-storage matrix.
- An *interchangeable stylus* held in the body.

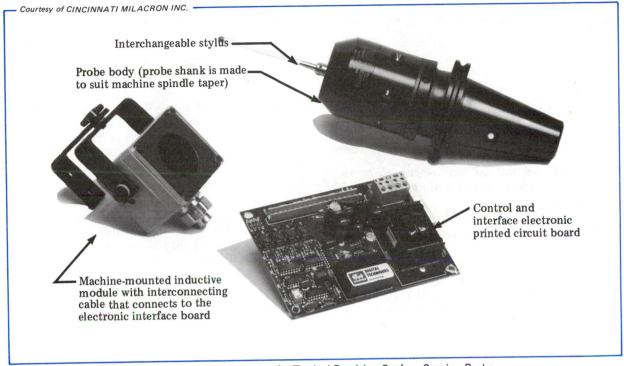

Courtesy of CINCINNATI MILACRON INC.

Interchangeable stylus

Probe body (probe shank is made to suit machine spindle taper)

Control and interface electronic printed circuit board

Machine-mounted inductive module with interconnecting cable that connects to the electronic interface board

Figure 64–2 Components of a Typical Precision Surface Sensing Probe

- A *probe-mounted inductive module* on the machine spindle connected to a *control circuit board* mounted in another location.

COMPONENTS OF A SURFACE SENSING PROBE

The major components of a precision surface sensing probe, generally used on machining centers, are shown in Figure 64–2. The probe is held in the storage matrix. After the workpiece is machined, the sensing probe is automatically transferred from the storage matrix onto the machine spindle (being interchangeable with the last used cutting tool). Any deflection of the probe stylus instantaneously records the surface location of a defect in machining or variation from an allowable tolerance.

Surface sensing probes serve these functions:

- Measuring and checking a particular surface,
- Detecting and compensating for size or form variations,

- Making in-process alignments, and
- Sensing any drift (moving away from a normal condition in machining) which may produce out-of-tolerance parts.

AUTOMATED FEED, SPEED, AND/OR ADAPTATIVE CONTROL SUBSYSTEMS

Adaptive control provides feed-back signals to a machine control unit (MCU) for purposes of adjusting feeds and speeds. Sensory circuits and computations are used to compare *cutting torque* with specific cutting torque limits. Limits are based on prevailing tooling, material, and machine tool operations.

The adaptive control unit senses such factors as torsional tool overload, cutter wear, a broken or damaged cutting tool, surface irregularities, and dimensional variations. Feed and speed rates are adjusted upward or downward automatically as machining conditions require.

Under these conditions, tool life is prolonged and there is maximum machining productivity.

FUNCTIONS OF ROBOTS

Robots are used to perform the following tasks in single- or multiple-machine manufacturing.

- Load and unload workpieces on or from transporter or work-positioning devices, like fixtures.
- Transfer workpieces from a conveyor at one or more levels to another processing level.
- Load and unload cutting tools for purposes of changing or replacing them in tool storage drums or matrices on machine tools.
- Perform programmed manufacturing tasks (for example, arc welding processes).

APPLICATION OF COMPUTER-CONTROLLED SIX-AXIS ROBOT

Figure 64–3 provides an application of a computer-controlled, six-axis robot in a *manufacturing cell.* Workpieces are brought to location on a transport car. From the car, the fingers of the robot hand are designed to transfer the material for a part to a preset location on a CNC lathe. Once machined, the finished part is removed and replaced.

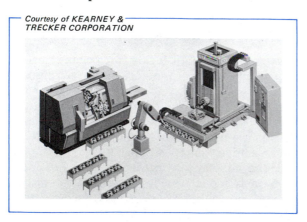

Courtesy of KEARNEY & TRECKER CORPORATION

Figure 64–3 Application of a Computer-Controlled Six-Axis Robot

FACTORS AFFECTING THE SELECTION OF AN INDUSTRIAL ROBOT

Stand-apart robots (Figure 64–3) serve multimachine functions. Major specifications used in selecting an appropriate robot follow.

Degrees of Freedom. Each robot is identified by a number of *degrees of freedom.* This means, for example, that a general-purpose robot with *six degrees of freedom has six axes of motion.*

Operating Range and Reach. The maximum reach and operating range for a *three-dimensional outer envelope.* The reach of this envelope is measured from the tool center point. Manufacturers also specify an *inner envelope.* It is within this inner envelope that the robot can handle a maximum load, move it at maximum speed, and position a workpiece with the greatest accuracy.

Load Capacity. The combined weight of the grippers and workpiece are identified as the load capacity of a robot.

Speed of Programming. This factor relates to the capability of the robot to be flexibly programmed for operating speeds. The robot operation may be controlled within an infinitely-variable speed range in which operating speeds are automatically increased or decreased.

Loading and Unloading Design Features. Design considerations are based on safe-carrying capacity, maximum movements, clearance of the arm and workpiece, maximum reach, positioning repeatability, and compliance. *Compliance* relates to the security with which the robot hand accommodates irregularities in the workpiece. Equally important is the fact that the workpiece is accurately and safely guided and piloted into position for machining, assembling, or other work processes.

Robot Hand (End Effector). Generally, a robot hand has two sets of *grippers (fingers).* One set is designed to accommodate an unmachined workpiece and to load it into a machine. After machining, the robot hand turns to permit the second set of fingers to grip and to remove the finished part.

In the case of flexible manufacturing systems where families-of-parts are produced, universal hands and fingers are designed to accommodate and to produce the full range of different parts in the family.

VERSATILITY OF ROBOTS IN PRODUCTION

The versatility of robots in production is illustrated in Figure 64–4. The *manipulator* (robot hand) in this application includes accessories used in combination with a laser unit.

The robot hand serves as a fixture to position and to rotate the workpiece in relation to a fixed position of the laser unit. Cutting is performed by laser. In all such applications, *cautions* must be strictly observed. All protective guards and safety equipment must be in place prior to operation. All *Warning* notices on machine tools must be followed.

The robot illustrated has been programmed *online* by the operator who manually moves the manipulator through each successive step in the cycle. Each step is recorded in memory in the robot controller unit for automatic programming during production.

Robots may also be programmed *offline* and tested for accuracy and reliability. The program may then be loaded onto the robot controller and further integrated into an FMS.

Courtesy of CINCINNATI MILACRON INC.

CAUTION: "Safety equipment may have been removed or opened to clearly illustrate products and must be in place prior to operation."

Figure 64–4 Application of a Robot in Performing Arc Welding Processes

CADD ROBOT MODEL

CADD has the following capabilities in relation to the design and development of robots.

- Storing a three-dimensional graphics model of a part in its *database*.

- Copying the model in the data base in order to design the work cell in which specified tasks are performed.

- Providing visual feedback against which design problems may be checked and corrected.

UNIT 64 REVIEW AND SELF-TEST

A. SUBSYSTEMS OF HIGHLY-AUTOMATED MANUFACTURING SYSTEMS

1. Describe briefly the functions served by (a) CADD, (b) GEN, and (c) FMS systems.

2. Identify any combination of four machining or fabricating processes which can be automated for computer-aided manufacturing.

3. List the functions served by four subsystems of a fully-automated flexible manufacturing system (FMS).

4. a. Identify one function served by automated in-process gaging.
   b. State one advantage of in-process and one of post-process gaging.

5. List three uses of automated surface sensing probes.

6. Explain briefly the principle on which an adaptive subsystem automatically regulates machining speeds and feeds.

B. ROBOTICS IN CONVENTIONAL AND MULTIMACHINE MANUFACTURING SYSTEMS

1. State four functions which are served by robots in multimachine manufacturing systems.

2. Describe each of the following design features of robots.
   a. Degree of freedom   c. Compliance
   b. Repeatability   d. End effector

APPENDIX

- **GLOSSARY OF TECHNICAL TERMS**

- **HANDBOOK TABLES**

- **INDEX**

Abrasive band A continuous fabric band coated with grains of aluminum oxide or silicon carbide. A band used for polishing a previously file finished surface.

Abrasive machining The faster removal of material to a specified finish and dimensional sizes, shapes, and tolerances than by using conventional cutting tools.

Adjustable floating-blade reamer A two-blade reamer for reaming large-diameter holes. Floating blades that align with a formed hole.

Amplifier An intermediate modifying stage. A device for increasing (amplifying) electronic signals. (When used with different heads, it measures length, thickness, diameter, flatness, taper, and concentricity.)

Annealing Heating and cooling metals for purposes of softening, producing a definite microstructure, removing internal stresses, or altering mechanical or physical properties.

Austempering, martempering, and isothermal quenching Interrupted methods of quenching. Interrupting the cooling that begins above the transformation point. Interrupted cooling at a temperature above the one at which martensite begins to form.

Automatic bar and chucking machine A turret lathe that may be programmed so that trip blocks actuate microswitches. (These switches control the functions of the carriage, cross slide, vertical slide, threading head, speed and feed changes, and forward and reverse directions.)

Automatic sizing (abrasive wheel) Design feature incorporated into grinding machines to compensate for abrasive wheel wear.

Axial runout The amount a shoulder varies from a true square in relation to its axis.

Bar turner A hexagon turret accessory that is used widely to control concentricity during an external turning process. A cutting tool assembly consisting of adjustable guide rolls that serve as a follower rest and a turning tool.

Baseline dimensioning Successive measurements that originate from one or more common finished surfaces. A series of continuous dimensions that originate from a fixed reference plane or point.

Bilateral tolerance A two-direction tolerance. A tolerance above (+) *and* below (−) a basic dimension.

Binary code A code in which each allowable position permits a choice between two alternatives.

Block address format An address code or system of identifying words that specifies the format as well as the meaning of the words in a block.

Broaching (scru-broaching) Forming a screw thread by using a multiple-tooth cutting tool. (The teeth are spiral-formed on a body that has a pilot end. The broach is turned by a feed screw that draws the workpiece into the cutting tool. Limited applications of scru-broaching are in the automotive field.)

Cam-lock adapter A device for quickly and securely locking and unlocking a cutting tool in an adapter. An eccentric stud and corresponding slotted cutter. A device for applying the force needed to seat and hold a cutter in an adapter.

Cam motion Three common forms of motion: uniform, parabolic, and harmonic movement. The motion imparted by a cam. Actuating another part a fixed distance during a specific interval through which a cam moves.

Cam rise, fall, and dwell Motion and rest imparted by a cam to operate the turret and front, rear, and vertical cross slides. Cam design features for rough, finish, form turning, and cutting-off operations. (Screw machines).

Carbide insert designation system An eight-position series of letters and numerical symbols. A system that provides full specifications for carbide inserts.

Casehardening A series of heat-treating processes to produce a hardenable outer case. The introduction into a steel surface of carbon, carbon-nitrogen, or nitrogen by penetration, diffusion, or absorption. Carburizing, carbonitriding, and nitriding processes affecting the surface hardness of metals.

Centerless grinder principle A variation in the speed of two turning bodies (grinding wheel and regulating wheel) that produces and controls the direction and speed (RPM and sfpm) a workpiece rotates.

Ceramic inserts Fine grains of aluminum oxide that are formed and sintered into regular shapes (inserts). Diamond, square, rectangular, triangular, and other round shapes of ceramic cutting tools.

Chamfer cutting edge The angular surface formed between the face and sides of a cutter. An angular-formed cutting edge on a milling cutter.

Chucking tooling setups The combination of tooling and positioning cutters and accessories at each turret position to perform required machining processes.

Classes of fit (holes) Standards accepted by manufacturers, designers, and engineers for mating parts. Specifications generally used by the craftsperson for sliding parts, forced fits, average machine fits, fine tool and work assemblies, and others.

Climb (down) milling A machine cutting technique. Feeding a workpiece in the same direction that the milling cutter rotates. Milling a chip that is full thickness (equal to the feed) at the beginning of the cut. (The chip thickness diminishes to almost no thickness as the cutter tooth rotates to the depth of the cut.)

Closed loop system A NC system that provides output feedback for comparison to input command.

Compound gearing (milling) Gears used to form a gear train between the milling machine table feed screw, intermediate studs, and the worm shaft of a dividing head for purposes of milling spiral grooves.

Compound sine plate Two modified sine blocks that are adjustable on a single base. (The angle of each block lies in a different plane. The setting of the two angles produces the required compound angle.)

Concentricity The trueness of the periphery of a cylinder in relation to its axis.

Conditioning diamond and CBN wheels Steps dealing with mounting, testing, and correcting any runout of diamond or cubic boron nitride wheels.

Continuous-path system (contour control) The continuous and independent control of two or more instantaneous tool motions.

Continuous and segmental chips Chips that flow over a cutter as a continuous spiral (continuous chip) or as a welded file-tooth-shaped ribbon (segmental chip).

Control setting chart (butt welder) Manufacturer's chart for jaw gap and jaw pressure to use in butt welding different widths, thicknesses (gage), and saw band materials. A chart guide that gives the settings for a jaw pressure selector and jaw gap control knob.

Control signal The application of energy to actuate the device that makes corrective changes.

Coordinate chart A table accompanying a part drawing. A table that gives dimensions to locate each feature with respect to zero coordinates. (The chart contains other feature specifications.)

Critical points of steel Temperatures at which alpha and gamma structures are formed in steels. Lower and upper temperatures at which transformations of grain structure occur. Upper and lower temperatures affecting hardness and other properties of steels.

Cross rail A horizontal machine element upon which one or more rail heads are mounted. The horizontal ways on which rail heads may be moved transversely.

Crush dressing Rolling the form of a slowly turning work roll (with great force) into a grinding wheel. Using a hardened high-speed steel or carbide preformed work or reference roll to impress the required form (in reverse) in a grinding wheel. Forming a grinding wheel that has properties appropriate to withstand great force and to efficiently maintain its shape in form grinding.

Cutter clearance angle measurement Measuring the angle at which tool or cutter relief or clearance angles are ground. The use of a dial indicator or a cutter clearance angle gage.

Cutting off (parting, turning) The severing of a section by cutting a groove through a revolving workpiece. The process of using a mounted parting tool to cut through a workpiece.

Cutting tool deflection An undesirable machining condition. Springing of a cutting tool away from the surface being machined. A tool movement usually produced by a dull cutting edge, too fast a speed, or too great a feed.

Datum (datum plane) An exact line, surface, or reference point.

Digital input data Pulses, digits, or other coding elements that supply information to a machine control.

Digital readout An attachment that visually displays table movements along X and Y axes in + or − directions and in inch or metric units of measure.

Dimensional stability (precision instruments) The ability of an instrument to provide the same precise measurements over a long period of service. The relief of stresses within an instrument to ensure continuing accuracy of measurement.

Dimensional tolerance The permissible variation of a workpiece from its nominal dimension. (This variation may be the result of unilateral or bilateral tolerances.)

Direct indexing The process of positioning a part for an angular dimension by direct movement of a spindle. Rotating a workpiece a required angular distance without any additional gearing. Positioning a workpiece by using a plate with equally spaced holes in a circle.

Discontinuous surface A plane surface that is interrupted by a slot, groove, ridge, hole, or other cutaway section.

Drawfiling A process for graining a workpiece. Drawing a single-cut file or a second-cut mill file across a work surface and cutting on both the forward and return stroke.

Dressing traverse rate The speed with which a dressing tool is moved across the face of a grinding wheel. A factor in controlling the sharpness of abrasive grains and the quality of surface finish.

Electroband machining A band machine cutting process requiring a low-voltage, high-amperage current to be fed into a saw band. The discharge from a saw band producing an arc that disintegrates the material to be cut at the same instant the arc is cooled by a flow of coolant.

Electronic comparator An electronic measurement unit consisting primarily of a gage amplifier and one or more gaging heads. A device using solid-state electronic components to highly magnify and precisely measure linear dimensions.

End milling Milling processes using the end and/or face of a cutter. Cutting with a solid-shank or adapter-held shell end mill.

Expansion hand reamer A hand reamer having flutes. (These flutes may be expanded by a tapered screw to enlarge the reamer size slightly.)

Extended-nose collet A machinable collet that extends the normal length of a collet to permit deeper counterboring and tool clearance.

Feed cycle change gears A gear train that permits gear combinations to be set up to control the rate of production. Gear combinations that control the time (in seconds) to machine one workpiece. (Screw machines).

Flat, round-nose, gouge, cape, and diamond-point chisels Five common types of cold chisels. (Cold chisels are named according to the shape of their cutting edge or the chip, surface, or groove that is produced.)

Floating zero The movement of the zero reference point on an axis. Establishing the zero point at any point in the travel.

Flood, through-the-wheel, and mist cooling systems Three basic grinding machine coolant systems. (The coolant floods over a specific work-wheel area, or is fed through the wheel, or a mist of air and coolant is supplied.)

Fluted reamer A machine reamer with narrow lands backed off the entire length. A rough finishing reamer with a greater number of cutting teeth than a rose reamer. (Cutting is done with cutting teeth that are rounded or beveled and are relieved.)

Form dressing Process of forming a wheel in a reverse contour to the form to be ground. Preparation of a grinding wheel to grind grooves, slots of varying shapes, and other profiles.

Friction sawing A sawing process where heat is instantly generated ahead of the saw teeth. The scooping out of metal at forging point temperature.

Gage blocks Hardened, precision-ground and -finished rectangular steel blocks used to lay out work and make measurements. Working to accuracies in ranges of 0.000008″ to 0.000002″ and equivalent metric values in decimal parts of millimeters.

Granite precision measurement products   Instruments and gaging products used in precision measurement.   Granite surface plates and other layout and measurement accessories. Angle plates, universal right angles, straight edges, parallels, sine bars, and sine plates made of black granite.

Grinding wheel markings (composition)   The use of five basic symbols.   (Each symbol designates a design feature such as type of abrasive grain, grain size, grade, structure, and bond.)

Grooving (necking)   The process of turning a cutaway surface of a particular shape and size.

Hardenability, machinability, ductility   Physical properties of industrial manufacturing and cutting tools.   Factors to consider in selecting material for a workpiece or a cutting tool.   Characteristics of materials for which engineering and design data establish conditions of use.

Hardening temperature   The temperature at which a steel part is heated or soaked before quenching.   A temperature range that depends on carbon content and composition (including alloying elements).

Hardness testing scales   Numerical values related to depth or area penetration of a hardness tester penetrator for a series of fixed loads, according to material.   Bhn, Rockwell, Vickers, Knoop, scleroscope, microhardness and other hardness testing measurement systems.   Numerical values that may be converted from one system to the equivalent in another system.

Headstock spindle tooling   Primarily work-holding devices for bar and chuck work on turret lathes.   (Collets, step chucks, regular chucks, and faceplates are mounted in the turret lathe spindle.)

Helix angle   The angle formed by the angular path of the helix and the centerline of the part containing the helix.

High amplification comparators   A series of comparative measuring instruments.   Instruments in which dimensional variations are multiplied.   Instruments for amplifying measurements for purposes of obtaining a high degree of accuracy.

High-speed steel and cobalt high-speed steel drills   A twist drill capable of cutting at high speed without changing the temper, softening the cutting edges, or reducing the cutting efficiency.   (The addition of cobalt to high-speed steel produces a cutting material that is capable of drilling through tough metals and other compositions.)

Hobbing (worm gear teeth)   Use of a formed multiple-width tool cutter to track in gashed teeth.   Milling worm gear teeth to shape and size by using a hobbing cutter. Machining as a hob automatically engages each tooth on a free-to-rotate gashed gear.

Incremental cuts (milling)   Machining a cam form by taking a series of cuts. Cuts that are stepped off to conform to the required cam shape at different angular positions of the cam.   A process of adjusting the cam being milled in relation to the cutter in order to produce the cam form.

In-feed, through-feed, and end-feed   Three basic techniques of feeding parts for centerless grinding.   Feeding techniques to accommodate a wide variety of cylindrical bars, shouldered forms, and other shapes requiring cylindrical grinding.

Infinitely variable spindle speeds   Versatility to set the spindle speed at any RPM within the minimum and maximum range of speeds.

Interchangeability The control of surface finish and all features of a part. (Regardless of the geographic location of manufacture, each part meets dimensional specifications exactly. Each part may be interchanged with every other similar part to serve an intended function.)

Interference fit A negative fit. The machining of a part to a dimension that requires the mating parts to be forced together.

Interlocking cutters Milling cutters with sides recessed and alternate teeth ground back on the inside faces. Two cutters that are meshed. (The teeth overlap and permit adjustment of a required width.)

International (ISO) metric thread system A metric thread system accepted as an international standard. A thread form having a 60° thread angle. (The crests and roots are flattened at 1/8th of the depth. A rounded root is recommended. Drawings are dimensioned with metric thread specifications.)

ISO metric thread designation Specifications of a metric thread giving the outside diameter and pitch in millimeters—for example, M20–2.5 means an outside diameter of 20mm and a pitch of 2.5mm.

Keyseat milling Machining a groove in a shaft to receive one-half a square or rectangular key.

Knurling The raising of a series of diamond-shaped or straight-line patterned surfaces. Causing a material to flow by exerting force on a revolving workpiece with a set of specially formed rolls.

Leader and follower threading-chasing attachment An attachment to the cross slide apron. A mechanism in which a half nut (follower) engages a lead screw (leader) to feed the cross slide according to the required thread pitch.

Leading and following side angles Thread angles resulting from the slope of a thread form around the periphery of the workpiece. (The leading side refers to the thread side of the thread-cutting tool that advances with the thread-cutting process. The following side relates to the opposite thread-cutting tool side.)

Left-hand helix (reamer) A reamer with flutes cut at an angle to the body and counter-clockwise. The left-hand direction of the flutes. (This direction is the reverse of the cutting direction.)

Lever (probe) gaging head A small lever arm mounted to an electronic unit for magnifying small movements of the contact point. A lever arm that may probe into small openings for purposes of measurement.

Limiting diameters Major, pitch, and minor diameters applied to screw threads.

Linear pitch The distance on a rack equal to the circular pitch of a gear.

Longitudinal (back taper) relief (machine reamers) Grinding the margins of a machine reamer to a slight taper of 0.0002" (0.005mm) per inch. A slight clearance from the cutting end to the shank end of the reamer margin (land).

Magnesium, chromium, molybdenum, tungsten, and cobalt Metal elements added to steel to form alloys that possess specific properties. Metal elements that increase physical properties of steels and alloys such as wear resistance, toughness, red hardness, and and hardenability.

Magnetic sine plate A work-holding accessory having two hinged sections at right angles to each other. A sine plate with a magnetic top plate. An accessory that is set at precise angles by using gage block combinations.

Microfinishing The removal of limited amounts of material to produce a finely finished and dimensionally precise surface. As treated in the text, honing, lapping, and superfinishing processes and machines.

Microinch A simplified form of stating a value in millionths of an inch. (For example, 11.3 microinches refers to 0.0000113", "eleven point three millionths of an inch").

Mode The positioning of the components of a machine tool to permit another combination of processes to be performed—for example, changing from the vertical mode to the horizontal mode on a vertical milling machine. Conversion of a vertical milling machine (by using attachments) from basic vertical milling processes to serve also as a horizontal milling machine.

Monochromatic light A light source that consists of one wavelength (or color) only. A practical ray of light that penetrates an optical flat. Light emitted from a monochromatic lamp having a wavelength of 23.1 microinches (0.0005867mm).

Multiple-spindle automatic screw machine Four, six, eight, or more work-rotating spindles on one machine. An automatic screw machine that permits operations to be performed simultaneously on the workpiece on each spindle.

National Acme thread form A 29° thread form with a flat crest and root. A series of standard thread pitches for specific diameters.

National Coarse and National Fine Two basic thread series built upon the American National Form. (Fine series (NF) threads are used on parts and instruments requiring finer, more precise threads.)

Nominal grade The actual grade of a grinding wheel. (The cutting condition of the nominal grade is affected by depth of cut, rate of feed, sfpm as the wheel wears, and other factors.)

Nominal size The size used for general thread identification. (For example, the nominal size of a 1"—8NC thread is 1". By contrast, the basic size is 1.000"—8NC.)

Off-line data processing Computations of a geometric and mathematical nature, tool offsets, feed codes, paths, and others; miscellaneous and preparatory function tasks. Generation of a control tape.

On-line data processing Translation of the information on a control tape into input signals on a NC machine tool. (On-line data processing is repeated each time the control tape is used to produce a part.)

Open-side shaper-planer A planer with one vertical column and a horizontal cross rail. A planer adapted to machining wider workpieces than can be accommodated by a double-housing planer. A manufacturer's designation of a popular size (42" width × 42" height, or approximately 1m × 1m) open-side planer.

Optical comparator (contour projector) A comparative measuring instrument. An instrument on which an enlarged object is projected on a screen. A master form against which a projected object is examined for contour conformance and accuracy of measurement.

Optical flat   A quartz, glass, sapphire, or other transparent very nearly perfect plane surface.

Overhead turning   Supporting one or more cutters in a multiple turning head on the hexagon turret. A method used for heavy stock removal.

Pantograph dresser   A surface grinder accessory incorporating a tracer linkage system for reproducing a template profile to form dress a grinding wheel.

Parallax error   A condition resulting in a dimensional reading error. An error caused by changing the viewing position of a measurement.

Peripheral grinding   Grinding with the grains at the periphery of the grinding wheel.

Permanent form relief   A design characteristic for relieving form teeth so that the tooth profile remains unchanged after numerous sharpenings. Grinding the face of a form cutter at a constant radial rake and maintaining the same tooth form.

Pitch diameter   An imaginary diameter midway between the major and minor diameters of a thread. The diameter where the thread and thread groove dimensions are the same. An important thread diameter from which tolerances are measured to provide a specified fit.

Plunge grinding (cylindrical grinding)   Forming a flat or other contour cylindrical surface by in-feeding the wheel head and grinding wheel into a workpiece without moving the table longitudinally.

Point-to-point positioning   A positioning control system in which the controlled motion requires no path control in moving from one end point to the next.

Positive, zero, and negative radial rake angle   An angle formed by a tooth face in relation to the centerline of the cutter. The included angle formed by the cutter tooth and a centerline that extends from the cutting edge to the center of the cutter.

Power down-feed attachment   A mechanism for controlling the up- and down-feed rates of the quill.

Quality control   Techniques and measurement systems used in manufacturing to ensure conformity with standards. Dimensional and materials control obtained through sampling plans. Maintenance of quality and the improvement of product resulting from continuous inspection.

Quick-change tooling system   A wide selection of cutter-holding devices and adapters that permits maximum tool changing flexibility. Holders and adapters to accommodate a wide variety of tools, tool sizes, and shank designs.

Radius dressing (truing) device   A mechanism for holding a diamond dressing tool to generate an accurate radius form in a grinding wheel. A fixture for holding a diamond dressing tool in a fixed radius relationship to permit an accurate concave or convex radius form to be produced.

Rapid traverse   A mechanism for increasing feeds by moving the rapid traverse positioning lever. A rapid feed for speeding up the process of bringing a cutter and workpiece together.

Rate of feed (ipm or mm/min)   The distance a workpiece feeds into a cutter in one minute.

Rate of feed (i³pm)   The volume (mass) of material removed in one minute. The number of cubic inches of material machined in one minute. A production milling term.

Readout (command) A display showing the absolute position. An absolute position derived from a position command. Readout information taken directly from the dimension storage command or as a summation of command departures.

Releasing type (external threading) holder A chucking device with a tapered nose in which a solid adjustable spring die is held. A die holder that permits a die to release at a preset depth and torque.

Reversing tap driver A tapping device used on a reversible spindle drill press. A mechanism for both driving and removing a machine tap.

Right-angle attachment A spindle head attachment used in confined spaces for angular, horizontal, slotting, and other milling or drilling processes. A device for machining internal or external surfaces in minimum working areas.

Rose reamer A reamer with teeth beveled on the end and provided with a clearance. A machine reamer on which the cutting takes place on the beveled end-cutting teeth instead of the sides. A reamer with teeth that have a slight back taper (0.001″ per inch) along the full length of each flute.

Secondary clearance machining setup (milling) Positioning the workpiece by indexing to permit a plain milling cutter to mill clearance between a flute and a land.

Sequencing of cuts Planning the tooling for each internal and/or external cut and the position of each tool and accessory on the turret or cross slide. Step-by-step scheduling of cuts to produce a desired part.

Serrated tool bit (planer work) A replaceable, specially formed cutting tool with a serrated bottom face. A tool bit with serrations to match the serrations of the seat.

Solid, pin, taper, center, and hollow punches Basic types of hand punches. (The names partially describe the shape and function of each punch. Punches are obtainable in sets of various diameters.)

Sparking out Uniformly traversing the ground surface of a workpiece after the last cut without further down-feed.

Spiral-edge saw band A circular saw band formed with a continuous spiral cutting edge. A round, spiral-cutting saw band for sawing a precise contour with a minimum radius.

Standard steep machine tapers A steep-angle taper system of 16°36′, or 3 1/2″ taper per foot. A self-releasing taper. A system of designating milling machine spindle nose tapers (#30, #40, #50, and so on).

Start drilling The process of centering or drilling a hole with minimum deviation from the axis. Using a short, rigid start drill to spot a cone shape in a workpiece.

Straddle milling The process of milling two parallel surfaces or steps at the same time. The application of two side milling cutters to machine two vertical surfaces at the same time. Milling two parallel steps in one operation.

Style A, B, and C arbors Three general designs of milling machine arbors. (Style A is a long arbor supported on the end of a pilot. Style B is an arbor that is steadied by a bearing sleeve and an arbor support. Style C is an arbor to which shell end mills and other cutting tools are attached directly.)

Synthetic cutting fluids A mixture of chemical additives that improve the wetting and cooling properties of water.

Tab sequential format A method of identifying a word by the number of characters in the block preceding the word. (The first character in each word is a tab character.)

Tempering A transformation temperature range below which a hardened or normalized steel part is reheated, followed by cooling at a specified rate.

Tenthset boring head A compactly designed boring head having a direct vernier reading. A boring head that permits boring diameter adjustments of 0.0001" (0.002mm).

Three-wire method of measuring threads The use of three precision-ground wires of a special size to measure the pitch diameter of a screw thread.

Tool circle Positioning the tooling on a turret to approximately the same starting point. Setups on the turret that provide for uniform starting positions and permit the operator to develop an easy machining rhythm and to minimize fatigue.

Tooth rest blade forms (cutter and tool grinding) Plain, rounded, offset, hooked, and inverted V-tooth formed blades. A variation of blade shapes to accommodate different cutter teeth forms.

Turning center (system) A numerically controlled universal turret lathe with a greater range of speeds and feeds and number of tool stations than a bar or chucking machine or combination.

Type D-1 cam-lock spindle chuck adapter A short, steep-angle, taper-bored adapter plate. A chuck plate bored to a 3.000" T_{pf}. An adapter plate designed to secure a chuck on a cam-lock spindle nose.

Type-L spindle nose A steep-angle taper lathe spindle nose. A spindle nose with an American Standard taper of 3.500" T_{pf}. A tapered lathe spindle nose having a key and a lock ring. (The lock ring draws and holds a chuck securely on the taper and against the shoulder of the spindle.)

Undercutting The removing of material from the corner of a shoulder. Cutting a groove deeper than the intersecting surface.

Unified Miniature Screw Thread Series An American Standard series of 60° form threads that are particularly adapted to instruments and microminiaturized mechanisms. A thread form that is compatible with the Unified inch-standard and the ISO metric-standard screw thread systems. Threads designated on drawings as UNM.

Universal bar equipment (ram- or saddle-type turret lathe) A combination of standard cross slide and turret tools. A setup of tools established by manufacturers after years of extensive experience. A tooling setup that may be left permanently on turret lathes for short-run production jobs requiring basic internal and/or external cuts.

Wheel head (cylindrical grinding) A mechanism for housing, driving, and controlling the grinding wheels for external and internal grinding. A grinding spindle head mounted on a carriage to permit swiveling the wheel at an angle to the axis of the workpiece.

Zero coordinate A starting point from which all dimensions along a specific axis originate. A fixed reference point for all dimensions on a plane line.

HANDBOOK TABLES

Table A–1 Decimal Equivalents of Fractional, Wire Gage (Number), Letter, and Metric Sizes of Drills

Decimal	Inch	Wire	mm
.0059		97	.15
.0063		96	.16
.0067		95	.17
.0071		94	.18
.0075		93	.19
.0079		92	.20
.0083		91	.21
.0087		90	.22
.0091		89	.23
.0095		88	.24
.0098			.25
.0100		87	
.0102			.26
.0105		86	
.0106			.27
.0110		85	.28
.0114			.29
.0115		84	
.0118			.30
.0120		83	
.0122			.31
.0125		82	
.0126			.32
.0130		81	.33
.0134			.34
.0135		80	
.0138			.35
.0145		79	
.0156	1/64		
.0158			.40
.0160		78	
.0177			.45
.0180		77	
.0197			.50
.0200		76	
.0210		75	
.0217			.55
.0225		74	
.0236			.60
.0240		73	
.0250		72	
.0256			.65
.0260		71	
.0276			.70
.0280		70	
.0292		69	
.0295			.75
.0310		68	
.0312	1/32		
.0315			.80
.0320		67	
.0330		66	
.0335			.85
.0350		65	
.0354			.90
.0360		64	
.0370		63	
.0374			.95
.0380		62	
.0390		61	
.0394			1.00
.0400		60	
.0410		59	
.0413			1.05
.0420		58	
.0430		57	
.0433			1.10
.0453			1.15
.0465		56	
.0469	3/64		
.0472			1.20
.0492			1.25
.0512			1.30
.0520		55	
.0532			1.35
.0550		54	
.0551			1.40
.0571			1.45
.0591			1.50
.0595		53	
.0610			1.55
.0625	1/16		
.0630			1.60
.0635		52	
.0650			1.65
.0669			1.70
.0670		51	
.0689			1.75
.0700		50	
.0709			1.80
.0728			1.85
.0730		49	
.0748			1.90
.0760		48	
.0768			1.95
.0781	5/64		
.0785		47	
.0787			2.00
.0807			2.05
.0810		46	
.0820		45	
.0827			2.10
.0847			2.15
.0860		44	
.0866			2.20
.0886			2.25
.0890		43	
.0906			2.30
.0925			2.35
.0935		42	
.0938	3/32		
.0945			2.40
.0960		41	
.0965			2.45
.0980		40	
.0984			2.50
.0995		39	
.1015		38	
.1024			2.60
.1040		37	
.1063			2.70
.1065		36	
.1083			2.75
.1094	7/64		
.1100		35	
.1102			2.80
.1110		34	
.1130		33	
.1142			2.90
.1160		32	
.1181			3.00
.1200		31	
.1221			3.10
.1250	1/8		
.1260			3.20
.1280			3.25
.1285		30	
.1299			3.30
.1339			3.40
.1360		29	
.1378			3.50
.1405		28	
.1406	9/64		
.1417			3.60
.1440		27	
.1457			3.70
.1470		26	
.1476			3.75
.1495		25	
.1496			3.80
.1520		24	
.1535			3.90
.1540		23	
.1562	5/32		
.1570		22	
.1575			4.00
.1590		21	
.1610		20	
.1614			4.10
.1654			4.20
.1660		19	
.1673			4.25
.1693			4.30
.1695		18	
.1719	11/64		
.1730		17	
.1732			4.40
.1770		16	
.1772			4.50
.1800		15	
.1811			4.60
.1820		14	
.1850		13	4.70
.1870			4.75
.1875	3/16		
.1890		12	4.80
.1910		11	
.1929			4.90
.1935		10	
.1960		9	
.1969			5.00
.1990		8	
.2008			5.10
.2010		7	
.2031	13/64		

Decimal	Inch	Wire	mm
.2040		6	
.2047			5.20
.2055		5	
.2067			5.25
.2087			5.30
.2090		4	
.2126			5.40
.2130		3	
.2165			5.50
.2188	7/32		
.2205			5.60
.2210		2	
.2244			5.70
.2264			5.75
.2280		1	
.2284			5.80
.2323			5.90
.2340		A	
.2344	15/64		
.2362			6.00
.2380		B	
.2402			6.10
.2420		C	
.2441			6.20
.2460		D	
.2461			6.25
.2480			6.30
.2500	1/4	E	
.2520			6.40
.2559			6.50
.2570		F	
.2598			6.60
.2610		G	
.2638			6.70
.2656	17/64		
.2658			6.75
.2660		H	
.2677			6.80

Decimal	Inch	Letter	mm
.2717			6.90
.2720		I	
.2756			7.00
.2770		J	
.2795			7.10
.2810		K	
.2812	9/32		
.2835			7.20
.2854			7.25
.2874			7.30
.2900		L	
.2913			7.40
.2950		M	
.2953			7.50
.2969	19/64		
.2992		N	
.3020			7.60
.3032			7.70
.3051			7.75
.3071			7.80
.3110			7.90
.3125	5/16		
.3150			8.00
.3160		O	
.3189			8.10
.3228			8.20
.3230		P	
.3248			8.25
.3268			8.30
.3281	21/64		
.3307			8.40
.3320		Q	
.3347			8.50
.3386			8.60
.3390		R	
.3425			8.70
.3438	11/32		

Decimal	Inch	Letter	mm
.3445			8.75
.3465			8.80
.3480		S	
.3504			8.90
.3543			9.00
.3580		T	
.3583			9.10
.3594	23/64		
.3622			9.20
.3642			9.25
.3661			9.30
.3680		U	
.3701			9.40
.3740			9.50
.3750	3/8		
.3770		V	
.3780			9.60
.3819			9.70
.3839			9.75
.3858			9.80
.3860		W	
.3898			9.90
.3906	25/64		
.3937			10.00
.3970		X	
.4040		Y	
.4062	13/32		
.4130		Z	

Decimal	Inch	mm
.4134		10.50
.4219	27/64	
.4331		11.00
.4375	7/16	
.4528		11.50
.4531	29/64	
.4688	15/32	
.4724		12.00
.4844	31/64	
.4921		12.50
.5000	1/2	
.5118		13.00
.5156	33/64	
.5312	17/32	
.5315		13.50
.5469	35/64	
.5512		14.00
.5625	9/16	
.5709		14.50
.5781	37/64	
.5906		15.00
.5938	19/32	
.6094	39/64	
.6102		15.50
.6250	5/8	
.6299		16.00
.6406	41/64	
.6496		16.50
.6562	21/32	
.6693		17.00
.6719	43/64	
.6875	11/16	
.6890		17.50
.7031	45/64	
.7087		18.00
.7188	23/32	
.7283		18.50
.7344	47/64	
.7480		19.00

Decimal	Inch	mm
.7500	3/4	
.7656	49/64	
.7677		19.50
.7812	25/32	
.7874		20.00
.7969	51/64	
.8071		20.50
.8125	13/16	
.8268		21.00
.8281	53/64	
.8438	27/32	
.8465		21.50
.8594	55/64	
.8661		22.00
.8750	7/8	
.8858		22.50
.8906	57/64	
.9055		23.00
.9062	29/32	
.9219	59/64	
.9252		23.50
.9375	15/16	
.9449		24.00
.9531	61/64	
.9646		24.50
.9688	31/32	
.9843		25.00
.9844	63/64	
1.0000	1	

Table A–2 Conversion of Metric to Inch-Standard Units of Measure

mm Value	Inch (decimal) Equivalent	mm Value	Inch (decimal) Equivalent	mm Value	Inch (decimal) Equivalent	mm Value	Inch (decimal) Equivalent
.01	.00039	.34	.01339	.67	.02638	1	.03937
.02	.00079	.35	.01378	.68	.02677	2	.07874
.03	.00118	.36	.01417	.69	.02717	3	.11811
.04	.00157	.37	.01457	.70	.02756	4	.15748
.05	.00197	.38	.01496	.71	.02795	5	.19685
.06	.00236	.39	.01535	.72	.02835	6	.23622
.07	.00276	.40	.01575	.73	.02874	7	.27559
.08	.00315	.41	.01614	.74	.02913	8	.31496
.09	.00354	.42	.01654	.75	.02953	9	.35433
.10	.00394	.43	.01693	.76	.02992	10	.39370
.11	.00433	.44	.01732	.77	.03032	11	.43307
.12	.00472	.45	.01772	.78	.03071	12	.47244
.13	.00512	.46	.01811	.79	.03110	13	.51181
.14	.00551	.47	.01850	.80	.03150	14	.55118
.15	.00591	.48	.01890	.81	.03189	15	.59055
.16	.00630	.49	.01929	.82	.03228	16	.62992
.17	.00669	.50	.01969	.83	.03268	17	.66929
.18	.00709	.51	.02008	.84	.03307	18	.70866
.19	.00748	.52	.02047	.85	.03346	19	.74803
.20	.00787	.53	.02087	.86	.03386	20	.78740
.21	.00827	.54	.02126	.87	.03425	21	.82677
.22	.00866	.55	.02165	.88	.03465	22	.86614
.23	.00906	.56	.02205	.89	.03504	23	.90551
.24	.00945	.57	.02244	.90	.03543	24	.94488
.25	.00984	.58	.02283	.91	.03583	25	.98425
.26	.01024	.59	.02323	.92	.03622	26	1.02362
.27	.01063	.60	.02362	.93	.03661	27	1.06299
.28	.01102	.61	.02402	.94	.03701	28	1.10236
.29	.01142	.62	.02441	.95	.03740	29	1.14173
.30	.01181	.63	.02480	.96	.03780	30	1.18110
.31	.01220	.64	.02520	.97	.03819		
.32	.01260	.65	.02559	.98	.03858		
.33	.01299	.66	.02598	.99	.03898		

Table A–3 Conversion of Fractional Inch Values to Metric Units of Measure

Fractional Inch	mm Equivalent	Fractional Inch	mm Equivalent	Fractional Inch	mm Equivalent	Fractional Inch	mm Equivalent
1/64	0.397	17/64	6.747	33/64	13.097	49/64	19.447
1/32	0.794	9/32	7.144	17/32	13.494	25/32	19.844
3/64	1.191	19/64	7.541	35/64	13.890	51/64	20.240
1/16	1.587	5/16	7.937	9/16	14.287	13/16	20.637
5/64	1.984	21/64	8.334	37/64	14.684	53/64	21.034
3/32	2.381	11/32	8.731	19/32	15.081	27/32	21.431
7/64	2.778	23/64	9.128	39/64	15.478	55/64	21.828
1/8	3.175	3/8	9.525	5/8	15.875	7/8	22.225
9/64	3.572	25/64	9.922	41/64	16.272	57/64	22.622
5/32	3.969	13/32	10.319	21/32	16.669	29/32	23.019
11/64	4.366	27/64	10.716	43/64	17.065	59/64	23.415
3/16	4.762	7/16	11.113	11/16	17.462	15/16	23.812
13/64	5.159	29/64	11.509	45/64	17.859	61/64	24.209
7/32	5.556	15/32	11.906	23/32	18.256	31/32	24.606
15/64	5.953	31/64	12.303	47/64	18.653	63/64	25.003
1/4	6.350	1/2	12.700	3/4	19.050	1	25.400

Table A–4 Machinability, Hardness, and Tensile Strength of Common Steels and Other Metals and Alloys

SAE Number	AISI Number	Tensile Strength (psi)	Hardness (Brinell)	Machinability Rating (percent)	SAE Number	AISI Number	Tensile Strength (psi)	Hardness (Brinell)	Machinability Rating (percent)
Carbon Steels					*Molybdenum Steels (continued)*				
1015	C1015	65,000	137	50	4140	A4140	90,000	187	56
1020	C1020	67,000	137	52	4150	A4150	105,000	220	54
×1020	C1022	69,000	143	62	×4340	A4340	115,000	235	58
1025	C1025	70,000	130	58	4615	A4615	82,000	167	58
1030	C1030	75,000	138	60	4640	A4640	100,000	201	69
1035	C1035	88,000	175	60	4815	A4815	105,000	212	55
1040	C1040	93,000	190	60					
1045	C1045	99,000	200	55					
1095	C1095	100,000	201	45	*Chromium Steels*				
					5120	A5120	73,000	143	50
Free-Cutting Steels					5140	A5140		174–229	60
×1113	B1113	83,000	193	120–140	52100	E52101	109,000	235	45
1112	B1112	67,000	140	100					
. . .	C1120	69,000	117	80	*Chromium-Vanadium Steels*				
					6120	A6120		179–217	50
Manganese Steels					6150	A6150	103,000	217	50
×1314	. . .	71,000	135	94					
×1335	A1335	95,000	185	70	*Other Alloys and Metals*				
					Aluminum (11S)		49,000	95	300–2,000
Nickel Steels					Brass, Leaded		55,000	RF 100	150–600
2315	A2317	85,000	163	50	Brass, Red or Yellow		25–35,000	40–55	200
2330	A2330	98,000	207	45	Bronze, Lead-Bearing		22–32,000	30–65	200–500
2340	A2340	110,000	225	40	Cast Iron, Hard		45,000	220–240	50
2345	A2345	108,000	235	50	Cast Iron, Medium		40,000	193–220	65
					Cast Iron, Soft		30,000	160–193	80
Nickel-Chromium Steels					Cast Steel (0.35 C)		86,000	170–212	70
3120	A3120	75,000	151	50	Copper (F.M.)		35,000	RF 85	65
3130	A3130	100,000	212	45	Low-Alloy, High-				
3140	A3140	96,000	195	57	Strength Steel		98,000	187	80
3150	A3150	104,000	229	50	Magnesium Alloys				500–2,000
3250	. . .	107,000	217	44	Malleable Iron				
					Standard		53–60,000	110–145	120
Molybdenum Steels					Pearlitic		80,000	180–200	90
4119	. . .	91,000	179	60	Pearlitic		97,000	227	80
×4130	A4130	89,000	179	58	Stainless Steel (12% Cr F.M.)		120,000	207	70

Table A–6 Recommended Drill Point and Lip Clearance Angles for Selected Materials

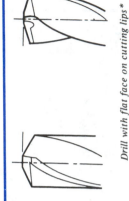

Drill with flat face on cutting lips*

Workpiece Material	Included Drill Point Angle	Lip Clearance Angle
—General-purpose, plain carbon steels		
—Gray cast iron	118°	8–12°
—Annealed alloy steels		
—Medium-hardness, pearlitic, malleable iron		
—Stainless steel		
—Heat-treated steel	125°	12°
—Hard-to-cut materials		
—Alloy steel		
—Drop Forgings		
—Soft cast iron		
—Aluminum		
—Wood	90° to 100°	15°
—Ferritic, malleable iron		
—Plastics		
—Hard rubber		
—Soft aluminum		
—Die casting	60°	20°
—Plastics		
—Wood		
—Soft and medium copper		
—Bronze*	100° to 118°	15°
—Brass*		
—Magnesium alloys*	60° to 118°	10°
—7 to 13 percent manganese steel*		
—Armor plate		
—Tough alloy steels	136° to 150°	10°
—Extra-hard materials		

*Grind flat face on cutting lips.

Table A–5 Recommended Saw Blade Pitches, Cutting Speeds, and Feeds for Power Hacksawing Ferrous and Nonferrous Metals*

Material	Pitch (teeth per inch)	Cutting Speed (feet per minute)	Feed (force in pounds)
Ferrous Metals			
Iron			
Cast	6 to 10	120	125
Malleable	6 to 10	90	125
Steel			
Drill rod	10	90	125
Forging stock, mild	3, 4, or 6	120	125
Alloy	4 to 6	90	125
Carbon tool	6 to 10	75	125
High speed	6 to 10	60	125
Machine	6 to 10	120	150
Stainless	6 to 10	75	125
Structural	6 to 10	120	125
Pipe	10 to 14	120	125
Tubing			
Thick-wall	6 to 10	120	90
Thin-wall	14	120	60
Nonferrous Metals			
Aluminum			
Alloy	4 to 6	150	60
Heat-treated	4 to 10	120	60
Soft	4 to 6	180	60
Brass			
Hard	10 to 14	90	60
Free-machining	6 to 10	150	60
Bronze			
Castings	4–6 to 10	110	125
Manganese	6 to 10	80	60
Copper			
Bars	3, 4, or 6	130	150
Tubing	10	120	60

*Coarser pitches are used for sizes over two inches and for thick-walled sections.

Table A–7 Conversion of Cutting Speeds and RPM for All Machining Operations
(metric and inch-standard diameters)

Cutting Speeds																		
Meters per Minute*	6.1	9.1	12.2	15.2	18.2	21.3	24.4	27.3	30.4	38	45.6	60.8	91.2	121.6	152	182.4	243.2	304
Surface Feet per Minute	20	30	40	50	60	70	80	90	100	125	150	200	300	400	500	600	800	1000

Diameter / **Revolutions per Minute**

mm	*Inches																		
1.6	1/16	1222	1833	2445	3056	3667	4278	4889	5500	6112	7639	9167							
3.1	1/8	611	917	1222	1528	1833	2139	2445	2750	3056	3820	4584	6112	9167					
4.7	3/16	407	611	815	1019	1222	1426	1630	1833	2037	2546	3056	4074	6112	8149				
6.3	1/4	306	458	611	764	917	1070	1222	1375	1528	1910	2292	3056	4584	6112	7639	9167		
7.8	5/16	244	367	489	611	733	856	978	1100	1222	1528	1833	2445	3667	4889	6112	7334	9778	
9.4	3/8	204	306	407	509	611	713	815	917	1019	1273	1528	2037	3056	4074	5093	6112	8149	
10.9	7/16	175	262	349	437	524	611	698	786	873	1091	1310	1746	2619	3492	4365	5238	6985	8731
12.7	1/2	153	229	306	382	458	535	611	688	764	955	1146	1528	2292	3056	3820	4584	6112	7639
15.6	5/8	122	183	244	306	367	428	489	550	611	764	917	1222	1833	2445	3056	3667	4889	6112
18.8	3/4	102	153	204	255	306	357	407	458	509	637	764	1019	1528	2037	2546	3056	4074	5093
21.9	7/8	87	131	175	218	262	306	349	393	437	546	655	873	1310	1746	2183	2619	3492	4365
25.4	1	76	115	153	191	229	267	306	344	382	477	573	764	1146	1528	1910	2292	3056	3820
28.1	1-1/8	68	102	136	170	204	238	272	306	340	424	509	679	1019	1358	1698	2037	2716	3395
31.7	1-1/4	61	92	122	153	183	214	244	275	306	382	458	611	917	1222	1528	1833	2445	3056
34.8	1-3/8	56	83	111	139	167	194	222	250	278	347	417	556	833	1111	1389	1667	2222	2778
38.1	1-1/2	51	76	102	127	153	178	204	229	255	318	382	509	764	1019	1273	1528	2037	2546
41.0	1-5/8	47	71	94	118	141	165	188	212	235	294	353	470	705	940	1175	1410	1880	2351
44.2	1-3/4	44	66	87	109	131	153	175	196	218	273	327	437	655	873	1091	1310	1746	2183
47.3	1-7/8	41	61	82	102	122	143	163	183	204	255	306	407	611	815	1019	1222	1630	2037
50.8	2	38	57	76	96	115	134	153	172	191	239	286	382	573	764	955	1146	1528	1910
63.5	2-1/2	31	46	61	76	92	107	122	138	153	191	229	306	458	611	764	917	1222	1528

*The meters per minute and mm diameters are approximately equivalent to the corresponding sfpm and the diameters given in inches.

Table A–8 Recommended Cutting Speeds for Drilling with High-Speed Drills*

Material	Hardness (Brinell)	Cutting Speed (sfpm)	Material	Hardness (Brinell)	Cutting Speed (sfpm)
Plain Carbon Steels			**Tool Steels**		
AISI–1019, 1020	120–150	80–120	Water-Hardening	150–250	70–80
1030, 1040, 1050	150–170	70–90	Cold-Work	200–250	20–40
1060, 1070, 1080, 1090	170–190	60–80	Shock-Resisting	175–225	40–50
	190–220	50–70	Mold	100–150	60–70
	220–280	40–50		150–200	50–60
	280–350	30–40	High-Speed Steel	200–250	30–40
	350–425	15–30		250–275	15–30
			Cast Iron		
Alloy Steels			Gray	110–140	90–140
AISI–1320, 2317, 2515	125–175	60–80		150–190	80–100
3120, 3316, 4012, 4020	175–225	50–70		190–220	60–80
4120, 4128, 4320, 4620	225–275	45–60		220–260	50–70
4720, 4820, 5020, 5120	275–325	35–55		260–320	30–40
6120, 6325, 6415, 8620	325–375	30–40			
8720, 9315	375–425	15–30	**Malleable Iron**		
			Ferritic	110–160	120–140
Alloy Steels			Pearlitic	160–200	90–110
AISI–1330, 1340, 2330	175–225	50–70		200–240	60–90
2340, 3130, 3140, 3150	225–275	40–60		240–280	50–60
4030, 4063, 4130, 4140	275–325	30–50			
4150, 4340, 4640, 5130	325–375	25–40	**Aluminum Alloys**		
5140, 5160, 5210, 6150,	375–425	15–30	Cast		
6180, 6240, 6290, 6340,			Nonheat-treated		200–300
6380, 8640, 8660, 8740,			Heat-treated		150–250
9260, 9445, 9840, 9850			Wrought		
			Cold-Drawn		150–300
Stainless Steel			Heat-treated		140–300
Standard Grades					
Austenitic, annealed	135–185	40–50	**Brass, Bronze (ordinary)**		150–300
Cold-drawn, austenitic	225–275	30–40			
Ferritic	135–185	50–60	**Bronze (high-strength)**		30–100
Martensitic, annealed	135–175	55–70			
	175–225	50–60			
Tempered, quenched	275–325	30–40			
	375–425	15–30			
Free Machining Grades					
Annealed, austenitic	135–185	80–100			
Cold-drawn, austenitic	225–275	60–90			
Ferritic	135–185	100–200			
Martensitic, annealed	135–185	100–130			
Martensitic, cold-drawn	185–240	90–120			
Martensitic, tempered	275–325	50–60			
Quenched	375–425	30–40			

*Use cutting speeds of 50% to 65% of speeds given in the table for high-speed steel reamers.

Table A–9 General NC Tap Sizes and Recommended Tap Drills (NC standard threads)

Size (outside diameter inch)	Threads per Inch	Tap Drill Size* (75% thread depth)
1/4	20	#7
5/16	18	"F"
3/8	16	5/16
7/16	14	"U"
1/2	13	27/64
9/16	12	31/64
5/8	11	17/32
11/16	11NS	19/32
3/4	10	21/32
13/16	10NS	23/32
7/8	9	49/64
15/16	9NS	53/64
1	8	7/8
1 1/8	7	63/64
1 1/4	7	1 7/64
1 3/8	6	1 13/64
1 1/2	6	1 11/32
1 5/8	5 1/2NS	1 29/64
1 3/4	5	1 35/64
1 7/8	5NS	1 11/16
2	4 1/2	1 25/32

*Nearest commercial drill size to produce a 75% thread depth.

Table A–10 General NF Tap Sizes and Recommended Tap Drills (NF standard threads)

Size (outside diameter inch)	Threads per Inch	Tap Drill Size* (75% thread depth)
1/4	28	#3
5/16	24	"I"
3/8	24	"Q"
7/16	20	"W"
1/2	20	29/64
9/16	18	33/64
5/8	18	37/64
11/16	11NS	19/32
3/4	16	11/16
13/16	10NS	23/32
7/8	14	13/16
15/16	9NS	53/64
1	12	59/64
1	14NS	15/16
1 1/8	12	1 3/64
1 1/4	12	1 11/64
1 3/8	12	1 19/64
1 1/2	12	1 27/64

*Nearest commercial drill size to produce a 75% thread depth.

Table A–11 General Metric Standard Tap Sizes and Recommended Tap Drills (metric-standard threads)

Metric Thread Size (nominal outside diameter and pitch) (mm)	Recommended Tap Drill Size			
	Metric Series Drills		Inch Series Drills	
	Size (mm)	Equivalent (inch)	Nominal Size	Diameter (inch)
M1.6×0.35	1.25	.0492		
M2×0.4	1.60	.0630	#52	.0635
M2.5×0.45	2.05	.0807	#45	.0820
M3×0.5	2.50	.0984	#39	.0995
M3.5×0.6	2.90	.1142	#32	.1160
M4×0.7	3.30	.1299	#30	.1285
M5×0.8	4.20	.1654	#19	.1660
M6×1	5.00	.1968	#8	.1990
M8×1.25	6.80	.2677	"H"	.2660
M10×1.5	8.50	.3346	"Q"	.3320
M12×1.75	10.25	.4035	13/32	.4062
M14×2	12.00	.4724	15/32	.4688
M16×2	14.00	.5512	35/64	.5469
M20×2.5	17.50	.6890	11/16	.6875
M24×3	21.00	.8268	53/64	.8281
M30×3.5	26.50	1.0433	1 3/64	1.0469
M36×4	32.00	1.2598	1 1/4	1.2500

Table A–12 Suggested Tapping Speeds (fpm) and Cutting Fluids for Commonly Used Taps and Materials

Material	Carbon Steel Taps				High-Speed Steel Taps				Cutting Fluid
	Hand Tap		Gun Tap		Hand Tap		Gun Tap		
	Thread System								
	Coarse	Fine	Coarse	Fine	Coarse	Fine	Coarse	Fine	
	Feet per Minute								
Stainless steel					15	20	20	25	Sulphur-base oil
Drop forgings	20	25	20	25	50	55	60	65	Soluble oil or sulphur-base oil
Aluminum and die castings					70	90	80	100	Kerosene and lard oil
Mild steel	35	40	35	40	70	80	90	100	Sulphur-base oil
Cast iron	35	40	35	40	80	90	100	110	Dry or soluble oil
Brass	70	80	70	80	160	180	200	220	Soluble or light oil

Table A–13 Recommended Cutting Speeds for Reaming Common Materials (High-Speed Steel Machine Reamers)

Material	Cutting Speed Range	
	sfpm	m/min*
Steel		
Machinery .2C to .3C	50–70	15–22
Annealed .4C to .5C	40–50	12–15
Tool, 1.2C	35–40	10–12
Alloy	35–40	10–12
Automotive forgings	35–40	10–12
Alloy 300–400 brinell	20–30	6–9
Free-machining stainless	40–50	12–15
Hard Stainless	20–30	6–9
Monel Metal	25–35	8–12
Cast Iron		
Soft	70–100	22–30
Hard	50–70	15–22
Chilled	20–30	6–9
Malleable	50–60	15–18
Brass/Bronze		
Ordinary	130–200	40–60
High tensile	50–70	15–22
Bakelite	70–100	20–30
Aluminum and its alloys	130–200	40–60
Magnesium and its alloys	170–270	50–80

*m/min: meters per minute.

Table A–14 Recommended Cutting Speeds (in sfpm) for Carbide and Ceramic Cutting Tools

Material	Carbide Cutting Tools (sfpm)		Ceramic Cutting Tools (sfpm)	
	Rough Cuts	Finish Cuts	Rough Cuts	Finish Cuts
Nonferrous				
Cast Iron	200–250	350–450	200–800	200–2000
Machine Steel	400–500	700–1000	250–1200	400–1800
Tool Steel	300–400	500–750	300–1500	600–2000
Stainless Steel	250–300	375–500	300–1000	400–1200
Ferrous				
Aluminum	300–450	700–1000	40–2000	600–3000
Brass	500–600	700–800	400–800	600–1200
Bronze	500–600	700–800	150–800	200–1000

Table A-15 Recommended Cutting Speeds and Feeds (in sfpm) for Selected Materials and Lathe Processes (High-Speed Steel Cutting Tools and Knurls)

| | Turning, Boring, Facing | | | | | | |
| | Rough Cut | | Finish Cut | | | | |
Material	RPM	Feed (")	RPM	Feed (")	Threading	Parting	Knurling
Ferrous							
Cast Iron	70	.015 to .025	80	.005 to .012	20	80	30
Machine Steel	100	.010 to .020	110	.003 to .010	40	80	30
Tool Steel	70	.010 to .020	80	.003 to .010	25	70	30
Stainless Steel	75–90	.015 to .030	100–130	.005 to .010	35	70	20
Nonferrous							
Aluminum	200	.015 to .030	300	.005 to .010	50	200	40
Brass	180	.015 to .025	220	.003 to .010	45	150	40
Bronze	180	.015 to .025	220	.003 to .010	45	150	40

Table A-16 Recommended Cutting Speeds and Feeds for Various Depths of Cut on Common Metals (Single-Point Carbide Cutting Tools)

Metal	Depth of Cut (inches)	Feed per Revolution (inches)	Cutting Speed (sfpm)
Aluminum	.005–.015	.002–.005	700–1000
	.020–.090	.005–.015	450–700
	.100–.200	.015–.030	300–450
	.300–.700	.030–.090	100–200
Brass, Bronze	.005–.015	.002–.005	700–800
	.020–.090	.005–.015	600–700
	.100–.200	.015–.030	500–600
	.300–.700	.030–.090	200–400
Cast Iron (gray)	.005–.015	.002–.005	350–400
	.020–.090	.005–.015	250–350
	.100–.200	.015–.030	200–250
	.300–.700	.030–.090	75–150
Machine Steel	.005–.015	.002–.005	700–1000
	.020–.090	.005–.015	550–700
	.100–.200	.015–.030	400–550
	.300–.700	.030–.090	150–300
Tool Steel	.005–.015	.002–.005	500–700
	.020–.090	.005–.015	400–500
	.100–.200	.015–.030	300–400
	.300–.700	.030–.090	100–300
Stainless Steel	.005–.015	.002–.005	375–500
	.020–.090	.005–.015	300–375
	.100–.200	.015–.030	250–300
	.300–.700	.030–.090	75–175

Table A-17 Average Cutting Tool Angles and Cutting Speeds for Single-Point, High-Speed Steel Cutting Tools

Material to Be Machined	Side[1] Relief	End[1] Relief	Side Rake[2]	True Back[3] Rake	Suggested Cutting Speeds[4]	
					sfpm	m/min
Steel						
Free-machining	10°	10°	10°-22°	16°	160-350	50-110
Low-carbon (.05%-.30%C)	10°	10°	10°-14°	16°	90-100	27-30
Medium-carbon (.30%-.60%C)	10°	10°	10°-14°	12°	70-90	21-27
High-carbon tool (.60-1.70%C)	8°	8°	8°-12°	8°	50-70	15-21
Tough alloy	8°	8°	8°-12°	8°	50-70	15-21
Stainless	8°	8°	5°-10°	8°	40-70	12-21
Stainless, free-machining	10°	10°	5°-10°	16°	80-140	24-42
Cast Iron						
Soft	8°	8°	10°	8°	50-80	15-24
Hard	8°	8°	8°	5°	30-50	9-15
Malleable	8°	8°	10°	8°	80-100	24-30
Nonferrous						
Aluminum	10°	10°	10°-20°	35°	200-1500	60-460
Copper	10°	10°	10°-20°	16°	100-120	30-36
Brass	10°	8°	0°	0°	150-300	45-90
Bronze	10°	8°	0°	0°	90-100	27-30
Plastics						
Molded	10°	12°	0°	0°	150-300	45-90
Acrylics	15°	15°	0°	0°	60-70	18-21
Fiber	15°	15°	0°	0°	80-100	24-30

[1]The side- and end-relief angles are usually ground to the same angle for general machining operations. Average 8° end-relief and 10° side-relief angles are standard for turning most metals. The end- and side-relief angles for shaper and planer cutting tools generally range from 3° to 5°.
[2]The smaller angle is used without a chip breaker; the larger angle, with a chip breaker.

[3]The rake angles are measured from horizontal and vertical planes.
[4]The slower speeds are used for roughing cuts, machining dry. The faster speeds are used for finishing cuts, using cutting fluids. Cutting speeds for cast alloy bits are 50% to 70% faster; for cemented carbide tips and inserts, two to four times faster.

Table A-18 Recommended Cutting Speeds (sfpm) of Common Cutter Materials for Milling Selected Metals

Material to Be Milled	Milling Cutter Material				
	High-Speed Steel	Super High-Speed Steel	Stellite	Tantalum Carbide	Tungsten Carbide
Nonferrous					
Aluminum	500-1000	. . . .	800-1500	. . . .	1000-2000
Brass	70-175	. . . .	150-200	. . . .	350-600
Hard Bronze	65-130	. . . .	100-160	. . . .	200-425
Very Hard Bronze	30-50	. . . .	50-70	. . . .	125-200
Cast Iron					
Soft	50-80	60-115	90-130	. . . .	250-325
Hard	30-50	40-70	60-90	. . . .	150-200
Chilled	. . . .	30-50	40-60	. . . .	100-200
Malleable	70-100	80-125	115-150	. . . .	350-370
Steel					
Soft (low-carbon)	60-90	70-100	. . . .	150-250	300-550
Medium-carbon	50-80	60-90	. . . .	125-200	225-400
High-carbon	40-70	50-80	. . . .	100-175	150-250
Stainless	30-80	40-90	. . . .	75-200	100-300

Table A–19 Recommended Cutting Fluids for Ferrous and Nonferrous Metals

	Ferrous Metals				Nonferrous Metals	
Group	I	II	III	IV	V	VI
Machinability:*	Above 70%	50–70%	40–50%	Below 40%	Above 100%	Below 100%
Materials	Low-Carbon Steels High-Carbon Steels		Stainless Steels		Aluminum and Alloys Brasses and Bronzes	
Type of Machining Operation)**	Malleable Iron Cast Steel Stainless Steel	Cast Iron	Ingot Iron Wrought Iron	Tool Steels High-Speed Steels	Magnesium and Alloys Zinc	Copper Nickel Inconel Monel
1 Broaching, internal	Em Sul	Sul Em	Sul Em	Sul Em	MO Em	Sul ML
2 Broaching, surface	Em Sul	Em Sul	Sul Em	Sul Em	MO Em	Sul ML
Threading, pipe	Sul	Sul ML	Sul	Sul		Sul †
3 Tapping, plain	Sul	Sul	Sul	Sul	Em Dry	Sul ML
Threading, plain	Sul	Sul	Sul	Sul	Em Sul	Sul †
Gear shaving	Sul L	Sul L	Sul L	Sul L		
4 Reaming, plain	ML Sul	ML Sul	ML Sul	ML Sul	ML MO Em	ML MO Sul
Gear cutting	Sul ML Em	Sul	Sul	Sul ML		Sul ML
5 Drilling, deep	Em ML	Em Sul	Sul	Sul	MO ML Em	Sul ML
6 Milling — Plain	Em ML Sul	Em	Em	Sul	Em MO Dry	Sul Em
Milling — Multiple cutter	ML	Sul	Sul	Sul ML	Em MO Dry	Sul Em
Boring, multiple-head	Sul Em	Sul HDS	Sul HDS	Sul Em	K Dry Em	Sul Em
7 Multiple-spindle automatic screw machines and turret lathes: drilling, forming, turning, reaming, cutting off, tapping, threading	Sul Em ML	Sul Em ML	Sul Em ML HDS	Sul ML Em HDS	Em Dry ML	Sul
8 High-speed, light-feed screw machines: drilling, forming, tapping, threading, turning, reaming, box milling, cutting off	Sul Em ML	Sul Em ML	Sul Em ML	Sul ML Em	Em Dry ML	Sul
Drilling	Em	Em	Em	Em Sul	Em Dry	Em
9 Planing, shaping	Em Sul ML	Em Sul ML	Sul Em	Em Sul	Em Dry	Em
Turning, single-point tool, form tools	Em Sul ML	Em Sul ML	Em Sul ML	Em Sul ML	Em Dry ML	Em Sul
10 Sawing, circular, hack	Sul ML Em	Sul Em ML	Sul Em ML	Sul Em ML	Dry MO Em	Sul Em ML
Grinding — Plain	Em	Em	Em	Em	Em	Em
Grinding — Form (thread, etc.)	Sul	Sul	Sul	Sul	MO Sul	Sul

CUTTING FLUID SYMBOL: K = Kerosene; L = Lard Oil; MO = Mineral oils; ML = Mineral-lard oils; Sul = Sulphurized oils, with or without chlorine; Em = Soluble or emulsifiable oils and compounds; Dry = No cutting fluid needed; HDS = Heavy-duty soluble oil.

*Machinability rating based on 100% for cold-drawn Bessemer screw stock (B1112).

**Operations are ranked according to severity; 1 is the greatest and 10 is the least.

†Palm oil is frequently used to thread copper.

Compiled from *Metals Handbook, Machinery's Handbook,* and *AISI Steel Products Manual.*

Table A–20 Natural Trigonometric Functions (1° to 90°)

Angle	Sine	Cosine	Tangent	Angle	Sine	Cosine	Tangent
1°	.0175	.9998	.0175	46°	.7193	.6947	1.0355
2°	.0349	.9994	.0349	47°	.7314	.6820	1.0724
3°	.0523	.9986	.0524	48°	.7431	.6691	1.1106
4°	.0698	.9976	.0699	49°	.7547	.6561	1.1504
5°	.0872	.9962	.0875	50°	.7660	.6428	1.1918
6°	.1045	.9945	.1051	51°	.7771	.6293	1.2349
7°	.1219	.9925	.1228	52°	.7880	.6157	1.2799
8°	.1392	.9903	.1405	53°	.7986	.6018	1.3270
9°	.1564	.9877	.1584	54°	.8090	.5878	1.3764
10°	.1736	.9848	.1763	55°	.8192	.5736	1.4281
11°	.1908	.9816	.1944	56°	.8290	.5592	1.4826
12°	.2079	.9781	.2126	57°	.8387	.5446	1.5399
13°	.2250	.9744	.2309	58°	.8480	.5299	1.6003
14°	.2419	.9703	.2493	59°	.8572	.5150	1.6643
15°	.2588	.9659	.2679	60°	.8660	.5000	1.7321
16°	.2756	.9613	.2867	61°	.8746	.4848	1.8040
17°	.2924	.9563	.3057	62°	.8829	.4695	1.8807
18°	.3090	.9511	.3249	63°	.8910	.4540	1.9626
19°	.3256	.9455	.3443	64°	.8988	.4384	2.0503
20°	.3420	.9397	.3640	65°	.9063	.4226	2.1445
21°	.3584	.9336	.3839	66°	.9135	.4067	2.2460
22°	.3746	.9272	.4040	67°	.9205	.3907	2.3559
23°	.3907	.9205	.4245	68°	.9272	.3746	2.4751
24°	.4067	.9135	.4452	69°	.9336	.3584	2.6051
25°	.4226	.9063	.4663	70°	.9397	.3420	2.7475
26°	.4384	.8988	.4877	71°	.9455	.3256	2.9042
27°	.4540	.8910	.5095	72°	.9511	.3090	3.0777
28°	.4695	.8829	.5317	73°	.9563	.2924	3.2709
29°	.4848	.8746	.5543	74°	.9613	.2756	3.4874
30°	.5000	.8660	.5774	75°	.9659	.2588	3.7321
31°	.5150	.8572	.6009	76°	.9703	.2419	4.0108
32°	.5299	.8480	.6249	77°	.9744	.2250	4.3315
33°	.5446	.8387	.6494	78°	.9781	.2079	4.7046
34°	.5592	.8290	.6745	79°	.9816	.1908	5.1446
35°	.5736	.8192	.7002	80°	.9848	.1736	5.6713
36°	.5878	.8090	.7265	81°	.9877	.1564	6.3138
37°	.6018	.7986	.7536	82°	.9903	.1392	7.1154
38°	.6157	.7880	.7813	83°	.9925	.1219	8.1443
39°	.6293	.7771	.8098	84°	.9945	.1045	9.5144
40°	.6428	.7660	.8391	85°	.9962	.0872	11.4301
41°	.6561	.7547	.8693	86°	.9976	.0698	14.3007
42°	.6691	.7431	.9004	87°	.9986	.0523	19.0811
43°	.6820	.7314	.9325	88°	.9994	.0349	28.6363
44°	.6947	.7193	.9657	89°	.9998	.0175	57.2900
45°	.7071	.7071	1.0000	90°	1.0000	.0000	

Table A–22 Conversion Table for Optical Flat Fringe Bands in Inch and SI Metric Standard Units of Measure

Number of Fringe Bands	Equivalent Measurement Value*		
	Microinches	Inches	Millimeters
0.1	1.2	0.0000012	0.000029
0.2	2.3	0.0000023	0.000059
0.3	3.5	0.0000035	0.000088
0.4	4.6	0.0000046	0.000118
0.5	5.8	0.0000058	0.000147
0.6	6.9	0.0000069	0.000176
0.7	8.1	0.0000081	0.000206
0.8	9.3	0.0000093	0.000235
0.9	10.4	0.0000104	0.000264
1.0	11.6	0.0000116	0.000294
2.0	23.1	0.0000231	0.000588
3.0	34.7	0.0000347	0.000881
4.0	46.3	0.0000463	0.001175
5.0	57.8	0.0000578	0.001469
6.0	69.4	0.0000694	0.001763
7.0	81.0	0.0000810	0.002056
8.0	92.5	0.0000925	0.002350
9.0	104.1	0.0001041	0.002644
10.0	115.7	0.0001157	0.002938
11.0	127.2	0.0001272	0.003232
12.0	138.8	0.0001388	0.003525
13.0	150.4	0.0001504	0.003819
14.0	161.9	0.0001619	0.004113
15.0	173.5	0.0001735	0.004407
16.0	185.1	0.0001851	0.004700
17.0	196.6	0.0001966	0.004994
18.0	208.2	0.0002082	0.005288
19.0	219.8	0.0002198	0.005582
20.0	231.3	0.0002313	0.005876

*For a general approximation, one band may be considered to be one microinch or 0.0003mm.

Table A–21 Tapers and Included Angles

Taper per Foot	Taper per Inch	Included Angle		
		Deg.	Min.	Sec.
1/8	0.010416	0	35	48
3/16	0.015625	0	53	44
1/4	0.020833	1	11	36
5/16	0.026042	1	29	30
3/8	0.031250	1	47	24
7/16	0.036458	2	5	18
1/2	0.416667	2	23	10
9/16	0.046875	2	41	4
5/8	0.052084	2	59	42
11/16	0.057292	3	16	54
3/4	0.06250	3	34	44
13/16	0.067708	3	52	38
7/8	0.072917	4	10	32
15/16	0.078125	4	28	24
1	0.083330	4	46	18
1 1/4	0.104666	5	57	48
1 1/2	0.125000	7	9	10
1 3/4	0.145833	8	20	26
2	0.166666	9	31	36
2 1/2	0.208333	11	53	36
3	0.250000	14	15	0
3 1/2	0.291666	16	35	40
4	0.333333	18	55	28
4 1/2	0.375000	21	14	2
5	0.416666	23	32	12
6	0.500000	28	4	2

Table A–23 Microinch and Micrometer (μm) Ranges of Surface Roughness for Selected Manufacturing Processes

Roughness Height in Microinches and Micrometers (μm)*

Manufacturing Process	4000 (101.60)	3000 (76.20)	2000 (50.80)	1000 (25.40)	500 (12.70)	250 (6.35)	125 (3.18)	63 (1.60)	32 (0.81)	16 (0.41)	8 (0.20)	4 (0.10)	2 (0.05)	1 (0.03)	0.5 (0.01)
Flame cutting															
Snagging															
Sawing															
Planing, shaping															
Drilling															
Electrical discharge machining															
Milling (chemical)															
Milling (rough)															
Broaching															
Reaming															
Boring, turning (finish)															
Turning (rough)															
Barrel finishing															
Electrolytic grinding															
Burnishing (roller)															
Grinding (commercial)															
Grinding (finish)															
Honing															
Polishing															
Lapping															
Superfinishing															
Sand casting															
Hot rolling															
Forging															
Mold casting (permanent)															
Extruding															
Cold rolling (drawing)															
Die casting															

Code ▬ General manufacturing (average) surface finish range

▒ Higher or lower range produced by using special processes

*Values rounded to nearest second place μm decimal

Table A–24 Constants for Setting a 5″ Sine Bar (0°1′ to 10°60′)*

Minutes	0°	1°	2°	3°	4°	5°	6°	7°	8°	9°	10°
0	0.00000	0.08725	0.17450	0.26170	0.34880	0.43580	0.52265	0.60935	0.69585	0.78215	0.86825
1	.00145	.08870	.17595	.26315	.35025	.43725	.52410	.61080	.69730	.78360	.86965
2	.00290	.09015	.17740	.26460	.35170	.43870	.52555	.61225	.69875	.78505	.87110
3	.00435	.09160	.17885	.26605	.35315	.44015	.52700	.61370	.70020	.78650	.87255
4	.00580	.09310	.18030	.26750	.35460	.44155	.52845	.61510	.70165	.78790	.87395
5	0.00725	0.09455	0.18175	0.26895	0.35605	0.44300	0.52985	0.61655	0.70305	0.78935	0.87540
6	.00875	.09600	.18320	.27040	.35750	.44445	.53130	.51800	.70450	.79080	.87685
7	.01020	.09745	.18465	.27185	.35895	.44590	.53275	.61945	.70595	.79225	.87825
8	.01165	.09890	.18615	.27330	.36040	.44735	.53420	.62090	.70740	.79365	.87970
9	.01310	.10035	.18760	.27475	.36185	.44880	.53565	.62235	.70885	.79510	.88115
10	0.01455	0.10180	0.18905	0.27620	0.36330	0.45025	0.53710	0.62380	0.71025	0.79655	0.88255
11	.01600	.10325	.19050	.27765	.36475	.45170	.53855	.62520	.71170	.79795	.88400
12	.01745	.10470	.19195	.27910	.36620	.45315	.54000	.62665	.71315	.79940	.88540
13	.01890	.10615	.19340	.28055	.36765	.45460	.54145	.62810	.71460	.80085	.88685
14	.02035	.10760	.19485	.28200	.36910	.45605	.54290	.62955	.71600	.80230	.88830
15	0.02180	0.10905	0.19630	0.28345	0.37055	0.45750	0.54435	0.63100	0.71745	0.80370	0.88970
16	.02325	.11055	.19775	.28490	.37200	.45895	.54580	.63245	.71890	.80515	.89115
17	.02475	.11200	.19920	.28635	.37345	.46040	.54725	.63390	.72035	.80660	.89260
18	.02620	.11345	.20065	.28780	.37490	.46185	.54865	.63530	.72180	.80800	.89400
19	.02765	.11490	.20210	.28925	.37635	.46330	.55010	.63675	.72320	.80945	.89545
20	0.02910	0.11635	0.20355	0.29070	0.37780	0.46475	0.55155	0.63820	0.72465	0.81090	0.89685
21	.03055	.11780	.20500	.29220	.37925	.46620	.55300	.63965	.72610	.81230	.89830
22	.03200	.11925	.20645	.29365	.38070	.46765	.55445	.64110	.72755	.81375	.89975
23	.03345	.12070	.20795	.29510	.38215	.46910	.55590	.64255	.72900	.81520	.90115
24	.03490	.12215	.20940	.29655	.38360	.47055	.55735	.64400	.73040	.81665	.90260
25	0.03635	0.12360	0.21085	0.29800	0.38505	0.47200	0.55880	0.64540	0.73185	0.81805	0.90405
26	.03780	.12505	.21230	.29945	.38650	.47345	.56025	.64685	.73330	.81950	.90545
27	.03925	.12650	.21375	.30090	.38795	.47490	.56170	.64830	.73475	.82095	.90690
28	.04070	.12800	.21520	.30235	.38940	.47635	.56315	.64975	.73615	.82235	.90830
29	.04220	.12945	.21665	.30380	.39085	.47780	.56455	.65120	.73760	.82380	.90975
30	0.04365	0.13090	0.21810	0.30525	0.39230	0.47925	0.56600	0.65265	0.73905	0.82525	0.91120
31	.04510	.13235	.21955	.30670	.39375	.48070	.56745	.65405	.74050	.82665	.91260
32	.04655	.13380	.22100	.30815	.39520	.48210	.56890	.65550	.74190	.82810	.91405
33	.04800	.13525	.22245	.30960	.39665	.48355	.57035	.65695	.74335	.82955	.91545
34	.04945	.13670	.22390	.31105	.39810	.48500	.57180	.65840	.74480	.83100	.91690
35	0.05090	0.13815	0.22535	0.31250	0.39955	0.48645	0.57325	0.65985	0.74625	0.83240	0.91835
36	.05235	.13960	.22680	.31395	.40100	.48790	.57470	.66130	.74770	.83385	.91975
37	.05380	.14105	.22825	.31540	.40245	.48935	.57615	.66270	.74910	.83530	.92120
38	.05525	.14250	.22970	.31685	.40390	.49080	.57760	.66415	.75055	.83670	.92260
39	.05670	.14395	.23115	.31830	.40535	.49225	.57900	.66560	.75200	.83815	.92405
40	0.05820	0.14540	0.23265	0.31975	0.40680	0.49370	0.58045	0.66705	0.75345	0.83960	0.92545
41	.05965	.14690	.23410	.32120	.40825	.49515	.58190	.66850	.75485	.84100	.92690
42	.06110	.14835	.23555	.32265	.40970	.49660	.58335	.66995	.75630	.84245	.92835
43	.06255	.14980	.23700	.32410	.41115	.49805	.58480	.67135	.75775	.84390	.92975
44	.06400	.15125	.23845	.32555	.41260	.49950	.58625	.67280	.75920	.84530	.93120
45	0.06545	0.15270	0.23990	0.32700	0.41405	0.50095	0.58770	0.67425	0.76060	0.84675	0.93200
46	.06690	.15415	.24135	.32845	.41550	.50240	.58915	.67570	.76205	.84820	.93405
47	.06835	.15560	.24280	.32990	.41695	.50385	.59060	.67715	.76350	.84960	.93550
48	.06980	.15705	.24425	.33135	.41840	.50530	.59200	.67860	.76495	.85105	.93690
49	.07125	.15850	.24570	.33280	.41985	.50675	.59345	.68000	.76635	.85250	.93835
50	0.07270	0.15995	0.24715	0.33425	0.42130	0.50820	0.59490	0.68145	0.76780	0.85390	0.93975
51	.07415	.16140	.24860	.33570	.42275	.50960	.59635	.68290	.76925	.85535	.94120
52	.07565	.16285	.25005	.33715	.42420	.51105	.59780	.68435	.77070	.85680	.94260
53	.07710	.16430	.25150	.33865	.42565	.51250	.59925	.68580	.77210	.85820	.94405
54	.07855	.16580	.25295	.34010	.42710	.51395	.60070	.68720	.77355	.85965	.94550
55	0.08000	0.16725	0.25440	0.34155	0.42855	0.51540	0.60215	0.68865	0.77500	0.86110	0.94690
56	.08145	.16870	.25585	.34300	.43000	.51685	.60355	.69010	.77645	.86250	.94835
57	.08290	.17015	.25730	.34445	.43145	.51830	.60500	.69155	.77785	.86395	.94975
58	.08435	.17160	.25875	.34590	.43290	.51975	.60645	.69300	.77930	.86540	.95120
59	.08580	.17305	.26028	.34735	.43435	.52120	.60790	.69445	.78075	.86680	.95260
60	0.08725	0.17450	0.26170	0.34880	0.43580	0.52265	0.60935	0.69585	0.78215	0.86825	0.95405

5″ sine bar applications of constant values to different sizes of sine bars, plates, and chucks

Sine Bar Size (Inches)	Constant for 5″ Sine Bar
10	Multiply by 2
20	Multiply by 4
2-1/2	Multiply by 0.5
3	Multiply by 0.6
4	Multiply by 0.8

*Complete tables of sine bar constants from 0°1′ to 59°60′ are included in engineering and trade handbooks.

Table A–25 Suggested Starting Speeds and Feeds for High-Speed Steel End Mill Applications

	Materials				
	Heat-Resistant Cobalt-Base Alloys, High Tensile Steels (50–55C)		Cast Iron, Mild Steel, Half-Hard Brass and Bronze		Aluminum, Plastics, Wood
	Types and Styles of End Mills				
	Premium Cobalt High-Speed Steel 2 or More Flutes		High-Speed Steel End Mills, 2 or More Flutes Surface Treatment Helpful in C. I. Applications		High-Speed Steel End Mills of High-Helix Type, 1 to 6 Flutes
Diameter of End Mills (in Inches)	Speed 5–10 sfpm RPM	Feed Chip Load per Tooth	Speed 10–15 sfpm RPM	Feed Chip Load per Tooth	Speed 15–20 sfpm RPM	Feed Chip Load per Tooth
1/16	*	*	*	*		
3/32	*	*	*	*	611–815	.0002–.0005
1/8	*	*	*	*	456–611	.0002–.0005
3/16	*	*	204–306	.0002–.0005	306–407	.0002–.0005
1/4	76–153	.0002–.001	153–230	.0002–.001	229–306	.0002–.001
5/16	61–122	.0002–.001	122–183	.0002–.001	183–244	.0002–.001
3/8	51–102	.0002–.001	102–153	.0002–.001	153–203	.0002–.001
7/16	44–88	.0005–.001	88–132	.0005–.001	131–175	.0005–.002
1/2	38–76	.0005–.001	76–115	.0005–.001	115–153	.0005–.002
9/16	34–68	.0005–.002	68–104	.0005–.002	104–136	.0005–.002
5/8	31–61	.0005–.002	61–92	.0005–.002	92–122	.0005–.002
11/16	28–56	.0005–.002	56–84	.0005–.002	84–111	.0005–.002
3/4	26–51	.0005–.002	51–76	.0005–.002	76–102	.001–.004
13/16	24–47	.001–.003	47–71	.001–.003	71–94	.001–.004
7/8	22–44	.001–.003	44–65	.001–.003	65–87	.001–.004
15/16	20–40	.001–.003	40–62	.001–.003	62–81	.001–.004
1	19–38	.001–.003	38–58	.001–.003	58–76	.001–.004
1 1/8	34	.0015–.004	34–51	.0015–.004	51–68	.0015–.005
1 1/4	31	.0015–.004	31–46	.0015–.004	46–61	.0015–.005
1 3/8	28	.0015–.004	28–42	.0015–.004	42–55	.0015–.005
1 1/2	26	.0015–.004	26–38	.0015–.004	38–51	.002 Up
1 5/8	24	.002 Up	35	.002 Up	35–47	.002 Up
1 3/4	22	.002 Up	32	.002 Up	32–43	.002 Up
1 7/8	20	.002 Up	30	.002 Up	30–40	.003 Up
2	19	.002 Up	29	.003 Up	29–38	.003 Up
2 1/8	18	.003 Up	28	.003 Up	36	.003 Up
2 1/4	17	.003 Up	26	.003 Up	34	.003 Up
2 3/8	16	.003 Up	25	.003 Up	32	.003 Up
2 1/2	15	.003 Up	23	.003 Up	30	.003 Up
2 5/8	15	.003 Up	22	.003 Up	29	.003 Up
2 3/4	14	.003 Up	21	.003 Up	28	.003 Up
2 7/8	14	.003 Up	20	.003 Up	27	.003 Up
3	13	.003 Up	19	.003 Up	26	.003 Up

*Use solid carbide end mills in small-diameter applications for materials that are harder than 46 C Rockwell.
Note: The speeds and feeds apply also to cavity and slotting cuts up to the depth equal to the diameter of the end mill. Deeper cuts require these values to be decreased.
Table adapted with permission from The Cleveland Twist Drill Co.

Table A–26 Allowances and Tolerances on Reamed Holes for General Classes of Fits

Class of Fit	*Allowances and Tolerances	*Nominal Diameters					
		Up to 1/2"	9/16" through 1"	1-1/16" through 2"	2-1/16" through 3"	3-1/16" through 4"	4-1/16" through 5"
A	High limit (+)	.0002	.0005	.0007	.0010	.0010	.0010
	Low limit (−)	.0002	.0002	.0002	.0005	.0005	.0005
	Tolerance	.0004	.0007	.0009	.0015	.0015	.0015
B	High limit (+)	.0005	.0007	.0010	.0012	.0015	.0017
	Low limit (−)	.0005	.0005	.0005	.0007	.0007	.0007
	Tolerance	.0010	.0012	.0015	.0019	.0022	.0024

Allowances and Tolerances for Forced Fits

Class of Fit	Allowances and Tolerances	Up to 1/2"	9/16" through 1"	1-1/16" through 2"	2-1/16" through 3"	3-1/16" through 4"	4-1/16" through 5"
F	High limit (+)	.0010	.0020	.0040	.0060	.0080	.0100
	Low limit (+)	.0005	.0015	.0030	.0045	.0060	.0080
	Tolerance	.0005	.0005	.0010	.0015	.0020	.0020

Allowances and Tolerances for Driving Fits

Class of Fit	Allowances and Tolerances	Up to 1/2"	9/16" through 1"	1-1/16" through 2"	2-1/16" through 3"	3-1/16" through 4"	4-1/16" through 5"
D	High limit (+)	.0005	.0010	.0015	.0025	.0030	.0035
	Low limit (+)	.0002	.0007	.0010	.0015	.0020	.0025
	Tolerance	.0003	.0003	.0005	.0010	.0010	.0010

Allowances and Tolerances for Push Fits

Class of Fit	Allowances and Tolerances	Up to 1/2"	9/16" through 1"	1-1/16" through 2"	2-1/16" through 3"	3-1/16" through 4"	4-1/16" through 5"
P	High limit (−)	.0002	.0002	.0002	.0005	.0005	.0005
	Low limit (−)	.0007	.0007	.0007	.0010	.0010	.0010
	Tolerance	.0005	.0005	.0005	.0005	.0005	.0005

Allowances and Tolerances for Running Fits

Class of Fit	Allowances and Tolerances	Up to 1/2"	9/16" through 1"	1-1/16" through 2"	2-1/16" through 3"	3-1/16" through 4"	4-1/16" through 5"
X	High limit (−)	.0010	.0012	.0017	.0020	.0025	.0030
	Low limit (−)	.0020	.0027	.0035	.0042	.0050	.0057
	Tolerance	.0010	.0015	.0018	.0022	.0025	.0027
Y (average machine work)	High limit (−)	.0007	.0010	.0012	.0015	.0020	.0022
	Low limit (−)	.0012	.0020	.0025	.0030	.0035	.0040
	Tolerance	.0005	.0010	.0013	.0015	.0015	.0018
Z (fine tool work)	High limit (−)	.0005	.0007	.0007	.0010	.0010	.0012
	Low limit (−)	.0007	.0012	.0015	.0020	.0022	.0025
	Tolerance	.0002	.0005	.0008	.0010	.0012	.0013

*These inch-standard measurements may be converted to equivalent metric-standard measurements.

Table A–27 ANSI Spur Gear Rules and Formulas for Required Features of 20° and 25° Full-Depth Involute Tooth Forms

Required Feature	Rule	Formula*
Diametral pitch (P_d)	Divide the number of teeth by the pitch diameter	$P_d = \dfrac{N}{D}$
	Add 2 to the number of teeth and divide by the outside diameter	$P_d = \dfrac{N+2}{O}$
	Divide 3.1416 by the circular pitch	$P_d = \dfrac{3.1416}{P_c}$
Number of teeth (N)	Multiply the diametral pitch by the pitch diameter	$N = P_d \times D$
	Multiply the diametral pitch by outside diameter and then subtract 2	$N = (P_d \times O) - 2$
Pitch diameter (D)	Divide the number of teeth by the diametral pitch	$D = \dfrac{N}{P_d}$
	Multiply the addendum by 2 and subtract the product from the outside diameter	$D = O - 2A$
Outside diameter (O)	Add 2 to the number of teeth and divide by the diametral pitch	$O = \dfrac{N+2}{P_d}$
	Add 2 to the number of teeth and divide by the quotient of the number of teeth divided by the pitch diameter	$O = \dfrac{N+2}{N/D}$
Circular pitch (P_c)	Divide 3.1416 by the diametral pitch	$P_c = \dfrac{3.1416}{P_d}$
	Divide the pitch diameter by the product of 0.3183 times the number of teeth	$P_c = \dfrac{D}{0.3183 \times N}$
Addendum (A)	Divide 1.000 by the diametral pitch	$A = \dfrac{1.000}{P_d}$
Working depth (W^1)	Divide 2.000 by the diametral pitch	$W^1 = \dfrac{2.000}{P_d}$
Clearance (S)	Divide 0.250 by the diametral pitch	$S = \dfrac{0.250}{P_d}$
Whole depth of tooth (W^2)	Divide 2.250 by the diametral pitch	$W^2 = \dfrac{2.250}{P_d}$
Thickness of tooth (T)	Divide 1.5708 by the diametral pitch	$T = \dfrac{1.5708}{P_d}$
Center distance (C)	Add the pitch diameters and divide the sum by 2	$C = \dfrac{D^1 + D^2}{2}$
	Divide one-half the sum of the number of teeth in both gears by the diametral pitch	$C = \dfrac{1/2\,(N^1 + N^2)}{P_d}$

*Dimensions are in inches.

Table A-28 Conversion of Surface Speeds (sfpm) to Spindle Speeds (RPM) for Various Diameters of Grinding Wheels (1″ to 20″ or 25.4mm to 508.0mm)

Wheel Diameter (mm)	Surface Speed in Feet per Minute (sfpm)										Wheel Diameter (Inches)
	4,000	4,500	5,500	6,500	7,500	9,500	12,500	14,200	16,000	17,000	
	Spindle Speeds in Revolutions per Minute (RPM)										
25.4	15,279	17,189	21,008	24,828	28,647	36,287	47,745	54,240	61,116	64,935	1
50.8	7,639	8,594	10,504	12,414	14,328	18,143	23,875	27,120	30,558	32,465	2
76.2	5,093	5,729	7,003	8,276	9,549	12,096	15,915	18,080	20,372	21,645	3
101.6	3,820	4,297	5,252	6,207	7,162	9,072	11,940	13,560	15,278	16,235	4
127.0	3,056	3,438	4,202	4,966	5,730	7,258	9,550	10,850	12,224	12,985	5
152.4	2,546	2,865	3,501	4,138	4,775	6,048	7,960	9,040	10,186	10,820	6
177.8	2,183	2,455	3,001	3,547	4,092	5,183	6,820	7.750	8,732	9,275	7
203.2	1,910	2,148	2,626	3,103	3,580	4,535	5,970	6,780	7,640	8,115	8
228.6	1,698	1,910	2,334	2,758	3,182	4,032	5,305	6,030	6,792	7,215	9
254.0	1,528	1,719	2,101	2,483	2,865	3,629	4,775	5,425	6,112	6,495	10
304.8	1,273	1,432	1,751	2,069	2,386	3,023	3,980	4,520	5,092	5,410	12
355.6	1,091	1,228	1,500	1,773	2,046	2,592	3,410	3,875	4,366	4,640	14
406.4	955	1,074	1,313	1,552	1,791	2,268	2,985	3,390	3,820	4,060	16
457.2	849	955	1,167	1,379	1,591	2,016	2,655	3,015	3,396	3,605	18
508.0	764	859	1,050	1,241	1,432	1,814	2,390	2,715	3,056	3,245	20

Table A-29 Recommended Heat Treatment Temperatures for Selected Grades and Kinds of Steel

AISI or SAE Number	Hardening Temperature Range		Annealing Temperature Range		Normalizing Temperature Range		Quenching Medium
	°F	°C	°F	°C	°F	°C	
1030	1550/1600	843/871	1525/1575	829/857	1625/1725	885/941	
1040	1500/1550	816/843	1475/1525	802/829	1600/1700	871/927	
1050	1475/1525	802/829	1450/1500	788/816	1550/1650	843/899	
1060	1450/1500	788/816	1425/1475	774/802	1500/1600	816/871	
1070	1450/1500	788/816	1425/1475	744/802	1500/1600	816/871	Water or brine
1080	1400/1450	760/788	1375/1425	746/774	1475/1575	802/857	
1090	1400/1450	760/788	1375/1425	746/774	1475/1575	802/857	
1132	1550/1600	843/871	1525/1575	829/857	1625/1725	885/941	
1140	1500/1550	816/843	1475/1525	802/829	1600/1700	871/927	
1151	1475/1525	802/829	1450/1500	788/816	1550/1650	843/899	
1330	1525/1575	829/857	1500/1550	816/843	1600/1700	871/927	Oil or water
1340	1500/1550	816/843	1475/1525	802/829	1550/1650	843/899	Oil or water
3140	1475/1525	802/829	1475/1525	802/829	1550/1650	843/899	Oil
4028	1550/1600	843/871	1525/1575	829/857	1600/1700	871/927	Oil or water
4042	1500/1550	816/843	1475/1525	802/829	1550/1650	843/899	Oil
4063	1475/1525	802/829	1450/1500	788/816	1550/1650	843/899	Oil
4130	1550/1600	843/871	1525/1575	829/857	1600/1700	871/927	Oil or water
4140	1525/1575	829/857	1500/1550	816/843	1600/1700	871/927	Oil
4150	1500/1550	816/843	1475/1525	802/829	1600/1700	871/927	Oil
4340	1500/1550	816/843	1500/1550	816/843	1600/1700	871/927	Oil

INDEX

A

Abrasive jet machining (AJM), 566
Abrasive machining (*See also* Grinding wheels, Cutter and tool, Surface, and Cylindrical grinders, 412–482.)
  characteristics of, 412–413
  cutter and tool grinding, 462–482
  cutting fluids, coolant systems, 435–437
  cylindrical grinding, 456–466
  grinding machines, 412–421
  grinding wheels, 422–437
  honing, lapping, superfinishing, 419–420
  microfinishing, 419
  polishing, buffing, surface finishing, 420
  surface grinding, 414–415, 438–455
  universal grinding, 415–418
  vibratory finishing, deburring, 418
Abrasives
  artificial, 100
  bonding, coating, backing, 101
  coated (structure), 99
  forms, 99–100
  grain sizes, 101
Acceleration, deceleration, NC/CNC, 499
Acceptable quality level (AQL), 63–66
Acme screw threads, 252–253
  stub Acme, 253–254
Adapter plates, magnetic chucks, 443
Adaptive controls, (FMS), speeds, feeds, 573
Adjustable chucking reamers, 160
Adjustable limit snap gage, 61
Adjustable thread ring gage, 62
AISI, SAE, steel classifications, 524–526
Allowances, tolerance limits, thread sizes, 243
Alloys, nonferrous, 530–531
Alloy steels, 524, 530
Aluminum Association (AA) designations, 526
Aluminum oxide abrasive wheels, 412
American system of manufacturing, 1
Angle plate sine bar, 447

Angles, single-point cutting tools, 186–194
  end-cutting, rake, relief angles, 188
  lead, angle of, 188
Annealing, normalizing, spheroidizing, 545–546
APT (Automatically Programmed Tools), manuscript preparation, 514–516
Arbor press, 238
Arbors, milling machines, 293
ASA Surface Texture Standards, measurement of, 66–69
Austempering, 541
Automated process systems, types of, NC/CNC, 494
Automatic bar and chucking machines, 256. *See also* Turret lathes
Automatic screw machines. *See* Screw machines

B

Balancing, grinding wheel, 434
Band grinding, polishing, 399–400
Band machines, horizontal, 382–391
  cutoff sawing, 126–131
  cutting action, fluids, 127–128
    straight, soluble oils, 127–128
    synthetic cutting fluids, 128
  design features, 382–386
  dry, wet, cutting, 131
  job selector, 129–131
    reading:
      coolant, feed, 130
      pitch, blade type, velocity, 131
  saw blade, forms:
    precision, buttress, claw, 127
    tungsten carbide, 128
Band machining processes, 387–401
  abrasive grinding, 399–400
    line band, 400
  band (saw) recommendations, 384
  blade: cutoff shearing, welding, 387
    electroband, 400
  filing, 398–399
  knife-edge band, 400–401
  sawing: external, internal, 392
    cutoff, requirements of, 129–130
    blade selecting for, 130

  coolants and feeds, 130
  friction, 395–397
  high-speed, 397
  problems, causes, corrective action (Table), 394
  slitting, 394–395
  slotting, 394
  spiral-band, 397
Bar (and chucking) machines. *See also* Turret lathes
Bar turners, 264–265, 272
  cutting feeds, speeds, (Table), 273
Basic oxygen steel manufacturing, 520–521
Baths, quenching, 540, 541
Bed-type, milling machine, 363
Bench comparator, 78
Bench work, hand tools, noncutting, 86–89
Bench work, layout, measuring tools:
  angle measuring, 20–21
  combination set, 19–20
  rules, steel, 15–18
    alignment of, discrimination, 15–16
    flexible steel tapes, 16
    measuring with, 17–18
  surface plates, granite, features of, 76
    angle, universal, right angle blocks, 77
    parallels, V-, right-angle blocks, 77
Bevel protractors, universal, vernier, 39
Binary code decimal (BCD) system, (notation), NC/CNC, 497
Blocks, parallel, precision gage, 44–45
Boring, defined, 170
  processes, mandrel work, 234–239
  tools, holders (lathe work), 234–235
    boring bar, 170
Boring, process, milling machine, 376–381
Boring head, set, 367
  offset, 376–381
Boring machines, numerically controlled, 510
  jig borer, NC, 510–511

C

Brinell Hardness testing, 554
British Standard Whitworth (BSW) thread, 242
Broaching, screw threads, 251
Brush analyzer, surface finish, 73

CADD:
  model, robotics, 575
  subsystem, (FMS), 570
Calipers and measurements, 22–24
Cams, 357–360
  milling of, uniform, irregular rise, 357–360
  motions, 357
  terminology, 358–359
  types of, 357–358
Cam-lock adapters, 296
Casehardening, 546–547
Cast alloy cutting tools, 531
Cast irons, kinds, properties, 527–529
  cupola, pearlitic, 529
Cemented carbide cutting tools, 531
Center drills, holes, 197
Centerless grinding, 417
Center work, setups for, 197–200
Ceramic cutting tools, 531–532
Chemical machining, 565
Chips, formation, control, thickness, 188
Chisels and chiseling, 92–94, 104
Chucking machines. See Turret lathes
Chucks and chucking, 224–229
  drill, Jacobs, 226–227
  four-jaw independent, 225–226
  Jacobs, spindle nose, 226–227
  mounting, truing work in, 227–228
  power-operated, self centering, 226
  spindle nose collets, 226–227
  universal three-jaw, 225
Circular interpolation, 499
Clamps, applications:
  bent, straight, finger, U-strap, 142–143
  column mounted (drill press), 142
  parallel, 90
Clearance, rake angles, flat cutting tools, 480
Closed loop NC system, 495
CNC basic components:
  presetting NC cutting tools, 518
  productivity, 517
  standard features, 517
  tape-controlled, programming, 516
Cobalt high-speed steel end mills, 371–372
Collets, chucks, turret lathe, 262
Combination drill and countersink, 168

Command pulse weight, 499
Comparators. See High amplification comparators
Compound rests, positioning, 201
Computer-assisted programming, NC, CNC, 512–513
Computer integrated manufacturing, 571
Computerized numerically controlled (CNC) machines. See NC/CNC machines, systems, programs.
Coolant systems. See Cutting fluids.
Coordinate system, rectangular, 495
Counterbores, counterboring, 169–170, 236
Countersinks, countersinking, 168, 170
Crush dressing, rolls, 451
Cubic boron nitride (CBN) grinding wheels, 432–434
Cutter and tool grinding, 467–482
  attachments:
    centering gage, 471
    cutter grinding arbor, 471
    cylindrical grinding, internal, 469
    gear cutter sharpening, 469
    small end mill, 470
    surface grinding, 468
    tooth rests, types, 471
      table mounted, 472
      wheel-head mounted, 473
  grinding fluids for:
    dry grinding, 436–437
    wet grinding, systems, 435–436
  milling cutter offsets, (Table), 474
  primary relief, secondary relief, setups, 473, 475, 477
    checking, dial indicator, 474
    clearance angles, (Table), 474
  tooth sharpening:
    clearance, rake, angles, 480
    flat cutting tools, 480
    end mills, side cutting edges, end teeth, 478–479
    form relieved cutters, 480
    reamer grinding, 479–480 (Table), 480
    single-, double-angle cutters, 479
    slitting saws, 476
    staggered-tooth milling cutters, 476
    taps, 480
  universal design features:
    work head, 467
    wheel head, 467–468
Cutting feeds, fluids, speeds. See separate machine tool applications
Cutting-off, (surface grinder), 453

Cutting oils:
  newly developed, 282
Cutting tools, toolholders. See also Machine tool applications.
  angles for relief, rake, side-cutting, 186
  cast nonferrous alloys, 189
  cemented carbides, 188–190
    specifications of, 191–192
  cutting-tool tools, holders, turret lathe, 267–268
  high-speed steels, 188–189
  industrial diamonds, 190
  tool circle, planned, 268
Cycling, constant, intermittent, NC/CNC, 495
Cylindrical grinders, grinding, 456–466. See also Grinding wheels.
  accessories, 459
  automated functions, 460–461
  external, small angle tapers, 463
  external, steep angle tapers, 463–464
  face, 464
  internal, straight (parallel), 464
    problems, 464
    small-, steep-angle tapers, 465
  operational data, 459
  plain, 456
  problems, causes, corrective action, 461–462; (Table) 461
  straight (parallel), 462
  supporting long workpieces, 463
  traverse, plunge, (Table), 460
  universal, 456–459
  wheel recommendations, (Table), 462

D

D-1 cam lock, spindle nose, 224
Decalescence point, 537
Degrees of freedom, robots, 574
Dial indicators, 41–42. See also Mechanical comparators, gages (indicating)
Diametral pitch, 347
  normal (circular), 343
Diamond cutting tools, 532
Diamond dressers, roll form, single-point, cluster, 430
  tools, 452
Die heads, 265–266
Dies, external thread cutting, 118–121
Digital:
  control, surface grinders, 438
  system, height gage, 50
  readouts:
    jig borer, 485
    vertical millers, 368
Dimensional measurement:
  characteristics, features, 9, 11

Dimensional tolerance, 58
Dividing heads. *See also* Indexing.
   accessories for, parts of, 333–334
   collet index fixture, 331
   direct indexing, 331
   rotary (circular) table, 331–332
     ultraprecise, 332
   standard, 332
   wide-range adapter, spiral, 332
   universal, 332
Double-housing planers, 406
Dovetail milling, 328–330
   characteristics of, 328–329
   measurement of, 329
Down-feed attachment, automatic,
    surface grinder, 440
   downfeeds, 442
Dressers, dressing:
   angle, radius, combination, 450
   microform, forming crush rolls,
    452–453
Drill drifts, 151
Drilling: 147–157
   calculating RPM, 153
   cutting feeds for 153; (Tables), 154
   cutting speeds, factors affecting, 155
   problems, causes, corrective action,
    (Table), 154
   processes, 137–138, 155–156, 276
Drilling machines, accessories, 137–146
   deep-hole, 140
   gang, 139–140
   heavy-duty upright, 139
   machines, numerically controlled,
    509
   multiple-spindle, 140
   radial, 140–141
   sensitive, 138–139
   standard upright, 139
   special hole-producing, 141–142
   turret, multiple station, 141
Drill jig, fixture, 144
Drill point gage, 151
Drill points, shapes, 149
Drills: 147–157
   carbide-tipped, 147–148
   carbon steel, 147–148
   cobalt high-speed steel, 147
   coolant feeding, 148
   measurement of, 153
   parts, function of, 148–149
   points, grinding, 152
   sizes of, 152–153
   split point, 150
   two-, three-, four-flute, 148
Drill shank adapters, 151

E

Electrical comparators, 58
Electrical discharge machining (EDM),
   562

accessories, 563
   cutting rates, factors affecting, 562
   traveling wire system, 563–564
Electric furnace, steel manufacturing,
   520
Electroband machining, 400
Electrochemical machining (ECM),
   564
   grinding (ECG), 565
Electromagnetic chucks, 445
Electron beam machining (EBM), 566
Electronic comparators, 58
   amplifier readout, 58
   gaging heads, 58–59
Electronic tool gage, 518
End effector, robots, 574
End mills, 370–374
   grinding, attachment, 478, 479
English module, gearing, 341
Eutectoid steels, 535
Expansion reamers, 160

F

Faceplate, driver plate, 177–178
Facing, lathe work, 202–203
Feeds. *See* separate machine tools
   and machining processes
Files:
   cuts of, names, 95
   features, 97: parts of, 94
   machinists: mill, round, half-round,
    pillar, 94–95
   needle-handle, 96
   special purpose, 96
   Swiss pattern, 95–96
Filing:
   cross, 97–98
   draw-, 98
   soft metal, 98
Fitch, Stephen, 2
Fits, major types, 63
Fixed-bed milling machine, 289
Fixed-size gages. *See* Gages
Flatness, checking for, 81, 85
Flexible manufacturing systems (FMS),
   570–576
   description of, 570
   display of, 571
   robotics, applications to, 574–575
   subsystems:
    adaptive controls, feeds, speeds,
     573
    CAD/CADD, 570
    CIM/CAM, CIM/GEN, 571
    functions of, 572
    in-process, post-process gaging,
     572
    materials handling, automated
     572
    tool-handling, automated, 572

Floating zero point, NC/CNC, 499–500
Formed relieved cutters, 480
Form dressing, grinding, 446–453
Furnaces:
   basic oxygen, steel-making, 521–522
   heat treating, 542

G

Gage blocks:
   angle measurements, computing
    (inch/metric combinations),
    78–79
   applications of, 44
   grades of, 43
   permanent magnetic sine plate,
    80–81
   precision, 78
   simple, compound sine plates, 80
   sine bar, 79
   sine block, 80
   Table of Constants (partial), 79
Gage, cutter clearance (cutter grinding), 475
Gage maker's tolerances (Table), 60
   basic gage designs, 61
Gages. *See also* High amplification
    comparators
   dimensional control, 59
   fixed-size, 60–62
   height, optical, 50
Gaging, in-process, post-process, FMS,
   572
Gear, burnishing, finishing, 338
Gear cutter sharpening attachment, 469
Gears and gear milling, 336–352
   bevel, computing dimensions of, 344
    design features, 343
    machining, 349–350
    straight, spiral teeth, 338
   grinding and lapping teeth, 338
   helical, machining, 348–349
   involute curve, 342
   involute cutters, sizes, 347
   materials for, 345
   modules: (English), 341
    (SI metric), 341–342
   spur:
    drawings (ANSI), 340
    formulas (ANSI), 342;
     (SI metric), 341
    machining, 347–348
    rack milling (indexing) attach-
     ment, 348
   worm:
    features, terminology, 345
    single-, multiple-thread, 344
    velocity ratio, 344
Gear production (nonmachining)
    processes, 337
   casting, hot rolling, extruding, 337
   powder metallurgy, 337–338

Gear shaping, 336–337
Gear shaving, 337
Gear tooth measurement, 338–340, 347
   vernier caliper, 339
   wire diameters, Van Keuren, 339–340
   two wire, 339–340
Gooseneck planer toolholder, 409
Grinding: *See also* Abrasive machining, Cylindrical, Surface, Tool and cutter grinding
   bench and floor, offhand, 132–136
     angular surfaces, 135
     features of machines, 132–133
     processes, defined, 132
   precision
     cylindrical, 456–466
     surface, 438–453
     tool and cutter, 467–482
Grinding coolants, fluids, 435–437
Grinding, machines, processes, tooling.
   See Abrasive machining
Grinding wheels, 133–136, 421–435
   See also Cylindrical grinding, Cutter and tool grinding, Surface grinding.
   abrasives, types, 421–422
   balancing and truing, 133–134
   balancing ways, parallel, overlapping disk, 434
   CBN, diamond wheels, conditioning, 432–434
   computing work (surface) speed of, 425
   cones, plugs, mounted, 425
   cylindrical grinder, 423
   form dressers, dressing, 430
     truing, dressing, cylindrical/surface grinder, 431–432
   problems, causes, corrective action, (Table), 426;
     selection factors, (Table), 427
   surface grinder, 423–425
     rate of travel, 425
   traverse dressing rate, effect on, 432

H

Hacksawing, power, metal cutting, 122–125. *See also* Horizontal band machines and Band machining
   blades:
     selection of pitches, 124
     set patterns for teeth, 124–125
   drives, machine operation, 122–123
   heavy-duty, production, 123
   major machine parts, 123
   speeds, feeds, 124
   stacking, 124
   types of, drives, machine operation, 122–123

Hardness testing, 552–558
   Brinell, 554–556
   general processes of, 552
   Knoop, 557
   Rockwell:
     features, penetrators, 552–553
     scales, testing factors, (Table), 554
     testing with, 553–554
   Scleroscope, Vickers, microhardness, 556
Heat treating, 537–551. *See also* Metallurgy
   annealing, 545–546
   baths, 540–541, quenching processes:
     austempering, martempering, 541
     isothermal, 541–542
   casehardening/carburizing, 546–548
   furnaces, processes of, 542–543
   hardening, critical points, 538–539
   hardening, flame, induction, 549
   hardening, Table, 544
   hardness stabilizing, subzero, 550
   nitriding, 548
   normalizing, spheroidizing, 546
   problems, causes, corrective action, 537, (Table), 538
   pyrometers (thermocouple), 542
   tempering, tungsten high-speed steels, 545
Height gages, 50
Helical milling, 353–360
   grooves, setups, 356–357
   machining, secondary clearance angles, 356–357
Helix, 353–355
   calculating angle of, 353
   change gears for machining, formulas, 353–355
     compound gearing, 354–355
   lead, angle, hand, 353
   short-lead, gearing for, 353
High amplification comparators, 52–58
   electrical, electronic, 58
   mechanical, dial indicator, 53–54
   mechanical, optical, 55
   pneumatic, column-, pressure-types, 57
Horizontal (cutoff sawing) band machine. *See* Band machines, horizontal
   dry and wet cutting, 131
   functions of, 126
   operating principles, 126–127
Horizontal milling. *See also* Milling: cutters, machines, processes
Horizontal milling, cutters, 299–310, 316
   angle, (single-, double-), 303
   ANSI, 299

   dovetail, single-angle, 328
   end, 303–304
     features of, 299–300
       solid end mills, 303–304
         cutting speeds, 318, 319
         feed rates (Table), 318
         holding, mounting, 319
       milling shoulders with, 304, 316
   face, general, 316
     cutter features, 316–317
     feed, cutter lines, 317
     lead angle, effect of, 317
   fly, 304
   formed tooth, 300
   heavy-duty, plain, 301
   historical:
     Howe, Frederick; Warner, Thomas, 3
   helical, plain, 301
   interlocking side, 321–323
   light-duty, plain, 300
   metal slitting saw, 299–300
   positive, negative, zero rake angles, 299–300
   screw slotting, (Table), 325–326
     work-holding setups, 326
   side, 301–302; applications, 321–324
     half-side, 301–302
     interlocking, 302, 322–323
     plain, 301
     staggered tooth, 302
     straddle, milling, 322
   slitting saws, 302–303, 325
     side-tooth, features of, 325
   standard tooth form, 299
   surface finish, factors affecting, 317
   T-slot and Woodruff keyseat, 330
Horizontal milling, processes, 299–310, 311–315
   climb (down), 311–312
   conventional (up), 311:
     computing ipm/mpm, 304–305
     feeds (rate) for, 304
   cutting fluids for, properties, 307
     kinds of, 308
     surface finishes, effect of, 307–308
     tool life, 307
   dovetails, 328–329
   end, 303–304
   face, 316–320
   flat surfaces, 312–313
   gang, 323
   indexing, direct (simple), 331–335
   keyseat, Woodruff:
     depth of cut, specifications, 327–328
     dimensions, (Table), 328
     setup, machining, 327–328

Horizontal milling *(continued)*
   parallel, right-angle surfaces, 313
   plain (peripheral or slab), problems,
      causes, corrective action
      (Table), 313–314
   sawing, slotting, 325, 326
   shoulder, 318, 320

I

Incremental spindle positioning, NC/
   CNC, 499
Indexing principles, applications,
   331–335
   direct, devices, 331–335
   rotary table, 331–332
      ultraprecise, optical, 332
   simple (plain), 334
      calculations, 335
      index plates, 335
   standard dividing head, 332
   universal dividing head, 332
   vibration damping unit, 314
Indicators. *See* Dial test indicators
Industrial magnifiers, 50–51
Industrial materials, 520–533. *See*
   Metals and alloys
Interchangeability:
   American National, Unified, screw
      thread forms, 63
   North, Simeon, 3
Interpolation, linear, circular, 499
Involute gear cutter, 347
Iron, alpha, gamma, 535
Iron-carbon phase diagram, 537

J

Jack (screw), 144
Jig borers/boring:
   machine tools and accessories,
      483–486
      boring chuck, bars, chucks,
         collets, 484
      digital readout and printer, 485
      locating microscopes, 485–486
      table mounting: rotary/micro-
         sine tables, index center, 485
      tapping heads, 484–485
Jig grinders/grinding,
   attachments:
      dressing, angle, spherical socket,
         pantograph, 488
      heads for, 487, 488
      slot grinding, 487
   material allowances for, 488–489
   processes:
      out-feed, 486; wipe, plunge,
         taper, shoulder, 487
      setting up for, 486
   wheel and diamond charged man-
      drels, speed of, 488

K

Keyseat milling, 326
Keyway cutting, Woodruff, 327
Keyway, slot, end milling, 374
Knife-edge bands, machining, 400–401
Knoop hardness testing, 557
Knurling and knurls:
   process of, 214
   roll sizes and holders, 213

L

Laminated parallels, magnetic, 440
Laser beam machining (LBM), 567–568
Lathes (engine):
   carriage, parts of, 175–176
   centers, 177, 198
      alignment of, checking, 182
   center work, setups, 197–200
   chucks, chucking, 224–229
      chucks:
         three-jaw, four-jaw, universal,
            225–226
         power-operated, self-centering,
            226
      collets, spindle nose, Jacobs,
         226–227
      mounting spindle accessories, 228
      work-truing methods, 227–228
   compound rest, 176
   controls for, 181–184
   cutting tools, 186–196:
      design features, materials, 188
      chip:
         formation, control, breakers,
            192
      grinding angles, cutting tools:
         end-cutting edge relief, 195
         side- and back-rake, 195
         side-cutting edge relief,
            194–195
      lead angle, chip thickness, force,
         188
   floturn, 180
   gear box, quick-change, 175, 181
   geared head of, variable speed drive,
      174
   headstock, 173–174
      controls, 181
      drives: step cone, back gear,
         variable, 181
      historical: Blanchard, Thomas,
         Maudslay, Henry, Wilkinson,
         David, 1–3
   machining problems, causes, correc-
      tive action, 184
   maintenance of, 182–183
   spindle nose types, 174–175,
      224–225
   tailstock, 176
   taper attachment, 178
   thread cutting, 178
   toolmaker's lathe, 180
Lathe work, processes:
   angle turning:
      compound rest, computing,
         221–222
      shoulder, 206–207
   arbor press, 238
   boring:
      bars, tools, holders,
         234–235
      mandrel work, 234–239
      straight holes, 236–237
      taper, 236
   center drilling, 197–198,
      199
   counterboring, recessing, 236
   cutting off, 209–211
   drilling, 230–232
   facing:
      process, 202–203
      tools, setups, 202
   filing:
      files, types of, 212
      turned surface, 212–213
   grooving, form turning, cutting off,
      208–211
   knurling, knurls, 213–215
   mandrels, types, 237–238
   polishing, 213
   reaming, 230–231
   rounded end, turning, 207
   shoulder turning, 205–206
      radius gage, use, 205
   straight turning, 201
      compound rest, positioning for,
         201
      depth of cut, setting, 201
   taper turning/tapers:
      American Standard Taper Pins,
         Jacobs, Morse, Brown &
         Sharpe, self-releasing steep
         tapers, 217–218
      definitions, formulas, 218–219
      dimensioning, measuring, gaging,
         219–220
      process of, 220–221
Layout and inspection practices, sur-
   face plate work, 76–85
Layout and transfer measuring tools,
   18–25
Lay, patterns, symbols, 68
Lead screw, lathes, 241–242
Linear interpolation, 499
Linear measurement:
   precision height gage, 82
Line grinding, band, 400
Liquid baths, 540
Locating microscopes, 485–486
Lost motion, 182

M

Machine control unit (MCU), 512–513
Machining centers, numerically controlled, NC/CNC, 511–512
Mandrels:
    expansion, 237
    gang, 238
    solid, 237
    taper-shank, 238
    threaded, 238
Malleability, steel, properties, 523–524
Malleable iron castings, 528
Martempering, 541
Martensite, characteristics, effects of, 537
Materials handling, FMS, automated, 572
Maximum waviness height, 66
Measurement:
    bases of, measured point, 11
    conversion tables for, 14
    errors in reading:
        manipulation, parallax, worker bias, 17
    industrial drawings, applications of, 11
    systems of (British, decimal-inch, metric), 13
Metallurgy, 534–539. *See also* Heat treating, Hardness testing
    carbon steels:
        cementite, ferrite, eutectoid, 535
        critical points, heating/cooling rates, 535
        iron-carbon phase diagram, 535–539
        martensite, characteristics of, 537
        rapid/slow cooling, effects of, 536–537
        temperature/time, effects of, 535
Metals and alloys, 520–533
    aluminum designation system, 526
    cast alloy cutting tools, 531
    cast iron:
        castings, kinds, characteristics, 527–529
    diamond (polycrystalline) cutting tools, 532
    nonferrous alloys, 530–531
    steel:
        alloy, 524
        carbon-, plain, medium/high carbon, 523–524
        castings, carbon, alloy, 529–530
        chemical, mechanical, properties of, 522–524
        manufacturing processes, 520–522
        SAE, AISI classifications, 524–526
    tool and die, coding of, 526
    Unified numbering system (and alloys), 526
    visible identification of, 527
Microform dresser, 452–453
Microhardness testing, 556
Micrometers:
    depth, 32–33
        measuring with, 33–34
    dial indicating, 42–43
    functions, advantages, 27
    inch standard, 28
    inside, 34
        measuring with, 29–30
    metric, 30–31
    principles for using, 28
    screw thread, 243–244
    vernier, 30–31
        scale graduations, 34–35
Microscopes:
    shop, application of, 49
    surface finish, 72
    toolmaker's measuring, 49
Micro-tapping attachments, 164
Milling machines. *See* Horizontal and Vertical milling machines
    two-, three-, four-axes, NC, 510
Mirror image, programming, 505–506
Module system of gearing, 341
Multiple-spindle automatic screw machines, 280–283
Multiple-start screw threads, 254

N

National Acme, Stub Acme, Thread Forms, 252–254
National Coarse, Fine, Thread series, 242
National Screw Thread Standards, 242
NC/CNC milling machines, 363
Nitriding, salt bath, gas, 548
Nodular cast iron, 527
Nonreinforced, cutoff wheels, 453
Nontraditional machines/Machining, 559–569
    energy groups, (Table), 559–560
    machining processes:
        electrical discharge (EDM)
            accessories, 563
            cutting rates, factors affecting, 562
            traveling wire system, 563–564
        electrochemical (ECM), 564
        (CHM), (ECG), (ECD), (USM), 565
        laser beam, (LBM), 567–568
        plasma arc (PAM), abrasive jet (AJM), electron beam (EBM), 566
    metal removal rates, (Table), 561
    surface integrity, 561
Normalizing, annealing, spheroidizing, 545–546
Numerical control, NC/CNC, machines, features, processes, 508–519
    CNC, components, features:
        presetting NC cutting tools, 518
        manuscript preparation, APT, 514–516
        productivity, 517
        standard features, 517
        tape-controlled programming, 516
NC, machines:
    basic control features, 508
    bar, chucking, 510
    boring, 510
    computer-assisted programming (CAP), 512
    drilling, 509
    jig boring, 510–511
    machining centers, 511–512
    machine control unit (MCU), 512–513
    milling, 1, 2, 3, 4 axes, 510
    NC data processing for, 512
    turning, two axes, continuous path, 509–510
    vertical millers, 363
Numerical control (NC), computer-assisted (CNC) machine tools, 492–519
NC programming, factors:
    acceleration, deceleration, cuts, 499
    command pulse weight, 499
    linear, circular interpolation, 499
    spindle positioning:
        floating zero point, 499–500
        incremental, coordinate, 499
NC systems, programming, 492–507
    advantages of, 492–494
    applications, standard machines, 494
    automated processes, types of, 494
    binary system, notation, (BCD), 497
    closed-, open-loop, systems, 495
    constant, intermittment cycling, 495
    NC rectangular coordinate system:
        quadrants, point values, 496
        X, Y, Z axes, reference point, rotational axes, 495
    programming information, 494
    tool changers, storage drum, 494
NC tape preparation:
    manual programming, 500
    tape block, rewind stop code, 500
NC TAB sequential format programs:
    description of, 501
    single-axis, single machining process, 501

speed programming, 502–503
    feed changes, milling, 503–504
    programming for, 504–505
  third-axis function, 503–504
  X, Y, Z axes and tool changes, 503
    programming, 503
NC word address format programs:
  continuous path machining, 505
  fixed block, 505
  mirror image, 505–506
  sample coded functions, 505
NC word language, 497
  dimension, nondimension words, 498
  selected NC functions, other, 498–499

O

Open-hearth steel, making, 520–522
Open-side shaper-planers, 406
Optical comparators:
  fringe bands, (Table), 52
  optical flats, 51
  reference and working, 51–52
Optical height gage, 56
Optical measuring system, vertical millers, 366
Optical thread comparators, 244
Overlapping disk balancing ways, 430
Oxygen steel manufacturing process, 520

P

Pantograph dressing attachment, 452
Parallel balancing ways, 430
Parallels, laminated, magnetic, 440
Permanent tool setups, turret lathes, 268–269
Pitch diameter, 110, 118
Planers, design features of, 406, 408
  adjusting cutting speeds, feed rate, 408
  cutting tool holders, 409
  planing flat surfaces, 408
  side head, planing with, 409
  tool bit, geometry of, 409–410
Plasma arc machining (PAM), 566
Plug gages, 60
  GO, NO-GO, 61
  Taperlock, cylindrical, 61
Pneumatic comparators, 55–56
  amplification, discrimination, (Table), 57
  flow, column type, 57
  gaging heads, 56
  pressure type, 57–58
Polycrystalline (diamond) cutting wheels, 432–433
Power hacksaws, metal cutting, 122–125

Precision inspection/layout (surface plate) accessories, 77
Precision grinding:
  centerless, 417
  cylindrical, 456–466
  form, 449–453
  plunge, 460
  surface, 438–453
  tool and cutter, 467–482
Processor, postprocessor, 512
Programming, computer, NC/CNC:
  APT, 514–516
  fixed block format, 505
  NC word language, 497–499
  single-, two-, three-, four axes, 501–505
  tab sequential format programs, 501–505
  word address format programs, 505–506
Primary, secondary relief, grinding of, 473, 475, 477
Punches, bench, hand, 94

Q

Quality control, 63–66
  acceptable quality level (AQL), 65
  charts for, 66
  normal frequency distribution curve, 63–64
  sampling plans, 65–66
  surface texture, 66–73. *See also* Surface texture and finish
Quenching baths, 540–541
Quick-change tooling system, 367

R

Rack milling, 348
Rack and pinion gears, 343
Radial relief angles, grinding, (Table), 477
Radius gage, 205
Rail and side heads (planer), 403, 409
Ram-saddle types, turret lathes, 257–260
Rapid traverse attachment, lathe, 179
Reamers and reaming, hand, 105–109
  adjustable expansion, 107
  cutting fluids, 108
  features, 105–106
  grinding, angles for, 479–480
  Morse, Brown & Sharpe, taper, 106–107
  solid, straight, tapered hole, 106
Reamers and reaming, machine, 158–162
  alignment, 161; process, 161–162
  cutting; feeds, fluids, 161; speeds, 160–161
  stock allowances for, Table of, 158

reaming, turret lathe, 276
reamers, characteristics, types of:
  chucking, adjustable, expansion, 160
  jobbers, rose, shell, 159
  step, taper, 160
Recalescence point, 539
Recessing, turret lathe, 276
Rectangular coordinate system, NC/ CNC, 495
Reed comparator, features of, 54
Right-angle attachment (milling), 366
Robotics/robots. *See also* Flexible manufacturing systems
  CADD model, 575
  degrees of freedom, range, capacity, end effector, 574
  functions of, 574
  selection factors, 574
  six-axis, computer controlled, 574
  versatility in production, 575
Rockwell hardness testing. *See* Metallurgy and Heat treating
Rotary cross slide milling head, 366
Rotary tables, 485
Roundness, checking for, 81, 85

S

SAE steel classifications, 524
Safe practices. *See* End-of unit personal, machine, product safety items
Sawing, hand, 90–92
  blades for, 91
  frames, types of, 90–91
Saw band machines:
  guide post, 385–387
  welding, 387–390
Sawing, vertical band machine, 382–402
Scleroscope hardness testing, 556
Screw extractor, 120
Screw threads, 240–255
  Acme, 252–253; stub Acme, 253–254
  allowances, tolerance limits, size, 243
  American Standard Unified Miniature, 251
  cutting tools:
    leading, following, side angles, 245
    single point, 240–241
  depth settings, $60°$ threads, 245–246
  forms, formulas:
    ANS, Whitworth, Unified, SI metric, 242
  lead screw, setting of, 241–242
  measurement of:
    center gage, thread gage, 240
    screw thread micrometer, 243–244
    systems for, 244

Screw threads (continued)
  multiple start, 254
  production cutting methods,
      249–251
  spindle speeds, cutting, 242
  square, cutting of, 251–252
  thread cutting:
      outside, RH, LH, 245–246
      internal, 247
      SI metric, 247
      taper, 247
Screw machines, automatic:
  accessories, 284–285
  cams for, designs of, 283
  classifications of, components, 280
  clutch, controls for, 281
  collapsing taps, 277
  core, spade, start drilling, 275
  cutting fluids for, 284
  end former, necking cutting block,
      274
  feed-cycle change gears, 280
  taper attachment, 273
  tooling for, 282
Screw thread micrometers, 243–244
Second machine operations, 284–285
Shank adapters, drill, 150
Shaper, horizontal, 403–404
  cutting tools, carbide, geometry, 406
  shaping, processes of:
      contour surfaces, dovetails, 405
      internal forms, 406
      simple and compound angles,
          404–405
      vertical surfaces, 404
Shaper (slotter), vertical, 403–404
Side angles, threading tools, 245
SI metrics:
  form, formulas, threads, 242
  screw threads, cutting of, 247
Sine bar, 79
Sine block, 80
Sine plates:
  simple, compound, 80
  permanent magnetic, 80–81
Sine value, angle measurements,
      computation of, 78–79
Single/double angle cutters, grinding, 479
Single-point thread cutting tools,
      240–241
Single-spindle automatic screw ma-
      chines, 280
Slide caliper, 24
Slide tools, turret lathe, 265–266
Slitting saws, 476
Snap gage, adjustable, limit, 61
Spheroidizing, annealing, normalizing,
      545–546
Spindle noses:
  threaded, 224–225

Type D–1 cam lock, 224–225
Type L, 224
Spindle speeds, feeds, milling, 362
Spindle speeds, thread cutting, 242
Spindle tap drivers, 163–167
Spiral-edge saw bands, 397
Spiral flutes, hand reamers, 106
Spiral point drill, 150–151
Spotfacing, defined, 169
Square, cylindrical, solid, 40
Staggered-tooth milling cutters,
      grinding, 476
Steady rest, follower rest, 179
Steel:
  annealing, 545–546
  carbon, 522–524
      iron-carbon phase diagram,
          535–539
  casehardening, 546–547
  classification of, 523–524
  critical points, 535
  hardening, 538–539
      baths, 540–542
  hardness testing, 552–557
  heat treatment,
      slow cooling, effects of, 536–537
  manufacturing processes, 520–522
  subzero treatment of, 550
  tool and die, 526
  tungsten high-speed, 545
Straight turning, 201
Strap, U-, clamps, 144
Subzero temperature stabilizing, 550
Surface comparators, 68–69
Surface finish analyzers, 69
Surface grinders, grinding, 438–453
  adapter plates, fine pole magnetic
      chucks, 443
  crush rolls, 451
  cutting-off processes, wheels:
      nonreinforced, reinforced, 453
  design features, 438
      automatic down feed, wheel
          dressing, 440
      digital controls, 439
      fully automatic, 438
  diamond dressing tools, 452
  down-, cross-feeds, 442
  form dressing, angles, 449
  form grinding, accessories for,
      449–450
  grinding; angular surfaces, setups,
      446–447
      edges, parallel, 442
      shoulder, undercutting for, 448
  microform dresser, forming crush
      rolls on, 452–453
  pantograph dressers, 452
  workholding accessories, magnetic,
      445

magna-vise clamps, vacuum
      chucks, 440
  problems encountered, 441–442
Surface plates, granite, 24–25
Surface plate work, layout, inspection,
      76–85
Surface sensing probes, automated
      systems, FMS, 572–573
Surface texture/finish, 66–72, 75
  comparator sepcimens, 69–70
  measuring, measurement values, 68;
      (Table), 69
  measuring machined surfaces, brush
      analyzer, profilometer
      (Table), 69–72
  roughness, 69–72
  specifying, terms, symbols
      wheel conditions affecting, 422,
      426

T
Table of Constants, sine bar, 79
TAB sequential format programs:
  description of, 501
  single-axis, single machining process,
      501
      programming, 502–503
  speed/feed changes, milling
      processes, 503–504
      programming for, 504–505
  third-axis function, 503–504
  X, Y axes, tool changes, 503
      programming for, 503
Tap drills, 114
Tap drivers:
  nonreversing, reversing, spindle,
      163
  torque-driven, 164
Tape blocks, NC/CNC, 500
Tape preparation, NC/CNC, 500
Taper turning, systems, 218–220
Tap extractors, 116
Tapered end mills, 371
Taper socket adapters:
  fitted socket, 151
  reamers for, 106
Taper thread cutting, turret lathe, 273
  setup for, lathe, 247
  turning attachment, turret lathe,
      261
Tapping, 110–117
  basic thread series, 110–111
      American National, Unified,
          110–111; ISO metric, 111
  extractors, 116
  hand, 115–116
  machine screw sizes, 112–113
  screw thread characteristics, 110
  threading, turret lathe, 276
  thread producing methods, 110

Taps, hand and machine types:
 drill sizes (Tables), 114–115
 features of, 165
 gun flute only, 113
 machine, 165
 pipe, Dryseal, straight, taper, 114
 sets, 112; serial sets, 113
 sharpening, 480
 spiral fluted, low angle, 113; high
  angle, 114
 wrenches for, 115
Telescoping (small hole) gages, 34
Test indicator, 40–41
Thread gage, 240
Thread ring gages, 62
 clearance, positive and negative, 62
Threads, screw:
 Acme, 252–253
  stub Acme, 253–254
 allowances, tolerances, limits, 243
 American Standard Unified Miniature, 251
 drawings of, 118
 external, die cut, 118–121
  hand cutting dies, 119–120
 internal, 247
 ISO metric:
  forms, formulas of, 242
  measurement of, pitch, 118
  setup for cutting, 247
 multiple-start, 254
 pitch gage, 118
 taper, setup for, 247
 Unified system, forms, formulas,
  242
Threads, screw: machine tapping, drill
  press, 163–167. See also
  Screw threads
 attachments:
  micro, heavy-duty; torque-
   driven, heads, 164
  reversing, nonreversing, spindle
   tap drivers, 163–167
 cutting fluids, 164; (Table) 165
 cutting speed, factors affecting, 164
 problems, causes, corrective action,
  (Table), 166
 processes of, 165–167
 threading, turret lathe, 274
Through-feed centerless grinding, 417
Tolerances, definition, 11
Tool changing, programming for,
  NC/CNC, 494
Tool and cutter grinder. See Cutter
  and tool grinding
Tool geometry, lathe, 186
Tool handling automated, FMS, 572
Toolholders,
 carbide inserts, 193
 heavy-duty, open-side, tool block, 194

quick-change systems, 193–194
 Types 1, 2, and 3, 176–177
 web-bar, 49–50
Tool length gage, electronic, 518
Toolmaker's microscope, 244
Tool slide (screw machine), 282
Tooth rests, grinding, 471–473
Tracer, milling machine, 362
Troubleshooting, automatic screw
  machines, 285
T-slot cutting, 375
Tungsten high-speed steels, heat treatment of, 545
Turning,
 angles, 206–207
 centers, universal, 256–257
 cutting off, 209–211
 facing, 202–203
 knurling, 213–215
 rounded corners, 207
 shoulder, 205–206
 straight, 201
 taper, 217–221
Turning machines. See Screw machines
Turret lathes: tooling, setups,
   processes:
 automated bar, chucking, machines,
  256
  classification of ram, saddle
   types, 257–260
  processes identified, 256
 bar turners, 264–265
 cutting-off tools, 267–268
 cutting tool holders, 267
 headstock spindle tooling:
  collets chucks, step chucks,
   closers, 262
  chucks, special fixtures, 263
  permanent setups, 263–265
 machine attachments, thread chasing, taper turning, 261
 machine cuts, kinds of, 262
 permanent setups, chucking tooling,
  268–269
 planned tool circle, 268
 slide tools for turrets:
  box, balance turning, die heads,
   265–266
 tool chatter problems, 269
 turning (universal) centers, 256–257
  ram, saddle types, 257–269
 turret, components of, 260
Turret lathes, work:
 bar, chucking, NC, 510
 bar turners, 272
  cutting speeds, feeds, 272
   (Table), 273
 turning cuts, external, 272:
  cutting off, 274
  facing, 273

forming, 274
 sequencing cuts, 274–275
 taper, 273
 threading, 274
 turning cuts, internal:
  boring, 276
  drilling, 276
  reaming, 276
  recessing, 276
  sequencing, 277–278
  threading, 276
Turrets, turning machines:
 mounted tools, 283
 multiple-, "eight" station, 2
 slide tools, box, balance turning,
  264–265
 two axes, NC, continuous path,
  509–510
Twist drills:
 functions, parts, 148–150
 points, shapes of, 149–151

U

Ultrasonic machining (USM), 565
Universal:
 dial indicator sets, 43
 measuring machines, 489–491
  bidirectional gaging system, 490
  TV microscopes, angular measuring instruments, 490
 milling machine, 288
 precision vise, 142

V

Vacuum chucks, 440
V-blocks, 142–143
Vernier:
 depth gage, 38
 height gage, 38–39
 metric, 37
 parts, scales, 35–36
 principle, 36–37; Brown, Joseph
  R., 2
 protractor, 39–40
 reading, 38
Vertical band machines. See Band
  machines and processes
Vertical/horizontal milling machines, 362
Vertical milling machines:
 bed type, 363
 combination vertical/horizontal, 362
 general purpose, 361
  features, components, sizes, 361–362
 NC/CNC, 363
 tracer, 362
 turret head, controls, 362
Vertical milling machines, accessories:
 boring head set, 367
 holders, cutting tool, 367

Vertical milling machines, accessories
    (continued)
  offset boring head, 376–381
  optical measuring system, 366
  quick-change tooling system, 367
  shaping tool set, 367
Vertical milling machines, attachments:
  digital readout, 368
  high-speed, 365
  right-angle, 366
  rotary cross-slide head, 366
  vertical shaper head, 365, 377–378
Vertical milling machines, processes:
  boring, straight, angle, radius,
      stepped, 376–381
  cutting speeds, feeds, 372–373
  end milling:
    angular surfaces, 373–374
    flat surfaces, 372–373

  keyways, slots, 374
  T-slot, 375
  end mills, types, features, 370–372
Vertical shaper head, 365, 377–378
Vertical slotter/shaper, 403–404
Vises:
  angle, standard, universal, 142
Vickers hardness testing, 556

W

Wavelengths, measurement of, 52
Welding, butt, band machines,
    387–390
Wheel forming accessories, 449–452
White cast iron, 527–529
Wipe grinding, 487
Woodruff keyseat, 330
Word address format programs:
  continuous path machining, 505

  fixed block, 505
  mirror image, programming for,
      505–506
  sample coded functions, 505
Work-holding devices/accessories:
  angle plate, 144
  column mounted, safety, 142
  drill jigs, fixtures, 144
  jack (screw), 144
  protecting strips, shims, 145
  strap clamps, T-head bolts,
      143–144
  universal swivel block, 145
  V-blocks, 142–143, 162–163
  Vises: standard, angle, universal,
      142
Work orders, production plan, 83
Worm gears, milling, 350–351
Worm thread, milling, 350